Advances in Intelligent and Soft Computing

Editor-in-Chief: J. Kacprzyk

Advances in Intelligent and Soft Computing

Editor-in-Chief

Prof. Janusz Kacprzyk
Systems Research Institute
Polish Academy of Sciences
ul. Newelska 6
01-447 Warsaw
Poland
E-mail: kacprzyk@ibspan.waw.pl

Further volumes of this series can be found on our homepage: springer.com

Vol. 41. P. Melin, O. Castillo,
E. Gómez Ramírez, J. Kacprzyk,
W. Pedrycz (Eds.)
Analysis and Design of Intelligent Systems Using Soft Computing Techniques, 2007
ISBN 978-3-540-72431-5

Vol. 42. O. Castillo, P. Melin,
O. Montiel Ross, R. Sepúlveda Cruz,
W. Pedrycz, J. Kacprzyk (Eds.)
Theoretical Advances and Applications of Fuzzy Logic and Soft Computing, 2007
ISBN 978-3-540-72433-9

Vol. 43. K.M. Węgrzyn-Wolska,
P.S. Szczepaniak (Eds.)
Advances in Intelligent Web Mastering, 2007
ISBN 978-3-540-72574-9

Vol. 44. E. Corchado, J.M. Corchado,
A. Abraham (Eds.)
Innovations in Hybrid Intelligent Systems, 2007
ISBN 978-3-540-74971-4

Vol. 45. M. Kurzynski, E. Puchala,
M. Wozniak, A. Zolnierek (Eds.)
Computer Recognition Systems 2, 2007
ISBN 978-3-540-75174-8

Vol. 46. V.-N. Huynh, Y. Nakamori,
H. Ono, J. Lawry,
V. Kreinovich, H.T. Nguyen (Eds.)
Interval / Probabilistic Uncertainty and Non-classical Logics, 2008
ISBN 978-3-540-77663-5

Vol. 47. E. Pietka, J. Kawa (Eds.)
Information Technologies in Biomedicine, 2008
ISBN 978-3-540-68167-0

Vol. 48. D. Dubois, M. Asunción Lubiano,
H. Prade, M. Ángeles Gil,
P. Grzegorzewski,
O. Hryniewicz (Eds.)
Soft Methods for Handling Variability and Imprecision, 2008
ISBN 978-3-540-85026-7

Vol. 49. J.M. Corchado, F. de Paz,
M.P. Rocha,
F. Fernández Riverola (Eds.)
2nd International Workshop on Practical Applications of Computational Biology and Bioinformatics (IWPACBB 2008), 2009
ISBN 978-3-540-85860-7

Vol. 50. J.M. Corchado, S. Rodriguez,
J. Llinas, J.M. Molina (Eds.)
International Symposium on Distributed Computing and Artificial Intelligence 2008 (DCAI 2008), 2009
ISBN 978-3-540-85862-1

Vol. 51. J.M. Corchado, D.I. Tapia,
J. Bravo (Eds.)
3rd Symposium of Ubiquitous Computing and Ambient Intelligence 2008, 2009
ISBN 978-3-540-85866-9

Vol. 52. E. Avineri, M. Köppen,
K. Dahal,
Y. Sunitiyoso, R. Roy (Eds.)
Applications of Soft Computing, 2009
ISBN 978-3-540-88078-3

Vol. 53. E. Corchado, R. Zunino,
P. Gastaldo, Á. Herrero (Eds.)
Proceedings of the International Workshop on Computational Intelligence in Security for Information Systems CISIS 2008, 2009
ISBN 978-3-540-88180-3

Vol. 54. B.-y. Cao, C.-y. Zhang,
T.-f. Li (Eds.)
Fuzzy Information and Engineering, 2009
ISBN 978-3-540-88913-7

Vol. 55. Y. Demazeau, J. Pavón,
J.M. Corchado, J. Bajo (Eds.)
7th International Conference on Practical Applications of Agents and Multi-Agent Systems (PAAMS 2009), 2009
ISBN 978-3-642-00486-5

Yves Demazeau, Juan Pavón, Juan M. Corchado,
Javier Bajo (Eds.)

7th International Conference on Practical Applications of Agents and Multi-Agent Systems (PAAMS 2009)

 Springer

Editors

Yves Demazeau
Laboratoire d'Informatique de Grenoble
Maison Jean Kuntzmann
Domaine Universitaire de
Saint Martin d'Heres
110 av. de la Chimie
38041 Grenoble
France
E-mail : Yves.Demazeau@imag.fr

Prof. Juan Pavón
Facultad de Informática de la
Universidad Complutense de Madrid
Avda. Complutense s/n
28040 Madrid
Spain
E-mail: jpavon@fdi.ucm.es

Prof. Juan M. Corchado
Departamento de Informática y Automática
Facultad de Ciencias
Universidad de Salamanca
Plaza de la Merced S/N
37008, Salamanca
Spain
E-mail: corchado@usal.es

Javier Bajo
Departamento de Informática y Automática
Facultad de Ciencias
Universidad de Salamanca
Plaza de la Merced S/N
37008, Salamanca
Spain
E-mail: jbajope@upsa.es

ISBN 978-3-642-00486-5   e-ISBN 978-3-642-00487-2

DOI 10.1007/978-3-642-004867-2

Advances in Intelligent and Soft Computing   ISSN 1867-5662
Library of Congress Control Number: 2009921170

©2009 Springer-Verlag Berlin Heidelberg

This work is subject to copyright. All rights are reserved, whether the whole or part of the material is concerned, specifically the rights of translation, reprinting, reuse of illustrations, recitation, broadcasting, reproduction on microfilm or in any other way, and storage in data banks. Duplication of this publication or parts thereof is permitted only under the provisions of the German Copyright Law of September 9, 1965, in its current version, and permission for use must always be obtained from Springer. Violations are liable for prosecution under the German Copyright Law.

The use of general descriptive names, registered names, trademarks, etc. in this publication does not imply, even in the absence of a specific statement, that such names are exempt from the relevant protective laws and regulations and therefore free for general use.

Typeset & Cover Design: Scientific Publishing Services Pvt. Ltd., Chennai, India.

Printed in acid-free paper

5 4 3 2 1 0

springer.com

Preface

Research on Agents and Multi-Agent Systems has matured during the last decade and many effective applications of this technology are now deployed. An international forum to present and discuss the latest scientific developments and their effective applications, to assess the impact of the approach, and to facilitate technology transfer, has become a necessity.

PAAMS, the International Conference on Practical Applications of Agents and Multi-Agent Systems is an evolution of the International Workshop on Practical Applications of Agents and Multi-Agent Systems. PAAMS is an international yearly tribune to present, to discuss, and to disseminate the latest developments and the most important outcomes related to real-world applications. It provides a unique opportunity to bring multi-disciplinary experts, academics and practitioners together to exchange their experience in the development of Agents and Multi-Agent Systems.

This volume presents the papers that have been accepted for the 2009 edition. These articles capture the most innovative results and this year's trends: Assisted Cognition, E-Commerce, Grid Computing, Human Modelling, Information Systems, Knowledge Management, Agent-Based Simulation, Software Development, Transports, Trust and Security. Each paper has been reviewed by three different reviewers, from an international committee composed of 64 members from 20 different countries. From the 92 submissions received, 35 were selected for full presentation at the conference, and 26 were accepted as posters.

We would like to thank all the contributing authors, as well as the members of the Program Committee and the Organizing Committee for their hard and highly valuable work. Their work has helped to contribute to the success of the PAAMS 2009 event. Thanks for your help, PAAMS 2009 wouldn't exist without your contribution.

Yves Demazeau	Juan Manuel Corchado
Juan Pavón	Javier Bajo
PAAMS 2009 Program Co-chairs	PAAMS 2009 Organizing Co-chairs

Organization

General Co-chairs

Yves Demazeau	Centre National de la Recherche Scientifique (France)
Juan Pavón	Universidad Complutense de Madrid (Spain)
Juan M. Corchado	University of Salamanca (Spain)
Javier Bajo	Pontifical University of Salamanca (Spain)

Program Committee

Yves Demazeau (Co-chairman)	Centre National de la Recherche Scientifique (France)
Juan Pavón	Universidad Complutense de Madrid (Spain)
Analia Amandi	ISISTAN (Argentina)
Luis Antunes	University of Lisbon (Portugal)
Ana Bazzan	Universidade Federal de Rio Grande do Sul (Brazil)
Olivier Boissier	Ecole Nationale Superieure des Mines de Saint Etienne (France)
Magnus Boman	Royal Institute of Technology (Sweden)
Juan A. Botía	University of Murcia (Spain)
Vicente Botti	Polytechnic University of Valencia (Spain)
Bernard Burg	Panasonic Ltd (USA)
Sven Brueckner	NewVectors (USA)
Monique Calisti	Whitestein Technologies (Switzerland)
Valerie Camps	University Paul Sabatier (France)
Javier Carbó	University Carlos III of Madrid (Spain)
Helder Coelho	University of Lisbon (Portugal)
Emilio Corchado	University of Burgos (Spain)
Juan M. Corchado	University of Salamanca (Spain)

Rafael Corchuelo	University of Sevilla (Spain)
Keith Decker	University of Delaware (USA)
Alexis Drogoul	IRD (Institut de Recherche pour le Development) (Vietnam)
Julie Dugdale	University Pierre Mendes France (France)
Edmund Durfee	University of Michigan (USA)
Torsten Eymann	University of Bayreuth (Germany)
Klaus Fischer	DFKI (Germany)
Rubén Fuentes	Complutense University of Madrid (Spain)
Francisco Garijo	Telefónica I+D (Spain)
Khaled Ghedira	National School of Computer Sciences (Tunisia)
Sylvain Giroux	Unversity of Sherbrooke (Canada)
Marie-Pierre Gleizes	University Paul Sabatier (France)
Jorge J. Gómez-Sanz	Complutense University of Madrid (Spain)
Vladimir Gorodetski	University of Saint Petersburg (Russia)
Dominic Greenwood	Whitestein Technologies (Switzerland)
David Hales	Delft University of Technology (The Netherlands)
Toru Ishida	University of Kyoto (Japan)
Kasper Hallenborg	University of Southern Denmark (Denmark)
Vicente Julián	Polytechnic University of Valencia (Spain)
Achilles Kameas	University of Patras (Greece)
Franziska Kluegl	University of Örebro (Sweden)
Matthias Klusch	DFKI (Germany)
Beatriz López	University of Gerona (Spain)
Adolfo López Paredes	University of Valladolid (Spain)
Rene Mandiau	University of Valenciennes (France)
Philippe Mathieu	University of Lille (France)
Fabien Michel	University of Reims (France)
José M. Molina	University Carlos III of Madrid (Spain)
Bernard Moulin	University Laval (Canada)
Jörg Müller	Clausthal University of Technology (Germany)
Eugenio Oliveira	University of Porto (Portugal)
Andrea Omicini	University of Bologna (Italy)
Sascha Ossowski	University of Rey Juan Carlos (Spain)
Van Parunak	New Vectors (USA)
Michal Pechoucek	Czech Technical University in Prague (Czech Republic)
Paolo Petta	University of Vienna (Austria)
Jeremy Pitt	Imperial College of London (UK)
Alessandro Ricci	University of Bologna (Italy)
Antonio Rocha Costa	Catholic University of Pelotas (Brazil)
Nicolas Sabouret	LIP6 (France)
Munindar Singh	North Carolina State University (USA)
Kostas Stathis	Royal Holloway University of London (UK)
Paolo Torroni	University of Bologna (Italy)
Domenico Ursino	University of Reggio Calabria (Italy)

José R. Villar   University of Oviedo (Spain)
Danny Weyns   Catholic University of Leuven (Belgium)
Franco Zambonelli   University of Modena (Italy)

Organizing Committee

Juan M. Corchado
  (Co-chairman)   University of Salamanca (Spain)
Javier Bajo (Co-chairman) Pontifical University of Salamanca (Spain)
Juan F. De Paz   University of Salamanca (Spain)
Sara Rodríguez   University of Salamanca (Spain)
Dante I. Tapia   University of Salamanca (Spain)
M.A. Pellicer   University of Salamanca (Spain)

Contents

A Holonic Approach to Warehouse Control
Hristina Moneva, Jurjen Caarls, Jacques Verriet 1

Developing Home Care Intelligent Environments: From Theory to Practice
J.A. Fraile Nieto, M.E. Beato Gutiérrez, B. Pérez Lancho 11

Distributing Functionalities in a SOA-Based Multi-agent Architecture
Dante I. Tapia, Javier Bajo, Juan M. Corchado 20

Mobile Agents for Critical Medical Information Retrieving from the Emergency Scene
Abraham Martín-Campillo, Ramon Martí, Sergi Robles, Carles Martínez-García 30

INGENIAS Development Assisted with Model Transformation By-Example: A Practical Case
Iván García-Magariño, Jorge Gómez-Sanz, Rubén Fuentes-Fernández 40

GENESETFINDER: A Multiagent Architecture for Gathering Biological Information
Daniel Glez-Peña, Julia Glez-Dopazo, Reyes Pavón, Rosalía Laza, Florentino Fdez-Riverola 50

Agent Design Using Model Driven Development
Jorge Agüero, Miguel Rebollo, Carlos Carrascosa, Vicente Julián 60

A Tool for Generating Model Transformations By-Example in Multi-Agent Systems
Iván García-Magariño, Sylvain Rougemaille, Rubén Fuentes-Fernández, Frédéric Migeon, Marie-Pierre Gleizes, Jorge Gómez-Sanz 70

Modelling Trust into an Agent-Based Simulation Tool to
Support the Formation and Configuration of Work Teams
Juan Martínez-Miranda, Juan Pavón 80

Multi-agent Simulation of Investor Cognitive Behavior in
Stock Market
Zahra Kodia, Lamjed Ben Said 90

How to Avoid Biases in Reactive Simulations
Yoann Kubera, Philippe Mathieu, Sébastien Picault 100

Generating Various and Consistent Behaviors in Simulations
Benoit Lacroix, Philippe Mathieu, Andras Kemeny 110

Foreseeing Cooperation Behaviors in Collaborative Grid
Environments
Mauricio Paletta, Pilar Herrero 120

MAMSY: A Management Tool for Multi-Agent Systems
*Victor Sanchez-Anguix, Agustin Espinosa, Luis Hernandez,
Ana García-Fornes* 130

A Multi-Agent System Approach for Algorithm Parameter
Tuning
R. Pavón, D. Glez-Peña, R. Laza, F. Díaz, M.V. Luzón 140

Relative Information in Grid Information Service and Grid
Monitoring Using Mobile Agents
Carlos Borrego, Sergi Robles 150

A Multi-Agent System for Airline Operations Control
Antonio J.M. Castro, Eugenio Oliveira 159

Agent-Based Approach to the Dynamic Vehicle Routing
Problem
Dariusz Barbucha, Piotr Jędrzejowicz 169

The Undirected Rural Postman Problem Solved by the
MAX-MIN Ant System
María Luisa Pérez-Delgado 179

Performance Visualization of a Transport Multi-agent
Application
Hussein Joumaa, Yves Demazeau, Jean-Marc Vincent 188

A Software Architecture for an Argumentation-Oriented
Multi-Agent System
Andrés Muñoz, Ana Sánchez, Juan A. Botía 197

A SOMAgent for Identification of Semantic Classes and Word Disambiguation
Vivian F. López, Luis Alonso, María Moreno 207

Multiagent Systems in Expression Analysis
Juan F. De Paz, Sara Rodríguez, Javier Bajo 217

Intentions in BDI Agents: From Theory to Implementation
S. Bonura, V. Morreale, G. Francaviglia, A. Marguglio,
G. Cammarata, M. Puccio 227

An Intrusion Detection and Prevention Model Based on Intelligent Multi-Agent Systems, Signatures and Reaction Rules Ontologies
Gustavo A. Isaza, Andrés G. Castillo, Néstor D. Duque 237

An Attack Detection Mechanism Based on a Distributed Hierarchical Multi-agent Architecture for Protecting Databases
Cristian Pinzón, Yanira de Paz, Rosa Cano, Manuel P. Rubio 246

Trusted Computing: The Cornerstone in the Secure Migration Library for Agents
Antonio Muñoz, Antonio Maña, Daniel Serrano 256

Negotiation of Network Security Policy by Means of Agents
Pablo Martin, Agustin Orfila, Javier Carbo 266

A Contingency Response Multi-agent System for Oil Spills
Aitor Mata, Dante I. Tapia, Angélica González, Belén Pérez 274

V-MAS: A Video Conference Multiagent System
Alma Gómez-Rodríguez, Juan C. González-Moreno,
Loxo Lueiro-Astray, Rubén Romero-González 284

Online Scheduling in Multi-project Environments: A Multi-agent Approach
José Alberto Arauzo, José Manuel Galán, Javier Pajares,
Adolfo López-Paredes 293

Experiencing Self-adaptive MAS for Real-Time Decision Support Systems
Jean-Pierre Georgé, Sylvain Peyruqueou, Christine Régis,
Pierre Glize 302

Induced Cultural Globalization by an External Vector Field in an Enhanced Axelrod Model
Arezky H. Rodríguez, M. del Castillo-Mussot, G.J. Vázquez 310

Towards the Implementation of a Normative Reasoning Process
Natalia Criado, Vicente Julián, Estefania Argente 319

Negotiation Exploiting Reasoning by Projections
Toni Mancini 329

A JADE-Based Framework for Developing Evolutionary Multi-Agent Systems
*Bertha Guijarro-Berdiñas, Amparo Alonso-Betanzos,
Silvia López-López, Santiago Fernández-Lorenzo,
David Alonso-Ríos* 339

A Multi-agent Approach for Web Adaptation
A. Jorge Morais 349

A Multi-tiered Approach to Context and Information Sharing in Intelligent Agent Communities
Russell Brasser, Csaba Egyhazy 356

A Multiagent Distributed Design System
Ewa Grabska, Barbara Strug, Grażyna Ślusarczyk 364

A Realistic Approach to Solve the Nash Welfare
A. Nongaillard, P. Mathieu, B. Jaumard 374

A Study of Bio-inspired Communication Scheme in Swarm Robotics
P.N. Stamatis, I.D. Zaharakis, A.D. Kameas 383

Artificial Intelligence for Picking Up Recycling Bins: A Practical Application
Maria Luisa Pérez-Delgado, Juan C. Matos-Franco 392

An Access Control Scheme for Multi-agent Systems over Multi-Domain Environments
*C. Martínez-García, G. Navarro-Arribas, J. Borrell,
A. Martín-Campillo* 401

Bridging the Gap between the Logical and the PhysicalWorlds
*Francisco García-Sánchez, Renato Vidoni,
Rodrigo Martínez-Béjar, Alessandro Gasparetto,
Rafael Valencia-García, Jesualdo T. Fernández-Breis* 411

Building Service-Based Applications for the iPhone Using RDF: A Tourism Application
Javier Palanca, Gustavo Aranda, Ana García-Fornes 421

Designing a Visual Sensor Network Using a Multi-agent Architecture
Federico Castanedo, Jesús García, Miguel A. Patricio,
José M. Molina 430

Designing Virtual Organizations
N. Criado, E. Argente, V. Julián, V. Botti 440

Dynamic Orchestration of Distributed Services on Interactive Community Displays: The ALIVE Approach
I. Gómez-Sebastià, Manel Palau, Juan Carlos Nieves,
Javier Vázquez-Salceda, Luigi Ceccaroni 450

Efficiency in Electrical Heating Systems: An MAS Real World Application
José R. Villar, Roberto Pérez, Enrique de la Cal, Javier Sedano 460

Hardware Protection of Agents in Ubiquitous and Ambient Intelligence Environments
Antonio Maña, Antonio Muñoz, Daniel Serrano 470

Management System for Manufacturing Components Aligned with the Organisation IT Systems
Diego Marcos-Jorquera, Francisco Maciá-Pérez, Virgilio Gilart-Iglesias,
Jorge Gea-Martínez, Antonio Ferrándiz-Colmeiro 480

MASITS – A Tool for Multi-Agent Based Intelligent Tutoring System Development
Egons Lavendelis, Janis Grundspenkis 490

Multi-agent Reasoning Based on Distributed CSP Using Sessions: DBS
Pierre Monier, Sylvain Piechowiak, René Mandiau 501

Natural Interface for Sketch Recognition
D.G. Fernández-Pacheco, N. Aleixos, J. Conesa, M. Contero 510

Performance of an Open Multi-Agent Remote Sensing Architecture Based on XML-RPC in Low-Profile Embedded Systems
Guillermo Glez. de Rivera, Ricardo Ribalda, Angel de Castro,
Javier Garrido 520

Privacy Preservation in a Decentralized Calendar System
Ludivine Crépin, Yves Demazeau, Olivier Boissier,
François Jacquenet 529

Protected Computing Approach: Towards the Mutual Protection of Agent Computing
Antonio Maña, Antonio Muñoz, Daniel Serrano 538

Toward a Conceptual Framework for Multi-points of View Analysis in Complex System Modeling: OREA Model
Mahamadou Belem, Jean-Pierre Müller 548

Using Hitchhiker Mobile Agents for Environment Monitoring
Oscar Urra, Sergio Ilarri, Eduardo Mena, Thierry Delot 557

Using Multiagent Systems and Genetic Algorithms to Deal with Problems of Staggering
Arnoldo Uber Junior, Ricardo Azambuja Silveira 567

VisualChord: A Personal Tutor for Guitar Learners
Alberto Romero, Ana-Belén Gil, Ana de Luis 576

Author Index 587

A Holonic Approach to Warehouse Control

Hristina Moneva[1,*], Jurjen Caarls[2], and Jacques Verriet[1]

[1] Embedded Systems Institute, P.O. Box 513, 5600 MB Eindhoven, The Netherlands
 `hristina.moneva@topic.nl, jacques.verriet@esi.nl`
[2] Dynamics and Control Group, Department of Mechanical Engineering, Eindhoven University of Technology, P.O. Box 513, 5600 MB Eindhoven, The Netherlands
 `j.caarls@tue.nl`

Abstract. Warehouses play a critical role in the distribution of products of many suppliers to many customers. Traditionally, warehouse operations are controlled by centralised control systems. Because of increasing customer demands, the complexity of such systems becomes too large to respond optimally to all warehouse events. Holonic control systems try to overcome this limitation. This paper presents a framework that supports the development of holonic warehouse control systems. The framework, which is built on top of JADE middleware, allows a holonic control system to be generated from a warehouse layout model and a library of reusable behaviour components. The architecture of the framework and an example of its application to an existing warehouse are presented.

Keywords: warehouse control systems, holonic control, agent technology, JADE, automated system generation.

1 Introduction

Warehouses are critical links in supply chains: they receive goods from many different suppliers, provide temporary storage of these goods, repack them, and distribute them to many different customers. Nowadays, it is not uncommon for a warehouse to deliver goods to different types of customers, such as other warehouses, various types of shops and Internet customers [1]. Each type of customer has its own delivery requirements. For instance, shop customers place large orders that have to be delivered in such a way that the shop customers can replenish their shelves in an efficient manner (i.e. similar products must be stored in the same container). On the other hand, Internet orders are very small orders that have to be delivered as a mail parcel.

A warehouse control system is responsible for controlling the operations needed to fulfil all customer requirements. Traditionally, warehouse control systems are centralised systems responsible for planning, scheduling and execution of all warehouse operations. These operations include normal operations like receiving, storage, picking, packing and shipping of goods, and frequently occurring exceptions like equipment failures. Because of increasing customer requirements [1], it has become almost impossible to control all (normal and exceptional) operations in an optimal manner using a centralised warehouse control system.

---

[*] Current affiliation: Topic Embedded Systems, P.O. Box 440, 5680 AK Best, The Netherlands.

An alternative to a centralised warehouse control system would be a fully decentralised warehouse control system. Such a system consists of a collection of agents, each controlling the operations in a limited part of the warehouse without any system-level coordination. Although fully decentralised warehouse control systems can adapt themselves to changing circumstances, it is unclear whether such systems are feasible. The desired system qualities, like performance and robustness, have to emerge from the individual qualities of the agents and the interaction of the agents. Engineering the complex warehouse flows without any system-level coordination will prove to be a very challenging task, especially for shop customers and their difficult delivery requirements.

Holonic Control. A control paradigm that attempts to overcome the limitations of fully centralised and fully decentralised control is holonic control, which can be seen as a hybrid form of both types of control. A holonic control system consists of dynamic hierarchies of autonomous agents, called *holons*, which individually attempt to reach their own objectives and collectively try to reach a common objective. The low-level holons are responsible for the basic operations and provide local adaptivity and the high-level holons coordinate the product flows to achieve the required system performance. In other words, holonic control has the flexibility and adaptivity of fully decentralised control and the performance and predictability of fully centralised control.

Holonic control has mainly been applied in the manufacturing domain. There are many publications describing the benefits of holonic manufacturing systems. An example is the holonic engine manufacturing line described by Fleetwood et al. [6]. They show that their holonic system has a better performance and is more flexible than its centralised counterpart. Fletcher [7] presents holonic control systems of a packing plant and a lumber mill and describes how these cope with normal scenarios and with exceptional scenarios like equipment breakdowns and rush orders. Other applications of holonic control systems include baggage handling systems [3] and transportation scheduling [5].

Warehouse control using agents or holons has hardly been studied and is generally limited to transport. Exceptions are Kachornvitaya-Wan [10] and Kim [11], who consider holonic control systems for two warehouses. They both propose holonic control systems, in which the high-level holons are responsible for providing an initial schedule. The low-level holons start with such a schedule, but constantly try to find a schedule that best fits the real-time conditions. If they find such a schedule, they negotiate with the high-level holons for a schedule change. Experiments show that the constant re-planning positively influences the system performance, especially in case of exceptions: when equipment breaks down, the adapted schedules greatly outperform the initial schedules.

Many applications of holonic control systems are very specific for the application. However, a few reference architectures have been proposed for holonic manufacturing systems. A well-known holonic reference architecture is PROSA [15]. PROSA is built upon three types of basic holons: resource, product and order holons. Resource holons represent physical production resources, product holons hold process and production knowledge, and order holons represent tasks to be performed. PROSA has been applied to a security glass production facility [4], a tile factory [9] and a car painting plant [16]. ADACOR [13], a reference architecture similar to PROSA, has been applied to a robotised production system [12].

Outline. In this paper, we go one step further than a holonic reference architecture. This paper presents a holonic control framework that supports warehouse design. From warehouse models, it generates a holonic control system that can be used for simulation. Traditionally, strict time constraints do not allow the analysis of many warehouse concepts during warehouse design. The facilities of our framework allow the exploration of a larger part of the warehouse design space, because it enables control experiments to be set up and performed quickly.

The paper is organised as follows. Chapter 2 explains the background of our framework's holonic warehouse control systems. The architecture and implementation of the framework is described in Chapter 3. In Chapter 4, we present the results of the first experiments performed with the framework. Chapter 5 provides a summary of this paper and an outlook into the future.

2 Warehouse Holons

The basis for our framework is formed by three basic holons. We distinguish resource holons, order holons, and logic holons. *Resource holons* represent the equipment in a warehouse, *order holons* represent the tasks to be performed and *logic holons* are service directories where holons can register services, which can be requested by other holons. Besides the basic holons, there are *staff holons* that provide functionality, like negotiation, which can be used by the basic holons.

The basic holons are similar to PROSA's basic holons [15]. The main difference involves the logic holon, which replaces PROSA's product holon: where the product holon only holds production knowledge, our logic holon holds information for complex tasks in general, making it applicable for more general purposes.

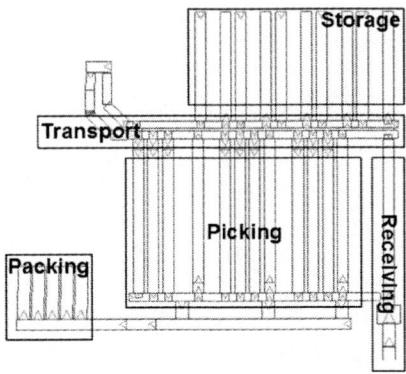

Fig. 1. Warehouse layout

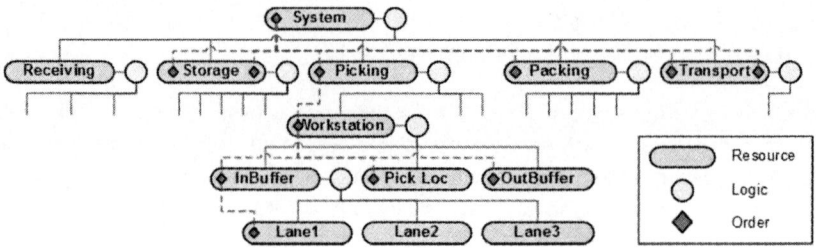

Fig. 2. Warehouse holarchies

Resource Holons. Warehouses can be seen as hierarchies of functional building blocks, each type with unique responsibilities. This hierarchical structure is used to create a hierarchy of resource holons, a *holarchy*. This is illustrated using Figure 1, which shows the layout of a retail warehouse built by our industrial partner Vanderlande Industries. The warehouse system contains areas for receiving, storage, picking, packing and transport. The receiving area is responsible for warehouse replenishment: it receives pallets as input and puts their content in storage bins, which are stored in the storage area. The picking area is responsible for fulfilling customer orders: items are picked from storage bins and placed in order bins. The packing area receives order bins and groups these onto roll containers, which are shipped to the customer. The transport area transports storage bins between the functional areas. Each area is represented by a resource holon.

The areas consist of workstations that are responsible for the actual execution of tasks. The receiving area has three workstations (not shown in Figure 1), the storage area has five aisles, the picking area has three workstations, the packing area has five workstations, and the transport area consists of a loop between the receiving, storage and picking area. Like the functional areas, the workstations are represented by resource holons. Workstations generally consist of elements that together provide the workstation's execution capabilities. In such situations, the workstation holon has child holons, resulting in a hierarchy of resource holons.

An important part of the resource holarchy for the warehouse in Figure 1 is shown in Figure 2. It shows a system resource holon and four area holons. Figure 2 also shows the resource holons of a picking workstation and its elements, an input buffer with three buffer lanes, a pick location, and an output buffer.

Logic Holons. The logic holons can be seen as service directories: holons can register their services at a logic holon, making these services available for other holons. Because equipment may break down, holons can also unregister their services at a logic holon, making their services unavailable for other holons. Holons having to perform a (complex) task can consult a logic holon to obtain a process plan that allows the task to be executed. For example, the process plan of a replenishment order involves filling, transporting and storing storage bins.

Because subtasks in a process plan can only be executed by a resource or its child resources, we decided to couple logic holons and resource holons. A resource holon has an associated logic holon if it has child resources or if it has tasks with a process plan. Child resource holons can use their parents' logic holon to register and unregister their services. The resulting hierarchy of logic holons resembles the distributed directory facilitator of López Orozco and Martinez Lastra [14], but does not involve propagation of services in the hierarchy.

Figure 2 shows the hierarchy of logic holons for the warehouse of Figure 1. On the top level, Figure 2 shows a logic holon associated to the system resource holon. The area holons can register their services at this logic holon. Similarly, there is a logic holon associated to every area holon. The workstation holons use these logic holons to register their services. Similarly, the picking workstation and its input buffer have associated logic holons which are used for the registration of their children's services.

Order Holons. Order holons are created for the tasks to be performed by the resource holons. If a resource holon is not able to execute a task, it creates an order holon, which ask a logic holon for a process plan. Using this process plan, the task is divided into subtasks to be executed by child resources. The order holon is responsible for the coordination of the execution of these subtasks.

On the top level, order holons represent customer orders and replenishment orders assigned to the system resource holon. These orders are first broken down into tasks for the functional areas. Next, these tasks are decomposed in subtasks for the workstations and their elements, resulting in a hierarchy of order holons.

Figure 2 shows the order holarchy for a customer order for a single item; it shows a customer order holon assigned to the system resource holon. The order is decomposed into a retrieval task, a picking task, a storage task, a packing task and two transportation tasks. These tasks are assigned to the storage area, the picking area, the storage area, packing area and the transportation area, respectively. These tasks are further divided into tasks for the workstation holons. This is shown for the picking task, including the division into tasks for the workstation's elements.

3 Framework Architecture

As holons are much alike agents [8], we opted to use existing agent technology to implement them. The main difference between agents and holons is the holons' recursive nature. We handled this difference by using hierarchies of agents and considering an agent as the representative of all agents hierarchically below it. Amongst various agent frameworks, we selected JADE [2], the Java Agent DEvelopment framework, because it is being actively developed, has a good performance, is well documented and supports multiple hosts.

In JADE, agents manage collections of behaviours. Our framework extends JADE's agent and behaviour classes: Figure 3 shows the class hierarchy for our basic holons. The logic holon is special, as it has to keep track of available services. Since JADE provides this functionality in a Directory Facilitator (DF) agent, it was natural to extend the DF agent for the logic holon.

Figure 4 shows the AgentX (Agent eXtended) framework embedded in the simulation environment of Vanderlande Industries. The framework's Gateway agent uses a standardised JLinc interface to interact with running simulations.

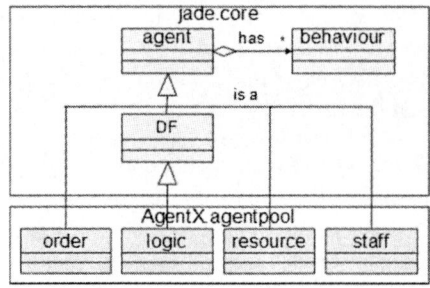

Fig. 3. Class hierarchy of the basic holons implemented in JADE

The AgentX framework has four inputs: an XML layout file, an agent library, an initialisation file and a scenarios file. The agent library contains predefined agent behaviours that extend JADE's agent and behaviour classes. The XML layout file contains a description of the warehouse resources, including their behaviours. The initialisation and scenarios files are used to run simulations. The initialisation file contains the initial storage bins and their locations. The scenario input is used to mimic customers placing specified orders at specified times.

XML Binding. Each resource specified in the XML layout file has a field tagged "agentX". Figure 5 shows an example of such a field; it contains an excerpt of an XML layout file for a picking workstation of the warehouse shown in Figure 1. The excerpt describes a picking workstation of the agent type resource with id 2.1.1. To allow other holons to use its services, the workstation has to register its service pickingWS at the logic holon with id 2.1. This service registration, which is done when the workstation holon is started, creates a hierarchy of resource and logic holons similar to the one in Figure 2.

Tasks are implemented as JADE behaviours. We distinguish two types of tasks: ones that are always active and ones that are received via JADE messages. The former type of tasks is specified with the parameter tag "behaviour", the latter type with the tag "task". The example workstation has a task pickingWS.main that is always active and a behaviour pickingWS.PickOrder that is executed when a PickOrder task has been assigned to it.

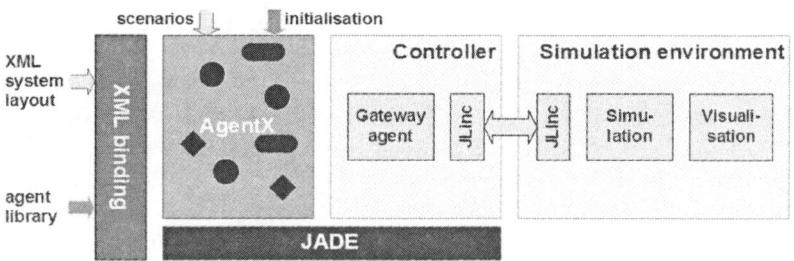

Fig. 4. AgentX framework architecture

```xml
<agentX>
  <resource>
    <parameter name="id" value="2.1.1."/>
    <parameter name="agent" value="BasicHolon.resource"/>
    <parameter name="behaviour" value="pickingWS.main"/>
    <service name="pickingWS" registerAt="2.1."/>
    <sensor name="EmergencyStop" value="2.1.1.0.2"
      behaviour="pickingWS.handleEmergencyStop"/>
    <task name="PickOrder" behaviour="pickingWS.PickOrder"/>
    <logic>
      <parameter name="agent" value="BasicHolon.logic"/>
      <task name="PickFromTote" behaviour="pickingWS.PickFromTote"/>
    </logic>
  </resource>
</agentX>
```

Fig. 5. Excerpt from an XML layout file

A task may have a process plan dividing the task into subtasks. These are specified using the tag "logic". It will cause the creation of a logic holon responsible for providing process plans. In the example in Figure 5, the behaviour pickingWS.PickFromTote provides the process plan for the task PickFromTote. This behaviour is not the process plan itself; it handles the request to provide the process plan. This allows dynamic changes of process plans.

The "sensor" tag is used to decouple sensors needed by the resource holon (EmergencyStop) and the actual sensor (2.1.1.0.2). The resource holon will subscribe to the sensor's events via the framework's Gateway agent. If needed, a behaviour (pickingWS.handleEmergencyStop) can be specified to handle those events. A similar method is used to find actuators.

4 Experimental Results

One of the first experiments we performed involves the retail warehouse shown in Figure 1. This experiment focuses on the operations of the picking workstations and uses the holarchies shown in Figure 2.

System Operation. The system resource holon receives customer orders. Fulfilling these orders involves two steps: reservation and picking. During reservation, the system resource informs storage of the items to be picked to fulfil a customer order. Storage responds by reserving storage bins holding the ordered products and sending the list of reserved bins to the system resource. This list is assigned to the picking area, which forwards it to one of its workstations. During picking, the reserved bins are transported from storage to this picking workstation, where items are picked from storage bins into order bins, after which the reserved bins are transported back to storage. The experiment ignores packing of order bins.

Figure 6 shows the running experiment; it shows storage bins travelling from the storage area via a picking workstation back to the storage area. A bin's colour indicates the customer order to which it belongs.

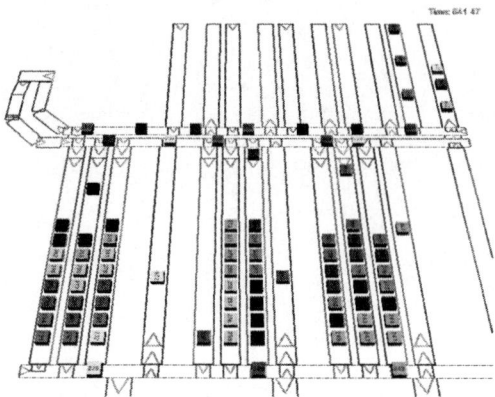

Fig. 6. Demonstration of the AgentX framework

To fulfil the customer's delivery requirements, all picks of a customer order have to be performed consecutively by a single picking workstation. The workstations' input buffers are responsible for fulfilling this requirement: they should receive storage bins in such a manner that the bins of one order can be released consecutively.

Buffer Behaviour. After a customer order has been assigned to a workstation, the input buffer receives the list of storage bins reserved for this order. The input buffer uses its logic holon to find the available buffer lanes. When there is space in one of these lanes, the input buffer builds a list containing the bins that it can receive without violating the delivery requirements. The buffer will communicate this list to its workstation. The workstation will send the list to the storage area holon, which selects the bins to come to the workstation based on the priorities specified by the workstation. The input buffer will specify the buffer lane in which these bins are to be stored.

Figure 6 shows that the picking workstations use different strategies for buffering storage bins. The middle workstation puts all bins of a customer order in the same buffer lane, whereas the other workstations spread the bins of an order over all available buffer lanes. Both strategies have their strengths and weaknesses: the former allows rush orders to overtake normal orders, whereas the latter avoids customer orders from being blocked by full buffer lanes.

The lane selection strategy is actually a behaviour of the input buffer, which can be set at start-up using the XML layout file, but which can also be adapted at runtime. A change of circumstances can necessitate a change of strategy; a message sent to the buffer holon will make it change its lane selection behaviour.

A related parameter is the number of available buffer lanes, which need not be constant. If a buffer lane breaks down, it will unregister its services at the input buffer's logic holon. The lane selection behaviour will automatically adapt to the new situation: it will act as if there are only two buffer lanes. In fact, the lane selection algorithm knows how to cope with this situation as it is designed to deal with a dynamic layout. In this way, exceptions such as breakdowns of buffer lanes are incorporated as normal behaviour.

5 Conclusion and Outlook

This paper has presented a framework that allows the automatic generation of holonic warehouse control systems from a warehouse layout file and a library of predefined behaviours. The framework has been built on top of JADE middleware and embedded in an industrial simulation environment.

The applicability of the framework was demonstrated using an experiment for an existing retail warehouse: the generated control system was capable of fulfilling the warehouse's complex delivery requirements.

The performed experiment demonstrates the adaptivity of our framework. An important source of the framework's adaptivity is the XML layout file. This file specifies hierarchies of holons and the initial behaviours of these holons. Because of the use of an XML layout file, the framework supports both changes of the hierarchies and changes of the behaviours.

The framework's adaptivity is also due to the logic holons. Since broken-down equipment will unregister its services, a holonic warehouse control system will automatically adapt itself to the new situation: the services of broken-down equipment will not be available until the equipment has been repaired.

A third way in which the framework allows the system to adapt to changing circumstances involves holons changing their behaviours for handling tasks. If a holon is informed of a change in its environment, then it can respond by selecting a behaviour from the agent library that best fits the new situation.

An important goal of the framework, which has not been considered in this paper, is the exploration of the warehouse design space. Since simulations with holonic control systems can be initiated and performed quickly, our framework can greatly reduce simulation effort and therefore allow many design alternatives to be analysed. We plan to study the usefulness of our framework for the design of a warehouse, in which shuttles handle the transport of bins. We intend to exploit the hybrid nature of holonic control systems by comparing a range of control systems varying from fully centralised to fully decentralised systems.

Acknowledgments. This work has been carried out as part of the FALCON project under the responsibility of the Embedded Systems Institute with Vanderlande Industries as the industrial partner. This project is partially supported by the Netherlands Ministry of Economic Affairs under the Embedded Systems Institute (BSIK03021) program.

References

1. Angel, B.W.F., van Damme, D.A., Ivanovskaia, A., Lenders, R.J.M., Veldhuijzen, R.S.: Warehousing space in Europe: meeting tomorrow's demand. Capgemini (2006)
2. Bellifemine, F., Caire, G., Greenwood, D.: Developing Multi-Agent Systems with JADE. John Wiley & Sons Ltd., Chichester (2007)
3. Black, G., Vyatkin, V.: On practical implementation of holonic control principles in baggage handling systems using IEC 61499. In: Mařík, V., Vyatkin, V., Colombo, A.W. (eds.) HoloMAS 2007. LNCS (LNAI), vol. 4659, pp. 137–148. Springer, Heidelberg (2007)

4. Blanc, P., Demongodin, I., Castagna, P.: A holonic approach for manufacturing execution system design: An industrial application. Engineering Applications of Artificial Intelligence 21, 315–330 (2008)
5. Bürckert, H.-J., Fischer, K., Vierke, G.: Holonic transport scheduling with TeleTruck. Applied Artificial Intelligence 14, 697–725 (2000)
6. Fleetwood, M., Kotak, D.B., Wu, S., Tamoto, H.: Holonic System Architecture for Scalable Infrastructures. In: Proceedings of the 2003 IEEE International Conference on Systems, Man and Cybernetics, pp. 1469–1474 (2003)
7. Fletcher, M.: Holonic Manufacturing Systems: Some Scenarios and Issues. In: Proceedings of the Workshop Intelligent Methods for Quality Improvement in Industrial Practice (2002)
8. Giret, A., Botti, V.: Holons and agents. Journal of Intelligent Manufacturing 15, 645–659 (2004)
9. Giret, A., Botti, V.: From system requirements to holonic manufacturing system analysis. International Journal of Production Research 44, 3917–3928 (2006)
10. Kachornvitaya-Wan, V.: Complex Integrated Order Fulfillment Technology: A Holonic Approach. PhD Thesis, Dartmouth College (2006)
11. Kim, B.I.: Intelligent Agent Based Planning, Scheduling, and Control: Warehouse Management Application. PhD Thesis, Rensselaer Polytechnic Institute (2002)
12. Leitaõ, P.J.P.: An Agile and Adaptive Holonic Architecture for Manufacturing Control. Dissertation, University of Porto (2004)
13. Leitaõ, P., Restivo, F.: ADACOR: A holonic architecture for agile and adaptive manufacturing control. Computers in Industry 57, 121–130 (2006)
14. López Orozco, O.J., Martinez Lastra, J.L.: Distributed director facilitator in a multiagent platform for networked embedded controllers. In: Mařík, V., Vyatkin, V., Colombo, A.W. (eds.) HoloMAS 2007. LNCS(LNAI), vol. 4659, pp. 137–148. Springer, Heidelberg (2007)
15. Van Brussel, H., Wyns, J., Valckenaers, P., Bongaerts, L., Peeters, P.: Reference Architecture for Holonic Manufacturing Systems: PROSA. Computers in Industry 37, 255–276 (1998)
16. Wyns, J.: Reference Architecture for Holonic Manufacturing Systems - the key to support evolution and reconfiguration. PhD Thesis, KU Leuven (1999)

Developing Home Care Intelligent Environments: From Theory to Practice

J.A. Fraile Nieto[1], M.E. Beato Gutiérrez[1], and B. Pérez Lancho[2]

[1] Pontifical University of Salamanca, c/ Compañía 5, 37002 Salamanca, Spain
{jafraileni,ebeatogu}@upsa.es
[2] University of Salamanca, Plaza de la Merced s/n, 37008 Salamanca, Spain
lancho@usal.es

Abstract. One of the main aims of the pervasive systems is to be able to adapt themselves in execution time to the changes in the number of resources available, the mobility of the users, variability in the needs of the users and failures of the system. This work presents HoCa, a multi-agent based architecture designed to facilitate the development of pervasive systems. HoCa presents a new model where multi-agent systems and service oriented architectures are integrated to facilitate compatible services. HoCa has been applied to case study in a real scenario, aimed to provide automatic assistance to dependent people at their home, and the results obtained are presented in this paper.

Keywords: Dependent environments, Pervasive Systems, Multiagent Systems, Home Care.

1 Introduction

Nowadays, there is a considerable growth in the development of automation technologies as well as intelligent environments as demonstrated by the relevance acquired by the Pervasive Systems. One of the main objectives of these systems is to look after the user's well-being at their home, at work, etc. [2] [16]. Pervasive systems demand their integration with computational applications in a non intrusive way, facilitating intelligent interfaces characterized to be natural, simple and transparent [16]. There are several benefits provided by Pervasive System [12]: users get easier access to services that are contracted, access to services is independent of the terminal which is used and use of services is simpler, and allowing a rapid assimilation by the user. In addition, users can receive personalized services entirely, so they have quick access to what they call their personal needs.

Home Care requires the improvement of the services offered to the users as well as the way they can be accessed. Moreover, it is necessary to adopt the trends already tested and proven in technological environments [1]. Intelligent environments are focused on the user, since the user is the centre of the new technological facilities and demands access to unified services [2]. In this sense, multi-agent systems can facilitate the development of pervasive home care environments.

The importance acquired by the dependency people sector has dramatically increased the need for new home care solutions [6]. Besides, the commitments that have been acquired to meet the needs of this sector, suggest that it is necessary to modernize the current systems. Multiagent systems [16], and intelligent devices-based architectures have been recently explored as supervisor systems for health care scenarios [1] for elderly people and for Alzheimer patients [6]. These systems allow providing constant care in the daily life of dependent patients [5], predicting potentially dangerous situations and facilitating a cognitive and physical support for the dependent patient [2]. Taken into account these solutions, it is possible to think that multi-agent systems facilitate the design and development of pervasive environments [7] and improve the services currently available, incorporating new functionalities. Multi-agent systems add a high level of abstraction regarding to the traditional distributed computing solutions.

The main objective of this paper is to define a hybrid Multi-Agent architecture, HoCa, for the control and supervision of pervasive environments [7] [8]. The architecture provides innovative mechanisms to integrate multi-agent systems with service oriented architectures and intelligent interfaces to obtain context-aware information. The architecture incorporates technologies for automatic identification, location, alarms management and movement tracking. These technologies facilitate the monitoring and management of dependent patients at their home in a ubiquitous way. One of the main contributions of the HoCa architecture is the use of both reactive and deliberative BDI agents [7], specialized in the interaction with sensors and the distribution of complex tasks respectively. Moreover an alert system based on SMS and MMS technologies allows a quick response to dangerous situations requiring medical attention. Finally, advanced location technologies based on RFID (Radio Frequency IDentification) [13] and video surveillance provide location and tracking of patients, and Java Card [15] technology allows automatic identification.

The rest of the paper is structured as follows: Section 2 describes the proposed architecture and in Section 3 the preliminary results obtained is presented and the conclusions are discussed.

2 HoCa Architecture

The HoCa multi-agent architecture uses a series of components to offer a solution that includes all levels of service for various systems. It accomplishes this by incorporating intelligent agents, identification and localization technology, wireless networks and mobile devices. Additionally, it provides access mechanisms to multi-agent system services, through mobile devices, such as mobiles phones or PDA. The architecture integrates two types of agents, each of which behaves differently for specific tasks. The first group of agents is made up of deliberative BDI agents, which are in charge of the management and coordination of all system applications and services. The second group of agents is made up of reactive agents responsible for handling information and offer services in run time. The communications between agents within the platforms follows the FIPA ACL (Agent Communication Language) standard. The protocol for communication between agents and services is based on the SOAP standard. Access is provided via Wi-Fi wireless networks, a notification and alarm management module based on SMS and MMS technologies, and user identification and localization system based on Java Card and RFID technologies. This system is dynamic, flexible, robust and very adaptable to changes of context. For all these reasons the proposed architecture is very appropriate for pervasive environments. Pervasive systems are characterized by the complexity of its service, the number of devices and software components that manage and obligation to facilitate cooperation between all these elements. The HoCa architecture facilitates the integration and management of all devices integrated into a pervasive system. All this makes it an open system, easy to integrate into complex environments that it does not dependent of a specific platform.

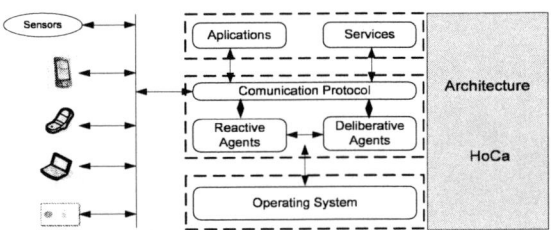

Fig. 1. HoCa Framework

HoCa architecture describes four basic blocks that can be seen in Figure 1 Applications, Services, Agents Platform and Communication Protocol. These blocks constitute the whole functionality of the architecture. HoCa allows the use of devices such as PDAs, mobile phones, laptops, sensors or Java Card chip with RFID technology. The agent's platform consists of deliberative agents and reactive agents. Reactive agents are responsible for monitoring sensors and actuators and deliberative agents are responsible for monitoring services and applications architecture. Reactive agents perform actions that you ask the agents deliberative and also provide information to agents deliberative. The agent platform is installed on the operating system of the architecture. This union is the architecture basis. The agents control the security, communications and each of the features offered HoCa. This architecture design allows us to be able to add new agents at any time with new features to the system. It follows mounting flexible and adaptable architecture to dynamic pervasive environments.

In the following subsections are the four basic building blocks of architecture and localization, identification and alerts systems that incorporate HoCa. In subsection 2.1 describes the agent's platform and the agent's types that interact in HoCa. In subsection 2.2 explains the communication protocol. Subsection 2.3 describes the location and identification system that is integrated into HoCa and subsection 2.4 describes the alerts system.

2.1 Agents

The Agents platform is the core of the architecture and integrates two types of agents, each of which behaves differently for specific tasks, as shown in Figure 2. The first group of agents is made up of deliberative BDI agents, who are in charge of the management and coordination of all system applications and services [6]. These agents are able to modify their behaviour according to the preferences and knowledge acquired in previous experiences, thus making them capable of choosing the best solution [4].

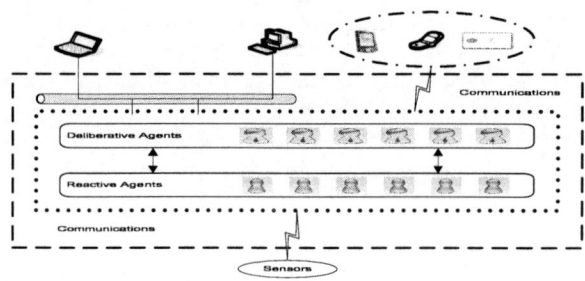

Fig. 2. Agents platform structure in the HoCa architecture

Deliberative agents constantly deal with information and knowledge [9]. Because they can be executed on mobile devices, they are always available and they provide ubiquitous access for the users [3]. There are different kinds of agents in the architecture, each one with specific roles, capabilities and characteristics. This fact facilitates the flexibility of the architecture to incorporate new agents.

However, there are pre-defined agents which provide the basic functionalities of the architecture:

- CoAp Agent: This agent is responsible for all communications between applications and the platform.
- CoSe Agent: It is responsible for all communications between services and the platform.
- Directory Agent. Manages the list of services that can be used by the system.
- Supervisor Agent. This agent supervises the correct functioning of the agents in the system.
- Security Agent. This agent analyzes the structure and syntax of all incoming and outgoing XML messages.
- Manager Agent. Decides which agent must be called taking into account the QoS and users preferences.
- Interface Agent. This kind of agent has been designed to be embedded in users' applications. Interface agents communicate directly with the agents in HoCa so there is no need to employ the communication protocol, but FIPA ACL specification.

The figure 3 along with a simple example helps to understand the communication between different types of agents in the architecture. A patient is visited by the medical service due to a feverish that suffers by an infection. The medical service went to the house before the patient's explicit request made through the alerts system. The patient through an application has inserted the alert in the system. The CoAp agent is responsible for registering this information into the system and notifies the supervisor agent. The security agent confirmed the credentials of the user who enters the information and validates the information entered. The supervisor agent at the same time performs two tasks, requests the directory agent you select the service to run to launch the alert and through the interface agent informs the manager agent of operations performed. The CoSe agent runs the service that launches the alert and finally the alert is sent

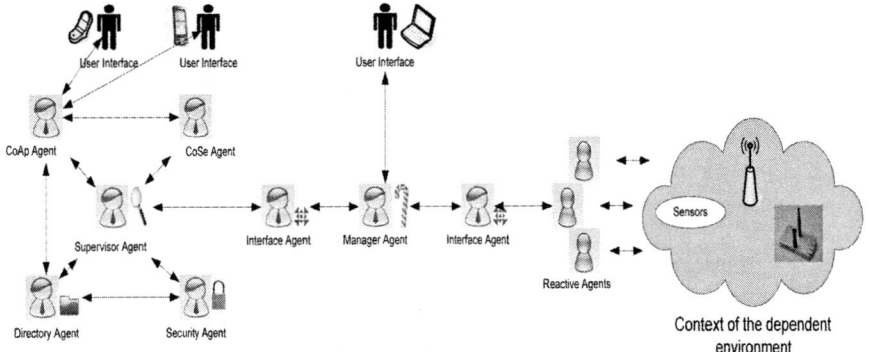

Fig. 3. Agents Workflow in the HoCa architecture

through the reactive agents of the architecture. At all times the manager agent is informed of the steps being taken in the system and is responsible for validating the alert sending through the interface agent to the corresponding reactive agent. Once the patient enters the information into the system, this process seems very laborious and slow is running in a few thousandths of a second.

The second group is made up of reactive agents. Most of the research conducted within the field of multi-agent pervasive systems focuses on designing architectures that incorporate complicated negotiation schemes as well as high level task resolution, but don't focus on temporal restrictions. In general, the multi-agent architectures assume a reliable channel of communication and, while some establish deadlines for the interaction processes, they don't provide solutions for limiting the time the system may take to react to events. It is possible to define a run-time agent as an agent with temporal restrictions for some of its responsibilities or tasks [11]. The use of run-time multi-agent system makes sense within an environment of temporal restrictions, where the system can be controlled by autonomous agents that need to communicate among themselves in order to improve the degree of system task completion. In this kind of environments every agent requires autonomy as well as certain cooperation skills to achieve a common goal.

2.2 HoCa Communication Protocol

Communication protocol allows applications, services and sensors to be connected directly to the platform agents. The protocol presented in this work is open and independent of programming languages. It is based on the SOAP (Simple Object Access Protocol) standard and allows messages to be exchanged between applications and services as shown in Figure 4.

SOAP is a standard protocol that defines how two objects in different processes can communicate through XML data exchange. For example, here are displayed as a HoCa user since the supervisor application, which can run on a PDA asks the agent CoAp (see Figure 3) the patient location in his home. The application requests the patient location with a SOAP message and the CoAp agent when has the patient location

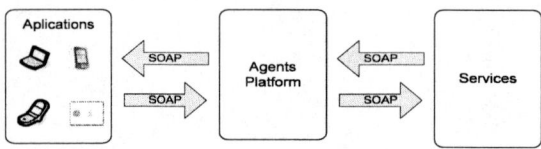

Fig. 4. Communication using SOAP messages in HoCa

gets to communicate with the application back in another SOAP message the information requested.

The communications between agents within the platforms follows the FIPA ACL (Agent Communication Language) standard. This way, the applications can use the platform to communicate directly with the agents. The agent's messages structure are key-value row. These rows are written in an agent's communications language as FIPA ACL. The messages include the names of the sender and the receiver and may contain other messages recursively. Moreover defining protocols for high-level interaction between the agents, called talks and it is possible to define new primitives from a core of primitive by composition. You can see a message ACL in Figure 5.

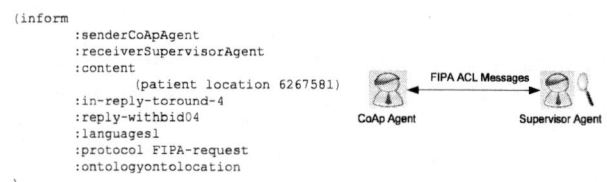

Fig. 5. Example of ACL message between two agents in HoCa

In Figure 5 shows the communication between two agents of the HoCa architecture. In this case the example of ACL message corresponds to the request for location of a patient seen in the example of SOAP messages. CoAp agent sends an ACL message to supervisor agent asking the location of a patient. Talks between agents tend to follow a certain pattern, typical sequence of messages called conversation protocols. An agent informs the protocol you want to use it with parameter ':protocol'. Basic protocols defined by FIPA are: (i) FIPA-request, (ii) FIPA-query, (iii) FIPA-request-when, (iv) FIPA-contract-net, (v) FIPA-iteraterated-contract-net and (vi) FIPA-auction-english.

2.3 Identification and Location Systems in the HoCa Architecture

This system incorporates Java Card and RFID technologies. The primary purpose of the system is to convey the identity of an object or person, as with a unique serial number, using radio waves. Java Card is a technology that permits small Java applications (applets) to be run safely in microchip smart cards and similar embedded devices.

The radio frequency identification (RFID) [10] is a wireless communication technology used to identify and receive information on humans, animals or objects in motion. The main applications of RFID technology have occurred in industrial environment, transport, and other sectors, including medicine, are becoming increasingly important [10]. RFID provides more information than other auto-identification technologies, speeds up processes without losing reliability, and requires no human intervention. The combination of these two technologies allows us to both identify the user or identifiable element, and to locate it, by means of sensors and actuators, within the environment, at which time we can act on it and provide services. The microchip, which contains the identification data of the object to which it is adhered, generates a radio frequency signal with this data. The signal can be picked up by an RFID reader, which is responsible for reading the information and sending it, in digital format, to the specific application.

2.4 Alert System in HoCa

The alert system is integrated into the HoCa architecture and uses mobile technology to inform users about alerts, warnings and information specific to the daily routine of the application environment. It gets so improve service quality of communication and control at all times the performance of applications that are implemented in architecture. This is a very configurable system that allows users to select the type of information they are interested, and to receive it immediately on their mobile phone or PDA.

The system is proactive, that is, users should not bother to monitor the environment in which the service is implemented, to see if there is information that interests them. It alerts the system itself which cares for users to get information immediately on their mobile devices, with the benefits that this entails, so they can get all this information without having to meet at her workplace.

3 Results and Conclusions

HoCa has been used to develop a multi-agent system for monitoring dependent patients at home. The main features of this system include reasoning and planning mechanisms, and alert and response management. Most of these responses are reactions in run time to certain stimuli, and represent the abilities that the reactive agents have in the HoCa architecture based platform. The technologies used to test the system include mobile technology for managing service alerts through PDA and mobile phones, and Java Card and RFID technology for identification and access control. HoCa improved security at home to dependents because the monitors, alerting relatives or other services to emergency situations and ensures they are properly. HoCa architectures proposes a model that goes a step further in designing systems for home care, providing generality and offering features that make it easily adaptable to any pervasive environment.

One of the main contributions of the HoCa architecture is the alert system. We implemented several test cases to evaluate the management of alerts integrated into the system. This allowed us to determine the response time for warnings generated by the users, for which the results were very satisfactory, with response times shorter than those obtained prior to the implementation of HoCa. The system studies the information collected, and applies a reasoning process which allows alerts to be automatically generated. For these alerts, the system does not only take response time into account,

Table 1. Comparison between the HoCa and the ALZ-MAS architectures

Factor	HoCa	ALZ-MAS
Average Response Time to Incidents(min.)	4 minutes	7 minutes
Assisted Incidents	12	17
Average number of daily planned tasks	12	10
Average number of services completed daily	46	32
Time employed by the medical staff to attend to an alert (min.)	25 minutes	35 minutes

but also the time elapsed between alerts, and the user's profile and reliability, in order to generalize reactions to common situations.

Table 1 presents the results obtained after comparing the HoCa architecture to the previously developed ALZ-MAS architecture [7] in a case study on medical care for patients at home. The ALZ-MAS architecture allows the monitoring of patients in geriatric residences, but home care is carried out through traditional methods. The case study presented in this work consisted of analysing the functioning of both architectures in a test environment. The HoCa architecture was implemented in the home of 5 patients and was tested for 30 days. The results were very promising.

The data shown in Table 1 are the results obtained from the test cases. They show that the alert system improved the communication between the user and the dependent care services providers, whose work performance improved, allowing them to avoid unnecessary movement such as travels and visits simply oriented to control or supervise the patient. The user identification and location system in conjunction with the alert system has helped to notably reduce the percentage of incidents in the environment under study. Moreover, in addition to a reduction in the number of incidents, the time elapsed between the generation of a warning and solution decreased significantly. Finally, due to the many improvements, the level of user satisfaction increased with the introduction of HoCa architecture since patients can live in their own homes with the same level of care as those offered at the residence.

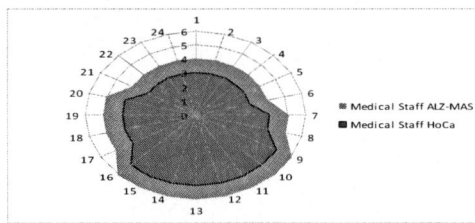

Fig. 6. Comparison of medical staff between the HoCa and the ALZ-MAS architectures

Figure 6 shows the comparison of the average medical personnel required in the two studies made with ALZ-MAS and HoCa. Figure 6 represents the number of people who are in need each hour of the day. The graph seen as the medical personnel necessary for both systems is lower during the hours of lesser activity of the patients. On the contrary during the hours of the morning and afternoon will need more health staff. It also looks like the HoCa system does not need as personal as ALZ-MAS.

As the result HoCa architecture creates an environment that facilitates intelligent and distributed and provides services to dependents at home. Automating tasks and patient monitoring improve the system security and efficiency of care to dependents. The use of RFID technology, JavaCard and mobiles with people provides a high level of interaction between users and patients through the system and is essential in building an intelligent environment. Moreover the good use of mobile devices can facilitate social interactions and knowledge transfer.

References

1. Angulo, C., Tellez, R.: Distributed Intelligence for smart home appliances. Tendencias de la minería de datos en España. Red Española de Minería de Datos. Barcelona, España (2004)
2. Augusto, J.C., McCullagh, P.: Ambient Intelligence: Concepts and Applications. Invited Paper by the International Journal on Computer Science and Information Systems 4(1), 1–28 (2007)
3. Bajo, J., et al.: Hybrid architecture for a reasoning planner agent. In: Apolloni, B., Howlett, R.J., Jain, L. (eds.) KES 2007, Part II. LNCS (LNAI), vol. 4693, pp. 461–468. Springer, Heidelberg (2007)
4. Bratman, M.E., Israel, D., Pollack, M.: Plans and resource-bounded practical reasoning. Computational Intelligence 4, 349–355 (1988)
5. Carrascosa, C., Bajo, J., Julian, V., Corchado, J.M., Botti, V.: Hybrid multi-agent architecture as a real-time problem-solving model. In: Expert Systems With Applications, vol. 34(1), pp. 2–17. Elsevier, Amsterdam (2008)
6. Corchado, J.M., y Laza, R.: Constructing Deliberative Agents with Case-based Reasoning Technology. International Journal of Intelligent Systems 18, 1227–1241 (2003)
7. Corchado, J.M., Bajo, J., de Paz, Y., Tapia, D.: Intelligent Environment for Monitoring Alzheimer Patients, Agent Technology for Health Care. Decision Support Systems 34(2), 382–396 (2008)
8. Corchado, J.M., Bajo, J., Abraham, A.: GERAmI: Improving the delivery of health care. IEEE Intelligent Systems (2008); Special Issue on Ambient Intelligence (March/April 2008)
9. Corchado, J.M., Gonzalez-Bedia, M., De Paz, Y., Bajo, J., y De Paz, J.F.: Replanning mechanism for deliberative agents in dynamic changing environments. Computational Intelligence 24(2), 77–107 (2008)
10. ITAA (2004), Radio Frequency Identification. RFID coming of age. Information Technology Association of America, http://www.itaa.org/rfid/docs/rfid.pdf
11. Jayaputera, G.T., Zaslavsky, A.B., Loke, S.W.: Enabling run-time composition and support for heterogeneous pervasive multi-agent systems. Journal of Systems and Software 80(12), 2039–2062 (2007)
12. Ramos, C., Augusto, J., Shapiro, D.: Special Issue on Ambient Intelligence. IEEE Intelligent Systems 23(2), 15–18 (2008)
13. Segarra, M.T., Thepaut, A., Keryell, R., Poichet, J., Plazaola, B., Peccatte, B.: Ametsa: Generic Home Control System Based on UPnP. In: Independent living for persons with disabilities and elderly people, pp. 73–80. IOS Press, Amsterdam (2003)
14. Tapia, D.I., Bajo, J., De Paz, F., Corchado, J.M.: Hybrid Multiagent System for Alzheimer Health Care. In: Rezende, S.O., da Silva Filho, A.C.R. (eds.) Proceedings of HAIS 2006. Ribeirao Preto, Brasil (2006)
15. Want, R.: You Are Your Cell Phone. IEEE Pervasive Computing 7(2), 2–4 (2008)
16. Weiser, M.: The Computer for the Twenty-First Century. Scientific American 265, 94–104 (1991)

Distributing Functionalities in a SOA-Based Multi-agent Architecture

Dante I. Tapia, Javier Bajo, and Juan M. Corchado

Departamento Informática y Automática
Universidad de Salamanca
Plaza de la Merced s/n, 37008, Salamanca, Spain
{dantetapia, jbajope, corchado}@usal.es

Abstract. This paper presents how functionalities are distributed my means of FUSION@, a SOA-based multi-agent architecture. FUSION@ introduces a new perspective for constructing multiagent systems, facilitating the integration with service-oriented architectures, and the agents act as coordinators and administrators of services. FUSION@ makes use of several mechanisms for managing and optimizing the services distribution. The results obtained demonstrate that FUSION@ can efficiently distribute functionalities in dynamic scenarios at execution time.

Keywords: Multi-Agent Systems, Services Oriented Architectures, Distributed Computing.

1 Introduction

Multi-agent systems are very appropriate for resolving problems in a distributed way [9]. Agents have a set of characteristics, such as autonomy, reasoning, reactivity, social abilities, pro-activity, mobility, organization, etc. which allow them to cover several needs for dynamic environments, especially ubiquitous communication and computing and adaptable interfaces. Agent and multi-agent systems have been successfully applied to several scenarios, such as education, culture, entertainment, medicine, robotics, etc. [6], [15]. Moreover, the continuous advancement in mobile computing makes it possible to obtain information about the context and also to react physically to it in more innovative ways [9]. Nevertheless, complex systems need higher adaptation, learning and autonomy levels than pure BDI model [3]. This can be achieved by modelling the agents' characteristics [21] to provide them with mechanisms that allow solving complex problems and autonomous learning [7]. However, excessive complex mechanisms (e.g. data mining, genetic algorithms, indexing, learning techniques, visualization, etc.) can cause malfunctioning and crashes in multi-agent systems developed with current technologies such as agent platforms. This is because developers tend to integrate all functionalities inside the agents' internal structure, creating agents with high computational requirements.

 The *Flexible User and ServIces Oriented multi-ageNt Architecture* (FUSION@) [18] tries to solve this problem. One of the most important characteristics in FUSION@ is the use of intelligent agents as the main components in employing a service oriented approach, focusing on distributing the majority of the systems' functionalities into remote and local services and applications. The architecture proposes a new and easier method of building distributed multi-agent systems, where the functionalities of the systems are not integrated into the structure of the agents; rather they are modelled as distributed services and applications which are invoked by the agents acting as controllers and coordinators. This approach optimizes usability and performance because it can be obtained lighter agents in terms of computational load.

FUSION@ has been previously presented in related papers [18], however, this paper focuses on describing how FUSION@ implements several mechanisms for optimizing and managing functionalities and resources in dynamic environments. Through these mechanisms, FUSION@ facilitates the development of distributed multi-agent systems and the agents have the ability to dynamically adapt their behaviour at execution time. FUSION@ provides an advanced flexibility and customization to easily add, modify or remove applications or services on demand, independently of the programming language. It also formalizes the integration of applications, services, communications and agents. The proposed approach has been applied to a real scenario to evaluate the performance in a multi-agent system for monitoring Alzheimer patients.

In the next section, the specific problem description that essentially motivated this research is presented. Section 3 describes the main characteristics of the FUSION@ architecture and briefly explains the mechanisms for distributing and optimizing functionalities and resources. Section 4 presents the results and conclusions obtained after testing the architecture in a real scenario.

2 Problem Description and Background

Excessive centralization of services negatively affects the systems' functionalities, overcharging or limiting their capabilities. Classical functional architectures are characterized by trying to find modularity and a structure oriented to the system itself. Modern functional architectures like Service-Oriented Architecture (SOA) consider integration and performance aspects that must be taken into account when functionalities are created outside the system. These architectures are aimed at the interoperability between different systems, distribution of resources, and the lack of dependency of programming languages [5]. Services are linked by means of standard communication protocols that must be used by applications in order to share resources in the services network [1]. The compatibility and management of messages that the services generate to provide their functionalities is an important and complex element in any of these approaches.

One of the most prevalent alternatives to these architectures is agents and multi-agent systems technology which can help to distribute resources and reduce the central unit tasks [1] [19]. A distributed agents-based architecture provides more flexible ways to move functions to where actions are needed, thus obtaining better responses at execution time, autonomy, services continuity, and superior levels of flexibility and scalability than centralized architectures [4]. Additionally, the programming effort is reduced because it is only necessary to specify global objectives so that agents cooperate in solving problems and reaching specific goals, thus giving the systems the ability to generate knowledge and experience.

Agent and multi-agent systems combine classical and modern functional architecture aspects. Multi-agent systems are structured by taking into account the modularity in the system, and by reuse, integration and performance. Nevertheless, integration is not always achieved because of the incompatibility among the agents' platforms (e.g. JADE agents with RETSINA or OAA agents). The integration and interoperability of agents and multi-agent systems with SOA and Web Services approaches has been recently explored [1] [9]. Some developments are centred on communication between these models, while others are centred on the integration of distributed services, especially Web Services, into the structure of the agents [11] [14] [16]. Although these developments provide an adequate background for developing distributed multi-agent systems integrating a service oriented approach, most of them are in early stages of development, with little systems (if any) developed upon them so it is not possible to actually know their potential in real scenarios. Please refer to [18] for a detailed comparison of these approaches.

3 FUSION@, A SOA-Based Multi-agent Architecture

FUSION@ [18] is a modular multi-agent architecture, where services and applications are managed and controlled by deliberative BDI (Belief, Desire, Intention) agents, [3], [13], which mainly follows the principles of the THOMAS architecture [12]. Deliberative BDI agents are able to cooperate, propose solutions on very dynamic environments, and face real problems, even when they have a limited description of the problem and few resources available [2], [8]. There are different kinds of agents in FUSION@, each one with specific roles, capabilities and characteristics. This fact facilitates the flexibility of the architecture in incorporating new agents. FUSION@ defines four basic blocks which provide all the functionalities of the architecture.

- Applications. Represent all the programs that can be used to exploit the system functionalities. They can be executed locally or remotely, even on mobile devices with limited processing capabilities, because computing tasks are largely delegated to the agents and services.
- Agents Platform. This is the core of FUSION@, integrating a set of agents, each one with special characteristics and behaviour. An important feature in this architecture is that the agents act as controllers and administrators for all applications and services, managing the adequate functioning of the system, from services, applications, communication and performance to reasoning and decision-making.
- Services. They are the bulk of the functionalities of the system at the processing, delivery and information acquisition levels. Services are designed to be invoked locally or remotely. Services can be organized as local services, web services, or even as individual stand alone services.
- Communication Protocol. This allows applications and services to communicate directly with the agents' platform. The protocol is completely open and independent of any programming language. This protocol is based on SOAP specification to capture all messages between the platform and the services and applications [5]. All external communications follow the same protocol, while the communication among agents in the platform follows the FIPA Agent Communication Language

(ACL) specification. Applications can make use of agents platforms to communicate directly (using FIPA ACL specification) with the agents in FUSION@, so while the communication protocol is not needed in all instances, it is absolutely required for all services.

There are pre-defined agents that provide the basic functionalities of FUSION@. *CommApp Agent* is the agent is responsible for all communications between applications and the platform. It manages the incoming requests from the applications to be processed by services. It also manages responses from services (via the platform) to applications. All messages are sent to *Security Agent* for their structure and syntax to be analyzed. *CommServ Agent* is responsible for all communications between services and the platform. The functionalities are similar to *CommApp Agent* but backwards. *Admin Agent* signals to *CommServ Agent* which service must be invoked. All messages are sent to *Security Agent* for their structure and syntax to be analyzed. This agent also periodically checks the status of all services to know if they are idle, busy, or crashed. *Directory Agent* manages the list of services that can be used by the system. For security reasons [17], FUSION@ does not include a service discovery mechanism, so applications must use only the services listed in the platform. However, services can be added, erased or modified dynamically. There is information that is constantly being modified: the service performance (average time to respond to requests), the number of executions, and the quality of the service (QoS). This last data is very important, as it assigns a value between 0 and 1 to all services. All new services have a quality of service (QoS) value set to 1. This value decreases when the service fails (e.g. service crashes, no service found, etc.) or has a subpar performance compared to similar past executions. QoS is increased each time the service efficiently processes the tasks assigned. *Supervisor Agent* supervises the correct functioning of the other agents in the system. *Supervisor Agent* periodically verifies the status of all agents registered in the architecture by sending ping messages. If there is no response, the Supervisor agent kills the agent and creates another instance of that agent. *Security Agent* analyzes the structure and syntax of all incoming and outgoing messages. If a message is not correct, the *Security Agent* informs the corresponding agent (*CommApp* or *CommServ*) that the message cannot be delivered. This agent also directs the problem to the *Directory Agent*, which modifies the QoS of the service where the message was sent. *Admin Agent* decides which agent must be called by taking into account the QoS and users' preferences. Users can explicitly invoke a service, or can let the *Admin Agent* decide which service is best to accomplish the requested task. This agent also checks if services are working properly. It requests the *CommServ Agent* to send ping messages to each service on a regular basis. If a service does not respond, CommServ informs *Admin Agent*, which tries to find an alternate service, and informs the *Directory Agent* to modify the respective QoS. *Interface Agent* was designed to be embedded in users' applications. *Interface agents* communicate directly with the agents in FUSION@ so there is no need to employ the communication protocol, rather the FIPA ACL specification. The requests are sent directly to the *Security Agent*, which analyzes the requests and sends them to the *Admin Agent*. These agents must be simple enough to allow them to be executed on mobile devices, such as cell phones or PDAs. All high demand processes must be delegated to services.

3.1 Services Management

FUSION@, and in particular the *Admin Agent*, employs a mechanism composed of a set of techniques that allows the architecture to select the most appropriate service to meet a request at any given time. The mechanism to assign the most appropriate service to respond to this request begins when a new request is received, taking into account the following parameters: the QoS value; the user preferences; and the estimated delivery time. The first two parameters are set in advance so it is not necessary to calculate by the *Admin Agent*. The execution time is estimated using a RBF (Radial Basis Function) Neural Network. The reason for using this type of network is the speed in the training phase, compared to a Multi Layer Perceptron (MLP). The neural network is made up of three layers: the input layer, an intermediate/hidden layer and an output layer. The number of neurons in the input layer is defined by the number of entries (i.e. parameters) to each service. In the case of arrays, each element counts as an entry to the service. The number of neurons in the middle layer is determined dynamically, making a cross-training and validation. The cross-validation is done through the method GCV (Generalized Cross-Validation) [19]. The variation in the number of neurons in the middle layer is made by the following algorithm.

First are initialized the list of errors $e = \{\}$ and the list of angles $\alpha = \{\}$. Then the training is conducted with a neuron in the middle layer and the training value is stored in e_1 in the values list $e = e \cup e_1$. Next, the number of neurons in the intermediate layer is initialized. If there is no prior training, it is initialized to $4n+1$ where n is the number of neurons in the input layer. However, if there is a prior training, it is initialized to $2n+1$, where n is the number of neurons of the prior training. r is the number of neurons in the current layer. The training is conducted for r intermediate neurons and the error in e_r $e = e \cup e_r$ is stored. Subsequently, the training is done for $r/2$ neurons and the error in $e_{r/2}$ $e = e \cup e_{r/2}$ is stored. The lines that pass through the points $r_{1,r/2} = (e_r e_{r/2})$ and $r_{1,r/2} = (e_1 e_{r/2})$ are calculated and the angle $\alpha_{r/2}$ formed by these lines is obtained. The angle $\alpha_{r/2}$ is introduced in the list of angles $\alpha = \alpha \cup_{r/2}$. If there is only one angle on the list of angles $\#\alpha = 1$ then the value of r is set to $r/4$ and the training is restarted. If $\#\alpha = 2$, being α_i the other existent value, then: If $i<r/2$ then $r = r/2$ and the training is restarted; else $r = r/2+r/4$ and the training is restarted. But selecting the two adjacent values to the left and right of $\alpha_{r/2}$ denoted by α_i and α_j so that:

- If $\alpha_i > \alpha_j$
    i. If $j=r/2+1$ the value of neurons is set to r
    ii. Else, set the new value of $r=r/2+(j-r/2)/2$
- Else
    i. If $i=r/2-1$ the value of neurons is set to r
    ii. Else, set the new value of $r=r/2-(r/2-i)/2$

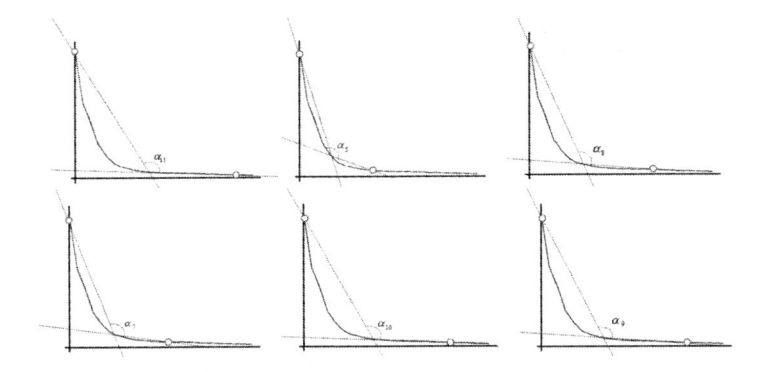

Fig. 1. Progression of the number of neurons in the hidden layer

Finally, the training is carried out and the value of neurons in the intermediate layer is set to r. Figure 1 shows the progress of the algorithm searching the lower angle α. The number of neurons in the middle layer is shown in the X axis and the cross validation is shown in the Y axis. The values correspond to a simulation, but for a real case there is only necessary to calculate the values that are obtained in the sets listed next. The sets of values studied at every moment are (23, 11, 1) (11, 5, 1) (17, 8, 1) (14, 7, 1) (20, 10, 1) (18, 9, 1). The end result is given by (18, 9, 1) therefore, the number of neurons in the final layer is 9. The output layer consists of a single neuron, whose output value is the time prediction.

Learning consists of an unsupervised learning phase for the middle layer and other supervised learning phase for the output layer. A cross-validation is carried out at the training phase by varying the number of neurons in the intermediate layer in case of no new results obtained. First, the training of the middle layer and the output layer are carried out. The retraining of the network is carried out when the average time is k times larger or smaller than the estimated time for the last n executions. These values are defined in advance for each system. A cross-validation is done through the GVC method in order to determine the completion of the training.

The available services can be assigned once the execution time of the requests in the queue is estimated. The assignment tries to maximize the performance in terms of the time required to respond to all requests. This assignment must be efficient and dynamically adaptable because it must be carried out in execution time. For this reason, a FIFO model has been chosen for each service. Each service has a queue associated. There is a common queue where all requests are queued and managed by the *Admin Agent*. The allocation to the queue of each service is done taking into account the requests assigned to each queue and the estimated performance given by (1). The request is assigned to the queue in order to minimize the cumulative performance and the estimation time.

$$r_i = (1 - QoS_i) \cdot t \cdot (1 - p_i) \qquad (1)$$

Where: r_i is the estimated performance of the service i; QoS_i is the quality of service i which corresponds to a value between 0.00 and 1.00; T is the estimated time of duration; and P_i are the preferences for the service i with a value between 0 and 1. Being n

the number of services, R_i the cumulative performance for the queue i, a new request s with performance estimations for the service j r_{ij} will be assigned to the queue R_i when satisfying the following condition. If the value of the preference is equal to 1, only the queues with that value for the preference are taken into account.

$$\min\{R_1 + r_{i1}, ..., R_n + r_{in}\} \tag{2}$$

Assuming that it is assigned to the k queue, the new cumulated performance for the k queue will be $R_k + r_{ik}$ and the queuing mechanism continues for the next service. Figure 2 shows a representation of the queuing mechanism. There is a primary queue associated with all services. Each request has an estimated performance r_{ij} and is associated with a queue until all queues are occupied.

In this way, it is possible to assign the requests entering into the system to each of the available services. This leads to a better performance and optimization of resources.

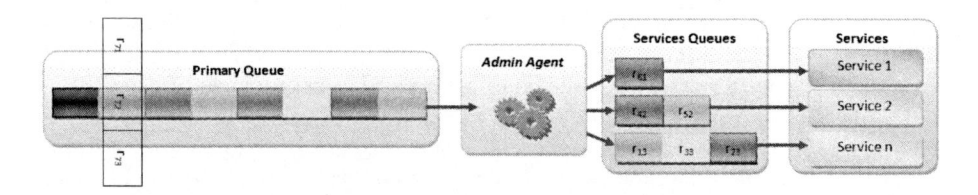

Fig. 2. Services assignation through the queuing mechanism

4 Results and Conclusions

Several tests have been done to demonstrate if the mechanisms in FUSION@ are appropriate to distribute resources and optimize the performance of multi-agent systems. Most of these tests basically consist on the comparison of two simple configurations (System A and System B) with the same functionalities. These systems are specifically designed to schedule a set of tasks using a planning mechanism [6]. System A integrates this mechanism into a deliberative BDI agent, while System B implements FUSION@, modelling the planning mechanism as a service. Table 1 shows an example of the results delivered by the planning mechanism for both systems. An agenda is a set of non organized tasks that must be scheduled by means of the planning mechanism or the planner service. There were 30 defined agendas each with 50 tasks. Tasks had different priorities and orders on each agenda. Tests were carried out on 7 different test groups, with 1, 5, 10, 15, 20, 25 and 30 simultaneous agendas to be processed by the planning mechanism. 50 runs for each test group were performed, all of them on machines with equal characteristics. Several data have been obtained from these tests, focusing on the average time to accomplish the plans. For System B five planner services with exactly the same characteristics were replicated.

Table 1. Example of the results delivered by the planning mechanism

Time	Activity
19:21	Exercise
20:17	Walk
22:00	Dinner

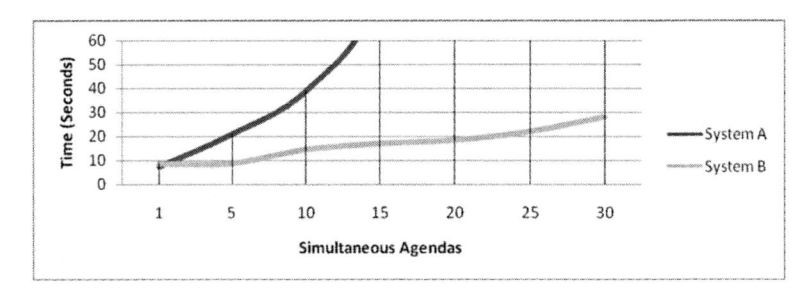

Fig. 3. Time needed for both systems to schedule simultaneous agendas

Figure 3 shows the average time needed by both systems to generate the paths for a fixed number of simultaneous agendas. System A was unable to handle 15 simultaneous agendas and time increased to infinite because it was impossible to perform those requests. However, System B had 5 replicated services available, so the workflow was distributed, and allowed the system to complete the plans for 30 simultaneous agendas. Another important data is that although the System A performed slightly faster when processing a single agenda, performance was constantly reduced when new simultaneous agendas were added. This fact demonstrates that the overall performance of System B is better when handling distributed and simultaneous tasks (e.g. agendas), instead of single tasks.

The FUSION@ architecture proposes an alternative where agents act as controllers and coordinators. The mechanisms implemented in FUSION@ can distribute resources, exploiting the agents' characteristics to provide a robust, flexible, modular and adaptable solution that covers most of the requirements of a wide diversity of projects. All functionalities, including those of the agents, are modelled as distributed services and applications. By means of the agents, the systems are able to modify their behaviour and functionalities at execution time. Developers can create their own functionalities with no dependency on any specific programming language or operating system.

Initial results demonstrate that FUSION@ is adequate for distributing composite services and optimizing performance for multi-agent systems. The *Admin Agent* learns and reason, which facilitates the optimum distribution of tasks and reduces the processing for the rest of the agents in the system. Future work consists on applying this architecture into composite multi-agent systems, as well as extending the experiments to obtain more decisive data from applications that consume multiple services with different capabilities in heterogeneous scenarios.

Acknowledgements. This work has been partially supported by the TIN2006-14630-C03-03 and the IMSERSO 137/07 projects. Special thanks to Tulecom for the technology provided and the know-how supported.

References

1. Ardissono, L., Petrone, G., Segnan, M.: A conversational approach to the interaction with Web Services. In: Computational Intelligence, vol. 20, pp. 693–709. Blackwell Publishing, Malden (2004)
2. Bratman, M.E.: Intentions, plans and practical reason. Harvard University Press, Cambridge (1987)
3. Bratman, M.E., Israel, D., Pollack, M.E.: Plans and resource-bounded practical reasoning. In: Computational Intelligence, vol. 4, pp. 349–355. Blackwell Publishing, Malden (1988)
4. Camarinha-Matos, L.M., Afsarmanesh, H.: A Comprehensive Modeling Framework for Collaborative Networked Organizations. Journal of Intelligent Manufacturing 18(5), 529–542 (2007)
5. Cerami, E.: Web Services Essentials Distributed Applications with XML-RPC, SOAP, UDDI & WSDL, 1st edn. O'Reilly & Associates, Inc., Sebastopol (2002)
6. Corchado, J.M., Bajo, J., Abraham, A.: GERAmI: Improving the delivery of health care. IEEE Intelligent Systems, Special Issue on Ambient Intelligence 23(2), 19–25 (2008)
7. Corchado, J.M., Bajo, J., De Paz, Y., Tapia, D.I.: Intelligent Environment for Monitoring Alzheimer Patients, Agent Technology for Health Care. In: Decision Support Systems. Eslevier, Amsterdam (in press, 2008)
8. Georgeff, M., Rao, A.: Rational software agents: from theory to practice. In: Jennings, N.R., Wooldridge, M.J. (eds.) Agent Technology: Foundations, Applications, and Markets. Springer, New York (1998)
9. Greenwood, D., Lyell, M., Mallya, A., Suguri, H.: The IEEE FIPA approach to integrating software agents and web services. In: Proceedings of the 6th international Joint Conference on Autonomous Agents and Multiagent Systems, AAMAS 2007, Honolulu, Hawaii, pp. 1–7. ACM, New York (2007)
10. Jayaputera, G.T., Zaslavsky, A.B., Loke, S.W.: Enabling run-time composition and support for heterogeneous pervasive multi-agent systems. Journal of Systems and Software 80(12), 2039–2062 (2007)
11. Li, Y., Shen, W., Ghenniwa, H.: Agent-Based Web Services Framework and Development Environment. In: Computational Intelligence, vol. 20(4), pp. 678–692. Blackwell Publishing, Malden (2004)
12. Ossowski, S., Julián, V., Bajo, J., Billhardt, H., Botti, V., Corchado Rodríguez, J.M.: Open Issues in Open MAS: An abstract architecture proposal. In: 12th Conferencia de la Asociación Española para la Inteligencia Artificial (CAEPIA 2007), Salamanca, Spain, vol. 2, pp. 151–160 (2007)
13. Pokahr, A., Braubach, L., Lamersdorf, W.: Jadex: Implementing a BDI-Infrastructure for JADE Agents. In: EXP - in search of innovation (Special Issue on JADE), Department of Informatics, University of Hamburg, Germany, pp. 76–85 (2003)
14. Ricci, A., Buda, C., Zaghini, N.: An agent-oriented programming model for SOA & web services. In: 5th IEEE International Conference on Industrial Informatics (INDIN 2007), Vienna, Austria, pp. 1059–1064 (2007)
15. Schön, B., O'Hare, G.M.P., Duffy, B.R., Martin, A.N., Bradley, J.F.: Agent Assistance for 3D World Navigation. LNCS, vol. 1, pp. 499–499. Springer, Heidelberg (2005)

16. Shafiq, M.O., Ding, Y., Fensel, D.: Bridging Multi Agent Systems and Web Services: towards interoperability between Software Agents and Semantic Web Services. In: Proceedings of the 10th IEEE International Enterprise Distributed Object Computing Conference (EDOC 2006), pp. 85–96. IEEE Computer Society, Washington (2006)
17. Snidaro, L., Foresti, G.L.: Knowledge representation for ambient security. In: Expert Systems, vol. 24(5), pp. 321–333. Blackwell Publishing, Malden (2007)
18. Tapia, D.I., Rodriguez, S., Bajo, J., Corchado, J.M.: FUSION@, A SOA-Based Multi-Agent Architecture. Advances in Soft Computing Series, vol. 50, pp. 99–107. Springer, Heidelberg (2008)
19. Tiwari, A.K., Shukla, K.K.: Implementation of generalized cross validation based image denoising in parallel virtual machine environment. Digital Signal Processing 14, 138–157 (2004)
20. Voos, H.: Agent-Based Distributed Resource Allocation in Technical Dynamic Systems. In: Proceedings of the IEEE Workshop on Distributed intelligent Systems: Collective intelligence and Its Applications (DIS 2006)., pp. 157–162. IEEE Computer Society, Washington (2006)
21. Wooldridge, M., Jennings, N.R.: Intelligent Agents: Theory and Practice. The Knowledge Engineering Review, vol. 10(2), pp. 115–152. Cambridge University Press, Cambridge (1995)

Mobile Agents for Critical Medical Information Retrieving from the Emergency Scene

Abraham Martín-Campillo, Ramon Martí, Sergi Robles, and Carles Martínez-García

Departament of Information and Communications Engineering, Universitat Autònoma de Barcelona, 08193 Cerdanyola del Vallès, Spain
{abraham.martin, ramon.marti.escale, sergi.robles, carlos.martinez}@uab.cat

Abstract. Lacking medical information about a victim in the aftermath of an emergency makes the early treatment and the efficient allocation of resources difficult. On the other hand, communication infrastructures are normally disrupted in these situations, thus hindering the gathering of the required information. This paper presents a new application of mobile agents for retrieving partial information of medical records upon request from the emergency scene. This solution fits well when mobile ad hoc networks are in use, and it is based on the asynchronous communication provided by mobile agents. By using the proposed system, it is possible to request remote hospitals for critical information about the victims, such as allergies or infectious diseases, thus facilitating more accurate diagnosis and bringing forward decision making. An implementation of the system has been developed, showing its feasibility.

1 Introduction

When an emergency occurs, especially in mass casualty incidents, lots of victims need medical attention. When the first responder medical personnel arrive to the scene, they triage the victims they find [11] [7]. The triage consists of a protocol to sort victims according to their medical status in order to prioritize their needs for receiving medical attention. When all the victims are triaged, the medical personnel begins to treat them following the color-based prioritization.

The objective of the first responder medical personnel is to stabilize the victim before they can be evacuated to an hospital. Only after stabilized, the victim is transferred to an hospital. Normally, while the victim is being stabilized or transferred, medical personnel search for victim's personal items that could identify them. Discovering who the victim is may not seem essential but it is; with this information, personnel from the hospital where the victim has been taken to can contact with their family. Those can provide doctors with important medical information about the patient. Without the help of their family it is probable that the hospital does not have all the important medical data of the victim unless it has implemented a Virtual Electronic Patient Medical Record (VEPMR) solution.

Some research has been done about electronic patient medical records systems, a hot topic nowadays [5] [13]. A VEPMR system makes available from any medical institution all the existing medical information about a patient. To achieve this purpose, medical institutions have to be part of the medical network where every member shares the information it has. As a consequence, every time a patient visits the doctor, each surgery and each test done (f.e. scans or x-rays) is recorded inside this distributed database to be available at the whole medical network. It is clear that having all the medical information about a patient, a faster and better diagnosis is obtained, thus a better treatment may be given [6].

Our motivation is making VEPMRs available in the emergency scene. This could improve the medical assistance of the victim. Seems obvious that the knowledge of some data like blood type, chronic and contagious diseases, without the need for doing tests, could save more lives as the medical team saves time and money. Furthermore, this knowledge could prepare the hospital where the victim is assigned if special attentions are needed. The possibility to have access to a VEPMR repository in the emergency scene is not a reality nowadays.

When incidents like hurricanes, floodings or tsunamis occurs, most communication networks are normally disrupted, being an obvious handicap for quick and coordinated assistance. Some projects propose deploying antennas to have communication networks in all the emergency scene [10]. Others propose using sensor or ad-hoc networks for communication inside the emergency scene [8][12].

We could use any of these systems to make true the VEPMR retrieval from the emergency scene. But, our goal is going a step forward proposing the use of mobile agents on ad-hoc networks to forward requests of VEPMRs to the medical network. This allows the retrieval of the patient's medical record in the place where the emergency has occurred, even in the worst case scenarios that others proposals have not taken into account. As a result, our proposal does not need the deployment of a network infrastructure. Only using personal information about the victims, found in the emergency scene, our system is able to identify the victims to make the request of their VEPMR.

The rest of the paper is structured as follows. Section 2 introduces the reader to the emergency world, describing an emergency scene and its actions and actors. Section 3 presents our system proposal, its description and implementation. Finally, Section 4 contains a discussion comparing our systems with different alternatives and complementary work. Section 5 concludes the paper.

2 Emergency Scene

This paper is centred in the worst scenario emergencies which usually are mass casualty incidents (MCIs). The main characteristic of MCIs is the big number of victims. In these cases, the triage of victims is needed to sort injured people into groups based on their need for immediate medical treatment. This triage is done by the first medical personnel that arrive at the emergency scene. Consequently, the medical personnel arriving latter know those victims who need more urgent attention. The victims are stabilized and prepared, in triage color order, to be evacuated to the hospital where they will be treated widely.

The triage process usually creates four groups of victims based on their condition. The first group, from worst to best condition order, is the black one. Those victims triaged as black group are dead or in very bad condition, impossible for the medical

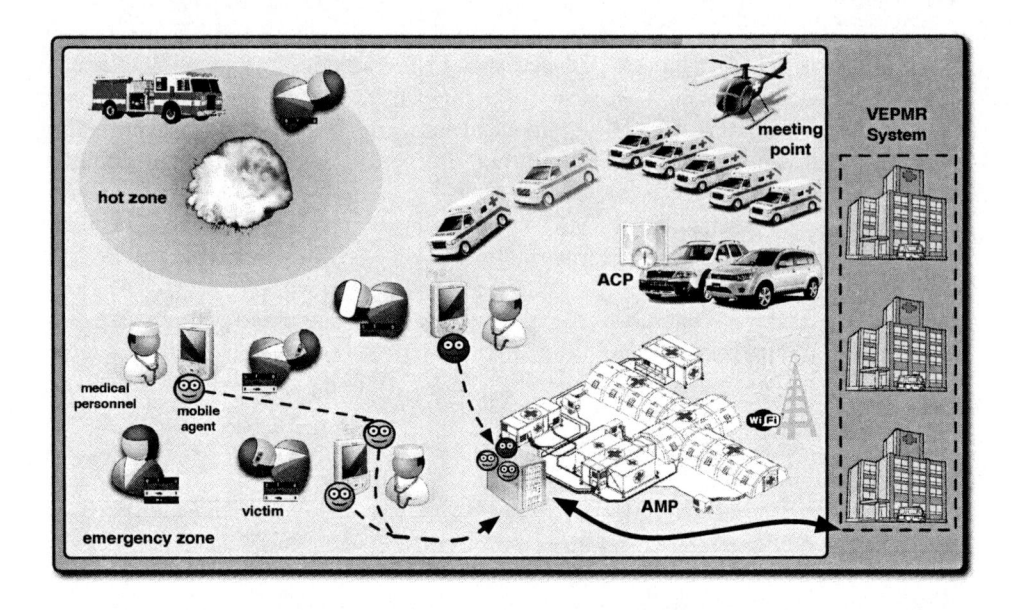

Fig. 1. Emergency scene and VEPMRE retrieval scheme

team to do something to save their live. The second group, red, are victims who need immediate attention. The victims in the third one, yellow, do not need for immediate but urgent medical attention so can wait for a short period of time. And finally, the green group, with victims with minor injuries who need help less urgently.

The first and foremost step is the triage because it focuses the medical attention in those victims in worst conditions. Once the triage is complete, rescue teams extract those victims who are trapped or cannot move from the hot spot to a safe place. The hot spot is a dangerous area where the emergency has happened. In this area the medical personnel cannot work because it may exists risk of danger (as explosion or contamination). Because of this, it is important for everybody to evacuate this area and for the rescue teams to extract the victims who cannot move. While rescue teams are doing their job, the medical personnel treat those victims in the red group that have been already evacuated and are in a safe place.

If an Advanced Medical Post (AMP) is installed, the victims are evacuated to this place (see Fig. 1). The AMP is a mobile hospital to treat the victims before they can be transferred to an hospital. In mass casualties, where is necessary to treat lots of victims in seriously condition, the AMP is a must that have to be installed near but in reasonable distance from the emergency, or hot spot, to be a safe place.

The main objective of the medical personnel regarding the victims in the red group is to stabilized them, to make them be ready to be transferred. Once the stabilization is done, the ambulance, helicopter or rescue vehicle in general are called to pick up each victim. Afterward, victims are transferred to the hospital assigned by the coordinator team. The coordinator team manage which victims are assigned to which hospitals.

Once all the red victims have been treated it is time for the yellow ones, that follow the same process as the red ones. The green patients are arranged together and then transfered as low priority to other hospitals or medical institutions using any available transportation.

The Advanced Command Post (ACP) is where the coordination team is. From this site, all the decisions about actions to be carried out by rescue and medical teams are taken.

It is necessary to consider that in a big emergency more than one hot spot or local emergency can exist. These local emergencies, for instance, in a hurricane each house devastated or vehicle crashed, can share the meeting point, AMP and/or ACP or have its own for each one if they are big enough. Usually, if the emergency has multiple hot spots or local emergencies, a crisis committee is created to manage and coordinate all the emergency in collaboration with different ACPs installed.

2.1 Communications in the Emergency Scene

Talking about communications in the emergency scene, on past days, it was only matter of walkie-talkie communications, but nowadays it is getting more and more important. This is due to the greater use of Internet enabled devices or mobile phones by the emergency personnel, that require mobile networks such as mobile phone network (3G) or WiMAX. In the great majority of emergency cases, hurricanes, terrorist attacks, floodings, etc, these networks become unstable, unaccessible, overused or even destroyed. As a consequence, emergency personnel cannot use existing network infrastructure and may deploy and use their own, or simply use wireless mobile ad-hoc networks (MANETs) or wireless mesh networks. These networks create routes by request of the nodes that are maintained as long as they are needed or the link is available.

Anyway, all these solutions have the same lack. If the area of emergency is big enough, could be possible that the ad-hoc network created by the medical personnel's devices is not fully connected. As a result, an attempt to communicate from one point of the network, for instance, a first responder, to another point of the network, for example, the AMP, could be unsuccessful.

The AMP and ACP always have Internet connection even if the network infrastructures are destroyed or unusable. They use their own deployed network infrastructure, for instance, satellite connections. For the AMP and the ACP, it is very important to have Internet connection for coordination or information communications (f.e. with another coordination point or with hospitals assigned to victims).

3 Critical Medical Information Retrieving System

We propose a mechanism to retrieve VEPMRs from the emergency scene. Each member of the triage and medical personnel is provided with a mobile device supporting mobile agents. With this mobile device, any member of the triage or medical personnel, anywhere in the emergency zone, can create a mobile agent with a VEPMR request when any personal document identification of the victim is found.

3.1 Retrieving System

Our proposal takes into account the three possibles scenarios in the emergency scene. The first one is the existence of a network infrastructure with Internet connection, for instance, mobile phones network infrastructure (GSM, EDGE or 3G). The second and third possible situation lack of a network infrastructure with Internet connection. In the second case, a MANET is created using all the neighbor devices of the medical and triage personnel to solve the lack problem. In this situation, all the mobile devices are close enough each other to create a fully connected MANET. Furthermore, some of the mobile devices that creates the MANET are near the AMP or ACP. As a result, all of them can access to Internet, using the network infrastructure of the AMP or ACP, and routing the packet to them using usual ad-hoc networks routing protocols. Consequently, from any part of the network it is possible to send directly the VEPMR request to the medical network. The third possible situation is similar to second one but the MANET is not fully connected, so clusters of devices are created, hence, not all the mobile devices can access to Internet directly.

Let us begin with the third situation. Our proposal is based on creating a mobile agent each time a VEPMR request is done. This mobile agent contains the identification of the victim whose VEPMR is requested. As the MANET is not fully connected, the request cannot be sent directly, thus, the mobile agent have to jump from device to device or cluster to cluster of devices until it reaches the ACP or AMP. Once the mobile agent has arrived, it can communicate the request to the VEPMR system of some hospital of the medical network.

Delay Tolerant Networks (DNTs) are one solution for this problem but the routing process in DTNs is an open issue nowadays. We wanted to go one step further and make possible a more complex routing protocol and an easy-to-change routing policy. As a consequence, we decided to use mobile agents and create a mobile agents based MANET. This MANET uses agent platforms and its services as routers for the mobile agents that arc treated as packets in traditional networks. Therefore, the routing process is done in the application layer. The complex routing decision is provided by the mobile agent itself, accessing and using attributes and values available in the application layer. Furthermore, it is possible to use different routing protocols depending on the situation of the mobile agent, or the network, or even allows the user to select which routing policy wants to use.

Using this method it is possible to route a mobile agent from the point where the request is created thought the whole, not fully connected, MANET to the AMP or ACP. In traditional networks this cannot be done because a fully connected network from the origin point to the destiny point is needed. Using mobile agents based MANET, the agents can wait for a connection if there is not, in any part of the network, thus, they can cover a part of the route. Furthermore, they can use different dynamic routing protocols to reach the destiny point. The dynamic routing protocol decides if it is better for the mobile agent to jump to another platform or to stay in the platform where it is, and also deciding the best platform to jump.

Different dynamic routing protocols can be used in the mobile agents based MANET. In our case, the objective of all mobile agents is to arrive to some point where Internet connections exists. This is usually the AMP or ACP if there is no network infrastructure. So, the objective of the routing protocol is to offer the best route spending the minimum time to arrive to the AMP or ACP. Our proposal aims to use the time that the medical personnel expects to stay in the emergency scene to calculate the routing. This is an approach and one solution for the problem, but other routing policies can be used to arrive to the ACP or AMP as soon as possible. When a first responder leaves the meeting point they has to state when they will come back. This is for security reasons because they will be in a emergency scene where some disaster has occurred and it is not fully safe. Therefore, they foresee a "time to return" (TTR) that they will put into their device. So, each device stores the time when the personnel who carries it expects to return to the AMP. The mobile agent asks for this information for the service's platform of all the neighbor devices and jumps to the one that has the smaller return time value. In this way, the mobile agents arrives to the AMP as soon as possible.

Some special devices, for instance those installed in an emergency vehicle, can store an especial "time to return" value. This is especially useful when an emergency vehicle has to go to the emergency scene and come back to the AMP or ACP, or the first responder has finished their job in the emergency scene and is coming back. In these situations the "time to return" will have a low value. Therefore, all the mobile agents near the platform with low TTR will jump into it because is the fastest way to get access to Internet.

It would wrongly seem that agents are not useful in the first situation, where mobile devices can connect directly to Internet using network infrastructure available in the emergency scene, and in the second situation, where the mobile devices can access to Internet through the fully connected MANET. However, in emergency situations the network infrastructures are usually unstable or over-saturated, thus the situation can change from the first one to the second one. In the second situation it would wrongly seem that mobile agents are also not needed. Moreover, as a mobile ad-hoc network, in any moment the situation can change from a fully connected to not fully connected network, hence the third situation. In this case, we do need mobile agents to make the requests work. In emergency scenes the situations are intermittent as they can change in any moment from first to second or third, and vice versa. For this reason we always have to focus in the worst case scenario, this is, the third situation. Therefore, the mobile agents based MANET scheme is always used regardless of in what situation the emergency scene is. In conclusion, mobile agents are always launched when a requests is done and they try to reach the AMP or ACP to make the request from there, where the network infrastructure is more stable.

When the VEPMR is retrieved, a copy is saved in the AMP and ACP. The mobile agent that has made the request, looks from the ACP or AMP for the mobile device that has launched it. If this mobile device is connected to the same MANET cluster than the ACP or AMP, the mobile agent can jump to its agent platform. As a consequence, if the victim is being treated in the emergency zone or in the AMP, their VEPMR will be always available for the medical personnel treating them.

Security has also been taken into account in our system. VEPMRs consist of sensible private medical data. For this reason, this information has to be dealt with carefully

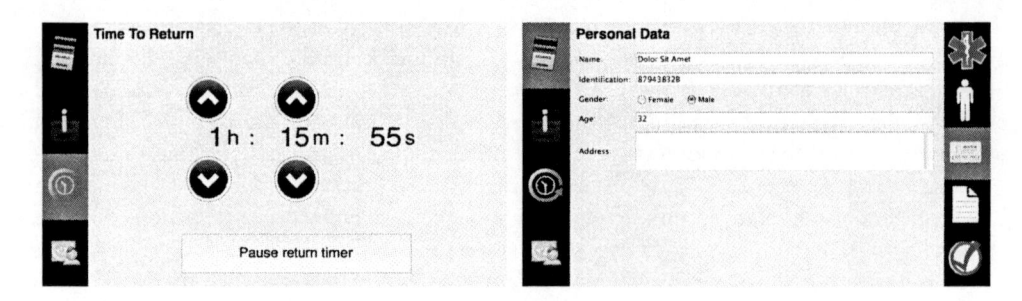

Fig. 2. Implementation screenshots

using strong security mechanisms, and the communication between ACP, or AMP, and the VEPRM system has to be secure. The problem of security in ad-hoc networks is not fully solved nowadays. Even thought, this is not an issue for our system because the very agents protect the data they carry [1]. Thus, the system can be used in insecure networks.

3.2 VEPMR for Emergencies

We also propose the creation of a special VEPMR containing only the relevant information for an emergency case. This VEPMR for emergencies (VEPMRE) contains the blood type, hepatitis, AIDS, allergies and more basic information to treat the patient in an emergency case. Thanks to this, the VEPMRE has a smaller size so it uses less network bandwidth.

An important issue is how a VEPMRE is associated with the victim. An AMP may receive many VEPMRE from requests launched by the mobile agents, so how do the medical personnel know what VEPMRE corresponds to which victim? Our proposal is adding the id number of the triage tag with each VEPMRE request inside the mobile agent. In this way, once the VEPMRE is received it is possible to identify the victim, since the triage tag is always carried by them.

3.3 Implementation

An implementation has been developed as a proof of concept. A Nokia n810 touch screen-based mobile device with a MAEMO linux distribution has been used as a hardware platform. The programming language used is Java with JADE Framework [2] as agents platform, together with its FIPA-compilant mobility services [3].

IEEE 802.11g has been used as a network interface with the handhelds. When a device leaves the AMP or ACP, a TTR value has to be added as the screenshot in the figure 2A shows. When a first responder finds an ID of the victim, it can be introduced in the system and a VEPMRE request is sent (figure 2B). At this moment a mobile agent is created containing the ID of victim and with a routing policy. The mobile agents tries to move from platform to platform following the routing policy until it reaches the AMP or ACP.

4 Discussion

Some other works have proposed some other ideas to implement in the emergency scene [10][9][8][4]. On the other hand, some researchers have proposed VEPMR solutions that complement our work [13].

"A Situation-Aware Mobile System to Support Fire Brigades in Emergency Situations" [10] proposes a network infrastructure deployment mechanism for communication between the fire brigades members. The firefighters are equipped with PDAs, for coordination, and sensors to read their vital signs. PDAs transmit data, voice over IP and video streaming through a mesh network, while the sensors create sensor networks to transport the data obtained to the PDA. The proposal for the mesh network is based on the deployment of new nodes where the coverage is lost. One or more of these nodes may be connected to Internet, allowing access to it to all PDAs. Our proposal does not deal with the deployment of new nodes in the infrastructure, but to use the mobile devices in asynchronous way in order to get VEPMREs.

The Wireless Internet Information System for Medical Response in Disasters (WIISARD) [9] project proposes a set of systems to use in the emergency scene. They provide triage personnel with PDAs to introduce the triage information of the victim [8]. The identification of the victim is done through a barcode reader. WIISARD also proposes a system in the ACP, able to receive all the triage information from the PDAs [4]. A decision making system is also included, using the triage information together with additional information about the emergency, like resources available at hospitals. Again, the deployment of a network, the existence of a network infrastructure or a fully connected MANET is needed.

Marques et al. [13] propose VEPMR mobile agents as a solution for the problems of heterogeneity and dispersion of electronic patient records in different healthcare institutions. Their paper describes an information-gathering system for securely integrating distributed, inter-institutional medical data using security-enhanced mobile agent technology into a single VEPMR. In this case, mobile agents move inside the infrastructure network in order to visit all medical institutions and departments to gather the required records. Our proposal is a step-forward, since it aims to create a significant reduced version of the VEPMR adapted to emergency situations, and use the mobility and intelligence of mobile agents to move asynchronously inside a MANET in order to reach the AMP as the entry point to the Internet.

5 Conclusions

Coordination and information gathering are hard chores to achieve during an emergency situation, especially when they involve scattered personnel and hospitals. It is particularly relevant the issue of getting critical information about a victim before arriving to a hospital for treatment. The proposal described in this paper provides a practical application of mobile agents to solve this particular problem, making it possible to asynchronously retrieve significant information, such as allergies or infectious diseases, from the patient record while the victim is still in transit. This information can be used to make a better resource allocation, such as operating rooms reservations, improve diagnosis, or to avoid hazardous treatments, for instance in the case of diabetics, or drug allergic victims. The proposed system can be used in mobile ad hoc networks, and is tolerant to network disruptions and high delays, which are common during the aftermath of an emergency. The early acquirement of this information might save lives in some cases. The system has been implemented as a proof of concept using real handheld devices, and has shown to be feasible.

There are aspects of the system that can be complemented with other technologies. As a future work, we aim to integrate this solution with a fault tolerance mechanism which is a must when sensible information is to be transmitted. Other additional functionalities we have planned to add are using the agent for bearing triage information, or inserting geographical position (GPS) of the victim's whereabouts in the agent.

Acknowledgements

This work has been funded by the Spanish Ministry of Science and Innovation through the project TSI2006-03481.

References

1. Ametller, J., Robles, S., Ortega-Ruiz, J.A.: Self-protected mobile agents. In: 3rd International Conference on Autonomous Agents and Multi Agents Systems, vol. 1, pp. 362–367. ACM Press, New York (2004)
2. Caire, G.: Jade: The new kernel and last developments. Technical report, Telecom Italia (2004), http://jade.tilab.com/papers/Jade-the-services-architecture.pdf
3. Cucurull, J., Martí, R., Robles, S., Borrell, J., Navarro-Arribas, G.: FIPA-based interoperable agent mobility. In: Burkhard, H.-D., Lindemann, G., Verbrugge, R., Varga, L.Z. (eds.) CEEMAS 2007. LNCS (LNAI), vol. 4696, pp. 319–321. Springer, Heidelberg (2007)
4. Demchak, B., Chan, T.C., Griswold, W.G., Lenert, L.A.: Situational awareness during mass-casualty events: command and control. In: AMIA Annu. Symp. Proc., p. 905 (2006) (Times Cited: 0)
5. Eden, K.B., Messina, R., Li, H., Osterweil, P., Henderson, C.R., Guise, J.-M.: Examining the value of electronic health records on labor and delivery. American Journal of Obstetrics and Gynecology, 199(3), 307.e1–e307.e9 (2008)
6. Hippisley-Cox, J., Pringle, M., Cater, R., Wynn, A., Hammersley, V., Coupland, C., Hapgood, R., Horsfield, P., Teasdale, S., Johnson, C.: The electronic patient record in primary care regression or progression? a cross sectional study. British Medical Journal 326, 1439–1443 (2003)
7. Kennedy, K., Aghababian, R.V., Gans, L., Lewis, C.P.: Triage: Techniques and applications in decisionmaking. Annals of Emergency Medicine 28(2), 136–144 (1996)
8. Killeen, J.P., Chan, T.C., Buono, C., Griswold, W.G., Lenert, L.A.: A wireless first responder handheld device for rapid triage, patient assessment and documentation during mass casualty incidents. In: AMIA Annu. Symp. Proc., pp. 429–33 (2006) (Times Cited: 0)
9. Lenert, L., Chan, T.C., Griswold, W., Killeen, J., Palmer, D., Kirsh, D., Mishra, R., Rao, R.: Wireless internet information system for medical response in disasters (wiisard). In: AMIA Annual Symposium Proceedings, pp. 429–433 (2006)

10. Luyten, K., Winters, F., Coninx, K., Naudts, D., Moerman, I.: A situation-aware mobile system to support fire brigades in emergency situations. In: Meersman, R., Tari, Z., Herrero, P. (eds.) OTM 2006 Workshops. LNCS, vol. 4278, pp. 1966–1975. Springer, Heidelberg (2006)
11. National Association of Emergency Medical Technicians U.S. and American College of Surgeons. PHTLS–basic and advanced prehospital trauma life support, 4th edn. Mosby, St. Louis (1999)
12. Portmann, M., Pirzada, A.A.: Wireless mesh networks for public safety and crisis management applications. IEEE Internet Computing 12(1), 18–25 (2008)
13. Vieira-Marques, P.M., Robles, S., Cucurull, J., Cruz-Correia, R., Navarro, G., Martí, R.: Secure integration of distributed medical data using mobile agents. IEEE Intelligent Systems 21(6), 47–54 (2006)

INGENIAS Development Assisted with Model Transformation By-Example: A Practical Case

Iván García-Magariño, Jorge Gómez-Sanz, and Rubén Fuentes-Fernández

Universidad Complutense de Madrid
ivan_gmg@fdi.ucm.es, jjgomez@sip.ucm.es, ruben@fdi.ucm.es

Abstract. INGENIAS is a methodology for the development of multi-agent systems. INGENIAS support tools has recently incorporated a plug-in called the *MTGenerator*, which has been developed to facilitate the creation of model transformations by-example from INGENIAS models. The *MTGenerator* tool overcomes some of the limitations of similar tools about the creation of many-to-many transformation rules. This paper introduces the practical application of the tool to a complete development case study made with INGENIAS, showing the role and benefits of such tools.

Keywords: Model-Driven Development, Model Transformation, Transformation By-Example, Multi-agent Systems, INGENIAS.

1 Introduction

INGENIAS [6] is a methodology for the development of multi-agent systems (MAS). Since its inception, the development process of INGENIAS has followed the principles of Model-driven Development (MDD) [10]. It bases the development on the specification of the models of the system and the automated generation of other artifacts, like documentation or code, from these models. These activities are assisted by a support suite called the INGENIAS Development Kit (IDK) [7]. Despite these aids, the application of MDD principles is not complete since work in Model Transformations (MTs) is missing. A MT transforms a source model, following a source meta-model, into a target model, following a target meta-model. It was considered that capturing this transformation knowledge into a program would help clarifying the MAS development process itself. Having a non ambiguous list of changes to apply at any moment would be meaningful towards a further automatization of the MAS development process. Developing MTs usually requires from designers knowing in detail those meta-models and the MT language itself. In addition, the development of these MTs usually requires manually managing metamodels and transformations as code. This problem difficulties the wide application of transformations to a model-driven MAS development process.

In literature, MTs are proposed for the Tropos agent-oriented methodology: initial requirements [3] are transformed into detailed requirements with many-to-many rules; platform independent models [1] are transformed into platform specific models. In addition, ADELFE [2] uses MTs to transform platform-independent models (with AAMAS-ML language) into platform-specific models (with *micro-Architecture Description Language, μADL*). Moreover, Jung et al [9] propose a language with high-level of abstraction, which can be transformed into more specific models. However, the definition of these MTs is not facilitated by any specific tool, until our recent research.

In order to reduce manual coding of MTs in the context of MAS and gain experience in the use of MTs for agents, this work focuses in the generation of Model Transformations By-Example (MTBE) in the context of INGENIAS methodology. Varro [11] defines MTBE as the creation of MTs from pairs of source and target models that are used as prototypes. The common steps of MTBE are: the manual set-up of prototype mapping models, the automated derivation of the rules that constitute the MTs from these pairs, the manual refinement of rules if required, and the automated execution of MTs. Existent MTBE tools, like [11, 12, 13], are not suitable for their use in INGENIAS modeling since they do not support the creation of many-to-many rules . A many-to-many rule converts a set of modeling elements into another set of modeling elements. As our experience has shown, many of the required transformations in a MAS model-driven development are of this kind. For this reason, our research has recently created a MTBE tool called the *MTGenerator* that supports these many-to-many transformation rules.

This paper illustrates the use of MTBE, concretely the *MTGenerator* [5] tool, through a practical development with INGENIAS sustained by the MTs created with it. The algorithm of this tool simulates the input patterns of the many-to-many rules with constraints. The definition of the elements of the output pattern relies on the common capabilities of most existing MT languages. The algorithm recursively links these elements during their creation. In addition, it can create the appropriate mappings for attributes, so the generated MT propagates the information from the source to the target model. This *matching-control mechanism* defines identifiers in the source model that are then referred from the target model. The tool called the *MTGenerator* implements this algorithm for the ATL (*ATLAS Transformation Language*) [8] and *INGENIAS* models. The *MTGenerator* tool and the IDK are available from the Grasia website[1]. Fig. 1 shows the elements of the *INGENIAS* notation used in the case study of this paper.

The project considers the development of a MAS to evaluate documents following the Delphi method [4]. This development focuses on the use cases and workflow diagrams of INGENIAS to create the other parts of the specifications using MTs. The paper presents a sequence of steps to define these MTs with the tool. As a result of the work done, a set of MTs have been constructed which can be reused in other MAS developments in a similar fashion.

---

[1] http://grasia.fdi.ucm.es (software section).

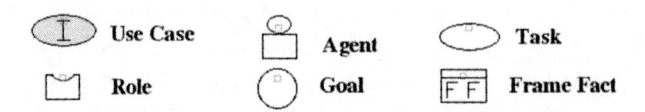

Fig. 1. Some relevant concepts of the *INGENIAS* notation

The rest of the paper is organized as follows. The case study that applies the *MTGenerator* with INGENIAS is the subject of Section 2. That section has a sub-section for each step of the MAS development assisted with MTBE. Finally, Section 3 discusses conclusions about the tool and the development with it, and future lines of research.

2 INGENIAS Modeling with MTBE Support

This section describes a process to model in INGENIAS with the support of MTs. The MTs guide and assist the designer in the specification, although the process gives the possibility of manual edition if necessary. The process reported here tries to be general in two ways: MTs can be reused in other similar INGENIAS developments; designers can define MTs with our MTBE as it has been done in the current project to support additional needs.

The case study where these MTs have been applied is an implementation of the Delphi process with MAS [4]. This method dates back to the fifties. It was created by the RAND corporation in Santa Monica, California. The method is made of structured surveys. It plans several rounds of questionaries which are sent to the different involved experts. The results collected can be included partially in a new round of questionaires, but respecting the anonymity of the participants.

2.1 Planning the Chain of Model Transformations

An INGENIAS Development based on MTBE begins choosing which models are the seeds to generate partially other models. The Delphi case study considered here regards the use case and workflow models as these sources. Use cases provide in INGENIAS the most abstract modeling level for requirements and specify their initial elicitation. Moreover, this domain problem is centered in the workflow models for the questionnaire rounds of the Delphi method. The use cases and the workflow of this case study are respectively presented in Fig. 2 and Fig. 5. From these types of source models, a chain of transformations generates several modeling artifacts for the development in INGENIAS. The steps are the following:

- Definition of the use cases (first group of source models).
- A MT creates the *Role* definitions.
- Definition of the workflow (second group of source models).
- A MT creates the *Tasks*.

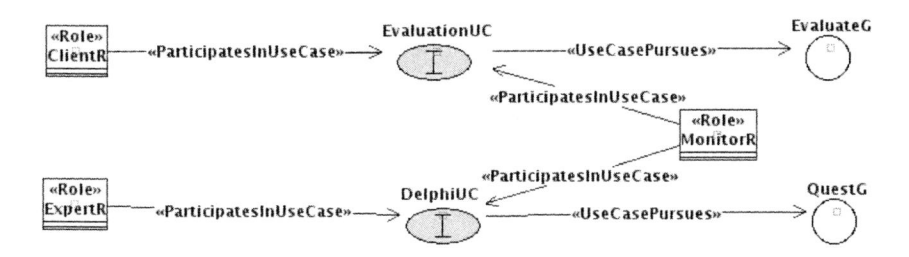

Fig. 2. Use cases for the Delphi MAS

- For each *Task*, a MT creates an INGENIAS *Code Component*.
- A MT creates the *Interactions* according to the workflow.
- A MT creates the agent deployments.
- A MT generates the tests.

All the MTs required to generate these elements are created in this case with MTBE using the *MTGenerator*. The remaining subsections show the generation of these MTs from their pairs of source and target examples.

2.2 MT for Generating Roles from Use Cases

In INGENIAS, the use case diagrams (see an example in Fig. 2) usually include roles, use cases where they participate, and goals satisfied by these cases. A later step must link roles and the goals they pursue. In addition, there must be at least one agent to play each role (see Fig. 4). Since these tasks are common in INGENIAS development, this work proposes to automatically create the role and agent definitions with a MT.

The model pairs in Fig. 3 can generate this MT. In the first pair, the source model contains a role, and the target model an agent that plays this role. The agent identifier is composed from the role identifier using the matching-control mechanism provided by the tool. In the second pair, the source model contains a goal connected to a role through a use case, and the target model has the same goal directly linked with the role. In this way, the MT links roles with their goals according to the existent use cases. For example, the MT generates the role definitions of Fig. 4 from the use cases of Fig. 2.

2.3 MT to Get the Tasks Related with a Workflow

The initial definition of a workflow in INGENIAS regards it as a chain of tasks (see an example in Fig. 5) connected with *WFConnects* relationships. Designers refine this definition giving the inputs and outputs of each task in different diagrams. Since the workflow concept is widely present in most domains, we expect a high reusability of the resulting MTs. In particular, the first task of a workflow is usually triggered by an event launched by some user's application. Fig. 6 shows the MT. Its source model constains the initial task of the workflow,

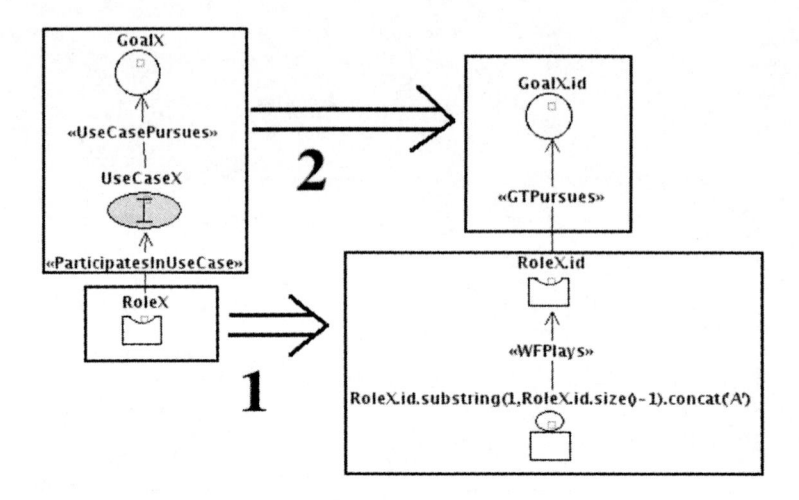

Fig. 3. MT from use cases into role and agent definitions. Each arrow represent a pairs of a source and a target models. The numbers in rules indicate the order.

and the target model shows the task consuming an event from an internal application that represents the user's application. The initial task of the workflow is the only one that is not preceded by any other task. However, this cannot be specified within the source model of the pair. Designers of the MT must define a constraint for the pair of models according to the INGENIAS metamodel with the Object Constraint Language (OCL). The expression of the constraint is:

```
MMA!WFConnects.allInstances()->select(e|WFConnectstarget.
     WFConnectstargettask.id=cin.id).isEmpty()
```

The constraint declares that there is no other task connected with this one with a *WFConnects* relationship. Therefore, it is the first of the workflow. This

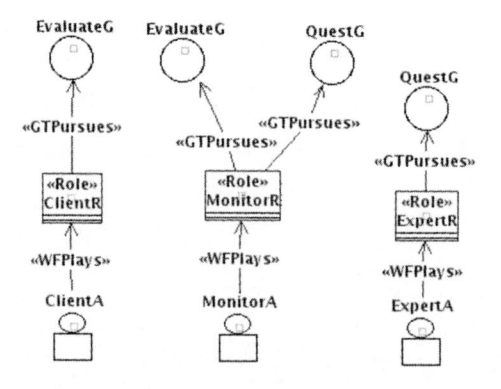

Fig. 4. *Roles* and *Agents* of Delphi

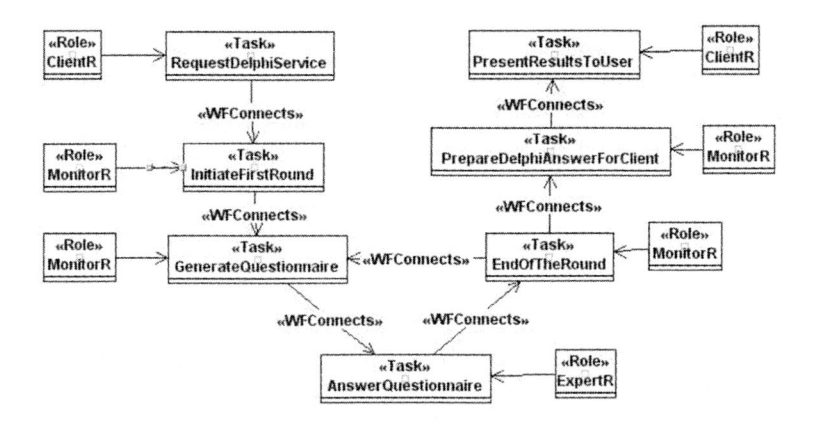

Fig. 5. Workflow for the Delphi MAS

constraint should not need to be defined if the MTBE tool included negative examples. The Grasia group is planning to include this feature in a next release of the *MTGenerator* tool.

The non-initial tasks of a workflow usually consume a *frame fact* produced by the previous task, and produce a *frame fact* to be consumed by the next task. Fig. 7 presents the MT that creates these frame facts and their connections with their related tasks. The first rule creates the frame fact and the *WFConsumes* relation, and the second one creates the *WFProduces* relation. The MT propagates the names in the source models to the target models by means of the matching mechanism. The name of each new frame fact is the concatenation of the names of the producer task and the consumer task with a suffix. The first rule of this MT is not applied to the initial task because it is not preceded by any other task. These rules can be applied several times to the same task if it precedes or is preceded by several tasks. For instance, the example in Fig. 5

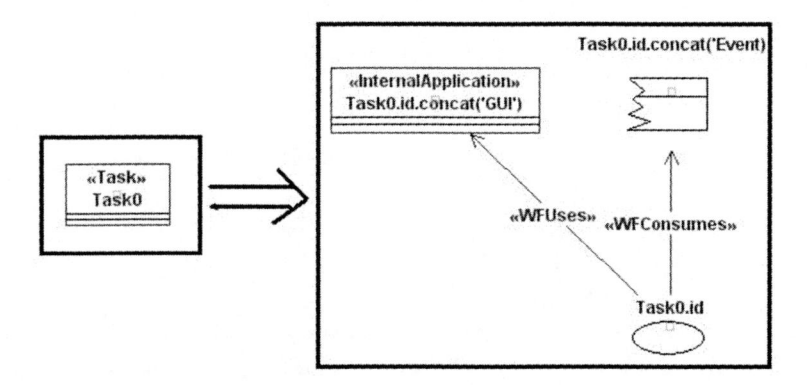

Fig. 6. MT for the initial task of a workflow

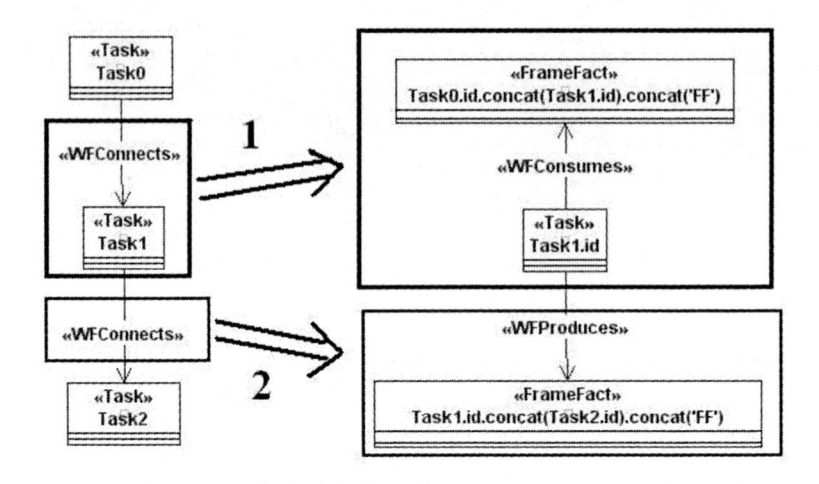

Fig. 7. MT for non-initial tasks of a workflow

includes the *EndOfRound* task that precedes two different tasks, so the second rule is applied two times for this task.

A INGENIAS specification also associates a code component to each task. The MT specified in Fig. 8 creates this code components for tasks.

2.4 MT for Creating Interaction Units

The *interaction units* of INGENIAS transfer information (i.e. *frame facts*) among agents. Tasks capture the way agents produce and consume these *frame facts*. If two different agents are responsible of two consecutive tasks in a workflow, then the *frame fact* that connects these tasks has to be transferred within an *interaction unit*. The MT specified in Fig. 9 receives as input two consecutive tasks and generates the corresponding interaction unit with its frame fact. The names of the new elements are the result of the concatenation of the two involved tasks and a suffix.

The MT generated from the specification in Fig. 9 is triggered even if the two agents responsible of the tasks are the same agent, although in this case it is

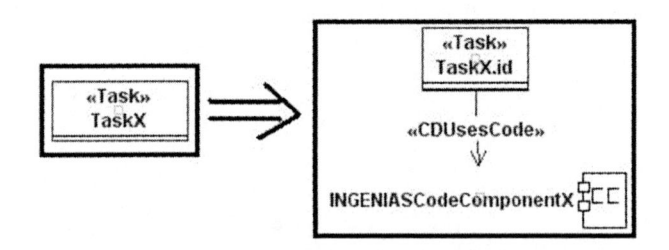

Fig. 8. MT to generate of the *Code Components* for *Tasks*

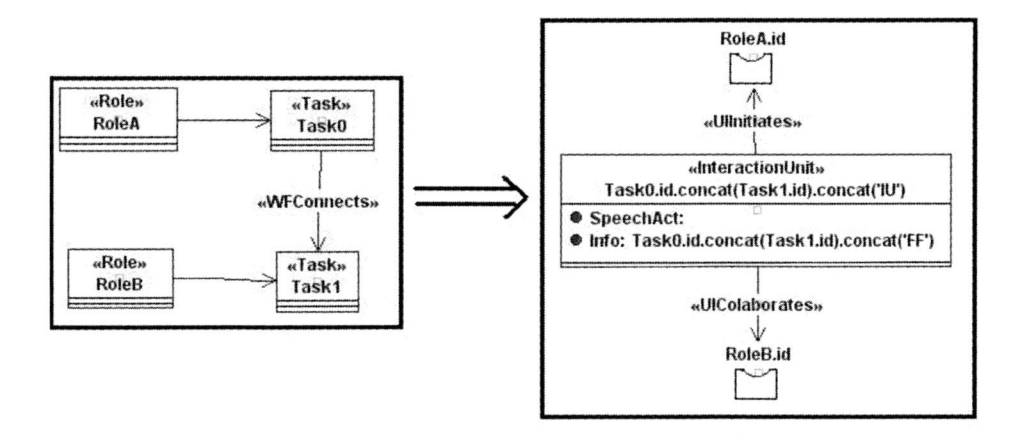

Fig. 9. MT to generate the *Interaction Unit* that communicates two tasks

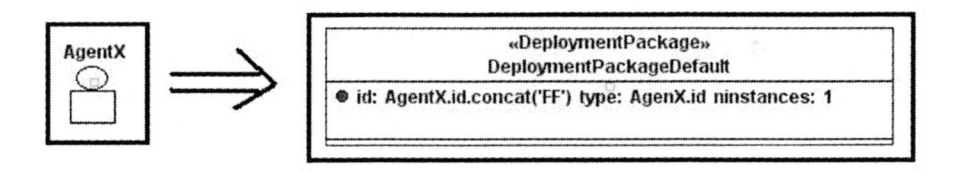

Fig. 10. MT to create *Deployment Packages*

not necessary to create the *interaction unit*. To avoid this, designers can add a constraint establishing that the two agents must be different:
not(RoleA.id=RoleB.id)

2.5 Generation of Deployments and Tests

At the end of modeling with INGENIAS, designers specify the number of agents instantiated for the MAS using deployments. The MT specified with Fig. 10 creates a deployment with one instance of each agent. This default deployment can be altered afterwards by designers. The INGENIAS methodology also recommends defining a test configuration for each deployment. The MT specified in Fig. 11 can create the required modeling elements for this purpose.

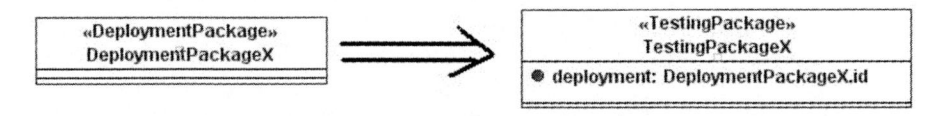

Fig. 11. MT to create *Testing Packages*

3 Conclusions and Future Work

This paper has introduced practical examples of pairs of models suitable for the application of MTBE within the INGENIAS methodology. In fact, they were used as input for the *MTGenerator* tool for MTBE. As a result, several MTs were generated as candidates to be reused for other domain problems.

The current experimentation with the *MTGenerator* has shown some open issues where the tool and its underlying algorithm can be improved. The tool needs to consider negative examples to reduce the need of specifying text constraints. Finally, further experimentation and measures are required to evaluate the gain in productivity produced by the use of the *MTGenerator* in INGENIAS developments.

Acknowledgments

This work has been supported by the Spanish Council for Science and Technology under grants TIN2005-08501-C03-01, and by the Region of Madrid (Comunidad de Madrid) and the Universidad Complutense Madrid as the Research Group 921354 and the Santander/UCM Project PR24/07 - 15865.

References

1. Amor, M., Fuentes, L., Vallecillo, A.: Bridging the Gap Between Agent-Oriented Design and Implementation Using MDA. In: Odell, J.J., Giorgini, P., Müller, J.P. (eds.) AOSE 2004. LNCS, vol. 3382, pp. 93–108. Springer, Heidelberg (2005)
2. Bernon, C., Camps, V., Gleizes, M.-P., Picard, G.: Engineering Adaptive Multi-Agent Systems: The ADELFE Methodology . In: Henderson-Sellers, B., Giorgini, P. (eds.) Agent-Oriented Methodologies, pp. 172–202. Idea Group Pub., NY (2005)
3. Bresciani, P., Perini, A., Giorgini, P., Giunchiglia, F., Mylopoulos, J.: Modeling Early Requirements in Tropos: A Transformation Based Approach. LNCS, pp. 151–168. Springer, Heidelberg (2002)
4. García-Magariño, I., Gómez-Sanz, J.J., Agüera, J.R.P.: A Multi-Agent Based Implementation of a Delphi Process. In: The Seventh International Conference on Autonomous Agents and Multiagent Systems, AAMAS 2008, Estoril Portugal, May 12-16, 2008, pp. 1543–1546 (2008)
5. García-Magariño, I., Rougemaille, S., Fuentes-Fernández, R., Migeon, F., Gleizes, M.-P., Gómez-Sanz, J.J.: A Tool for Generating Model Transformations By-Example in Multi-Agent Systems. In: 7th International Conference on Practical Applications of Agents and Multi-Agent Systems (PAAMS 2009) (2009)
6. Gomez-Sanz, J.J., Fuentes, R., Pavon, J.: The INGENIAS Methodology and Tools. In: Agent-oriented Methodologies, pp. 236–276. Idea Group Publishing (2005)
7. Gómez-Sanz, J.J., Fuentes-Fernández, R., Pavón, J., García-Magariño, I.: INGENIAS Development Kit: a visual multi-agent system development environment (BEST ACADEMIC DEMO OF AAMAS 2008). In: The Seventh International Conference on Autonomous Agents and Multiagent Systems, AAMAS 2008, Estoril Portugal, May 12-16, 2008, pp. 1675–1676 (2008)

8. Jouault, F., Kurtev, I.: Transforming Models with ATL. In: Bruel, J.-M. (ed.) MoDELS 2005. LNCS, vol. 3844, pp. 128–138. Springer, Heidelberg (2006)
9. Jung, Y., Lee, J., Kim, M.: Multi-agent based community computing system development with the model driven architecture. In: AAMAS 2006: Proceedings of the fifth international joint conference on Autonomous agents and multiagent systems, pp. 1329–1331. ACM, New York (2006)
10. Schmidt, D.: Model-Driven Engineering. IEEE Computer 39(2), 25–31 (2006)
11. Varro, D.: Model transformation by example. LNCS, pp. 410–424. Springer, Heidelberg (2006)
12. Varró, D., Balogh, Z.: Automating model transformation by example using inductive logic progamming. In: Proceedings of the 2007 ACM symposium on Appiled computing, pp. 978–984 (2007)
13. Wimmer, M., Strommer, M., Kargl, H., Kramler, G.: Towards Model Transformation By-Example. In: Proceedings of the 40th Annual Hawaii International Conference on System Sciences (2007)

GENESETFINDER: A Multiagent Architecture for Gathering Biological Information

Daniel Glez-Peña, Julia Glez-Dopazo, Reyes Pavón, Rosalía Laza, and Florentino Fdez-Riverola

Escuela Superior de Ingeniería Informática
Universidad de Vigo
Edificio Politécnico, Campus Universitario As Lagoas s/n, 32004, Ourense, Spain
dgpena@uvigo.es, jgdopazo@correo.ei.uvigo.es, pavon@uvigo.es,
rlaza@uvigo.es, riverola@uvigo.es

Abstract. The past decade has seen a tremendous growth in the amount of experimental and computational biomedical data, specifically in the areas of genomics and proteomics. In this context, the immense volume of data resulting from DNA microarray experiment presents a major data analysis challenge. Current methods for genome-wide analysis of expression data typically rely on biologically relevant gene sets instead of individual genes. This has translated into a need for sophisticated tools to mine, integrate and prioritize massive amounts of information. In this work we report the development of a multiagent architecture that gives support to the construction of gene sets coming from multiple heterogeneous data sources. The proposed architecture is the base of a publicly available web portal in which final users are able to extract lists of genes from multiple heterogeneous data sources.

Keywords: MAS architecture, integrative data sources, gene set construction.

1 Introduction

Since the introduction of microarray technology in 1995 [1], it now has become a consolidated methodology widely used in health care, drug discovery and basic biological research studies. The main advantage of this technology is that it enables studying the molecular bases of interactions on a gene-scale that is impossible using conventional analysis. The field of gene expression data analysis has grown in the past few years from being purely data-centric to integrative, aiming at complementing microarray analysis with data and knowledge from diverse available sources. The challenge no longer lies in obtaining gene expression profiles, but rather in interpreting the results to gain insights into biological mechanisms.

In this context, recently developed genes set analysis (GSA) methods evaluate differential expression patterns of gene groups instead of those of individual genes [2]. From a biological perspective, GSA methods are promising because functionally related genes often display a coordinated expression to accomplish their roles in the cell. In GSA, as important as the available tools are the gene sets. They are manually arranged using diverse sources of biological knowledge such as the gene ontology information, cytogenetic bands, pathways such as KEGG, Gen MAPP and Biocarta, cis-acting regulatory motifs and co-regulated genes in a microarray study [3]. Through the logical operations of gene sets (e.g. intersection, union, etc.) between different functional classifications, gene sets can be separated into more specific and smaller groups of genes, which facilitates a much more detailed analysis of expression patterns. This has translated into a need for sophisticated tools to mine, integrate and prioritize massive amounts of information [4, 5].

In this work, we report the development of a multiagent architecture that gives support to the whichgenes project [6]. Whichgenes is a simple but powerful gene set building tool, allowing the user to extract lists of genes from multiple heterogeneous data sources in order to use them for further experiments. In the proposed architecture, we exemplify the use of cooperative agents in an information retrieval environment.

This paper has been structured as follows: Section 2 presents theoretical background aspects of the intelligent information agents used in our work. Section 3 focuses in the description of the platform, addressing its architecture, interaction protocols and the use of ontologies. Finally, Section 4 summarizes the main conclusions and discusses future work.

2 Intelligent Agents

In order to solve complex problems in heterogeneous environments, individual information agents must cooperate with other agents [7]. From this cooperation point of view (the agents may collaborate between them at the execution of tasks) we can distinguish two kinds of agents: cooperative and non-cooperative agents. A set of cooperative agents which communicate to execute a given task constitute a multiagent system (MAS). This cooperation allows solving problems that are beyond the individual capabilities of each individual information agent. Basically, the advantages of an approach based on cooperative agents are: simplicity, flexibility, robustness, scalability and the integration of existing legacy systems [8].

2.1 Non-cooperative Agents

Enterprises and investigation groups have developed a high number of non-cooperative information agents. We have distinguished several kinds of non-cooperative agents:

- *Search agents* (Bullseye, Copernic) that help the user to retrieve information from a close list of heterogeneous and distributed sources.
- *Monitor agents* (Mind-It, Informant) control the changes in different information sources (e.g., changes in web structures, updated news in a newspaper, etc.).
- *Filter agents* (InfoScan, BotBox news) reduce the incoming information keeping only the relevant data from the user point of view.
- *Browser agents* (Interquick, Letizia) help the user in navigation through the Web.
- *Agents for electronic commerce* (MySimon, Pricerunner) offer commercial services in order to save time and money.

2.2 Cooperative MAS

The cooperation determines the agent behaviour, this means that it closely defines the principal characteristics of the system design. Therefore, the complexity of system, and its functionality, depends on whether it has agents that make tasks without the collaboration of other agents (without communication) or it has several of them making tasks co-ordinately [9].

The main advantages of an approach based on cooperative agents are: (*i*) *simplicity*, agents are a natural way of modeling complex systems; (*ii*) *flexibility*, MAS provide solutions in situations where knowledge is spatially and/or temporally distributed for obtaining a global solution; (*iii*) *robustness*, MAS distribute computer resources and task execution capabilities among a net of interconnected agents; (*iv*) *scalability*, new agents can be easily introduced in the system for integrating new entities and (*v*) *legacy systems*, MAS allow for the interconnecting an interoperation of multiple existing legacy systems.

The previous idea shows that building a cooperative system is very complex because it is necessary to define its interactions and to implement protocols and methods of cooperation like tasks delegation, contracts or negotiation between the autonomous agents that conform the system.

Furthermore, current investigation related to cooperative information agents is focussed on the two main functionalities that must be provided for the correct behaviour of the whole system: (*i*) a mechanism for linking the different agents and (*ii*) a way of solving the heterogeneity of the information managed.

2.3 Intermediation

Taking into account the role played by the agents, two main classes can be distinguished [10]: *Provider agents* (servers) offer their capabilities to users and other agents and *Requester agents* (clients) use information and services offered by provider agents.

In a very simple multiagent system, the easiest method of coordination among agents is the use of agent-to-agent messages. To make possible this coordination method, all providers and their available services have to be known in advance by every agent.

If the multiagent system is desired to be open (components/participants may enter and exit from the system at any time) this knowledge will not be managed by every agent. In conclusion, the problem of finding agents who might have, or produce, the information or other services needed by requesters is a complex problem.

Certainly, in an open multiagent information system it is necessary the presence of especial agents that mediate between providers and requesters. This kind of agents is called *middle agents* [11, 12]. The process of mediation carried out by middle agents is based on the following steps: (*i*) Provider agents advertise their capabilities to one or more middle agents, describing the service they provide; (*ii*) middle agents store all these advertisement; (*iii*) a requester agent asks for locating and connecting to provider agents, which offer a desired service; (*iv*) middle agents, using the stored advertisements, return the result.

Depending on the result type returned, two types of middle agents can be distinguished:

- *Matchmaker agent.* The result is an ordered list of provider agents, which offer the requested service. Once the result is received by a requester agent, it is the responsible for contacting the provider agent, negotiate and perform the transaction.
- *Broker agent.* In contrast with matchmaker, it performs the complete transaction. This means that there is no direct communication between provider and requester agent, because all the operations go through the broker agent. The main tasks of these agents are contacting to appropriate provider agents, negotiating, performing and controlling the transaction and giving back the results of service to requester agent.

Given the fact that different types of middle agents provide different performance results, deciding what types of middle agents are appropriate depends on the application [10]. In addition, deciding between the use of matchmaking or brokering techniques to solve the connection problem yields to performance tradeoffs along a number of dimensions both quantitative (such as the time needed to fulfil a request) and qualitative (such as the robustness and adaptively of the system to the failure or addition of agents). In our system has been defined a matchmaker agent.

In the area of computational trust one common classification distinguishes between 'probabilistic' and 'non-probabilistic' models. While non-probabilistic systems vary considerably and need further classification (e.g., as social networks or cognitive), probabilistic systems usually have common objectives and structure. In such models, trust information is based on the past behaviour (history). Histories are used to estimate the probability of potential outcomes arising in a next interaction with an entity. Probabilistic models, called 'game-theoretical' by Sabater and Sierra [13], rely on Gambetta's view of trust [14]: *"...trust is a particular level of the subjective probability with which an agent assesses that another agent or group of agents will perform a particular action, both before he can monitor such action (or independently of his capacity ever to be able to monitor it) and in a context in which it affects his own action"*.

2.4 Ontology

As it has been said, one of the main problems for cooperation in an agent's society is the semantic heterogeneity of information that the agents must manage. Semantic heterogeneity considers the content of an information item and its 'intended' meaning. In order to manage this semantic heterogeneity, the meaning of the interchanged information has to be understood across the agent's society. This task can be done with the use of ontologies.

Ontology are defined as a explicit specification of a conceptualisation [15], that is, a representation (with a set of concepts and the relationships among them) of an abstract and simplified view of the world. In agent's society, ontologies can be used to describe the semantics of the request and service descriptions and to make explicit the content of the different information sources. They also reduce conceptual and terminological conflicts providing a unified framework.

Middle agent overcomes the semantic heterogeneity by means of a knowledge-based process, which relays on using ontologies. That is, the use of ontologies enables shared understanding among different agents with different aims and different viewpoints of the global system. In our system it has been defined two kinds of ontology agents.

2.5 Interaction Protocols

All agents need to cooperate in order to accomplish their objectives. Usually they utilize a special type of an ACL (*Agent Communication Language*) to exchange knowledge and to cooperate effectively. Therefore, they have to maintain conversations (groups of interrelated messages) among them. These conversations between agents often fall into typical patterns called interaction protocols [16].

Many standard interaction protocols have been defined by the FIPA (*Foundation for Intelligent Physical Agents*). In our system, all complex interactions among agents have been defined using a reduced set of the interaction protocols defined by FIPA. In this work, KQML (*Knowledge Query and Manipulation Language*) is used. KQML provides a basic architecture for knowledge sharing through a special class of agent called *communication facilitators* which coordinate the interactions of other agents.

3 System Architecture

The main goal of the proposed system is to provide uniform access to a set of heterogeneous sources of information, which in this case are sources of biological knowledge. Based in a set of cooperative agents, the system will allow the user to make a request, which will be distributed among the available databases in order to achieve the results.

Gene sources are connected to the multiagent system via wrapper agents coping with the heterogeneous technologies and protocols that the different databases offer. For example, the KEGG pathway database [17] publishes a SOAP based API to extract its pathways and their related genes. In other cases, like in GeneCards [18], the wrapper agent accesses directly to the web page, fills a query form and parses the HTML output to obtain the resulting gene list. Wrapper agents register themselves in a matchmaker agent whose main objective is to keep a list of available sources.

Figure 1 presents the system architecture which is composed of several kinds of agents. The characterisation of each agent is introduced below:

- *Interface agent*. Each user interacts with the system through its own interface agent. It is responsible for showing the user a request interface for entering his request. Interface agent captures and translates the request before sending it to an intermediary agent. Moreover, it presents to the user the results obtained and facilitates the navigation through the collection of genes. In the basic system usage, a list of available gene sources is displayed to the user. Currently, there are two types of sources: (*i*) free-query sources, which extract genes from a free text query like a disease name and (*ii*) constrained-query sources, which can search genes given a set of controlled terms (typically database identifiers) such as pathways, GO terms, etc. These two types of query are defined in a global ontology carried out by the *Global Ontology* agent.

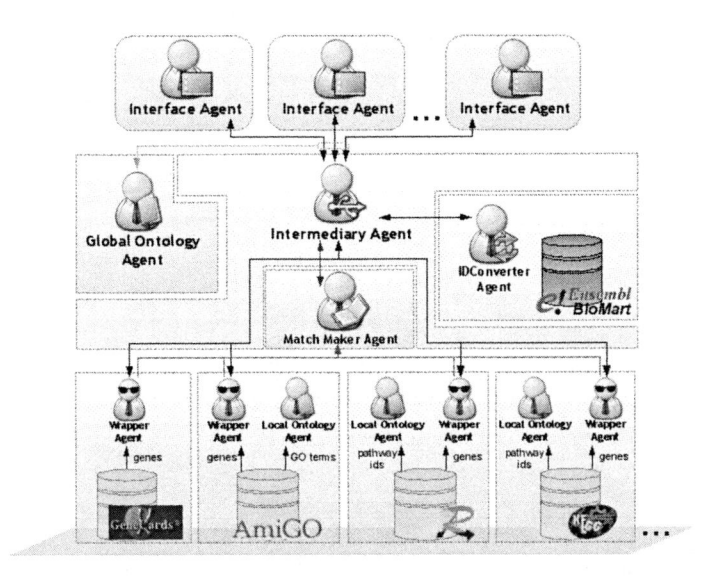

Fig. 1. GENESETFINDER system architecture

- *Intermediary agent.* It is the core of the interrogation process to the different databases. It receives the request from the interface agent, asks the matchmaker for the suitable wrappers, integrates the existing results from the different sources and gives them back to the interface agent.
- *Matchmaker agent.* It is a facilitator that carries out several communication services (in this case, providing 'matchmaking' between information providers and clients) [19]. It works as a yellow pages service, giving back to the intermediary agent the list of wrapper agents it has to contact to.
- *Wrapper agent.* There is a wrapper agent for each publicly available knowledge database. Wrapper agents implementing constrained-query sources make use of local ontology agents which basically keeps information about the source database specific identifiers. The interface agent interacts with an intermediary agent who (*i*) offers the list of available sources (coming from the mathmaker agent) (*ii*) redirects the query and gets the response from the corresponding wrapper agent and (*iii*) normalizes the gene list output to a same namespace by the assistance of the ID-Converter agent. Currently GENESETFINDER displays HUGO IDs in the case of human and MGI names in the case of the mus musculus specie.

3.1 Intermediation Process

In our developed system provider agents are implemented by wrapper agents who offer their capabilities to interface agents. Nevertheless some middle agents are introduced to mediate between wrappers and interface agents. These middle agents are the intermediary agent and the matchmaker agent.

The process of mediation implies the following interactions:

- (I1) *Interface-Intermediary*. Intermediary agent accepts the request from interface agent.
- (I2) *Intermediary-Matchmaker*. Intermediary agent asks matchmaker for a list of wrapper agents with sources containing these concepts.
- (I3) *Intermediary-Wrapper*. Once intermediary agent obtains the list of wrappers distributes the request to the appropriate wrapper.
- *Wrapper-Intermediary*. The answers obtained from different wrapper agents are integrated by the intermediary agent before giving them back to the interface agent. This process may be not as straightforward as merging the results and may require, in some cases, some processing by the intermediary agent. This interaction is the answer in step number (I3).
- *Intermediary-Interface*. The results integrated in the previous step are sent to the interface agent who started the request process. This interaction is part of number (I1).

3.2 Ontologies

Ontology agents are used to deal with semantic heterogeneity among sources. In order to achieve this objective two kinds of ontology agents can be distinguished in our system: (*i*) global ontology agent and (*ii*) local ontology agent.

Global ontology agent has a conceptual model at higher abstraction level than the data models from each of the participant sources. This means that it provides concepts over which the user can make the request, and the semantic relationships among concepts. In addition, each source has its own local ontology agent. This agent relates each concept defined in the global ontology to its particular representation in the data source.

These agents collaborate with the rest of the system agents for solving a user request in the following way:

- (O1) *Interface-Global ontology*. The two types of supported queries (free and constrained) are defined in a global ontology carried out by the global ontology agent.
- (O2) *Wrapper-Local ontology*. Wrapper agent asks local ontology in order to translate each concept contained in the request received from intermediary agent to the corresponding concept in the source. It keeps information about the source database specific identifiers.

When a new source of information is added or removed from the system, a process of actualization is required. This process implies a first step which is done manually and consists of defining the local ontology and wrapper associated to the source. Next, the new local ontology agent communicates to global ontology agent the concepts the source is able to deal with, and to matchmaker the list of concepts associated with the new source.

3.3 Interaction Protocols

Figure 2 shows a sequence diagram which represents the underlying interaction protocols of the steps of the intermediation process (Section 3.1) and the ontology query process (Section 3.2). Each step implies one or more interaction protocols which are characterised in the graph with a rectangle labelled with the same name.

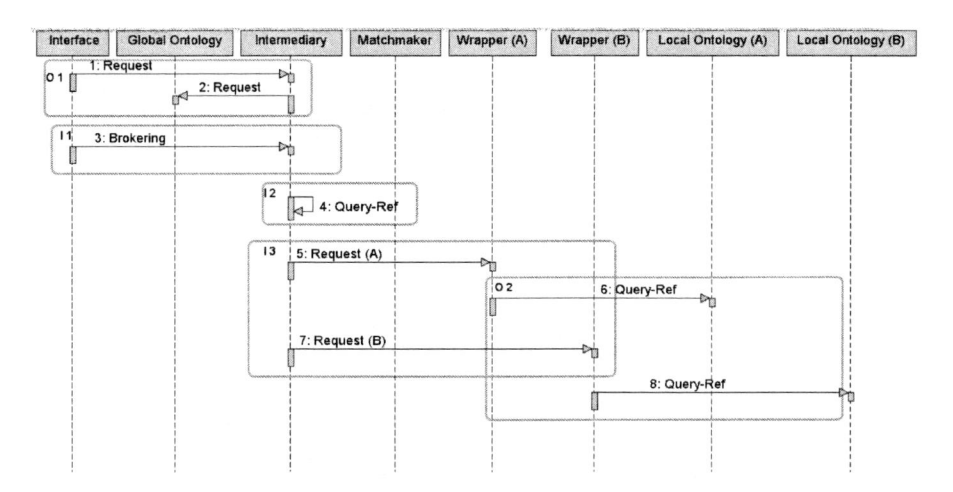

Fig. 2. FIPA interaction protocols used in GENESETFINDER

Each box at the top represents a set of agents with the same role or functionality. A special case is the wrapper and local ontology agents because they are subdivided in two boxes in order to emphasize that their communications are dependent on their underlying sources and related knowledge. Each arrow indicates a FIPA interaction protocol [16, 20].

In this example the user makes a request about two different sets of concepts, represented by A and B. This question implies the following interaction protocols:

- *O1*. The interface agent *speaks* with the intermediary agent using the FIPA request protocol. This interaction protocol allows the intermediary agent (initiator) to request the global ontology (participant) to perform the provide action of the domain knowledge.
- *I1*. With the FIPA brokering interaction protocol the interface agent translates the user request (about A and B) to intermediary agent. The intermediary will determine a set of appropriate wrapper agents (helped by the matchmaker) for answering the request. It will send the request to those agents and will give their answers back to the interface agent.
- *I2*. The intermediary agent asks the matchmaker about the relevant wrappers for a specific user request (A or B). They use for this task the FIPA query-ref interaction protocol. Interface agent uses the *subscribe* performative to request that matchmaker monitor for the truth of request. If wrapper subsequently informs matchmaker that it believes request to be true, then matchmaker can in turn inform interface agent [19].
- *I3*. When the intermediary agent knows the list of wrapper agents, the request is sent (FIPA request protocol). This interaction will take place with different groups of wrapper agents depending on the concepts that they manage.
- *O2*. Each wrapper agent, in order to obtain the information of its underlying source, asks the local ontology agent with a FIPA query-ref interaction protocol about the requested concepts.

4 Results and Conclusions

Many advantages in using cooperative agents for information retrieval in heterogeneous environments are noticed: simplicity, robustness and scalability. This work emphasizes three important aspects that must be taken into account in the construction of cooperative agents systems: (*i*) the model of intermediation, (*ii*) the interaction protocols among the different agents and (*iii*) how ontologies are used to deal with the heterogeneity of information.

GENESETFINDER uses two middle agents, one of them does a matchmaking process, while the other (an intermediary agent) operates as a broker taking part into the request process. In order to solve semantic conflicts among the sources of information two kinds of ontology agents are introduced.

The system is able to easily grow. When a new source is added, an associated wrapper agent and a local ontology agent are created and the global ontology agent is actualised by means of incorporating the new semantics. At the moment this process is done manually by a human with a deep knowledge of the source.

For future research in the domain of ontology, we propose to study the way of automating the process for the actualisation of global ontology and matchmaker agents. This process will take into account the information available from the local ontology agent and will be done using the features of a description logic system.

In order to guarantee the robustness of the system, new analysis must be done about the possibility of replicating intermediary, matchmaker and global ontology agents for improving the overall availability of the system. This copy may also increase the performance of the system with a high number of sources.

Other important line of research is the definition of methodological aspects in the development. This includes the description of agents and roles, the association of interactions and roles, diagrams of different levels of abstraction, description of knowledge transference, etc.

Finally, the developed tool can be freely accessed and tested at http://www.whichgenes.org/.

Acknowledgements. We thank Gonzalo Gómez (CNIO, http://www.cnio.es/) for valuable discussion in early versions of the manuscript as well as for acting as system evaluator. This work is partly funded by the research project BioTools (ref. 2008-INOU-2) from University of Vigo. The work of DGP is supported by a 'Maria Barbeito' research contract from Xunta de Galicia.

References

1. Schena, M., Shalon, D., Davis, R.W., Brown, P.O.: Quantitative monitoring of gene expression patterns with a complementary DNA microarray. Science 270(5235), 467–470 (1995)
2. Nam, D., Kim, S.-Y.: Gene-set approach for expression pattern analysis. Briefings in Bioinformatics 9(3), 189–197 (2008)
3. Khatri, P., Drăghici, S.: Ontological analysis of gene expression data: current tools, limitations, and open problems. Bioinformatics 21(18), 3587–3595 (2005)

4. Ball, C.A., Sherlock, G., Brazma, A.: Funding high-throughput data sharing. Nature Biotechnology 22(9), 1179–1183 (2004)
5. Kanehisa, M., Bork, P.: Bioinformatics in the post-sequence era. Nature Genetics 33(Suppl.), 305–310 (2003)
6. Wichgenes, http://www.wichgenes.org/
7. Jennings, N.R.: Agent-Based Computing: Promise and Perils. In: Proc. 16th International Joint Conference on Artificial Intelligence, pp. 1429–1436 (1999)
8. Wooldridge, M., Jennings, N.R.: Intelligent Agents: Theory and Practice. The Knowledge Engineering Review 10(2), 115–152 (1995)
9. Sycara, K., Paolucci, M., van Velsen, M., Giampapa, J.: The RETSINA MAS Infrastructure. Robotics Institute Technical Report # CMU-RI-TR-01-05 (2001)
10. Klusch, M., Sycara, K.: Brokering and matchmaking for coordination of agent societies: a survey. In: Omicini, A. (ed.) Coordination of Internet Agents. Models, Technologies, and Applications, pp. 197–224 (2001)
11. Decker, K., Sycara, K., Williamson, M.: Middle-Agents for the Internet. In: Proc. 15th International Joint Conference on Artificial Intelligence, vol. 1, pp. 578–583 (1997)
12. Huhns, M.N., Singh, M.P.: All agents are not created equal. IEEE Internet Computing 2(3), 94–96 (1998)
13. Sabater, J., Sierra, C.: Review on Computational Trust and Reputation Models. Artiff. Intell. Rev. 24(1), 33–60 (2005)
14. Gambetta, D.: Can we trust trust? In: Gambetta, D. (ed.) Trust: Making and Breaking Cooperative Relations, pp. 213–237. University of Oxford, Department of Sociology (2000)
15. Gruber, T.: Toward principles for the design of ontologies used for knowledge sharing. International Journal on Human Computer Systems 43, 907–928 (1994)
16. FIPA organization, http://www.fipa.org/
17. Kanehisa, M., Araki, M., Goto, S., Hattori, M., Hirakawa, M., Itoh, M., Katayama, T., Kawashima, S., Okuda, S., Tokimatsu, T., Yamanishi, Y.: KEGG for linking genomes to life and the environment. Nucleic Acids Res. 36(database issue), D480–D484 (2008)
18. Rebhan, M., Chalifa-Caspi, V., Prilusky, J., Lancet, D.: GeneCards: integrating information about genes, proteins and diseases. Trends in Genetics 13(4), 163 (1997)
19. Finin, T., Fritzson, R., McKay, D., McEntire, R.: KQML as an agent communication language. In: Proc. 3rd International Conference on Information and Knowledge Management, pp. 456–463 (1994)
20. FIPA Interaction Protocol Lib. Specification, http://www.fipa.org/specs/fipa00025/

Agent Design Using Model Driven Development

Jorge Agüero, Miguel Rebollo, Carlos Carrascosa, and Vicente Julián

Departamento de Sistemas Informáticos y Computación
Universidad Politécnica de Valencia
Camino de Vera S/N 46022 Valencia, Spain
{jaguero, mrebollo, carrasco, vinglada}@dsic.upv.es

Abstract. Object-oriented software development methodologies have adopted a *model-driven* approach to analysis and design phases. Currently, a similar approach is being adopted for Multi-Agent Systems to improve the development process and the quality of the agent-based software. Model-Driven Development is a technique that allows to obtain executable code by means of transformations from models and meta-models. This work presents a transformation process that allows to generate automatically the code of an agent over his execution platform. That is, an agent is developed under the MDD approach in an easy and transparent way for the user. The code obtained from the transformations is executed over ANDROMEDA and JADE-Leap embedded agent platforms.

1 Introduction

Though Multi-agent Systems (MAS) are a powerful technology with relevant applications, to develop a MAS requires, currently, a great experience in one or more design platforms. A big challenge for the design of MAS is to provide efficient tools to be used by any kind of user, even not expert ones. But the view provided by the Model-Driven Development approach (MDD) can facilitate and simplify the design process and the quality of the agent-based software, because it allows to re-use software and transformations between models[11].

Basically, MDD proposes to automatically generate code from the models and some specific platform details using transformations. Recently, some proposals for applying MDD ideas and technicals to MAS have been proposed [9, 13, 10]. But these previous approaches have meaningful differences with the one presented here: Perini and Susi[13] proposal is based on using meta-models that are specific of a methodology; Guessoum and Jarraya[9] do not propose to obtain executable code of the agent; and Hans et al.[10] do not take into account models of agents which platforms work specifically in systems with little computational resources, as embedded agent platforms.

To sum up, this work presents a transformation process to generate an agent executable code. This is made applying the basic ideas from MDD to agent design, that is, to design agents using models or abstract concepts forgetting implementation and platform details. After that, the proper agent for the platform desired to his execution is generated by means of transformation models. This allows to reduce the gap between the design and the implementation of agents. To test the proposal feasibility, the executable code obtained is implemented in two embedded agent platforms: ANDROMEDA[2] and JADE-Leap.

This paper is structured as follows: section 2 explains the model-driven development and how to apply it to agent design. Section 3 describes the concepts of the agent metamodel. ANDROMEDA platform description is presented in section 4. Section 5 explains how to design an agent from the MDD viewpoint. Section 6 shows an application example, proving the feasibility of developing agents using MDD. Finally, some conclusions of the current work are presented in section 7.

2 Model-Driven Development

The MDD is a quite new resource in the software engineering field. The main point in MDD is to define a process guided by using models (model-driven), in which visual modeling languages are used to integrate a huge diversity of technologies applied in the computing system design [11]. Currently, Model-Driven Architecture (MDA[1]) is an approximation for developing software based on the MDD that has been defined and standardized for the OMG (Object Management Group).

MDA proposes to move the philosophy of object-oriented programming to one based on models, changing the software artifacts of the programming technology: from "all is an object" to "all is a model"[4]. MDA approximation uses and creates different models at different abstraction levels, to fusion and combine them when it is needed to implement the application. When the abstraction levels are too high, these models are known as meta-models. A meta-model is just a model of a modeling language, defining the structure, semantics and restrictions for a family of models. In MDA, MOF[12] (Meta Object Facility) is the language facilitating the meta-models creation. MDA considers different kinds of models. These models have different names according to their abstraction level, as the Computation Independent Model (CIM) that details the system's requirements in a model that is independent of the computation, the Platform Independent Model (PIM) that represents the system's functionalities without considering the final platform where it is going to be implemented), and the Platform Specific Model (PSM), obtained from combining the PIM model with the specific details of the selected platform. One fundamental aspect is the definition of the transformation model, which allows to automatically convert the models. Transformations allow to pass from a model with an abstraction level to another one with a different level. Transformations are relational entities describing mapping rules between concepts of different meta-models.

From the viewpoint of the MAS design, different methodologies have identified a set of models to specify the different features of a system. These models can be fitted or re-flexed in different MDA meta-models specifying the concepts describing the MAS, as: roles, behaviors, tasks, interactions or protocols. These models can be used to model multi-agent systems in a very abstract way, without focusing in the specific details or requirements of the platform, as a platform independent model[9, 13, 10]. Meta-models can be specified from different points of view and they can be represented in different abstraction levels. After that, it is possible to develop transformations models from the agent independent meta-models (PIM) to platform specific models (PSM)[10]. Figure 1 shows the possible relations between the concepts of different MDA models and

---
[1] http://www.omg.org/mda/

the possible transformations between them (code transformations assume that there are three platforms to execute agents).

3 Agent Meta-model

The main problem to define a platform-independent agent model is to select the appropriated concepts that will be included in the model and that will be used to build the different features and classes of agents. At the moment, there is a large amount of agent models that provide a high-level description of their components and their functionalities, but they need to be changed and manually implemented when applied to specific agent platforms. To define the agent model presented in this paper, some of the most used and complete agent model proposals have been studied. The purpose of this study was to extract their common features and adapt them to the current proposal. In this way, Tropos [6], Gaia [14], Opera [7] and Ingenias [8] have been considered. So, the proposed process allows to do the analysis and design of the system according to different well-known methodologies (corresponding to the CIM). Then, the obtained design will be transformed in terms of our agent model (PIM). Furthermore, an automatic transformation into different agent platforms can be obtained according to the MDD process. In this case: JADE-Leap and ANDROMEDA platforms (PSMs). The process used in the proposal presented can be seen in Figure 1.

The main components and basic concepts employed in the meta-model are summarized in Table 1. This meta-model is called *agent-π* (agent-PI: agent Platform Independent). For a more detailed explanation of the model refer to Agüero et al.[1]

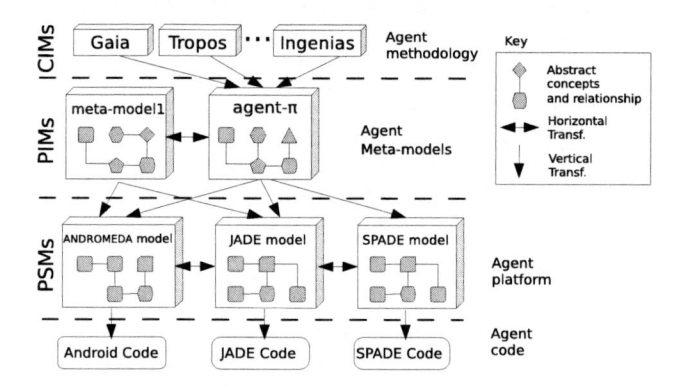

Fig. 1. MDA meta-models and their transformation schemes

4 The ANDROMEDA Platform

ANDROMEDA (ANDROid eMbeddED Agent platform)[2, 1] is an agent platform specifically oriented to embedded agents over the *Android*[2] operating system. The

---
[2] Android System, http://code.google.com/android/

Table 1. Main concepts employed in the *agent-π* meta-model

agent-π concepts	Description
Agent	The entity *agent* represented in any methodology
Behaviour	It encapsulates a set of capabilities active in specific circumstances, it represents the abstract concept of role
Capability	It represents an event-driven approach to solve a specific problem
Task	The know-how respect a specific problem
Event	It is employed to activate capabilities inside the agent
Environment	It is the way to model the external world
BeliefContainer	Abstraction employed to represent the agent knowledge

agents developed inside this platform are based on the *agent-π* meta-model. *Android* can be seen as a software system specifically designed for mobile devices which includes an operating system, a middleware and key applications. Developers can create applications for the platform using the Android SDK. Applications are written using the Java programming language and they run on Dalvik (the *Android* Virtual Machine), a custom virtual machine designed for embedded use, which runs on top of a Linux kernel.

The proposed ANDROMEDA platform includes all the abstract concepts of the *agent-π* meta-model. The implementation was done using the main API components of *Android* (SDK 1.0, Release 1). The correspondence between the *Android* components and the main *agent-π* abstract concepts are shown in Table 2.

Table 2. The *Android* components used in the *agent-π* meta-model

agent-π Components	Android Components	Overloaded methods
Agent	Service	onCreate(), onStart(), onDestroy()
Behaviour	BroadcastReceiver	registerReceiver(), onReceive()
Capability	BroadcastReceiver	registerReceiver(), onReceive()
Task	Service	onStart(), onDestroy()
Events	Intents	IntentFilter()
Beliefs	Contentprovider	–

5 Designing an Agent with a Model Driven Approach

According to the proposal of this paper, the agent design will have an MDA approach, that is, it will be model-driven. The first step consists in obtaining an agent representation using the abstract concepts of the proposed model. After this, each agent modeled in the system could be transformed into code according to its specific agent platform where the agent will be executed. In this work, the proposal is centered in the study of the transformation of the agent models into the agent platforms ANDROMEDA and JADE-Leap.

According to Figure 1, it is necessary to apply two transformation processes. The first one is model-to-model (PIM to PSM applying mapping rules) and the second one is model-to-text (PSM to code applying templates). In the presented case, it is needed

Table 3. Transformation rules of the *agent-π* meta-model to ANDROMEDA PSM

Rule	Concept	Transformation
1	Agent	Agent-π.Agent ⇒ ANDROMEDA.Agent
2	Behaviour	agent-π.Behaviour ⇒ ANDROMEDA.Behaviour
3	Capability	agent-π.Capability ⇒ ANDROMEDA.Capability
4	Task	agent-π.Task ⇒ ANDROMEDA.Task
5	Events	agent-π.Event ⇒ ANDROMEDA.Event
6	Beliefs	agent-π.BeliefContainer ⇒ ANDROMEDA.BeliefContainer
7	Environment	agent-π.Environment ⇒ ANDROMEDA.Environment

to apply a model-to-model transformation by doing correspondence between the concepts of the *agent-π* meta-model (PIM) and the concepts employed in the target agent platform (PSM of the ANDROMEDA or JADE-Leap platforms). After this, a model-to-text transformation is needed, translating the obtained PSM of the selected target agent platform into executable code of the same platform. Next sections explain these transformations for each one of the selected agent platforms, the which is based in the *EMF*[5] project (Eclipse Modelling Framework).

5.1 Implementing Agent-π Models in ANDROMEDA

The ANDROMEDA platform has been designed using the same concepts of the *agent-π* model. So, an agent in ANDROMEDA is implemented using the same concepts employed in the abstract model. This design greatly simplifies the automatic transformation between PIM and PSM, because the ANDROMEDA PSM is very similar to the *agent-π* model. Accordingly, the first step of the MDA process is not necessary. In this case, the needed transformation rules are mainly model-to-text, generating directly ANDROMEDA code which can be combined with additional user code. A selected set of the defined transformation rules are shown in Table 3. Then, the two transformation processes of the MDA approach are merged in only one step.

For reasons of brevity only the main rules have been shown. It's good to clarify that in **Rule 3**, a *Task* can be a simple or complex task. For example, if a simple task exists in the model (OneShot task), this task is transformed according to the transformation rule *agent-π.Task.OneShot* ⇒ ANDROMEDA.*Task.OneShot*.

It is important to mention that these transformations are done using *MOFScript*[3] for model to text transformations (code skeletons or code templates generation), which mainly: (i) changes the agent concepts or models into the Java classes used in ANDROMEDA, and (ii) defines how to obtain the needed attributes from the model. Figure 2 illustrates how these rules are implemented using *MOFScript*, part of the code of the **Rule 1** (which transforms the *agent* concept).

5.2 Implementation of the Agent-π Model in JADE-Leap

JADE-Leap was chosen as a target agent platform for the presented MDA proposal, because it is a revision of the most used agent platform, JADE[3], to be employed in limited devices as PDA or mobile phones and differ from JADE is at the interfaces.

---
[3] http://jade.tilab.com/

```
texttransformation UMLAGENT2ANDROMEDA (in myAgentModel:uml2)
...
//Rule1: Agent transformation
uml.Package::mapPackage () {
  self.ownedMember->forEach(c:uml.Class)
     if (c.name != null) if (c.name = Agent) c.outputGeneralization()
}
uml.Class::outputGeneralization(){
   file (package_dir + self.name + ext)
   self.classPackage()
   self.standardClassImport ()
   self.standardClassHeaderComment ()
   <% public class %> self.name <% extends Agent { %>
      self.classConstructor()
      <% // Attributes    %>
      self.ownedAttribute->forEach(p : uml.Property) {
          p.classPrivateAttribute()
      } newline(2)
   <%}%>  ...
}
```

Fig. 2. Transformation of the *agent* concept using MOFScript

The transformation from agent-π models to JADE-Leap code must be done employing the two phases previously commented: (i) the first phase translates from the PIM model to JADE-Leap PSM mode, and allows to obtain a correspondence among the abstract concepts of the agent-π model to JADE-Leap concepts; (ii) the second phase allows to translate the JADE-Leap PSM models obtained in the previous phase into executable JADE-Leap code. The JADE-Leap Model (PSM model) will be called JADELM.

The transformation needed in the first phase is done using the ATL toolkit [3], a plug-in of the Eclipse IDE[4], which automatically translates between the two agent models. ATL employs model-to-model transformation rules. A subset of the transformation rules needed in this phase is shown in Table 4, and the ones related to the main agent concepts are detailed next.

Table 4. Transformation rules for the *agent-π* meta-model to JADE-Leap

Rule	Concept	Transformation
8	Agent	agent-π.Agent $\Rightarrow$ JADELM.Agent
9	Behaviour	agent-π.Behaviour $\Rightarrow$ JADELM.ParallelBehaviour
10	Capability	agent-π.Capability $\Rightarrow$ JADELM.OneShotBehaviour
11	Task	agent-π.Task $\Rightarrow$ JADELM.Behaviour
12	Events	agent-π.Event $\Rightarrow$ JADELM.ACLMessage
13	Beliefs	agent-π.BeliefContainer $\Rightarrow$ JADELM.Schema
14	Environment	agent-π.Environment $\Rightarrow$ JADELM-Java.ports

The conversion in **Rule 8** is direct because our agent model matches with the JADE-Leap agent model. After the transformation, the methods have to be reviewed to check

---
[4] http://www.eclipse.org/

that the JADE-Leap agent works properly. One of the most important methods to be derived is init() because this method contains the code executed by the agent. Then, the init() method of the agent-π is moved into the setup() method of JADELM, i.e. init() → setup(). Other methods are also derived: the method to destroy the agent destroy() → takeDown() and the method to add behaviors addBeh() → addBehaviour().

For **Rule 9**, a *Behaviour* in this agent model is a set of actions that can be executed. To make possible to launch several actions, a **Behaviour** correspond with a CompositeBehaviour in JADELM. Specifically, for each *Behaviour* referenced in agent-π a ParallelBehaviour must be added in JADELM. These ParallelBehaviour will be empty at first, but new *Behaviour* will be added for each task in the model when the *capability* and *Task* of agent-π will be transformed.

In **Rule 10** a *capability* is a component that can launch an activity or not depending on the arrival of the corresponding event, that is, the capability launch a task if its trigger event has arrived (event-driven). To emulate this behaviour, each *capability* correspond with a JADE-Leap simpleBehaviour, which goal is to verify the arrival of an event and, if the event is the correct, then the activity will be launched.

For **Rule 11**, a *Task* in our agent model can be a simple or a complex action. The type of *Task* establishes a specific transformation to a SimpleBehaviour or a CompositeBehaviour; in general, to a *Behaviour*. For example, if there is a cyclic task in agent-π, a CyclicBehaviour() have to be added in JADE-Leap. In short, for each *Task* in agent-π a type of *Behaviour* must be added in JADELM. But this type of *Behaviour* must be contained into a OneShotBehaviour (which emulates the agent-π *capability*) and controls its triggering (depending on the event). After that, the OneShotBehaviour must be added to a ParallelBehaviour (created empty in the beginning). *Tasks* will be added using addsubBehaviour() over the mentioned ParallelBehaviour. A *Task* is the place where users write their code. So it is important to define how to it to JADELM.

This can be done by translating the doing() method of agent-π to the action() method in JADE-Leap. There are more transformation rules, one for each abstract concept of the agent, but for reasons of space they are not expounded in this paper. The PSM model is transformed into code (PSM-to-code) using *MOFScript*.

6 Transformation Example

To validate the proposal, transformations to develop an agent on two different platforms are going to be evaluated. There is a set of users who use a system that makes electronic auctions, in the style of the classic 'eBay'. Users have available an agent to represent them in the auction system. Agents will make all the transactions (bids and sales) on behalf of their users.

The agent can play two different roles: auctioneer or bidder, depending on it is selling or buying products. Users have to establish the conditions for the auction (minimun or maximum price or availability period). Auctioneer role is in charge of advertising new products to the platform and processing the sale after the process is completed and the

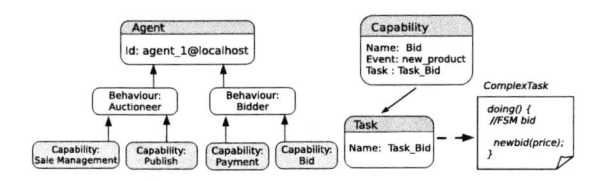

Fig. 3. Platform independent model (PIM) of the agent

auction has a winner. Bidder role bids for articles in which it is interested and completes the payment when an auction is won.

The development of the agent begins using the agent-π model (PIM). Each agent role is modeled as a behaviour, resulting an auctioneer and a bidder behaviour, each one of them with two capabilities: *publish* and *sale management* for the auctioneer role and *bid* and *payment* for the bidder role.

Each capability executes its associated action once the triggering event is received. This event is different for each capacity. When the event is received, the capability executes the actions corresponding to its main task. The above model is shown in Figure 3. The next step is to apply the transformation rules to the PIM to obtain the corresponding PSMs.

In the case of the ANDROMEDA platform, the transformation is straightforward as previously commented: beginning with rule 1 to obtain the agent, then the rule 2 to behaviors and after that rules 3 and 4 for the tasks and capabilities respectively. The obtained model can be seen in Figure 4.

For the JADE-Leap platform, rule 8 must be applied to get an agent and the agent in the model JADE-Leap. After that, rule 9 generates a *ParallelBehaviour* for each

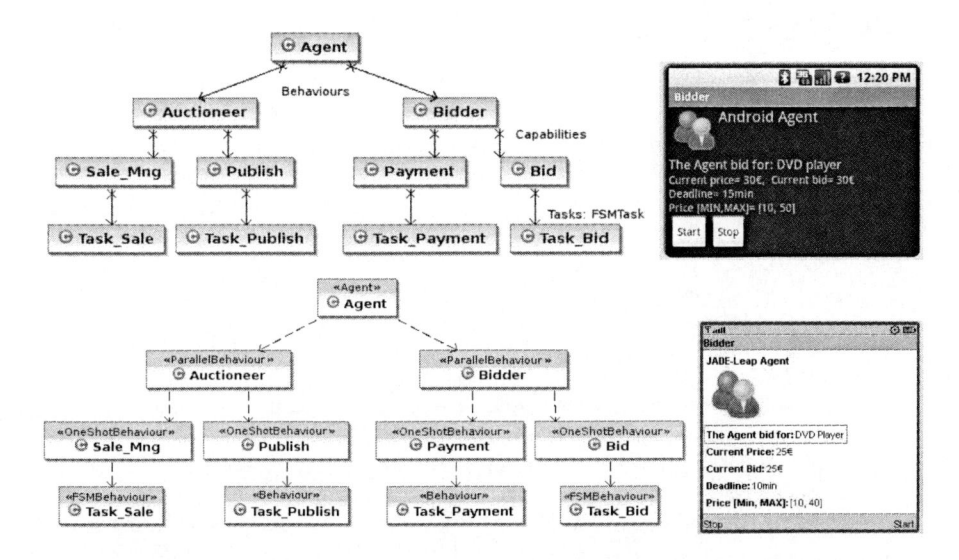

Fig. 4. PSM agent models and sample execution screen over ANDROMEDA and JADE-Leap

behavior (role). Then, rule 10 generates two *OneShotBehaviour* (one for each capability) and they are included into the *ParallelBehaviour*. Finally, rule 11 extracts the *Behaviours* that executes the agent actions. For example, in the case of the bidding behavior, they will be *bid* (with a *CompositeBehaviour*) and *payment* (a *SimpleBehaviour*). In Figure 4 can be seen the obtained model.

Finally, the proper PSM-to-code transformation is applied to obtain the executable instances of the agents. The code generated by the tool can be edited by the user to complete the development of the agent with specific details of the running platform. The result is a set of agents running as Figure 4 shows.

7 Conclusions

This paper presents a transformation process for embedded agent platforms JADE-Leap and ANDROMEDA, using the ideas proposed by the MDA of OMG, although the use of the MDA is concerned principally with the methodologies for the development of object-oriented software. It has been verified that this approach can be efficiently adopted in agent-oriented software development, simplifying to the users the process of agent design as it hides as much as possible implementation details. Besides automating the design process, it reduces the human intrusion in the development of the agent. It also can verify the inter-operability between platforms as a single agent model can be run on different platforms thanks to the automatic transformations.

The future work of this research line will focus on: (i) developing new rules for conversion of agents to other platforms, and (ii) increasing the concepts or components that are used to model the agent, e.g the inclusion of organizations.

Acknowledgment

This work was partially supported by the Spanish government under grant CSD2007-00022 and under FEDER grant TIN2006-14630-C0301 project and in part by the Valencian Government under grant PROMETEO 2008/051.

References

1. Agüero, J., Rebollo, M., Carrascosa, C., Julián, V.: Does android dream with intelligent agents? In: International Symposium on Distributed Computing and Artificial Intelligence 2008 (DCAI 2008), vol. 50, pp. 194–204 (2008) ISBN: 978-3-540-85862-1
2. Agüero, J., Rebollo, M., Carrascosa, C., Julián, V.: Towards on embedded agent model for android mobiles. In: The Fifth Annual International Conference on Mobile and Ubiquitous Systems (Mobiquitous 2008). CD Press (2008) ISBN: 978-963-9799-21-9
3. Allilaire, F., Bézivin, J., Jouault, F., Kurtev, I.: Atl: Eclipse support for model transformation. In: European Conference on Object-Oriented Programming (ECOOP 2006) (2006)
4. Bézivin, J.: On the unification power of models. Software and Systems Modeling 4(2), 171–188 (2005)
5. Budinsky, F., Brodsky, S., Merks, E.: Eclipse Modeling Framework. Pearson Education, London (2003)

6. Castro, J., Kolp, M., Mylopoulos, J.: A requirements-driven development methodology. In: Dittrich, K.R., Geppert, A., Norrie, M.C. (eds.) CAiSE 2001. LNCS, vol. 2068, pp. 108–123. Springer, Heidelberg (2001)
7. Dignum, V.: A model for organizational interaction: based on agents, founded in logic. Phd dissertation, Utrecht University (2003)
8. Gomez Sanz, J.: Modelado de sistemas multi-agente. Phd thesis, Universidad Complutense de Madrid, Spain (2002)
9. Guessoum, Z., Jarraya, T.: Meta-models & model-driven architectures. In: Contribution to the AOSE TFG AgentLink3 meeting (2005)
10. Hahn, C., Madrigal-Mora, C., Fischer, K.: A platform-independent metamodel for multiagent systems. In: Autonomous Agents and Multi-Agent Systems, vol. 16 (2008)
11. Kleppe, A., Warmer, J.B., Bast, W.: MDA Explained: The Model Driven Architecture: Practice and Promise. Addison-Wesley Professional, Reading (2003)
12. (OMG): Object management group. meta object facility (mof) 2.0 core specification (October 2004), http://www.omg.org/docs/ptc/04-10-15.pdf
13. Perini, A., Susi, A.: Automating model transformations in agent-oriented modelling. In: Müller, J.P., Zambonelli, F. (eds.) AOSE 2005. LNCS, vol. 3950, pp. 167–178. Springer, Heidelberg (2006)
14. Zambonelli, F., Jennings, N.R., Wooldridge, M.: Developing multiagent systems: The gaia methodology. ACM Trans. Softw. Eng. Methodol. 12(3), 317–370 (2003)

A Tool for Generating Model Transformations By-Example in Multi-Agent Systems

Iván García-Magariño[1], Sylvain Rougemaille[2], Rubén Fuentes Fernández[1], Frédéric Migeon[2], Marie-Pierre Gleizes[2], and Jorge Gómez-Sanz[1]

[1] D. Software Engineering and Artificial intelligence
   Facultad de Informatica - Univesidad Complutense Madrid, Spain
   {ivan_gmg,ruben}@fdi.ucm.es, jjgomez@sip.ucm.es
[2] Institut de Recherche en Informatique de Toulouse
   SMAC Research Group
   Université de Toulouse, France
   firstname.name@irit.fr

Abstract. Many Multi-Agent Systems (MAS) methodologies incorporate a model-driven approach. Model Driven Engineering is based on three main ideas: models are the "first-class citizens", meta-models define modelling languages that are used to specify models and models are transformed during the development. However, model transformation is still a challenging issue in MAS. At first, MAS designers are not necessarily familiar with existing model transformation languages or tools. Secondly, existing tools for creating model transformations do not satisfy the necessities of agent-oriented software engineering, since they focused on coding with little support for developers. This paper proposes a tool for the creation of model transformations that is based on the generation of model transformations by-example. This tool overcomes the limitations of other similar tools in the sense that it can generate many-to-many transformation rules. The tool application is exemplified with two MAS methodologies, *INGENIAS* and *ADELFE*.

1 Introduction

Developing *Multi-Agent Systems* (MAS) in the scope of modern information systems is a demanding activity. MAS are usually related to distributed applications in changing environments, where knowledge is partial and there are requirements of a flexible behaviour. Different proposals have been made to alleviate the designer's work, mainly through methodologies and their support tools. This work gathers the experience of two different research groups that developed the agent-oriented methodologies *INGENIAS* [10] and *ADELFE* [1]. These two methodologies use Model Driven Engineering (MDE) principles to carry out the development process. One of the backbones of these approaches are model transformations which allow the automation of tasks and code generation. *ADELFE* proposes model-to-model transformations to integrate several modelling languages and separate concerns into different models [15]. *INGENIAS* also raises the need of model-to-model transformations for designer assistance and model refinement.

The specification of these transformations is still a critical point since only MDE specialists are able to do it. In this paper, we propose a further step in the application of MDE in the scope of MAS development. Model Transformation By Example (MTBE) proposes to generate model-to-model transformations from representative source and target models so that, transformation specification task would no more be assigned to specialists. However, existing tools adopting this approach do not satisfy yet the requirements of model transformations in MAS. The work presented here describes a tool for the creation of model transformations based on MTBE. It overcomes the limitations of other similar tools since it can generate many-to-many transformation rules.

This paper introduces some practical uses of MTBE for model refactoring purpose, firstly in the scope of *INGENIAS*, for which was originally design the MTBE tool, and secondly to the *ADELFE* methodology. The *INGENIAS* methodology [11] is devoted to the development of multi-agent systems, it was based on the use of meta-modelling techniques. It covers the whole development cycle, from analysis to implementation and provides tool support with the *INGENIAS Development Kit (IDK)* [5], which follows the principles of *Model-Driven Development*. *INGENIAS* and the IDK have been applied successfully in several areas; for instance, in surveillance [12], mobile tourist guide [9] and social simulation [8]. *ADELFE*[1] [1] is an agent-oriented methodology for designing Adaptive Multi-Agent System (AMAS) [4]. It proposes specific modelling languages such as *AMAS-ML* (AMAS Modelling Language) and integrates a model-driven implementation phase based on model transformations [14].

From the experience acquired during the definition of these two methodologies, we can state that MDE are specially appropriate to increase the productivity of MAS development. MTBE and the proposed tool can strengthen the whole development process, because it provides means for MAS designers to define their own model transformations.

The paper focuses on the application of MTBE in the scope of MAS. The next section depicts the principles and steps of the model transformations generation process. Section 2 introduces the interests of the proposed MTBE tool as well as its implementation. Section 4 presents two practical model refinement applications of MTBE, these examples are taken from both *INGENIAS* and *ADELFE* methodologies. In Section 5 we present some other works that deal with MTBE as well as MAS methodologies that could take advantage from it. Finally, the last section presents the conclusions we can draw from these experiments and proposes some specific further issues that have to be coped with in the application of MDE to MAS development.

2 Principles of Model Transformation by Example

As a recent initiative, Model transformation By-Example (MTBE) [16] is defined as the automatic generation of transformations from source and target model pairs. The common steps of MTBE are the following:

1. *Manual set-up of prototype mapping models*. The transformation designer assembles an initial set of interrelated source and target model pairs.
2. *Automated derivation of rules*. Based upon the available prototype mapping models, the transformation framework should synthesise (see Figure 1(a)) the set of model transformation rules. These rules must correctly transform (see Figure 1(b)) at least the prototypical source models into their target equivalents.

---
[1] ADELFE is a French acronym for "Atelier de Développement de Logiciels à Fonctionnalité Emergente".

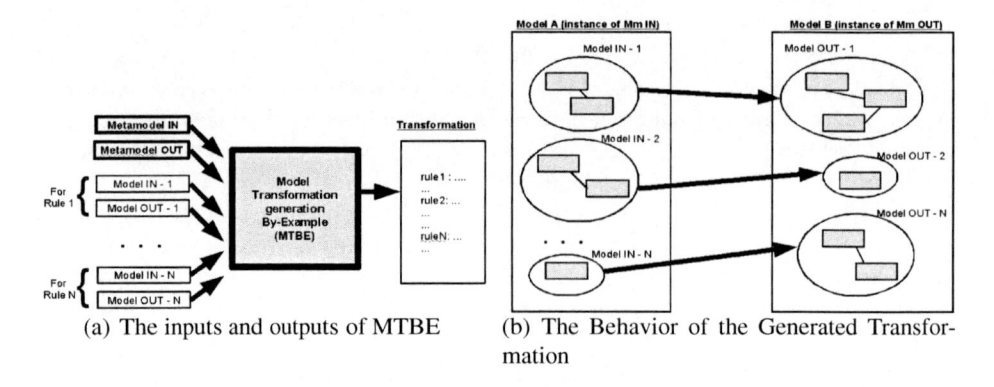

Fig. 1. Description of the Model Transformation By-Example (MTBE)

3. *Manual refinement of rules*. The transformation designer can refine the rules manually at any time. However, MTBE recommends these modifications to be included in the pairs of models, so the alterations are not overwritten the next time the transformation is generated.
4. *Automated execution of transformation rules*. The transformation designer validates the correctness of the synthesised rules by executing them on additional source-target model pairs as test cases.

The MTBE approach avoids the hard-coding of transformations, which frequently hinders the principles of MDE. MTBE follows MDE principles because its main products are models and transformations. In addition, transformation designers in MTBE do not need to learn a new model transformation language; instead they only use the concepts of the source and target modelling languages.

3 Tool for Model Transformation by Example in MAS

Existing MTBE algorithms and tools [18, 16, 17] are only able to generate one-to-one transformations. For this reason, Grasia research group has defined an algorithm for MTBE that overcomes this limitation. This algorithm can generate many-to-many transformation rules. The input patterns of the many-to-many rules are simulated with constraints. The elements of the output pattern are defined directly with the transformation language and the connection among these elements are recursively defined by the algorithm. In addition, it can create the appropriate mapping of attributes so the generated transformation propagates the information from the source to the target. This *matching-control mechanism* relies on the use of identifiers in the source model that are referred in the target model. A prototype tool (see Figure 2) implements this algorithm for generating ATL (Atlas Transformation Language) transformations [7] from *INGENIAS* model; this tool is called *MTGenerator*.

The tool provides an interface (GUI) in which the user can select the input and output meta-models of the transformation. The user must define the meta-models with the ECore language and select the corresponding location paths in the top-left area of the

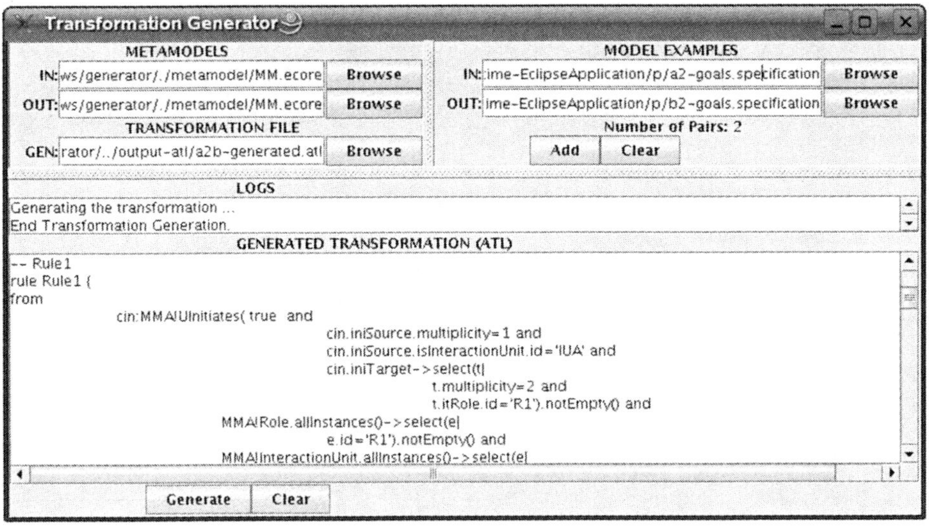

Fig. 2. Model-Transformation Generator Tool

GUI. The user can add the pairs of model with the top-right area of the generator tool, by selecting the corresponding location paths and adding them. After the automatic generation, the tool shows some logs in the *Logs* text area, confirming that the generation has finished successfully. The generated model transformation is shown in the bottom text area of Figure 2. In this manner, the user can examine the generated transformation. In brief, the presented MTGenerator tool automatically generates a model transformation. Even if the user wants to manually improve the generated transformation, the tool saves time for the user because it provides a generated transformation as a basis for the final model transformation.

4 Application of MTBE in MAS

MAS meta-models usually involves concepts semantically rich that have to be specified in terms of several meta-classes. Models conforming to MAS meta-models contain many instances of these meta-classes, as a consequence, their refinement or translation involve complex patterns of modelling elements. Therefore, MAS model driven development can profit considerably of many-to-many model transformation generation.

Furthermore, MAS meta-models used to be less stable than others, the concepts they define are still evolving as no consensus has been reached in the agent community. This implies that models need to be updated each time their meta-models undergo modifications. MTBE can help this upgrading task for models which meta-models have been modified. Source models are the models to upgrade, thus the only task left is the description of target models. The evolution of MAS concepts implies that this upgrade process is potentially more frequent in MAS methodologies. MTBE constitutes a powerful means to reduce this upgrade time, as it allows MAS designers to generate the required transformations without having to assimilate transformation languages.

One application of MTBE in MAS is the automation of repetitive mandatory tasks which are often related to phase transition in the development process, as for instance the translation of requirements models in the beginning of *ADELFE* analysis phase. MTBE can also be used to improve the quality of models by defining specific automatic refactoring that prevent designers from potential mistakes. As MTBE helps designers to define their own transformations, they could embody easily their pragmatic knowledge of the application domain and improve the process. The following sections presents two examples of MTBE practical uses in the *INGENIAS* and *ADELFE* methodologies.

4.1 MTBE in INGENIAS

In *INGENIAS*, the MTBE is applied to create model transformations for assisting the designers in the creation of the model specification. There are several processes for modelling a whole MAS with *INGENIAS*. Most of the processes start with the definition of the use cases. For this reason, this paper presents the generation of a model transformation that creates the definition of roles from the specification of the use cases. In *INGENIAS*, the use case diagrams usually include roles and goals, which must be linked afterwards for defining the roles. In addition, at least, an agent must be created for playing each role. In this example we propose to automatically create the role and agent definitions with a model transformations.

In particular, Figure 3 shows the pairs of models, from which the model transformation was generated. In the first pair, for each role in the source model, the target model

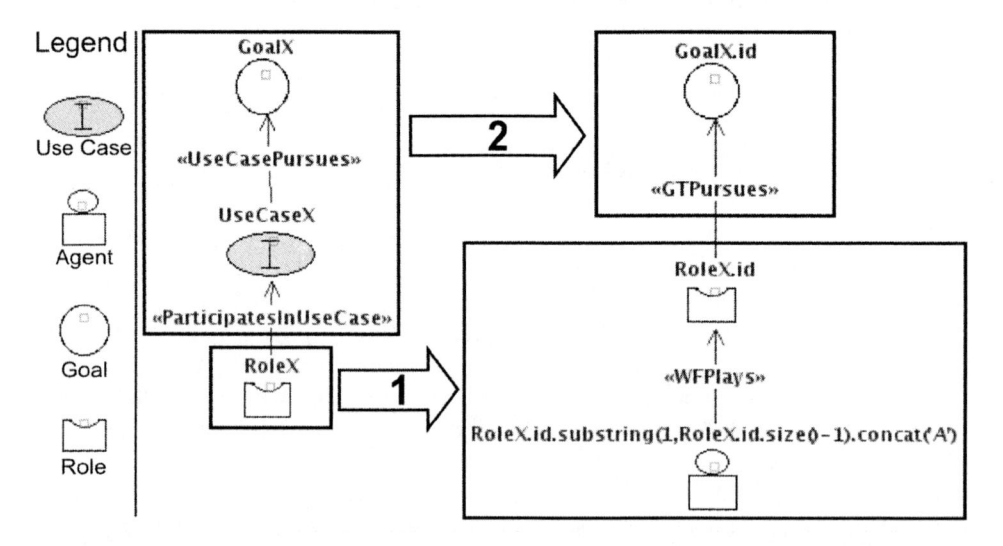

Fig. 3. Model transformations for generating the Specification of Roles. Each square represent a model example for MTBE. Each pair of models is related with an arrow and a number, in which the source and target models are respectively situated at the left and right sides of the arrow.

contains the role and an agent playing this role. The identifier of the role is copied whereas the identifier of the agent is an alteration of the role identifier. The expression of the agent identifier in the target model is defined according to the matching-control mechanism provided by our tool (see Section 2). In the second pair, the source model contains a goal connected to a role through a use case. The target model has the same goal that is directly linked with the role. In this manner, the model transformation link roles with the goals according to the existent use cases.

As one can observe, this example needs to transform patterns of several modelling elements; thus, one-to-one transformation rules do not satisfy the requirements of this transformation.

4.2 MTBE Use in ADELFE

One task of the *ADELFE* methodology design phase is the specification of agent interactions (direct communication between agents). The UML 2.0 sequence diagram is used to achieve this task. A model-to-model transformation has been specified to translate UML sequences of messages into AMAS-ML *Cooperative Interaction Protocols*. However, once the protocols are integrated to the AMAS-ML model messages emitting and reception have to be declared in the agents involved in these protocols. This example proposes to assist the designer by automating this process. The model-to-model transformation is created applying the MTBE principles (see Section 2).

Figure 4 shows the way the MTBE is specified via a meta-object notation (instances of AMAS-ML meta-classes). The source example model is figured on the left hand side. It presents a message (*m1*) that is owned by a protocol (*protocol1*) and sent by the *agent1* to the *agent2*. On the right hand side the figure presents the wished result of the transformation. The idea is to add a communication action (*cAm1*) to the *agent1* action

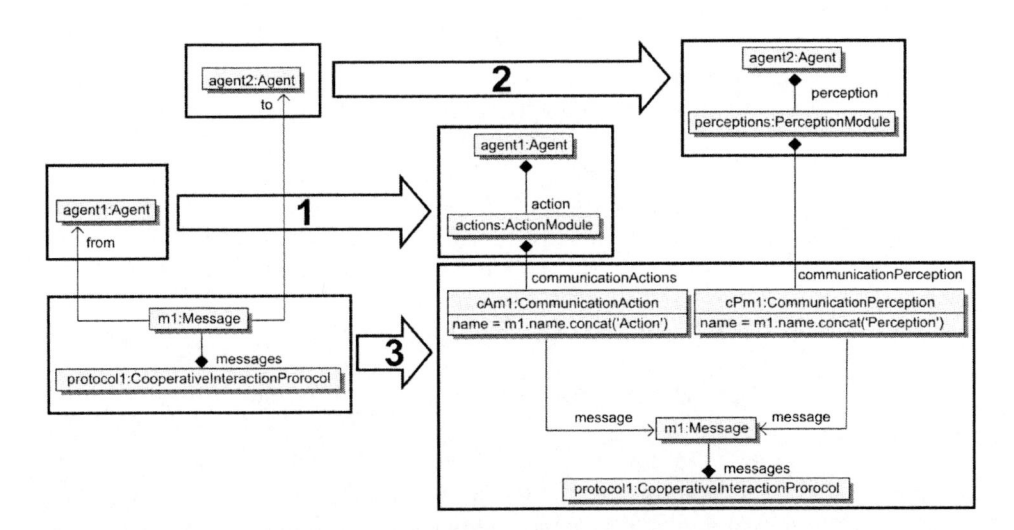

Fig. 4. Model refactoring example in ADELFE using MTBE : creating communication action and perception in respective modules for each message specified in AMAS-ML interaction protocols

module (*actions*) and the respective communication perception (*cPm1*) to the *agent2* perception module (*perceptions*). Both the created communication action and perception are related to the *m1* message (*cAm1.message* and *cPm1.message*) in conformance to the AMAS-ML meta-model. Furthermore, each numbered arrows represents a transformation rule. As a matter of fact, the result of the generation will be separated in three different parts (ATL rules). One (*arrow number 1*) to add the action module (*actions*) to the source agent (*agent1*), another (*number 2*) to do the same with the perception module (*perceptions*) of the target agent (*agent2*) and a last one (*number 3*) to integrate the communication action (*cAm1*) and perception (*cPm1*) to their respective modules and links them to the message (*m1*).

Although this process could be achieve by AMAS designers, we advocate that their productivity should be improved by a transformation that abstains them to perform these quite repetitive actions. Considering the set of protocols and messages that are usually defined in AMAS-ML models, the automation of this process could save a precious time during the design phase. In addition, this transformation could strengthen the design phase by avoiding errors while treating each messages from each protocols by hand.

5 Related Work

First of all, there are other MTBE tools. For instance, Wimmer et al.[18] present another MTBE tool which uses the same model transformation language: ATL. However, Wimmer et al. generate simple ATL rules that transforms only one isolated element into another one. Moreover, Varro and Balogh[16, 17] use inductive logic programming to derive the transformation rules with a MTBE approach. An innovation of this work is the learning of negative constraints from *negative examples*. A *negative examples* is a context of elements for which a rule do not have to be applied. In addition, this work carries what Varro and Balogh call *connective analysis*. In the connective analysis, the references among modelling elements are analysed. However this analysis is only successfully executed in the rule outputs. Their approach only generates one-to-one rules.

The great advantage of the work presented here over Wimmer's, Varro and Balogh's is the generation of many-to-many rules, by means of OCL constraints within the input side of the rules. In other words, the rules generated by the tool presented in Section 2 allow one to transform patterns of modelling elements into other patterns of modelling elements.

The most relevant features of the existing tools and the one we have presented are compared in Table 5

Besides *INGENIAS* and *ADELFE*, there are other MAS methodologies that use refinement model transformations such as Tropos[2] [3]. It is associated with a design tool called TAOM4E (Tool for visual Agent Oriented Modelling for the Eclipse platform) [2]. TAOM4E uses a specific modelling language and introduces a model driven approach. Perini et al. [13] presents the different types of transformation which were

---

[2] http://www.troposproject.org/

Table 1. Comparison of existent MTBE tools with the presented tool

Features of MTBE	Varro and Balogh	Wimmer et al	Our Technique
Mapping of attributes	yes	yes	yes
Propagation of links	yes	yes	yes
Negative Examples	yes	no	no
Generation of Constraints	no	yes	yes
Explicit Allocation of Target Elements	yes	no	yes
Limit number of input elements of rules	1	1	no-limit
Limit number of output elements of rules	1	1	no-limit

implemented in Tropos (model refinement and translation of preliminary UML2.0 models). In the same scope, *MDAD* (Model Driven Agent Development) applies MDE principles for the development of MAS [6]. It defines UML based meta-models (profile), covering aspects such as the domain, the agents and the organisation. In *MDAD* the preliminary abstract model is transformed into a platform specific model conforming to the INteractive Agent Framework (INAF) meta-model.

The Tropos and MDAD approaches as well as all MAS methodologies using models, can benefit from the presented MTBE tool. The MTGenerator tool (see section 2) provides means to assist MAS designers in defining many-to-many model transformation rules with an user friendly GUI. This kind of transformations is especially useful in MAS development.

6 Conclusions and Future Work

This paper presents a tool based on the MTBE principles. This tool can generate model transformations that satisfy the fundamental requirements in the scope of MAS. MTBE facilitates the task of the MAS designers and reduces design time by providing transformations generation. Moreover, MTBE speeds up the transformation specification as the process just consists of defining two models and using the tool to generate the transformation that relates them. MTBE brings a real improvement specially considering refining transformations (same source and target meta-models). MAS designers can easily defines new refactoring or refinement transformations as they are used to manipulate the concepts from both source and target models (they are MAS concepts). For the experimentation, this work includes two examples from two agent oriented methodologies: *INGENIAS* and *ADELFE*.

This particular aspect can be exploited to enhance development processes. Each transition from task to task or from phase to phase that needs specific treatments could be automated. Designers practical knowledge can be embodied into transformations that they are able to define thanks to MTBE. We foresee to automate most of the the *INGENIAS* and *ADELFE* process workflow definition with the help of the presented MTBE tool.

Future work with the generator tool includes that the user can define model-to-model transformations for a wider range of meta-models. Another future direction is to apply the presented tool for exogenous model-to-model transformations (different source and target meta-models). Finally, the use of negative examples can be incorporated in the tool to facilitate the expression of complex constraints over source model elements.

References

1. Bernon, C., Camps, V., Gleizes, M.P., Picard, G.: Engineering Adaptive Multi-Agent Systems: The ADELFE Methodology. In: Henderson-Sellers, B., Giorgini, P. (eds.) Agent-Oriented Methodologies, pp. 172–202. Idea Group Pub., NY (2005)
2. Bertolini, D., Delpero, L., Mylopoulos, J., Novikau, A., Orler, A., Penserini, L., Perini, A., Susi, A., Tomasi, B.: A tropos model-driven development environment. In: Boudjlida, N., Cheng, D., Guelfi, N. (eds.) CAiSE Forum. CEUR Workshop Proceedings, vol. 231. CEUR-WS.org. (2006)
3. Bresciani, P., Perini, A., Giorgini, P., Giunchiglia, F., Mylopoulos, J.: Tropos: An agent-oriented software development methodology. Autonomous Agents and Multi-Agent Systems 8(3), 203–236 (2004)
4. Capera, D., Georgé, J.P., Gleizes, M.P., Glize, P.: The AMAS Theory for Complex Problem Solving Based on Self-organizing Cooperative Agents. In: TAPOCS 2003 at WETICE 2003, Linz, Austria. IEEE CS, Los Alamitos (2003)
5. Gómez-Sanz, J.J., Fuentes, R., Pavón, J., García-Magariño, I.: INGENIAS development kit: a visual multi-agent system development environment. In: AAMAS (Demos), pp. 1675–1676. IFAAMAS (2008)
6. Jarraya, T., Guessoum, Z.: Towards a model driven process for multi-agent system. In: Burkhard, H.-D., Lindemann, G., Verbrugge, R., Varga, L.Z. (eds.) CEEMAS 2007. LNCS, vol. 4696, pp. 256–265. Springer, Heidelberg (2007)
7. Jouault, F., Kurtev, I.: Transforming Models with ATL. In: Bruel, J.-M. (ed.) MoDELS 2005. LNCS, vol. 3844, pp. 128–138. Springer, Heidelberg (2006)
8. Pavon, J., Arroyo, M., Hassan, S., Sansores, C.: Agent-based modelling and simulation for the analysis of social patterns. Pattern Recognition Letters 29(8), 1039–1048 (2008)
9. Pavón, J., Corchado, J., Gómez-Sanz, J., Ossa, L.: Mobile Tourist Guide Services with Software Agents. LNCS, pp. 322–330. Springer, Heidelberg (2004)
10. Pavón, J., Gómez-Sanz, J.: Agent Oriented Software Engineering with INGENIAS. In: Mařík, V., Müller, J.P., Pěchouček, M. (eds.) CEEMAS 2003. LNCS, vol. 2691, pp. 394–403. Springer, Heidelberg (2003)
11. Pavón, J., Gómez-Sanz, J., Fuentes, R.: Model Driven Development of Multi-Agent Systems. In: Rensink, A., Warmer, J. (eds.) ECMDA-FA 2006. LNCS, vol. 4066, pp. 284–298. Springer, Heidelberg (2006)
12. Pavón, J., Gómez-Sanz, J.J., Fernández-Caballero, A., Valencia-Jiménez, J.J.: Development of intelligent multisensor surveillance systems with agents. Robotics and Autonomous Systems 55(12), 892–903 (2007)
13. Perini, A., Susi, A.: Automating model transformations in agent-oriented modelling. In: Müller, J.P., Zambonelli, F. (eds.) AOSE 2005. LNCS, vol. 3950, pp. 167–178. Springer, Heidelberg (2006)
14. Rougemaille, S., Arcangeli, J.P., Gleizes, M.P., Migeon, F.: ADELFE Design, AMAS-ML in Action. In: International Workshop on Engineering Societies in the Agents World (ESAW), Saint-Etienne, mai 2008, pp. 213–224. Springer, Heidelberg (2008), http://www.springerlink.com/
15. Rougemaille, S., Migeon, F., Maurel, C., Gleizes, M.P.: Conception d'applications adaptatives basées sur l'IDM (Accepted). In: Artikis, A., O'Hare, G.M.P., Stathis, K., Vouros, G. (eds.) ESAW 2007. LNCS, vol. 4995, pp. 318–332. Springer, Heidelberg (2008)

16. Varro, D.: Model transformation by example. In: Nierstrasz, O., Whittle, J., Harel, D., Reggio, G. (eds.) MoDELS 2006. LNCS, vol. 4199, pp. 410–424. Springer, Heidelberg (2006)
17. Varró, D., Balogh, Z.: Automating model transformation by example using inductive logic programming. In: Proceedings of the 2007 ACM symposium on Applied computing, pp. 978–984 (2007)
18. Wimmer, M., Strommer, M., Kargl, H., Kramler, G.: Towards Model Transformation By-Example. In: Proceedings of the 40th Annual Hawaii International Conference on System Sciences, vol. 40(10), p. 4770 (2007)

Modelling Trust into an Agent-Based Simulation Tool to Support the Formation and Configuration of Work Teams

Juan Martínez-Miranda and Juan Pavón

Dep. Ingeniería del Software e Inteligencia Artificial
Universidad Complutense de Madrid
Ciudad Universitaria s/n, 28040, Madrid, Spain
jmartinez@microart.cat, jpavon@fdi.ucm.es

Abstract. One important factor that contributes to create good or bad relationships between individuals inside human societies is the notion of trust. In particular, some research works have proved the influence of trust in the performance of the activities that team-members perform jointly. This paper presents our initial theoretical work to include the trust factor into our TEAKS (TEAm Knowledge-based Structuring) model. TEAKS is an agent-based model to simulate the interaction between individuals when working together in the development of a project. Each team-member is represented through a set of pre-selected human characteristics: the emotional state, social characteristics, cognitive abilities, and personality types. The main outcome of the TEAKS simulation is statistical information about the possible performance at the individual and team levels. In this context we use two (emotional state and personality traits) of the four modelled human characteristics to introduce a model of trust into TEAKS to analyse the impact of trust in the results of the team.

Keywords: Agent-Based Simulations, Trust, Human Performance Modelling.

1 Introduction

The concept of trust has been widely discussed during the last decade, especially within the Organisational Behaviour discipline [10] [6]. This increasing interest in trust within organisations could be explained as there are more and more large companies or consortiums where several people need to work together from different geographically locations. New theories and hypotheses about the thinking and functioning of organisations have been replacing traditional aspects of management by collaborative approaches emphasising ideas of coordination, sharing of responsibilities and risk taking [23]. Nevertheless, the study of trust has also been addressed from other disciplines such as Psychology [12] [20], Sociology [2] [7] and Economic Sciences [1]. More recently and with the great development of applications in Internet, the interest in the study of trust has grown up and some research works put efforts towards the modelling of trust and reputation concepts addressed mainly to e-Commerce applications (for a good review of some existent models of trust and reputation see [22]). Most of these models of trust and reputation use software agents as the entities where the relationship of trust takes place and is represented using specific characteristics of each model. Some other models and studies have been developed in the Human Resources and Management disciplines to analyse the importance of trust within work teams and how it is related with performance effectiveness [8] [5].

All these works show the evidence that, in some way, the trust relationship between the members of a work team affects the performance of the team over its tasks or activities. Given this evidence, we are interested in the inclusion of a trust model into our previous work called TEAKS, which is an agent-based simulation tool to help in the formation and configuration of work teams [18]. The initial inclusion of a theoretical model of trust into TEAKS is described in this paper.

2 Overview of the TEAKS Model

The formation and configuration of work teams is not a trivial task, which is typically performed by managers based on past experience and available (though frequently scarce, uncertain and dynamic) information about the cognitive, personal and social characteristics of the potential team members. To support this decision-making process we have developed the TEAKS (TEAm Knowledge-based Structuring) simulation tool [18]. TEAKS is an agent-based model where a virtual team can be configured using some human characteristics of the real candidates to form the team, and given a set of tasks, the model generates the possible performance of the team-members. A set of selected human characteristics are used to observe how the possible individual behaviour within the work team can be obtained from the combined values in this agent's internal state, which is affected by the interaction with the internal state of the other agents and by the characteristics of the assigned task(s). The human characteristics currently included into TEAKS agents are described in the following subsections.

2.1 Personality Traits

During the last years, several studies have shown the importance of personality traits in job performance [13] [3]. We have taken from [19] a set of four personality traits to model the internal state of the agents:

Amiable: they are dependable, loyal and easygoing. They like things that are non-threatening and friendly. They hate dealing with impersonal details and cold hard facts. Often described as a warm person and sensitive to the feelings of others but at the same time wishy-washy.

Expressive: very outgoing and enthusiastic, with a high energy level. They are also great idea generators, but usually do not have the ability to see the idea through to completion. They enjoy helping others and are particularly fond of socialising.

Analytical: these people are known for being systematic, well organized and deliberate. These individuals appreciate facts and information presented in a logical manner as documentation of truth. Others may see them at times as someone who does things too much 'by the book'.

Driver: they thrive on the thrill of the challenge and the internal motivation to succeed. Drivers are practical folks who focus on getting results. They can do a lot in a very short time. Often viewed as decisive, direct and pragmatic.

2.2 Emotional State

Research works have proved that emotions influence the behaviour of a person at work place [11], and a critical decision for the TEAKS model has been the selection of the set of basic emotions to include into the internal state of the agents. From the extended psychology classification of the basic emotions, we have selected a set of four basic emotions to model the agents' emotional state at work. Two of them are considered positive emotions: *desire* and *interest* of the person to develop a specific task in a given moment and the other two are negative emotions: *disgust* and *anxiety* generated by the context (the characteristics of the assigned tasks and characteristics of the team-mates in that task) at specific moment.

We made this selection given the context of application and thinking in the most common emotions produced by the activities of a person at work. In a project it is more common that one specific task produces the *interest* or *desire* (while developing that task) positive emotions in a worker than the *happiness* or *joy* emotions (most commonly identified during personal life situations). On the other hand, the negative emotions *disgust* and *anxiety* produced by specific work activities are more common than *fear*, *pain* or *sadness*, even though that there are some special circumstances through the development work activities that can produce these emotions (such as dangerous or high risk tasks), but currently we concentrate on projects where these type of tasks do not frequently appear.

2.3 Social Characteristics and Cognitive Capabilities

The characteristics of human relations in groups and teams (such as competence, trust, and co-operation, among others) are the main topic of research in areas such as social psychology, social sciences and organisational behaviour. Modelling all these characteristics together is out of the scope of the TEAKS model and initially only two parameters, representing social characteristics, were considered: *introverted / extroverted*, and *prefers to work alone / in team*: when a person at work prefers to perform alone a specific task(s), even though this person can be extroverted, and vice versa. The work presented in this paper is focused in the introduction of *trust* as a new social characteristic to the model.

Regarding the cognitive capabilities, the TEAKS model takes into account the degree of expertise of a person in a particular domain. This is given by the role within the work team: *Project Manager:* a person in charge of managing the project and assigning tasks to the other team-members. *Co-ordinator:* person in charge of specialised tasks, re-configuration of tasks and re-location of resources. *Specialist:* person in charge of complex and specialised tasks. *Technician:* Person who can deal with technical and non-specialised tasks. *Assistant:* Person in charge of not complex, routine and repetitive tasks. In addition of each person's role, every team-member has two other independent-role parameters to represent his/her cognitive capabilities: *experience level* and *creativity level*.

2.4 Modelling the Team Performance

The resulting team-members and team performance are calculated by taking into account the internal characteristics of the agents, the interaction with its assigned task and the interaction with the rest of team-members. We have considered five parameters to evaluate the expected final performance of each team member: *goals achievement, timeliness* and *quality* of the performed task; *level of collaboration*, and *level of required supervision* during each one of the assigned tasks.

Additionally, the TEAKS model includes the representation of a project in which the agents simulate to work. Every project is represented by its division into tasks and every task is represented by the following parameters: *number of participants* in the task, *estimated duration* (measured in days), *sequence* (two or more tasks could be executed sequentially or in parallel), *priority* within the project, *estimated cost, task description, level of difficulty, type* (required specialisation level), *timeliness* and *quality*.

The behaviour of every agent modelled in the virtual team (i.e. the simulation of its performance within the project individually and as a team-member), is generated by the combination of above mentioned human characteristics. We use fuzzy values to assess the parameters in the agent's internal state and two of the tasks parameters (difficulty and type). Using these fuzzy values and a set of pre-defined fuzzy rules, the simulated individual behaviour of each virtual team member is obtained in a three-step cyclical process.

In the first step, the initial values of the emotional parameters of the agents are modified as they are influenced by the characteristics of the task assigned to the agent, the personality traits of its team-mates and its own personality. In the second step, each emotional parameter of the agent is defuzzified to introduce random values in the model. This randomness represents the non-deterministic nature of human emotions: the reaction of the same person under similar circumstances, in front of similar circumstances will not always be exactly the same. During the third step, the crisp values in the emotional state of the agent are fuzzified again to generate the fuzzy values for each one of the agent's performance parameters (please refer to [18] for details of this cyclical process).

3 Introducing a Trust Model into TEAKS

3.1 Foundations

As we introduced in Section 1, some computational models of trust have already been developed [17] [4] and some of them also include the concept of reputation, mainly for their use in e-Commerce [21] [24]. The focus in the TEAKS model is on the concept of trust, following the point of view described in [23], where *trusting attitudes are reinforced each time an outcome meets expectations. The outcome becomes part of the history of the relationship, increasing the chance that partners will have positive expectations about joint actions in the future and the increased trust reduces the sense of risk for these future actions.* Although this approach of trust is related to inter-organisational collaboration, we adopt this vision in terms of trust at the individual level where the outcome results obtained in every joint developed task influence the *trust level* between the team-mates. Then for the TEAKS model we assume that:

1. *Good results in the developed task(s) increase trust and bad results decrease trust among the participants in the task(s).*

Additionally, since we are considering the internal state of each agent as fundamental in the generation of the individual and team behaviour, we also include the influence of this internal state in the trust attitudes of the agents. Specifically, we use the set of emotions and the personality traits already modelled in TEAKS as additional influences on the increasing or decreasing of trust.

To include the influence of emotions on trust into TEAKS, we use the findings reported in [9] where five studies were developed to analyse this kind of influence. In these studies they used six emotions: *anger, gratitude, happiness, pride, guilt* and *sadness*. One of the results they obtained suggests that happiness and gratitude, emotions with positive valence, increase trust, and anger, an emotion with negative valence, decreases trust. Nevertheless, the generalisation that all negative-valence emotions would decrease trust and that all positive-valence emotions would increase trust must not be assumed given the appraisal control of the emotions. In one of the studies, presented in [9], four emotions were characterized by either positive or negative valence and either appraisals of other-person control (anger and gratitude) or appraisals of personal control (pride and guilt). In this study they found that *"emotions with appraisals of other-person control influenced trust in a manner consistent with the emotion's valence; anger decreased trust and gratitude increased trust. Emotions with personal control influenced trust significantly less than did emotions with other-person control; participants in the gratitude condition were more trusting than were participants in the pride condition, and participants in the anger condition were less trusting than were participants in the guilt condition"*.

From the four modelled emotions in TEAKS (anxiety, disgust, interest and desire) any of them can be considered as emotion with a personal control appraisal because each of them are caused by the context of the agent (the characteristics of the assigned tasks and the characteristics of the other team-members) and not directly by the agent's actions (although the obtained performance also influence the emotions, in TEAKS we are not considering emotions such as pride or guilty that can be directly influenced by the simulated actions of the agents). Then for the TEAKS model we assume that:

2. *Anxiety and disgust influence negatively the trust behaviour of an agent, while interest and desire influence positively the trust behaviour of an agent.*

The other human characteristic that we model as an additional influence on trust is the personality type. There are few studies about how personality traits affect interpersonal trust within work teams, and we use the results reported in [15], which identifies types of personality that induce more trust than others. Although the goal of this study was to analyse the trust of people when shopping online, we use these findings to set a variable that we call *trust tendency* used as the tendency of a team-member to trust in its teammates when no previous interaction with them has occurred (see next subsection).

The study developed in [15] considers four personality traits to analyse how trust is affected. These four personality traits were taken from [14] and include:

Popular Sanguine: the extrovert, talker, and optimist. Individuals with this personality type are generally appealing to others. They are enthusiastic and expressive and live life in the present.

Perfect Melancholy: The introvert, thinker and pessimist. Individuals with this personality type are generally deep, thoughtful and *analytical*.

Powerful Choleric: the extrovert, doer and optimist. Individuals with this personality type are independent and self-sufficient.

Peaceful Phlegmatic: the introvert, watcher and pessimist. Individuals in this category tend to be easy going and agreeable or *amiable*.

Although the names used to identify the personality types differ from our four modelled personalities, the traits they describe are the same and even some of the titles of our own modelled personalities are used as characteristics to describe these traits. Then we matched our modelled personalities to each one of the personalities used in the study as: *Popular Sanguine* → *Expressive*; *Perfect Melancholy* → *Analytical*; *Powerful Choleric* → *Driver* and *Peaceful Phlegmatic* → *Amiable*.

According to the results of [15] *popular sanguine* personalities are optimists who focus on the details of a 'story', they were the most trusting of the respondents. The assessment of trustworthiness of *perfect melancholy* personalities was the lowest and yet they attribute the highest importance to trust triggers. Next to popular sanguine personalities, the optimistic *powerful choleric* personalities were the most trusting of the respondents. Finally, the pessimistic *peaceful phlegmatic* personalities got lower trustworthiness ratings and they appear to have attributed relatively high importance ratings to trust triggers.

Using these results for the TEAKS model we assume that:

> 3. Agents with Expressive and Driver personalities have a higher tendency to trust in their team-mates that agents with Analytical and Amiable personalities.

From the above described foundations, and using the three numbered assumptions, a model of trust is introduced into TEAKS, which is described in the next subsection.

3.2 Implementation

In real life it is difficult to accept that numerical values could be used to measure the intensity of either human emotion, experience or creativity level. That is the main reason behind the use of Fuzzy Logic in TEAKS to represent most of the modelled human characteristics, the performance parameters and some of the tasks parameters. In the same sense, it is difficult to use quantitative values to represent how much one person trusts in his/her co-worker and how much this trust can be increased/decreased during all the project life. Similarly to other TEAKS parameters, we have defined fuzzy sets to represent the values of trust and the values for the increment/decrement of the level of trust in each agent (see Figure 1). These fuzzy values are introduced in the three-step cyclical process for the generation of the possible agent's behaviour described in the Section 2.4.

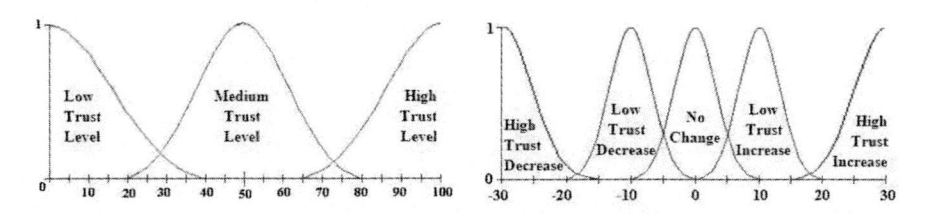

Fig. 1. Fuzzy sets defined for the values in trust (left shape) and for the trust change values (right shape)

During the first step of this cyclical process (when the agent "discovers" the characteristics of its assigned task and the personalities of its team-mates on that task) the level of trust is not affected directly. Following the first assumption presented in Section 3.1, the trust level is only increased/decreased depending on the obtained results in the task. So since during this first step there are no task's results yet involved, we have introduced the concept of *trust tendency* as the tendency of an agent to trust in its team-mates and for each one of these team-mates the trust tendency is obtained in the following way:

1. If there is no previous interaction with a team-mate, then the value of the trust tendency concerned to that team-mate is obtained from the current internal state (the value in each one of the emotions) and from the personality of the agent. Values in trust tendency are also fuzzy values obtained from fuzzy sets similar to those sets used for the trust parameter.
2. If there is a previous interaction between the agent and its team-mates (i.e. the agent has already participated in a previous task with the other agents) then the trust tendency value is only influenced by its current emotional state.

We must say that in the current TEAKS model there is no representation of possible past interaction between the agents in former projects, i.e. at the beginning of each simulation the trust level of each agent is only obtained from its emotional state and personality.

The second step of the cyclical getting behaviour process is where randomness is introduced into the model (see [18] for details) and at this step the level of trust in each agent is not affected. In the third step is where the values in each one of the performance parameters of every agent is obtained and the task's result parameters (*quality* and *timeliness*) are also acquired as the result of the performance of all the agents involved in the task. At this stage is where the trust level of each agent concerned with all of its team-mates is modified using four inputs:

1. *The current emotional state* that was updated according to the values obtained in its performance parameters.
2. *The trust tendency* that was previously set when the task was assigned to the agent.
3. *The task results* obtained after the simulated interaction between the agents and the parameters of the specific task.
4. *Elapsed simulation time*, i.e. the position of the task within the project used to represent that changes in the trust level are more frequent at the beginning than at the end of the project (e.g. when a good/bad relationship of trust between two people has been built during the first and medium stages of the project, it is more difficult that this level of trust decreases/increases at final stages of the project – unless something really bad/good occurs– than when a trust relationship is bad/good since the beginning of the project).

After this step, the trust level of the agent regarding each one of its co-workers on the already finished task is stored as the *trust history* to be used in future interactions

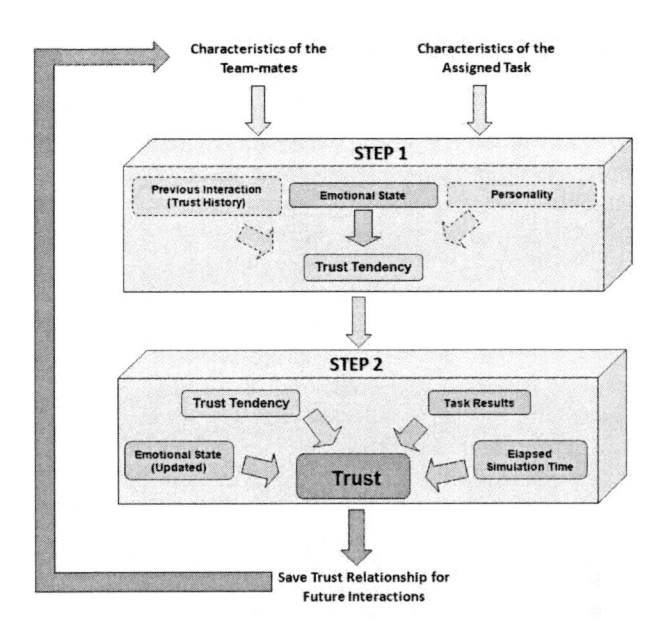

Fig. 2. Model of trust in TEAKS

when the two agents will possible work again together in a task. Figure 2 summarises through a graphical representation how the trust is modelled in TEAKS.

The change in both, trust tendency and trust parameters, will be done through the definition of a pre-defined set of fuzzy rules similarly as it is already implemented for the emotional parameters. Examples of these fuzzy rules are:

IF *anxiety* of A is *high* THEN
  The *trust_tendency* regarding $B_1...B_n$ has a *high decrease*
IF *current level of trust* in B_1 is *high* THEN
  The *trust_tendency* regarding B_1 has a *low increase*
IF the personality trait *driver* of A is *high* THEN
  The *trust_tendency* regarding $B_1...B_n$ has a *low increase*
...
IF *trust_tendency* of A regarding B_1 is *low* THEN
  The *trust* regarding B_1 has *low increase*
IF the obtained *quality* of T_x is *satisfactory* THEN
  The *trust* regarding $B_1...B_n$ has a *low increase*
...

In these examples of fuzzy rules, A represents the agent who is calculating its level of trust, and ($B_1...B_n$) is the set of A's team-mates in a specific task (T_x) of the project. For the matching rule we use the Mamdani fuzzy rule-based model [16] (the minimum operator) to represent the "AND" in the premise and the implication. The example shows that the level of trust and trust tendency may increase or decrease according to the fired fuzzy rules.

4 Conclusions and Future Work

We have presented our initial efforts towards the modelling of trust into TEAKS, an agent-based simulation tool to support the formation and configuration or work teams. The current implementation of TEAKS (without the model of trust) has been tested and validated in a real scenario obtaining good results when comparing the TEAKS results versus a real work team as it is described in [18]. The immediate next steps in this work (currently in progress) is the implementation of the proposed theoretical model through the complete definition of the fuzzy rules, the implementation and integration of these rules into the TEAKS prototype (using the FuzzyJess engine) and the most important task: the study of the obtained results to analyse if the inclusion of the notion of *trust* enhance our previous results.

Introducing trust in the model we then can analyse some observed situations in real projects that were not explained with the previous model, such as how the familiarity between the team-members influence the performance of the team or the situation observed in the real work team that was used as validation for our model: when some tasks were developed with almost the same set of people, the time duration of those tasks were relatively lower than the same type of tasks developed by people with no much previous interaction during the project.

Acknowledgements. This work has been developed with support of the program "Grupos UCM-Comunidad de Madrid" with grant CCG07-UCM/TIC-2765, and the project TIN2005-08501-C03-01, funded by the Spanish Council for Science and Technology.

References

1. Burchell, B., Wilkinson, F.: Trust, Business Relationships, and the Contractual Environment. Cambridge Journal of Economics 21, 217–237 (1997)
2. Buskens, V.: The Social Structure of Trust. Social Networks (20), 265–298 (1998)
3. Code, S., Langdan-Fox, J.: Motivation, cognitions and traits: predicting occupational health, well-being and performance. Stress and Health J. 17(3), 159–174 (2001)
4. Castelfranchi, C., Falcone, R.: Principles of Trust for MAS: Cognitive Anatomy, Social Importance, and Quantification. In: Proc. of ICMAS 1998, pp. 72–79 (1998)
5. Costa, A.C.: Work team trust and effectiveness. Personnel Review 32(5), 605–622 (2003)
6. Das, T.K., Teng, B.G.: Between trust and control: Developing confidence in partner cooperation in alliances. Academy of Management Review 23, 491–512 (1998)
7. di Luzio, G.: A Sociological Concept of Client Trust. Current Sociology 54(4), 549–564 (2006)
8. Dirks, K.T.: The Effects of Interpersonal Trust on Work Group Performance. Journal of Applied Psychology (84), 445–455 (1999)
9. Dunn, J.R., Schweitzer, M.E.: Feeling and believing: The influence of emotion on trust. Journal of Personality and Social Psychology 88(5), 736–748 (2005)
10. Elangovan, A.R., Shapiro, D.L.: Betrayal of trust in organizations. Academy of Management Review 23, 547–566 (1998)
11. Fisher, C.D.: Mood and emotions while working: missing pieces of job satisfaction? Journal of Organisational Behaviour 21(2), 185–202 (2000)

12. Gurtman, M.B.: Trust, distrust, and interpersonal problems: A circumplex analysis. Journal of Personality and Social Psychology 62(6), 989–1002 (1992)
13. Heller, D., Judge, T.A., Watson, D.: The confounding role of personality and trait affectivity in the relationship between job and life satisfaction. Journal of Organisational Behaviour 23(7), 815–835 (2002)
14. Littauer, F.: Personality Plus: How to Understand Others by Understanding Yourself, Grand Rapids. Fleming H. Revell Publishing, Michigan (2005)
15. Lumsden, J., Mackay, L.: How does personality affect trust in B2C e-commerce? In: Proc. of the 8th Int. Conf. on Electronic Commerce (ICEC 2006), pp. 471–481 (2006)
16. Mamdani, E.H., Assilian, S.: An experiment in linguistic synthesis with a fuzzy logic controller. International Journal Machine Stud. 7(1) (1975)
17. Marsh, S.: Formalising Trust as a Computational Concept. Ph.D. thesis, Department of Mathematics and Computer Science, University of Stirling (1994)
18. Martínez-Miranda, J., Pavón, J.: An Agent-Based Simulation Tool to Support Work Teams Formation. Advances in Soft Computing 50(2009), 80–89 (2009)
19. Merrill, D.W., Reid, R.H.: Personal Styles & Effective Performance. CRC Press, Boca Raton (1981)
20. Pillutla, M.M., Malhotra, D., Murnighan, J.K.: Attributions of trust and the calculus of reciprocity. Journal of Experimental Social Psychology 39(5), 448–455 (2003)
21. Sabater, J., Sierra, C.: REGRET: A reputation model for gregarious societies. In: Proc. of the 4th Workshop on Deception, Fraud and Trust in Agent Societies, pp. 61–69 (2001)
22. Sabater, J., Sierra, C.: Review on Computational Trust and Reputation Models. Artificial Intelligence Review 24(1), 33–60 (2005)
23. Vangen, S., Huxham, C.: Nurturing Collaborative Relations: Building Trust in Interorganizational Collaboration. J. of App. Behavioral Science 39(1), 5–31 (2003)
24. Yu, B., Singh, M.P.: Distributed Reputation Management for Electronic Commerce. Computational Intelligence 18(4), 535–549 (2002)

Multi-agent Simulation of Investor Cognitive Behavior in Stock Market

Zahra Kodia and Lamjed Ben Said

Laboratoire d'Ingénierie Informatique Intelligente (LI3)
Institut supérieur de Gestion de Tunis
41, Rue de la Liberté, Cité Bouchoucha
2000 Le Bardo, Tunis, Tunisia
{Zahra.Kodia,Lamjed.BenSaid}@isg.rnu.tn

Abstract. In this paper, we introduce a new model of Investor cognitive behavior in stock market. This model describes the behavioral and cognitive attitudes of the Investor at the micro level and explains their effects on his decision making. A theoretical framework is discussed in order to integrate a set of multidisciplinary concepts. A Multi-Agent Based Simulation (MABS) is used to: (1) validate our model, (2) build an artificial stock market: SiSMar and (3) study the emergence of certain phenomena relative to the stock market dynamics at the macro level. The proposed simulator is composed of heterogeneous Investor agents with a behavioral cognitive model, an Intermediary agent and the CentralMarket agent matching buying and selling orders. Our artificial stock market is implemented using distributed artificial intelligence techniques. The resulting simulator is a tool able to numerically simulate financial market operations in a realistic way. Preliminary results show that representing the micro level led us to build the stock market dynamics, and to observe emergent socio-economic phenomena at the macro level.

Keywords: Multi-agent based simulation, Cognitive and behavioral modeling, Stock market.

1 Introduction

The complexity of the financial rules governing the stock market and their confrontation with Investors' activities make the explication of observed global behavior very difficult to understand. Empirical and numerical analyses are powerful; nevertheless they still insufficient [10]. We notice that the existing mathematical and statistical models have shown critical limits by failing to explain and to anticipate the extreme perturbation we are living during the last months of 2008.

Previous researches, such as in [13] and [8], are generally based on the hypothesis of rational behavior. However, recently, emerged evidences show that stock markets could not be only studied with a rational paradigm such as in [1] and [17]. In the last decade, we complete the description of theoretical phenomena by many aspects based on the individual's behavior and their interactions. We distinguish an evolution of approaches used to study the stock market: (1) numerical approach during the eighties: [13] and [8], (2) multi-agent based systems during the nineties: [15], [6], [16], [17], [1] and [21] and (3) behavioral multi-agent based systems in recent years: [11]

and [10]. This evolution shows that the multi-agent based simulation is a promising approach to study the stock market. An excellent overview of former and significant models in agent-based computational finance is given by [14]. These models range from relatively simple models like the one of [15] to very complicated models like the Santa Fe Artificial stock market [2].

This paper is structured as follows. In the second section, we introduce our integrative theoretical framework. This framework exposes concepts considered in our model. In the third section, we design a new conceptual model of the Investor which is based on rational and cognitive mechanisms. Finally, in the fourth section, we present our simulator named SiSMar: Simulation Stock Market. We expose the experiments led with this simulator in order to observe the impact of behavioral attitudes on the investor decision and the emergence of socio-economic phenomena at the macro level.

2 Integrative Theoretical Framework

We focus on behavioral finance. It uses social psychology and sociological insights to clarify phenomena. However, we can not pass over the numerical and the economic approaches.

- Behavioral finance

Behavioral finance is the paradigm where financial markets are studied using models that are wider than those based on Von-Neumann Morgenstern expected utility theory and arbitrage assumptions [20]. Psychologists have shown that the real behavior of investors can not be explained only by the basic economic assumptions [12]. The decision of these actors is influenced by behavioral factors such as speculation, pessimism, caution and self control.

- Economic sociology

The goal of economic sociology is to study the social interactions in markets and how social structures are created and are evolved. According to [23],"economic sociology suggests that it is necessary to meet the economic and sociological theories in order to provide better explanations of economic concepts". It considers the social circumstances for economic change, and the effects of these arrangements upon social dissimilarity and well-being [7]. This discipline seeks to understand how the modern economy could be integrated into the social institutions. It involves the micro and macro organizational behavior of investor under risk.

- Cognitive psychology

Cognitive psychology is the study of empirical process of information involved in human behavior. It aims to explore how individuals perceive their environment (neighborhood), how they understand, diagnostic and solve problems. It also intends to analyze how they store information about the location outside or themselves and interpret this information. This discipline examines internal mental processes such as problem solving, perception, short-term memory, long-term memory, habits, and anxiety [18].

Cognitive psychology

Problem solving, Short-term memory,	Caution	Speculation
Trust, Rationality, Micro-organizational Behavior	Self control, Perception	Pessimism, Optimism, Rumor
Imitation, Social neighbor, Social institutions, Macro-organizational Behavior	Risk, Leadership	Supply and demand, Volatility, Price, Efficient market theory

Behavioral finance (top right)

Economic sociology (bottom left) — **Financial economics** (bottom right)

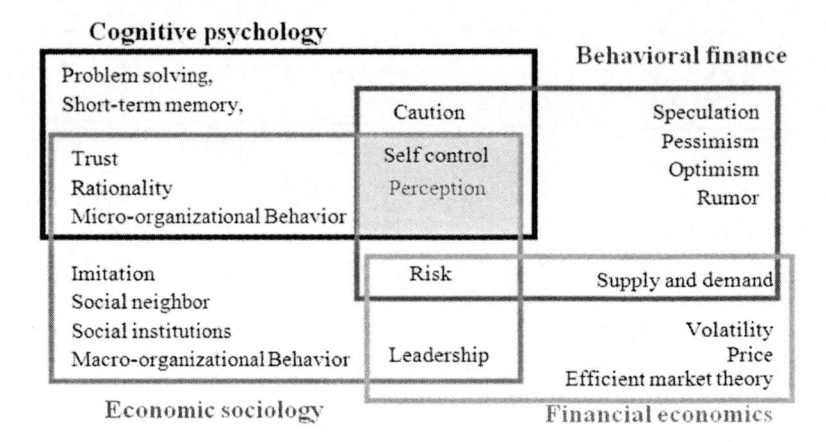

Fig. 1. Integrative theoretical framework of our model

- Financial economics

Financial economics determines a set of rules used by investors and managers in decision-making process [22]. This discipline examines the price formation under risk and inspects volatility under the supply and demand mechanism.

A fundamental assumption in financial economics theory is the linear relationship between risk and return. An issue of this assumption is the efficient market hypothesis [5]. This hypothesis assumes that investors act permently in a rational way and attempt to maximize the expected utility of their risk and return decisions.

Figure 1 represents concepts and their related domains which constitute our integrative theoretical framework. This framework underlines the multidisciplinary character of certain concepts such as trust, risk and perception.

According to the behavioral finance, we interpret the stock market dynamics as a direct result of the confrontation of the demand for capital by companies and the bid capital from novice and expert investors. In fact, in this paper, we identify mainly three kinds of actors: novice investors, expert investors and market intermediary. In reality, all the transactions happen through an intermediary who manages the supply and the demand and seeks to ensure a balance.

3 Stock Market Model

We describe now the investor behavior while taking the decision of selling or buying and the interactions between the stock market actors. Our model includes two granularity levels: the micro and the macro level. At the micro level, it describes the cognitive behavior of investors. In addition, it represents actors' interactions. These interactions participate in the emergence and stabilization of socio-economic phenomena observed at the macro level which influence reciprocally behavior and individual interactions. Figure 2 shows the global stock market dynamics integrating micro and macro levels. It specifies at the micro level the five sequential steps for the transaction realization.

We introduce mainly two hypotheses for the constructing of our model. We assume that the stock market is represented by a social network where information

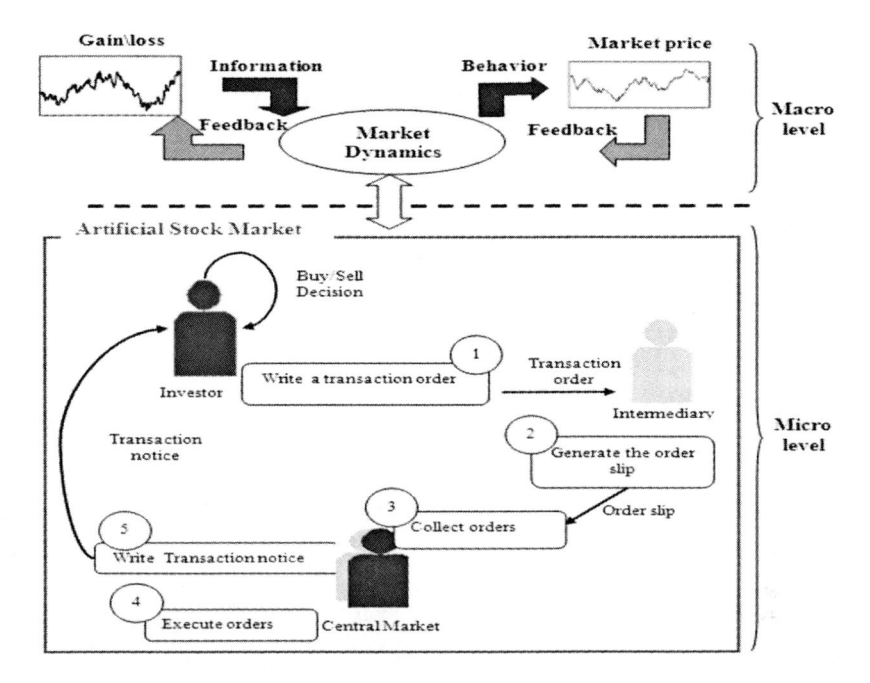

Fig. 2. The micro/macro level dynamics of the stock market

circulates randomly among heterogeneous set of investors. This information concerning stocks and indexes is available permanently for all investors. Their interactions described in figure 2 form the stock price. We neglect the external factors as financial crushes due to crisis, wars and climatic disasters. In addition, we admit that the buying and selling are accomplished immediately and closed out the same day. More explicitly, if an investor gives a purchase order, he must own in advance the capital corresponding to the amount of his purchase. In opposition, if he gives a sell order, he must own in advance the corresponding stocks.

Our artificial stock market is composed of: (1) a set of agents corresponding to the considered three types of actors: ExpertInvestor agents, NoviceInvestor agents and Intermediary agents, (2) a CentralMarket agent responsible of conducting transactions and controlling the dynamics of the stock market. The novice investor acquires a summary knowledge of financial analysis, while expert investor is mainly based on a more complex analysis. The cognitive behavior model of the investor describes his perceptual, informational and decisional processes. It includes the behavioral attitudes and the social profile which influence these processes.

3.1 Rational Analysis

The rational analysis is composed of fundamental analysis and chart analysis.

- *Fundamental analysis*
For our modeling, we adapt the two complementary approaches necessary to accomplish fundamental analysis. The first is the *Stock Evaluation* which is

based on the constant-growth model (known as Gordon Shapiro model). This model only requires data from one period and an average growth rate which can be found from past financial statements [19]. The second approach is *Performance Measuring* which is based on calculating rates of return and risk. In fact, the rate of total return on shares is composed of the rate of overall performance in dividends and the return of capital. We used the notion of systematic risk (or market risk), which is indicated by a given coefficient β. This coefficient indicates how the expected return of a stock or portfolio is correlated to the return of the financial market as a whole. It is calculated by the CentralMarket agent.

- *Chart analysis*

Chart analysis is based on the hypothesis that the past development of a financial asset provides better information about its own future. The trend of our artificial stock market is determined by the CentralMarket agent from trends calculated by all expert investors. If the trend is upward, it implies that the stock value should continue to rise. The market presents an uptrend. If investors realize that the following value outside the market price, they notice a sign of rising. Otherwise, it announced a reversal of trend (downtrend) and causes a signal to sell. In our model, we are guided by the Points and Figure Charting (PFC) [4]. This method represents the changes of the stock price.

3.2 Behavioral Analysis

We introduce in our model three pairs of behavioral attitudes: *optimism/pessimism, speculation/caution and imitation/leadership*. For the representation of behavior attitudes, we adopt the generic approach introduced in [3]. This approach is based on the specification of a set of inhibitor and triggering thresholds. In fact, each Investor agent receives various kinds of qualitative stimuli (experts' opinion, prediction and advice) and quantitative stimuli (market trend and market price). The stimuli affect the Investor agent decisions. Their effects are weighted according to the Investor agent behavioral profile described through the three remaining pairs of behavioral attitudes. These attitudes play the role of reactive modulators that filter and weight the effect of external stimuli.

- *Optimism/pessimism attitudes*

In our model, this behavioral component plays a crucial role in determining the estimated rates. The optimistic Investor agent, which has a confidence in the outcome, does not react the same way as the pessimistic Investor agent. These attitudes affect the rational analysis and more specifically the evaluating performance. Furthermore, in the informational process of our model, optimistic Investor agent overestimates rates while the pessimistic Investor agent under estimates them.

- *Speculation/caution attitudes*

A speculator Investor agent decides to conduct a transaction (buy or sell) by accepting the risk of losing in order to gain maximum benefits. Besides, a cautious Investor agent proceeds with prudence and prefers to take every detail into account before buying or selling. This difference influences the rational analysis. Indeed, the speculator Investor agent decides to buy a stock even if it presents a high rate of risk, something unacceptable by the cautious Investor agent. The parameters taken into account in PFC method related to the chart analysis are also influenced by this behavioral component. Besides, we consider that these attitudes affect the perceptual process. In fact, a speculator investor presents a lower threshold of acceptance information than a cautious investor.

- *Imitation/leadership attitudes*

An imitator Investor agent reproduces unconsciously the reaction of his entourage of investors. It follows and is aligned with the overall trend of the market. Whereas, the leader holds the dominant market position and take initiatives to buy or sell stocks. These behavioral attitudes influence the perceptual process which is more extended for the leader than the imitator investor. Indeed, the number of persons composing the confidence network (called TrustNet) of an imitator investor is larger compared to the leader one who has confidence in a few number of investors.

4 Simulation and Experimental Results

4.1 SiSMar: The Artificial Stock Market

SiSMar (Simulation Stock Market) is an artificial stock market composed of a set of Investor agents having diversified behavioral attitudes and social profiles. The purpose of this simulation is to understand the influence of psychological character of an investor and its neighborhood on its decision-making and their impact on the market in terms of price fluctuations. Our simulator is implemented using the MadKit platform [9] and is written in Java.

The Investor agent is in a direct relationship with other agents. In our simulation, the neighborhood is not physical but it is a neighborly relationship (trust, privacy, etc.). An investor can make and receive advice or opinion of its neighbors. However, the agent takes into account the message received if it is filtered through the filter of privacy and / or the filter of confidence. On the privacy filter, if the sender of the message is part of trust network of the receiver, the message is accepted. Otherwise, the message is refused. On the confidence filter, if the acceptance threshold is lower than the information certainty threshold, this information is taken into account. Otherwise, the receiver refuses the message and ignores it. We assume that the cardinal of TrustNet is less than or equal to six investors for novices and it does not exceed two for experts. This network is dynamic, since if the chart analysis of the agent does not coincide with the advice of a member of his TrustNet, the information is ignored and the investor eliminates the sender of this message and randomly chooses another one. At each simulation step, every Investor agent may behave in several ways, depending on his state. He may be inactive if the number of simulation steps designated by t is not a multiple of its periodicity. Once active, Investor agents interact on the market. The ExpertInvestor agents' decision-making takes place after processing four tests relative to: stock evaluation, risk measuring, dividend rate measuring and chart analysis. Each test gives out a signal to buy, sell or do nothing respectively designed by $d_{StockEval}$, $d_{RiskMeasure}$, $d_{RDividend}$ and $d_{ChartAnalysis}$. The final decision is calculated as follows:

$$D = \alpha * d_{StockEval} + \gamma * d_{RiskMeasure} + \lambda * d_{RDividend} + \delta * d_{ChartAnalysis}$$

With: $\alpha + \gamma + \lambda + \delta = 1$.

The NoviceInvestor agent anticipates changes of the stock price using only chart analysis and filtered information diffused by its neighborhood.

4.2 The Market Price Volatility Evolution

The purpose of this experiment is to observe the magnitude of changes in stock prices. Figure 3 represents the evolution of four stocks in a bull market and its evolution when the trend is downward. In figure 3 (a), we notice that the upward movements are longer than those of downward. This reflects a resistance to downtrend pressure on the bull market. In addition, changes in the prices of the four studied stocks are very considerable which shows that a large number of transactions were executed. Contrarily, in figure 3 (b), we observe some positive stock price fluctuations however dominated by a global downtrend. We can conclude that there is resistance to the increase in the bear market. These two results are in line with what happens in reality on a stock market and coincide with the principles of Dow Theory. In this experiment, we observe the change in stock price in a market where the trend is not determined from the outset. We assume that Investors' experts, as the market leaders, calculate at every

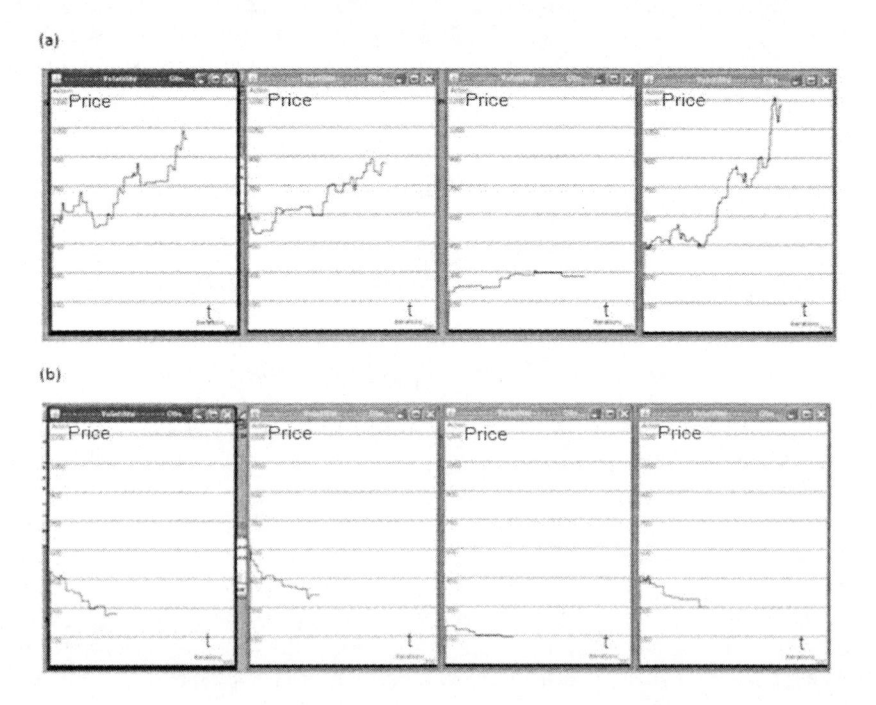

Fig. 3. Analysis of market trend of four stocks: (a) uptrend market, (b) downtrend market

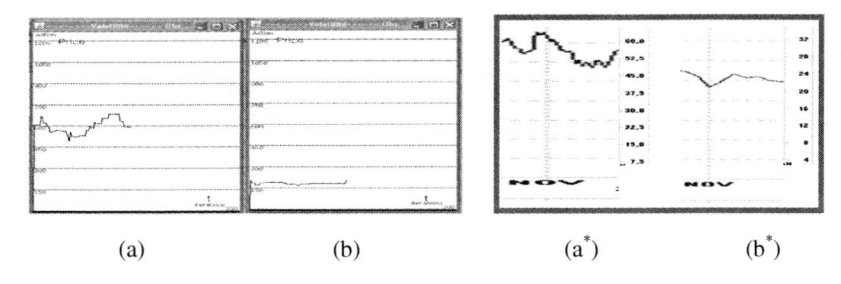

Fig. 4. Comparative of the stock volatility: (a) Peugeot volatility overall SiSMar, (b) Air-France volatility overall SiSMar, (a*) Peugeot volatility overall EuroNext Paris (November 2007), (b*) Air-France volatility overall EuroNext Paris (November 2007)

step of simulation the tendency of stocks and send it to the Central Market. The last one determines the overall trend.

We take the example of these two stocks[1] presented in figure 4. The table 1 resumes the characteristics taken into consideration in this experiment, with a population composed of 50 ExpertInvestor agents and 80 NoviceInvestor agents.

We observe a very significant volatility of the first stock figure 4 (a) compared with that in the figure 4 (b) despite that the price of the second is much lower than the first and they all possess both an upward trend.

We explain this fact by the coefficient β which influences the rate of risk and it is considered by the behavioral profile of our Investor agents. More than β is near to 1, more the stock is risky. This risk is visible by the measure of volatility. In addition, the annual dividend offered by a stock improves the number of transactions of this stock. This fact increases his volatility. We notice also that the volatility of these two stocks overall SiSMar is very close to the values extracted from reality (figure 4 (a*) and (b*)). What we are emphasizing here is that it is possible to reproduce realistic market evolutions using the SiSMar behavioral model.

Table 1. The characteristics considered of two stocks: Peugeot and Air-France

Name	Peugeot	Air-France
Acquisition Price	60.7	20.2
Annual dividend	3.1	0.9
Trend	1	1
B	0.6	0.2

5 Conclusion and Future Works

In this paper, we introduced a new model of stock market dynamics. Our research focuses on the modeling of the stock market and particularly modeling the behavior and

---

[1] We notice that data relative to these two real stocks, Peugeot and Air-France, are extracted from EuroNext Paris (November 2007) and used to calibrate the parameters of our simulator.

decision making of two types of Investors: novice and expert Investor. Our contribution is to consider the stock market as a social organization of autonomous actors with dependents heterogeneous beliefs and different behavioral attitudes. Different perspectives can be considered in our work. The first is to refine the model and enrich its implementation with including new cognitive concepts at the micro level. Learning techniques and fuzzy logic can be used respectively to explore the memory effect and the uncertainty at various levels in the stock market. Whereas, the perceptual process may include components which are based on fuzzy sets such as threshold triggers and inhibitor sets.

References

1. Arifovic, J.: The behavior of the exchange rate in the genetic algorithm and experimental economies. Journal of Political Economy 104, 510–541 (1996)
2. Arthur, W., Durlauf, S., Lane, D.: The economy as an evolving complex system. Santa Fe Institute, vol. 27, pp. 15–44. Addison-Wesley, Reading (1997)
3. Ben Said, L., Bouron, T.: Multi-agent simulation of virtual consumer populations in a competitive market. In: SCAI 2001, pp. 31–43. IOS Press, Denmark (2001)
4. Dorsey, T.: Point & figure charting: The essential application for forecasting and tracking market prices, 3rd edn. Wiley Trading, Chichester (2007)
5. Fama, E.: The behavior of stock prices. Journal of Business (January 1965)
6. Gode, D., Sunder, S.: Allocative efficiency of markets with zero intelligence traders. Journal Of Political Economy 101(1), 119–137 (1993)
7. Granovetter, M.: Les institutions économiques comme constructions sociales. Orléan André édition, Analyse économique des conventions, Paris, Presses Universitaires de France, chapitre, vol. 3, pp. 119–134 (2004)
8. Grossman, S., Stiglitz, J.: On the impossibility of informationally efficient markets. American Economic Review 70, 393–408 (1980)
9. Gutknecht, O., Ferber, J.: The madkit agent platform architecture. In: Agents Workshop on Infrastructure for Multi-Agent Systems, pp. 48–55 (2000)
10. Hoffmann, A., Jager, W.: The effect of different needs, decision-making processes and network-structures on simulating stock-market dynamics: a framework for simulation experiments. In: ISSA 2004 (2004)
11. Hoffmann, A., Jager, W., Eije, J.V.: Social simulation of stock markets: Taking it to the next. Level Journal of Artificial Societies and Social Simulation 10(2) (2007)
12. Kahneman, D., Tversky, A.: Prospect theory: An analysis of decision under risk. Econometrica 47(2), 263–292 (1979)
13. Karaken, J., Wallace, N.: On the indeterminacy of equilibrium exchange rates. Quarterly journal of economics 96, 207–222 (1981)
14. LeBaron, B.: Agent-based computational finance: suggested readings and early research. Journal of Economic Dynamics and Control 24, 679–702 (2000)
15. Lettau, M.: Explaining the facts with adaptive agents: the case of mutual fund flows. Journal of Economic Dynamics and Control 21, 1117–1148 (1997)
16. Lux, T., Marchesi, M.: Volatility clustering in financial markets: A micro-simulation of interactive agents. In: 3rd Workshop on Economics and Interacting Agents, Ancona (1998)
17. Margarita, S., Beltratti, A.: Stock prices and volume in an artificial adaptive stock market. In: Mira, J., Cabestany, J., Prieto, A.G. (eds.) IWANN 1993. LNCS, vol. 686, pp. 714–719. Springer, Heidelberg (1993)

18. Neisser, U.: Cognitive psychology, New York. Appleton-Century-Crofts (1967)
19. Peyrard, J.: La Bourse. Edition Vuibert, collection Etreprise, 9 ème édition (2001)
20. Ricciardi, V.: Risk: Traditional Finance versus Behavioral Finance. John Wiley & Sons, Chichester (2008)
21. Routledge, B.: Artificial selection: genetic algorithms and learning in a rational expectations model. Technical report, GSIA, Carnegie Mellon, Pittsburgh, Penn (1994)
22. Sharpe, W.: Corporate pension funding policy. Journal of Financial Economics 3, 183–193 (1976)
23. Steiner, P.: The sociology of economic knowledge. European Journal of Social Theory, 443–458 (2001)

How to Avoid Biases in Reactive Simulations

Yoann Kubera, Philippe Mathieu, and Sébastien Picault

Laboratoire d'informatique Fondamentale de Lille, University of Lille, France
forename.surname@lifl.fr

Abstract. In order to ensure simulations reproducibility, particular attention must be payed to the specification of its model. This requires adequate design methodologies, that enlightens modelers on possible implementation ambiguities – and biases – their model might have. Yet, because of not adapted knowledge representation, current reactive simulation design methodologies lack specifications concerning interaction selection, especially in stochastic behaviors. Thanks to the interaction-oriented methodology IODA – which knowledge representation is fit to handle such problems – this paper provides simple guidelines to describe interaction selection. These guidelines use a subsumption like-structure, and focus the design of interaction selection on two points : how the selection takes place – for instance first select the interaction, and then select the partner of the interaction, or first a partner and then an interaction – and the nature of each selection – for instance at random, or with a utility function. This provides a valuable communication support between modelers and computer scientists, that makes the interpretation of the model and its implementation clearer, and the identification of ambiguities and biases easier.

1 Introduction

Any Multi-Agent-Based Simulation (MABS) – and more generally any kind of simulation – is implemented according to a model defined by domain specialists. These specialists are not always fully aware of implementation requirements. As a result, computer scientists have to make implementation choices, that may lead to biased results. Even worse, because of programming habits and too permissive methodologies and frameworks, these choices might be implicit. For instance in Epstein and Axtell ecosystem simulation [5], a bias occurred in results because the interactions in which an agent might participate at the same time were not specified.

To ensure simulation reproduction – *i.e.* obtain similar results with implementations of the model made by different persons – and to hedge against ambiguities and biases, domain specialists have to consider the most exhaustive set of questions about what they want or expect of their model. Indeed, their answers elicit implementation choices, and remove ambiguities that may lead to different implementations. Moreover, it makes sure that choices – including the choice to not answer some questions – are made willingly, and aware of the biases they may introduce.

In this paper, "bias" means "erroneous/distorted simulation outcomes". Thus, biases are the result of either defective means such as faulty random number generators, or of wrong implementation choices. This paper focuses on this second point.

Our goal is to provide a generic and domain independent simulation design methodology called *IODA* and framework called *JEDI* [8]. This paper participates in that effort by defining a particular question – and its answer – that all simulation methodologies should consider to prevent implementation biases: *"how agents select the action or interaction they perform among their perceived affordances*[1] *?"* [13]. Since this aspect is well specified for cognitive agents, the focus of this paper is reactive agents. Nevertheless, our proposition remains valid for cognitive ones.

In order to clarify precisely this point, at least two properties are required:

- *knowledge of agents* – what they are able to do – *has to be defined separately from action/interaction selection* – what an agent chooses to do. This separation has to be made even for reactive simulations;
- *interaction* – a notion underlying any simulation – *has to appear explicitly in the methodology as well as in the implementation, as a software entity*.

Yet, simulation methodologies do not meet the requirements of the last point (see section 3), and thus remain ambiguous on how action/interaction selection is handled in reactive simulations. Indeed, agents define only how they select the action or interaction they perform [3, 15], but do not provide guidelines on how target agents are selected (see section 2), even though these processes are deeply bound together.

This paper aims at filling this gap by first specifying an architecture that underlies any kind of multi-agent-based simulation, and that is fit to enlighten modelers on the problem mentioned above (see section 2). Then, a solution of this problem is presented in section 4. In this solution, the modeler elicits the action/interaction selection process of agents in two parts. First he has to specify how selection takes place among three recurrent patterns met in simulations – *first interaction then target* selection, or *first target then interaction* selection or *tuple* selection – and then the nature of every selection among three ones – either *random*, *by preference* or *weighted*. We uphold that such specifications provide a valuable communication support between modelers and computer scientists, that makes the interpretation of the model and its implementation clearer, and that makes the identification of model ambiguities and possible biases easier. We illustrate this solution on a modeling example (see section 5), that shows the importance of such a specification.

2 Separation in Functional Units

Even if the application domains of multi-agent simulations are heterogeneous, they can be split into different and weakly dependent functional units [4, 16]. We consider a particular functional decomposition that underlies any kind of simulation. This decomposition is done in three main units (see Fig. 1), called ACTIVATION UNIT, DEFINITION UNIT and SELECTION UNIT (see [7]).

Because the design of simulations implies crucial choices about those three units, we claim that it is important to make this separation clear, even in reactive simulations, in order to make modeling choices explicit.

---

[1] What an agent knows it can perform in a given context.

```
┌─→ ACTIVATION UNIT
│   selects the next agent a that will behave.
│              ↓
│   DEFINITION UNIT
│   provides the informations required to build a's per-
│   ceived affordances.
│              ↓
│   SELECTION UNIT
└── selects from perceived affordances of a what action
    or interaction a initiates.
```

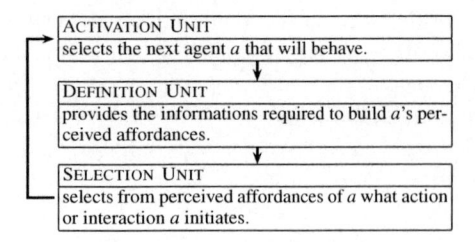

Fig. 1. The three main functional units of a multi-agent simulation described in [7]

The significance of the DEFINITION UNIT specification and generic representation has been addressed in [8], and its relevance to elicit model ambiguities and possible biases is demonstrated in [7] and in this paper. The impact of implementation choices of the ACTIVATION UNIT was dealt with in [7], and studies possible answers to the questions *"when agents trigger their behavior ?"* and *"in which interactions an agent may participate simultaneously ?"*. Thus, we focus in this paper on the latest unit, the SELECTION UNIT.

3 Related Works

The space of implementation choices is really wide. To guide modelers in the hard task of eliciting modeling and implementation choices, many agent-based simulation design methodologies exist and claim to handle this issue.

Some of them are all purpose design methodologies – like VOLCANO [14]. On the opposite, many are specific to particular subsets of simulations – like DESIRE [15] that designs reasoning agents, or ADELFE [1] that designs adaptative agents. Because they are developed for particular use, they target more specific problems, and thus provide a more exhaustive specification of implementation choices for it.

Reactive simulation design methodologies have a particular status among these last. Indeed, even if they claim to be methodologies, many just consist in writing the simulation in the agent language or architecture they provide. Thus, unless the structure of the architecture forces to make choices, there is no guidelines to build behaviors. Some methodologies and frameworks make the separation between knowledge of agents and action/interaction selection, and provide guidelines to build reactive agents behavior. This is the case of component based frameworks like MALEVA [2], or of hybrid frameworks like InteRRap [11] and PRS-based ones [6].

Nevertheless this separation is a necessary but not a sufficient condition to avoid biases coming from action/interaction selection. Indeed, agents action/interaction selection is the art of selecting the next action or *interaction* it will initiate. The underlying problem is that agents have to consider which interactions they will initiate *and with which other agent it will be performed* (*i.e.* the target agent). Sadly, target agent choice process remains unspecified in such methodologies.

To elicit such issues, we uphold that separation must be made between knowledge declaration, perceived affordances listing and action/interaction selection process. Thanks to that, different patterns of action/interaction selection were identified. The modeler has to take into account these last to ensure that his model will be understood and implemented as it was firstly thought.

4 Unit Specification Proposal

To specify clearly the SELECTION UNIT, we center the action/interaction selection process of agents on the notion of interaction and perceived affordances – *i.e.* the set of all actions and interactions an agent might perform in a particular context.

Perceived affordances construction requires a specific representation of actions and interactions. We use the one of the *IODA* methodology [8], that reifies interactions and perceived affordances even at implementation. *IODA* provides advanced methodological tools to design interactions in MABS. Since we do not need all refinements it provides, we use a simplified version of [8] definitions.

4.1 Knowledge and Affordances Representation

To make the difference between the abstract concept of agent (for instance Wolves), and agent instances (a particular Wolf), we use the notion of **agent families** as abstract concept of agent. Thus, the word **agent** refers to an agent instance.

Definition 1. *An **agent family** is an abstract set of agent instances, which share all or part of their attributes and behavior.*

Definition 2. *An **interaction** is a structured set of actions involving simultaneously a fixed number of agents instances that can occur only if some conditions are met.*

An interaction is represented as a couple (*conditions, actions*), where *condition* is a boolean function and *action* is a procedure. Both have agent instances as parameters. Agents that are involved in an interaction play different roles. We make a difference between **Source** agents that may initiate the interaction (in general the one selected by the ACTIVATION UNIT) and **Target** agents that may undergo it.

Definition 3. *Let $\mathscr{S} \in \mathbb{F}$ and $\mathscr{T} \in \mathbb{F}$ be agent families. We note $a_{\mathscr{S}/\mathscr{T}}$ the **set of all interactions** that an instance of the $\mathscr{S}$ agent family is able to initiate with an instance of the $\mathscr{T}$ agent family as a target.*

Thanks to these definitions, we can specify the knowledge of an agent family $\mathscr{S} \in \mathbb{F}$ as the set $\bigcup_{\mathscr{T} \in \mathbb{F}} a_{\mathscr{S}/\mathscr{T}}$, which contains every interactions it is able to initiate as source with any agent family as target.

To unify knowledge, actions are considered as interactions that occur with no target. We do not add this to our notations, please see [8] for more informations.

The definition of perceived affordances uses the notion of realizable interaction, in order to determine if two agents can participate in an interaction.

Definition 4. *Let I be an interaction, and $x \prec \mathscr{S}$, $y \prec \mathscr{T}$ two agents. The tuple (I, x, y) is **realizable** (written $r(I, x, y)$) if and only if :*

- $I \in a_{\mathscr{S}/\mathscr{T}}$, i.e. agents of $\mathscr{S}$ family are able to perform I with agents of $\mathscr{T}$ family;
- the conditions of I hold true with x as source and y as target.

A realizable tuple represents one interaction that an agent can initiate with a particular target agent. Moreover, an agents perceived affordances are the set of all interactions it can initiate in a given context. Thus, at a time t, the perceived affordances of the x agent are the set of all realizable tuples that x may perform.

Definition 5. *Let $\mathbb{A}_t$ be the set of all agents in the simulation at a time t, and $x \in \mathbb{A}_t$. Then, the **perceived affordances** $\mathbb{R}_t(x)$ that x may perform at time t is the set:*

$$\mathbb{R}_t(x) = \bigcup_{y \in \mathbb{A}_t} \bigcup_{I \in a_{x/y}} \{(I,x,y) | r(I,x,y)\}$$

4.2 Selection Unit

In reactive simulation, agents try in general to perform actions and interactions sequentially until a realizable one is found – *i.e.* they use nested if/else structures. We propose to use a similar principle in the SELECTION UNIT (see Fig. 2) : every possible interaction I between a source $\mathscr{S}$ and a target agent family $\mathscr{T}$ is assigned a priority $p(I, \mathscr{S}, \mathscr{T})$, just as [12] did for classical actions (that do not involve any target). Selection takes place on interactions in decreasing order.

4.2.1 Interaction Selection Policies

Thanks to the interaction-oriented study of experiments, different policies used to select a tuple from a set of realizable tuples $\mathbb{R}_t^p(x)$ were identified. An interaction selection policy is decomposed in two parts : the nature of the selection, and on which elements the selection takes place. Indeed, the SELECTION UNIT can:

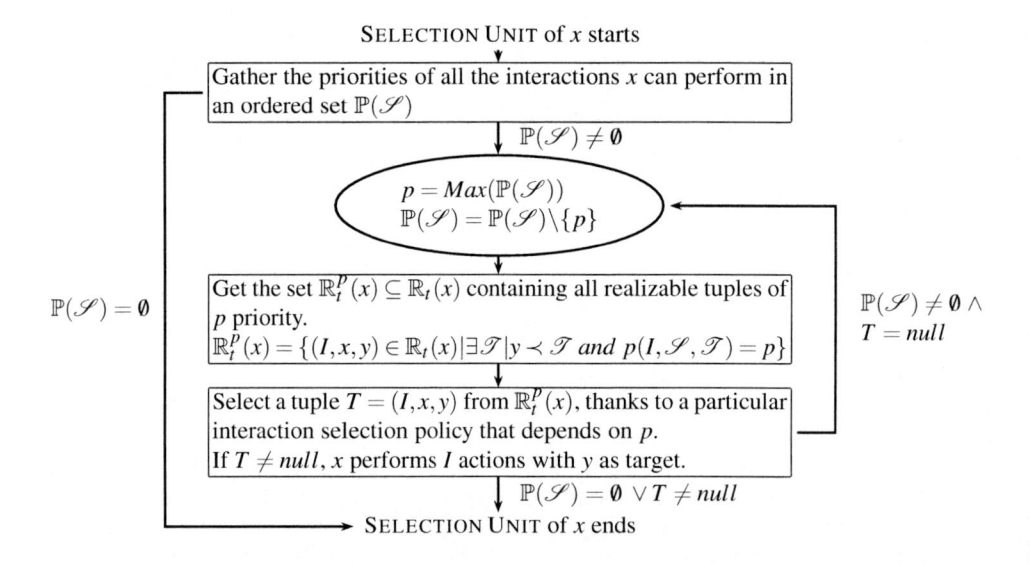

Fig. 2. Generic description of a reactive agent's (named x) SELECTION UNIT

- **First** select the **interaction** that will occur, and **then** select its **target**. If the selected interaction is degenerate (*i.e.* is an action), no target selection takes place;
- **First** select the **target** on which an interaction will occur, and **then** select the **interaction** that will occur. This selection cannot involve degenerate interactions;
- Directly select a **tuple** (*Interaction/Target*). If an interaction is degenerate, the corresponding tuple is only (*Interaction*).

The selection of each element – interaction, target or tuple – has one nature chosen among three different ones:

- the element is selected at **random**;
- every element is given a **preference** value. The selected element is the one with the highest preference value. If more than one have this value, one of them is selected at random. This selection is intensively used in cognitive agents;
- every element e is given a **weight** $w(e) \in [0,1[$, and an interval $\mathscr{W}(e) \subset [0,1[$ of length $w(e)$ such that intervals are pairwise not intersecting. A number $r \in [0,1[$ is chosen at random. The selected element e is the element such that $r \in \mathscr{W}(e)$.

4.2.2 Design Guidelines of the SELECTION UNIT

To design an SELECTION UNIT containing fewer ambiguities, the modeler has:

- to provide priorities to every interaction an agent may perform;
- to provide for each couple (source agent family, priority):

  – on what element the selection is made (either *interaction then target*, or *target then interaction*, or *tuple*);
  – the nature of each selection (either *random*, *by preference* or *weighted*);
  – how preference and weights are computed.

Obviously, he has to understand what his choices imply. This kind of specification is possible only if interactions are at center of simulation, like in IODA [8].

5 Illustration on a Modeling Problem

Reactive MABS application fields widen everyday, and tackle very different problems. Among these appears chemistry, for which MABS provide more realistic diffusion behaviors than in numerical simulations. In this application field, one of the most difficult issue of multi-agent programing has to be tackled: defining the behavior of agents according to macroscopic rules. These rules use probabilities, and thus require to define stochastic behaviors for agents. These kind of behaviors introduce issues that do not appear in non-stochastic multi-agent simulations. Consequently, biases may occur in situations that might seem correct for regular simulations design methodologies. We illustrate this point on a modeling example, and show how our solution provides guidelines that leads modelers to identify biases.

5.1 The Modeling Problem

We consider simulations that describe chemical reactions. In those kinds of simulations, the behavior of agents is almost completely summarized in reaction rules. The modeling problem we consider is the implementation of the rules:

$$A + B \xrightarrow{k_1} C \qquad (R1)$$
$$A + D \xrightarrow{k_2} E \qquad (R2)$$

The reaction rule R1 means that an agent of A family can react with an agent of B family in order to from a new agent of C family. The two agents of A and B family are then destroyed. This reaction occurs only at a particular reaction rate k_1.

This rate is deeply bound with the probability that a R1 reaction occurs. For convenience, we consider that k_1 is the probability that the reaction R1 occurs[2]. The same goes for k_2 and R2.

Due to the lack of space, we focus only on the definition of A chemical species behavior. Moreover, since this is not the topic of this paper, we do not describe how A, B, C, D and E agents move in the environment.

This modeling problem is common in chemical reaction modeling, since chemical species are often involved in many different chemical reactions.

5.2 First Encountered Problem

To implement such agents, the reaction rate of R1 and R2 have to be integrated into their behavior. Many different implementations of this behavior can be made (see [9]). These implementations correspond to different interpretations of reaction rates. Indeed, k_1 can:

- Either represent the probability that the reaction R1 occurs if an agent of A family is close to at least one agent of B family. In that case, the probability $\mathscr{P}(R1)$ that an agent of A family executes R1 depends only on the presence of one nearby agent of B family. Thus, if at least one agent of B family is close-by, $\mathscr{P}(R1) = k_1$;
- Or represent the probability that the reaction R1 occurs with one particular agent of B family. In that case, the probability that an agent of A family performs R1 depends on the number of nearby agents of B family. Thus, if there is n_b close-by agents of B family, $\mathscr{P}(R1) = 1 - (1 - k_1)^{n_b}$;

With our solution, the two interpretation correspond to two different SELECTION UNIT, where R1 and R2 have the same priority:

1. Either a *First interaction then target* selection, with a *weighted* selection for interactions (where $w(R1) = k_1$ and $w(R2) = k_2$), and a *random* one for targets;
2. Or a *Tuple* selection, where a tuple is realizable only if it meets the probability : the stochastic factor is tried in the condition of R1 and R2. The performed tuple is selected at random among realizable tuples.

The different interpretations appear clearly in the SELECTION UNIT. Indeed the use of the *First interaction then target* selection policy implies that the probability to trigger a

---

[2] Usually, this probability is obtained through computations, and is different from k_1.

reaction is independent from the number of neighboring agents. The use of the *tuples* selection policy implies that the probability to trigger a reaction is proportional to the number of neighboring agents.

5.3 Second Encountered Problem

Let us consider the second implementation, where the probability that an agent of A family performs R1 depends on the number of neighboring agents of B family.

Because an agent of A family that performs R1 disappears from the environment, such simulations are sometimes written like in figure 3.

```
ask every agent of A family [
    ;; List in its perceived affordances realizable R1
    ;; reactions with close-by B agents.
    ;; If at least one such tuple exists, the agent performs
    ;; one of them, and disappears from the environment.
]
ask the remaining agents of A family [
    ;; List in its perceived affordances realizable R2
    ;; reactions with close-by D agents.
    ;; If at least one such tuple exists, the agent performs
    ;; one of them
]
```

Fig. 3. An implementation example of our modeling problem

This kind of implementation provides biased results. Indeed, agents of A family perform R2 only if they failed to perform R1. Thus, conditional probabilities are introduced: $\mathscr{P}(R2) = (1 - \mathscr{P}(R1)) \times (1 - (1-k_2)^{n_d})$

The greater k_1 or the density of agents of B family are, the greater the bias coming from conditional probabilities becomes. Thus, if the simulation is verified with low densities – or with a low reaction probability k_1 – the error has a weak impact on simulation results, and simulation seems unbiased.

Even if this bias seems obvious, it exists in real implementations, for instance in the Netlogo [17] implementation of Henry-Michaelis-Menten kinetics [10].

If the reaction rates raise, experiments showed that a huge gap appeared between the reaction speed in biased implementations and the reaction speed in the unbiased ones. Because the reaction speed is at the center of many chemical reactions (like in Henry-Michaelis-Menten kinetics), such a bias is not acceptable. Thus, particular attention must be payed to this point.

With our solution, to obtain such a bias, different priorities must be given to R1 and R2. Thus, the fact that an agent of A family performs an R2 interaction only if it could not perform an R1 interaction appears explicitly in the SELECTION UNIT.

6 Conclusion

Designing simulations implies making implementation choices. These choices have a deep impact on simulation results, and might even introduce biases in them. To avoid this problem, modelers have to provide a precise description of implementation choices, to ensure the reproducibility of the model. This is only possible if the modeling methodology they use provides guidelines that elicits all these choices.

Current reactive MABS design methodologies do not specify clearly how the target of interactions are selected, because they do not provide both the separation between knowledge and action/interaction selection process of agents, and the reification of interactions.

Thanks to the IODA methodology – that meets the requirements mentioned above – we built guidelines to design the behavior of agents in order to solve this issue. The guidelines consist in providing knowingly a priority to every interaction an agent may perform, and then specifying for every priority:

- on what element the selection is made (either *first interaction then target*, *first target then interaction*, or *tuple*);
- the nature of each selection (either *random*, *by preference* or *weighted*);
- how preference and weights are computed.

The importance of such guidelines was illustrated in the case of chemical reactions simulations. It avoids the misuse of probabilities, that could introduce critical biases in results.

We uphold that such specifications provide a valuable communication support between modelers and computer scientists for the design of any kind of reactive simulations. It makes the interpretation of the model and its implementation clearer – and thus avoids ambiguities in the model – and the identification of possible biases easier.

Acknowledgement. This research is supported by the FEDER and the "Contrat-Plan État Région TAC" of Nord-Pas de Calais.

References

1. Bernon, C., Camps, V., Gleizes, M.-P., Picard, G.: Engineering Adaptive Multi-Agent Systems: The ADELFE Methodology. In: Agent-Oriented Methodologies (2005)
2. Briot, J.-P., Meurisse, T.: An experience in using components to construct and compose agent behaviors for agent-based simulation. In: Proceedings of IMSM 2007 (2007)
3. Brooks, R.A.: A robust layered control system for a mobile robot. IEEE journal of robotics and automation 2(1) (1986)
4. Demazeau, Y.: From interactions to collective behaviour in agent-based systems. In: Proceedings of ECCS 1995, Saint-Malo, France (1995)
5. Epstein, J., Axtell, R.: Growing Artificial Societies. Brookings Institution Press, Washington (1996)
6. Georgeff, M.P., Lansky, A.L.: Reactive reasoning and planning. In: Proceedings of AAAI 1987, Seattle, WA (1987)
7. Kubera, Y., Mathieu, P., Picault, S.: Biases in multi-agent based simulations : An experimental study. In: Proceedings of ESAW 2008, St Etienne, France (2008)
8. Kubera, Y., Mathieu, P., Picault, S.: Interaction-oriented agent simulations : From theory to implementation. In: Proceedings of ECAI 2008, Patras Greece (2008)
9. Kubera, Y., Mathieu, P., Picault, S.: Interaction selection ambiguities in multi-agent systems. In: Proceedings of IAT 2008, Sydney, Australia (2008)

10. Michaelis, L., Menten, M.L.: Die kinetik der invertinwirkung. Biochemische Zeitschrift 49 (1913)
11. Müller, J.P., Pischel, M.: Modelling interacting agents in dynamic environments. In: Proceedings of ECAI 1994, Amsterdam, The Netherlands (1994)
12. Nilsson, N.J.: Teleo-reactive programs for agent control. Journal of Artificial Intelligence Research 1 (1994)
13. Norman, D.A.: The Psychology of Everyday Things. Basic Books (1988)
14. Ricordel, P.-M., Demazeau, Y.: Volcano, a vowels-oriented multi-agent platform. In: Dunin-Keplicz, B., Nawarecki, E. (eds.) CEEMAS 2001. LNCS, vol. 2296, p. 253. Springer, Heidelberg (2002)
15. van Langevelde, I., Philipsen, A., Treur, J.: Formal specification of compositional architectures. In: Proceedings of ECAI 1992 (1992)
16. Weyns, D., Parunak, H., Michel, F., Holvoet, T., Ferber, J.: Environments for multiagent systems: State-of-the-art and research challenges. In: Environments for multiagent systems, New York, NY, USA (2004)
17. Wilensky, U.: Netlogo. Center for Connected Learning and Computer-Based Modeling. Northwestern University, Evanston, IL (1999), http://ccl.northwestern.edu/netlogo

Generating Various and Consistent Behaviors in Simulations

Benoit Lacroix[1], Philippe Mathieu[2], and Andras Kemeny[3]

[1] Renault, Technical Center for Simulation and LIFL, University of Lille, France
benoit.lacroix@renault.com
[2] LIFL, University of Lille, Cite Scientifique Bat M3, 59655 Villeneuve d'Ascq, France
philippe.mathieu@lifl.fr
[3] Renault, Technical Center for Simulation, 1 av. du Golf, 78288 Guyancourt, France
andras.kemeny@renault.com

Abstract. In multi-agent based simulations, providing various and consistent behaviors for the agents is an important issue to produce realistic and valid results. However, it is difficult for the simulations users to manage simultaneously these two elements, especially when the exact influence of each behaviorial parameter remains unknown. We propose in this paper a generic model designed to deal with this issue: easily generate various and consistent behaviors for the agents. The behaviors are described using a normative approach, which allows increasing the variety by introducing violations. The generation engine controls the determinism of the creation process, and a mechanism based on unsupervised learning allows managing the behaviors consistency. The model has been applied to traffic simulation with the driving simulation software used at Renault, SCANeR© II, and experimental results are presented to demonstrate its validity.

1 Introduction

A typical application of multi-agent based simulations is the reproduction of real world situations. In such cases, the simulations validity is first assessed by reproducing known situations, before studying how new designs or behaviors influence the results. The variety and the consistency of the behaviors are fundamental in these applications. The variety is often obtained by providing the agents with individual characteristics [3]; the consistency is essential for the results validity, which will be questioned if aberrant behaviors appear. However, these issues are often not specifically considered. A model designed to ease the work of the designers has to consider them and to include different characteristics: provide a high-level representation of the behaviors, allowing abstracting the domain specificities; be flexible and generic to adapt easily to various users needs; and finally, allow users to check that the produced behaviors remain in the limits they wish.

The paper is organized as follows. We introduce in section 2 the different characteristics of the proposed model. The behaviors are described using a normative approach, taking advantage of the descriptive capacities of norms, rather than the more usual prescriptive way typically used in multi-agent systems. The creation of the behaviors is based on nondeterministic principles, and unsupervised classification is used to control the potential violating behaviors and the norms evolution. In section 3 we describe how the model allows producing variety and consistency, and illustrate the genericity of the approach. Section 4 describes the application of the model in the traffic simulation field, and the implementation in the driving simulation software used at Renault, SCANeR© II. Finally, section 5 presents experimental results demonstrating the model validity.

2 Description of the Model

2.1 Normative Representation of the Behaviors

In multi-agent based simulations, normative systems are usually used to introduce regulation possibilities in the environment, and to add cooperation and coordination mechanisms [1, 14]. For instance, Electronic Institutions [10] exploit them to regulate agents interactions: the institution describes the conventions governing agents interactions, and the norms assess actions consequences within the scope of the institution. Normative models are applied in various fields: disaster management, market monitoring...

In simulations where the behaviors rely upon many parameters of different kinds (discrete, continuous...), controlling their values and their consistency is a complex issue. The notion of norm, which presents an intuitive description means of the parameters sets, has been proposed to answer it [7]. Norms are used to describe agents behaviors, using the following metaphor: the institution handles parameters limits, and norms behaviors families. The institutional elements are defined as follows.

Definition 1. *An Institution is a tuple $\langle P, D_P, P_i, P_e \rangle$ where: P is a finite set of parameters; $D_P = \{d_p, \forall p \in P\}$ is a set of definition domains; P_i is a set of institutional properties; and P_e is a set of environmental properties.*

The institution provides a fixed reference for the norms. It handles a finite set of parameters. A definition domain is associated to each of them, to provide limits for the parameters. Finally, sets of institutional and environmental values link the institution to its context. Application or domain specificities are taken into account this way.

Definition 2. *A Norm is a tuple $\langle I, P_n, D_{P_n}, \Gamma_P, P_{n_d}, P_{n_i}, P_{n_e} \rangle$ where: I is the institution the norm refers to; $P_n \subset P$ is the subset of parameters associated to the norm; $D_{P_n} \subset D_P$ is the subset of definition domains; $\Gamma_{P_n} = \{\gamma_{p_n} : d_{p_n} \to \mathbf{R}, \forall p_n \in P_n\}$ is a set of distance functions; $P_{n_d} = \{p_{n_d}, \forall p_n \in P_n\}$ is a set of default values of the parameters; P_{n_i} is a set of institutional properties; and P_{n_e} is a set of environmental properties.*

A norm is defined as a subset of the institution parameters, associated to a subset of the definition domains. For each parameter, a distance function is specified, which provides a metric allowing quantifying the final parameter value regarding its original definition domain. A set of default values for the parameters is defined. Finally, the norm handles a set of institutional and environmental properties, which can specialize the institution's ones. Conflicting norms are allowed, several norms can be defined for the same environment, and norms can have non-empty intersections.

Definition 3. *A Behavior is a tuple $\langle N, P_b, V_{P_b} \rangle$ where: N is a reference to the instantiated norm; P_b is a subset of the set of parameters defined in the instantiated norm; and V_{P_b} is the set of values associated to the parameters.*

A behavior is the instantiation of a norm. Each element of the behavior is described by a parameter taken from the corresponding norm, and a value associated to this parameter. This value can be taken in or outside the definition domain, but has to remain within the institution's one. A behavior having at least one of its parameters values outside the definition domain specified in the norm is in violation.

2.2 Generation Engine

The instantiation from norm to behavior uses a generation engine build on nondeterminism principles, like applied in some displacement models [13]. It does not include domain specific elements, to preserve genericity and flexibility. Its main characteristic is to manage the determinism of the creation process: this way, users are able to guarantee the simulations reproducibility when needed, while allowing the creation of unexpected behaviors in other cases.

Each agent is associated to a finite set O of available objects. All objects in O are balanced with a factor p_o ($\forall o \in O$, $p_o \in [0, 1]$ and $\sum_{o \in O} p_o = 1$). A deterministic process d is associated to the agent to select the next object. Let p be a random parameter, $p \in \,]0, 1]$ (uniform distribution). At each time step t, $O_t \subset O$ is the set of objects which can be selected. Using the probability $1/p$, the agent uses randomly one of the objects in O_t, else it uses the deterministic process d to choose it. p itself is randomly chosen with a probability q. If $q = 0$, the probability to choose randomly an object is null: the resulting behavior of the agent is deterministic. If $q = 1$, the object is randomly chosen at each step, and the behavior is purely non-deterministic. When $0 < q < 1$, the randomization level of the process changes depending on the user choice: the number of random drawings increases with q.

The engine runs in three steps (Algorithm 1). First, the parameter q is chosen by the user or loaded from the configuration. Then, p is computed at each time step, and used to select the next object o. Finally, the agent applies o according to its own factor p_o. This allows managing the determinism's level: either follow the provided deterministic process, or easily introduce nondeterminism. The engine is used to instantiate the

---

Algorithm 1. Generation engine

Require: a an agent associated to a set of objects $O = \{(o_i, p_{o_i}), i \in I \subset N\}$; $q \in [0, 1]$ global randomization parameter; at time t: $O_t \subset O$ available actions
1:   $p \leftarrow random(\,]0, 1])$
2:   **if** $p < q$ **then**
3:      **for all** $o_i \in O_t$ **do**
4:         **if** $\sum_0^i p_{o_k} < p \leq \sum_0^{i+1} p_{o_k}$ **then**
5:            $o_{selected} \leftarrow o_i$ {select o_i}
6:            break
7:         **end if**
8:      **end for**
9:   **else**
10:      $o_{selected} \leftarrow d(O_t)$ {select the object using the deterministic process d}
11:   **end if**
12:   apply($o_{selected}$, $p_{o_{selected}}$)

behaviors. To do so, it is applied at the norm level: $O = P_n$. $O_t = P_b \subset P_n$ is the set of available parameters for the processed Behavior. Finally, the parameter's value v_{p_b} is selected by the model from the corresponding norm's definition domain d_{p_n}.

2.3 Control Mechanism

In order to manage the evolution of agents' behaviors, a control mechanism based on unsupervised learning is used. The data we classify are behaviors, which can be put under a vectorial form (vectors of parameters values). Among the various classical algorithms [5], Kohonen neural networks offer the characteristics we are looking for [6]. The classification is first used to study the agents' behavior evolution during the simulation: they can be classified to check if their behavior remains in the original norm, or matches another one. The second use is to control the evolutions of the norm set. When agents evolve during the simulation, new behaviors emerge. The study of these behaviors can lead to observe new norms (i.e. sets of similar behaviors), which are characterized using the classification mechanism. This can help users to improve their design and analyze the simulations results. Finally, the network can be trained on real data sets: the norms produced can then be used to recreate behaviors similar to the input data.

To handle these applications, a reference network is trained during the simulation using the instantiated behaviors, before the simulation could modify their characteristics. Only behaviors respecting the norms are used for this training. The study of the behaviors evolution is done by classifying them with the reference network. The evolution of the norm set is observed by training another network using the current behaviors, and comparing the two networks.

3 Variety and Consistency

The normative data structure provides two different ways to produce various agents' behaviors. The first method takes advantage of the norms description capabilities: any norms can be defined, each one using a different set of parameters, associated to different definition domains. A large potential of behaviors is thus made available, and very specific behaviors can be created by restricting the domains to single values. Unusual behaviors may be built this way, while keeping full controls over the simulation: the behavior is still the instantiation of a norm, and will never be produced again if this norm is deleted. Note that within the limits of each norm, the generation engine introduces the desired behavioral variety. The second method is based on norms violations. The parameters and associated values are determined during the instantiation, using the generation engine (Algorithm 2): if $q \neq 0$, violating behaviors can appear. This possibility to create violations allows the appearance of unspecified behaviors, which increases the variety.

As for the behaviors consistency, the limits set with the norms guarantees it if violations are forbidden. To allow reproducible simulations, we only have to store the seeds used by the generation engine. When violations are allowed, the consistency criteria can no more be guarantied. However, the model definition allows quantifying the violations,

Algorithm 2. Instantiating violating behaviors

Require: a an agent applying norm N; $q \in [0, 1]$ the global randomization parameter
1: **for all** $p_n \in P_N$ **do**
2:     $p \leftarrow random([0,1])$
3:     **if** $p < q$ **then** {violation allowed}
4:         $r \leftarrow random([0,1])$
5:         **if** $r < p$ **then**
6:             $v_{p_n} \leftarrow random(D_P)$ {take the value in institution's domain}
7:         **else**
8:             $v_{p_n} \leftarrow random(D_N)$ {take the value in norm's domain}
9:         **end if**
10:    **else**
11:        $v_{p_n} \leftarrow random(D_N)$ {no violation allowed: use the norm's domain}
12:    **end if**
13: **end for**

using the distance functions specified in the norms. These functions provide quantified values of the deviations, and allows excluding too deviant behaviors.

Finally, once the content of the normative model has been defined, the execution can be done without any modification to the engine. The model is generic, and can be applied in various fields. For instance, if we consider soccer player agents with a single action "shoot" where they have to choose the direction[1], we have an institution, using one parameter *direction* of definition domain $[0, 2\pi]$. Norms can describe players always shooting right or left $D_{P_n} = \{-\pi/4, \pi/4\}$, shooting in front of them ($D_{P_n} = [0, \pi]$)... In artificial economics, market agents are characterized by a direction, a price and a quantity, and apply different norms on the market: zero-intelligent traders, chartists, fundamentalists and speculators can be observed. They can be created with the model, using the same method as presented in section 4.

4 Application to Traffic Simulation

In order to illustrate our approach, the model was applied in the context of traffic simulation with the driving simulation software used at Renault, SCANeR© II[2]. SCANeR© IIis dedicated to run a wide range of driving simulators, for various applications: ergonomics of the driver's cab, design, validation of car lightings... In these applications, real humans drive a simulator, immersed in a traffic of autonomous vehicles. In most experiments, the behaviors of the vehicles have to be as realistic as possible, to allow the immersion of the users in the simulation and ensure the validity of the results. However, specific behaviors are sometimes needed, to simulate for instance drunk drivers and study the influence on the reactions of drivers without endangering them.

Various traffic management strategies exist, which use different decision models to simulate drivers' behaviors[11]. They often take into account individual characteristics,

---
[1] http://www2.lifl.fr/SMAC/projects/cocoa/football.html
[2] http://www.scaner2.com

Fig. 1. The SCANeR© II 2D and 3D visual outputs during the experiment presented in Section 5

including several psychological factors [3]: personality, emotion, motivation and social behavior. In SCANeR© II, the autonomous vehicles use a perception – decision – action architecture as reasoning basis [8]. First, the perception phase identifies the various elements which may interact with the vehicle: roads, lanes, other vehicles, road signs and pedestrians. Then the decision phase is built on three levels: strategic, tactical, and operational. The strategic level plans the itinerary, the tactical level selects the next maneuver to be executed using a finite state automaton, and the operational level computes the acceleration and wheel angles resulting from the chosen maneuver. Finally, the action phase computes the next position, using a dynamic model of the vehicle.

Different pseudo-psychological parameters are used during the decision phase. The "maximal speed" parameter is the maximal acceptable speed for the driver. "safety time" describes the security distance it will adopt, depending on its speed. "overtaking risk" represents the risk a driver will accept to overtake, function of the available gaps with oncoming vehicles. The "speed limit risk" allows it bypassing speed limits, and "observe priority" and "observe signs" are boolean rules regarding the respect of signalization and priorities. These parameters influence the resulting behaviors, and are adapted inputs to the traffic model, so we chose in this work to apply the proposed differentiation model directly on them. The description of the whole set of available parameters constitute the institution: $P = \{$ maximal speed, safety time, overtaking risk, speed limit risk, observe signs, observe priority $\}$; $D_P = \{[0,300],[0,100],[-1,2],[0,10],\{true, false\},\{true, false\}\}$; P_i and P_e are empty sets.

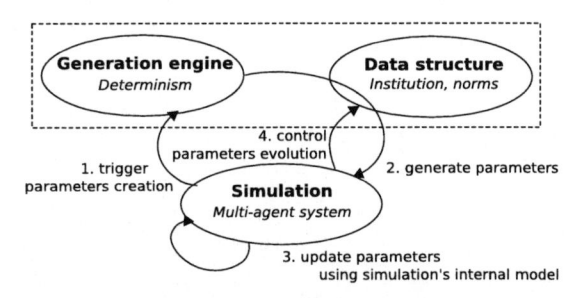

Fig. 2. Model implementation. The simulation keeps using its internal decision model, and only requires the differentiation model for parameters creation.

Finally, the model was implemented as an external tool providing input parameters to the traffic simulation model, as presented in Figure 2. It was thus easily introduced in the pre-existing application, as it remains non-intrusive.

5 Experimental Results

Based on this institution, three norms were introduced, describing normal, cautious and aggressive drivers (Table 1). They reproduce the behaviors humans are able to distinguish among a simulated traffic [15]. The simulation was done on a database representing a highway, on a 11 km long section (Figure 1). The vehicles were generated at the beginning of the section, using a traffic demand of 3000 veh/h, during 2h30. The generation function was a uniform distribution. Three detector were placed on the highway, to record vehicles data at kilometer 2.2, 6 and 10.8.

The vehicles were created using three distinct sets of norms. In the first one, *no norms*, all the vehicles are created with the same parameters (the behavioral differentiation model is deactivated). In the second one, *normal only*, only the norm "normal driver" is used. In the third case, *all norms*, the three norms are used, and the norm instantiated to create a new vehicle is chosen with a probabilistic law (cautious 10 %, normal 80 %, aggressive 10 %). After the initial creation by the differentiation model, the traffic model of the application handles all the vehicles.

The Figure 3 represents the distributions of vehicles speeds at kilometer 6. When no norm is used, the recorded speeds are either low (70 to 90 km/h, 46 % of the vehicles), or high (130 km/h, 40 % of the vehicles). The left lane remains slow, the right

Table 1. Norms parameters. The definition domain are truncated normal distribution, presented as following: [minimal value, maximal value], (mean μ, variance σ.

parameter	cautious driver	normal driver	aggressive driver
maximal speed	$[90, 125]$, $(115, 10)$	$[100, 140]$, $(125, 10)$	$[140, 160]$, $(150, 5)$
safety time	$[1.5, 5.0]$, $(2.0, 0.5)$	$[0.8, 5.0]$, $(1.5, 0.5)$	$[0.1, 1.2]$, $(0.8, 0.4)$
overtaking risk	$[-0.5, 0.5]$, $(0.0, 0.25)$	$[-0.5, 0.5]$, $(0.0, 0.25)$	$[1.0, 2.0]$, $(1.5, 0.5)$
speed limit risk	$[0.0, 1.1]$, $(1.0, 0.05)$	$[0.0, 1.1]$, $(1.0, 0.05)$	$[1.0, 10.0]$, $(1.5, 0.25)$

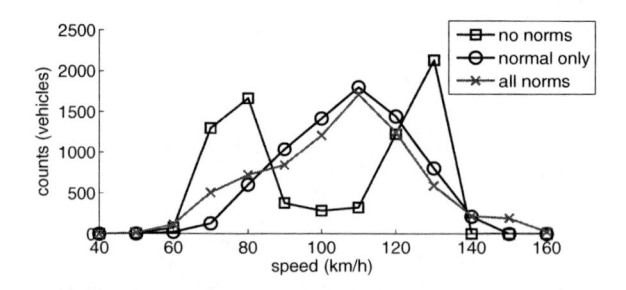

Fig. 3. Distribution of vehicles speeds at kilometer 6 (total vehicles crossing the detector during the simulation: 7400)

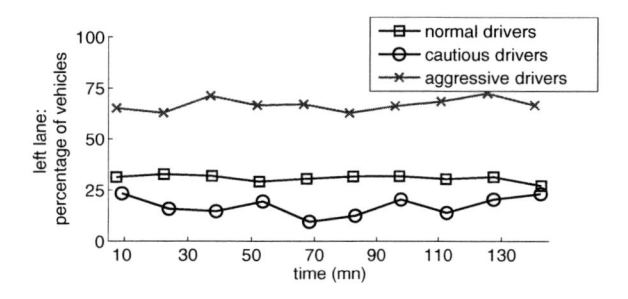

Fig. 4. Vehicles repartition on the highway lanes, function of their norm

one fast, and we observe few lane changes or overtaking. With one norm, 60 % of the speeds are between 90 and 115 km/h, 30 % between 115 and 140 km/h: the resulting distribution is more balanced. In the last case, the distribution presents a similar shape, widened because of the increased variety of maximal speeds. To evaluate the behaviors consistency, we studied the percentage of each type of drivers on the left and right lane (Figure 4). The set of norms used is the *all norms* case. Most of the aggressive drivers are on the left lane (71 %), when cautious ones stays on the right lane (82 %). In addition, the flow on the left lane represents only 34 % of the total flow. These results are consistent with real world situations. With the addition of norms, an increased variety of behaviors is observed in the simulation, and their consistency is guaranteed by the choice of the definition domains.

Different elements can be discussed. Firstly, the values used to define the norms have been chosen empirically: an important improvement would be using calibration with real data, which is currently under work. Secondly, the use of statistical data conceals some of the traffic characteristics. The visual observation of the simulation shows an important increase of behaviors variety (overtaking, speed choices...): we need to introduce indicators quantifying these elements. Finally, violating behaviors were not exploited during these simulations. They will be introduced in further experiments, to simulate for instance loss of control or drunk drivers.

6 Related Works and Conclusion

The generation of various behaviors in multi-agent systems has been approached from different perspectives. In [9], the authors increase the variety by automatically modifying characteristics of virtual humans in crowds, like clothes or accessories. In [15], virtual personalities are used to improve the behavioral variety in traffic simulation, which contribute to the subjective realism felt by the users. However, these approaches do not handle the issue of behaviors consistency, and the mechanisms remain domains specific. In [12], parameters settings for simulations models are automatically generated using a method based on bayesian networks, without user supervision. However, the consistency issue faced with when dealing with agents behaviors is not taken into account.

Some works have explored the use of normative approaches for traffic simulation. Bou et al. [2] study how traffic control strategies can be improved by extending Electronic Institutions with autonomic capabilities. The system dynamically adapts to maximize the respect of traffic law. In [4], non-normative behaviors are introduced to improve autonomous vehicles behaviors in intersections, by allowing agents to break some of the rules of the road. However, these works focus on the regulation possibilities of norms, when we take advantage of their description capabilities.

In this paper, we have presented a generic model designed to manage the variety and the consistency of agents' behaviors in multi-agents based simulations. These two elements are crucial for the simulations realism, and often not specifically taken into account. The behaviors are described using a normative approach: norms offer an intuitive way to define them, and intrinsically handle the notion of violations. Their creation is computed using a generation engine managing the determinism of the process. Control capabilities have been added, based on Kohonen neural networks. Their purpose is to check the behaviors deviations against their initial norms, and offer a tool to study the system evolutions. The behavioral variety is achieved by two different ways: with the norms definitions, by creating multiple and/or large definition domains, and by taking advantage of the violation possibilities. In the first case, the consistency is guaranteed within the limits of the norms. In the second case, it has to be checked by the control mechanism, using the possibility to quantify the potential deviations. Finally, the model was applied in the traffic simulation field. It was used to improve the realism of autonomous vehicles behaviors, and experimental results demonstrated the validity of the approach. Introducing various normative behaviors in the traffic increased the traffic dynamicity, while providing consistent drivers behaviors.

References

1. Boella, G., van der Torre, L.: An architecture of a normative system: count-as conditionals, obligations and permissions. In: Int. Conf. AAMAS, New-York, USA, pp. 229–231 (2006)
2. Bou, E., López-Sánchez, M., Rodríguez-Aguilar, J.A.: Adaptation of autonomic electronic institutions through norms and institutional agents. In: O'Hare, G.M.P., Ricci, A., O'Grady, M.J., Dikenelli, O. (eds.) ESAW 2006. LNCS, vol. 4457, pp. 300–319. Springer, Heidelberg (2007)
3. Dewar, R.: Individual Differences. In: Human Factors in Traffic Safety, pp. 111–142 (2002)
4. Doniec, A., Espié, S., Mandiau, R., Piechowiak, S.: Non-normative behaviour in multi-agent system: Some experiments in traffic simulation. In: IEEE/WIC/ACM Int. Conf. on Intelligent Agent Technology, Hong Kong, China, pp. 30–36 (2006)
5. Gallant, S.I.: Neural Network Learning and Expert Systems. MIT Press, Cambridge (1993)
6. Kohonen, T.: Self-Organizing Maps. Springer, Heidelberg (1995)
7. Lacroix, B., Mathieu, P., Kemeny, A.: A normative model for behavioral differentiation. In: IEEE/WIC/ACM Int. Conf. on Intelligent Agent Technology, Sydney, Australia, pp. 96–99 (2008)
8. Lacroix, B., Mathieu, P., Rouelle, V., Chaplier, J., Gallée, G., Kemeny, A.: Towards traffic generation with individual driver behavior model based vehicles. In: Driving Simulation Conference North America, Iowa City, USA, pp. 144–154 (2007)
9. Maim, J., Yersin, B., Thalmann, D.: Unique instances for crowds. In: Computer Graphics and Applications (2008)

10. Noriega, P.: Agent mediated auctions: The Fishmarket Metaphor. PhD thesis, Univ. de Barcelona (1997)
11. Olstam, J.: Simulation of rural road traffic for driving simulators. In: Transportation Research Board, Whashigton, USA (2005)
12. Pavón, R., Díaz, F., Luzón, V.: A model for parameter setting based on bayesian networks. Engineering Applications of Artificial Intelligence 21, 14–25 (2008)
13. Reynolds, C.W.: Steering behaviors for autonomous characters. In: Game Developers Conference, San Francisco, California, pp. 763–782 (1999)
14. Vázuez-Salceda, J., Dignum, V., Dignum, F.: Organizing multiagent systems. J. of Autonomous Agents and Multi-Agents Systems 11, 307–360 (2005)
15. Wright, S., Ward, N.J., Cohn, A.G.: Enhanced presence in driving simulators using autonomous traffic with virtual personalities. Presence 11, 578–590 (2002)

Foreseeing Cooperation Behaviors in Collaborative Grid Environments

Mauricio Paletta[1] and Pilar Herrero[2]

[1] Departamento de Ciencia y Tecnología
 Universidad Nacional Experimental de Guayana (UNEG)
 Av. Atlántico, Ciudad Guayana, 8050, Bolívar, Venezuela
 mpaletta@uneg.edu.ve
[2] Facultad de Informática
 Universidad Politécnica de Madrid (UPM)
 Campus de Montegancedo S/N. 28.660, Spain
 pherrero@fi.upm.es

Abstract. Balancing the load of the computational nodes is one important problem in the management of distributed computing, specifically collaborative grid environments. Some models were developed to deal with this problem. Most of them have in common a lack of ability to learn from previous experience in order to make predictions about future collaborations in the environment. Therefore improving the way in which equilibrium/balance is done, and to optimize efforts, resources, and time spent. This paper presents CoB-ForeSeer (Cooperation Behavior Foreseer), a new learning strategy proposal to solve the particular problem presented above. This strategy is based on neural network technology, specifically on Radial Based Function Network (RBFN). The paper also presents the way in which this learning strategy is properly configured and its corresponding evaluation. Results show that CoB-ForeSeer can successfully learn (because it reaches an acceptable average error) previous cooperation in the grid, and use this knowledge to improve new scenarios where collaboration is needed.

Keywords: Learning, prediction, collaboration, RBFN, Grid Computing.

1 Introduction

AMBLE (Awareness Model for Balancing the Load in Collaborative Grid Environments) [3] is an example of a model developed to equilibrate the computational nodes load in collaborative Grid Computing (GC) [1], [2]. These models do not take advantage of the experience on the collaborations taken in the past, in order to make new and better collaborations in the future. To undertake this situation, the collaborations that were carried out in the environment could be stored in a historic repository. This data could be potentially used to predict the best way to equilibrate the load of the computational nodes for the models, therefore a learning strategy based on this data is necessary to deal with this matter.

 This paper presents a new learning strategy to deal with the problem previously mentioned. This strategy named CoB-ForeSeer (Cooperation Behavior Foreseer) is based on Artificial Neural Networks (ANNs) [4], [5], more specifically on the Radial

Based Function Network (RBFN) model [11]. Results presented in this work show that CoB-ForeSeer learns from historical data of collaboration/ cooperation to improve cover load balancing delivery.

The rest of this paper is organized as follows. Section 2 presents a brief explanation of the RBFN model. The CoB-ForeSeer learning strategy is presented in section 3, including how data, relate to the collaborations in grid environments, is represented. Section 4 presents results of experimental tests done. Some existing related work in the area is presented in section 5. Finally, section 6 exposes the paper conclusions as well as the future work related with this research.

2 Radial Based Function Network (RBFN) Background

ANNs are arrays of processing units (artificial neurons) that are interconnected with one another, forming the network. Any edge has a weight that makes the role of synapses in the neurophysiologic system. Through an appropriate adjustment of these synaptic weights, the ANNs are able to "learn" the association between a characteristic set of input (incentives) and output (response) vectors regarding a specific problem.

Moreover, the operation of an ANN is given by three basic properties: 1) a nonlinear transfer function associated with the processing unit; 2) topology interconnection units and 3) the learning or weights adjustment law (algorithm). The specific description of these elements makes an ANN model different from another, and RBFN is one example.

RBFs (Radial Based Functions) were originally developed by Hardy to discuss problems involving the adaptation of irregular topographic contours through a series of geographic data [9], [10]. ANN's based on this technique (RBFN) are among the best choices in models spread forward as an alternative to achieve excellent results in alignment of data caused either by stochastic or deterministic functions [11].

This is a model consisting of three layers of processing units: the input layer, the output layer and a layer of hidden units. Information flows in one direction, from the input layer (elements sensors) to a layer of units that shows the system output. On the way, the information is processed in part by the intermediate (hidden) units.

The learning process is supervised and provides a method to adjust the synaptic weights. Therefore the network learns $(\vec{y}, \vec{s})$ correspondence, being $\vec{y}$ the input vector and $\vec{s}$ the associated output vector. The basis of this algorithm is the method of gradient descent that is used to minimize a function of quality. The most commonly used is mean-squared error, whose expression can be seen in (1), where: O - output units; s - desired outputs; i - units index for output layer; μ - learning patterns index.

$$E(\vec{w}) = \frac{1}{2} \sum_{\mu i} (s_i^\mu - O_i^\mu)^2 \qquad (1)$$

Expressions (2) and (3) are used to calculate the spread of hidden units and output units respectively. Where: V - hidden layer unit; y - input unit; j - hidden units index; k - input units index; f - activation function; σ - standard deviation.

$$V_j = f(-\sum_k \frac{(y_k^\mu - w_{jk})^2}{2\sigma^2}) \tag{2}$$

$$O_i = \sum_j W_{ij} V_j \tag{3}$$

$$f(x) = \exp(x) \tag{4}$$

Taking into account expressions (2) to (4) as well as the adjustment weights expressions below, this model has its foundation in the Gaussian, hence the use of the standard deviation. Expressions for adjusting weights between the output layer and the hidden layer are shown in (5). Expressions (6) and (7) are used to adjust the weights and standard deviations between the hidden layer and the input layer.

$$\begin{aligned}\Delta W_{ij} &= \eta \sum_\mu (s_i^\mu - O_i^\mu)(\sum_j W_{ij} V_j) V_j^\mu \\ \Delta W_{ij} &= \eta \sum_\mu \delta_i^\mu V_j^\mu \; ; \; \delta_i^\mu = (\sum_j W_{ij} V_j)(s_i^\mu - O_i^\mu)\end{aligned} \tag{5}$$

$$\Delta w_{jk} = \lambda \sum_{\mu i}(s_i^\mu - O_i^\mu)(\sum_j W_{ij} V_j) W_{ij}(\frac{y_k - w_{jk}}{\sigma_{jk}^2}) f(-\sum_k \frac{(y_k^\mu - w_{jk})^2}{2\sigma_{jk}^2}) \tag{6}$$

$$\Delta\sigma_{jk} = \gamma \sum_{\mu i}(s_i^\mu - O_i^\mu)(\sum_j W_{ij} V_j) W_{ij}(\frac{(y_k - w_{jk})^2}{\sigma^3}) f(-\sum_k \frac{(y_k^\mu - w_{jk})^2}{2\sigma^2}) \tag{7}$$

Data consists of real values in the interval [0, 1]. Any data outside this range must be adequately normalized. η, λ and γ are the corresponding learning factors to each weight matrix. Next section presents the way in which RBFN is used in CoB-ForeSeer to achieve its main purpose: learn data of collaboration/cooperation to properly cover load balancing delivery in collaborative grid environments.

3 CoB-ForeSeer: Cooperation Behavior Foreseer in GC

3.1 Representing Collaborations in Grid Environments

In order to represent and store collaborations, we use part of the specifications adopted in [3]. It is an extension and reinterpretation of the Spatial Model of Interaction (SMI) [6], an awareness model designed for Computer Supported Cooperative Work (CSCW). This reinterpretation, open and flexible enough, merges all the OGSA [7] featured to create a collaborative and cooperative environment within which it is possible to manage different levels of awareness.

Given a distributed environment E containing a set of resources R_i ($1 \leq i \leq N$), and a task T which needs to be solved in this environment (being required by one of the R_i), the objective is to solve T in a collaborative way. T is a set of P tuples (p_j, rq_j) ($1 \leq j \leq P$). p_j are the processes needed to solve the task in the system, and rq_j are requirements (such as power, disk space, data and/or applications) needed to solve each of these p_j processes. By extending and reinterpreting the SMI's key concepts in the context of grid environments we have:

Foreseeing Cooperation Behaviors in Collaborative Grid Environments   123

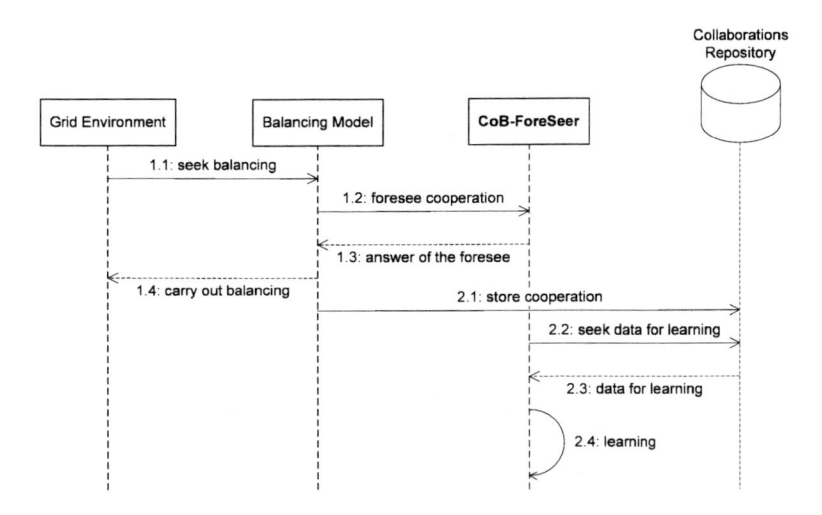

Fig. 1. CoB-ForeSeer consultation process (between 1.1 to 1.4) and learning process (between 2.1 to 2.4)

- *Focus(R_i)*: It can be interpreted as the subset of the space in which the user has focused his attention aiming of interacting with it. This selection will be based on different parameters and characteristics (such as power, disk space, data and/or applications). Given a resource R_i in the system, its focus *Focus(R_i)* would contain, at least, the subset of resources that are composing the Virtual Organization (VO) [8] in which this resource is involved. It could be modified and oriented towards any other VO, if needed.
- *Nimbus(R_i) = (NimbusState(R_i), NimbusSpace(R_i))*: It contains information about the current state of each resource R_i in the environment E.
- *NimbusState(R_i)*: The state of R_i in a given time. It could have three possible values: *Null, Medium* or *Maximum*. If the load of R_i is not high, and this node could receive some processes (p_j) or even the whole task T, then *NimbusState(R_i)* will get the *Maximum* value. If there are not receptor nodes in the system and R_i can accept, at least, a process, then *NimbusState(R_i)* would be *Medium*. Finally, if R_i is overload, its *NimbusState* would be *Null*.
- *NimbusSpace(R_i)*: The subset of the space where R_i is present. It will determine those machines that could be taking into account in the collaborative/cooperative process.
- *TaskResolution(R_i, T) = \{(p_1, s_1), ..., (p_P, s_P)\}*: Determines if there is a service in the resource R_i, being *NimbusState(R_i) ≠ Null*, such that could be useful to execute T (or at least a part of it). s_j ($1 \leq j \leq P$) is the "score" to carry out p_j in the resource R_i, being its value within the range $[0, \infty)$: The lower the value of s_j is (closer to 0), R_i fulfils all the minimum requirements to carry out the process p_j. Otherwise, the higher is the surplus over these requirements. This concept would also complement the *Nimbus* concept, because the *NimbusSpace* only determines those machines that could be taking into account in the assignment process, because they are not overloaded yet.

However, the *TaskResolution* determines which of these machines can contribute effectively to solve T or, at least, a part of this task.

As we said earlier, the basic idea of this proposal is, one, to keep a historic data with all those collaborations that were carried out in the environment. And second, this data will be used by CoB-ForeSeer to train its RBFN model to foresee next collaborations needed. Details of the two processes involved in the use of CoB-ForeSeer are presented in Fig. 1: 1) to make an inquiry aiming to predict the collaboration; and 2) to train the RBFN with the data stored.

3.2 Learning Collaborations in Grid Environments

Learning cooperation in this context means to learn the association between the situation grid environment in a given moment ($E + T$), and the response given to that specific situation (*TaskResolution*(R_i, T) for each resource R_i). Therefore, the RBFN topology has to be defined so as to receive $E + T$ in the input layer and to produce the corresponding *TaskResolution*(R_i, T) by the output layer units.

According to the model presented in Section 3.1, to obtain the current state of E it is necessary to have at least the following information, for each resource R_i: 1) A N-vector Ns representing the *NimbusSpace*(R_i) $\Rightarrow Ns_b$ ($1 \leq b \leq N$) = $1 \Leftrightarrow R_b \in$ *NimbusSpace*(R_i); 0 otherwise; 2) A value representing the *NimbusState*(R_i) $\in$ {*Null, Medium, Maximum*}.

Due to the fact that Ns (one for each resource) are N-bits vectors, and with the intention to reduce the number of input units in the RBFN, a transformation function *VtoR* (see (8)) has been defined. The idea is to use this function to obtain a different single real value, in the range of [0, 1], associated to any particular N-bits vector. This means that, for each vector of N bits, a single real value is required reducing, therefore, the complexity in the network topology.

$$VtoR([b_o, b_1, \ldots, b_{N-1}]) = \frac{\sum_{i=0}^{N-1} b_i \times 2^i}{2^N - 1} \in [0,1] \quad (8)$$

On the other hand, to represent task T, and the information related with the resource R_t in which T can be executed, the following information is required: 1) a N-bits vector F representing the *Focus*(R_t) $\Rightarrow F_a$ ($1 \leq a \leq N$) = $1 \Leftrightarrow R_a \in$ *Focus*(R_t); 0 otherwise; where R_t is the resource that receives the task T, and R_a is the resource in which R_t focus its attention to execute T; 2) a value representing R_t; 3) a P-vector Tp that represents $T \Rightarrow Tp_c$ ($1 \leq c \leq P$) is the requirement rq_c associated with the execution of the process p_c (being p_c a T's process). The F vector is also reduced to a single real value by using (8).

For the input, at least 2 input units (*NimbusSpace*(R_i) and *NimbusState*(R_i)) for each resource R_i are required to represent E, and $1 + 1 + P$ units are required to represent T. Related to the output units, it is necessary to represent the P-vector *TaskResolution*(R_i, T) for each resource. Therefore, the RBFN topology to deal with this problem has $2 \times (N + 1) + P$ units in the input layer and $N \times P$ units in the output layer.

The RBFN training patterns Γ have to be prepared according to the following aspects to get the more suitable learning process:

1) *NimbusSpace*(R_i) is expressed as a set of resources, so that the corresponding N-bits vector $Ns(R_i)$ is defined according to (9). Therefore, and as was explained above,

the corresponding component in Γ is the real value given by $VtoR(Ns(R_i))$. For example: $N = 8$ and $NimbusSpace(R_2) = \{R_1, R_2, R_3, R_5\} \Rightarrow Ns(R_2) = (1, 1, 1, 0, 1, 0, 0, 0)$ $\Rightarrow VtoR(Ns(R_2)) = 0.090196$.

$$Ns_b(R_i) = \begin{cases} 1, R_b \in NimbusSpace(R_i) \\ 0, otherwise \end{cases}; \quad 1 \leq b \leq N \qquad (9)$$

2) $NimbusState(R_i)$ is expressed as one of the following possible values {*Null*, *Medium*, *Maximum*}; in Γ is represented by the real value *Nst* expressed as it is indicated in (10).

$$Nst(R_i) = \begin{cases} 1, NimbusState(R_i) \text{ is } Maximum \\ 0.5, NimbusState(R_i) \text{ is } Medium \\ 0, NimbusState(R_i) \text{ is } Null \end{cases} \qquad (10)$$

3) $Focus(R_t)$ is expressed as a set of resources, so that the corresponding N-bits vector $F(R_t)$ is defined according to (11). Therefore, the corresponding component in Γ is the real value given by $VtoR(F(R_t))$. For example: $N = 8$ and $Focus(R_5) = \{R_2, R_3\}$ $\Rightarrow F(R_5) = (0, 1, 1, 0, 0, 0, 0, 0) \Rightarrow VtoR(F(R_5)) = 0.023529$.

$$F_a(R_t) = \begin{cases} 1, R_a \in Focus(R_t) \\ 0, otherwise \end{cases}; \quad 1 \leq a \leq N \qquad (11)$$

4) R_t is expressed as a value between 1 and N. As R_t is outside the range [0, 1] the corresponding value R_{tn} in Γ has to be normalized to fit this range. Expression (12) is used to obtain the valid [0, 1]-based R_{tn} value with the corresponding [1, N]-based R_t value. For example: $R_t = 5$ and $N = 8 \Rightarrow R_{tn} = 0.625$.

$$R_{tn} = \frac{R_t}{N} \qquad (12)$$

5) T is expressed as a set of P tuples (p_j, rq_j) $(1 \leq j \leq P)$. In Γ is represented as the P-vector T_P calculated as it is indicated in (13), where Q is the number of requirements, and rq_j is the requirement associated to the p_j process. For example: $P = 4$, $T = \{(p_1, 1), (p_2, 2), (p_3, 3), (p_4, 4)\}$, and $Q = 4 \Rightarrow T_P = (0.25, 0.5, 0.75, 1)$.

$$T_{Pc} = \frac{rq_c}{Q}; \quad (1 \leq c \leq P) \qquad (13)$$

6) $TaskResolution(R_i, T)$ is expressed as a set of P tuples (p_j, s_j) $(1 \leq j \leq P)$ with $s_j \in [0, \infty)$. This information is represented in Γ as the P-vector T_R where $T_{Rj} \in [0, 1]$. For example: $P = 4$ and $TaskResolution(R_9, T) = \{(p_2, 0.8), (p_3, 1), (p_4, 0.5)\} \Rightarrow T_R = (0.0, 0.8, 1.0, 0.5)$.

Fig. 2 shows a diagram related to RBFN learning. Left hand side corresponds to input data ($E + T$), and right hand side is the output information ($TaskResolution(R_i)$). Data for training has to be previously selected from those collaborations that were carried out in the environment, and stored in the historic repository. Thanks to one of the RBFN features, a network previously trained can be re-trained with more data. Therefore, the learning process can be continuously reinforced, improving the quality of response (by reducing the average error).

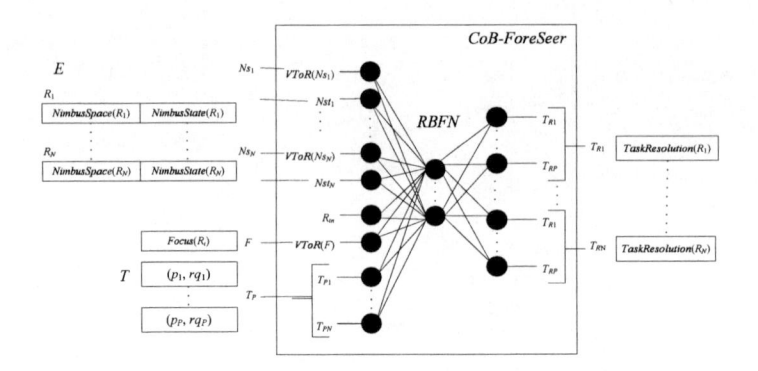

Fig. 2. RBFN topology and the input/output specifications

By using CoB-ForeSeer, the balancing model can predict an appropriate collaboration/cooperation scenario by estimating the "score" to indicate how convenient the processes to resources assignments are. It is made based on how E is in a given moment and a new task T needs to be executed. Next section presents details of the way in which this proposal was evaluated. Some results and statistical values are included.

4 Evaluation

RBFN is very susceptible to the initial conditions, overall in the initial values associated to the weights and standard deviations. Additionally, it is necessary to find appropriate values for the learning factors (η, λ and γ), and the number of units of the hidden layer. To deal with this situation, it was necessary to design an Evolutionary Algorithm (EA) [12] to find the right combination of all these parameters.

On the other hand, historical data used for testing CoB-ForeSeer was generated with the AMBLE model. Several experiments were made with 10, 25 and 50 patterns obtained from this data base. In all cases acceptable average errors in the range of 0.001 were achieved. For case of 10 patterns 2500 training cycles were carried out. On 25 patterns 10000 cycles were required. All experiments were conducted with data generated with a scenario of 8 resources (N), a maximum of 4 processes for task (P) and a maximum of 4 types of requirements (Q). Therefore, the RBFN topology for these tests (see Section 3.2) has 22 input units and 32 output units. The number of units of the hidden layer depended on results of EA.

For space reasons only the results related with the most complicated case (50 patterns) will be presented in this paper. The RBFN obtained has 17 units in the hidden layer and was trained with 10000 cycles achieving an average error of 0.0846. Fig. 3-a shows a chart in which the difference between real *TaskResolution* "scores" (clearer lines) and those given by the network after training (darker lines) can be appreciate. Since there are 32 values for each pattern, the numbers shown in the chart correspond to the average of these 32 values. Therefore, it appears the comparison of 50 results (one per pattern).

As it can be appreciated in these figures, the training results are over 90% efficiency (properly trained in this context), having some patterns with a short error (patterns: 10, 21, 46, 48, 49 and 50). These results can be improved by modifying conditions, both the EA as the RBFN.

(a) (b)

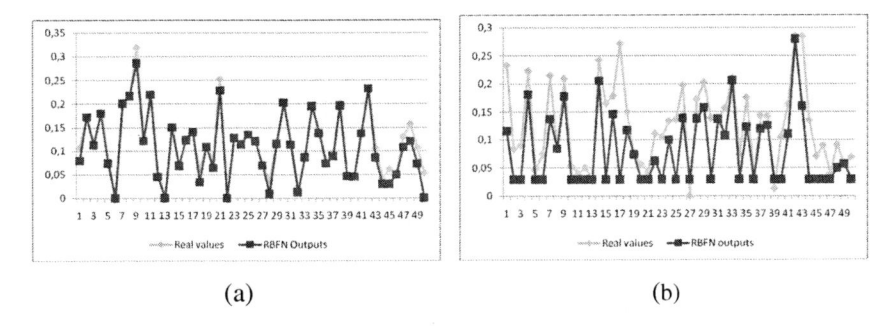

Fig. 3. a) Results of training CoB-ForeSeer with 50 patterns; b) results of testing CoB-ForeSeer previously trained

After the training process, CoB-ForeSeer was tested with 50 different patterns randomly generated to prove the capacity of the strategy as well as to give an appropriate answer. Fig. 3-b presents the raw results. It is important to highlight that even although the margin of error is high, the network tends to converge properly according to the data. This means that an adjustment in the training process is needed to reduce the error, and therefore to improve the quality of the answer.

5 Related Work

Cooperative learning on grid systems has been addressed in several approaches [13], [14], [15]. Authors in [13] examine a simple algorithm for distributed resource allocation in a simplified grid environment. They propose a system consisting of a large number of heterogeneous reinforcement learning agents that share common resources for their computational needs.

In [14] authors propose a model based on the reinforcement learning and the multi-level organization learning. The objective is to improve the efficiency of resources used in dynamic network environment. This research includes a series of formal definitions: Dynamic Network Grid (DNG), computing agent, cooperation computing team and relations among them.

Finally, authors in [15] propose a dynamic rule mechanism of agents for fitting the state changes of idle computing resources during the computing and migration processes. They are focusing on the idle computational resources of CSCW environment that is composed of computer clusters.

Even though these researches address the topic of cooperative learning on grid systems, which have in common the use of rule based systems as the agent learning technique, as far as we know, there are not any ANN based system for learning cooperation on a grid system. Moreover, there are approaches that combine GC and ANN techniques aiming to configure neural networks on grid-based environments [16].

On the other hand, RBFN has been used in several researches and applications. Recent examples could be consulted in [17], [18]. The use of RBFN is the only thing these researches have in common with this work. The final goal is completely different.

6 Conclusion and Future Work

This paper presents CoB-ForeSeer, a RBFN based learning strategy to provide collaborative grid environments with learning load balancing capabilities. The environment learns by using historical data collected from previous collaborations. CoB-ForeSeer can estimate a suitable "score" to indicate how convenient the resources to solve processes efficiently are. The higher is the score, the most suitable collaboration/cooperation is. Therefore, CoB-ForeSeer improves cover load balanced delivery resources on a grid environment.

We are working on the validation of the algorithm over real grid environment scenarios to improve the VO management. We are also working on improving the RBFN configuration to optimize the training time as well as reducing the error gotten. We are also evaluating others neural models, with different activation function than the Gaussian, with the purpose mentioned above.

References

1. Buyya, R., Abramson, D., Giddy, J., Stockinger, H.: Economic Models for Resource Management and Scheduling in Grid Computing. Journal of Concurrency and Computation: Practice and Experience 14(13-15), 1507–1542 (2002)
2. Foster, I., Kesselman, C., Tuecke, S.: The Anatomy of the Grid: Enabling Scalable Virtual Organizations. International Journal of Supercomputer applications and High Performance Computing 15(3), 200–222 (2001)
3. Herrero, P., Bosque, J.L., Salvadores, M., Pérez, M.S.: AMBLE: An Awareness Model for Balancing the Load in collaborative grid Environments. In: Proc. The 7th IEEE/ACM International Conference on Grid Computing, Barcelona, Spain, pp. 246–253 (2006)
4. Muller, B., Reinhardt, J., Strickland, M.T.: Neural Networks An Introduction, 2nd edn. Springer, Heidelberg (1995)
5. Ritter, H., Martinetz, T., Schulten, K.: Neural Networks. ISBN: 3–89319–131–3 (1991)
6. Benford, S.D., Fahlén, L.E.: A Spatial Model of Interaction in Large Virtual Environments. In: Proc. of the Third European Conference on Computer Supported Cooperative Work, Milano, Italy, pp. 109–124. Kluwer Academic Publishers, Dordrecht (1993)
7. Foster, I., Kesselman, C., Nick, J., Tuecke, S.: The Physiology of the Grid: An Open Grid Services Architecture for Distributed Systems Integration. Globus Project (2002)
8. Panteli, N., Dibben, M.R.: Revisiting the nature of virtual organizations: reflections on mobile communication systems. Futures 33(5), 379–391 (2001)
9. Lingireddy, S., Ormsbee, L.E.: Neural Networks in Optimal Calibration of Water Distribution Systems. Artificial Neural Networks for Civil Engineers: Advanced Features and Applications, 53–76 (1998)
10. Shahsavand, A., Ahmadpour, A.: Application of Optimal Rbf Neural Networks for Optimization and Characterization of Porous arterials. Computers and Chemical Engineering 29, 2134–2143 (2005)
11. Jin, R., Chen, W., Simpson, T.W.: Comparative Studies of Metamodelling Techniques under Multiple Modeling Criteria. Struct Multidiscip Optim. 23, 1–13 (2001)
12. Eiben, A.E., Smith, J.E.: Introduction to Evolutionary Computing, 9th edn. Springer, Heidelberg (2003)
13. Galstyan, A., Czajkowski, K., Lerman, K.: Resource Allocation in the Grid with Learning Agents. Journal of Grid Computing 3, 91–100 (2005)

14. Qingkui, C., Lichun, N.: Multiagent Learning Model in Grid. International Journal of Computer Science and Network Security (IJCSNS) 6(8B), 54–59 (2006)
15. Qingkui, C.: Cooperative Learning Model of Agents in Multi-cluster Grid. In: Proc. 11th International Conference on Computer Supported Cooperative Work in Design (CSCWD), pp. 418–423 (2007) 10.1109/CSCWD.2007.4281472
16. Schikuta, E., Weishaupl, T.: N2Grid: neural networks in the grid. In: Proc. IEEE International Joint Conference on Neural Networks, vol. 2, pp. 1409–1414 (2004)
17. Panda, S.S., Chakraborty, D., Pal, S.K.: Prediction of Drill Flank Wear Using Radial Basis Function Neural Network. Applied Soft Computing 8(2), 858–871 (2008)
18. Wang, D., Zhou, Y., He, X.: Radial basis function neural network-based model predictive control for freeway traffic systems. International Journal of Intelligent Systems Technologies and Applications 2(4), 370–388 (2007)

MAMSY: A Management Tool for Multi-Agent Systems[*]

Victor Sanchez-Anguix, Agustin Espinosa, Luis Hernandez, and Ana Garcia-Fornes

Universidad Politecnica de Valencia, Departamento de Sistemas Informaticos y Computacion,
Grupo Tecnologia Informatica Inteligencia Artificial, Valencia Camino de Vera 46022, Spain
{sanguix, aespinos, lhernand, agarcia}@dsic.upv.es

Abstract. Multi-agent systems may pose a real challenge to management since resources like hosts, agents or agent societies may have different owners and thus different interests. Management tools for multi-agent systems should take into account these inconveniences in order to apply a proper management. We detected in our study that most current management tools only show information about current state. We also think that information about the history of the system can supply a wealth of additional information to system administrators. In this paper we present MAMSY, a management tool for multi-agent systems aimed at showing and controlling current state and past related information.

Keywords: Multi-agent management, Resource management, Multi-agent platform tools.

1 Introduction

Over the last few years multi-agent systems have proved to be one of the most productive areas for Artificial Intelligence researchers. There are still several subjects that are under research, i.e security, agent programming languages and core enhancements for efficient and scalable systems. Although some advancements have been made in this area, most multi-agent systems still remain under the *closed multi-agent system* category, where the whole system is owned by a private user or organization. One of the main goals in multi-agent research is producing *open multi-agent systems*. According to [15], an *open multi-agent system* is a system where elements can be removed or can be added during the runtime. We are more interested in a specialization of such systems: *open multi-agent systems* where the whole system may not be owned by a private user or organization. From now on we will refer to this system category.

As people extend the system, they will increasingly become interested in monitoring and controlling their part of the system. This is where the term *management* comes in, having been commonly used in software engineering where management systems are the most frequent type of application. The term *management* acquires a special meaning when it is applied to multi-agent systems. In multi-agent system management, the items that are managed are resources. To our knowledge extent, there is no formal definition for the term *resource* in multi-agent systems thus we decided to provide one.

---
[*] This work was supported by CONSOLIDER-INGENIO 2010 under grant CSD2007-00022 and Spanish goverment and FEDER funds under TIN2008-04446 project.

Definition 1. *A resource is a physical or virtual object which has an owner, and therefore needs to be administered according to its owner's goals.*

Physical resources are mainly *node machines or nodes* where the multi-agent system is being is executed. Obviously, administrators are interested in seeing how their machines are being used, which agents are being executed, current system overload and so forth. Administrators may want to modify their state when it is not desirable. It is also normal practice to associate the term *resource* with something physical but in multi-agent systems it is also possible to find resources which are not physical. These are called virtual resources. From the administrator's point of view, agents are virtual resources that he himself has placed in order to achieve certain goals.

Since research into multi-agent system management is still in its infancy stages, there is neither an agreed definition for the term *multi-agent system management*. It is possible to find some works about management related issues at [12], but no formal definition has been given. In fact, to our knowledge extent, [12] is the only work related to multi-agent management. In accordance with our research and in order to help future works we define the term *multi-agent system management*.

Definition 2. *Multi-agent system management consists in the visualization and control of the physical and virtual resources located in a multi-agent system.*

As we stated before, resources in an *open multi-agent system* may have different owners with perhaps different objectives and interests. It is obvious that owners will be interested in what actions can be performed remotely over their resources by other entities or users. A user is an owner who places virtual or physical resources into the multi-agent system. Since our goal is *open multi-agent systems* it is possible to find in the system users with opposing goals and even untrusted users which only goal is to attack the system. It is not advised to deploy an open multi-agent system which has not treated properly these issues. Serious problems arise with the authentication of users in these highly dynamic and open systems, so associating authenticated owners to resources is still a problem. In fact, authentication is one of the main lines of research currently available in multi-agent systems. Once we have authenticated owners and resources, it is necessary to set permissions for these resources, always bearing in mind the highly variable set of users we may find in a multi-agent system.

But permissions and security are not the only problems management in multi-agent systems may face. For instance, the amount of information and data in *open multi-agent systems* may vary from a small set to very large sets of data. Management tools should show data in a friendly and clear way so users take full advantage of the developed tools and perform their tasks efficiently. User interface research has already provided solutions for most of these inconveniences and their application here is highly encouraged.

MAMSY is a management tool for multi-agent systems hosted by Magentix[23][5] platforms. In the following sections we will will analyze current management tools briefly in order to point their common features. After that we will describe Magentix platform briefly and then give a more detailed description of MAMSY architecture and its main features. Finally, we will give some conclusions and describe our future lines of work in management for multi-agent systems.

2 Current Management Tools

For the purpose of developing MAMSY, a previous study was carried out on current management possibilities in available platforms. The targets of the study were 3APL[10], AgentBuilder[4], AgentScape[8], Aglets[18], AGlobe[22], CometWay Agent Kernel (AK)[1], Cougaar[14], Decaf[17], Diet[16], Hive[19], Jade[7], Jadex[6], Madkit[13], Open Agent Architecture[9], Semoa[2], Tryllian Agent Development Kit [3] and Zeus[20]. The objective of the analysis was to detect which functionalities were implemented in current management tools and to take into account clear visualization methods. We designed an ideal taxonomy of services which should be included in every multi-agent management tool. This taxonomy and results of the analysis on current management tools can be observed in Table 1.

It was observed that some management tools [1][14][16][3] do not include modifying functionalities for platform current state. They usually only offer a way of configuring the multi-agent system before it is started or visualization functionalities. Management tools that are able to perform modifying functionalities offer a heterogeneous set of actions which deal with nodes and agents in the current system. Although some functionalities are common, they differ in the way they are implemented. The largest set of actions avaliable was found in [7] and related tools like [6].

The visualization methods are similar but they are not prepared to handle big volumes of data since most of them do not offer search and filter mechanisms, [14] being the only management tool that had such services. Although current state management offers a large set of actions, current management tools do not offer services for looking through platform past. In [11] it is described the problems a human faces when observing and monitoring a set of distributed concurrent processes. It is also pointed that global-time-stamped information is the only realiable picture for such systems. We are convinced that this past information could be used in debugging tasks in the way [11] proposed.

3 Magentix: A Multiagent Platform Integrated in Unix

Magentix is a multi-agent platform developed by GTI-IA, a research group located at Universidad Politecnica de Valencia . It has been developed entirely using the C programming language and makes considerable use of Unix's core calls. So the platform is currently only available on Unix-like systems. One of the biggest problems faced in multi-agent systems is platform core performance when the number of agents and nodes is large. Through the use of a non-interpreted language like C and Unix's core calls, Magentix aims for a multi-agent platform that solves scalability and performance issues.

3.1 Agent Messaging

Older versions of Magentix [23][5] used plain text as a way of exchanging information between platform agents. Recently, a new version of Magentix was developed using RDF as a way of structuring messages. RDF is a resource description language recommended by W3C which expresses information as a triplet. Each triplet has a subject, a predicate and an object. Subjects and predicates content can only be expressed as

Table 1. Current management tools analysis

	1	2	3	4	5	6	7	8	9	10	11	12	13	14	15
3APL						x		x	x	x			x		x
AgentBuilder						x			x						x
AgentScape	x					x		x	x						
Aglets						x		x	x	x	x	x			
AGlobe						x		x	x						x
AK															
Cougaar	x	x	x	x		x	x							x	
Decaf						x			x				x		x
Diet						x								x	
Hive	x			x		x		x	x						
Jade	x		x	x		x		x	x	x	x	x	x		x
Jadex	x		x	x		x		x	x	x	x	x	x		x
Madkit	x			x		x		x	x					x	x
OAA	x					x			x						
Semoa	x					x		x	x						
Tryllian															
Zeus						x								x	x
MAMSY	x		x	x		x	x	x	x					x	x

1:Visualize nodes: List instances of platform kernel in our system and related information – **2:Filter/Search nodes:** Narrow the list of nodes according to our interests – **3:Add node:** Add a new node to the current system – **4:Remove node:** Stop the activity of a system node and delete it from the system – **5:Stop/Restart node:** Stop the activity of a system node and its agents/Restart the activity of a stopped node, recovering the state it had before it was stopped. – **6:Visualize agents:** List agents in our system and related information – **7:Filter/Search agents:** Narrow the list of agents according to our interests – **8:Add agent:** Add a new agent to our system – **9:Remove agent:** Delete an agent from our system – **10:Stop/Restart agent:** Stop the activity of an agent/Restart the activity of a stopped agent, recovering the state it had before it was stopped – **11:Migrate agent:** Stop an agent and restart its execution in another node – **12:Clone agent:** Copy an agent and start its execution in a node – **13:Send message to agent** – **14:Organizational support:** Visualization of agent organizations and operations that deal with organizations – **15:Communication analyzer tools:** Tools that let us observe messages exchanged between agents.

a URI (Universal resource identificator) while objects may be a URI or a literal string. Knowledge graphs can be formed if we consider subjects and objects as nodes and predicates as named links. Using RDF graphs as a way of structuring information generates an extra overload of data in communications but in exchange it allows developers to use richer and more structured knowledge representation. Magentix messages are RDF graphs which are serialized and then sent to agents. A Magentix simplified agent message can be observed in Fig.1. Round graph nodes symbolize URIs whereas box graph nodex symbolize literal values. Those round graph node which content in enclosed by parentheses symbolize a special unique URI node which is called *blank*. The figure shows a *Hello world* message in Magentix which is sent to agent *mgx_history*.

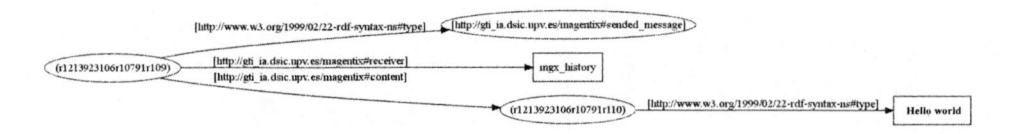

Fig. 1. Hello World! sample message

4 MAMSY: A Tool for Multi-Agent System Management

MAMSY stands for *Magentix Management System* and is a tool designed to help system administrators monitor and control their resources in a Magentix multi-agent system. Furthermore MAMSY aims to solve some of the management problems explained in the introductory section.

Functionalities offered to users are a subset of our ideal service taxonomy, but since MAMSY and Magentix are still in the development phase, more services are expected to be offered to system administrators. However, we include a new type of management called *past time management* which allows users to analyze and observe past events and messages.

Due to the fact that one of MAMSY's goals is to be user friendly, the design team decided to enhance the tool with a desktop graphical interface. Selected libraries were Trolltech's Qt C++ APIs.

4.1 Functionalities

MAMSY's services and functionalities may be classified into two types, according to the time period they act on or visualize. Firstly we have present time functionalities, services which allow users to visualize and modify the platform's current state. Lastly, we can also find past time functionalities that are focused on visualizing events and messages in a selectable time range.

4.1.1 Present Time Functionalities
Three different resources may be managed in current state: Nodes, agents and units. As we stated, nodes are physical hosts that form the multi-agent system, each one executing a magentix core instance. Agents are the main subject in MAMSY's management capabilities and units only allow visualization since unit management is relegated to agents. We will describe these functionalities briefly below. A short matching between our proposed taxonomy of present services and MAMSY's current services can be found in Table 1. We can observe MAMSY showing present time information in Fig.2.

Node Managing

-Add host: Adds a new node to the current Magentix system. It requires that Magentix is installed on the target host.
-Remove host: A host and its agents are deleted from the system.

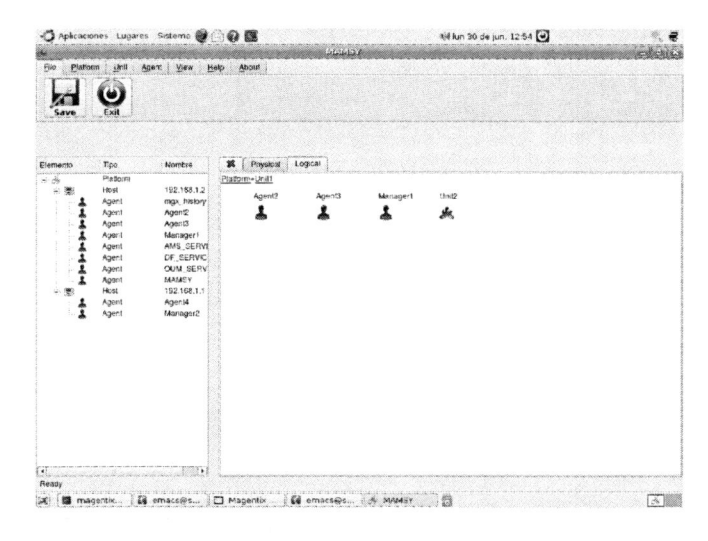

Fig. 2. MAMSY showing present information

-Stop Platform: The system permanently stops its activities.
-Node visualization: Nodes appear in a hierarchical view with their agents nested.

Agent Managing

-Add agent: An agent may be added to any node.
-Remove agent: A local or remote agent is deleted from the system.
-Search agent: A search engine is included so users can search agents by name.
-Agent visualization: Agents can appear along with the node they are executing in or with the organizations they are a member of.

Agent Organization Managing

-Search organization: A search engine is included so users can search for organizations by name.
-Organization visualization: Units only appear in 'Logical View', showing what members they have or even which organizations they are a member of.

4.1.2 Past Time Functionalities

One of the most important features implemented in MAMSY is past time related functionalities. No similar features were found when analyzing current management tools. Whereas current systems just deal with the very near past, more specifically a very limited time window, our management tool can store a wide time range for later visualization. The information we are interested in storing for later visualization is platform events, i.e agent creation/deletion, new host, host deletion, unit creation/deletion, unit member addition/deletion and messages exchanged between users. We will review in section 4.2 who is in charge of storing and distributing this information.

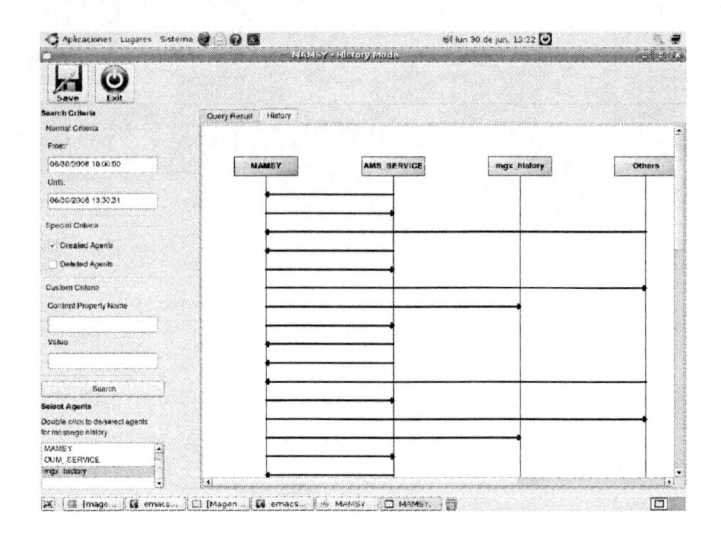

Fig. 3. MAMSY showing past information

We used a sniffer-like visualization method to show past messages and events to system the administrator, due to the fact that it is a commonly spread application in the world of multi-agents. Since the amount of data may be quite large, we decided to include several filters that allow users to narrow the information they are interested in. More specifically, we allow through an user interface to filter agents created/deleted in a time range, agents present in a time range and agents that sent/received messages with a concrete property and value specified by the user. This visualization method and filters can be observed at Fig.3.

Information gathered and showed to users was considered extremely interesting to system administrators because of its applications to error detecting tasks and debugging. Furthermore, stored information also contains messages exchanged between agents so further analysis may highlight interesting information regarding agent behaviour, agent likes and dislikes and so forth.

4.2 Architecture

One of the key factors in the design of MAMSY was its placement in the multi-agent system. The final design determined that MAMSY would consist of two main components. Each of these components was placed as a special agent in the multi-agent system.

The first agent is commonly referred to as a history agent and its main task consists of capturing events and messages produced in the system. The second agent is called the MAMSY agent and its task consists of capturing user requests and showing requested information to the user. However MAMSY is really the set of the two types of agents. When the history agent is present in the system, platform services send a special RDF notification to the history agent with the event that has been produced. Then the history agent looks for the MAMSY agent instances and sends them RDF notifications with the event produced, thus they can update their state and look. Arriving events and messages are properly stored into a database once they have been processed. Magentix RDF graphs are stored into a database using Redland API. The chosen database was open source PostgreSQL although it is possible to use other free database engines like MySQL and SQLite. When MAMSY agents desire information about the past or history agent needs to query for information, they send a SparQL query, a RDF query language specified by the W3C, which is executed on history agent database. MAMSY agent users do not input the SparQL query, but instead they fill a form which is translated by the MAMSY agent to SparQL once the user has decided to execute the query.

4.3 Security Features

It has been mentioned that one of the main problems in multi-agent management is security, more specifically user authentication and permission policies. MAMSY solves part of these inconveniences using SSH protocol as a secure channel for sending commands to be performed over resources. Using SSH forces users to have an appropiate username and password for the host machine they want to act on. Consequently, when a user performs an action on a Magentix resource it is almost certain that they are trusted users.

5 Conclusions and Future Work

In this paper we have described some problems faced when dealing with multi-agent system management. Even though security is a big problem, tool designers should not forget good user interface practices due to the large amount of data a multi-agent system may generate.

One of the conclusions extracted from the study of current management tools is that they provide a heterogeneous set of management actions. We also detected the lack of visualization about past time information which can be useful for a wide range of applications.

In this paper we presented MAMSY which is our first attempt at addressing management of Magentix multi-agent systems. We have described its current functionalities, its architecture and security features. It offers services for current state management and a new service for visualization of past messages and events. Our current tool has been used in order to help the development of a Tourism Application[21].

However MAMSY and Magentix are not finished works. Both of them are still in development phase and gradually including new features. Some of the most important functionalities in future MAMSY versions will include:

-Add more actions for current state management
-Explore new security models which enhance authentication and permissions issues:
    Unfortunately, the current model still lacks flexibility due to the fact that there are situations where SSH is not suitable. For instance, in an ideal system there may be hosts whose owners let other users execute their agents freely. This is not accomplished using only the SSH security model.

- Explore new visualization methods for past events: More visualization methods which give a clearer view about platform past.
- Intelligent and autonomous management: Avoid system overloads through autonomous and intelligent management.

References

1. Cometway agent kernel, http://www.agentkernel.com/
2. Semoa, http://semoa.sourceforge.net/
3. Tryllian agent development kit, http://www.tryllian.org/
4. Acronymics, Inc.: AgentBuilder User's Guide (2004)
5. Alberola, J.M., Mulet, L., Such, J.M., Garca-Fornes, A., Espinosa, A., Botti, V.: Operating system aware multiagent platform design. In: Proceedings of Fifth European Workshop On Multi-Agent Systems (EUMAS 2007), pp. 658–667. Association Tunisienne D'Intelligence Artificielle (2007)
6. Alexander Pokahr Lars Braubach, W.L.: Jadex: Implementing a bdi-infrastructure for jade agents. EXP - In Search of Innovation (Special Issue on JADE) 3(3), 76–85 (2003)
7. Bellifemine, F., Poggi, A., Rimassa, G.: JADE - A FIPA-compliant agent framework. In: Proceedings of the Practical Applications of Intelligent Agents (1999)
8. Brazier, F.M.T., van Steen, M., Wijngaards, N.J.E.: Distributed shared agent representations. In: Krose, B., de Rijke, M., Schreiber, A.T., van Someren, M. (eds.) Proceedings of the 13th Dutch-Belgian AI Conference, p. 77 (2001); Extended abstract of AEMAS 2001 publication
9. Cheyer, A., Martin, D.: The open agent architecture. Autonomous Agents and Multi-Agent Systems 4(1-2), 143–148 (2001)
10. Dastani, M., van Riemsdijk, M.B., Dignum, F.P.M., Meyer, J.-J.C.: A programming language for cognitive agents goal directed 3APL. In: Dastani, M., Dix, J., El Fallah-Seghrouchni, A. (eds.) PROMAS 2003. LNCS, vol. 3067, pp. 111–130. Springer, Heidelberg (2004)
11. Garcia-Molina, H., Germano Jr., F., Kohler, W.H.: Debugging a distributed computing system. IEEE Trans. Software Eng. 10(2), 210–219 (1984)
12. Giampapa, J.A., Juarez-Espinosa, O.H., Sycara, K.P.: Configuration management for multi-agent systems. In: AGENTS 2001: Proceedings of the fifth international conference on Autonomous agents, pp. 230–231. ACM, New York (2001)
13. Gutknecht, O., Ferber, J.: The madkit agent platform architecture. In: Agents Workshop on Infrastructure for Multi-Agent Systems, pp. 48–55 (2000)
14. Helsinger, A., Wright, T.: Cougaar: A robust configurable multi agent platform. In: Aerospace Conference, pp. 1–10. IEEE, Los Alamitos (2005)
15. Hewitt, C.: Open information systems semantics for distributed artificial intelligence. Artif. Intell. 47(1-3), 79–106 (1991)
16. Hoile, C., Wang, F., Bonsma, E., Marrow, P.: Core specification and experiments in diet: A decentralised ecosystem-inspired mobile agent system. In: Proc. 1st Int. Conf. Autonomous Agents and Multi-Agent Systems (AAMAS 2002), 2002, pp. 623–630. ACM Press, New York (2002)
17. Graham, J.R., Decker, K.S., Mersic, M.: Decaf - a flexible multi agent system architecture. Autonomous Agents and Multi-Agent Systems 7, 7–27 (2003)
18. Karjoth, G., Lange, D., Oshima, M.: A security model for aglets. IEEE Internet Computing 1(4), 68–77 (1997)
19. Minar, N., Gray, M., Roup, O., Krikorian, R., Maes, P.: Hive: Distributed agents for networking things. In: International Symposium on Mobile Agents, Systems and Applications, p. 118 (1999)

20. Nwana, H.S., Ndumu, D.T., Lee, L.C., Collis, J.C., Heath, M.: Zeus: A collaborative agents tool-kit
21. Palanca, J., Aranda, G., Garcia-Fornes, A.: Building service-based applications for the iphone using rdf: a tourism application. In: Proceedings of the International Conference on Practical Applications of Agents and Multiagent Systems (2009)
22. Šišlák, D., Rollo, M., Pěchouček, M.: A-globe: Agent platform with inaccessibility and mobility support. In: Klusch, M., Ossowski, S., Kashyap, V., Unland, R. (eds.) CIA 2004. LNCS (LNAI), vol. 3191, pp. 199–214. Springer, Heidelberg (2004)
23. Such, J.M., Alberola, J.M., Garca-Fornes, A., Espinosa, A., Botti, V.: Kerberos-based secure multiagent platform. In: Sixth International Workshop on Programming Multi-Agent Systems (ProMAS 2008), pp. 173–186 (2008)

A Multi-Agent System Approach for Algorithm Parameter Tuning

R. Pavón[1], D. Glez-Peña[1], R. Laza[1], F. Díaz[2], and M.V. Luzón[3]

[1] Computer Science, University of Vigo, Spain
{pavon,dgpena,rlaza}@uvigo.es
[2] Computer Science, University of Valladolid, Spain
fdiaz@infor.uva.es
[3] Computer Science, University of Granada, Spain
luzon@ugr.es

Abstract. The parameter setting of an algorithm that will result in optimal performance is a tedious task for users who spend a lot of time fine-tuning algorithms for their specific problem domains. This paper presents a *multi-agent tuning system* as a framework to set the parameters of a given algorithm which solves a specific problem. Besides, such a configuration is generated taking into account the current problem instance to be solved. We empirically evaluate our *multi-agent tuning system* using the configuration of a genetic algorithm applied to the root identification problem. The experimental results show the validity of the proposed model.

1 Introduction

Parametric algorithms are generally employed by users interested in solving an specific problem, and frequently such users are experienced in the problem domain to solve, but not in the selected algorithm. They typically configure their algorithms following an iterative process on the basis of some runs of different configurations that are felt as promising. However, tuning algorithms parameters for robust and high performance is a tedious, repetitive and time-consuming task. Moreover, the optimal parameter configuration to use may differ considerably across different problem instances, so one needs some experience to choose the most effective parameter values.

It is the reason why it will be interesting to have a system that, for a given problem instance, it automatically calculates the right values for the parameters associated with the algorithm employed for solving it.

In recent years there have been a number of different approaches to automatic identification of the best default parameter configuration in a single algorithm. To sum up, three approaches can be distinguished: the *evolutionary* approach [8, 18], the *model selection* approach [13, 1], and the *statistical* approach [11, 4]. However few approaches have been developed that take into account the characteristics of the problem instance at hand and past experience of similar instances.

In this paper, we describe a multi-agent system that assists the user when configuring the parameters of an algorithm, taking into account the current problem instance to be solved. The proposed approach is based on estimating the best parameter configuration that maximizes the algorithm performance when it solves an instance-specific domain problem. Such a multi-agent system incorporates the following features:

- The system does not require *a priori* knowledge about the mode of operation of the algorithm. In absence of this explicit knowledge, the multi-agent system only needs data about previous algorithm runs solving specific instances of the problem.
- The parameter tuning will be particular for each instance of the problem, in terms of its relevant features that can be supplied by the expert in the problem domain.
- The multi-agent system will be able to adapt its behaviour at the same time as it captures more knowledge about the problem.

The design of the multi-agent system is mainly based on a set of cooperative agents, where every agent is specialized in a problem instance and models its knowledge about the parametric algorithm using a Bayesian Network (BN). These agents adapt their knowledge and behaviour over time, due to one collaborative approach with other agents of the system, becoming gradually more effective at achieving their goals.

As an example, we empirically evaluate our proposal tuning a genetic algorithm (GA) that solves the *root identification problem* [3, 14]. The *multi-agent tuning system* automatically will supply the right values for the control parameters associated with the GA, in terms of the features of the geometric instance to be solved, resulting in a practical application for computer aided design.

The rest of the paper is structured as follows. Section 2 is devoted to describing the architecture of the proposed multi-agent system, focusing the attention on agents capabilities and communication. Section 3 describes the problem used throughout the paper to illustrate the application of the proposed system and a summary of some experiments carried out as well as the results attained. Finally, in Section 4 some conclusions and future work are introduced.

2 System Overview

As previously mentioned, the purpose of our *multi-agent tuning system* is to assist the user when he needs to configure an algorithm that solves a given problem instance. For this purpose, we have developed a system that consists of four kinds of agents integrated in an organization whose architecture can be observed in Figure 1.

In brief, the *multi-agent tuning system* has an *interface agent*, which interacts with the user which wants to configure the algorithm for solving the problem instance. *Interface agent* captures the request of configuration and sends the features of the specific problem instance to *broker agent*. *Broker agent* asks

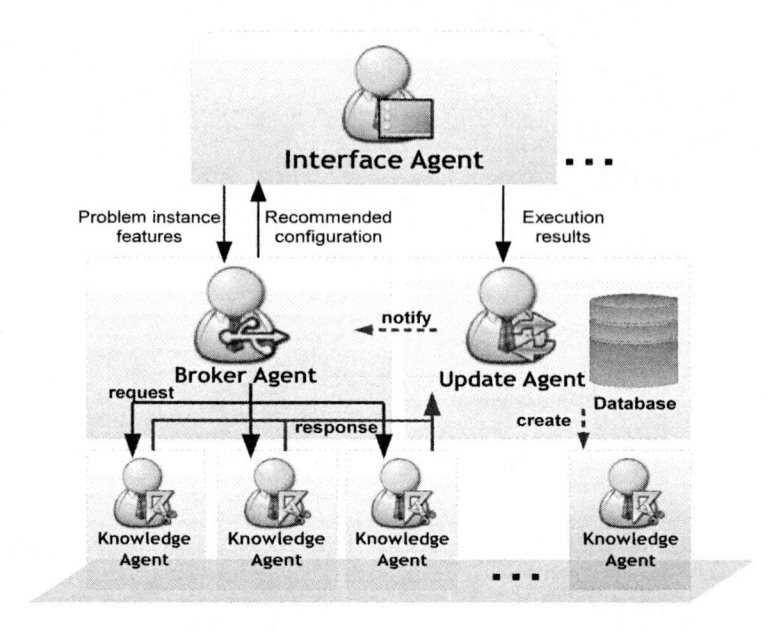

Fig. 1. Architecture at the *multi-agent tuning system*

knowledge agents for the most probable configuration, integrates the different results and gives them back to the *interface agent*. Next, the *interface agent* solves the problem instance with the recommended configuration and stores the obtained results in a database. Registered data are used by the *update agent* in order to gradually acquire more knowledge about the problem domain and the algorithm to configure remodelling the multi-agent system to assist the user more efficiently.

In the following subsections, we briefly describe existing agents focusing attention on their capabilities.

2.1 Update Agent

In order to help users to configure their algorithms, our system needs some knowledge about the task it has to perform. Initially, the multi-agent system needs a minimum of background knowledge about the algorithm to configure $A(\mathbf{X})$ and the problem instance to solve P_i, where $\mathbf{X}$ denotes the set of algorithm parameters that the user must fix. Initial knowledge is obtained running the algorithm (with several parameter configurations) for solving different problem instances. The data gathered are stored in a local database and they are related with (i) the features of the problem instance solved, (ii) the parameter configuration used and (iii) the efficiency measurements obtained.

From this data, the *update agent* will create a set of *knowledge agents*, every one specialized in configuring the algorithm for a specific problem instance. Moreover, the *update agent* will notify to *broker agent* the *knowledge agents* incorporated into the *multi-agent tuning system*.

2.2 Knowledge Agent

There is a *knowledge agent* for each problem instance. Each *knowledge agent* observes the data runs recorded into the initial database for a specific problem instance P_i (characterized by a set of attributes) and induces a model in the form of a bayesian network. The domain variables of each bayesian network are the algorithm parameters **X** that the user must set up as well as the parameters that measures the performance of the algorithm. So, each *knowledge agent* induces a bayesian network which models dependence/independence relationships between algorithm parameters as well as efficiency measures of the algorithm, for each problem instance.

BNs have been proposed as the model which encodes data stored into the database because they take into account the following features: they store explicit knowledge (both qualitative and quantitative) about domain variables; they provide a mechanism that permits the model to carry out reasoning under uncertain conditions; lastly, they offer a sound and practical methodology for automatically discovering probabilistic knowledge in databases, updating it when new data are presented [9, 12]. Each *knowledge agent* uses bayesian reasoning strategies to encode and update knowledge.

Once the system is working and new data runs are recorded by the *update agent*, it will notify the corresponding *knowledge agent* that refines the structure and local probability distributions of its bayesian network to better reflect the experience gathered. Sequential update in *knowledge agents* represents a crucial capability for making our system adaptable and for overcoming errors. Once the update process is carried out, the data used are removed from the database.

2.3 Broker Agent

In our proposal, we have defined a *broker agent* to receive the request from *interface agent* and to ask *knowledge agents* for an efficient parameter configuration. *Broker agent* is able to store information about the problem instance that each *knowledge agent* can broach. In this way, when the user needs to solve a problem instance P_i using a parametric algorithm, the *broker agent* has to determine the appropriate *knowledge agent*. The selected *knowledge agent* will send to the *broker agent* the model that encodes its knowledge. This model will serve to the *broker agent* for estimating the best parameter configuration for the algorithm used for solving the problem instance P_i. To do this, it carries out abductive inference [16] over the bayesian network and it infers the most probable configurations (MPE) in order to achieve a "good" performance for the algorithm.

However, there is not always a *knowledge agent* suitable to the problem instance to solve. In this case, the *broker agent* has the task of selecting the best *knowledge agent* having into account both the similarity between the problem instances and the reliability of the bayesian networks induced. Sometimes, if several *knowledge agents* are selected as suitable, the *broker agent* has to fusion several bayesian networks, representing several sources of information, in order to unify the knowledge [17].

In this last scenario, after *interface agent* receives a configuration and executes the algorithm with the recommended configuration, the data generated and used like feedback to the multi-agent system can not be used to update the bayesian network of any existing *knowledge agent*. Instead of this, the results of the executions are stored in the database. When a sufficient number of runs for the new specific problem instance have been gathered into the local database, the *update agent* will create a new *knowledge agent* specific for this problem instance, and with a bayesian network induced from these executions. Moreover, the *update agent* will notify to *broker agent* that a new *knowledge agent* is incorporated into the multi-agent system.

2.4 Interface Agent

Finally, the *interface agent* captures the request of configuration of the user being able to identify the relevant features of the problem instance to solve and to deliver it to the *broker agent*. Moreover, it presents to the user the recommended configuration and it executes the algorithm, storing into the local database the resulting data. This data will be used by *knowledge agents* to revise their models or by the *update agent* to increment the number of *knowledge agents* of the system.

2.5 Communication

All agents need to cooperate in order to accomplish their objectives. Usually they utilize a special type of an ACL (*Agent Communication Language*) to exchange knowledge and to cooperate effectively. Therefore, they have to maintain conversations (group of interrelated messages) among them. These conversations between agents often fall into typical patterns. These typical patterns of message exchange are called interaction protocols [6].

Many standard interaction protocols have been defined by the Foundation for Intelligent Physical Agents (FIPA). In this work all complex interactions among agents have been defined using a reduced set of the interaction protocols defined by FIPA.

Figure 2 shows a sequence diagram which represents the underlying interaction protocols of the steps of the intermediation process. Each arrow indicates a FIPA interaction protocol, that is, it represents the start of the conversation, the subsequent control messages and the final response [7]. In the example illustrated there is not a *knowledge agent* suitable to the problem instance to solve. This question implies the following interaction protocols:

A The *interface agent* captures the request of configuration and it speaks with the *broker agent* using the FIPA request protocol. This interaction protocol allows the *interface agent* to request the broker to perform the provide action of the domain knowledge.

B With the FIPA brokering interaction protocol the *interface agent* translates the user request to the *broker agent*. The broker will determine a set of appropriate *knowledge agents* for answering the request and these agents will give their answers back to the *broker agent*.

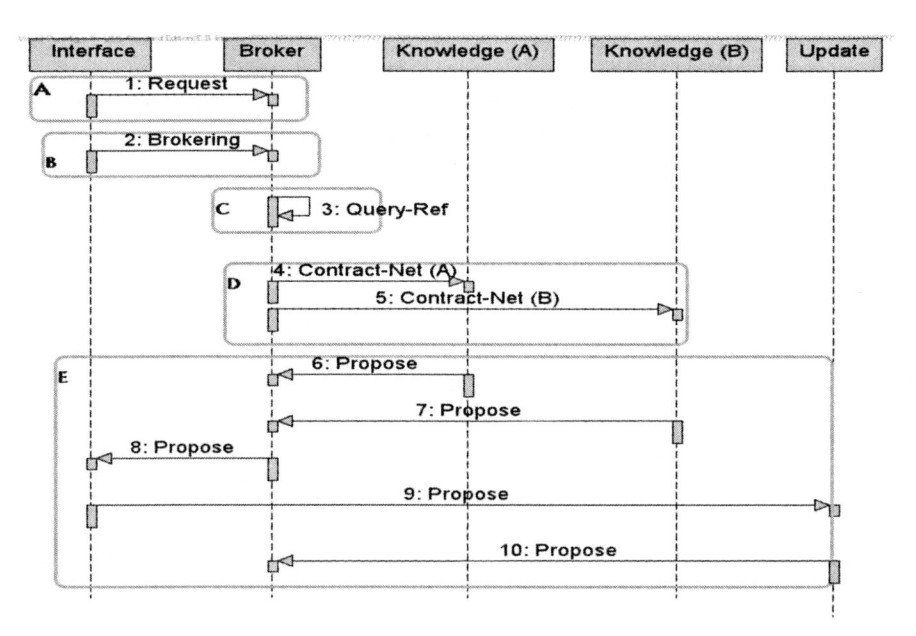

Fig. 2. FIPA interaction protocols used in the proposed system

C The *broker agent* asks itself about the relevant *knowledge agents* for a specific user request. They use for this task the FIPA query-ref interaction protocol.

D When the *broker agent* knows the list of *knowledge agents*, the negotiation between them starts (FIPA contract-net protocol). This interaction will take place with different groups of *knowledge agents*. The variable under negotiation is the similarity between problem instances and reliability of bayesian networks induced.

E The *interface agent* receives a configuration (propose interaction protocol) and executes the algorithm with the recommended configuration in order to solve the specific problem instance. The results of executions are stored in the local database. The *update agent* will create a new *knowledge agent*. Moreover, the *update agent* will notify to *broker agent* that a new *knowledge agent* is incorporated into the multi-agent system (propose interaction protocol).

3 Example of Use and Experimental Results

In order to visualize the functionality of the proposed multi-agent system, this section describes an example in which a GA, that solves the *root identification problem* [10], is automatically adjusted in function of the specific geometric problem that it solves.

Several approaches to the geometric constraint solving problem have been reported in the literature, but the *constructive* technique is one of the most promising approaches [10]. However, in this class of constraint solvers, a well-constrained geometric constraint problem has an exponential number of solutions while the user is only interested in one instance. This instance solution is called the *intended solution*. In this context, finding the intended solution (named the *root identification problem*) means finding an instance solution in the search space [5] that represents the geometric object drawn up by the user.

In [14], GAs were proved as a suitable and effective searching mechanism for solving the *root identification problem*. Nevertheless, in order to apply a particular GA, a number of parameters must be specified by the user and to choose the right parameter values is a hard task. Because of this, we have applied the *multi-agent tuning system*, which automatically supplies the right values for the control parameters of the GA used for solving a given geometric constraint problem.

In the example we are developing, we have limited the parameters configurable of the GA, **X**, to the population size N, the crossover rate P_c and the mutation rate P_m. In relation to the performance measure used to evaluate the goodness of the proposed configurations, we have used the expectation of the evaluations (see Expression 1), that stands for the number of evaluations that the GA needs to carry out in order to find a valid solution, where n is the number of runs, k is the number of successful runs, $eval_{max}$ represents the termination condition and $eval_i$ means the number of evaluations in the $i-th$ run.

$$E = \frac{1}{k}\sum_{i=1}^{k} eval_i + \frac{n-k}{k} \times eval_{max} \qquad (1)$$

In a typical working scenario of the *multi-agent tuning system*, the operation starts when the user defines a rough sketch, annotated with dimensions and constraints. From this sketch, *interface agent* captures the set of features that identify this geometric problem and sends them to *broker agent*. The features relevant for each geometric problem are as follows: the size of the search space; the number of points; the number of distances, angles, tangents, arches and perpendicular from a point to a segment; the percentage of outlier points with additional constraints and the average of additional constraints per points defined by the user.

Broker agent receives the features values of geometric problem at hand and it has the ability to request a configuration solution from the appropriate *knowledge agent* of the system. Each *knowledge agent* in the system has a BN that models the relationships between the control parameters (N, P_c, P_m) and the performance measure E reached by the GA for its specific geometric problem. These relationships are initially induced by the *knowledge agent* from the database of executions of GA solving this specific geometric problem using the PC algorithm [19].

Once the *broker agent* receives the best model solution in the form of a BN, its objective is to infer the most probable configuration $\{p_m, p_c, n\}$ that minimizes the variable E. To do this, the *broker agent* carries out abductive inference [15, 16] over the BN. Next, the most probable configuration $\{p_m, p_c, n\}$ is sent to the *interface agent*. The *interface agent* shows the configuration to the user

Table 1. Comparison of GA performance obtained with validation figures between configurations recommended by the *multi-agent tuning system* and Barreiro

	GA performance		Mann-Whitney test	
Cross	*Multi-agent*	Barreiro	Statistic	p-value
CV 1	6402,35	5473,89	59	0,5205
CV 2	3383,88	2949,44	62	0,3846
CV 3	2399,57	2205,60	53	0,8098
CV 4	2259,47	1984,70	60	0,6761
CV 5	3327,20	2685,44	72	0,2962
CV 6	7192,82	5601,27	64	0,3074

and executes the GA with such a recommended configuration, solving the specific geometric problem sketched. Both the configuration data and the performance measure are stored in the local database of the system. These run data are managed by the *update agent*, which will require to create new *knowledge agents* or will ask *knowledge agents* involved for revising their BNs (sequential learning).

The experiments carried out to test our *multi-agent tuning system* have used 30 different geometric problems. Due to this limited number, we have decided to employ a methodology of 6-fold cross validation of equal size. Each of these folds is used as validation set in turns, with the rest of figures (25) as training set, represented in the system by 25 *knowledge agents*, each one with its BN. With this initial system we have solved each figure of the validation set.

The results of this experiment are given in Table 1, grouped by cross. The second column shows the performance (on average) attained by the GA when it solves the validation figures using the configuration recommended by the multi-agent system. In order to validate the results obtained, a third column presents the GA performance obtained by Barreiro [2] for the same problem, which is considered as the reference model or golden standard for the comparison of the other models, since he carried out a complete statistical study using all information available. Table 1 shows in the fourth and fifth columns results of a comparative analysis between both solutions using the Mann-Whitney test. This test revealed that there is no statistically significant difference between GA performances and it highlights the effective behaviour of the *multi-agent tuning system* in a dynamic context. Also, the *multi-agent tuning system* does not require all available knowledge. So, at the end of the experiment, the number of executions encoded by the BN of each *knowledge agent* are on average 3000, against the 75600 used by Barreiro.

Therefore, it has been possible to confirm experimentally that the *multi-agent tuning system* solves figures proposing configurations almost as efficiently as Barreiro in his study, but it has the advantage of not needing a complete state of information/runs. Experiments have clearly shown the utility of the *multi-agent tuning system* for solving new figures in terms of their features, which in turn demonstrates the learning capacity of the system.

4 Conclusions and Future Work

In this paper we have described a multi-agent system with the capability of tuning a parametric algorithm according to the features of problem instance that it solves. The proposed system is based mainly on the cooperation among a set of agents which use the formalism of bayesian network to model the knowledge about the algorithm. The main advantages obtained are scalability, robustness and flexibility, due to the integration of intelligent agents.

The multi-agent system has been described in a generic way, but later it has been applied with success to configure the GA that solves the root identification problem. Although good results have been obtained, for future work it would be desirable to study the behaviour of the multi-agent system with other parameterised algorithms and more complex problems, since we think that the generality of the approach presented could be limited by the kind of problem domain to solve, affecting the functionality of *interface agent*.

References

1. Akaike, H.: A new look at statistical model identification. IEEE Transactions on Automatic Control 19, 716–723 (1974)
2. Barreiro, E.: Modelización y optimización de algoritmos genéticos para la selección de la solución deseada en resolución construtiva de restricciones geométricas. PhD thesis, Dpto. de Informática. Universidade de Vigo., Marzo (2006)
3. Bouma, W., Fudos, I., Hoffmann, C., Cai, J., Paige, R.: Geometric constraint solver. Computer-Aided Design 27(6), 487–501 (1995)
4. DeGroot, M.H.: Optimal Statistical Decisions. McGraw-Hill, New York (1970)
5. Essert-Villard, C., Schreck, P., Dufourd, J.F.: Sketch-based pruning of a solution space within a formal geometric constraint solver. Artificial Intelligence Journal 1(124), 139–159 (2000)
6. Document Title Fipa. Fipa interaction protocol library specification, http://www.fipa.org/specs/fipa00025/
7. http://www.fipa.org
8. Goldberg, D.E.: Genetic Algorithms in Search, Optimization and Machine Learning. Addison-Wesley, Reading (1989)
9. Heckerman, D.: A tutorial on learning with bayesian networks. Technical Report MSR-TR-95-06, Microsoft Research, Redmond, Washington (1995)
10. Hoffmann, C.M., Joan-Arinyo, R.: A brief on constraint solving. Computer-Aided Design and Applications 2(5), 655–663 (2005)
11. Jeffreys, H.: Theory of Probability. Oxford University Press, Oxford (1983)
12. Jensen, F.V.: An Introduction to Bayesian Networks. Springer, Heidelberg (1996)
13. Kohavi, R., John, G.: Automatic parameter selection by minimizing estimated error. In: Prieditis, A., Russell, S. (eds.) Proceedings of the 12th International Conference on Machine Learning, Lake Tahoe, CA, pp. 304–312. Morgan Kaufmann, San Francisco (1995)
14. Luzón, M.V.: Resolución de Restricciones geométricas. Selección de la Solución Deseada. PhD thesis, Dpto. de Informática. Universidade de Vigo., Septiembre (2001)

15. Nilsson, D.: An efficient algorithm for finding the m most probable configurations in bayesian networks. Statistics and Computing 8(2), 159–173 (1998)
16. Pearl, J.: Probabilistic Reasoning in Intelligent Systems: Networks of Plausible Inference. Morgan Kaufmann, San Mateo (1998)
17. Sagrado, J.D.: Fusión topológica y cuantitativa de redes causales. Universidad de Almería. Servicio de publicaciones (2000)
18. Schwefel, H.P.: Evolution and Optimum Seeking. John Wiley and Sons, Chichester (1995)
19. Spirtes, P., Glymour, C., Scheines, R.: Causation, Prediction and Search. Lectures Notes in Statistics, vol. 81. Springer, New York (1993)

Relative Information in Grid Information Service and Grid Monitoring Using Mobile Agents

Carlos Borrego[1] and Sergi Robles[2]

[1] High Energy Physics Institute (IFAE), Campus UAB Edifici Cn. Facultat Ciències E-08193 Bellaterra
`cborrego@ifae.es`
[2] Department of Information and Communications Engineering (dEIC), Campus UAB Edifici Q, E-08193 Bellaterra
`sergi.robles@deic.uab.es`

Abstract. Grid computing is consolidated as a technology capable of solving scientific projects of our century. These projects' needs include complex computational and large data storage resources. The goal of grid Computing is to share these resources among different institutes and virtual organizations across high-speed networks. The more resources there are the more difficult it gets to monitor them and to assure users find the best resources they are looking for. We introduce the concept of relative information to enhance grid information and grid monitoring services. Resource relative information is information that is not only gathered from the resource itself but also that takes into account other resources of the same type. The resource is not described in terms of the resource local information but also relative to other resources. We present a framework based on mobile agents that enables to publish relative information in the grid information service and another framework to monitor grid resources using relative criteria.

1 Introduction

Service and resource discovery in grid systems [1, 8] becomes a challenging issue when resources increase enormously. These resources may be classified into different types which include computational services, storage services, databases, catalogs, special-purpose services, etc. An increase of the number of grid services and resources makes it harder to guarantee that a user will find the service or resource they are looking for. The more resources and services there are the easier it will be to come across resources that will not match user needs even if the resources claim to do so. On the other hand, with this increase it is also harder to monitor efficiently grid services and resources. There is a need to monitor grid services and resources in a more global way and not just performing the traditional service oriented monitoring.

We present in this paper an infrastructure based on mobile agents that allows grid resources to publish relative information and to be monitored as well using relative criteria. Resource relative information is information that is not only gathered from the very resource itself, but also taking into account other resources of the same type. The resource is not only described in terms of the resource local information but also relative to other resources. This concept can be applied both to the grid information service and the monitoring service.

Mobile agents [16] are a good option to create this relative information. They are able to migrate their code from machine to machine and continue their execution on the destination machine. The entities on which mobile agents run are called platforms. A platform can be installed on a grid node so mobile agents can visit them in order to retrieve local information and conceive the relative information. This is a strong requirement for our proposal. Without a platform running on the grid node, mobile agents will not be able to migrate to the node.

The rest of the paper is organized as follows: Section 2 explains how to create this relative information. Section 3 presents an infrastructure to publish relative information in grid services using mobile agents. Section 4 describes another infrastructure that monitors grid services using relative criteria and mobile agents. Section 5 presents the results we have achieved, and finally section 6 describes the conclusions we have reached and identifies open issues for future work.

2 Obtaining Relative Information in Grid Services Using Mobile Agents

In this section we explain why mobile agents are useful to monitor grid resources as well as to retrieve data for the information system. Other working groups like Tomarchio et al. [13, 15], Chen et al. [10] and Dong et al. [9] have defined different grid monitoring systems using mobile agents but without using relative information.

If a resource needs to be described in a relative way, besides the local values it will need to collect external data from other resources. In order to collect relative information in grid services, it must firstly consider whether a new framework is needed or a current grid framework can be used. There are three possibilities here:

- Letting the resources contact rest of the resources asking for their local information. This is not efficient due to the number of messages exchanged and could flood the system.
- Publishing local information to a central node and then provide a relative information service. Comparing information inside a central node where we can find all resource's information is a trivial issue. As described in [18], grid services tend to be decentralized because of scalability issues for being a single point of failure. Therefore, this would create an unnecessary bottleneck.
- Letting a software entity *visit* every resource, extract the local information and create the relative information after all these visits have been made. This means obtaining relative information from the source that is from the resource itself.

The goal of this paper is to create an infrastructure based on mobile agents that allows obtaining relative information in grid services. In order to achieve this, we will install a platform in every grid node so mobile agents can visit the node and obtain local information. Once the mobile agent has gone through all resources of the same type it will have a global view of these resources and will be able to construct the relative information. The mobile agent behaves as a finite state machine. This machine is depicted in Figure 1.

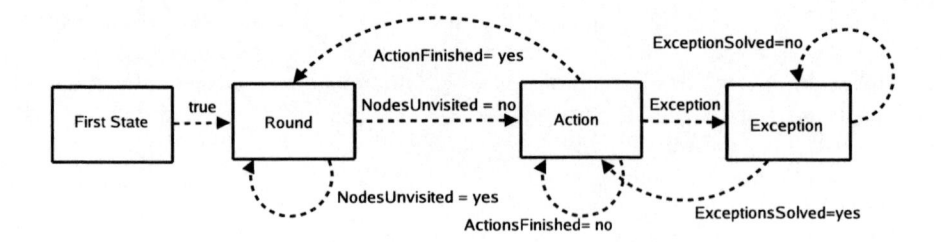

Fig. 1. Agent's finite state machine

The platform we have chosen to install on every grid node is JADE/IPMS [3, 7]. This infrastructure provides platform-to-platform mobility for JADE agents. The benefit of using this infrastructure is that we do not have a centralized point which manages every agent migration. Grid services must be independent enough to assure flexibility. Other infrastructures which also allow this flexibility, as seen in [7], do not follow the IEEE-FIPA standards.

On the *First state*, the mobile agent will query the information service to check for all the grid nodes of the type it is trying to relativize filtering by those having a platform running on it. This is what we call the itinerary of the agent. This query can be done to the grid information system or to an alternative information system in which platforms announce their existence. This latter option is more efficient in terms of avoiding bottlenecks.

After creating the itinerary, it will change to *Round* state visiting every grid node and obtaining the local information. Once all the grid nodes are visited the mobile agent will be able to obtain the relative information since it has a global view of all the resources. After this, it can change to the *Action* state, use this relative information and start again. If an error happens, the agent will stay in an *Exception* state until this error is solved. Comparing this proposal with others like [2] that have a central node, there are three advantages:

- It is not necessary to introduce any change in any central node which are normally critical services with risky update processes.
- The user that will use the relative information may be not the developer of implementation of the relative criteria. A change in the criteria just implies an upgrade of the mobile agent's code. Its deployment is immediate since we just need the agent to make a whole round to obtain the relative information with the new criteria. This is a substantial benefit from other centralized services that need to wait for all grid administrators to update sensors in the grid resources.
- And finally, using mobile agents to gather the resource's local information assures that we obtain the information from the different resources in the same way. This is really important since our goal is to compare them.

We propose two different applications of our proposal to show how useful it is, as well as its feasibility. These applications are:

- An application that allows publishing relative information in the grid information service using mobile agents.
- An application that monitors grid services using relative criteria and mobile agents.

Sections 3 and 4 describe these applications.

The mobile agent following the general finite state machine of Figure 1, will first of all create the itinerary querying the information service for the type of grid node which it wants to relativize and has a platform running. Then it will migrate from node to node gathering from each one the local value for information we want to relativize. The mobile agent keeps at all times a global view of the information that has been collected. Once the mobile agent has finished the itinerary, it will be capable of creating the relative information. It will change its state to *Action* by migrating to the nodes that need to be updated. Should a problem arise while updating the node, the mobile agent will change its state to *Exception* and will not exit from this state until it has solved the incoherence.

The agent will carry information about the resources it has visited. This information can be as detailed as the agent programmer wishes to be. The more detailed will be this information, the more accurate the global view the agent has, but on the contrary, it will be less efficient for the agent transport. The global view can be the result of a statistical function of local values, as a mean for example. In this case, this is extremely light for the agent to carry. The agent will gather local information on each node visit and will compare it with the global view, whereupon the agent will update the latter.

On the other hand, the global view can be represented using a round robin structure in which local values and/or statistical information are stored. This will allow us to have a historical information of local values and global views. This option makes the agent data increase, but, since the round robin structure has a fixed size, this size can be controlled by the agent programmer. An interesting application from this option is that we will be capable of discarding peak local values which may not be representative of the service or the resource.

3 Publishing Relative Information in the Grid Information Service

The grid information service describes resources by publishing information about them. We present an infrastructure based on mobile agents that allows publishing relative information in the grid information service. As an example of its functionality we present a general use case based on grid's service computing element. The main functionally of a computing element is to manage grid jobs, that is to say, job submission and job control. The objective of this system is to publish under a group of computing elements which are the best two computing elements given a criteria. The criteria could be, for instance, the efficiency in grid jobs, i.e., the percentage of jobs that have successfully finished.

At all times the mobile agent keeps which are the best two computing elements, concerning the efficiency. Once the mobile agent has finished the itinerary it will be capable of knowing which are the best two. It will change its state to *Action*, it will migrate to these computing elements and change the information service. In fact, it is not the mobile agent itself that changes the information system. It just leaves information that the dynamic plug-ins of the information system will retrieve to publish the relative information. Finally, the mobile agent will return to the *Action* state and start over again.

To implement this infrastructure using gLite [2] we have added two attributes to the schema [4]. The first one, *GlueCEPlatformUp*, indicates whether the computing element has a platform running or not. This is crucial for the agent's itinerary creation that will be defined querying the information service. To fill in this attribute we define a dynamic plugin that will check if the platform is running in the computing element. The other attribute, *GlueCEInfoIsTop2*, represents whether or not the computing element is among the best two computing elements taking into account the efficiency criteria. To complete this attribute we implement a dynamic plugin that will collect the information left by the mobile agent during the *Action* state.

As described in section 2 we have installed a running Jade/IPMS platform [3, 7] in every computing element. The mobile agent will jump from computing element to computing element following the behaviour defined in chapter 2. The final goal is to let a user include in his job requirements the variable *GlueCEInfoIsTop2* set to "yes". This will force his jobs to be run in any of the two best computing elements following the chosen criteria. This allows the user to be detached from the implementation of such criteria.

4 Monitoring Grid Resources Using Relative Criteria

Grid services are monitored to check their functionality. In grid computing it is normal to have plenty of identical nodes that normally have the same behaviour in terms of the different criteria we monitor. We would like to monitor different aspects of grid services but some of these aspects imply relative criteria. There is a need to compare local sensors with a global view of other nodes of the same type. The goal is to point out possible critical nodes comparing the node itself with the rest of the nodes of the same type as a whole. We have defined the agent *WMSMonitAgent* which monitors gLite's resource broker, the grid service which is in charge of job distribution over the grid. This service is called WMS [14], which stands for Workload Management System. The monitoring will be looking at the content of its information supermarket. The information supermarket is the WMS component which contains all the resources the WMS knows about at a given time. It is fed by querying the information system, while working in pull mode or by resources themselves when services run on push mode. At a given time these resources should be similar on every WMS in the grid. We have developed a system that points out failing WMS by comparing them with the rest of the WMS. As seen in section 2, we will install a Jade/IPMS platform in every WMS so the *WMSMonitAgent* can run on it. The agent will jump from WMS to WMS checking the content of the information supermarket. If the content differs from the global view the agent has, the agent will consider to flag the WMS as problematic. There is a grace period so we can avoid false positives created by resources in push mode that update a single WMS. After this period the WMS will be marked as true problematic. Recovery actions launched by agents have been also implemented.

5 Results

Tests performed in a test environment using virtual machines show that we have achieved our goal to allow users to include relative requirements in their jobs. The resources publish relative information while the users are able to match these resources using relative criteria as if it were standard requirements as seen in Figure 2.

Computing elements in Figure 2 are publishing relative information. The user can search for computing elements which have the relative attribute *other.GlueCEInfoIPMSisTop2* set to "yes".

The following code is the result of an experiment of job matching using gLite [2]:

```
$ cat ipms.jdl
Executable = "/usr/bin/whoami";
Requirements = other.GlueCEInfoIsTop2 == "1" ;
$edg-job-list-match --vo atlas ipms.jdl
testce2.ifae.es:2119/jobmanager-lcgpbs-atlas
testce1.ifae.es:2119/jobmanager-lcgpbs-atlas
```

The file *ipms.jdl* contains the description and the requirements of the user's grid job. One of these requirements could be to run the job in those Computing Elements which have the variable *other.GlueCEInfoIsTop2* set to *1*. Using the command *edg-job-list-match* the user can ask the WMS for a list of Computing Elements which match the user's jdl. As it can be seen above, the result are just two Computing Elements which are the best two following the criteria explained before.

Concerning grid resources monitoring using relative criteria, we have shown how our system based on mobile agents points out malfunctioning services. Several WMS are instanced as virtual machines. On every WMS we run a Jade/IPMS platform. As seen in Figure 3, we simulate the situation in which a WMS loses connection with its information service. Hence, its information supermarket becomes obsolete. After several rounds, the mobile agents finds this problem and flags the WMS as problematic.

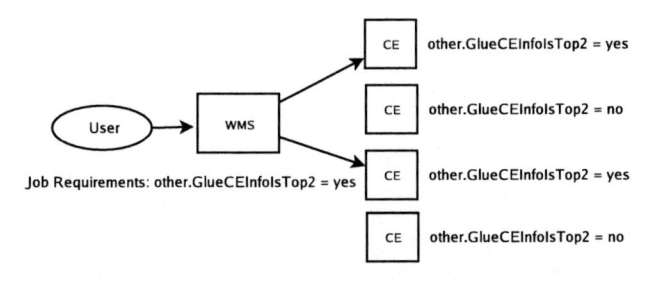

Fig. 2. Job match making with relative requirements

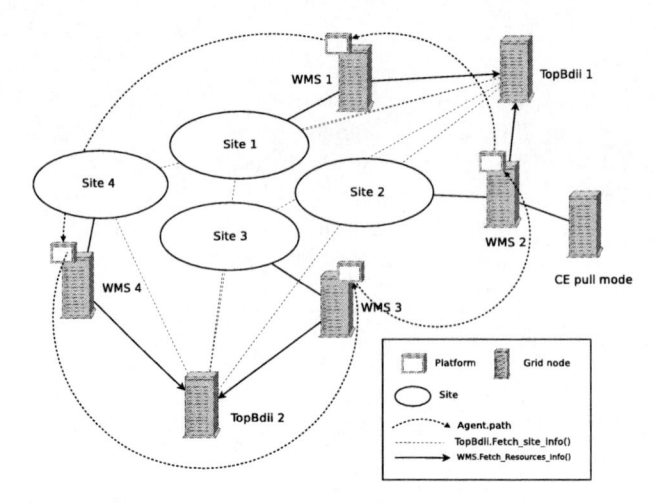

Fig. 3. Monitoring WMS. Test scenario

Simulations performed in virtual machines show, as expected, that round trip time increases linearly with the number of platforms and the size of the agent's code as seen in figures 4 and 5. We are aware that 10 platforms is not representative enough of a real grid, but we believe watching these graphs, that it could still scale linearly when arriving to hundreds of nodes.

When a mobile agent with an average code size is run, round trip time does not exceed 1 second/platform. Services like the ones mentioned before, Computing Element and WMS, in the largest grids like EGEE [11] do not outnumber 500 nodes. In this case the longest round trip time would be around 7 minutes. We believe this time is short enough to let the agent being up-to-date when it finishes the round trip. Moreover, this is an overhead that can be assumed by most applications.

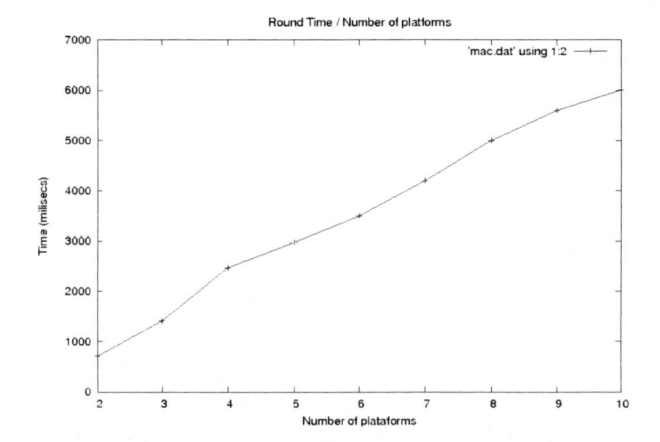

Fig. 4. Round trip time as a function of the number of platforms

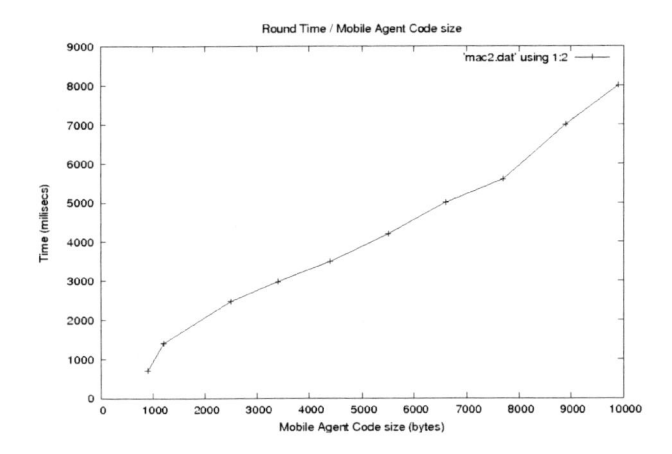

Fig. 5. Round trip time as a function of the size of the agent's code with fixed number of platforms

6 Conclusions and Future Work

We have shown that it is feasible and useful to publish relative information and monitor grid services using relative criteria. We have seen that relative information gives a more complete view of grid services, and it helps the user to choose the best resources for their needs. Monitoring grid services with relative criteria is a good way to find malfunctioning services. In order to flag a service as problematic our proposal does not take into account just local sensors of the service but as well sensors values extracted from the rest of the services of the same type. Mobile agents are a very good option to use as an infrastructure to go through with these implementations. This infrastructure proposed is more flexible than traditional ones because of the sensors deployment. The update of the way local information is extracted becomes a trivial matter while using mobile agents, even if we have a large number of different services. There is an inconvenience though: Every grid node must have installed a mobile agent platform where the agents will run. This requirement does not significantly affect performance. since mobile platforms do not consume much CPU time.

Security, at the moment, is focused on the mobile agent platforms. As a future work, we could move security from the platform to the agent following the ideas in [6]. Mobile agents inside the platform could act as gateways to local resources, enhancing security as well as making it more flexible. Besides, the number of platforms could be minimized from one platform per grid node to just one platform per site. This implies some studies on how local information from grid nodes can be gathered from just one node on a site. Finally, a more ambitious research could be to define a complete information service and monitoring service using mobile agents capable of creating relative information and replacing traditional communications.

References

1. Foster, I., Kesselman, C., Tuecke, S.: The Anatomy of the Grid: Enabling Scalable Virtual Organizations. International Journal of High Performance Computing Applications 15(3), 200–222 (2001)
2. Laure, E., Fisher, S.M., Frohner, A., Grandi, C., Kunszt, P.: Programming the Grid with gLite. Computational Methods in Science and Technology 12(1), 33–45 (2006)
3. Bellifemine, F., Poggi, A., Rimassa, G.: JADE. A FIPA compliant agent framework. In: Proceedings of PAAM, pp. 97–108 (1999)
4. Glue Schema (access date April 15, 2008), http://globus.org/toolkit/mds/glueschemalink.html
5. Kuang, H., Bic, L.F., Dillencourt, M.B.: Iterative grid-based computing using mobile agents. In: International Conference on Parallel Processing (ICPP 2002), p. 109 (2002)
6. Ametller, J., Robles, S., Ortega-Ruiz, J.A.: Self-protected mobile agents. In: Proceedings of the Third International Joint Conference on Autonomous Agents and Multiagent Systems, pp. 362–367 (2004)
7. Cucurull, J.: Efficient mobility and interoperability of software agents Phdthesis, Autonomous University of Barcelona (2008)
8. Czajkowski, K., Fitzgerald, S., Foster, I., Kesselman, C.: Grid Information Services for Distributed Resource Sharing. In: 10th IEEE International Symposium on High Performance 2001, pp. 181–194 (2001)
9. Dong, G., Tong, W.: A Mobile Agent-based Grid Monitor Architecture. In: IEEE/ACS International Conference on Computer Systems and Applications, AICCSA 2007, pp. 293–300 (2007)
10. Chen, A., Tang, Y., Liu, Y., Li, Y.: MAGMS: Mobile Agent-Based Grid Monitoring System, Frontiers of WWW Research and Development. LNCS, pp. 739–744. Springer, Heidelberg (2006)
11. Gagliardi1, F., Jones, B., Grey, F., Bgin, M., Heikkurinen: Building an infrastructure for scientific Grid computing: status and goals of the EGEE project. Philosophical Transactions: Mathematical, Physical and Engineering Sciences 363(1833), 1729–1742 (2005)
12. Iamnitchi, A., Foster, I., Nurmi, D.: A peer-to-peer approach to resource discovery in grid environments. In: Proceedings of the 11th Symposium on High Performance Distributed Computing, pp. 340–348 (2002)
13. Tomarchio, O.: Active Monitoring In Grid Environments Using Mobile Agent Technology. In: Proceedings of the 2nd Workshop on Active Middleware Services (2000)
14. Andreetto, P., et al.: Practical approaches to grid workload and resource management in the EGEE project. In: Proceedings of CHEP 2004, Interlaken, CH (September 2004)
15. Tomarchio, L.V.: On the use of mobile code technology for monitoring Grid system. In: First IEEE/ACM International Symposium on Cluster Computing and the Grid, 2001. Proceedings, pp. 450–455 (2001)
16. White, J.E.: Mobile agents make a network an open platform for third-partydevelopers. Computer 27(11), 89–90 (1994)
17. Cai, M., Frank, M., Chen, J., Szekely, P.: A Multi-Attribute Addressable Network for Grid Information Services. In: Fourth International Workshop on Grid Computing, Information Sciences Institute, University of Southern California, p. 184 (2004)
18. Iamnitchi, A., Foster, I.: On fully decentralized resource discovery in grid environments. In: Lee, C.A. (ed.) GRID 2001. LNCS, vol. 2242, pp. 51–62. Springer, Heidelberg (2001)

A Multi-Agent System for Airline Operations Control

Antonio J.M. Castro and Eugenio Oliveira

Informatics Engineering Department and LIACC/NIAD&R
Faculty of Engineering, University of Porto
Rua Dr. Roberto Frias s/n, 4200-465, Porto, Portugal
ajmc@fe.up.pt, eco@fe.up.pt

Abstract. The Airline Operations Control Center (AOCC) tries to solve unexpected problems that might occur during the airline operation. Problems related to aircrafts, crewmembers and passengers are common and the actions towards the solution of these problems are usually known as operations recovery. In this paper we present the implementation of a Distributed Multi-Agent System (MAS) representing the existing roles in an AOCC. This MAS has several specialized software agents that implement different algorithms, competing to find the best solution for each problem and that include not only operational costs but, also, quality costs so that passenger satisfaction can be considered in the final decision. We present a real case study where a crew recovery problem is solved. We show that it is possible to find valid solutions, with better passenger satisfaction and, in certain conditions, without increasing significantly the operational costs.

Keywords: Airline Operations Control, Operations Recovery, Disruption Management, Multi-agent system, Software Agents, Operational Costs, Quality Costs.

1 Introduction

Operations control is one of the most important areas in an airline company. Through operations control mechanisms the airline company monitors all the flights checking if they follow the schedule that was previously defined by other areas of the company. Unfortunately, some problems arise during this phase [7]. Those problems are related to crewmembers, aircrafts and passengers. The Airline Operations Control Centre (AOCC) is composed by teams of people specialized in solving the above problems under the supervision of an operation control manager. Each team has a specific goal contributing to the common and general goal of having the airline operation running with few problems as possible. The process of solving these problems is known as Disruption Management [8] or Operations Recovery.

To be able to choose the best solution to a specific problem, it is necessary to include the correct costs on the decision process. It is possible to separate the costs in two groups: Operational Costs (easily quantifiable costs) and Quality Costs (less easily quantifiable costs). The operational costs are, for example, crew costs (salaries, hotel, extra-crew travel, etc.) and aircraft/flights costs (fuel, approach and route taxes, handling services, line maintenance, etc.). The quality costs that we are interested in calculating in the AOCC domain are, usually, related to passenger satisfaction.

Specifically, we want to include on the decision process the cost of delaying or cancelling a flight from the passenger point of view, that is, in terms of the importance that a delay will have to the passenger.

Starting from the work presented in [6] and based on our observations we have done on an AOCC of a real airline company we hypothesize that the inclusion of quality costs in the decision process will increase the customer satisfaction (a fairly obvious prediction) without increasing significantly (or nothing at all) the operational costs of the solutions in a given period. Basically, we expect to find valid alternate solutions within the same operational cost but with a better impact on the passenger satisfaction.

The rest of the paper is organized as follows. In section 2 we present some work of other authors regarding operations recovery. Section 3 shows how we arrived at the formulas we have used to express the importance of the flight delay, from the passenger point of view. Section 4 shows how we have updated the MAS presented in [6] to include quality costs, including the MAS architecture and the algorithm used to choose the best solution. In section 5 we present the scenario used to evaluate the system as well as the results of the evaluation. Finally, we discuss and conclude our work in section 6.

2 Related Work

The paper [2] gives an overview of OR applications in the air transport industry. We will present here the most recent published papers according to [7]. We divided the papers in three areas: crew recovery, aircraft recovery and integrated recovery.

Aircraft Recovery. The most recent paper considering the case of aircraft recovery is [10]. They formulate the problem as a Set Partitioning master problem and a route generating procedure. The goal is to minimize the cost of cancellation and retiming, and it is the responsibility of the controllers to define the parameters accordingly. It is included in the paper a testing using SimAir [11] simulating 500 days of operations for three fleets ranging in size from 32 to 96 aircraft servicing 139-407 flights. Although the authors do try to minimize the flights delays, nothing is included regarding the use of quality costs.

Crew Recovery. In [1] the flight crew recovery problem for an airline with a hub-and-spoke network structure is addressed. The paper details and sub-divides the recovery problem into four categories: misplacement problems, rest problems, duty problems, and unassigned problems. The proposed model is an assignment model with side constraints. Due to the stepwise approach, the proposed solution is suboptimal. Results are presented for a situation from a US airline with 18 problems. This work omits the use of quality costs.

Integrated Recovery. In [4] the author presents two models that considers aircraft and crew recovery and through the objective function focuses on passenger recovery. They include delay costs that capture relevant hotel costs and ticket costs if passengers are recovered by other airlines. According to the authors, it is possible to include, although hard to estimate, estimations of delay costs to passengers and costs of future lost ticket sales. To test the models an AOCC simulator was developed, simulating domestic operations of a major US airline. It involves 302 aircrafts divided into 4 fleets, 74 airports and 3 hubs. Furthermore, 83869 passengers on 9925 different passengers' itineraries per day are used. For all scenarios are generated solutions with reductions in passenger delays and disruptions. The difference regarding our proposal is that we use the opinion of the passengers when calculating the importance of the delay.

In [6] the author presents a Multi-Agent System (MAS) to solve airline operations problems, using specialized agents in each of the three usual dimensions of this problem: crew, aircraft and passengers. However, in the examples presented, the authors ignore the impact that a delay in the flight might have on the decision process and only use operational costs to make the best decision. That is the biggest difference regarding the work we present in this paper.

3 How to Quantify Quality Costs

Overview. The Airline Operations Control Center (AOCC) has the mission of controlling the execution of the airline schedule and, when a disruption happens (aircraft malfunction, crewmember missing, etc.) of finding the best solution to the problem. It is generally accepted that, the best solution, is the one that does not delay the flight and has the minimum operational cost. In summary, it is the solution that is nearest to the schedule, assuming that the schedule is the optimal one. Unfortunately, due to several reasons (see [9] for several examples), it is very rare to have available solutions that do not delay a flight and/or do not increase the operational cost. From the observations we have done in a real AOCC, most of the times, the team of specialists has to choose between available solutions that delay the flight and increase the operational costs. Reasonable, they choose the one that minimize these two values.

The Perception of Quality Costs. In our observations, we found that some of the teams in the AOCC, used some kind of rule of thumb or hidden knowledge and, in some cases, they did not choose the solutions that minimize the delays and/or the operational costs. For example, suppose that they have disruptions for flight A and B with similar schedule departure times. The best solution to flight A would cause a delay of 30 minutes and the best solution to flight B would cause a delay of 15 minutes. Sometimes, and when technically possible, they would prefer to delay flight A in 15 minutes and flight B in 30 minutes or more if necessary. We can state that flights with several business passengers, VIP's or for business destinations correspond to the profile of flight A in the above example. In our understanding this means that they are using some kind of quality costs when taking the decisions, although not quantified and based on personal experience. In our opinion this makes the decision less reliable but that knowledge represents an important part in the decision process and should be included on it.

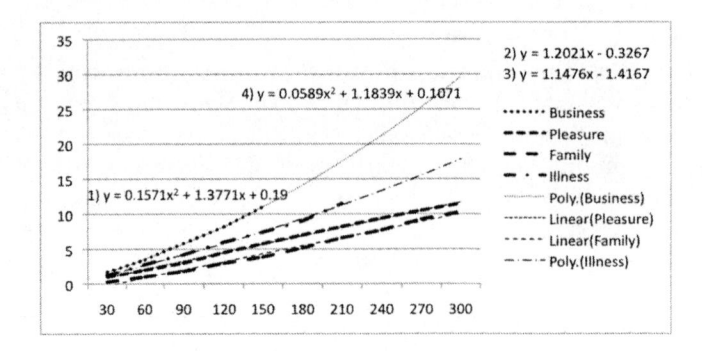

[Fig. 1. Delay time versus importance]

Quantifying Quality Costs. To be able to use this information in a reliable decision process we need to find a way of quantifying it. What we are interested to know is how the delay time and the importance of that delay to the passenger are related in a specific flight. It is reasonable to assume that, for all passengers in a flight, less delay is good and more is bad. However, when not delaying is not an opinion and the AOCC has to choose between different delays to different flights, which ones should they choose? To be able to quantify this information, we have done a survey to several passengers on flights of an airline company. Besides asking in what class they were seated and the reason for flying in that specific flight, we asked them to evaluate from 1 to 10 (1 – not important, 10 very important) the following delay ranges (in minutes): less that 30, between 30 and 60, between 60 and 120, more than 120 and flight cancellation. From the results we found four passenger profiles: Business, Pleasure (travelling in vacations), Family (usually immigrants visiting their families) and Illness (travelling due to medical care). To be able to get the information that characterizes each profile we used the airline company database, including relevant database fields like number of business and tourist class passengers, frequent flyer passengers, passengers with special needs, etc. The important information that we want to get from the survey data is the trend of each profile, regarding delay time/importance to the passenger. Plotting the data and the trend we got the graph in figure 1 (x – axis is the delay time and y – axis the importance).

From the graph in figure 1 it is possible to see the equations that define the trend of each profile. If we apply these formulas as is, we would get quality costs for flights that do not delay. Because of that we re-wrote the formulas. The final formulas that express the importance of the delay time for each passenger profile are presented in table 1. We point out that these formulas should be updated frequently to express any change on the airline company passenger's profiles.

Table 1. Final quality costs formulas

Profile	Formula
Business	$y = 0.16 \ast x^2 + 1.38 \ast x$
Pleasure	$y = 1.20 \ast x$
Family	$y = 1.15 \ast x$
Illness	$y = 0.06 \ast x^2 + 1.19 \ast x$

4 Using Quality Costs in Operations Recovery

Overview. The MAS we used is a modification of the one used in [6] and represents the Airline Operations Control Center (AOCC). The development followed the methodology presented in [5]. Some of the Agent/MAS characteristics that make us adopt this paradigm are the following [12]:

Autonomy. MAS models problems in terms of autonomous interacting component-agents, which are a more natural way of representing task allocation, team planning, and user preferences, among others.

Distribution of resources. With a MAS we can distribute the computational resources and capabilities across a network of interconnected agents avoiding problems associated with centralized systems.

Scalability. A MAS is extensible, scalable, robust, maintainable, flexible and promotes reuse. These characteristics are very important in systems of this dimension and complexity.

A high-level graphical representation of the MAS architecture is presented in figure 2. The square labeled BASE A shows the part of the MAS that is installed in each operational base of the airline company. Each operational base has its own resources that are represented in the environment. Each operational base has also software agents that represent roles in the AOCC. The *Crew Recovery*, *Aircraft Recovery* and *Pax Recovery* are sub-organizations of the MAS dedicated to solve crew, aircraft and passengers problems, respectively. The *Apply Solution Agent* applies the solution found and authorized in the resources of the operational base.

Architecture and Specialized Agents. The MAS sub-organizations have their own architecture with their specialized agents. Figure 3 shows the architecture for *Crew Recovery* in a UML diagram according to the notation expressed in [5]. The architecture for *Aircraft Recovery* and *Pax Recovery* are very similar.

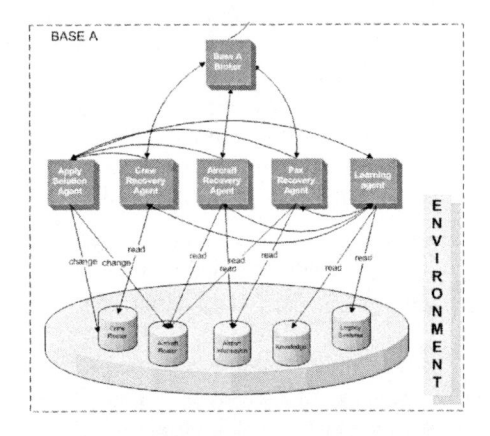

Fig. 2. MAS Architecture (one Base)

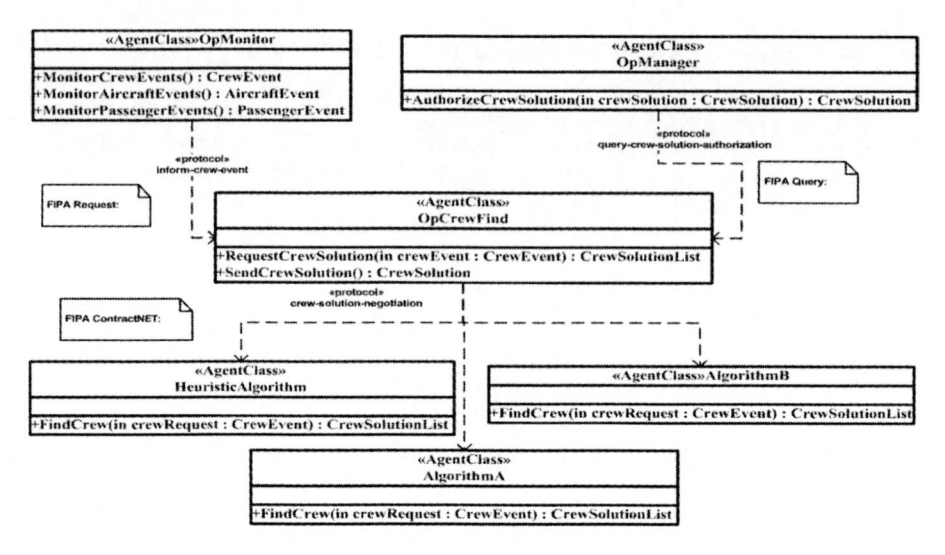

Fig. 3. Crew recovery architecture

The agent class *OpMonitor* is responsible for monitoring any crew events, for example, crewmembers that did not report for duty or duties with open positions, that is, without any crewmember assigned to a specific role on board (e.g., captain or flight attendant). When an event is detected, the service *MonitorCrewEvents* will initiate the protocol inform-crew-event (FIPA Request) informing the *OpCrewFind* agent. The message will include the information necessary to characterize the event. This information is passed as a serializable object of the type *CrewEvent*. The *OpCrewFind* agent detects the message and will start a CFP (call for proposal) through the *crew-solution-negotiation* protocol (FIPA contractNET) requesting to the specialized agents *HeuristicAlgorithm*, *AlgorithmA* and *AlgorithmB* (or any other that is implemented and deployed in the MAS) of any operational base of the airline company, a list of solutions for the problem. Each agent implements a different algorithm specific for this type of problem. When a solution is found a serializable object of the type *CrewSolutionList* is returned in the message as an answer to the CFP. The *OpCrewFind* agent collects all the proposals received and chooses the best one according to the algorithm in Table 2. This algorithm is implemented in the service *SendCrewSolution* and produces a list ordered by total cost that each solution represents. The computed values in the algorithm in Table 2 are the following:

***TotalDuty*:** Monthly duty minutes of the crewmember after the new duty.
***CredDuty*:** Minutes to be paid case the crewmember exceeds the duty limit.
***DutyPay*:** Cost of duty computed according to the hour salary of the crewmember.
***PerdiemDays*:** Days of work for the specific duty.
***PerdiemPay*:** Cost of duty computed according to the *perdiem* value
***BaseFactor*:** If the crew belongs to the same operational base where the problem happened, the value is one. Otherwise, it will have a value greater than one.
***OperCost*:** The operational cost of the solution.
***PaxBus*:** The total of passengers in the *business* profile on the disrupted flight.

PaxFam: The total of passengers in the *family* profile on the disrupted flight.
PaxIll: The total of passengers in the *illness* profile on the disrupted flight.
PaxPlea: The total of passengers in the *pleasure* profile on the disrupted flight.
BusPfCost: The importance of the delay for passengers of the *business* profile.
IllPfCost: The importance of the delay for passengers of the *illness* profile.
PleaPfCost: The importance of the delay for passengers of the *pleasure* profile.
FamPfCost: The importance of the delay for passengers of the *family* profile.
QualCost: The quality cost of the solution.

It is important to point out the use of coefficient C1 in the quality cost formula. The goal of this coefficient is to give a value to the quality costs in the same unit of the operational costs. Operational costs are expressed in monetary units (Euros, Dollars, etc.) because they are direct and real costs. On the other hand, quality costs are not real costs and express a level of satisfaction of the passengers. Besides transforming the quality costs into a monetary unit, airline companies can also use this coefficient to express the importance that this type of cost has in the decision process, by increasing its value.

The *SendCrewSolution* service initiates the protocol *query-crew-solution-authorization* (FIPA Query) querying the *OpManager* agent for authorization. The message includes the serializable object of the type *CrewSolution*.

Table 2. Selection algorithm

foreach item in *CrewSolution* list
   totalDuty = monthDuty+credMins
   if (totalDuty-dutyLimit) > 0
     credDuty = totalDuty-dutyLimit
   else
     credDuty = 0
   end if
   perdiemDays = (endDateTime-dutyDateTime
   perdiemPay = perdiemDays*perdiemValue
   dutyPay = credDuty*(hourSalaryValue/60)
   operCost = (dutyPay+perdiemPay)*baseFactor
   paxBus = cPax+vipPax+fflyerPax+paxTot*busDest
   paxFam = yPax+paxTot+imigDest
   paxIll = illPax
   paxPlea = yPax+paxTot+vacDest
   busPfCost = $0.16*fltDelay^2+1.38*fltDelay$
   illPfCost = $0.06*fltDelay^2+1.19*fltDelay$
   PleaPfCost = 1.2*fltDelay
   famPfCost = 1.15*fltDelay
   qualCost = C1*(busPfCost*paxBus+illPfCost*paxIll +
                PleaPfCost*paxPlea+famPfCost*paxFam)
   totalCost = operCost+qualCost
end foreach
order all items by *totalCost* **desc**
select first item on the list

5 Scenario and Experiments

Scenario. To evaluate our MAS we have setup the same scenario used by the authors in [6] that include 3 operational bases (A, B and C). Each base includes their crew-members each one with a specific roster. The data used corresponded to the real operation of June 2006 of base A. After setting-up the scenario we found the solutions for each crew event using our Crew Recovery Architecture and Specialized Agents of our MAS. As a final step, the solutions found by our MAS were presented to AOCC users to be validated.

Results. Table 3 presents the results that compare our method (method B) with the one used by the authors in [6], updated with quality costs for a better comparison (method A). We point out that in method A the quality costs were not used to find the best solution. From the results obtained we can see that in average, method B produced solutions that decreased flight delays in 36%. Method B took 26 seconds to find a solution and method A took 25, a 3% increase.

Table 3. Comparison of the results

	Method A Total	%	Method B Total	%	A/B %
Delay (avg):	11	100	7	64	-36
Time (avg)	25	100	26	103	3
Total Costs:	11628	100	8912	77	-23
Oper. Costs:	3839	100	4130	108	8
Qual. Costs:	7789	100	4782	61	-39

Regarding the total costs (operational + quality), the method B has a total cost of 8912 and method A a total cost of 11628. Method B is, in average 3% slower than method A in finding a solution and produces solutions that represent a decrease of 23% on the total costs. Regarding operational costs, method A has a cost of 3839 and method B a cost of 4130. Method B is 8% more expensive regarding operational costs. Regarding quality costs, method A has a cost of 7789 and method B a cost of 4782. Method B is 38% less expensive regarding quality costs.

6 Discussion and Conclusions

Regarding our first hypothesis we were expecting that the inclusion of quality costs would increase customer satisfaction. This is a fairly obvious conclusion. The quality costs we present here measure the importance of flight delays to the passengers and this is one of the most important quality items in this industry. If we decrease delays we are increasing passenger satisfaction.

Regarding hypothesis two we were expecting to increase the passenger satisfaction without increasing significantly (or nothing at all) the operational costs in a given period. From the results in table 3 we can see that operational costs increased 8% when comparing with the method used by [6]. If we read this number as is we have to say that our hypothesis is false. An 8% increased on operational costs can represent a lot of money. However, we should read this number together with the flight delay figure. As we can see, although method B increased the operational costs in 8% it was able to choose solutions that decrease, in average, 36% of the flight delays. This means that, when there are multiple solutions to the same problem, our method is able to choose the one with less operational cost, less quality costs (hence, better passenger satisfaction) and, because of the relation between quality costs and flight delays, the solution that produces less flight delays.

From this conclusion, one can argue that if we just include the operational costs and the expected flight delay, minimizing both values, the same results can be achieved having all passengers happy. In general, this assumption might be true. However, when we have to choose between two solutions with impact on other flights, which one should we choose? In our opinion, the answer depends on the profile of the passengers of each flight and on the importance they give to the delays, and not only in minimizing the flight delays. Our method takes into consideration this important information when taking decisions.

It is fair to say that we cannot conclude that our MAS will always have this behavior. For that we need to evaluate much more situations, in different times of the year (we might have seasoned behaviors) and, then, find an average value. From the results we can also obtain other interesting conclusions. As in the previous MAS [6] the cooperation between different operational bases has increased. The reason is the same, we evaluate all the solutions found (including the ones from different operational bases where the event happened) and we select the one with less cost. This cooperation is also possible to be inferred from the costs by base.

This paper has presented an improved version of the distributed multi-agent system in [6] as a possible solution to solve airline operations recovery problems, including sub-organizations with specialized agents, dedicated to solve crew, aircraft and passenger recovery problems that take into consideration the passenger satisfaction in the decision process. We have detailed the architecture of our MAS regarding the sub-organization dedicated to solve crew recovery problems, including agents, services and protocols. We have introduced a process of calculating the quality costs that, in our opinion, represent the importance that passengers give to flight delays. We show how, through a passenger survey, we build four types of passenger profiles and, for each one of these profiles, how we calculate a formula to represent that information. We have introduced an updated multi-criteria algorithm for selecting the solution with less cost from those proposed as part of the negotiation process, taking into consideration the quality costs. A case study, taken from a real scenario in an airline company where we tested our method was presented and we discuss the results obtained. We have shown that our method is able to choose solutions that contribute to a better passenger satisfaction and that produce less flight delays when compared with a method that only minimizes operational costs.

References

1. Abdelgahny, A., Ekollu, G., Narisimhan, R., Abdelgahny, K.: A Proactive Crew Recovery Decision Support Tool for Commercial Airlines during Irregular Operations. Annals of Operations Research 127, 309–331 (2004)
2. Barnhart, C., Belobaba, P., Odoni, A.: Applications of Operations Research in the Air Transport Industry. Transportation Science 37, 368–391 (2003)

3. Bellifemine, F., Caire, G., Trucco, T., Rimassa, G.: JADE Programmer's Guide. JADE 3.3 TILab S.p.A (2004)
4. Bratu, S., Barnhart, C.: Flight Operations Recovery: New Approaches Considering Passenger Recovery. Journal of Scheduling 9(3), 279–298 (2006)
5. Castro, A., Oliveira, E.: The rationale behind the development of an airline operations control centre using Gaia-based methodology. Int. J. Agent-Oriented Software Engineering 2(3), 350–377 (2008)
6. Castro, A.J.M., Oliveira, E.: Using Specialized Agents in a Distributed MAS to Solve Airline Operations Problems: a Case Study. In: Proceedings of IAT 2007 (Intelligent Agent Technology Conference), Silicon Valley, California, USA, pp. 473–476. IEEE Computer Society, Los Alamitos (2007)
7. Clausen, J., Larsen, A., Larsen, J.: Disruption Management in the Airline Industry – Concepts, Models and Methods. Technical Report, 2005-01, Informatics and Mathematical Modeling, Technical University of Denmark, DTU (2005)
8. Kohl, N., Larsen, A., Larsen, J., Ross, A., Tiourline, S.: Airline Disruption Management – Perspectives, Experiences and Outlook. Technical Report, CRTR-0407, Carmen Research (2004)
9. Kohl, N., Karish, S.: Airline Crew Rostering: Problem Types, Modeling and Optimization. Annals of Operations Research 127, 223–257 (2004)
10. Rosenberger, J., Johnson, E., Nemhauser, G.: Rerouting aircraft for airline recovery. Technical Report, TLI-LEC 01-04, Georgia Institute of Technology (2001)
11. Rosenberger, J., Schaefer, A., Goldsmans, D., Johnson, E., Kleywegt, A., Nemhauser, G.: A Stochastic Model of Airline Operations. Transportation Science 36(4), 357–377 (2002)
12. Wooldridge, M.: When is an Agent-Based Solution Appropriate? Introduction to Multiagent Systems, pp. 225–226. John Wiley & Sons, Ltd., West Sussex (2002)

Agent-Based Approach to the Dynamic Vehicle Routing Problem

Dariusz Barbucha and Piotr Jędrzejowicz

Dept. of Information Systems, Gdynia Maritime University, Morska 83, 81-225 Gdynia, Poland
{barbucha, pj}@am.gdynia.pl

Abstract. The term dynamic transportation problems refers to a wide range of problems where the required information is not given a priori to the decision maker but is revealed concurrently with the decision-making process. Among the most important problems belonging to this group are routing problems, which involve dynamic decision making with respect to vehicle routing in response to the flow of customer demands. The goal of such routing is to provide the required transportation with minimal service cost subject to various constraints. The paper proposes an approach to the dynamic vehicle routing problem based on multi-agent paradigm.

1 Introduction

One of the important group of transportation problems are vehicles routing problems (VRP), where a set of customers is to be served by the fleet of capacited vehicles in order to minimize the service cost and satisfying the set of given constraints. The VRP is said to be *static* if all its input data do not depend explicitly on time, otherwise it is *dynamic*.

The goal of the paper is to present a multi-agent approach to solving the dynamic vehicle routing problem (DVRP). Since DVRP can be viewed as a distributed problem it is proposed to search for solutions using intelligent software agents. To support search process a multi-agent platform simulating activities of the transportation company has been implemented.

The paper is organized as follows. Section 2 includes a formulation of the dynamic vehicle routing problem and introduces measures of the degree of its dynamism. Section 3 describes main features of the multi-agent approach proposed. Section 4 reports on the results of the computational experiment. Finally, Section 5 contains conclusions and suggestions for future research.

2 Formulation of the Dynamic Vehicle Routing Problem

The problem considered in the paper is modelled as an undirected graph $G = (V, E)$, where $V = \{0, 1, \ldots, N\}$ is the set of nodes and E is a set of edges. Node 0 is a central depot with NV identical vehicles of capacity W and each other node $i \in V \setminus \{0\}$ represents customer (with its request). Each customer (denoted as $cust(i)$) is characterized by coordinates in Euclidean space $(x(i), y(i))$ and a non-negative demand $d(i)$ $(i = 0 \ldots, N)$. Each link (i, j) between two customers denotes the shortest path from customer i to j and is described by the cost $c(i, j)$ of travel from i to j by shortest path $(i, j = 1 \ldots, N)$ and $t(i, j)$ $(i, j = 1 \ldots, N)$ - the travel time for each edge (i, j). It is assumed that $c(i, j) = c(j, i)$ and $t(i, j) = t(j, i)$.

Let $R = \{R(1), R(2), \ldots, R(NV)\}$ be a partition of V into NV routes of vehicles that cover all customers. Denote a length of the route $R(i)$ by $len(R(i))$ and a cost (or travel distance) by $cost(R(i))$, where $i = 1, \ldots, NV$.

The goal is to find vehicle routes which minimize the total cost of travel - $cost(R) = \sum_{i=1}^{NV} cost(R(i))$ and such that each route starts and ends at the depot, each customer is serviced exactly once by a single vehicle, and the total load on any vehicle associated with a given route does not exceed the vehicle capacity.

In the dynamic version of the problem defined above, it is also assumed that certain number of customers' requests are available in advance and the remaining requests arrive in sequence while the system is already running. Let us assume that the planning horizon starts at time 0 and ends at time T. Let $t(i) \in [0, T]$, where $i = 1, \ldots, N$ denotes the time when the $i - th$ customer request is submitted. Let N_s denotes the number of *static* (i.e. submitted in advance) requests available in $t(i) = 0$, where $i = 1, \ldots, N_s$ and N_d - the number of *dynamic* requests arriving within the $(0, T]$ interval. Of course $N_s + N_d = N$.

There are a few measures of degree of dynamism. In the model considered in the paper it has been used the one given by Lund et al. [8] and Larsen [7] who defined the *degree of dynamism (dod)* as a proportion of the number of dynamic requests to the number of all requests $(dod = N_d/N)$. It is easy to see that $dod \in [0, 1]$. According to the formula, the problem is more dynamic, if the above proportion is much closer to 1. If $dod = 0$, then the problem is static, and if $dod = 1$, the problem is fully dynamic.

During the recent years there have been many important advances in the field of static VRP. Because of the fact that this problem is computationally difficult, most of them are based on heuristics or metaheuristics [6]. Definitely much less works have been done with respect to solving dynamic VRP (see for example [4]).

3 Multi-agent Approach for DVRP

Among the methods for solving DVRP, the approaches based on intelligent software agents seems to be promising. In recent years only few approaches based on using intelligent agents for solving some transportation problems have been proposed. Some of them, rather simple, refer direct to the solving one of the problem from VRP class. Such approaches are presented for example in [10] where agent-based architecture is proposed for solving classical VRP, and in [5], where the authors consider Dynamic Pickup and Delivery Problem with Time Windows and propose an agent-based approach to solve it. Much complex multi-agent system developed to simulate planning and scheduling in a shipping company is presented for example in [2]. It solves the dynamic scheduling problem using a set of heterogeneous agents (drivers, trucks, trailers, containers) represented as holonic agents. Each holonic agent (or holon) consists of a set of subagents and co-ordinates and controls the activities of its subagents.

A multi-agent approach for solving DVRP presented in this paper is based on a specially designed multi-agent platform developed to simulate a transportation company activities. The platform is based on JABAT middleware, originally developed for solving difficult combinatorial optimization problems [1].

3.1 Main Features of the JABAT Middleware

JABAT is a middleware supporting design and development of the population-based applications intended to solve different computational problems. It produces solutions to combinatorial optimization problems using a set of intelligent optimising agents, each representing an improvement algorithm. The process of solving of a single task (i.e. a problem instance) in JABAT consists of several steps. At first an initial population of solutions is generated. Then, the individuals from the population are, at the subsequent computation stages, improved by independently acting optimization agents, thus increasing chances for reaching a global optimum. Finally, when the stopping criterion is met, the best solution in the population is taken as the result.

This functionality is realized mainly by two types of agents: *SolutionManagers* and *OptiAgents*. Each *SolutionManager* maintains a population of solutions and is responsible for finding the best solution of a single instance of the problem. *OptiAgents*, each representing a single optimizing algorithm, are used in process of finding/improving the solution of the problem. The agents of both types act in parallel and communicate with each other exchanging solutions that are either to be improved (when solutions are sent to *OptiAgent*) or stored back (when solutions are sent to *SolutionManager*).

Apart from *OptiAgents* and *SolutionManagers* there are also other agents working within the system which are responsible for initialising, organising the process of migrations between different platforms, writing down the results during the process of searching the best solution, and monitoring unexpected behaviors of the system.

More detailed description of JABAT environment were described in [1].

3.2 JABAT Implementation of the DVRP Simulator

The proposed implementation is based on the assumptions that population of solutions consist of a single individual, and only one optimization agent is responsible for improving a solution.

The structure of the *OptiAgent* designed for solving DVRP, viewed as a part of JABAT environment, is presented in Fig. 1. The specialized *OptiAgent*, called *DVRP-OptiAgent*, which reflects a transport company and typical elements of it, is itself a set of the following types of agents:

- *ACompany (AC)* - an agent which runs first and initializes all others agents.
- *ARequestGenerator (ARG)* - an agent which generates (or reads) new orders and sends them to the *ARequestManager* agent.
- *ARequestManager (ARM)* - an agent which manages the list of requests received from *ARG*. After receiving the new request, *ARM* announces it to each *AVehicle* agent and chooses the best from offers returned by *AVehicle* agents.
- *AVehicle (AV)* - an agent that represents a vehicle and is described by the capacity of the vehicle, actual route assigned to this vehicle, actual cost of the route, actual available space and the vehicle state. Periodically *AV* receives customer's request from the *ARM* one at a time, tries to assign it to the existing route in order to minimize the cost, and sends back its offer (i.e. calculated cost of insertion) to the *ARM*. If the offer turns out to be the best, the respective request is added to the actual route. Most

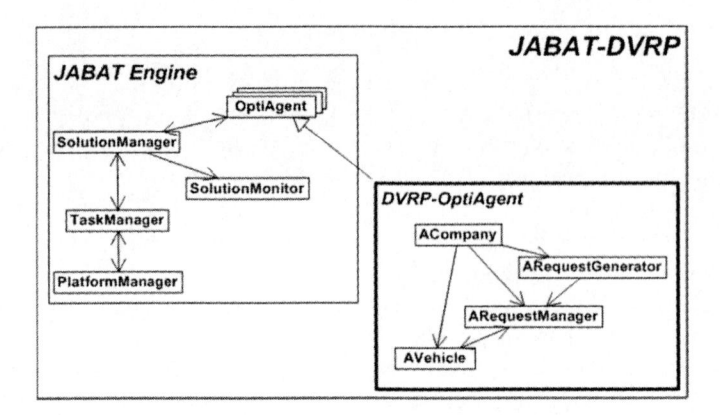

Fig. 1. Structure of the DVRP implementation based on JABAT

of its lifetime, a vehicle spends serving requests. It starts after receiving and accepting the first request. After reaching the nearest customer it goes to the next customer belonging to the route. In the model considered in the paper it is assumed that if the vehicle reaches the last customer on the current route, it waits in this location until a new arriving request is assigned to it or until the end of pool of requests is reached. In the first case the vehicle breaks waiting and moves to the new assigned customer, in the second case the waiting vehicle returns back to the depot.

Although the development of software agents representing elements of such company is not a new idea (similar approach is presented for example in [5], where three types of agents are proposed: agent-customer, agent-company and agent-vehicle), it seems to be very natural, taking into account a distributed nature of the problem.

The process of solving DVRP is divided into three phases:

1. Allocation of the pool of static requests to the available vehicles,
2. Allocation of the new dynamic requests to the fleet of vehicles,
3. Improvement of the current solution by *intra-route* and *inter-routes* operations.

The **first phase** is in fact the process of solving the static VRP for N_s requests. The solution obtained by procedure used in the system and described below gives the initial solution to the DVRP.

The initial solution is generated basing on polar representation of each vertex (customer with its request) in graph G, which uses the idea originated from *split* phase of the *sweep* algorithm of Gillett and Miller [3]. First, each vertex (customer) $i \in V$ is transformed from cartesian coordinates to polar coordinates $(\theta_i; \rho_i)$, where θ_i is the angle and ρ_i is the ray length.

Generation of individual (solution) starts from randomly choosing an arbitrary vertex i^* and assigning a value θ_{i^*} to it. Next, the remaining angles centered at 0 from the initial ray $(0; \rho_{i^*})$ are computed and the vertices (customers) are ranked in increasing order of their θ_i value. Resulting ranking determines the order in which requests are assigned to the available vehicles.

The process of assignment vertices to vehicles starts from the first unrouted vertex having the smallest angle, and next assigning vertices to the first vehicle as long as its capacity is not exceeded. If the capacity of the vehicle is exhausted, the vertices are assigned to the next vehicle. The whole process is repeated until the end of pool of static requests is reached.

Assignment of requests to vehicles is carried out in the form of messages exchange between *ARequestManager* and *AVehicle* agents.

After assigning all static requests to the available vehicles, all vehicles with the requests assigned to them start moving, and in loop the system is waiting for an event. Taking into account the objective of the problem, the most important event is a new request event.

The **second phase** includes assigning new dynamic requests to available vehicles and is realized in dynamically changing environment, where all vehicles are serving the customers already assigned to their routes.

The main steps of this phase are:

1. *ARG* reads (or generates) a new dynamic request and sends it to the *ARM*.
2. After receiving a new request from the *ARG*, *ARM* initializes a session using the *Contract Net Protocol (CNP)* [9] and starts communication between *ARM* and *AV* agents. As *Initiator* it announces the request to each *AV* agent (moving and waiting vehicles) sending around a call for proposal (cfp) message. *AV* (as *Participants* or *Contractors*) are viewed as potential contractors.
3. Each *AV* agent after receiving the request (with customer data) from the *ARM*, calculates the cost of inserting a new customer into the existing route. If an insertion of a new customer into the existing route does not violate the vehicle's capacity, the calculated cost of insertion is sent back (as propose message) to the *ARM*. Otherwise, the *AV* sends back the rejection (reject) message.
4. *ARM* after receiving proposals from all *AV* agents, chooses the one with the lowest cost of insertion. Next, it sends the accept-proposal message to the *AV* which is awarded and the reject-proposal to the others.
5. *AV* which receives the accept-proposal message, inserts the customer into the current route and sends the inform-done message if the operation is performed successfully and failure message, otherwise.
6. The above process is repeated for each new request.

Additionally, two kinds of improvement procedures are defined for *ARM* and *AV* agents and performed in the **third phase** of the process of solving DVRP. Each *AV* agent executes a set of operations that aim at improving the cost of its route (*intra-route* operations which operate on one selected route). In addition, the *ARM* agent also periodically performs global moves that aim at improving the global solution (*inter-routes* operations which operate on at least two selected routes).

Three *intra-route* operations include:

- *v1_2opt* - The implementation of *2-opt* algorithm where the sequence of customers visited by the vehicle on route $R(p)$ ($p \in \{1,\ldots,NV\}$) is changed by eliminating two edges and reconnecting the two resulting paths in a different way to obtain a new route.
- *v1_relocate* - The sequence of customers visited by the vehicle on route $R(p)$ is changed by moving one customer $cust(i) \in R(p)$, $i = 1,\ldots,len(R(p))$, $p \in \{1,\ldots,NV\}$ from its current position to another one.

- *v1_exchange* - Two selected customers from current route $R(p)$ are swapped, i.e. for each pair of customers $cust(i), cust(j) \in R(p)$ ($i \neq j$, $i, j = 1, \ldots, len(R(p))$, $p \in \{1, \ldots, NV\}$), $cust(i)$ moves to position occupied by $cust(j)$ and $cust(j)$ moves to the position occupied by $cust(i)$.

All possible moves are considered in above operations and moves that shorten the current route are accepted. The resulting route with the greatest reduction of the total cost is accepted as a new tour of the vehicle.

The above *intra-route* operations could be initialized by the *ARM* agent and next performed by *AV* agent or directly performed by *AV* agent. In the first case, *ARM* decides whether and when use the operation and sends the proper message to the particular *AV*. In the second case, *AV* autonomously decides about performing the operation.

Two *inter-routes* operations are proposed and implemented in the system:

- *v2_relocate* - One selected customer $cust(i)$ from one route $R(p)$ is moved to the second route $R(q)$ ($i = 1, \ldots, len(R(p))$, $p, q \in \{1, \ldots, NV\}$),
- *v2_exchange* - Two selected customers from two different routes ($cust(i) \in R(p)$ and $cust(j) \in R(q)$) are selected and swapped ($i = 1, \ldots, len(R(p))$, $j = 1, \ldots, len(R(q))$, $p, q \in \{1, \ldots, NV\}$).

As it is easy to see, during the process of assigning each new request to the available vehicles, each *AV* agent competes with others in order to get the request to servicing but during the execution improvement operations, agents cooperate in order to improve the current solution.

To sum up, the whole algorithm based on multiple agents for solving DVRP is presented in form of pseudocode as follows.

```
1. Allocate static requests
2. All vehicles with the assigned requests start moving
3. In loop system is waiting for an event
   IF (request_event)
     CASE "new request":
       allocate request to available vehicles
     CASE "end of requests":
       waiting vehicles return to the depot
   ELSE IF (vehicle_event)
     CASE "vehicle v(i) reached the location p":
       IF (NOT location p is the last location)
         vehicle v(i) proceeds to the next location
       ELSE
         IF (NOT end_of_requests)
           vehicle v(i) waits in location p
         ELSE
           vehicle v(i) is moving to the depot
   ELSE
     between events do intra-route and inter-routes operations
4. If all vehicles returned to the depot then STOP.
```

4 Computational Experiment

To validate effectiveness of the approach computational experiment has been carried out. For static cases, solutions (global cost of serving all request) obtained by the proposed approach were compared with the best known solutions using the mean relative error (MRE) from the best known solution. Additionally, for dynamic cases, the influence of the degree of dynamism of the problem and frequency of the customer requests arrivals on the solution have been observed.

The proposed agent-based approach was tested on classical VRP dataset transformed into its dynamic version, in which not all requests are known in advance. The experiment involved 5 benchmark instances (*vrnpc1* - *vrnpc5*) available from *OR-Library* benchmark set [11]. Each benchmark set includes information about number of customers, capacity of vehicles, coordinates of depot and coordinates and demands of customers. The selected problems contain 50-199 customers located randomly over the plane and have only capacity restriction.

In the experiment arrivals of the dynamic requests have been generated using the Poisson distribution with λ parameter denoting the mean number of requests occurring in the unit of time (1 hour in our experiment).

The proposed simulation model was run for the number of dynamic requests (N_d) varying from 0% (pure static problem) to 100% of all requests with step equal 20%. Additionally, for each positive value of the degree of dynamism, it has been assumed that dynamic requests may arrive with various frequencies. For the purpose of experiment λ was set to 3, 4, 5, 6, 10, 15, 20.

Additionally, it has been assumed that the vehicle speed was set at 60 km/h.

The above assumptions produced 36 test instances (1 static and 35 dynamic) for each dataset, giving in total 180 test instances. Moreover, each test problem was repeatedly executed five times and mean results from these runs were recorded.

The experiment results are presented in Tables 1-2.

Table 1 shows mean relative errors averages over all runs for each tested static instance of VRP from the OR-Library. Together with the problem names the header of the table includes the number of customers in the brackets. In rows of the table the average (Avg), minimum (Min) and maximum (Max) values of errors for each instance are shown.

Table 2 shows values of the percentage increase in cost of allocating all dynamic requests for selected instances of the problem as compared to the cost of the best known solution to the static instance. The first two columns of the table include degree of dynamism (in %) and mean number of requests per hour. The remaining five columns

Table 1. Mean relative error from the best known solution for selected instances of DVRP with all static requests

	vrpnc1 (50)	vrpnc2 (75)	vrpnc3 (100)	vrpnc4 (150)	vrpnc5 (199)
Avg	0,53%	3,83%	3,86%	5,07%	4,63%
Min	0,00%	2,25%	1,63%	2,82%	3,91%
Max	2,20%	7,91%	5,77%	8,09%	5,78%

Table 2. The performance of the proposed agent-based approach (measured as dynamic/best known static cost) for selected instances of the DVRP

Degree of dynamism	Mean number of requests per hour	vrpnc1 (50)	vrpnc2 (75)	vrpnc3 (100)	vrpnc4 (150)	vrpnc5 (199)
20%	3	34,8%	12,3%	17,8%	22,4%	14,3%
	4	36,4%	12,1%	17,8%	21,3%	14,7%
	5	14,1%	11,2%	10,7%	12,8%	14,8%
	6	15,3%	10,4%	10,6%	10,1%	9,2%
	10	12,7%	9,5%	9,3%	9,9%	8,8%
	15	5,1%	9,4%	7,9%	3,9%	8,6%
	20	2,2%	7,9%	4,5%	2,6%	7,7%
40%	3	45,2%	37,0%	52,4%	53,7%	39,2%
	4	52,9%	24,4%	30,8%	45,6%	35,8%
	5	53,5%	22,3%	29,8%	39,8%	27,2%
	6	48,8%	20,9%	27,4%	34,2%	20,6%
	10	10,9%	17,6%	23,7%	30,7%	18,7%
	15	3,2%	15,9%	16,7%	18,2%	15,8%
	20	2,8%	14,3%	10,6%	16,3%	8,4%
60%	3	80,6%	61,7%	58,6%	90,5%	78,1%
	4	41,7%	45,7%	54,7%	82,1%	56,2%
	5	55,9%	27,8%	40,8%	80,4%	45,8%
	6	29,1%	34,6%	38,9%	73,2%	44,8%
	10	36%	37,9%	46,8%	48,7%	38,7%
	15	47,5%	33,4%	34,1%	47,7%	38,7%
	20	18,5%	25,2%	23,9%	44,2%	30,6%
80%	3	105,6%	84,7%	71,8%	111,1%	117,3%
	4	83,7%	40,1%	59,3%	88,6%	103,1%
	5	60%	42,7%	54,3%	68,0%	78,4%
	6	65,3%	44,6%	50,1%	66,9%	70,6%
	10	24,4%	39,7%	46,6%	65,7%	63,6%
	15	41,3%	40,1%	42,8%	53,1%	62,3%
	20	13,4%	38,1%	33,5%	51,4%	43,6%
100%	3	101,6%	98,5%	80,9%	128,7%	121,7%
	4	77,4%	89,5%	80,2%	94,8%	113,2%
	5	82,4%	65,6%	77,5%	82,2%	99,8%
	6	73,6%	68,8%	77,1%	73,2%	96,2%
	10	38,5%	60,9%	72,4%	56,1%	75,3%
	15	23,9%	60,7%	46,4%	53,6%	75,2%
	20	22,4%	51,0%	49,8%	50,4%	56,5%

show calculated values obtained by proposed agent-based approach for each tested instance of the problem.

Results obtained during the experiment and presented in Table 1 show that the proposed agent-based approach produces quite good solutions in case of static requests only. The average value of MRE is not greater than 5%, but it depends on the instance and for most instances is smaller. Minimal value of MRE observed during the experiment is equal to 0% for *vrpnc1* instance or close to 2-3% for most of the instances, which is only slightly worse than the results produced by other methods (see, for example [6]).

By analyzing the results presented in Table 2 it is easy to see, that overall cost depends strongly on the degree of dynamism and frequency of dynamic request arrivals for all tested instances. In most cases the total cost may be substantially higher than for the static case.

It should be noted that comparisons of obtained results to dynamic cases can not be directly compared with the approaches proposed by other authors mentioned in this paper. This is mainly due to differences in problem formulation and differences in datasets used for evaluating the results. However, in case of static instances the results are fully comparable.

5 Conclusions

The paper proposes a multi-agent approach to solving the dynamic vehicle routing problem. The approach is based on a multi-agent platform which can be used to simulate activities of the transport company and analyze various scenarios with respect to the dynamic routing of the fleet of vehicles.

Computational experiment proved that presented approach can offer good quality solutions for the static case (as compared with state of the art approaches). It also shows how dynamic nature of the problem can influence the total cost of realizing customer requests. The overall evaluation of the presented approach is positive thanks to several features typical for multiple agent systems, like autonomy of agents, ability to increase computational efficiency through parallelization and possibility of using distributed environment.

References

1. Barbucha, D., Czarnowski, I., Jędrzejowicz, P., Ratajczak, E., Wierzbowska, I.: An Implementation of the JADE-based A-Team Environment. International Transactions on Systems Science and Applications 3(4), 319–328 (2008)
2. Burckert, H.J., Fischer, K., Vierke, G.: Holonic Transport Scheduling with TeleTruck. Journal of Applied Artificial Intelligence 14, 697–725 (2000)
3. Gillett, B.E., Miller, L.R.: A heuristic algorithm for the vehicle dispatch problem. Operations Research 22, 240–349 (1974)
4. van Hentenryck, P.: Online Stochastic Combinatorial Optimization. MIT Press, Cambridge (2006)
5. Kozlak, J., Creput, J.C., Hilaire, V., Koukam, A.: Multi-agent environment for dynamic transport planning and scheduling. In: Bubak, M., van Albada, G.D., Sloot, P.M.A., Dongarra, J. (eds.) ICCS 2004. LNCS, vol. 3038, pp. 638–645. Springer, Heidelberg (2004)
6. Laporte, G., Gendreau, M., Potvin, J., Semet, F.: Classical and modern heuristics for the vehicle routing problem. International Transactions in Operational Research 7, 285–300 (2000)
7. Larsen, A.: The on-line vehicle routing problem. Ph.D. Thesis, Institute of Mathematical Modelling, Technical University of Denmark (2001)

8. Lund, K., Madsen, O.B.G., Rygaard, J.M.: Vehicle routing problems with varying degrees of dynamism. Technical report, Institute of Mathematical Modelling, Technical University of Denmark (1996)
9. Smith, R.G.: The Contract Net Protocol: High Level Communication and Control in a Distributed Problem Solver. IEEE Transactions on Computers 29(12), 1104–1113 (1980)
10. Thangiah, S.R., Shmygelska, O., Mennell, W.: An agent architecture for vehicle routing problem. In: Proc. of the ACM Symposium on Applied Computing (SAC 2001), Las Vegas, pp. 517–521 (2001)
11. OR-Library, http://people.brunel.ac.uk/~mastjjb/jeb/orlib/vrpinfo.html

The Undirected Rural Postman Problem Solved by the MAX-MIN Ant System

María Luisa Pérez-Delgado

Departamento Informática y Automática
Universidad de Salamanca
Escuela Politécnica Superior de Zamora. AV. Requejo, 33, 49022, Zamora, Spain
`mlperez@usal.es`

Abstract. This work describes the application of the MAX-MIN Ant System algorithm to solve the Undirected Rural Postman Problem. The results obtained when we apply the proposed solution to a data set used by other authors demonstrate that this approach is very good. Moreover, the method only requires the graph formulation of the problem, so that no complex mathematical formulation of the same is required.

Keywords: Rural Postman Problem, NP-hard problems, Max-Min Ant System.

1 Introduction

Ants are social insects that exhibit collective behaviour in performing tasks that cannot be carried out by an individual ant. In ant colonies, a chemical substance called pheromone is used as a way to communicate important information on global behaviour. By exploiting pheromone information, ants are capable of finding the shortest path from a food source to their nest without using visual cues [1]. Ants looking for food lay the way back to their nest with a specific type of pheromone. When they decide about a direction to go in, they probably choose paths marked by strong pheromone concentrations, contributing to this by accumulating more pheromone in order to make such paths more desirable. Pheromone evaporates over time, making the paths selected by fewer ants less desirable. This behaviour is the basis for a cooperative interaction which leads to the emergence of the shortest paths.

The application of algorithms based on the behaviour of natural ants arose from the thesis of Marco Dorigo [2]. The first algorithm of this type, called Ant-System, was first applied to solve the Traveling Salesman Problem (TSP). Given a set of n points or cities joined by weighted connections, the aim of the TSP is to find a closed path of minimum cost passing through all the cities once and only once.

The way in which the next stop in a path is selected and the pheromone trail is updated, generates several ant-based algorithms. For an overview of all ant-based algorithms, [3] can be consulted.

Artificial ants have been successfully applied to several NP-hard combinatorial optimization problems; for an overview, [4] can be consulted.

The algorithm that we will use in this work is called the MAX-MIN Ant System (MMAS). Our objective is to apply this algorithm to solve the Undirected Rural Postman Problem (URPP). We will prove that the proposed algorithm can find good solutions to the RPP, by using a small amount of computational time and without the use of a special mathematical formulation of the problem. Computational results show that this method outperforms other heuristics applied to solve the same problem.

2 The Undirected Rural Postman Problem

Let $G = (V, E)$ be an undirected graph, where V represents the set of points in the graph, and E represents the set of connections. The elements of E are of the form $E=\{(i,j) / i, j \in V\}$ and they have a nonnegative cost associated with them, defined by the cost function c. Let $F \subseteq E$, $F \neq \emptyset$, be a set of required connections. This set induces a graph $G_F=(V_F, F)$, formed by the connections of F and the end-points of such connections, therefore, $V_F \subseteq V$. The aim of the URPP is to find a closed path of minimum cost in G which contains each connection in F at least once, [5]. The problem is NP-hard, since the TSP can easily be transformed to it [6].

The problem appears in a variety of practical situations such as: the inspection of electrical lines; the reading of meters; waste collection; street sweeping; mail delivery; school bus routing; snow plowing; or the optimization of the movements of a plotter, [7].

There are several exact methods for solving the URPP ([8]-[11]). Because the problem is NP-hard, several approximate solutions have also been applied to it, including heuristic methods ([12]-[20]); and metaheuristics ([21]-[26]).

3 The MAX-MIN Ant System

The MMAS algorithm was proposed by Stützle and Hoos, [27]. It is an improvement over the first ant-based algorithm, called Ant-System, and it is one of the best performing ant-based algorithms. It was first applied to solve the TSP. This algorithm utilizes the graph representation of the problem to be solved.

Let $G = (V, E)$ be the graph associated with the TSP, where V is the set of n cities or nodes in the problem, E is the set of connections among the nodes, and d is a function that associates a cost to each element in E. To solve the problem, we consider a set of m ants cooperating to find a solution to the TSP (a tour). To each connection, (i, j), of the TSP graph, a value τ_{ij} (called pheromone) is associated with $\tau_{ij} \in (0, 1]$. The pheromone allows ants to communicate among themselves, contributing in this way to the solution of the problem. At each iteration of the algorithm, each one of the m ants looks for a solution to the problem (a tour). When all the ants in the set have constructed their solution, the pheromone trail associated with the graph is updated, thus contributing to make the connections pertaining to the best solution more desirable for the ants in the next iteration. The process is repeated until the solution converges or until completing the maximum number of iterations allowed for the algorithm to be performed.

To allow each ant to build a valid solution to the problem, visiting each city once and only once, each ant has an associated data structure called the tabu list, which stores the cities that have already been visited by the ant. When the ant begins the search for a new solution, its tabu list is empty. Each time an ant visits a city, it is added to its tabu list. When it has completed the path, all the cities will be in such a list.

Each ant generates a complete tour starting at a randomly selected city and choosing the next city of its path as a function of the probabilistic state transition rule (1), which defines the probability with which ant k chooses to move from city i to city j at iteration t.

$$p_{ij}^k(t) = \frac{[\tau_{ij}(t)]^\alpha [\eta_{ij}(t)]^\beta}{\sum_{l \in N_i^k} [\tau_{lj}(t)]^\alpha [\eta_{lj}(t)]^\beta} \quad \text{if } j \in N_i^k \qquad (1)$$

where $\tau_{ij}(t)$ is the pheromone associated to the connection (i, j) at time t, $\eta_{ij}(t)$ is called visibility of the connection (i, j). N_i^k is the feasible neighborhood for ant k, whereas the parameters α and β determine the relative influence of the pheromone and the visibility, respectively. For the TSP, the visibility of a connection is a fixed value equal to the inverse of the distance associated to such a connection. The feasible neighborhood for ant k, now placed on city i, N_i^k, is the set of cities not yet visited by ant k and accessible from city i.

If $j \notin N_i^k$ we have $p_{ij}^k(t) = 0$.

The state transition rule (1) indicates that ants prefer to move to cities near the present one and connected to it with arcs or edges having a high amount of pheromone.

To update the pheromone, the expression (2) is applied to all the connections in the graph.

$$\tau_{ij}(t+1) = (1-\rho)\tau_{ij}(t) + \Delta\tau_{ij}^{best} \qquad (2)$$

where ρ is a parameter called evaporation rate of the pheromone, $0 < \rho \leq 1$, which determines the fraction of pheromone eliminated from each connection. It represents the effect of pheromone evaporation over time observed in natural ants. $\Delta\tau_{ij}^{best}$ is the amount of pheromone deposited on the connections belonging to the path of the best ant, whose value is given by expression (3).

$$\Delta\tau_{ij}^{best} = \begin{cases} \dfrac{1}{L_{best}} & \text{if } (i,j) \text{ is part of the tour of the best ant} \\ 0 & \text{otherwise} \end{cases} \qquad (3)$$

with L_{best} being the length of the tour of the best ant.

Therefore, first a fraction of the pheromone associated with each connection is evaporated, which avoids an unlimited buildup of it; moreover, this represents the phenomenon observed in natural ant colonies. Next, the best ant deposits an amount of pheromone on the connections of its tour, proportional to the length of that tour.

To select the best ant we can take the iteration-best ant (the one that generated the best solution at the present iteration) or the global-best ant (the one that generated the best solution so far).

To avoid search stagnation, the MMAS algorithm limits the pheromone trails to the interval [τ_{min}, τ_{max}], being $\tau_{min} > 0$. Both values must be determined for each particular problem, the first one being more critical. In [28] it is shown why these values can be calculated in a heuristic way.

Before starting the search for a solution, the pheromone of all connections is set to the value τ_{max}, which permits a greater exploration of the search space at the beginning of the algorithm. Moreover, when we apply the update rule for the trial, the pheromone remains on the connections of the better solutions with high values, and it is reduced on the bad ones. When pheromone is updated, the values are forced to the interval indicated, so that each pheromone trail greater than τ_{max} is set to the maximum, and those lower than τ_{min} equal the minimum.

While iterations proceed, if search stagnation is detected, a re-initialization is applied, setting all the pheromone trails to τ_{max} again. Stagnation occurs when all the ants follow the same path and construct the same tour.

The algorithm is usually combined with some improvement heuristic, such as 2-OPT, 3-OPT, which usually improves the results. This technique is commonly used in all ant-based algorithms, as in other metaheuristics.

4 The Proposed Solution

To solve the RPP problem using artificial ants, we first transform the RPP graph into a TSP graph. The resulting problem is solved by the MMAS algorithm, and finally the TSP solution is transformed into a RPP solution. The first and the last operations were described in [24].

Table 1. Pseudo-code of the algorithm applied to solve the RPP

```
PROCEDURE RPPMMAS
    Transform the RPP graph into a TSP graph
    Compute the initial pheromone matrix for the TSP graph
    Set Lg_best = ∞, iterate = TRUE, i = 0
    While iterate= TRUE
        Set i = i+1
        For h = 1 to m
            Set tabu_h = ∅
            Select a city, c₀, randomly as the starting point of the path
            Add c₀ to tabu_h
            For j = 1 to n−1
                Select the next city, c, according to equation (1)
                Add c to tabu_h
            end-for
            Improve the solution by applying 2-OPT exchange
            Compute the length of the path, L(h)
        end-for
        Select the best solution among the m calculated, Lb
        If Lg_best > Lb
            Set Lg_best = Lb
        end-if
        If Lg_best has not been improved during the last 15 iterations
            Set iterate= FALSE
        end-if
        If Lx = Ly ∀x, y / x ≠y, 1≤x, y ≤ m
            Reset the pheromone trails to the value τmax
        else
            Update the pheromone matrix according to expression (2)
        end-if
    end-while
    Transform the TSP solution into a RPP solution
END
```

All the steps in the solution process are outlined in table 1.

If the global-best solution does not improve during 15 consecutive iterations of the algorithm, we stop the search, taking this solution to be the best one.

If all the m solutions found by the ants at some iteration have the same length, we apply a re-initialization of the pheromone trail, setting the pheromone matrix to the maximum value, τ_{max}. The search for a solution will proceed, considering the global-best solution found until this moment, but discarding the values calculated for the pheromone trail.

5 Computational Results

The algorithm has been coded in C language. The tests have been performed on a personal computer with Intel Centrino Core 2 Duo processor, 2.2 GHz, with 2G RAM memory and working on Linux Operating System.

Table 2. Results considering $\alpha=1$, $\beta=2$, m=10, $\tau_{min}=0.01$, $\tau_{max}=1$, $\rho=\{0.01, 0.02\}$

	$\rho=0.01$				$\rho=0.02$			
	MIN	AV	SD	T	MIN	AV	SD	T
p01	76	76.00	0.00	0.05	76	76.00	0.00	0.00
p02	152	152.00	0.00	0.05	152	152.00	0.00	0.05
p03	102	102.30	0.92	0.55	102	102.15	0.67	0.55
p04	86	86.00	0.00	0.25	86	86.00	0.00	0.25
p05	124	124.00	0.00	0.05	124	124.00	0.00	0.10
p06	102	102.20	0.62	0.25	102	102.00	0.00	0.20
p07	130	130.65	1.18	0.35	130	130.15	0.67	0.30
p08	122	122.20	0.70	0.35	122	122.00	0.00	0.35
p09	83	83.00	0.00	0.05	83	83.00	0.00	0.05
p10	84	84.00	0.00	0.00	84	84.00	0.00	0.05
p11	23	23.00	0.00	0.05	23	23.00	0.00	0.00
p12	19	19.00	0.00	0.00	19	19.00	0.00	0.00
p13	35	35.00	0.00	0.00	35	35.00	0.00	0.00
p14	202	203.95	1.93	0.70	202	203.85	1.81	0.70
p15	441	441.00	0.00	0.15	441	441.35	1.57	0.15
p16	203	203.00	0.00	0.95	203	203.10	0.45	1.00
p17	112	112.00	0.00	0.10	112	112.00	0.00	0.10
p18	146	146.30	0.92	0.10	146	146.85	2.56	0.15
p19	257	259.50	3.56	0.65	257	259.25	3.61	0.80
p20	398	399.70	1.87	6.20	398	400.60	2.16	5.80
p21	366	370.90	3.16	8.50	366	370.95	5.20	7.45
p22	621	622.90	3.16	10.90	621	624.40	3.89	10.45
p23	475	478.35	3.12	11.05	475	477.35	2.03	11.30
p24	405	408.35	4.56	4.20	405	406.50	1.96	4.10
albaidaa	10943	11278.85	163.13	29.85	10947	11235.50	133.32	25.75
albaidab	8629	8855.55	156.75	18.25	8629	8867.05	120.52	16.90

Table 3. Results considering $\alpha=1$, $\beta=2$, m=10, $\tau_{min}=0.01$, $\tau_{max}=1$, $\rho=\{0.1, 0.2\}$

	$\rho=0.1$				$\rho=0.2$			
	MIN	AV	SD	T	MIN	AV	SD	T
p01	76	76.00	0.00	0.00	76	76.00	0.00	0.00
p02	152	152.00	0.00	0.05	152	152.00	0.00	0.10
p03	102	102.00	0.00	0.60	102	102.00	0.00	0.60
p04	86	86.00	0.00	0.25	84	85.44	0.87	0.20
p05	124	124.00	0.00	0.10	124	124.00	0.00	0.10
p06	102	102.10	0.45	0.20	102	102.00	0.00	0.20
p07	130	130.30	0.92	0.35	130	130.30	0.92	0.40
p08	122	122.10	0.31	0.50	122	122.00	0.00	0.40
p09	83	83.00	0.00	0.05	83	83.00	0.00	0.05
p10	84	84.00	0.00	0.00	84	84.00	0.00	0.05
p11	23	23.00	0.00	0.00	23	23.00	0.00	0.00
p12	19	19.00	0.00	0.00	19	19.00	0.00	0.00
p13	35	35.00	0.00	0.00	35	35.00	0.00	0.00
p14	202	203.70	1.84	0.75	202	203.10	1.80	0.95
p15	441	441.35	1.57	0.15	441	441.35	1.57	0.15
p16	203	203.00	0.00	0.90	203	203.00	0.00	1.05
p17	112	112.00	0.00	0.10	112	112.00	0.00	0.15
p18	146	146.15	0.67	0.10	146	146.00	0.00	0.10
p19	257	259.15	2.58	0.70	257	257.80	1.51	0.90
p20	398	400.40	2.30	6.45	398	399.10	1.52	6.95
p21	366	370.50	4.11	8.85	366	372.45	6.44	7.90
p22	621	625.15	4.80	11.20	621	623.30	3.45	10.25
p23	475	478.00	2.34	16.85	475	478.60	3.44	11.05
p24	405	408.65	4.17	4.35	405	406.05	2.19	5.00
albaidaa	10947	11192.30	135.65	32.95	10732	11165.85	161.78	32.90
albaidab	8657	8907.00	142.73	18.15	8629	8811.90	110.31	23.40

To check the proposed solution, we consider the set of problems proposed in [8] and [9].

In the tests performed, we always considered $\alpha=1$, $\beta=2$, m=10, $\tau_{min}=0.01$, $\tau_{max}=1$, as proposed in [4]. We considered several values for the evaporation rate: {0.01, 0.02, 0.1, 0.2}, to observe which of these values generated better results. We always considered the iteration-best ant to update the pheromone trail. Tables 2 and 3 summarize the results obtained. The first column identifies every problem to be solved. For each value of ρ, the table shows the best solution reached (MIN), the average solution (AV), the standard deviation (SD), and the average time in seconds to reach a solution (T).

We observe that 24 out of the 26 problems are solved to optimality. Problems p10 and albaidaA are never solved to optimality. The best solution for problem albaidaA is reached when $\rho=0.2$.

Table 4 compares the best solution reached by applying MMAS, with the solutions reported by several authors who have applied approximate methods. The first column shows the name of the problem, the second one shows the optimum for each problem, and in the remaining columns the percentage over the optimum of the best solution reached by several authors is shown: in S1 a shortest spanning tree is computed, [8]; in S2 Monte Carlo methods are used, [15]; S3 shows the results obtained by Frederickson's heuristic, a heuristic similar to the one proposed by Christofides to solve the TSP, [16]; S4 and S5 combine Frederickson's heuristic with 2-OPT and 3-OPT, respectively, [18]; S6 and S7 show the results obtained by applying two local search heuristics to improve the solution obtained by Frederickson, [19]; S8 combines the GRASP metaheuristic and genetic algorithms, [21]; S9 combines the GRASP metaheuristic, genetic algorithms and memetic algorithms, [22]. The column labeled ACS shows the results we obtained by applying Ant Colony System, [24]. The last column shows the best result obtained when MMAS was applied. We must remark that cells without a value in the table correspond to data not reported by the authors.

Regarding problem p18, there are some inconsistencies in its reported optimal value. In [15] an optimal cost solution is reported with a value of 148; in [19] it is reported that the optimum is 147; whereas in [18] a feasible solution is found with 146. We will initially consider 148 to be the optimum value for the problem because it was reached by exact methods.

Table 4. Percentage over the optimum for the set of test problems by applying several approximate solutions optimum – (calculated as 100(best-OPT)/OPT)

	OPT	S1	S2	S3	S4	S5	S6	S7	S8	S9	ACS	MMAS
p01	76	0	0	0	0	0	0	0	0	0	0	0
p02	152	7.90	7.24	1.97	0.66	0	0	0	7.24	7.24	0	0
p03	102	0	0	2.94	0.98	0.98	0	0	0	0	0	0
p04	84	0	2.38	0	0	0	0	0	0	0	0	0
p05	124	8.87	4.03	4.84	0	0	0	0	4.03	4.03	0	0
p06	102	4.90	0	4.90	4.90	0	0	0	0	0	0	0
p07	130	0	0	0	0	0	0	0	0	0	0	0
p08	122	0	0	0	0	0	0	0	0	0	0	0
p09	83	1.21	0	0	0	0	0	0	0	0	0	0
p10	80	0	5.00	0	0	0	0	0	0	0	0	5.00
p11	23	0	0	13.04	0	0	0	0	0	0	0	0
p12	19	15.79	10.53	15.79	0	0	0	0	10.53	10.53	0	0
p13	35	8.57	8.57	0	0	0	0	0	8.57	8.57	0	0
p14	202	4.95	3.47	2.48	0.99	0	0	0	3.47	3.47	0	0
p15	441	0.91	0.91	0.91	0	0	0	0	0.91	0.91	0	0
p16	203	0	0	5.91	0.99	0	0	0	0	0	0	0
p17	112	3.57	0	3.57	0	0	0	0	0	0	0	0
p18	146	1.37	1.37	0.69	-	-	0.69	0.69	1.37	1.37	0	0
p19	257	8.95	2.34	6.62	5.45	3.50	0	0	2.34	2.34	1.56	0
p20	398	0.50	0.25	1.01	0.50	0.50	0.50	0	0	0	6.53	0
p21	366	1.64	0.55	1.64	1.64	1.64	0	0	1.64	0	7.92	0
p22	621	1.77	0	1.93	0.16	0.16	0.16	0	2.42	0	5.96	0
p23	475	1.05	2.95	0.84	0.42	0.42	0	0	2.53	1.05	7.58	0
p24	405	1.48	0	1.48	0	0	0	0	0	0	2.47	0
albaidaa	10599	-	1.75	0	0	0	0	0	3.74	0.12	-	1.25
albaidab	8629	-	1.07	0	0	0	0	0	2.94	0	-	0

We can observe that with the MMAS algorithm the optimum is reached for 24 out of the 26 test problems. This is a better result than the one obtained by other methods reported in the table and was only improved by one of the solutions given by Hertz. The percentage over the optimum of the best solution reached for problems p10 and albaidaA, not solved to optimality, is: 5 and 1.25, respectively. For the problem labeled p18, the best solution obtained by MMAS is 146, the smallest of the three values reported in several works.

When we compare the two ant-based algorithms applied to this problem (MMAS and ACS), we can see that the results are significantly better when MMAS is applied. Nevertheless, we observe that problem labeled as p10 is solved by ACS, but not by MMAS.

6 Conclusions

We have proven that MMAS can be applied to solve the URPP. Computational results show that this algorithm can obtain a good solution (the optimal in many cases) with low computational cost. Moreover, this technique outperforms many existing heuristic solutions.

Our method does not need a linear programming formulation of the problem, but uses its graph representation. To apply artificial ants, we transform one NP-hard problem into another one. Although this transformation increases the number of nodes of the problem, it is possible to solve it in a reasonable amount of time.

Although we have tested the method on undirected problems, we must remark that the proposed solution can also be applied to directed problems.

In the future, we will work on improvements of the basic algorithm that allow us to reach better results, such us the use of candidate lists, or the application of other exchange heuristics (3-OPT and Lin-Kernighan). Another interesting line of future work is the application of artificial neural networks to define good values for the parameters involved in the algorithm.

Acknowledgements. We thank Angel Corberán for sending us the data set used in his experiments, which we have used in this work to compare the proposed solution with other well-known solutions.

This work has been partially financed by the Samuel Solórzano Barruso Memorial Foundation, of the University of Salamanca.

References

1. Deneubourg, J.L., Aron, S., Goss, S., Pasteels, J.M.: The self-organizing exploratory pattern of the argentine ant. J. Insect Behav. 3, 159–168 (1990)
2. Dorigo, M.: Optimization, learning and natural algorithms. Ph.D. thesis, Dip. Elettronica, Politecnico di Milano, Italy (1992)
3. Dorigo, M., Blum, C.: Ant colony optimization: a survey. Theorical Computer Science 344, 243–278 (2005)
4. Dorigo, M., Stützle, T.: Ant Colony Optimization. MIT Press, Cambridge (2004)
5. Orloff, C.S.: A fundamental problem in vehicle routing. Networks 4, 35–64 (1974)
6. Lenstra, J.K., Rinnooy-Kan, A.H.G.: On general routing problems. Networks 6(3), 273–280 (1976)
7. Eiselt, H.A., Gendreau, M., Laporte, G.: Arc routing problems, part II: The rural postman problem. Oper. Res. 43(3), 399–414 (1995)

8. Christofides, N., Campos, V., Corberán, A., Mota, E.: An algorithm for the rural postman problem. Tech. Rep. IC-O.R.-81-5, Imperial College, London, UK (1981)
9. Corberán, A., Sanchís, J.M.: A polyhedral approach to the rural postman problem. Eur. J. Oper. Res. 79, 95–114 (1994)
10. Letchford, A.N.: Polyhedral results for some constrained arc routing problems. Ph.D. thesis, Lancaster University, Lancaster (1996)
11. Ghiani, G.: A branch-and-cut algorithm for the undirected rural postman problem. Math. Programming 87(3), 467–481 (2000)
12. Chistofides, N., Campos, V., Corberán, A., Mota, E.: An algorithm for the rural postman problem on a directed graph. Math. Programming Stud. 26, 155–166 (1986)
13. Chistofides, N., Mingozzi, A., Toth, P.: Exact algorithms for the vehicle routing problem based on spanning tree and shortest path relaxations. Math. Programming 20(1), 255–282 (1986)
14. Fernández, E., Meza, O., Garfinkel, R., Ortega, M.: On the undirected rural postman problem: Tight bounds based on a new formulation. Oper. Res. 51(2), 281–291 (2003)
15. Fernández, P., García, L.M., Sanchis, J.M.: A heuristic algorithm based on Monte Carlo methods for the rural postman problem. Comput. Oper. Res. 25(12), 1097–1106 (1998)
16. Frederickson, G.N.: Approximation algorithms for some postman problems. J. ACM 26(3), 538–554 (1979)
17. Ghiani, G., Laganà, D., Musmanno, R.: A constructive heuristic for the undirected rural postman problem. Comput. Oper. Res. 33(12), 3450–3457 (2006)
18. Groves, G.W., van Vuure, J.H.: Efficient heuristics for the rural postman problem. Orion 21(1), 33–51 (2005)
19. Hertz, A., Laporte, G., Nanchen-Hugo, P.: Improvement procedures for the undirected rural postman problem. INFORMS J. Comput. 11(1), 53–62 (1999)
20. Pearn, W.L., Wu, T.C.: Algorithms for the rural postman problem. Comput. Oper. Res. 22(8), 819–828 (1995)
21. Baldoquín, M.G.: Heuristics and metaheuristics approaches used to solve the rural postman problem: a comparative case study. In: Proc. Fourth Internat. ICSC Symposium on Engineering of Intelligent Systems (EIS 2004), Maderia, Portugal (2004)
22. Baldoquín, M.G., Ryan, G., Rodriguez, R., Castellini, A.: Un enfoque híbrido basado en metaheurísticas para el problema del cartero rural. In: Proc. of XI CLAIO, Concepci'on de Chile, Chile (2002)
23. Kang, M.J., Han, C.G.: Solving the rural postman problem using a genetic algorithm with a graph transformation. Tech. rep., Dept. of Computer Engineering, Kyung Hee University (1998)
24. Pérez-Delgado, M.L.: A solution to the rural postman problem based on artificial ant colonies. In: Borrajo, D., Castillo, L., Corchado, J.M. (eds.) CAEPIA 2007. LNCS (LNAI), vol. 4788, pp. 220–228. Springer, Heidelberg (2007)
25. Pérez-Delgado, M.L., Matos-Franco, J.C.: Self-organizing feature maps to solve the undirected rural postman problem. In: Moreno Díaz, R., Pichler, F., Quesada Arencibia, A. (eds.) EUROCAST 2007. LNCS, vol. 4739, pp. 804–811. Springer, Heidelberg (2007)
26. Rodrigues, A.M., Ferreira, J.S.: Solving the rural postman problem by memetic algorithms. In: MIC 2001 - 4TH Metaheuristics Internat. Conf., Porto, Portugal (2001)
27. Stützle, T., Hoos, H.: The MAX-MIN Ant System and local search for the traveling salesman problem. In: Bäck, T., Michalewicz, Z., Yao, X. (eds.) Proc. IEEE Internat. Conf. on Evolutionary Computation, pp. 309–314 (1997)
28. Stützle, T., Dorigo, M.: A short convergence proof for a class of ant colony optimization algorithms. IEEE Trans. Evol. Comput. 6(4), 358–365 (2002)

Performance Visualization of a Transport Multi-agent Application

Hussein Joumaa, Yves Demazeau, and Jean-Marc Vincent

Laboratoire d'Informatique de Grenoble (LIG)
Maison Jean Kuntzmann, 110 avenue de la Chimie, Domaine Universitaire de Saint-Martin
d'Hères, 38041 Grenoble CEDEX9, France
Hussein.Joumaa@imag.fr, Yves.Demazeau@imag.fr,
Jean-Marc.Vincent@imag.fr

Abstract. Performance visualization is one of the performance evaluation techniques that can be used to perform a global analysis of a system's behaviour, from its internal point of view. In this work, we describe how a visualization tool, initially designed for classical distributed / parallel systems, has been adapted to visualize the internal behaviour of a multi-agent system. We present such an adaptation though a multi-agent application that can be considered as a typical example for analysis and performance study.

Keywords: Multi-agent systems, performance, evaluation, visualization, application.

1 Introduction

The analysis and understanding of computer systems behaviour is an important part in the development cycle. Performance visualization is one of the techniques that can be used to perform a global analysis of a system's behaviour. This technique supports performance debugging. It is essential to identify where and when the resources are consumed during running time. It provides the programmer performance indices and the internal behaviour of the entities in the system. Due to the recent interest in using multi-agent systems in a wide variety of application domains, the analysis and understanding of the internal complex characteristics (interaction, distribution of intelligence, cooperation, coordination, decentralization,...[1]) of multi-agent systems has become a priority. Usually, the behaviour of a multi-agent system is studied from a global point of view. The multi-agent system is viewed as a black box where entities interact one with the other in order to reach a global goal. Unfortunately, this approach does not provide sufficient information for a complete study of the multi-agent system behaviour. A study at the level of the entities composing the multi-agent system is necessary. The performance visualization of a multi-agent system application can be an approach to deal with this problem. In this article, we demonstrate the adaptation of a visualization tool, initially designed to visualize distributed / parallel systems performance, to visualize the internal behaviour of a multi-agent system application.

The classical application on which this adaptation is shown consists in the simulation of societies of robots collecting pieces of ore [3]. This application was selected as being an ideal candidate for analysis and performance study due to their characteristics, and the large variety of parameters that can be changed in order to study the different phenomena in multi-agent systems. The main scenario takes place at a particular planet, represented as a 2-dimensional torus grid world, which is known to have some amount of ore spread over the surface. In order to collect this ore, one or more spaceships land on the planet. A spaceship acts as a base for a number of exploration robots and transport robots. The explorers look for ore, which is moved back to the base by the transporters. The task must be performed under a number of constraints including vision range, communication scope and limited energy of the robots. Multiple bases on the same planet can be either competing or cooperative. The robots can perceive the environment including the other robots within their vision range. The communication between robots only happens within their respective communication scopes. Robots consume energy while acting. Their energy is limited and they need to recharge when running low on power to prevent them from dying. Robots are able to store a limited amount of coordinates internally. In each time step a given robot can perform one action at most, such as one message sending, a body movement, a perception or an ore sample picking. Each of these actions is associated with an energy cost. The process ends when all bases have filled their storage capacity with ore samples or when a given number of time steps has been reached.

The article is divided into 5 parts: The first part deals with visualization tools that are currently used to visualize the behaviour of parallel/distributed systems and how these tools may be adapted to multi-agent systems. We then describe the visualization tool that has been selected and adapted to visualize the behaviour of VOWELS multi-agent applications. The next two parts establish the links between multi-agent systems notions and the performance visualisation model. The fifth part presents some visualization patterns, focusing on interaction and resources management patterns. The final part discusses the current state of the art of the work and presents some perspectives.

2 Visualization Tools

A classical visualization process includes several phases: the data collection, data analysis and presentation of the performance indices to the programmer, usually by a visualization tools. We are interested in this latter phase. Different trace-based visualization tools for performance visualization of parallel applications have been developed: Paragraph [4, 5], Pablo [6, 7] and SvPablo [8]. These tools allow monitoring the execution of processors in large-scale PC clusters and offer a graphic representation of the behaviour of parallel applications. These tools propose a very large number of possible graphical representations. From the point of view of adapting these tools for multi-agent systems, the main problems with these visualization tools are the following:

- The ability to represent the execution of systems during long periods for large-sized systems;
- The possibility to add new functionalities or to extend existing ones without having to change the rest of the tool;
- The problem of displaying the entities that can be created and destroyed dynamically during the execution of the system.

The visualization tool Paje [9] has been created to face these problems. Paje provides an interactive interface - the space-time interface - and offers scalability in order to adapt to a large number of data when the number of nodes increases. This features, as well as its potential of Paje to adapt to multi-agent systems, have motivated us to choose it among other alternatives.

2.1 Presentation of Paje

A visualization constructed by Paje is composed of objects organized according to a tree type hierarchy. This organisation reflects the hierarchy of the programming model of the system to be visualized. Elementary objects, the leaves of the tree, are called *entities* while intermediate nodes of the tree are named *containers*. Entities are the objects that can be visualized in Paje's space-time diagram, while containers are the objects of the high level in the hierarchy, whose role is to organize the space of visualization and to gather the entities and other containers. There are four types of Paje entities, associated with their visualization in the space-time diagram:

- The *event* entity, to represent events that happen at a certain instant;
- The *state* entity - usually displayed as rectangles - to figure that a given container was in a certain state during a certain period of time;
- The *link* entity - displayed as an arrow - to visualize a relation between two containers that started at a certain instant and finished at a possibly different instant;
- The *variable* entity, used to represent the evolution in time of a certain value associated to a container - displayed as graph -.

Like every visualisation tool, the visualisation in Paje is based on a trace file. A trace file is composed of trace events. The trace events represent the definition of containers types, the definition of entities types, the start and the end of the visualized entities. A trace event can be seen as a table composed of named fields. Each trace event has designated fields, and each is composed of a name, a type, and a value. The visualization constructed by Paje results in a set of objects organized according to a tree type hierarchy containing containers and entities.

3 Agents Modelling

In this section, we specify the Paje objects in case of a multi-agent application. The discussion that follows is based on the four fundamental aspects [11] that characterize the multi-agent system: Agents, Environment, Interaction, and Organization.

The classic definitions [10, 11, 12, 13] of an agent consider it as a physical or virtual entity having a given mission. This mission is accomplished without the direct intervention of other entities but while cooperating with them. The behaviour state of the agent results from its observations, its knowledge, and its interaction with other agents. The distinction between the types of agents is based on the type of the attributed mission, so that for each type of agents a set of behaviour states can be characterized. In accordance to this analysis, the objects that can be considered as Paje entities, at the agent level, are the following:

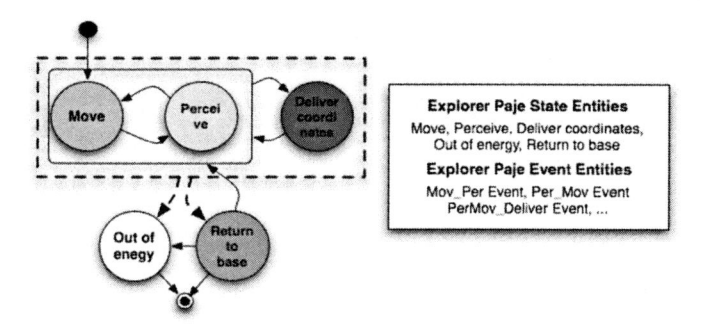

Fig. 1. State diagram and Paje entities associated to an explorer agent

- The behaviour state of an agent that can be visualised as a state;
- A transition between behaviour states can be considered as an event;
- A variable that considered as fundamental in the lifecycle of an agent can be visualized as a variable.

Regarding the specification of the robots simulation application [3], three types of agents can be modelled, according to their mission: the *explorer*, the *transporter*, and the *base*.

3.1 Explorer

The goal of an explorer agent is to identify ore. It has capabilities to perceive the environment in order to locate ore and to provide their coordinates to the agents that are able to capture these coordinates. The initial state is the *explore* state from which such agents move around and make perceptions at given intervals, according to the size of their perception scope, to locate ore. Whenever an explorer decides to inform other agents about its findings it moves into the *deliver ore coordinates* state in which it tries to communicate with others given its communication scope. At every time, the explorer can think that it needs to recharge, and turns to the *return to base* state. The figure 1 presents the state diagram of an explorer agent and the entities that are associated to

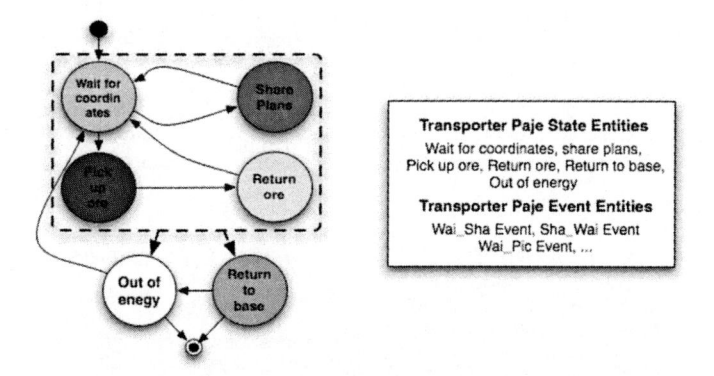

Fig. 2. State diagram and Paje entities associated to a transporter agent

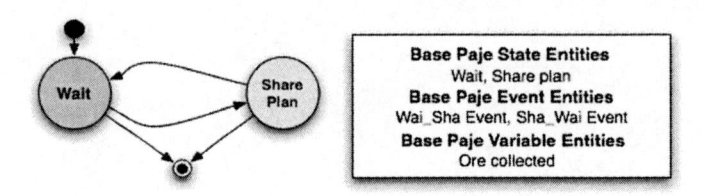

Fig. 3. State diagram and Paje entities associated to a base agent

an explorer. Other variables, such as the quantity of collected ore or the quantity of consumed energy, can be also visualized as Paje entities associated to an explorer.

3.2 Transporter

The goal of the transporters is to collect ore quickly at a minimal energy cost. They all start waiting at the base for coordinates. Once a transporter receives a number of coordinates it decides whether it will go ahead with collecting them or if it needs to share its partial plan with others in order to coordinate, to possibly derive together a complete plan for picking up the maximum of ore. Just as explorers, transporters need to eventually recharge at the base and can likewise run out of energy as visualized in figure 2.

3.3 Base

A base may plays the role of guaranteeing the communication from the explorers to transporters when explorers for one reason or another are not able to reach any transporters in order to deliver ore coordinates. It takes over these coordinates from explorers and then offers them to transporters at a later time. The state diagram and the events Paje associated to the base agent are showed in figure 3.

4 Infrastructure Modelling

4.1 Environment

The environment consists in the space shared by the set of agents, meaning the grid word in the case of our application. The environment is also the space containing the passive entities manipulated by the agents, as the ore in the application. In this work, the environment is considered as a passive entity, it reacts to the actions the agents exert on it, and responds to their request. The model used for the environment is presented

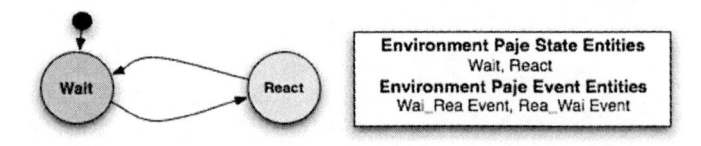

Fig. 4. State diagram and Paje entities associated to the environment

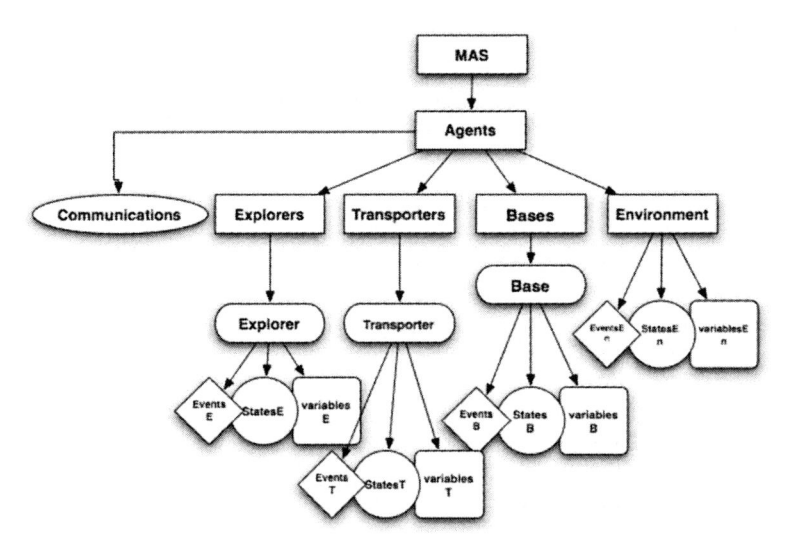

Fig. 5. Organisation diagram of the application and the related Paje entities

in figure 4. As for the agents, the objects visualized at level of the environment are the behaviour states and their changes. Some variables - the quantity of ore collected for example - representing the passive entities in the environment can also be modelled with Paje variables.

4.2 Interactions

A multi-agent system is distinct from a collection of independent entities by the fact that agents may interact together to accomplish a common task. The agents interact by communicating directly between them, or through another agent or by acting on the environment. The direct communication links between the different parts of a multi-agent system are presented by Paje link entities. These links are typed and differentiated in the visualization interface. Through this distinction between link types, we reflect the fact that the interactions inside the multi-agent system are typed, based on the content of the messages they transmit [14]. For example, the types of messages passed from explorers, are the following:

- *Ore pick up offer* with current coordinates in memory. This message is broadcasted to transporters or sent to the home base in case of unavailability of transporters at the given time.
- A direct *order to pick up ore* at chosen coordinates. This message is sent to an available transporter or for the base to take over the task of informing transporters about ore coordinates.

4.3 Organisation

We define the organisation as the set of relations between the several parts of a multi-agent system. It is also the way of modelling how to distribute the tasks among agents, how to share the information, and how to coordinate agents. In this work, we are interested only by the static organization that represents the relations between the different types of agents. This organization will be linked to the hierarchy imposed by Paje. The organisation of the visualized entities in the application is presented in the figure 5. This figure presents: the containers that represent the multi-agent system and the different types of agents, and the entities that refer the viewed events associated to these containers.

5 Multi-agent System Analysis

In this section, we present the Paje visualization of the multi-agent application. Figure 6 shows a global view of the execution. The sequences of states for each agent, the interaction links, and the events are displayed in the Paje space-time interface.

5.1 Patterns Analyse

To demonstrate the importance of visualizing agents' behaviour inside a multi-agent application, we focus here on two patterns that are extracted from the data presented by Paje. The first one shows how Paje may be used to debug a multi-agent application. The second one shows a possible metric that can be delivered from the Paje visualization. In Figure 7, we present an interaction diagram of an explorer trying to deliver ore coordinates. A positive response is received from the transporters T4 and T7. The transporter T8 ignores the offer because its state is *Pick up ore*.

The figure 8 visualizes a scenario developed when delivering ore coordinates. In order to relate this representation to the previously reported ones, the colours used in this figure to represent the states and the interactions are the same than in figures 1, 2 and 7.

Figure 9 presents the percentage of time spent in every state by two particular agents, E2 and E3. This figure puts in light the balance of tasks distribution for each of these

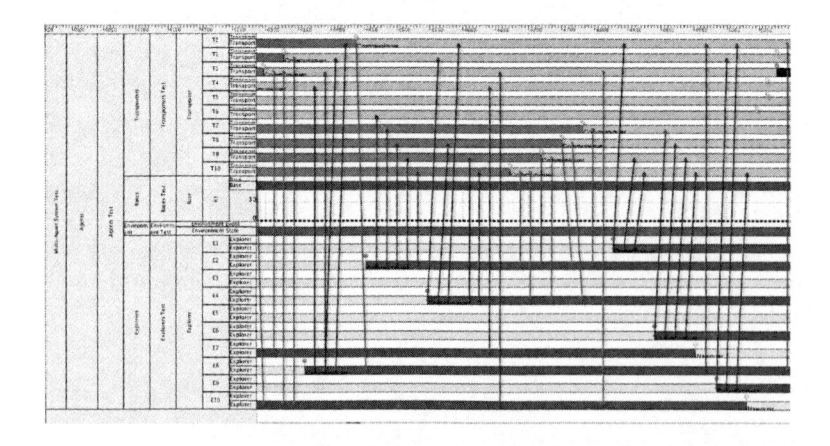

Fig. 6. Paje space-time interface

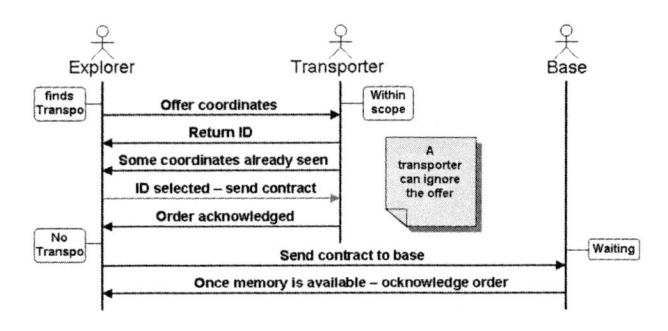

Fig. 7. Interaction diagram of delivering ore coordinates

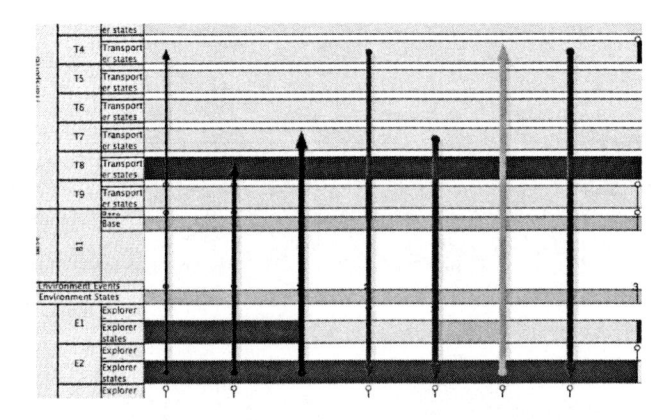

Fig. 8. Visualizing a scenario of delivering ore coordinates

two agents. This can be considered as a measure to understand the internal behaviour inside the multi-agent application, and to help the designer to better realize its design and the consequences of this design.

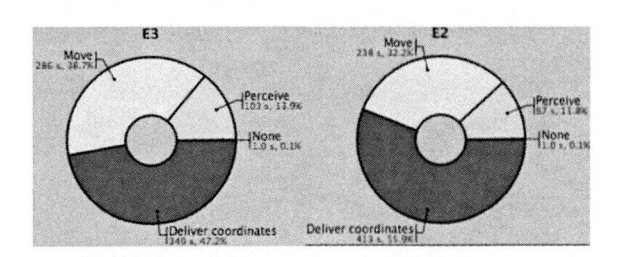

Fig. 9. The compared percentages of activities for different agents

6 Conclusion

In this article, we have presented a first trial to adapt Paje [9] to visualize VOWELS multi-agent systems [11]. We have demonstrated the usability of a visualization tool, designed initially for parallel/distributed systems, to visualize a VOWELS multi-agent system. The terms of the visualization tool have been adapted to the multi-agent application. This adaptation has been illustrated by the visualization of the execution of the explorer-transporter application [3]. The patterns presented in section 4 explicit the role of the visualisation in the performance debugging and in the system's behaviour presentation. This article has focused on the presentation of data aspects in the visualisation process. The results of the visualisation of the explorer-transporter application motivate us to pursue our efforts towards a general system for visualisation, MAS-Paje. MAS-Paje, which is under development, will achieve the visualization process of VOWELS multi-agent application, from the data collection step to the final delivering of a trace-file adapted to the Paje format.

References

1. Demazeau, Y. (ed.): Systémes Multi-Agents, ARAGO, OFTA, Lavoisier, vol. 29 (2004)
2. Chassin de Kergommeaux, J., Maillet, E., Vincent, J.-M.: Parallel program development for cluster computing: methodology, tools and integrated environments. Nova Science Publishers (2001)
3. Demazeau, Y.: MAS Examination Subject 2004, University South Denmark, Odense (2004)
4. Health, M.-T., Finger, J.E.: ParaGraph: A Tool for Visualizing Performance of Parallel Programs, Rapport technique No ORNL/TM-11813, Oak Ridge National Laboratory, Oak Ridge, TN (1991)
5. Heath, M.-T., Etheridge, J.-A.: Visualizing the Performances of Parallel Programs. IEEE Trans. Software Engineering 8(5) (1991)
6. Reed, D.-A.: An Overview of the Pablo Performance Analysis Environment, Rapport technique, Department of Computer Science, University of Illinois (1992)
7. Reed, D.-A.: Scalable Performance Analysis: The Pablo Performance Analysis Environment. In: Proceedings of the Scalable Parallel Libraries Conference. IEEE Computer Society, Los Alamitos (1993)
8. De Rose, L., Zhang, Y., Reed, D.-A.: SvPablo: A multi-language performance analysis system. LNCS. Springer, Heidelberg (1998)
9. Stein, B., Chassin de Kergommeaux, J.: Interactive visualization environment of multi-thread parallel programs. In: Parallel Computing: Fundamentals Applications and New Directions, Amsterdam (1998)
10. Demazeau, Y., Rocha Costa, A.: Populations and organisations in open multi-agent systems. In: First Symposium on Parallel and Distributed Artificial Intelligence (1996)
11. Demazeau, Y.: From Cognitive Interactions to Collective Behaviour in Agent-Based Systems. In: 1st European Conference on Cognitive Science, Saint-Malo, France, pp. 117–132 (1995)
12. Wooldridge, M., Jennings, N.: Intelligent agents: Theory and practice, the Knowledge Engineering Review (1995)
13. Wooldridge, M., Ciancarini, P.: Agent-oriented software engineering: The state of the art. In: Ciancarini, P., Wooldridge, M.J. (eds.) AOSE 2000. LNCS, vol. 1957, pp. 1–28. Springer, Heidelberg (2001)
14. Joumaa, H., Demazeau, Y., Vincent, J.-M.: Evaluation of Multi-Agent Systems: The case of Interaction. In: Proceedings of the 3rd International Conference on Information & Communication Technologies: from Theory to Applications. IEEE Computer Society, Los Alamitos (2008)

A Software Architecture for an Argumentation-Oriented Multi-Agent System

Andrés Muñoz, Ana Sánchez, and Juan A. Botía

Departamento de Ingeniería para la Información y las Comunicaciones
University of Murcia, Spain
{amunoz, anasanchez, juanbot}@um.es

Abstract. This paper proposes the materialization of a complete argumentation system ready to be built in conventional agent software platforms. In particular, an example for the Jadex agent platform is shown. The platform uses the BDI (Belief, Desire, Intention) model of agency. The main goal of this work is to foster usability of argumentation frameworks. The approach followed here to achieve it is the proposal of a formal representation for the dialogs that can occur, a specification of how to build the application domain model (i.e. the universe of discourse) and the necessary guidelines to manage these elements when building BDI agents.

1 Introduction

In the last years, knowledge-based multiagent systems (KBMAS) [6] have shown to be useful in complex scenarios such as traffic control or network administration. In these systems, cooperation and competition between agents to augment knowledge and to impose some beliefs, respectively, are the two main approaches. Both proposals need global techniques in order to detect and resolve potential *knowledge* or *opinion* conflicts on the different agents' beliefs.

Knowledge conflicts can be divided into two types: *semantic-independent* conflicts, so-called *contradictions*, which appear independently of the domain modeled, e.g. a positive and negative assertion on the same information; and *semantic-dependent* conflicts or *differences*, tightly related to a specific domain. For example, classifying an object as square and as rectangular is not *a priori* conflictive. However, the constraint that an object must be described with a unique shape could be elicited from the domain. In this case, the two classifications would be in conflict. In this work, a conflict-aware KBMAS is proposed in order to cope with both types of conflicts. This proposal consists of a software architecture that combines BDI agents, a formal knowledge representation model and a mechanism to solve conflicts based on argumentative persuasion dialogs. They are briefly introduced here.

Firstly, the universe of discourse (i.e. the domain knowledge) is described here by means of a formal and explicit model, based on Semantic Web [1] technologies. They provide an appealing vision about how to represent the agent's knowledge in distributed environments. More precisely, we refer to OWL ontologies [5], which offer a logical language to supply an exact description of any domain. These ontologies present a standardized means of exchanging and reusing knowledge, with a semantic enrichment of it. Due to the addition of this meta-information, agents can support reasoning operations, e.g. deductive processes or consistency checking.

Secondly, by integrating this formal knowledge into BDI (Belief-Desire-Intention) agents [11], they are provided with an explicit model where beliefs are related to the domain knowledge, desires amount to the motivations of the agent (e.g. to win an opinion discussion), and intentions are the agents' goals (e.g. to establish a belief in a competition). In particular, BDI agents are a specialization of *deliberative* agents, which allow for the definition of plans to achieve those goals. By means of the plans, the rationality of the agents can be extended with conflict management strategies.

Finally, one alternative to reach an agreement about the status of a conflict resides in establishing a *persuasion* dialog between agents [10]. Persuasion dialogs are one of the six basic types of dialogs defined by Walton and Krabbe [13]. They claim that each dialog tries to resolve an undesired situation (conflict of opinion, lack of information, etc.) by defining a set of protocols and rules. Specifically, persuasion dialogs consist of an exchange of opinions among agents that are for/against an issue, with the aim of clarifying which one is the most acceptable.

Argumentation [3] is considered as a promising materialization of persuasion dialogs in multi-agent systems. In this way, a negotiation protocol is defined via argumentation, that leads to a persuasion dialog in which an agent tries to convince others about a concrete proposal. But in this case, not only are the proposals exchanged. Furthermore, *arguments* (i.e. premises and rules used to derive those proposals) are also communicated. Thus, conflicts may be resolved more efficiently.

This work introduces an argumentation system based on ontologies (ASBO henceforth), which rests on the BDI agency model. The ASBO theoretical model has been published elsewhere [8], but enough details to introduce it are mentioned throughout the paper. An implementation of agents equipped with ASBO has been developed in a BDI agent system, the Jadex platform[1]. The particularity of this approach resides in an OWL-based domain representation.

The rest of the paper is structured as follows. Section 2 discusses the related work and our contribution to the state of art within argumentation literature. Section 3 introduces the argumentation system proposed here as a software architecture ready to be used. Its employment is illustrated in section 4 through a persuasion scenario in which ASBO is applied to build and exchange arguments in order to solve conflicts. Finally, section 5 summarizes our contribution and points out the future work.

2 Related Work

Argumentation deals with several fields in knowledge engineering [3]. Consequently, an abstract and generic framework for reasoning under incomplete and inconsistent information has been defined [2]. We have instantiated such a generic framework by using the Semantic Web information model approach. As a result, all the knowledge managed in ASBO, including arguments and rules themselves, is represented by means

---

[1] http://vsis-www.informatik.uni-hamburg.de/projects/Jadex/

of Semantic Web ontologies expressed in OWL. Thus, a researching line starts here, focusing on agents that automatically create and attack arguments, thanks to the semantic information obtained from OWL ontologies.

The Prakken's framework [10] allows specifying two-party persuasion dialogs which status is defined by exploiting a tree structure that is built as the dialog progresses. Termination is defined as the situation that a party is to move but has got no legal moves. The abstract protocol proposed is multi-move and multi-reply, and allows for all kinds of instantiations. This approach is used here to implement a persuasion dialog within ASBO, extending it to include argumentative attacks.

There are already some implementations of argumentation systems for a MAS. For example, Vreeswijk's *IACAS* [12], or P-DeLP [4], an extension to DeLP argumentation-based engine. Nevertheless, since all them are directed to a specific type of argumentation domains (e.g. legal topics), the possibility of achieving a general purpose argumentation system as ASBO is lost.

3   A Three-Layer Architecture in ASBO

This section describes the layered architecture of the conflict-aware KBMAS introduced in section 1. Agents that fulfill the architecture will be called *ASBO agents* henceforth. Figure 1 shows a representation of a couple of ASBO agents based on a block diagram. The top layer is related to argumentation tasks. Here, the argumentation system and the definition of the persuasive dialog are included (section 3.1). The middle layer constitutes the formal description model. It contains a common representation (i.e. OWL ontologies) for the domain-specific knowledge on which arguments are built (section 3.2). The bottom layer is the BDI layer. It covers details about the deliberative realization that agents must satisfy (section 3.3).

3.1   Argumentation Layer

This layer comprehends the core functionality that gives an agent the ability to persuade others through argumentation. Firstly, it allows ASBO agents to build arguments by using the domain knowledge obtained from the *ontology layer* and find attacks to opponents' arguments. Secondly, this layer defines a dialog for exchanging arguments

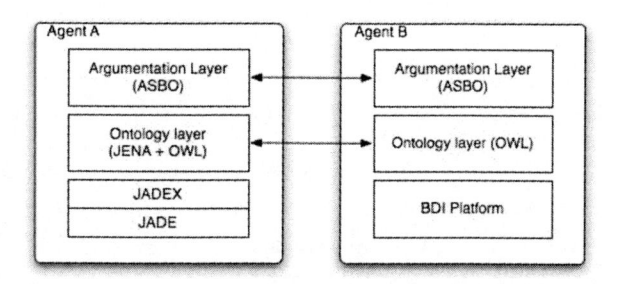

Fig. 1. Two ASBO agents showing the layered disposition of elements. The one on the right is an abstract representation meanwhile the one on the left shows a concrete implementation.

Table 1. Performatives available in ASBO persuasion dialogs

Performative	Informal semantics	Possible Responses	Illocutionary Effects
Claim φ	Assertion of proposition φ	Why φ, Concede φ	Publicly associates φ to the speaker
Why φ	Challenge of proposition φ, looking for reasons	Retract φ, φ since S (S is the argument for the proposition φ)	None
φ since S	Disclosure of the argument S that supports φ	Why v ($v \in S$), Concede v ($v \in S$), Claim $\neg v$ ($v \in S$), $\neg v$ since T ($v \in S$), Accept S	The speaker publicly associates to $v_i \in S$, i = 1..n
Concede φ	Assumption of proposition φ	None	The speaker publicly associates to φ
Retract φ	Rejection of proposition φ if a valid argument for it is not found	None	The speaker publicly withdraws from the proposition φ
Accept S	Acceptation of argument S. Update of the KB according to S	None	The speaker updates its KB according to the conclusions in S, regarding potential conflicts with previous knowledge
Question φ	Asks about agent's opinions on φ	Claim φ, Claim $\neg \varphi$	None

in order to resolve knowledge conflicts. Persuasion dialogs in this layer are specified by the following five elements: a communication language, an interaction protocol, and the effect, termination and result rules.

The **communication language**, denoted with L_{cm}, is formed by the performatives listed in Table 1. For each one, the semantics, possibles responses and illocutionary effects associated to them are given. Propositions φ, v are beliefs exchanged among agents, and S, T are arguments on which those propositions are based.

The **interaction protocol**, denoted with P_{ASBO}, is a formal specification of the legal persuasion dialogs that can occur in the system by using elements from L_{cm}. AUML has been used here to define it (see figure 2). P_{ASBO} starts with *claiming* a proposition by the dialog proponent. The opponent may respond *conceding* that proposition if agrees, or (the diamond symbol ◇ represents a decision node) asking for reasons through the performative *Why*. In this case, the proponent *retracts* the proposition if no argument can be built for it, or reacts with a valid argument (performative *Since*). Now, the opponent may *accept* the argument, or concede some premises from it together (the vertical black bar symbol represents a join node, in which several performatives are sent during the same speaking turn) attacking any other premise by claiming its negation, or attack a premise or rule giving the counterargument directly. The rest of the protocol can easily be followed in figure 2. Moreover, a *Question* performative may be sent at any time during the dialog, although it has not been included in the figure for simplicity (see section 4 for an example of persuasion dialog using L_{cm} and P_{ASBO}).

The set of **effect rules** expresses the illocutionary effects on the emitter and receiver of the performatives. They are given in the most right column of Table 1. The set of **termination rules** defines when a conversation ends. It occurs when an agent is in its turn of speaking and it can not make any legal movement according to P_{ASBO}. Termination of a dialog can be proved only if the agents' knowledge bases are finite. This topic falls out the scope of this paper and it is currently being studied as future work. From now on, it is assumed that KBs are finite. Finally, **result rules** determine the outcome

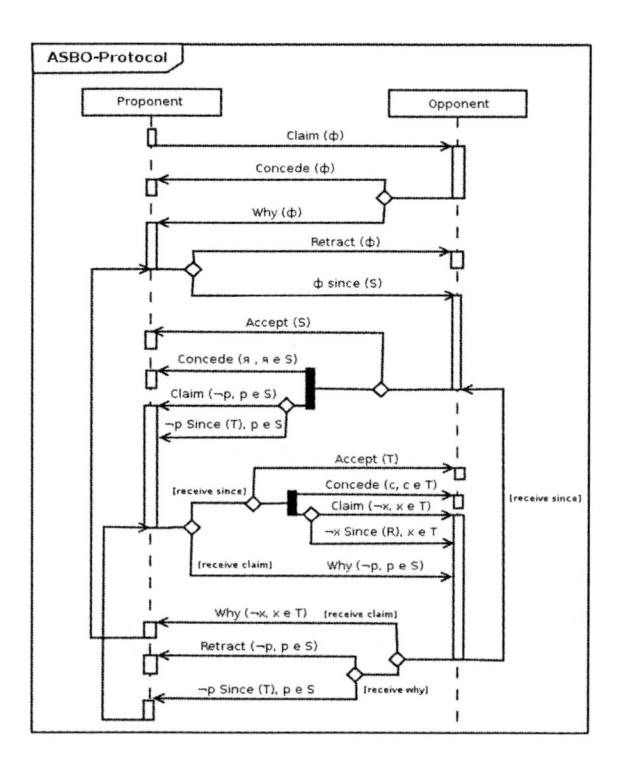

Fig. 2. AUML diagram of the ASBO persuasion dialog

of the dialog. Using them, an evaluation of the dialectical tree created during the dialog is performed so as to return which agent is the winner and which one is the looser. The evaluation process will be explained in section 4.

As a proof of concept, this persuasion dialog has been developed under the Jadex platform. The L_{cm} performatives have been included by extending the FIPA language, together with the protocol P_{ASBO}. This issue deserves a deeper discussion, but it is beyond the scope of this paper, just to remark that none of the L_{cm} performatives has the same semantics as the FIPA ones. Regarding arguments, they are of the form of

$$U = \{a, \langle s_{u,1}, \ldots s_{u,n} \rangle\},$$
$$s_{u,1} = \{p_{1,s_{u,1}}, \ldots, p_{m,s_{u,1}}, (b_{1,s_{u,1}} \wedge \ldots \wedge b_{m,s_{u,1}} \to c_{s_{u,1}})\} \vdash c_{s_{u,1}}$$
$$\ldots$$
$$s_{u,n} = \{p_{1,s_{u,n}}, \ldots, p_{m,s_{u,n}}, (b_{1,s_{u,n}} \wedge \ldots \wedge b_{m,s_{u,n}} \to a)\} \vdash a$$

where U is the argument, a is the conclusion derived from it, and the set of $s_{u,i}$ are the support elements of the argument. Each $s_{u,i}$ is composed of a set of premises $p_{j,s_{u,i}}$ and a domain knowledge rule $(b_{j_1,s_{u,i}} \wedge \ldots \wedge b_{j_m,s_{u,i}} \to c_{s_{u,i}})$ whose antecedents $b_{j_i,s_{u,i}}$ are variables instantiated by some of those premises, and the consequent $c_{s_{u,i}}$ is an intermediate conclusion. Note that the conclusion of the last support element $s_{u,n}$ must be the final conclusion a. The operator $\vdash$ represents the *modus ponens* inference rule in the deductive process. Premises in a $s_{u,i}$ are either initial facts from the KB or previous intermediate conclusions, $c_{s_{u,k}}, k < i$. Conflicts or attacks between arguments (i.e. *rebut*, or a conflict between two final argument conclusions; and *undercut*, or a conflict between a final conclusion and a $s_{u,i}$ premise, rule or intermediate conclusion), defeat relationships, and a global status of each argument are defined in this layer to establish which arguments are accepted and which ones are rejected in the dialog. The interested reader is referred to [8] for more details.

3.2 Ontology Layer

Due to the argumentation layer, arguments can be exchanged and attacked. The domain knowledge contained in the arguments (i.e. premises, rules, etc.) and the language used to define arguments are provided here. ASBO agents represent the domain knowledge and the arguments by means of a formal logic model, namely OWL ontologies. Some features of these ontologies are described here.

OWL [5] is a formal logic language for representing any knowledge through ontologies. The goal of OWL is twofold: On one hand, it represents knowledge as a set of concepts, properties, relationship among concepts, and instances of all these elements. On the other hand, it allows for a sound and computationally feasible deductive inference mechanism.

OWL ontologies are divided into two disjoint sets of statements: The terminological set *TBox*, which contains the definition of concepts, properties and relationships; and the assertional set *ABox*, which contains facts that instantiate those definitions, namely *individuals*. They are asserted as *concept assertions*, $C(a)$, indicating that the individual a belongs to the concept C, and as *role assertions*, $R(a,b)$, satisfying that individual a is related to b by the relation R. Some examples of these assertions are given in section 4.

In this proposal, not only is the domain knowledge represented by means of OWL ontologies. Arguments are themselves modeled after this idea. Concretely, the argument structure is expressed in OWL, whereas rules are defined using SWRL [7], an abstract rule language which extends OWL's syntax and semantics. Thus, the rebutting and undercutting attacks can automatically be discovered, since a consistency checking can be performed on arguments thanks to the OWL's formal features.

The Jena framework is employed here to manage OWL ontologies, together with ORE[2] [9], a rule editor tool. Due to this technique of knowledge representation and the argumentation layer explained in section 3.1, a generic purpose argumentation framework is offered, independent from the domain knowledge.

3.3 BDI Layer

The bottom layer is directly connected to an agent platform, in which the ASBO agents are developed. These agents are required to implement a BDI deliberative model (the Jadex platform is adopted here, although others such as JASON or 3APL may also be

---

[2] ORE is available at sourceforge.net/projects/ore

used). To this end, this layer is composed of three independent modules, gathering the minimal generic functionality that must be implemented.

The first module is the **Knowledge Module**. As explained in section 3.2, agents' knowledge is represented by OWL ontologies. These ontologies are managed by the Jena framework. Since it may have different configurations, an `OntologyHandler-Interface` has been created. This interface should be implemented by the agent through an ontology handler class with its preferred configuration.

The second module is the **Performative Processing Plan Module**(PPPM). Each agent determines here its attitude when receiving a P_{ASBO} performative. It has been developed using Jadex plans. They enable agents to have a modular implementation of the persuasion dialog, since each plan provides the adequate functionality to decide how to react when receiving a performative. An ASBO agent should implement at least one plan for each type of performative, and invoke it appropriately.

The last module is the **Plan Module**. ASBO agents define a set of goals in order to establish a group of dialogs that try to resolve opinion conflicts. When a goal is created, the plan associated to it is sent to the scheduler to be executed. There are two plans in this module: `ASBOInitiatePlan` and `ASBOResponderPlan`, to be executed depending on whether the agent acts as the dialog proponent or opponent, respectively. The initiator plan starts the dialog sending a *Claim* performative. The content of this message (i.e. the proposal) is received as a parameter from the goal associated to the plan. Then, the proponent waits for any ASBO protocol response and reacts according to its PPPM. The responder plan starts waiting for an ASBO protocol message and replies using its specific perfomative processing plans.

Note the benefits in implementing an argumentative agent in a modular manner, since either the knowledge base, the message processing or the conflict management strategy may be adapted to new requirements without changing the rest of modules. The only requisite is the implementation of each module to become an ASBO agent.

4 Using ASBO in a Real Argumentative Scenario

This section is devoted to illustrate the software architecture explained so far in a persuasion scenario. It has been developed in the ASBO framework, which contains specialized-knowledge agents. For this scenario, there are two significant agents: an agent *"E"* with a wide knowledge in environmental issues, and agent *"H"* specialized in handicapped people topics. They are discussing on the convenience of forbidding traffic in the city center.

The knowledge owned by each agent is shown below, using OWL's concept and role assertions. Agent E agrees with forbidding cars, as its goal is to protect the environment. It knows that cars emit fumes and are a threat to pedestrians, while buses represent an alternative to them. As counterpart, H opposes to forbid cars since they are a means of transport that minimizes the problems of impaired. R_{H1} states that an element X that helps a handicapped person Z has the quality of necessary. R_{H2} considers that an element X is not necessary if an alternative Y does exist, and moreover that element X is not preferred to the alternative Y:

Δ_E = {$ArtificialElem(Cars), ArtificialElem(Buses),$
$PollutionType(Fumes), PollutionType(Noise),$
$ProtectableElement(Pedestrians),$
$threat\_to(Cars, Pedestrians),$
$emit(Cars, Fumes), alternative\_to(Buses, Cars)$}

Δ_H = {$ArtificialElem(Cars), ArtificialElem(Buses),$
$Handicapped(impaired), help\_to(Cars, Impaired),$
$preferred\_to(Cars, Buses)$}

$R_{E1} = emit(X, Fumes) \Rightarrow PollutionElement(X)$
$R_{E2} = emit(X, Noise) \Rightarrow PollutionElement(X)$
$R_{E3} = threat\_to(X, Z) \Rightarrow dangerous\_to(X, Z)$
$R_{E4} = PollutionElement(X) \Rightarrow ForbiddenElement(X)$

$R_{H1} = Handicapped(Z) \wedge help\_to(X, Z)$
$\Rightarrow NecessaryElement(X)$
$R_{H2} = alternative\_to(Y, X) \wedge \neg preferred\_to(X, Y)$
$\Rightarrow \neg NecessaryElement(X)$
$R_{H3} = NecessaryElement(X) \Rightarrow \neg ForbiddenElement(X)$

Apart from the knowledge owned by each agent, there is a set of common rules shared by all them (both specific and common rules are written in SWRL). These rules represent general knowledge, and may be attacked for more specific agents' rules. R_{C1} states that if an element X represents a risk to Y, then the former must be forbidden. R_{C2} establishes that if an element Y offers an alternative to X, then the latter loses its "necessary" condition:

$$R_{C1} = dangerous\_to(X, Y) \Rightarrow ForbiddenElement(X)$$
$$R_{C2} = alternative\_to(Y, X) \Rightarrow \neg NecessaryElement(X)$$

A persuasion dialog obtained from the ASBO framework in this scenario is shown in figure 3. Following Prakken's terminology, each node of the tree represents a *move* in the dialog. They are evaluated to determine if the performative in them is accepted. Thus, a move is *in* when it has no attacks or all its attacking replies are *out*. A move is *out* whether it is a surrendered node (*Accept*, *Concede* and *Retract* moves) or it has a reply that is *in*. Hence, the dialog is won by the proponent if the root of the tree (the initial move) is *in*. Otherwise, the opponent wins the dialog. These are the result rules and the evaluation process mentioned in section 3.1.

As figure 3 shows clearly the dialog by following the numerical sequence (E_1, H_2,...,H_{18}), only the arguments built there are explained. The first argument, U_p, is disclosed in E_3 when agent E has to justify the proposal $ForbiddenElement(Cars)$:

$$U_p = \{ForbiddenElement(Cars), \langle s_{p_1}, s_{p_2} \rangle\},$$
$$s_{p_1} = \{emit(Cars, Fumes), R_{E1})\} \vdash PollutionElement(Cars),$$
$$s_{p_2} = \{PollutionElement(Cars), R_{E4}\} \vdash ForbiddenElement(Cars)$$

H needs to attack argument U_p. Since H does not possess any knowledge that counterattacks U_p premises, nor the argument's rules, it must assume them. Then, H tries to build an argument for $\neg ForbiddenElement(Cars)$. Looking into its knowledge, the argument $U_{\neg p}$ is built and revealed in H_7:

$$U_{\neg p} = \{\neg ForbiddenElement(Cars), \langle s_{\neg p_1}, s_{\neg p_2} \rangle\},$$
$$s_{\neg p_1} = \{Handicapped(Impaired), help\_to(Cars, Impaired), R_{H1}\} \vdash NecessaryElement(Cars),$$
$$s_{\neg p_2} = \{NecessaryElement(Cars), R_{H3}\} \vdash \neg ForbiddenElement(Cars)$$

E attacks now argument $U_{\neg p}$. It finds an attack on the $NecessaryElement(Cars)$ premise by taking the common rules into account. Thus, $U_{\neg n}$ undercuts $U_{\neg p}$ in E_{11}:

$$U_{\neg n} = \{\neg NecessaryElement(Cars), \langle s_{\neg n_1} \rangle\},$$
$$s_{\neg n_1} = \{alternative\_to(Buses, Cars), R_{C2}\} \vdash \neg NecessaryElement(Cars)$$

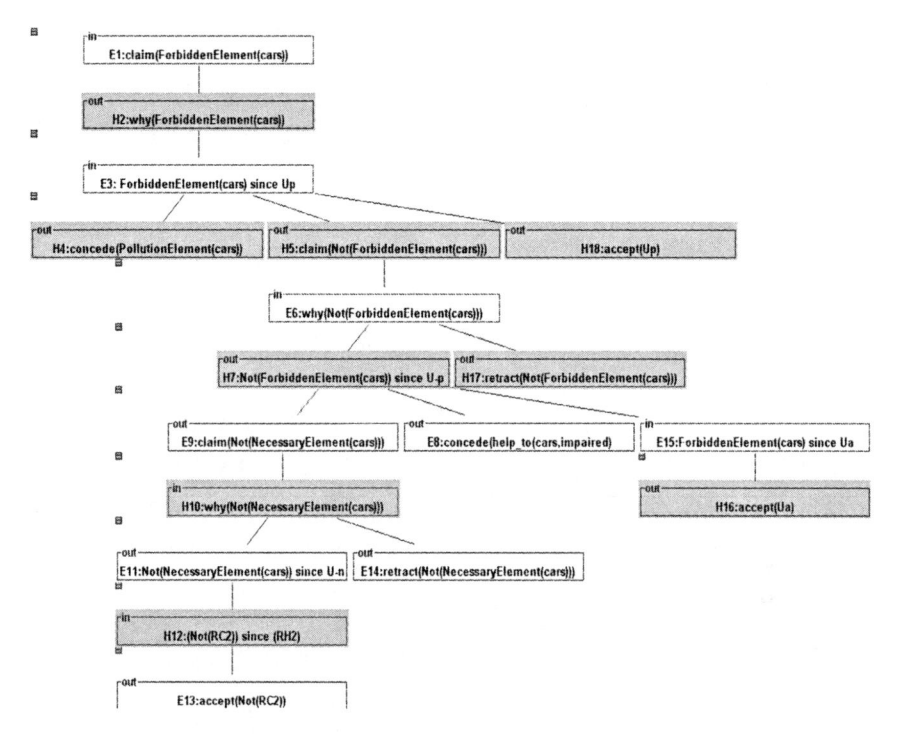

Fig. 3. ASBO snapshot of the dialog tree generated and evaluated in the persuasion scenario

Then, H detects that one of its specific rules, R_{H2}, has the same consequent that R_{C2} in $U_{\neg n}$. Furthermore, R_{H2} adds a new condition to obtain the conclusion $\neg Necessary$ $Element(Cars)$, namely that the element which an alternative is offered to must not be preferred to this alternative. Therefore, H can attack argument $U_{\neg n}$ by defeating R_{C2} with R_{H2} in H_{12}. This is a specific kind of attack defined in ASBO:

$$U_{\neg R_{C2}} = \{\neg R_{C2}, \langle R_{H2} \rangle\}$$

Now, E retracts that cars are not necessary (E_{14}) in response to $U_{\neg R_{C2}}$, as it can not attack that argument (E_{13}). Looking into common rules, E notices that R_{C1} opens a new way to support its goal. The argument U_a is sent to H in E_{15} as a new attack to H_7. Finally, H accepts U_a as it does not have any further knowledge for a new attacking move (H_{16}, H_{17}, H_{18}). Since the initial move is *in* in that moment, E becomes the winner of the dialog, and the proposal $ForbiddenElement(Cars)$ is accepted.

$$U_a = \{ForbiddenElement(Cars), \langle s_{a_1}, s_{a_2} \rangle\},$$
$$s_{a_1} = \{threat\_to(Cars, Pedestrians), R_{E3}\} \vdash dangerous\_to(Cars, Pedestrians),$$
$$s_{a_2} = \{dangerous\_to(Cars, Pedestrians), R_{C1}\} \vdash ForbiddenElement(Cars)$$

5 Conclusions and Future Work

This work has settled down the basis for new advances in argumentation within a MAS. As a result, a software architecture for building argumentation systems based on ontologies is developed and ready to be integrated into a MAS. Firstly, the agents' knowledge is represented by means of Semantic Web OWL ontologies. They allow for several types of automatic knowledge manipulation and reasoning processes. Secondly, the BDI agency introduces rationality in the agents, due to the goals and plans that process arguments and solve conflicts. Finally, a persuasion dialog is defined to exchange and attack arguments. This work has been implemented through the Jadex platform in a real persuasive scenario. Currently, the ASBO framework is extended to ease the argumentation debugging process.

Acknowledgement. The authors would like to thank the Spanish Ministerio de Ciencia e Innovación for sponsoring this work under the FPU grant AP2006-4154 and the Research Project TIN-2005-08501-C03-02, and also thank the Fundación Séneca for the grant 04552/GERM/06.

References

1. Berners-Lee, T., Hendler, J., Lassila, O.: The Semantic Web. Scientific American (2001)
2. Bondarenko, A., Dung, P.M., Kowalski, R.A., Toni, F.: An abstract, argumentation-theoretic approach to default reasoning. Artificial Intelligence 93(1-2), 63–101 (1997)
3. Carbogim, D.V., Robertson, D., Lee, J.: Argument-based applications to knowledge engineering. Knowledge Engineering Review 15(2), 119–149 (2000)
4. Chesnevar, C.I., Simari, G.R., Alsinet, T., Godo, L.: A logic programming framework for possibilistic argumentation with vague knowledge. In: AUAI 2004: Proceedings of the 20th conference on Uncertainty in Artificial Intelligence, pp. 76–84 (2004)
5. Dean, M., Connoll, D., van Harmelen, F., Hendler, J., Horrocks, I., McGuinness, D.L., Patel-Schneider, P.F., Stein, L.A.: Web ontology language (OWL). Tech. rep., W3C (2004)
6. Halpern, J.Y., Shore, R.A.: Reasoning about common knowledge with infinitely many agents. In: Logic in Computer Science, pp. 384–393 (1999)
7. Horrocks, I., Patel-Schneider, P.F., Boley, H., Tabet, S., Grosof, B., Dean, M.: SWRL: A semantic web rule language combining OWL and RuleML. Tech. rep., W3C (2004)
8. Munoz, A., Botia, J.A.: ASBO: Argumentation system based on ontologies. In: Klusch, M., Pěchouček, M., Polleres, A. (eds.) CIA 2008. LNCS (LNAI), vol. 5180, pp. 191–205. Springer, Heidelberg (2008)
9. Munoz, A., Vera, A., Botia, J.A., Skarmeta, A.F.G.: Defining basic behaviours in ambient intelligence environments by means of rule-based programming with visual tools. In: 1st Workshp of Artificial Intelligence Techniques for Ambient Intelligence. ECAI (2006)
10. Prakken, H.: Coherence and flexibility in dialogue games for argumentation. Journal Logic and Computation 15(6), 1009–1040 (2005)
11. Rao, A.S., Georgeff, M.P.: BDI-agents: from theory to practice. In: Proceedings of the First Intl. Conference on Multiagent Systems, San Francisco (1995)
12. Vreeswijk, G.: IACAS: an implementation of Chisholm's principles of knowledge. In: Dutch/German Workshop on Nonmonotonic Reasoning. Proceedings of the Second Workshop, pp. 225–234. Universiteit Utrecht (1995)
13. Walton, D.N., Krabbe, E.C.W.: Commitment in Dialogue: Basic Concepts of Interpersonal Reasoning. State University of New York Press, Albany (1995)

A SOMAgent for Identification of Semantic Classes and Word Disambiguation

Vivian F. López, Luis Alonso, and María Moreno

Departamento Informática y Automática
Universidad de Salamanca
Plaza de la Merced s/n, 37008, Salamanca, Spain
vivian@usal.es, lalonso@usal.es, mmg@usal.es

Abstract. This work describes a method that uses artificial neural networks, specially a Self-Organising Map (SOM), to determine the correct meaning of a word. By using a distributed architecture, we take advantages of the parallelism in the different levels of the Natural Language Processing system, for modeling a community of conceptually autonomous agents. Every agent has an individual representation of the environment, and they are related through the coordinating effect of communication between agents with partial autonomy. The aim of our linguistic agents is to participate in a society of entities with different skills, and to collaborate in the interpretation of natural language sentences in a prototype of an Automatic German-Spanish Translator.

Keywords: agent, natural language processing, semantic, syntactic, automatic translator, Kohonen Maps.

1 Introduction

As Timo Honkela indicates in [10], serious efforts to develop computerized systems for natural language understanding and machine translation have taken place for more than half a century. However, the more general the domain or complex the style of the text the more difficult it is to reach high quality translation. All systems need to deal with problems like ambiguity, lack of semantic coverage and pragmatic insight that also applies to natural language understanding. In the automatic understanding of language one needs to determine the concept that is described by certain words in a given context, that is called the Disambiguation of the Meaning of Words (DMV). Typically DMW has been applied in most of natural language processing system which use two fundamentals approaches [38]:

1. Methods such as Lesk and the Cottrell-Veronis-Ide (CVI) [36], that exploits the idea that if a pair of words in a sentence are semantically related, this relationship can be used to find the correct meaning in combination with their definitions.
2. Establishing the context, by starting with a given body of sentences, and taking into account the occurrence frequencies or global statistics of the words.

Many of these methods turn out to be impracticable in real systems because, in general, the information must be manually coded, and, especially in unrestricted text, the disambiguation mechanisms operates on very limited context and even offers incorrect answers in some special cases.

On the other hand most Natural Language Processing (NLP) system traditionally uses a sequential architecture that represents the classical linguistic levels [35]. To deal with this complex related information, and to make it available for text analysis, some previous studies have pointed to a distributed architecture. Some of them, as [34][33][1] report research on the possibilities of using multi-agent system [40] in NLP, to represent cooperation among distinct linguistic levels.

This paper describes a method that uses artificial neural networks, specifically a Self-Organising Map (SOM), to determine the correct meaning of a word used in context. This classifier is incorporated inside a prototype of an Automatic German-Spanish Translator, which can be trained as semantic agent (SOMAgent) that cooperates with other linguistic agents and that can improve the first choice of the translator. In future implementations, needing only training examples, coding effort for new languages could be reduced.

2 Self-Organizing Maps in Natural Language Processing

The Self-Organizing Maps due to Kohonen [15] [16] are used for the extraction of information from a primary multidimensional signal and its representation in two dimensions. Much of the formal and computational study of written language is centered on structural aspects, and not on semantics and pragmatics [8]. The SOM is used to resolve ambiguity in [31]. Scholtes has also used the SOM to parsing [32]. A model for lexical disambiguation is presented in [23].

The basic method for creating word category maps was introduced by Ritter and Kohonen [28][29]. Charniak [13] presented a scheme for grouping or clustering words into classes that reflect the commonality of some property. Pulkki [26] presents means for modeling ambiguity using the SOM.

Contextual information has widely been used in statistical analysis of natural language corpora [8]. During recent years, this statistical approach had considerable successes, based on the availability of large parallel corpora and some further methodological developments (consider e.g., [14][41]).

Tikkala et. al [37] have presented connectionist models for simulating both normal and disordered word production as well as child language acquisition. In [20] it is indicated that connectionist modeling of language acquisition has made significant progress since the Rumelhart and McClelland's pioneering model of the acquisition of the English past tense [30]. However, three major limitations need to be considered for the further development of neural network models of language acquisition (see models reviewed in [3][19][27]).

Previous work by [27][22], has shown that self-organizing neural networks, SOMs, are particularly suitable as models of the human lexicon. Various aspects of modeling translation and language use have been considered in [8]. There is recently a considerable interest on the models of language evolution (see, e.g., [2]).

2.1 The Semantic Memory of the SOMAgent Model

In [9], the main focus is in modeling communities of conceptually autonomous agents. An agent is conceptually autonomous if it learns its representation of the environment by itself, where a concept is taken to be simply a means of specifying a relationship between language and world.

Partial autonomy means a setting in which the learning process of an agent is influenced in some way by other agents. This influence can then serve as a basis for communication between agents. Thus, although each agent has an individual representation of the environment, the representations are related through the coordinating effect of communication between agents in situations where all agents have access to similar perceptions of the environment.

In our model, the environment consists of the context where the symbols (words) are represented during the process of learning, that is to say coded units should include a group of concurrent elements. In linguistics, the concept of the representation of the context is associated with a number of adjacent words. Thus similarity between words is a reflection of similarities of the context. The basic idea is to teach small *context maps* so that the SOMAgent can process the contextual information into clusters. Each model vector of the single-word maps corresponds to a particular *meaning* of the word.

Our agent implements a mechanism of class analysis, clustering, to represent and identify groups of meanings that are semantically associated. A class is a set of associated meanings with a central concept, whose members can be concepts or other classes. The agent has been used to choose the correct meaning from various candidates. Therefore the agent is conceptually autonomous, but it has partial autonomy: in cases where syntactic-semantic analysis is insufficient to solve a lexical ambiguity it must collaborate with other agents and take the context into account.

Let's assume that some sample data sets would be mapped onto an array that will be called the *map*. The set of input samples is described by a n-dimension real vector $x(t)$. Each unit in the map contains an n-dimension vector $m(t)$.

Let Xs be the vector which represents the symbolic expression of an element and Xc the representation of the context. The simplest neuronal model assumes that Xs and Xc are connected through the same neuronal unit, so that the vector X (the pattern) is formed by the concatenation of Xs and Xc:

$$X = \begin{bmatrix} Xs \\ Xc \end{bmatrix} = \begin{bmatrix} Xs \\ 0 \end{bmatrix} + \begin{bmatrix} 0 \\ Xc \end{bmatrix} \qquad (1)$$

The central foundation of the symbolic map is that the two parts have their owns weights during the process of self-organization; but the size of the context predominates, reflecting the metric relationships of the members of the set and implementing a spatial order that reflects semantic similarities. To find semantic relationships between words the semantic or conceptual space is explicitly modeled with the SOM algorithm [15].

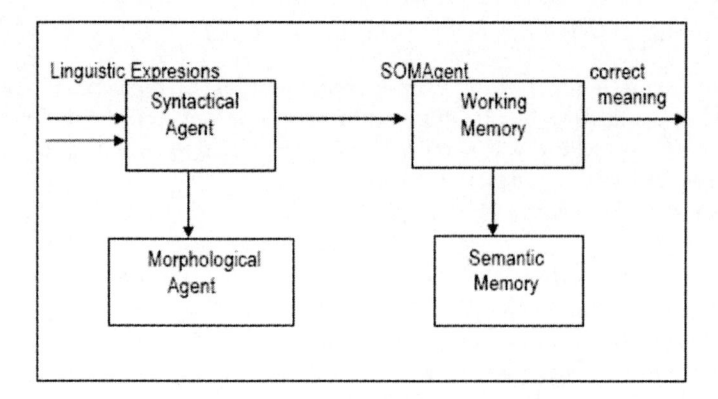

Fig. 1. Architecture of multi-agent system

The overall architecture of multi-agent system is presented in Figure 1. The SOMAgent receives perceptual inputs: linguistic expressions. There are potential actions: the agent can disambiguate an expression. The words perceptions are primarily stored in the working memory. The semantic memory associates contextual information and gives the correct meaning. Communication between the agents is motivated by the exchange of information related to linguistic expressions: morphological, syntactical and semantic information about the lexical items that are necessary for the resolution of specific tasks.

3 The Multi-Agent System

In this application it is intended to demonstrate that Kohonen Maps [16] can be applied to the organization of linguistic information to illustrate the idea of using the SOMAgent in finding a mapping between vocabularies of two different languages in machine translation. According to [10] the mapping between any two languages is based on an intermediate level of representation. This knowledge is embedded in specific autonomous agents in a multi-agent system [6]. In the following sections, we present the SOMAgent perspective from each of those agents.

3.1 Morphological Agent

Our lexical items are stored in a hierarchy of dictionaries: central dictionary or terms dictionary, idioms and terms that present lexical ambiguities. Terms description are realized by feature structure.

The terms dictionary was constructed in declarative form, associating with every term of source language its possible translations and corresponding morphological, syntactical and semantic information. The conjugations of every term are generated through morphologic rules.

Since the German language has a rich morphology inflexion there exist types of words that change their morphologic structure, the models allow the binding of termination groups to bases that perform identical inflexion behavior, thus reducing the number of both characters and descriptors stored in the dictionary. The words appear in the dictionary in degree zero and not their declines.

3.2 Syntactical Agent

Generalised Phrase Structure Grammar (GPSG) [12] is used in this point. GPSG further augments syntactic descriptions with semantic annotations that can be used to compute the compositional meaning of a sentence from its syntactic derivation tree.

In order to implement this model, grammar knowledge comprises the initial tree models, which represent the structure of German sentences and the lexicalization dictionary forming the syntactical agent knowledge. This agent can be seen as a subsociety [35], formed by agents handling simpler task or information associated with the features (e.g. complements) used in the parsing. This subsociety can be dynamically organized according to the problem it is expected to solve: to assist in German Spanish translation. One possible organization for this subsociety is a group of autonomous agent handling:

Agent 1, initial trees
Agent 2, auxiliary trees
Agent 3, lexicalization dictionary
Agent 4, formalism operation and organization of the working memory of the subsociety.
Agent 5, morphological and lexical transfer.

Taking as input German sentences the parsing is performed, giving as result a decorated abstract syntax tree (DAST). The dictionary agent gets the morphological information from the morphological agent (agent 5). The dictionary agent must negotiate with agents 1 to 3. It sends to agent 4 the set of trees that must be evaluated. Agent 4 tests all possible combinations with the received information, and set the values of the working memory.

4 Experiments

4.1 Lexicon

In this case a parallel vocabulary with all the words of the central dictionary is created, presenting lexical ambiguities that have to be determined from their context. This lexicon was integrated as part of the agent semantic memory. The vocabulary used for this specific example consists of nouns, ambiguous verbs and objects that will define the fundamental context as indicated by the number on the right hand side as shown in Table 1. The total number of contexts is 14.

The net is trained using the SOM Algorithm with a large set of sentences which reflects every type of context. In this way a trained net is created with the principal classes or with the active regions defined. A class is active if it contains any of the meanings included in the training for ambiguous cases. We take, for example, a subset of German verbs from the dictionary that have double meanings and whose right meaning can only be selected from their context.

Table 1. German lexicon for the experiments

Words	Cat	Words	Cat
Peter-Paul-Andreas	1	Klavier-Gitarre-Flöte	2
Fußball-Karten-Schach	3	Film-Szene	4
Draht-Rohr-Stange	5	Feuer-Licht-Kerze	6
Durst	7	Programm-Kasstette-Aufnahme	8
Schule-Kurs-Uni	9	Freund-Museum-Mutter	10
Spielt	11	dreht	12
Löscht	13	besucht	14

We consider the use of a subset of words in German, in a number of contexts of real-life situation. To illustrate the idea of using the SOMAgent in finding a mapping between ambiguous verbs of two different languages, we take, for example, the German verb *spielen* (to play) has two meanings represented by different Spanish verbs: either *tocar*, which appears in the context of playing musical instruments *Klavier, Gitarre, Flöte* (Cat=2), or *jugar* which appears in the context of games, *Fußball, Karten* or *Schach* (Cat=3).

4.2 Sentences Patterns

To study semantic relationships in their pure form, it is recognized that semantic value should not be inferred from any semantic pattern used for the encoding of individual words but only from the context where each word appears. Thus, the working memory in the simplest approach, takes all those words which occur in a certain "windows", represented by Xc and defined as inputs to the neural network. In this way vector inputs, X, to the network are created, using equation (1). In the self-organizing process, the inputs consist of sequences of three words selected from certain patterns of contexts. Such class patterns are defined off-line.

With sentence patterns generated based on the above contexts, sentences are created covering every possible context combination, for example, with the pattern 1-11-3, one takes a noun of context (1) the verb of context (11) and a game of the context (3). Sentences are created such as: *Peter spielt Fußball* (Peter plays football), *Peter spielt Karten* (Peter plays cards) or *Peter spielt Schach* (Peter plays chess).

4.3 Training Phase

The training phase consists of the sequential presentation of semantically correct sentences until the net converges. The set of parameters used as input to the training software was fixed by trial and error and each winning neuron was labeled according to its corresponding word from the vocabulary. After training, the network becomes topologically ordered, and it can be verified what units of the map are active for each input vector and are then labeled.

These sentences, following the steps of the general algorithm, form a file of input data vectors for doing the training, creating the semantic memory (a trained network) with the semantic classes specified in Table 2.

Table 2. Semantic class for the network

Class	Verbs in Spanish	Verbs in German
1	Tocar	Spielt
2	Jugar	Spielt
3	Filmar	dreht
4	torcer	dreht
5	apagar	löscht
6	quitar	löscht
7	borrar	löscht
8	ir	besucht
9	visita	besucht

Each class represents the meaning of the verbs according to the context. As it can be observed, nine classes were created. Corresponding to each meaning in Spanish, classes corresponding to the same verb in German are close on the maps.

4.4 Recognition Phase

The above algorithm was used to choose the correct meaning from various candidates, in a prototype tool to assist in German-Spanish translation in cases where syntactic-semantic analysis of the society of agents be insufficient to resolve a lexical ambiguity so that it should be solved by context reference. For those cases the semantic agent is called to collaborate in solving the ambiguity, the agent takes as its input the results of the previous agents: the semantic agent searches for meanings associated with each word, forming key sentences with the combination of words in German which could not disambiguate, these feed the network input, which should be able to classify it within the active classes, taking the best answer as the correct meaning.

For example for the sentence *Peter spielt Fußball* after training, the net find the true meaning of German verb *spielt*, classifying this entry inside the active classes, in this case, the class 2 (to play) whose meaning is "jugar" in Spanish. The following table (Table 3) shows how the sentence was translated by our multi-agent system:

Table 3. Results of the translation

German Sentence	Spanish Sentence
Peter spielt Fußball	Peter juega football

The system was implemented in C language under the UNIX operating system and using the Som-pack v3.1.10 software tool [17].

5 Conclusions

The scheme described in the paper considers a distributed architecture that takes advantages of the parallelism in the different levels of the NLP system, for modeling a community of conceptually autonomous agents. The agents have an individual representation of the environment, and they are related through the coordinating effect of communication between agents with partial autonomy. We consider that a distributed proposal is a better approach to NLP problems, since it allows the use of different formalisms at each level of linguistic processing. We developed a contextual environment using the Self-Organizing Map where we model a semantic agent (SOMAgent) that learns the correct meaning of a word used in context. In order to deal with specific phenomena such as ambiguity, we have given those regular agents the capacity to dynamically form subsocieties.

The aim of our linguistic agents is to participate in a society of entities with different skills, and to collaborate in the interpretation of natural language sentences in a prototype of an Automatic German-Spanish Translator system for natural language processing using the Self-Organizing Map, where the tasks to determinate the correct meaning of a word used in context emerges from the statistical properties of the training examples.

In the set of tests carried out, around 150 sentences, a correct classification was achieved in 97 percent of cases with a quantified error of less than 0.164. Thus the results can be considered satisfactory for this type of task.

This model too offers a methodology that illustrates the formation of a terminological mapping between two languages through an emergent conceptual space, and that can improve the first choice of the translator.

References

1. Balsa, J., Lopes, G.: A distributed approach for a robust and evolving NLP system. In: Christodoulakis, D.N. (ed.) NLP 2000. LNCS, vol. 1835, pp. 151–161. Springer, Heidelberg (2000)
2. Christiansen, M., Kirby, S.: Language Evolution. Oxford University Press, Oxford (2003)
3. Elman, J., Bates, A., Johnson, A., KarmiloffSmith, A., Parisi, D., Plunkett, K.: Rethinking innateness: A connectionist perspective on development. MIT Press, Cambridge (1996)
4. Fodor, J.A., Pylyshyn, Z.W.: Connectionism and cognitive architecture: A critical analysis. In: Connections and symbols. MIT Press, Cambridge (1998)
5. Gallant, S.I.: A practical approach for representing context and for performing word sense disambiguation using neural networks. ACM SIGIR Forum 3(3), 293–309 (1991)
6. Gärdenfors, P.: Conceptual Spaces. MIT Press, Cambridge (1999)
7. Hirst, G.: Semantic interpretation and the resolution of ambiguity. Cambridge University Press, Cambridge (1987)
8. Honkela, T.: Self-Organizing maps in natural language processing. Thesis for the degree of Doctor of Philosophy, Helsinki University of Technology, Department of Computer Science and Engineering. Public defense at 12th of December (1997)
9. Honkela, T., Winter, J.: Simulating language learning in community of agents using self-organizing maps. Technical report, Helsinky University of Technology, Computer and Information Science Report A71 (2003) ISBN 951-22-6881-7

10. Honkela, T.: Philosophical aspects of neural, probabilistic and fuzzy modeling of language use and translation. In: International joint conference on neural networks, IJCNN 2007, pp. 2881–2886 (2007)
11. Kaski, S.: Dimensionality reduction by random mapping: Fast similarity computation for clustering. In: Proceedings of IJCNN98, International Joint Conference on Neural Networks, vol. 1, pp. 413–418. IEEE Service Center, Piscataway (1998)
12. Klein, E., et al.: Generalized phrase structure grammar (1985)
13. Charniak, E.: Statistical language learning. MIT Press, Cambridge (1993)
14. Koehn, P., Och, F.J., Marcu, D.: Statistical phrase-based translation. In: NAACL 2003: Proceedings of the 2003 Conference of the North American Chapter of the Association for Computational Linguistics on Human Language Technology, pp. 48–54. Association for Computational Linguistics, Morristown (2003)
15. Kohonen, T.: Self-organized formation of topologically correct feature maps. Neurocomputing, 511–522 (1990)
16. Kohonen, T.: Self-organized maps. Proceedings of the IEEE 78(9), 1464–1480 (1990b)
17. Kohonen, T.: SOM-PAK: The self-Organizing map program package. Helsinki University of Technology Laboratory of Computer and Information Science, Finland (1995)
18. Koskenniemi, K.: Two-level morphology: A general computational model for word-form recognition and production. PhD thesis, University of Helsinki, Department of General Linguistics (1983)
19. Li, P.: Language acquisition in a self-organizing neural network model. In: Quinlan, P. (ed.) Connectionist models of development: Developmental processes in real and artificial neural networks, pp. 115–149. Psychology Press, New York (2003)
20. Li, P.: Farkas Early lexical development in a self-organizing neural network. Neural Networks 17, 1345–1362 (2004)
21. López, V.: Semantic disambiguation based on conexionist methods for a German Spanish problem of automatic translation. Thesis for the degree of Doctor in computer science. Valladolid University, Spain (1996)
22. MacWhinney, B.: Lexicalist connectionism. In: Broeder, P., Murre, J.M. (eds.) Models of language acquisition: Inductive and deductive approaches, pp. 342–932. Oxford University Press, Oxford (2001)
23. Mayberry III, M.R., Miikkulainen, R.: Lexical disambiguation based on distributed representations of context frequency. In: Proceedings of the 16th Annual Conference of the Cognitive Science Society (1994)
24. Morris, G., Hirst: Semantic interpretation and ambiguity. Artificial Intelligence 34, 131–177 (1998)
25. Pollack, J.B.: Recursive distributed representations. Artificial Intelligence 46, 77–105 (1990)
26. Pulkki, V.: Data averaging inside categories with the selforganizing map. Report A27, Helsinki University of Technology, Laboratory of Computer and Information Science, Espoo, Finland (1995)
27. Quinlan, P.: Modeling human development: In brief. In: Quinlan, P. (ed.) Connectionist models of development: Developmental processes in real and artificial neural networks, p. 112. Psychology Press, New York (2003)
28. Ritter, H., Kohonen, T.: Self-organizing semantic maps. Biological Cybernetics 61(4), 241–254 (1989)
29. Ritter, H., Kohonen, T.: Learning "semantotopic maps" from context. In: Proceedings of IJCNN-90-WASH-DC, International Joint Conference on Neural Networks, vol. I, pp. 23–26. Lawrence Erlbaum, Hillsdale (1990)

30. Rumelhart, D., McClelland, J.: On learning the past tenses of English verbs. In: Parallel Distributed Processing. Psychological and Biological Models, vol. 2, pp. 216–271. MIT Press, Cambridge (1986)
31. Scholtes, J.C.: Resolving linguistic ambiguities with a neural data-oriented parsing (DOP) system. In: Aleksander, I., Taylor, J. (eds.) Artificial Neural Networks, vol. 2, II, pp. 1347–1350. North-Holland, Amsterdam (1992)
32. Schütze, H.: Dimensions of meaning. In: Proceedings of Supercomputing, pp. 787–796 (1992)
33. Silva, J.L., Lima, V.: An alternative approach to lexical categorical disambiguation using a multi-agent system architecture. In: Proceedings of the RANLP 1997, Bulgaria, pp. 6–12 (1997)
34. Stefanini, M.H., Berrendonner, A., Lallich, G., Oquendo, F.: Talisman: Un système multi-agents gouverné par des lois linguistiques pour le traitement de la langue naturelle. In: COLING 1992, pp. 490–497 (1992)
35. Strube, V.L., Carneiro, P.R., Filho, I.: Distributing linguistic knowledge in a multi-agent natural language processing system: re-modelling the dictionary. In: Procesamiento del lenguaje natural (23), pp. 104–109 (1998) ISSN 1135-5948
36. Sutcliffe, R., Slater, B.: Word sense disambiguation of text by association methods: a comparative study. In: Proceedings of the SEPLN 1994 (1994)
37. Tikkala, A., et al.: The Production of finnish nouns: a psycholinguistically motivated connectionist model. Connect. Sci. 9(3), 295–314 (1997)
38. Veronis, J., et al.: Word sense disambiguation with very large neural networks extracted from machine readable. Dictionaries. In: COLING 1990, Helsinki (1990)
39. Wermter, S., Riloff, E., Scheler, G.: Connectionist, statistical and symbolic approaches to learning for natural language. Processing. Springer, New York (1996)
40. Wooldridge, M.: Intelligent Agents. In: Weiss, G. (ed.) Multiagent Systems. MIT Press, Cambridge (1999)
41. Zhang, R., Yamamoto, H., Paul, M., Okuma, H., Yasuda, K., Lepage, Y., Denoual, E., Mochihashi, D., Finch, A., Sumita, E.: The NiCTATR statistical machine translation system for IWSLT 2006. In: Proceedings of the International Workshop on Spoken Language Translation, Kyoto, Japan, pp. 83–90 (2006)

Multiagent Systems in Expression Analysis

Juan F. De Paz, Sara Rodríguez, and Javier Bajo

Departamento Informática y Automática
Universidad de Salamanca
Plaza de la Merced s/n, 37008, Salamanca, Spain
{fcofds, srg, jbajope}@usal.es

Abstract. This paper presents a multiagent system for decision support in the diagnosis of leukemia patients. The core of the system is a type of agent that integrates a novel strategy based on a case-based reasoning mechanism to classify leukemia patients. This agent is a variation of the CBP agents and proposes a new model of reasoning agent, where the complex processes are modeled as external services. The agents act as coordinators of Web services that implement the four stages of the case-based reasoning cycle. The multiagent system has been implemented in a real scenario, and the classification strategy includes a novel ESOINN neuronal network and statistics methods to analyze the patient's data. The results obtained are presented within this paper and demonstrate the effectiveness of the proposed agent model, as well as the appropriateness of using multiagent systems to resolve medical problems in a distributed way.

Keywords: Multiagent Systems, Case-Based Reasoning, microarray, neuronal network, ESOINN, Case-based planning.

1 Introduction

Currently, there exist many different systems aimed to provide decision support in medical environments [11] [12]. Cancer diagnosis is a field requiring novel automated solutions and tools, able to facilitate the early detection, even prediction, of cancerous patterns. The continuous growth of techniques for obtaining cancerous samples, specifically those using microarray technologies, provides a great amount of data. Microarray has become an essential tool in genomic research, making it possible to investigate global gene in all aspects of human disease [13]. Currently, there are several kinds of microarrays such as CGH arrays [16], expression arrays [17]. Expression arrays contain information about certain genes in patient's samples. Specifically, the HG U133 plus 2.0 [17] are chips used for this kind of analysis of expression. These chips analyze the expression level of over 47.000 transcripts and variants, including 38.500 well-characterized human genes. It is comprised of more than 54.000 probe sets and 1.300.000 distinct oligonucleotide feature. The great amount of data requiring analysis makes it necessary the use of data mining techniques in order to reduce the processing time. These data have a high dimensionality and require new powerful tools. Usually, the existing systems are focused on working with very concrete problems or diseases, with low dimensionality for the data, and it is very difficult to adapt them to new contexts for diagnosis of different diseases. Nowadays, there are different approximations as myGrid [23] [20] aimed to resolve this situation. They base their functionality in the creation of web service that are implemented following the OGSA (Open Grid Services Architecture) [14], but the main disadvantage is that the user must be the responsible of creating the sequence of actions that resolve a concrete problems. These systems provide methods for resolving complex problems in a distributed way through SOA [19] and Grid architectures, but lack of capacity for adaptation. On the other hand, there exist new research lines focused on reasoning mechanism with high capacity for learning and adaptation. Among these mechanisms highlights the case-based Reasoning (CBR) systems [2], which solve new problems taking into account the knowledge obtained in previous experiences [2], [15] and its integration within agents and multiagent systems. The inconvenience of the decision support systems based on CBR for microarray classification is the high dimensionality of the data and the corresponding complexity.

One alternative to the SOA architectures are multi-agent systems [18], that provide distributed entities with autonomous reasoning skills, called agents. Some proposals provide the agents with special capabilities for learning and adaptation by means of CBP (Case-Based Planning) mechanisms [3]. However, the CBP-BDI agents [3] present lacks when working with problems of high dimensionality and their efficiency is reduced. In [24] the incorporation of web services in multiagent architectures for implementing the agent's capabilities is studied. MAS and SOA architectures have been integrated and used in fields as gas turbine plant control [25] or hydrocarbure industries [26] to define the processes workflow.

This paper presents an innovative solution to model decision support Systems consisting of a multi-agent architecture which allows integration with Web services. In this sense it is possible to analyse data, for example from microarrays, in a distributed way. Moreover, the architecture incorporates CBP (Case-based planning) agents [3] specifically designed to act as coordinators of Web services. Thus, it is possible to reduce the computational load for the agents and expediting the classification process. The DASA architecture proposed within this paper has been applied to a case study, consisting of the classification of leukemia patients, and incorporates novel strategies for data analysis and classification. The process of studying a microarray is called expression analysis [1] and consists of a series of phases: data collection, data preprocessing, statistical analysis, and biological interpretation. These phases analysis consists basically of three stages: normalization and filtering; clustering and classification; and extraction of knowledge. In this work, a multiagent system based on the DASA architecture models the phases of the expression analysis by means of agents and incorporates innovative algorithms implemented as Web services, as filtering techniques based on statistical analysis, allowing a notable reduction of the data dimensionality and a classification technique based on a ESOINN [4] neural network. The core of the system are reasoning agents based on the CBP [3] mechanism. Furthermore, the system incorporates other agent types to accomplish complementary tasks required for the expression analysis.

Section 2 presents the DASA architecture. Section 3 describe a case study and finally,Section 4 presents the results and conclusions obtained.

2 DASA Architecture

DASA (Device, Agents and Service Architecture) is a multi-agent architecture that incorporates agents with skills to generate plans for analysis of large amounts of data. This is a novel mechanism for the implementation of the stages of CBP mechanisms through Web services. The architecture provides communication mechanisms that facilitate integration with SOA architectures.

DASA has been initially designed to facilitate the processing of data from expression arrays. To do this, DASA has been divided into three main blocks: devices modules, agents and services. The services are responsible for carrying out the processing of information by providing replication features and modularity. The agents act as coordinators and managers of services. Agents in the organization layer are available to run on different types of devices, so are created versions suitable to them.

The agents layer constitutes the core of the architecture, as can be seen in Figure 1. Figure 1 shows four groups of agents:

- Organization: The agents of the organization run on the user devices or on servers. The agents installed on the devices of the users make a bridge between the devices and agents of the system that perform data analysis. The agents installed on servers will be responsible for conducting the analysis of information on the model of reasoning CBP [3].
- Analysis: The agents in the analysis layer are responsible for selecting the configuration of the services that better suit the problem to solve. They communicate with Web services to generate results. The agents of this layer follow the model of reasoning CBP [3].
- Representation: These agents are in charge of generating the tables with the classification data and the graphics for the results.
- Import/Export: These agents are in charge of formatting the data in order to adjust them to the needs of agents and services.
- The Controller agent module manages the agents available in the analysis layer. It allows the registration of agents in the layer, as well as their use in the organization.

On the other hand, the services layer is divided into two groups:

- Analysis Services: The analysis services are services used by agents of analysis for carrying out different tasks. Within the analysis services are services for pre-processing, filtering, clustering and extraction of knowledge.
- Representation Services: They generate graphics and result tables.

Within the services layer, there is a service called Facilitator Directory that provides information on the various services available and manages the XML file for the UDDI (Universal Description Discovery and Integration). To facilitate communication between agents and services the architecture integrates a communication layer that provides support for the FIPA-ACL and SOAP protocols.

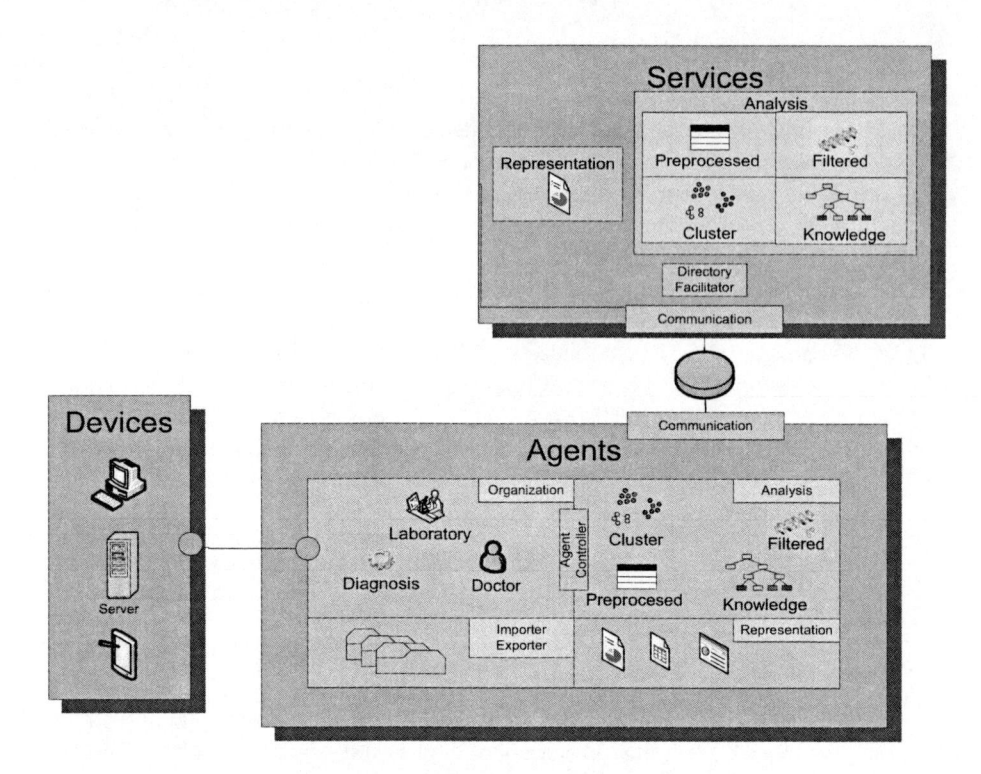

Fig. 1. Arquitectura de DASA

2.1 Coordinator Agent Based on CBR and CBP

The agents in the organization layer and the agents in the analysis layer have the capacity to learn from the analysis carried out in previous procedures. To do so, they adopt the model of reasoning CBP, a specialization of case-based reasoning (CBR) [2]. The primary concept when working with CBP's is the concept of case. A case can be defined as a past experience, and is composed of three elements: A problem description, which describes the initial problem; a solution, which provides the sequence of actions carried out in order to solve the problem; and the final state, which describes the state achieved once the solution was applied. A CBR manages cases (past experiences) to solve new problems. The way cases are managed is known as the CBR cycle, and consists of four sequential phases: retrieve, reuse, revise and retain.

Case-based planning (CBP) is the idea of planning as remembering [3]. In CBP, the solution proposed to solve a given problem is a plan, so this solution is generated taking into account the plans applied to solve similar problems in the past. The problems and their corresponding plans are stored in a plans memory.

A plan P is a tuple $<S,B,O,L>$:

- S is the set of plan actions.
- O is an ordering relation on S allowing to establish an order between the plan actions.

- *B* is a set that allows describing the bindings and forbidden bindings on the variables appearing in *P*.
- *L* is a set of casual links. That is, relations allowing to establish a link between plan actions.

3 Case Study: Decision Support for Leukemia Patients Diagnosis

The DASA multiagent architecture has been used to Developer a decision support system for the classification of leukemia patients. In framework of this research the system developed had available 212 samples from analyses performed on patients either through punctures in marrow or blood samples and affected by five types of leukemia (ALL, CLL, AML, MDS, CML). In this way, the analysis of data from microarrays is made in a distributed manner and certain tedious tasks are automated. The process consists of three steps: Initially the laboratory personnel hybridizes the samples, then the data analysis is made and finally a human expert interprets the results obtained. The multiagent system built from the DASA architecture reproduces this behaviour. The aim of the tests performed is to determine whether the system is able to classify new patients based on the previous cases analyzed and stored. Next, the developed agents and services are explained.

3.1 Services Layer

The services implement the algorithms that allow the analysis expression of the microarrays [1]. These services are invoked by the agents and present novel analysis techniques.

3.1.1 Preprocessing Service
This service implements the RMA algorithm and a novel control and errors technique. The RMA (R*obust Multi-array Average*) [5] algorithm is frequently used for preprocessing Affymetrix microarray data. RMA consists of three steps: (i) Background Correction; (ii) Quantile Normalization (the goal of which is to make the distribution of probe intensities the same for arrays); and (iii) Expression Calculation. During the Control and Errors phase, all probes used for testing hybridization are eliminated. Some few control points should contain the same values for all individuals. On occasion, some of the measures made during hybridization may be erroneous; not so with the control variables. In this case, the erroneous probes that were marked during the RMA must be eliminated.

3.1.2 Filtering Service
The filtering service eliminates the variables that do not allow classification of patients by reducing the dimensionality of the data. Three services are used for filtering:
Variability: Las variables con baja variabilidad no poseen valores similares para todos los individuos por tanto no son significativas a la hora de realizar clasificaciones. The first stage is to remove the probes that have low variability according to the following steps: Calculate the standard deviation for each of the probes, standardize the

above values, discard of probes for which the value of z meet the following condition: $z < \alpha$. **Uniform Distribution:** All remaining variables that follow a uniform distribution are eliminated. The variables that follow a uniform distribution will not allow the separation of individuals. The contrast of assumptions followed is explained below, using the Kolmogorov-Smirnov [6] test. H_0: The data follow a uniform distribution; H_1: The analyzed data do not follow a uniform distribution. **Correlations:** The linear correlation index of Pearson is calculated and correlated variables are removed so that only the independent variables remain.

3.1.3 Clustering Service
It addresses both the clustering and the association of a new individual to the group more appropriate. The services included in this layer are: the ESOINN neural network [4] (Enhanced self-organizing incremental neuronal network) y the NN clustering algorithm (Nearest Neighbor). Additional services in this layer for clustering are the Partition around medoids (PAM) [21] and demdograms [22].

The ESOINN [4] (Enhanced self-organizing incremental neuronal network) clustering technique is variation of neural network SOINN (self-organizing incremental neuronal network) [7]. ESOINN consists of a single layer, so it is not necessary to determine the manner in which the training of the first layer changes to the second. With a single layer, ESOINN is able to incorporate both the distribution process along the surface and the separation between low density groups. The initial phase could be understood as a phase of competition, while in a second phase, the network of nodes begins to expand just as with a NG. The classification is carried out bearing in mind the similarity of the new case using the NN cluster. The similarity measure used is as follows:

$$d(n,m) = \sum_{i=1}^{s} f(x_{ni}, x_{mi}) * w_i \qquad (1)$$

Where s is the total number variables, n and m the cases, w_i the value obtained in the uniform test and f the Minkowski [8] Distance that is given for the following equation.

$$f(x,y) = \sqrt[p]{\sum_i |x_i - y_j|^p} \quad \text{con } x_i, y_j \in R^p \qquad (2)$$

This dissimilarity measure weighs those probes that have a less uniform distribution, since these variables don't allow a separation.

3.1.4 Knowledge Extraction Service
The extraction of knowledge technique applied has been CART (Classification and Regression Tree) [9] algorithm is carried out. The CART algorithm is a non parametric test that allows extracting rules to explain the classification carried out. There are others techniques to generate the decision trees, as the methods based on ID3 trees [10], although the most used currently is CART.

3.2 Agents Layer

The agents in the analysis layer implement the CBP reasoning model and, for this, select the flow for services delivery and decide the value of different parameters based on previous plans made. A measure of efficiency is defined for each of the agents to determine the best course of the recovered for each phase of the analysis process.

In the analysis layer, at the stage Preprocessed only a service is available, so that the agent only selects the settings. The efficiency is calculated by the deviation in the microarray once have been preprocessed. All the cases are recovered and the configuration with greater efficiency is selected. At the stage of filtering, the efficiency of the plan p is calculated by the relationship between the proportion of probes and the resulting proportion of individuals falling ill.

$$e(p) = \frac{s}{N} + \frac{i'}{I} \tag{3}$$

Where s is the final number of variables, N is the initial number of probes, i' the number of individuals misclassified and I the total number of individuals. In the phase of clustering and classification the efficiency is determined by the number of individuals misclassified. The measure of similarity to retrieve the most similar plans is defined as the difference in the number of probes. Finally, in the process of extracting knowledge at the moment, CART has only been implemented for this task so that the agent responsible for conducting the selection method does not take into account the plans recovered, as in the previous phase Efficiency is determined by the number of individuals misclassified.

In the organization layer, the diagnosis agent is in charge of choosing the agents for the expression analysis [1]. The diagnosis agent establishes the number of plans to recover from the plans memory for each of the agents as well as the agents to select from the analysis layer. In a similar way the laboratory agent and the human expert select the agents from the organization layer that will be used. If the review at all stages is positive, and the human expert feels good result, the plan is stored in the memory of plans for further use.

4 Results and Conclusions

This paper has presented the DASA multiagent architecture and its application to a real problem. DASA facilitates tasks automation by means of intelligent agents capable of autonomously plan the stages of an expression analysis. Moreover, DASA facilitates the distributed execution of complex computational services, reducing the number of crashes in agents, since DASA separates the processing tasks from the agent architecture, and consequently, reduces the posibility of failure due to agents overload [27]. The multiagent system developed integrates within web services aimed to reduce the dimensionality of the original data set and a novel method of clustering for classifying patients. The multiagent perspective allow the system to works in a way similar to how human specialists operate in the laboratory, but is able to work with great amounts of data and make decisions automatically, thus reducing significantly both the time required to make a prediction, and the rate of human error due to confusion. The system focuses on identifying the important variables for each of the variants of blood cancer so that patients can be classified according to these variables.

Table 1. Plans of the filtering phase and plan of greater efficiency

Variability (z)	Uniform (α)	Correlation (α)	Probes	Errors	Efficiency
-1.0	0.25	0.95	2675	21	0.1485
-1.0	0.15	0.90	1341	23	0.1333
-1.0	0.15	0.95	1373	24	0.1386
-0.5	0.15	0.90	1263	24	0.1365
-0.5	0.15	0.95	1340	23	0.1333
-1.0	**0.1**	**0.95**	**785**	**24**	**0.1277**
-1.0	0.05	0.90	353	32	0.1574
-1.0	0.05	0.95	357	34	0.1669
-0.5	0.05	0.9	332	47	0.2278
-0.5	0.05	0.95	337	53	0.2562
-1.0	0.01	0.95	54	76	0.3594

In the study of leukaemia on the basis of data from microarrays, the process of filtering data acquires special importance. In the experiments reported in this paper, we worked with a database of bone marrow cases from 212 adult patients with five types of leukaemia. Table 1 shows the plans managed by the filtering agent for the analysis of the data from the HG U133 chip. Table 1 shows the values of the different parameters, significance levels for the tests and efficiency, and has been generated from the previous analysis of the 212 individuals.

Subsequently, we proceeded to repeat the test without pre-established settings, so that the agent automatically selected the plan of greater efficiency from equation (3). It has been highlighted in bold plan selected.

In Figure 2 it is possible to observe the performance of DASA for each of the agents of the organization and analysis layers. 11 tests were conducted based on manual planning and the results were compared with the automatic analysis provided by the multiagent system. Each of the agents of the organization layer selects the agents from the analysis layer as the previous plans and, in turn, each of these agents selects the services and configuration parameters. At the bottom of Figure 2 it can be seen the kind of agent of the analysis layer, and at the top the agent of the organization layer.

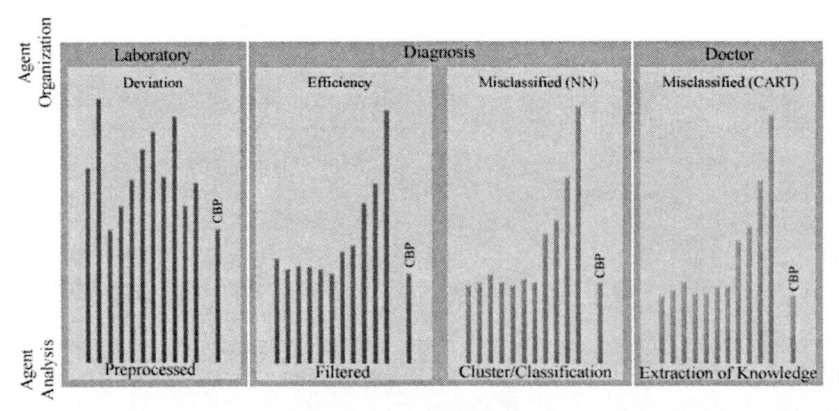

Fig. 2. DASA Architecture

In each chart the efficiency measure used is shown. The bar for the CBP agent is the highest efficiency according to the definitions applied.

DASA distributes the functionality among Web services, automatically calculates the expression analysis and allows the classification of leukaemia patients from the microarray data. DASA notably improves the performance provided by the manual procedures.

Acknowledgements. Special thanks to the Institute of Cancer of Salamanca for the information and technology provided. This work was supported in part by the projects MEC THOMAS TIN2006-14630-C03-03, IMSERSO 137/07 and JCYL SA071A08.

References

1. Lander, E., et al.: Initial sequencing and analysis of the human genome. Nature 409, 860–921 (2001)
2. Kolodner, J.: Case-Based Reasoning. Morgan Kaufmann, San Francisco (1993)
3. Glez-Bedia, M., Corchado, J.: A planning strategy based on variational calculus for deliberative agents. Computing and Information Systems Journal 10(1), 2–14 (2002)
4. Furao, S., Ogura, T., Hasegawa: An enhanced self-organizing incremental neural network for online unsupervised learning. Neural Networks 20, 893–903 (2007)
5. Irizarry, R., Hobbs, B., Collin, F., Beazer-Barclay, Y., Antonellis, K., Scherf, U., Speed, T.: Exploration, Normalization, and Summaries of High density Oligonucleotide Array Probe Level Data. Biostatistics 4, 249–264 (2003)
6. Brunelli, R.: Histogram Analysis for Image Retrieval. Pattern Recognition 34, 1625–1637 (2001)
7. Shen, F.: An algorithm for incremental unsupervised learning and topology representation. Ph.D. thesis. Tokyo Institute of Technology, Tokyo (2006)
8. Gariepy, R., Pepe, W.: On the Level sets of a Distance Function in a Minkowski Space. Proceedings of the American Mathematical Society 31(1), 255–259 (1972)
9. Breiman, L., Friedman, J., Olshen, A., Stone, C.: Classification and regression trees. In: Wadsworth International Group, Belmont, California (1984)
10. Quinlan, J.: Discovering rules by induction from large collections of examples. In: Michie, D. (ed.) Expert systems in the micro electronic age, pp. 168–201. Edinburgh University Press, Edinburgh (1979)
11. Chua, A., Ahna, H., Halwanb, B., Kalminc, B., Artifond, E., Barkune, A., Lagoudakisf, M., Kumar, A.: A decision support system to facilitate management of patients with acute gastrointestinal bleeding. Artificial Intelligence in Medicine 42(3), 247–259 (2008)
12. François, P., Cremilleux, B., Robert, C., Demongeot, J.: MENINGE: A medical consulting system for child's meningitis. Study on a series of consecutive cases. Artificial Intelligence in Medicine. 32(2), 281–292 (1992)
13. Quackenbush, J.: Computational analysis of microarray data. Nature Review Genetics 2(6), 418–427 (2001)
14. Foster, I., Kesselman, C., Nick, J., Tuecke, S.: The Physiology Of The Grid: An Open Grid Services Architecture For Distributed Systems Integration. Technical Report of the Global Grid Forum (2002)
15. Leake, D., Kendall-Morwick, J.: Towards case-based support for e-science workflow generation by mining provenance. In: Althoff, K.-D., Bergmann, R., Minor, M., Hanft, A. (eds.) ECCBR 2008. LNCS (LNAI), vol. 5239, pp. 269–283. Springer, Heidelberg (2008)

16. Shinawi1, M., Cheung, S.W.: The array CGHnext term and its clinical applications. Drug Discovery Today 13(17-18), 760–770 (2008)
17. Affymetrix, http://www.affymetrix.com/support/technical/datasheets/hgu133arrays_datasheet.pdf
18. Wooldridge, M., Jennings, N.: Agent Theories, Architectures, and Languages: a Survey. In: Wooldridge, Jennings (eds.) Intelligent Agents, pp. 1–22. Springer, Berlin (1995)
19. Erl, T.: Service-Oriented Architecture (SOA): Concepts, Technology, and Design. Prentice Hall PTR, Englewood Cliffs (2005)
20. Vittorini, P., Michettia, M., di Orio, F.: A SOA statistical engine for biomedical data. Computer Methods and Programs in Biomedicine 92(1), 144–153 (2008)
21. Saitou, N., Nie, M.: The neighbor-joining method: A new method for reconstructing phylogenetic trees. Molecular Biology and Evolution 4, 406–425 (1987)
22. Kaufman, L., Rousseeuw, P.: Finding Groups in Data: An Introduction to Cluster Analysis. Wiley Series in Probability and Statistics (1990)
23. Stevens, R., McEntireb, G.C., Greenwooda, M., Zhaoa, J., Wipatc, A., Lic, P.: MyGrid and the drug discovery process. Drug Discovery Today: BIOSILICO 2(4), 140–148 (2004)
24. Huhns, M., Singh, M.P.: Service-Oriented Computing: Key Concepts and Principles. Internet Computing 9(1), 75–81 (2005)
25. Arranz, A., Cruz, A., Sanz-Bobi, M.A., Ruíz, P., Coutiño, J.: DADICC: Intelligent system for anomaly detection in a combined cycle gas turbine plant. Expert Systems with Applications 34(4), 2267–2277 (2008)
26. Contreras, M., Sheremetov, L.: Industrial application integration using the unification approach to agent-enabled semantic SOA. Robotics and Computer-Integrated Manufacturing 24(5), 680–695 (2008)
27. Tapia, D.I., Rodriguez, S., Bajo, J., Corchado, J.M.: FUSION@, A SOA-Based Multiagent Architecture. In: International Symposium on Distributed Computing and Artificial Intelligence, Advances in Soft Computing, vol. 50, pp. 99–107 (2008)

Intentions in BDI Agents: From Theory to Implementation

S. Bonura, V. Morreale, G. Francaviglia, A. Marguglio, G. Cammarata, and M. Puccio

Intelligent Systems unit - R&D Laboratory - Engineering Ingegneria Informatica S.p.A.
{bonura,morreale,francaviglia,marguglio,cammarata,puccio}@eng.it

Abstract. In the context of the Belief Desire Intention (BDI) agent model and Bratman's theory, intentions play a primary role in reasoning towards actions. Indeed, intentions are supposed to be stable, constrain further deliberation, be conduct-controlling and influence beliefs about the future.

Thus, in this paper we present how PRACTIONIST, which is an integrated suite to develop BDI agent systems, embodies such properties of intentions. This allows to develop agents with the ability to know if desires are impossible, incompatible with other intentions and if intentions are achieved or no longer of interest.

We first give an overview of the PRACTIONIST deliberation process. Then the implementation of such properties is shown throughout a running example, i.e. the PSTS (PRACTIONIST Stock Trading System), which is aimed to monitor investors stock portfolio by managing risk and profit and supporting decisions for on-line stock trading, on the basis of investors trading rules and their risk attitude.

1 Introduction

The role that intentions play in action-directed reasoning has long been of interest to philosophers. Unlike most of them, Bratman [2] argues that intentions are not on a par with other basic mental attitudes, as beliefs and desires, but they play a stronger role in influencing action. More in detail, Bratman's theory of human practical reasoning (also referred to as Belief-Desire-Intention, or BDI) asserts that agents are influenced by their beliefs, desires and intentions in their reasoning on which action to perform. Practical reasoning consists of two distinct processes: (1) *deliberation*, to decide what states of affairs to achieve (the output of this process is intention); and (2) *means-ends reasoning*, to decide how to achieve these states of affairs (the output of this process is a plan to satisfy that intention). Thus intentions involve a much stronger commitment to action than other pro-attitudes (e.g. mere desires) and play in practical reasoning the following roles:

Intentions are Stable. If I formed an intention, to drop it without a good reason is not rational. A 'good reason' may be that I believe I have achieved my intention, or I cannot achieve it, or there is no reason anymore for having that intention. Such good reasons are formally described in literature on rational agents [12] by three levels of commitment strategies to intentions: *(i) Blind Commitment* - agents will continue to maintain an intention until they believe the intention has actually been achieved; *(ii) Single-minded commitment* - agents will continue to maintain an intention until they believe that either the intention has been achieved, or else that it is no longer possible to achieve the intention; *(iii) Open-minded commitment* - agents will maintain an intention as long as it is still believed possible.

Intentions Constrain Further Intentions. An intention to do an action this afternoon constrains the other intentions I will form during the day, by making my intentions consistent with them.

Intentions are Conduct-*controlling*. I will attempt to achieve my intentions, deciding means to achieve them, and if such means fails I will search for others. If this morning I formed the intention to have a pizza at lunch, as soon as lunchtime arrives I will simply proceed to try to meet my intention and order a pizza. My intention will not only inflence my conduct, it will control it too.

Intentions Influence Beliefs About the Future. If I have an intention, it is rational to plan the future on the basis of the premise that I believe that I will achieve that intention.

In the context of the Agent Oriented Software Engineering (AOSE), the BDI model and practical reasoning appear very attractive for dealing with the complexity of modern software applications, as they provide the essential abstractions necessary to cope with the real world issues [5]. But although it represents a well-established and fascinating theory its engineering involves several issues regarding the efficient implementation of the mental states, deliberation process and means-ends reasoning [13]. Most of existing BDI agent platforms (e.g. JACK [4], JAM [7]) generally use goals instead of desires. Moreover, the actual implementations of mental states differ from their original semantics: desires (or goals) are treated as event types (such as in AgentSpeak(L) [11]) or procedures (such as in 3APL [6]) and intentions are executing plans. Therefore the deliberation process and means-ends reasoning are not well separated, as being committed to an intention (ends) is the same as executing a plan (means). As a result, there is a gap between BDI theories and implementations [14].

In our research, our position is about considering intentions as related to the ends, being the plans related to the means to achieve those ends. Thus, the result of such a research is PRACTIONIST (PRACTIcal reasONIng syS Tem), which is an integrated suite providing a methodology, a design and development environment and a framework to develop BDI agent-based systems [10]. It adopts a goal-oriented approach to develop BDI agents and stresses the separation between the deliberation process and the means-ends reasoning. Agents can be programmed in terms of goals, which then will be related to either desires or intentions according to whether some specific conditions are satisfied or not. In PRACTIONIST the explicit declaration of goals and relations among them provides the agents with the ability to *know* if desires are impossible, incompatible with other intentions and if intentions are achieved, no longer of interest and so forth. This in turn supports the *commitment strategies* of agents and the capability to autonomously drop, reconsider, replace or pursue goals.

Some other BDI agent platforms deal with declarative goals. Indeed, in JADEX agent platform, goals are explicitly represented according to a generic model, enabling the agents to handle their lifecycle and reasoning about them [3]. Nevertheless, the model defined in JADEX does not deal with relations among goals.

In order to show how the abovementioned properties of intention are implemented and exploitable in PRACTIONIST, we chose to use the PSTS (PRACTIONIST Stock Trading System) as a running example throughout the paper, designed and developed by using PRACTIONIST Studio [8]. The PSTS is an intelligent system whose high-level goals are (i) monitoring investors' stocks portfolio in terms of risk and profit management and (ii) supplying a decision support for the on line stock trading, by considering investors'trading rules (i.e. stop loss, stop profit, profit target, tolerance, maximum budget to be inevested a week) e.g. degree of willingness to risk. Besides, if users wish the PSTS to sell stocks which are too risky or profitable, the PSTS autonomously does it by asking a broker (i.e. a Bank) to place the orders. At architectural level, the PSTS is mainly composed of two agents: the *HoldingStockManager*, which is the agent which monitors investors stock current prices and places sell order for that stocks resulting too profitable (if they have reached the profit target indicated by investor or the profit has descended below the stop profit of investors) or too risky (if their value has descended below the stop loss of investors); the *Trader*, which is the agent in charge of managing all kinds of order (i.e. market, limit and system orders) by asking the broker to place them.

The paper is organized as follows: we first give an overview of the Deliberation Process in PRACTIONIST in order to show how intentions are formed (section 2); then we describe how the properites of intentions according to the BDI model are satisfied in PRACTIONIST also throughout the PSTS example (sections 3, 4, 5, 6). Finally conclusions and further works are outlined in section 7.

2 Deliberation in PRACTIONIST

The fulfillment of the intention requirements stated above is achieved in PRACTIONIST through the abstraction of goal used to formally define both desires and intentions during the deliberation process. In other words, a PRACTIONIST agent will be provided with or autonomously figure out goals, which will be related to either desires or intentions according to whether some specific conditions are believed or not. Formally, a PRACTIONIST *goal g* is defined as a pair $g = \langle \sigma_g, \pi_g \rangle$, where σ_g is the *success condition* of the goal and π_g is the *possibility condition* stating whether g can be achieved or not. In the PRACTIONIST framework such elements are local properties of goals and are defined as operations that have to be implemented for each kind of goal. The *goal model* of PRACTIONIST agents contains the set of goals the agent could pursue and all existing relations among such goals. Such relations can be as follows [9]:

- a goal g_1 is *inconsistent* with a goal g_2 if and only if when g_1 succeeds, then g_2 must fail;
- a goal g_1 *entails* a goal g_2 (or equivalently g_2 is *entailed by* g_1) if and only if when g_1 succeeds, then also g_2 succeeds;
- a goal g_1 is a *precondition* of a goal g_2 if and only if g_1 must succeed, to be possible to pursue g_2;

- a goal g_1 *depends* on a goal g_2 if and only if g_2 is precondition of g_1 and g_2 must be successful while pursuing g_1.

Moreover, in case of inconsistency between two goals, it might be useful to specify that one is preferred to the other.

Thus, since in PRACTIONIST desires and intentions are mental attitudes towards goals, an agent can just *desire* to pursue a goal, which it is not committed to because of several possible reasons (e.g. it believes that the goal is not possible). On the other hand, a goal can be related to an *intention*, that is the agent is actually and actively committed to pursuing it. More accurately, suppose that an agent α *desires* to pursue the goal $g = \langle \sigma_g, \pi_g \rangle$, while working for achieving the set of active goals $\bar{G}$, because α has currently the intentions to achieve them. However, since an agent will not be able to achieve all its desires, α checks if it believes that the goal g is *possible* (i.e. if it believes that π_g is true) and not *inconsistent* with any active goal within $\bar{G}$. If both conditions hold, the desire to pursue g will be promoted to *intention* and the agent α will get to be committed to pursuing it. This will trigger tha means-ends reasoning with the aim of figuring out one or more plans (i.e. the means) to achieve g (i.e. the end).

For a detailed description of the deliberation process and goal model in PRACTIONIST, the reader can refer to [9, 10].

3 Intentions Are Stable

In PRACTIONIST the stability of intentions is assured by the adoption of the *single-minded intention commitment* strategy as a default. Indeed, if the agent α *intend*s to pursue g (i.e. g is *possible* and not *inconsistent* with active goals) and has already handled the means-ends reasoning by selecting a plan to meet g, α will continue to maintain the intention to achieve g, and consequently to execute that plan, until it believes that either such an intention has been achieved (i.e. the *success* condition of g is believed true) or it is no longer possible to achieve it (i.e. g does not turn out to be *possible* anymore). Moreover, the failure of a plan that has tried to achieve an intention will not result in dropping that intention. Indeed, the agent will try alternative means (see section 5).

With reference to the PSTS, let us suppose that the Trader's commitment strategy, whenever it has the intention to place a buy order on behalf of an investor (*PlaceSystemPurchaseOrder* goal), has to be that it maintains that intention until it believes that either the goal has been achieved or it is no longer possible (e.g. because that purchase is in conflict with the investor's trading rules).

In this instance, in order to implement such a commitment strategy, it is sufficient to model the goal's properties as shown in the Fig. 1. More in detail, the *PlaceSystemPurchaseOrder* goal was set such that it is possible (*applicable* according to the PRACTIONIST notation) when the agent believes that the user trading rules are verified and succeeded when the agent believes that the order was placed. Thus, while the Trader agent will be working to achieve the *PlaceSystemPurchaseOrde* goal, that is it will be executing the main plan to meet that goal, it will be tested if the intention to achieve such a goal is not in contrast with the user's trading rules (expressed as

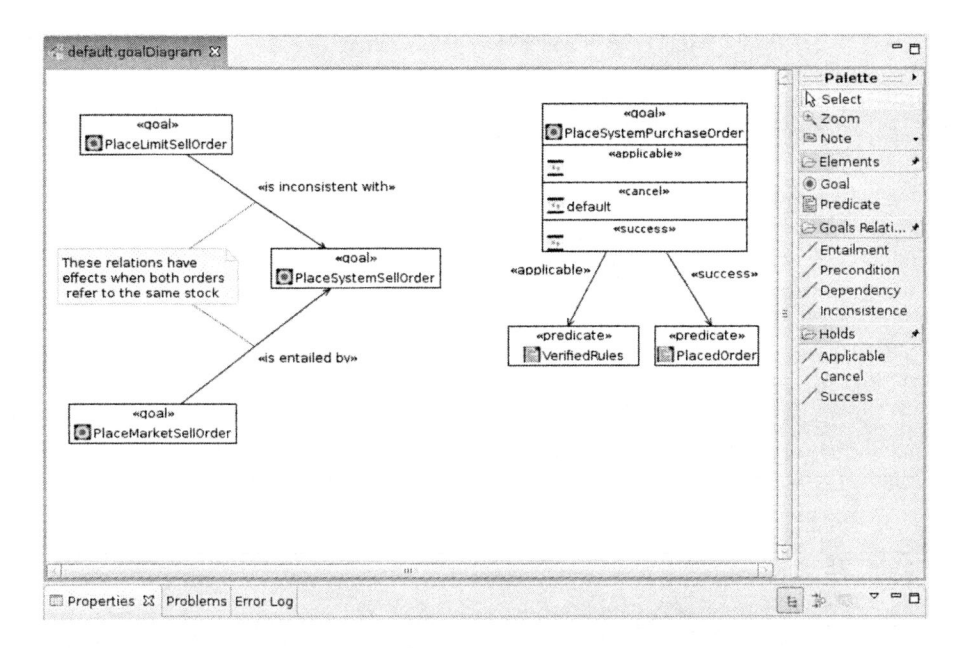

Fig. 1. Goal Diagram related to the HoldingStocksManager agent

conditions that can be believed or not by the agent). If at a certain point that checks don't pass, the goal becomes impossible and the Trader agent will give up working for that intention.

In order to model the applicability condition it was defined a *hold* relation between the goal and the *VerifiedRules* predicate asserting the goal is possible if the agent believes that the user trading rules hold; instead the success condition of the goal was modelled by putting a *hold* relation between the goal defined and the *PlacedOrder* predicate asserting that the goal succeeds when the agent has this belief.

A snippet of the *PlaceSystemPurchaseOrder* goal, coded by the code generator of the PRACTIONIST Studio, follows:

```
public class PlaceSystemPurchaseOrder implements Goal {
    private String id;
    private String uid;
    private String symbol;
    private String capital;
    private BeliefSet beliefSet = null;
    .....
    public boolean applicable(){
      return beliefSet.bel(AbsPredicateFactory.create(
        "verifiedRules(uid: %, symbol: %, capital: %)",
        uid, symbol, capital));}

    public boolean succeed(){
      return beliefSet.bel(AbsPredicateFactory.create(
          "placedOrder(id: %)", id)); }}
```

4 Intentions Constrain Further Intentions

Also the consistency between intentions is guaranteed in PRACTIONIST. Indeed, as already axplained, during the deliberation process, in order to decide if the *desire* to achieve a goal *g* can become an *intention*, the *inconsistency* relationship among *g* and some active goals is checked; thus the desire to pursue *g* will be promoted to an intention only if *g* is preferred to all inconsistent active goals (if any), which in turn will be dropped (i.e. α does not have the intention to pursue them anymore). This ability avoids that agents entertain goals that are inconsistent with the active ones.

Let us assume that the Trader agent is working to place a limit sell order (*PlaceLimitSellOrder* goal) and the HoldingStockManager agent, on the basis of the user trading rules and market prices, sends the Trader a request to place a market sell order for the same stock (*PlaceSystemSellOrder* goal) because it will be very risky to maintain that stock in the user portfolio; so, the Trader agent will drop the intention to place the limit order and commit to the goal to place that market order. Indeed, placing a market order on demand of the HoldingStockManager Agent and placing a limit order on demand of the investor are inconsistent goals, and the former is preferred to the latter as it implies a risky situation. Thus, in this scenario, the Trader agent drops its current intention in favour of the new intention (see Fig. 1).

In the diagram the inconsistency relation between the *PlaceLimitSellOrder* and *PlaceSystemSellOrder* goals for a stock was modelled, with a preference relation in favour of the second one (expressed in the diagram through the ending arrow). A code snippet of this relation follows:

```
public class GR_PlaceLimitSellOrder_PlaceSystemSellOrder
   implements InconsistencyRel {
   public Goal verifiesRel(SerializableGoal goal1, SerializableGoal goal2) {
      if (goal1 instanceof PlaceLimitSellOrder
         && goal2 instanceof PlaceSystemSellOrder) {
         String symbol1 = ((PlaceLimitSellOrder) goal1).getSymbol();
         String symbol2 = ((PlaceLimitSellOrder) goal2).getSymbol();
         if (symbol1.equals(symbol2))
            return goal2; }
      return null; }}
```

The goals *PlaceSystemSellOrder* and *PlaceLimitSellOrder* are inconsistent only if both refer to the same symbol of stock; besides, the former is the preferred one.

5 Intentions Are Conduct-Controlling

In PRACTIONIST agents, intentions control their conduct in several ways. Firslty, if an agent has formed an intention, it will try to achieve that intention, by deciding *how* to achieve it, by selecting a means and putting it in practice. If that particular means fails to achieve an intention, then it typically will attempt others.

More in detail, a PRACTIONIST agent has an event queue, containing among others, internal goal events, which are generated when the desire to pursue a given goal is promoted to an intention and some means-ends reasoning is required to figure out how to achieve such an intention. Thus, first of all, in PRACTIONIST intentions are

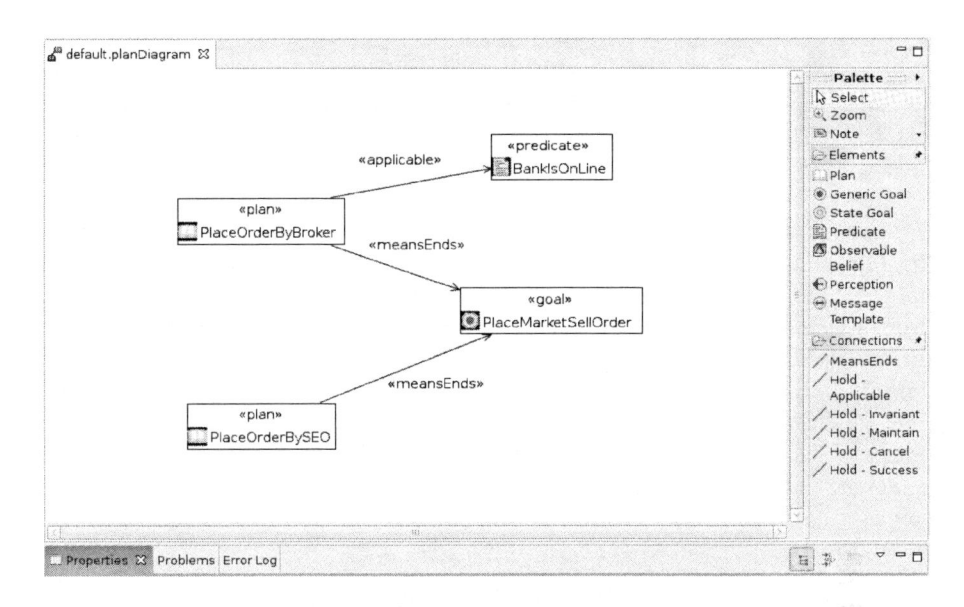

Fig. 2. Plan Diagram related to the Trader agent

conduct-controlling as they drive means-ends reasoning. Indeed, for each intention to achieve a goal, the agent performs the following *means-ends reasoning*: *(i)* it figures out the *practical* plans, which are those plans whose trigger event refers to a goal; *(ii)* among practical plans, the agent detects the *applicable* ones, which are those plan whose context is believed true, and selects one of them (which is called *main plan*); *(iii)* it builds the *intended means*, containing the main plan and the other alternative practical plans. This is the current set of means that the agent can exploit to achieve the selected intention.

When an event refers to an intention to pursue a goal, each new intented means is put on top of the stack containing the intended means that has generated the commitment to that intention. Thus, every intended means stack can contain several nested intended means, each able to achieve a goal and its subgoals, possibly through several alternative plans.

Secondly, the intentions are conduct-controlling as an actual strategy is followed to achieve them. Indeed, for each stack, the main plan of the topmost intended means is executed. Meanwhile, both success and cancel conditions of the plan are checked, in order to stop the execution (either with success or failure) before its normal completion.

Finally, intentions are conduct-controlling as several alternatives strategies are considered if necessary. Indeed, in order to be able to recover from *plan failures* and try other means to achieve an intention, if an executing main plan fails or is no longer appropriate to achieve the intention, then the agent selects one of the applicable *alternative plans* within the same intended means and executes it.

If none of the alternative plans is able to successfully pursue the goal g, the agent takes into consideration the non active goals that *entail* g. Thus the agent selects one of them and considers it as a desire, processing it in the way already described, from deliberation to means-ends reasoning. If there is no plan to pursue alternative goals, the achievement of the intention has failed, as the agent has not other ways to pursue its intention. Thus, according to agent's beliefs, the goal was *possible*, but the agent was no able to pursue it (i.e. there are no plans).

With reference to the PSTS, let us assume that the Trader agent has committed to the intention to place a market sell order; through the means-end reasoning it will find two practical plans by means of which the order will be placed by the bank or by a direct line with the Stock Exchange Office (SEO). In this scenario, if the Trader agent is not able to contact the bank (i.e. the corresponding plan is not applicable, while being practical), then it will try to place the order by contacting the SEO, which is applicable instead. The Fig. 2 shows the plans Trader agent may use to achieve its intention. *PlaceOrderByBroker* and *PlaceOrderBySEO* plans were modelled as two alternative ways to achieve the goal to sell a stock (indeed both plans have a means-end relation with the same goal, *PlaceMarketSellOrder*). The first plan models the set of acts needed to place an order by interacting with the broker, while the second one tries to place an order by interacting directly with the SEO.

6 Intention Influence Beliefs about the Future

In PRACTIONIST if the agent α has an intention, it will plan the future on the assumption that it will achieve that intention. Indeed, when a desire to pursue g is promoted to an *intention*, before starting the means-ends reasoning, the agent checks if the goal g *is entailed* by some of the current intentions (i.e. some other means is working to achieve a goal that entails g). If so, there is no reason to pursue the goal g and the agent does not need to make any means-ends reasoning to figure out how to pursue it. In other words, the agent believes that such current intention will be achieved and it will wait for its success as it implies the achievement of the intention to pursue g.

Let us assume that while the Trader agent is committed to place a market sell order (*PlaceSystemSellOrder* on demand of the HoldingStockManager Agent, the user sends it a request to place a market sell order for the same stock *PlaceMarketSellOrder*. In this case the Trader agent does not need to make any means-ends reasoning to figure out how to process the user's request because the achievement of the former goal entails the latter. Thus, in this scenario the Trader agent believes that the current intention will succeed, and only if the executing plan fails it will start the means-ends reasoning for directly pursuing the new intention.

The entailment relationship between the *PlaceSystemSellOrder* and the *PlaceMarketSellOrder* goals shown in the Fig 1 just reflects this reasoning.

7 Conclusions and Future Work

The BDI model of agency by Bratman [2] has been proving a fascinating theory to deal with the complexity of modern distributed applications. Indeed a lot of systems exist that draw inspiration from it or even implement it (e.g. [4, 1]). This model underpins the importance of intentions in action-directed reasoning, involving a much stronger commitment to action than mere desires.

In this paper we presented how our PRACTIONIST suite supports the development of agents endowed with the ability to believe if goals are impossible, succeeded, inconsistent with other goals and autonomously drop, reconsider, replace or pursue their intentions.

The benefits include the opportunity of not explicitly specifying end coding the desired behaviour of agents entailed by some statements about their goals as well as the relationships among them and with plans. Such behaviours are implicitly embodied within the agents, which only need to know those statements, being able to exploit them and performing rational choices during deliberation, means-ends reasoning and plan execution. Thus, through a running example, i.e. the PSTS, we showed how the properties of intentions are already implemented in the computational model of PRACTIONIST agents, and how it is easy to exploit them when modelling and developing intentional systems.

It is also worth mentioning that PRACTIONIST Studio is able to automatically generate code from the defined PAML models and diagrams.

As part of our future work, we aim at developing some other real-world applications by using PRACTIONIST to confirm and enhance the compliance with the practical reasoning approach and, on the other hand, evaluate the costs of such a coherence, in terms of time and computational resources.

Finally we are currently working to define a methodology that supports the development of applications using the PRACTIONIST framework and Studio.

References

1. Bellifemine, F., Poggi, A., Rimassa, G.: JADE - a FIPA-compliant agent framework. In: Proceedings of the Practical Applications of Intelligent Agents (1999)
2. Bratman, M.E.: Intention, Plans, and Practical Reason. Harvard University Press, Cambridge (1987)
3. Braubach, L., Pokahr, A., Lamersdorf, W., Moldt, D.: Goal representation for BDI agent systems. In: Second International Workshop on Programming Multiagent Systems: Languages and Tools, pp. 9–20 (July 2004)
4. Busetta, P., Rönnquist, R., Hodgson, A., Lucas, A.: JACK intelligent agents - components for intelligent agents in java. Agentlink News (January 1999)
5. Georgeff, M.P., Pell, B., Pollack, M.E., Tambe, M., Wooldridge, M.: The belief-desire-intention model of agency. In: Rao, A.S., Singh, M.P., Müller, J.P. (eds.) ATAL 1998. LNCS, vol. 1555, pp. 1–10. Springer, Heidelberg (1999)
6. Hindriks, K.V., De Boer, F.S., van der Wiebe, H., Jc Meyer, J.: Agent programming in 3APL. In: Autonomous Agents and Multi-Agent Systems, vol. 2(4), pp. 357–401. Kluwer Academic Publishers, Netherlands (1999)
7. Huber, M.J.: Jam: a bdi-theoretic mobile agent architecture. In: AGENTS 1999: Proceedings of the third annual conference on Autonomous Agents, pp. 236–243. ACM Press, New York (1999)
8. Marguglio, A., Cammarata, G., Bonura, S., Francaviglia, G., Puccio, M., Morreale, V.: Design and development of intentional systems with PRACTIONIST Studio. In: Proceedings of Joint Workshop From Objects to Agents, Palermo, Italy (2008)

9. Morreale, V., Bonura, S., Francaviglia, G., Centineo, F., Cossentino, M., Gaglio, S.: Goal-oriented development of BDI agents: the PRACTIONIST approach. In: Proceedings of Intelligent Agent Technology, Hong Kong, China. IEEE Computer Society Press, Los Alamitos (2006)
10. Morreale, V., Bonura, S., Francaviglia, G., Centineo, F., Puccio, M., Cossentino, M.: Developing intentional systems with the practionist framework. In: Proceedings of the 5th IEEE International Conference on Industrial Informatics (INDIN 2007) (July 2007)
11. Rao, A.S.: AgentSpeak(L): BDI agents speak out in a logical computable language. In: van Hoe, R. (ed.) Seventh European Workshop on Modelling Autonomous Agents in a Multi-Agent World, Eindhoven, The Netherlands (1996)
12. Rao, A.S., Georgeff, M.P.: Modeling rational agents within a BDI-architecture. In: Proceedings of the 2nd International Conference on Principles of Knowledge Representation and Reasoning, pp. 473–484. Morgan Kaufmann publishers Inc., San Francisco (1991)
13. Weiss, G. (ed.): Multiagent Systems: A Modern Approach to Distributed Artificial Intelligence. MIT Press, Cambridge (1999)
14. Winikoff, M., Padgham, L., Harland, J., Thangarajah, J.: Declarative & procedural goals in intelligent agent systems. In: Proceedings of the Eighth International Conference on Principles of Knowledge Representation and Reasoning, Toulouse, France, pp. 470–481 (2002)

An Intrusion Detection and Prevention Model Based on Intelligent Multi-Agent Systems, Signatures and Reaction Rules Ontologies

Gustavo A. Isaza[1], Andrés G. Castillo[2], and Néstor D. Duque[3]

[1] Departamento de Sistemas e Informática, Universidad de Caldas, Calle 65 # 26-10, Manizales, Colombia
gustavo.isaza@ucaldas.edu.co
[2] Departamento de Lenguajes y Sistemas Informáticos e Ingeniería del Software, Universidad Pontificia de Salamanca, Campus Madrid, Paseo Juan XXIII, 3, Madrid, Spain
andres.castillo@upsam.net
[3] Departamento de Administración de Sistemas, Universidad Nacional de Colombia, Sede Manizales, Campus la Nubia, Colombia
ndduqueme@unal.edu.co

Abstract. Distributed Intrusion Detection Systems (DIDS) have been integrated to other techniques to incorporate some degree of adaptability. For instance, IDS and intelligent techniques facilitate the automatic generation of new signatures that allow this hybrid approach to detect and prevent unknown attacks patterns. Additionally, agent based architectures offer capabilities such as autonomy, reactivity, pro-activity, mobility and rationality that are desirables in IDSs. This paper presents an intrusion detection and prevention model that integrates an intelligent multi-agent system. The knowledge model is designed and represented with ontological signature, ontology rule representation for intrusion detection and prevention, and event correlation.

Keywords: Multi-agent systems, Intrusion Prevention, Intrusion Detection Systems, Ontology, Intelligent Security, correlation alarms.

1 Introduction

Security Computing requires a permanent optimization in protection mechanisms and strategies that prevent attacks on the network and information systems. The event monitoring process that happens in a system or a network using patterns or signs is known as Intrusion Detection System (IDS). The trends in IDS are focused more on prevention models than correction models. These systems test traffic using a set of signs to detect malicious activities, report incidents o take corrective actions; but, any change inserted in the attack pattern can compromise the system, avoid the underlying technology and make insufficient the Intrusion Detection [1].

An Intrusion Prevention System (IPS) is a security component that has the ability to detect attacks (known and unknown), and prevent the malicious behaviour to succeed. Over the years different models based on Artificial Intelligence techniques have been considered to help the automatic signs and patterns generation without human intervention.

The main problems of using traditional architectures for intrusion detection systems are central console becoming a single point of failure, in such case; the network could go down without protection if this one fails, the scalability is limited, processing all the information in a node implies a limit in the network size that can monitor, reconfiguration difficulties and add capabilities to scale IDS signatures, the data analysis can be defective, the autonomous behaviour and learning capabilities are precarious.

This paper aims to present a progress in a PhD. thesis using an intrusion detection and prevention architecture based on multi-agents systems and representing their knowledge using ontology and semantic models, integrating hybrid intelligent techniques. The remainder of this paper is organized as follows: In the section 2 we introduce related works in this area, we present the multi-agent system for intrusion detection and the intelligence method applied in section 3, the ontology used for attacks signatures and reaction rules to support the prevention system integrated in the MAS is presented in section 4. Finally, we summarize our research work and discuss the future contributions in this investigation.

2 Related Work

Similar projects have used multi-agent systems (MAS) in the intrusion detection problem (IDP) [2], [3], [4], [5], [6], [7], in these investigations the MAS problem has been integrated with intelligent techniques like Neuronal Networks, Case Based Reasoning, genetic algorithms, among others. The Purdue *CERIAS (Center for Education and Research in Information Assurance and Security)* has contributed with great developments and advances using distributed intelligence and multi-agent systems in intrusion detection. In the AAFID model [8] the IDS nodes are organized in a hierarchical tree architecture composed by agents, transceivers and supervisors, every one with relevant functions and roles to manage the system.

On the other hand, different intelligent techniques have been integrated in the IDP, Probabilistic model [9], expert systems [10], supervised and non supervised models using Neural Networks (ANN)[11], [11, 12], [13], [14], genetic algorithms, decision trees, Bayesian networks and Petri networks[15], [16], [17], [18] to optimize the pattern recognition, and novel attacks. Other classification techniques such SVM (Support Vector Machines) [19], Data mining [20], [21], and hybrid methods such as Fuzzy Logic, Neuronal Networks and genetic algorithms [22], [23].

New approaches have been suggested using ontologies as a way to represent and understand the attacks domain knowledge, expressing the intrusion detection system much more in terms of their domain and performing intelligent reasoning. Projects such [24], [25] propose an ontology DAML-OIL (DARPA Agent Markup Language + Ontology Interface) target centric based on the traditional taxonomy classification migrated to semantic model, the investigation done in [26] integrates ontology entities and interactions captured by means of an centric-attack ontology which provides agents with a common interpretation of the environment signatures which are matched through a data structure based on the internals of the Snort network intrusion detection tool.

3 Intelligent Multi-Agent System

The multi-agent architecture was analyzed and designed using an agent software engineer methodology (UPSAM) proposed by one of the authors of this paper [27], this methodology raises to integrate the process and the specific agents models to analyze and design based on RUP (Rational Unified Process) oriented by cases uses, centered in the architecture and providing interdependence between models. It is iterative and incremental. Their main models are task model, architecture model, agent model, communication model, resource model and domain model. This methodology has been chosen because it provides a framework to integrate the models following a similar RUP process and allowing the refining of the models using AUML diagrams and the possible implementation approach using JADE. The main roles in the agents that participate in the MAS model are:

Sensor Agent
- *Intentions*: Capture packets from the network and send them to other agents to be analyzed and processed.

Analyzer Agent
- *Intentions*: Receive data from Sensor Agent, minimize false positives and false negatives, and compare signatures with predefined patterns. Integrate a hybrid intelligent technique combining neuronal networks and fuzzy logic, describing a linguistic representation to help the end user to interpret and understand the data.

Correlation Agent
- *Intentions*: Aims to the integration and correlation from events and alarms stored in the Signature Ontology model (*OntoIDPSMA.owl*). This agent uses correlation and classification methods using reasoning tools applied in ontological models.

Reaction Agent
- *Intentions*: Manage the events to generate alarms and to create a prevention model integrating reaction rules to reconfigure other network devices.

The multi-agent for IDS and IPS architecture (*OntoIDPSMA* – Ontological Intrusion Detection and Prevention Multi-agent system) is shown in Figure 1; this architecture illustrates the interaction and the OWL message exchange used by the agents; as well as the performative parameter ACL Message and action.

At the moment, the intelligent component behaviour has been probed using Artificial Neuronal Networks with a supervised learning. In this component a normalized process for packet captures with JPCAP has been used; the relevant fields were classified and coded to binary, the fields information used were ToS (Type of Service), Length, TTL (Time to Live), Sequence, ACK, Flags, TCP and Data content. The Neuronal network type is *MLP (Multilayer Perceptron)* these networks are trained using the *Backpropagation* algorithm, learning how to transform input data into a desired response, ideal for pattern classification.

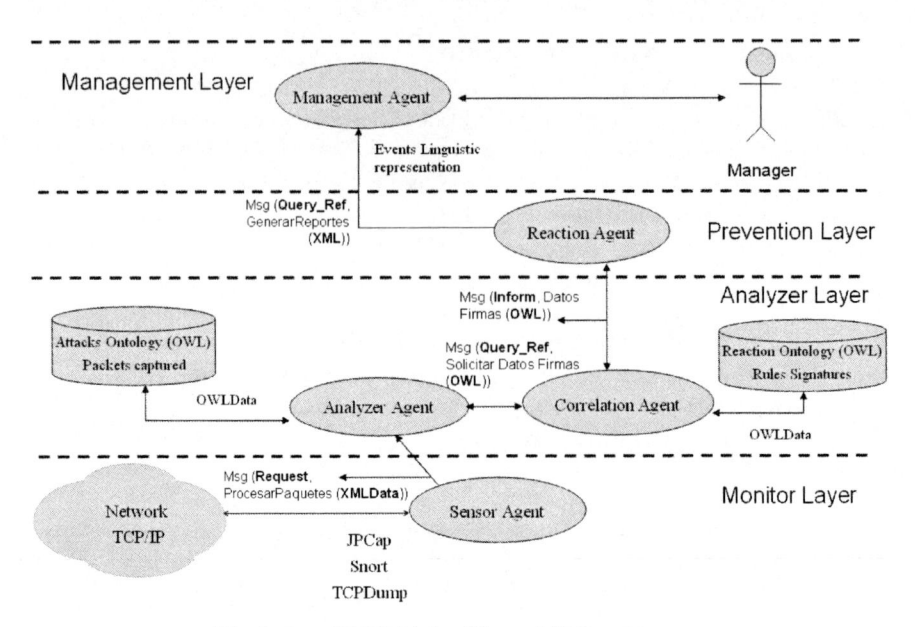

Fig. 1. OntoIDPSMA Intelligent MAS architecture

So far, there are *16* protocol fields and *393* content characters data, given 409 inputs for neuronal network. In the output layer there is only a neuron that expresses the result if a packet is normal or anomalous[28], [14]. This neuron contains values in a [0:1] range, classifying the values near zero as normal and near one as dangerous. The hidden layer has 450 neurons; therefore the topology tested at this moment is 409-450-1. The network topology, the training results and the SSE (Network performance function measuring performance according to the sum of squared errors) are depicted in Figure 2.

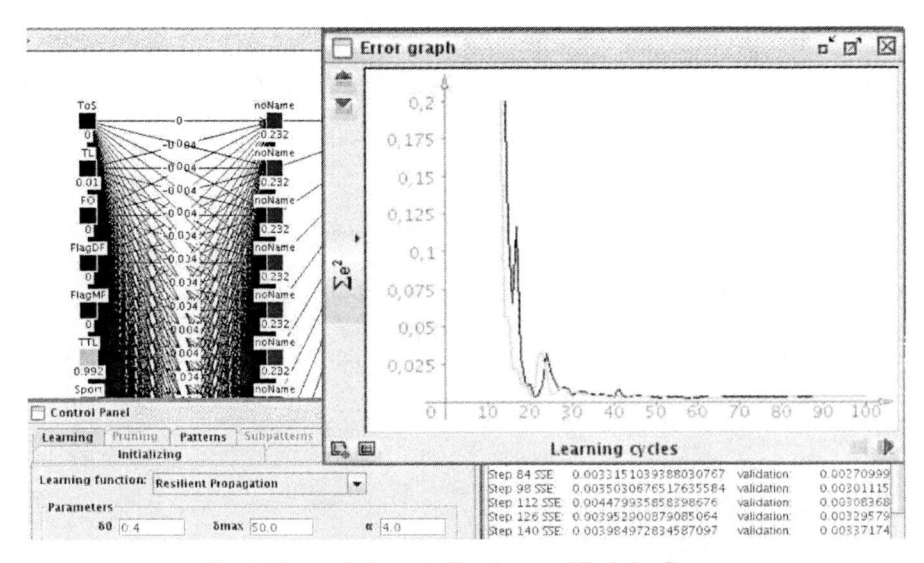

Fig. 2. Neuronal Network Topology and Training Process

Table 1. Performance comparison using multiple training algorithms

Algorithm	Training Parameters	Error SSE	Validation	Elapsed Time
Back Prop	n=0.2, max=0.1, cycles=100	22,73	0,97	312 sec
Resilient Prop	δ0=0.2,δmax=50.0,ά=4.0, cycles=140	0,93	0,132	285 sec
Quick Prop	n=0.5,μ=2.25,v=0.0001, cycles=300	19,42	1,58	128 sec

The simulation and training were developed using Matlab Neuronal Network Toolbox[29] and JavaNNS[30]. The input network data used to generate the knowledge base was de DARPA Data Sets Intrusion Detection Evaluation (two weeks captured) [31], complemented with the information gathered generating attacks with *IDSWakeUP*[32] and *Mucus* [33] synchronized with Snort (Open Source Network Intrusion Detection) and normal traffic captured using *libpcap*. The first tests using different training algorithms gave the results presented in Table 1. As it is demonstrated, the most efficient result having the pertinent values from SSE and MSE was Resilient Propagation.

After making multiple simulations and training with different configurations and transfer functions, the best performance can be seen in Table 2.

Table 2. Best Performance using different configurations

Algorithm	Tolerance	% trained successful	% normal traffic successful	% anomalous successful
traincgb	0,01	98,09	89,78	94,11
	0,001	99,94	90,09	85,29
traincgp	0,01	97,84	87,98	97,05
	0,001	100	91,89	91,17
Trainrp (Resilient Prop)	0,01	99,80	96,09	94,11
	0,001	99,94	97,89	97,05

The prototype for the Multi-agent system has been developed using *JADE* (Java Agent Development Framework, a software framework fully implemented in Java language It simplifies the implementation of multi-agent systems through a middleware that complies with the FIPA specifications) [34], we are integrating the intelligent technique with JOONE and JavaNNS.

4 Ontology

The ontology used allows represent the signatures for known attacks and novel attacks using the intelligent agent proposed; then an ontological model for reaction rules provide the prevention system. Using tools and algorithms based on inference rules and reasoning engine derivate and infer a semantic model. Integrating the API *Agent OWL* in our JADE implementation is possible including OWL language in the

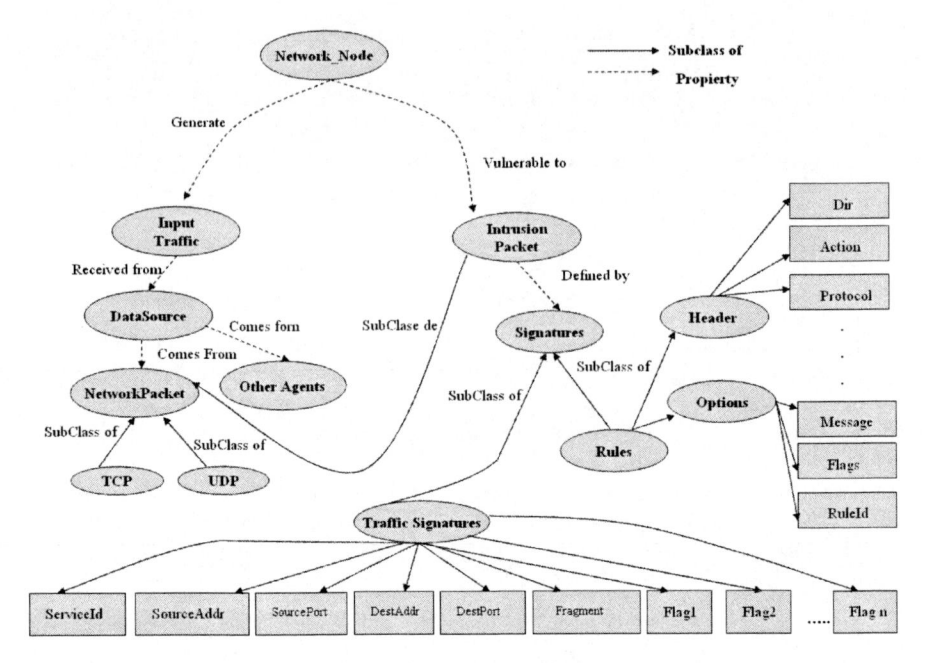

Fig. 3. Ontology OntoIDPSMA.owl

content data for ACL Messages exchange, providing the SPARQL query language for the ontology and including reasoning tools with the execution system.

The ontology defined that implements the intrusion detection and prevention knowledge is depicted in Figure 3 that presents a high level view. When constructing our ontology, we designed and implemented multiple classes and their interrelationships to define over 1800 attacks and intrusions.

Therefore we developed an Intrusion Prevention ontology characterized by network components, intrusion elements and classification defining traffic signatures and reaction rules classes and instances.

An ontological signature example using OWL is demonstrated in the following code that specifies the class for an exploit type attack:

```
<rdf:RDF xmlns:. . . . . . .
   <owl:Ontology rdf:about="Ontology OntoIDPSSMA">
   </owl:Ontology>
   <owl:class rdf:ID="Exploit Rule1.1.2">
   <rdfs:Subclass of>
     <owl class rdf:ID="Reglas">
     <TipoAlerta>Prevent</TipoAlerta>
     <pto_local> * </pto_local>
     <pto_remoto> 80 </pto_remoto>
     <nombre> Exploit HTTP </nombre>
        …..
     <type> str TS ES </type>
     <content> "\B4|B4|!|8B 83E9 04 8B|'"</content>
</rdf>
```

Compared with the standard IDMEF[35], that represents a specification for a standard alert format and a standard transport protocol, it becomes a syntactic representation based on XML and not an ontology that does not allow reasoning or inference model. In fact, we are developing and replacing simple taxonomies with semantic ontologies. The ontology OWL-DL is designed using Protégé and integrated to JADE using JENA[36] and *Agent OWL* to incorporate OWL language in ACL Messages content. At the moment we are working on an ontological domain subgroup representing SQL Injections, Web and *DoS* (Denial of Service) attacks.

5 Results, Conclusions and Future Work

The effect of the execution of the Multi-agent system on intrusion detection and prevention aims to decrease the bandwidth and the latency and also to obtain an autonomous, intelligent and cooperative model. The Ontology integration for the MAS in IDPS is an important data representation to solve heterogeneous problems in distributed environments and to allow building a scalable semantic model finding inferences using reasoning tools. The emerging intelligent agent paradigm is an efficient solution for the intrusion detection problem. Consequently, the present project implements a neural network in a supervised learning; further work will explore integration and hybrid technique combining fuzzy logic providing linguistic interpretation and information to the user and manager, on the other hand, the domain attacks and rules ontology will be extended to other classifications and will apply classification and correlation methods offering a semantic model with inferences and reasoning. The prevention architecture based on reaction rules generated by the intelligent and correlated component in our OntoIDPSMA creates new rules in other security tools; we are working on developing an integrated IDPS to support multiple network devices. So far, our project is in development, the whole system is being implemented and integrated; however the partial results shown in section 3 using neuronal networks in the intelligent agent demonstrate the optimization in identifying novel pattern of attacks and reducing false positives and negatives.

Comparing with a standard Open Source IDS, the partial evaluation of multi-agent system demonstrated the data presented in Table 3.

Table 3. Performance using MAS for Intrusion Detection and Prevention

Data Analyzed	Standard IDS	OntoIDPSMA
Traffic Processed	92,1 %	91,64 %
Packets Processed	10375	9847
Packets per second	11273	10714
Anomalous packets successful (detected)	89,3 %	90,4%
Average Detection Time	44,54 sec	33,21 sec

References

1. McHugh, J.: Intrusion and Intrusion Detection. International Journal of Information Security 1(1), 14–35 (2001)
2. Dasgupta, D., Gonzalez, F., Yallapu, K., Gomez, J., et al.: CIDS: An agent-based intrusion detection system. Computer and Security: Science Direct 24(5), 387–398 (2005)
3. Boukerche, A., Machado, R., Juc, K.: An agent based and biological inspired real-time intrusion detection and security model for computer network operations. Butterworth-Heinemann, 2649–2660 (2007)
4. Al-Hamami, A.H., Hashem, S.H.: A Proposed Multi-Agent System for Intrusion Detection System in a Complex Network. In: Information and Communication Technologies, ICTTA 2006, vol. 2, pp. 3552–3556 (2006)
5. Spafford, E., Zamboni, D.: Intrusion detection using autonomous agents. Computer Networks 34(4), 547–570 (2000)
6. Orfila, A., Carbo, J., Ribagorda, A.: Autonomous decision on intrusion detection with trained BDI agents. Butterworth-Heinemann, 1803–1813 (2008)
7. Herrero, A., Corchado, E., Pellicer, M., Abraham, A.: Hybrid Multi Agent-Neural Network Intrusion Detection with Mobile Visualization in Innovations in Hybrid Intelligent Systems, pp. 320–328. Springer, Heidelberg (2008)
8. Spafford, E.: Autonomous Agents for Intrusion Detection. Purdue CERIAS (Center for Education and Research in Information Assurance and Security. Consulted (2008), http://www.cerias.purdue.edu/about/history/coast/projects/aafid.php
9. Ning, P.: Probalistic states in Network Security. North Carolina State University (2003)
10. Eid, M.: A New Mobile Agent-Based Intrusion detection System Using distributed Sensors. In: Proceeding of FEASC, pp. 114–125 (2004)
11. Golovko, V., Kachurka, P., Vaitsekhovich, L.: Neural Network Ensembles for Intrusion Detection. In: 4th IEEE Workshop on Intelligent Data Acquisition and Advanced Computing Systems: Technology and Applications, IDAACS 2007, pp. 578–583 (2007)
12. Oksuz, A.: Phd Thesis Unsupervised Intrusion Detection System. Informatics and Mathematical Modelling, Technical University of Denmark (2007)
13. Laskov, P., Dussel, P., Schafer, C., Rieck, K.: Learning intrusion detection: Supervised or unsupervised? In: Roli, F., Vitulano, S. (eds.) ICIAP 2005. LNCS, vol. 3617, pp. 50–57. Springer, Heidelberg (2005)
14. Duque, N., Bonilla, C.M., Bohorquez, D., Isaza, G.: Sistema Neuronal de Detección de Intrusos. In: Zapata, C.M.y.G. (ed.) Tendencias en Ingeniería de Software e Inteligencia Artificial, G.M: Medellin (Colombia), vol. 2, pp. 99–105 (2008)
15. Abadeh, M., Habibi, J., Barzegar, Z., Sergi, M.: A parallel genetic local search algorithm for intrusion detection in computer networks, pp. 1058–1069. Pergamon Press, Inc., Oxford (2007)
16. Ye, N., Li, X., Emran, S.: Decision Tree for Signature Recognition and State Classification. In: IEEE Systems, Man, and Cybernetics Information Assurance and Security Workshop, West Point, New York, pp. 194–199 (2000)
17. Garcia, P.: Intensive Use of Bayesian Belief Networks for the Unified, Flexible and Adaptable Analysis of Misuses and Anomalies in Network Intrusion Detection and Prevention Systems. In: Proceedings of the 18th International Conference on Database and Expert Systems Applications. IEEE Computer Society, Los Alamitos (2007)
18. Kumar, S.: Classification and Detection of Computer Intrusions. Department of Computer Sciences. Purdue University, Purdue (1995)

19. Li, K., Teng, G.: Unsupervised SVM Based on p-kernels for Anomaly Detection. In: Proceedings of the First International Conference on Innovative Computing, Information and Control, vol. 2. IEEE Computer Society, Los Alamitos (2006)
20. Zurutuza, U., Uribeetxeberria, R., Fernández, I., Zamboni, D.: Un marco inteligente para el análisis de tráfico generado por gusanos en internet. In: XRECSI X Reunión Espanola sobre Criptología y Seguridad de la Información, Salamanca, pp. 607–618 (2008)
21. Zurutuza, U., Uribeetxeberria, R., Azketa, E., Gil, G., et al.: Combined Data Mining Approach for Intrusion Detection. In: International Conference on Security and Criptography, Barcelona, Spain (2008)
22. Mukkamala, S., Sung, A.H., Abraham, A.: Intrusion detection using an ensem-ble of intelligent paradigms. Journal of Network and Computer Applications 28(2), 167–182 (2005)
23. Tsang, C., Kwong, S., Wang, H.: Genetic-fuzzy rule mining approach and evaluation of feature selection techniques for anomaly intrusion detection, pp. 2373–2391. Elsevier Science Inc., Amsterdam (2007)
24. Undercoffer, J., Joshi, A., Pinkston, J.: Modeling Computer Attacks: An Ontology for Intrusion Detection. In: Vigna, G., Krügel, C., Jonsson, E. (eds.) RAID 2003. LNCS, vol. 2820, pp. 113–135. Springer, Heidelberg (2003)
25. Undercoffer, J., Finin, T., Joshi, A., Pinkston, J.: A target centric ontology for intrusion detection: using DAML+OIL to classify intrusive behaviors. In: Knowledge Engineering Review - Special Issue on Ontologies for Distributed Systems, pp. 2–22. Cambridge University Press, Cambridge (2005)
26. Mandujano, S., Galvan, A., Nolazco, J.: An ontology-based multiagent approach to outbound intrusion detection. In: The 3rd ACS/IEEE International Conference on Computer Systems and Applications, p. 94 (2005)
27. Castillo, A.: Modelos y Plataformas de Agentes Software Móviles e Inteligentes para Gestión del Conocimiento en el Contexto de las Tecnologías de la Información, Departamento de Informática, Universidad Pontificia de Salamanca, Madrid (2004)
28. Perez, C., Isaza, G., Brito, J.: Aplicación de Redes Neuronales para la detección de intrusos en redes y sistemas de información. Scientia et Technica XI(27), 225–230 (2005)
29. MathWorks. Neural Network Toolbox^{TM} 6.0 Design and simulate neural networks. Consulted: 2008 (2008), http://www.mathworks.com/products/neuralnet/
30. Fischer, I., Hennecke, F., Bannes, C., Zell, A.: User Manual, versión 1.1 of JAVA-NNS (Java Neural Network Simulator), University of Tübingen, Wilhelm-Schickard-Institute for Computer Science, Department of Computer Architecture (2002)
31. DARPA. DARPA Intrusion Detection Evaluation, The 1999 DARPA off-line intrusion detection evaluation, LINCOLN LABORATORY Massachusetts Institute of Technology. Consulted (2008), http://www.ll.mit.edu/IST/ideval/data/1999/1999_data_index.html
32. Herve, C.: IDSWakeUP. Consulted: 2008 (2002), http://www.hsc.fr/ressources/outils/idswakeup/index.html.en
33. Mutz, D., Vigna, G., Kemmerer, R.: An Experience Developing an IDS Stimulator for the Black-Box Testing of Network Intrusion Detection Systems, Department of Computer Science University of California, Santa Barbara (2003)
34. LuigiBellifemine, F., Caire, G., Greenwoo, D.: Developing Multi-Agent Systems with JADE. Wiley Series in Agent Technology, vol. 2008 (2007)
35. Curry, D.A., Debar, H., Feinstein, B.S.: Intrusion Detection Message Exchange Format. Intrusion Detection Working Group – Internet Engineering Task Force, Internet Draft (2004)
36. JENA. Jena – A Semantic Web Framework for Java. Consulted: Enero 2008 (2007), http://jena.sourceforge.net/

An Attack Detection Mechanism Based on a Distributed Hierarchical Multi-agent Architecture for Protecting Databases

Cristian Pinzón[1], Yanira de Paz[2], Rosa Cano[3], and Manuel P. Rubio[4]

[1] Universidad Tecnológica de Panamá, Av. Manuel Espinosa Batista, Panama
cristian.pinzon@utp.ac.pa
[2] Universidad Europea de Madrid, Tajo s/n 28670, Villaviciosa de Odón, Spain
yanirarosario.depaz@uem.es
[3] Instituto Tecnológico de Colima, Av. Tecnológico s/n, 28976, Mexico
rdegca@gmail.com
[4] Escuela Politécnica Superior de Zamora, Av. Cardenal Cisneros 34, 49022, Zamora, Spain
mprc@usal.es

Abstract. This paper presents an innovative approach to detect and classify SQL injection attacks. The existing approaches are centralized while this proposal is based on a distributed hierarchical architecture to provide a robust and dynamic strategy. The strategy for the classification and detection of SQL injection attacks uses a combination based on detection by anomalies and misuses. The detection by anomaly uses a case-based reasoning mechanism incorporating a mixture of neural networks. The approach has been tested and the results are presented in this paper.

Keywords: SQL injection, Security database, IDS, Multi-agent, case-based reasoning.

1 Introduction

A potential security problem on the database is a SQL injection attack. This attack seriously affects the database and it takes place when an original query is modified and is executed on the database by a hacker. The SQL injection attack has been addressed by the majority of the proposal from a centralized perspective [1] [2]. The main drawback of these approaches is that they solve the SQL injection attacks partially. Other solutions more sophisticated apply intrusion detection techniques [3] [4], but they have as drawback their large rate of cases poorly classified.

The proposal presented in this work tackles the SQL injection attack problem through a distributed hierarchical multi-agent architecture. Within the architecture are implemented strategies based on misuse and anomaly detection [5]. The key component of the architecture is a type of BDI (Belief, Desire and Intention) deliberative agent [6] which incorporates a based-case reasoning (CBR) mechanism [7]. The idea of a CBR mechanism is to exploit the experience gained from similar problems in the past and to adapt then successful solution to the current problem. This CBR-BDI type of agent [8] has been specially adapted to resolve the SQL injection attack problem. This agents use the CBR concept to gain autonomy and improve their problem-solving capabilities. In addition, it integrates a novel strategy of classification that lies in a mixture of neural networks which allows carrying out short term attack predictions. This work presents an entirely new approach in order to face the problem of SQL injection attack and describing an architecture that is unique in its conception.

The rest of the paper is structured as follows: section 2 presents the problem that has prompted most of this research work. Section 3 focuses on the details of the multiagent architecture, section 4 explains in detail the classification model integrated within the classifier agent. Finally, section 5 describes how the classifier agent has been tested inside a multi-agent system and presents the results obtained.

2 SQL Injection Attacks

A SQL injection attack takes place when a hacker changes the semantic or syntactic logic of a SQL text string by inserting SQL keywords or special symbols within the original SQL command that will be executed at the database layer of an application [1] [9]. The results of this attack can produce unauthorized handling of data, retrieval of confidential information, and in the worst possible case, taking over control of application server. The main problem for the detecting of SQL injection attack is the large number of variants. The detection of some SQL injection results trivial whereas that the detection of other result extremely complex due to large number of possible strategies.

Nowadays, this type of attack has been handled from distinct perspectives. The string analysis [10] has been the support of many others approaches such as [1] and [11], which carried out an analysis more complete applying a treatment dynamic and hybrid over the SQL string. In other cases, artificial intelligence techniques have been applied to face the SQL injection attack, such as [12] with WAVES (Web Application Vulnerability and Error Scanner). This proposal uses a black-box technique which includes a machine learning approach. Valeur [3] presented an IDS approach which uses a machine learning technique based on a dataset of legal transactions. These are used during the training phase prior to monitoring and classifying malicious accesses. Rietta [4] proposed an IDS at the application layer using an anomaly detection model. Finally, Skaruz [13] proposed the use of a recurrent neural network (RNN). The detection problem becomes a time serial prediction problem. Usually, many approaches present a large number of false positive and false negative. The proposals based on intrusion detection depend on database, which requires a continue updating in order to detect new attacks.

Our approach takes advantage of the multi-agent system to reanalyze the problem in a distributed mode. Moreover, intrusion detection technique based on misuse and anomaly has been incorporated at strategic level into the architecture. The detection by anomaly is built by means of a case-based reasoning (CBR) mechanism [7], whose characteristics do it especially suitable to tackle classification problems and this is reinforced with the predictive capacity of a mixture of neural network [14]. The capture of SQL queries is carried out through of distributed agents and the detection can be executed in a distributed mode. Moreover, the architecture presents a high scalability, flexibility and learning capacity that allows it a greater adaptation for distributed environments and new strategic of attacks.

3 Detection SQL Injection Based on Multi-agent Architecture

The agents are characterized through their capacities such as autonomy, reactivity, proactivity, social abilities, reasoning, learning and mobility [6]. One of the main features of agents is their ability to carry out cooperative and collaborative work, when they are grouped into multi-agent systems to solve problems in a distributed way [15], [16] [17] These features make to the agents suitable to face the SQL injection attack problem. A distributed hierarchical multi-agent presents a great capacity for the distribution of task and responsibilities, error recovering, adaptation to new changes and high level of learning. These factors are keys to achieve a robust and efficient solution. One main innovation of the architecture is the use of a CBR-BDI agent [8], which presents a great capacity of learning and adaptation. The agents BDI have a deliberative structure based on the BDI model [6]. Moreover, a BDI agent integrates a case-based reasoning mechanism [7] that allow it solve problems through the use de past experiences. As the core of the strategy for the classification of SQL queries is based on an anomaly detection technique, it seems appropriate to use a CBR mechanism [7] that leverage past experience to detect anomaly. This CBR mechanism additionally incorporates a mixture of neural networks [14] in its reuse phase. Using a mixture of neural networks improves the performance provided by other classification techniques such as Back-Propagation Neural Networks, Bayesian Forecasting Method, Exponential Regression, Polynomial Regression, Linear Regression, but also improves performance provided by the neural networks working individually. The number of cases where the classifier mechanism can not provide a decision is small and in few cases would be needed the intervention of a human expert.

An additional advantage provided by the architecture is the ability for executing agents on mobile devices. It is common to find SQL queries that can be originated from different mobile devices including assistant personals (PDA), mobile phones, notebook computers and workstations. Specialized agents (misuse and anomaly) can be organized in a distributed way to take advantage of available resources and improve the performance of the classification process, regardless of the nature of the physical devices. Finally, another advantage to use this type of agents is the ability to inform to security staff about events that are happening regardless of their physical location, sending alerts on mobile devices. All these advantages are achieved through an organizational design based on a hierarchical multi-agent architecture. These agents are distributed so when a classification starts, each type of agent knows its concrete tasks; the data required to carry out its job and where to send its results. The interaction and communication between the agents is crucial to achieve the goal of classification of the new SQL query:

Next, is described each type of agent within of the architecture:

- Sensor agents: Located in each of the devices accessing the database. They have 3 specific functions: a) The capture of datagrams launched by the devices. b) Order TCP fragments to extract the request's SQL string. c) Syntactic analysis of the request's SQL string. The duties of the agent Sensor end when the results (the SQL string transformed by the analysis, the result of the analysis of the SQL string and the user data) are sent to the next agent at the hierarchy of the classification process.

- FingerPrint agents: The numbers of agents FingerPrint depend on the workload at a given time. An agent FingerPrint receives the information of a Sensor agent and executes a pattern matching known attacks stored at a previously built database. The FingerPrint agent finishes its task when it sends its results to the Anomaly agent. The results of the FingerPrint agent consist of the SQL string transformed by the analysis, the result of the analysis of the SQL string, the user data and the results achieved by pattern matching.
- Pattern Agent: It is the responsible to save the new SQL string patterns in the database and search for patterns when the FingerPrint agent requests it.
- Anomaly agents: These agents are based on the CBR-BDI model. They are the key component within the classification process. Their strategy is based on a case-based reasoning mechanism that incorporates a mixture of neural networks. These agents retrieval those similar past cases to the new case of classification, training the neural networks with the recovered cases and generating the final classification. The numbers of Anomaly agents depend on the workload at a given time. The result of the classification is sent to the Manager agent for the evaluation.
- Loguser agent: This agent records the actions of the user and searching for the user profile (the historical profile and the user statistics) when it is requested by the Anomaly agent.
- Manager agent: It is the agent responsible for decision-making, evaluation and coordination of the overall operation of architecture. It evaluates the final decisions for classifications, manages alerts of attacks and coordinates the actions necessary when an attack is detected. It selects an Anomaly agent by means of a voting method.

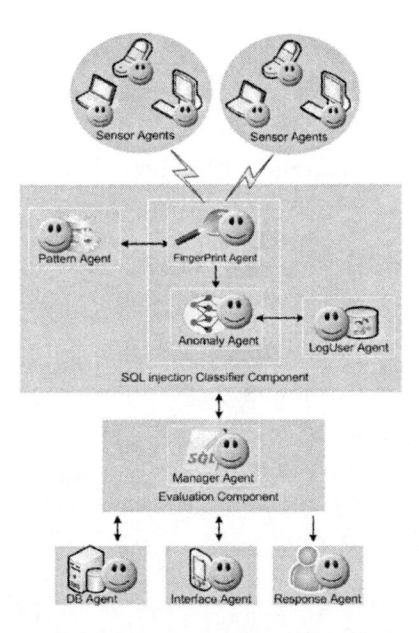

Fig. 1. Description of the hierarchical multi-agent architecture

- Interface agent: This agent allows the interaction of the user of the security system with the architecture. The interface agent communicates the details of an attack to the security personnel when an attack is detected. It has the ability to work on mobile devices. This capacity allows a ubiquitous communication to attend the alerts immediately.
- DB agent: It is in charge of executing the query in the database when the classification of the SQL query is legal, that is, the SQL query is not malicious.
- Response agent: This agent provides an answer to the user once obtained a solution of the classification. If the query has been classified as legal, the result of the query is sent to the user interface. Otherwise, if the query has been classified as illegal, it is sent to the user interface a warning message.

In figure 1 is presented the hierarchical multi-agent architecture showing different types of agents in charge of the classification of SQL queries.

4 Classifier Model of SQL Injection Attacks

The Anomaly agent type has been specially adapted to resolve the SQL injection attack problem. This agent is based on CBR-BDI model [8], which incorporates a case-based reasoning system that allows the detection and blocking of SQL injection attacks reusing previous experiences. This mechanism uses a prediction model based on neural networks, configured for short-term predictions of intrusions. To carry out this short-term prediction, the CBR mechanism uses a memory of cases which identifies past experiences with the corresponding indicators that characterize each of the attacks. A case is defined as a previous experience and is composed of three elements: a description of the problem; a solution; and the final state. To introduce a CBR motor into a BDI agent, we represent CBR system cases using BDI and implement a CBR cycle. This CBR cycle consists of four steps: retrieve, reuse, revise and retain. The elements of the SQL query classification problem are described as follows:

- Problem Description: It describes the initial information available for generating a classification. As can see in Table 1, the problem description consists of a case identification, user session and SQL query elements.
- Solution: Describes the action carried out to solve the problem description. As shown in Table 1, contains the case identification and the applied solution.
- Final State: Describes the achieved state after that the solution has been applied. It takes three possible values: attack, not attack o suspect. The Manager agent allows an expert to evaluate the classification.

The integration CBR-BDI [8] allows a BDI agent to use case-based reasoning to resolve the problem of classifying of a SQL query and blocking SQL injection attack. Regarding each state of a CBR system equivalent to a belief, the intention will be the plan that contains an ordered set of actions that the CBR-BDI agent should make to achieve the goals and each desire corresponds to one or more of the achieved final states in the past. The result of classifying a SQL query as attack, not attack or suspect is the desire that the agent seeks to achieve.

Table 1. Problem definition and solution for a case of SQL query classification

Problem Description fields		Solution fields	
IdCase	Integer	Idcase	Integer
Sesion	Session	Classification_Query	Integer
User	String		
IP_Adress	String		
Query_SQL	Query_SQL		
Affected_table	Integer		
Affected_field	Integer		
Command_type	Integer		
Word_GroupBy	Boolean		
Word_Having	Boolean		
Word_OrderBy	Boolean		
Numer_And	Integer		
Numer_Or	Integer		
Number_literals	Integer		
Length_SQL_String	Integer		
Cost_Time_CPU	Float		
Start_Time_Execution	Time		
End_Time_Execution	Time		
Query_Category	Integer		

The proposed mechanism is responsible for classifying SQL database requests made by users. When a user makes a new request, it is checked for matching well-known patterns of attack by a FingerPrint agent (misuse detection). These patterns are stored at a database that handles a significant number of signature not allowed on user level such as symbol combination, binary and hexadecimal encoding and reserved statement of language (union, execute, drop, revoke, concat, length, asc, chr among others). If the FingerPrint agent detects some known signature, it is automatically identified as an attack. In order to identify the rest of the SQL attacks, the Anomaly agent uses a CBR mechanism, which must have a memory of cases dating back at least 4 weeks, with the structure described in Table 1. The problem description of a case is obtained by means of a string analysis technique on the SQL query. This process can be understood easily through the following example: It is captured a SQL query with the following syntax: Select field1, field2, field3 from table1 where field1 = input1 and field2=input2. If we assume that the fields input1 and input2 are used to bypass the authentication mechanism with the following input data: Input1=' or 9876= 9876 -- and Input2= (blank). The result of these input data would alter the SQL string as follows: Select field1, field2, field3 from table1 where field1 ='' or 9876 = 9876 -- 'and field2=''

A syntactical analysis of the SQL string would generate the result presented in the Table 2 with the following fields: Affected_table[c1], Affected_field[c2], Command_type[c3], Word_GroupBy[c4], Word_Having[c5], Word_OrderBy[c6], Numer_And[c7], Numer_Or[c8], Number_literals[c9], Length_SQL_String[c10], Cost_Time_CPU[c11]. The fields Command_type and Query_Category have been encoding with the following nomenclature Command_Type: 0=select, 1=insert, 2=update, 3=delete; Query_Category: -1= suspect, 0=illegal, 1=legal.

Table 2. SQL String transformed through a syntactical analysis technique

c1	c2	c3	c4	c5	c6	c7	c8	c9	c10	c11	c12
1	3	0	0	0	0	1	1	2	81	0,3	0

The first phase of the CBR cycle consists of a retrieval past experience from the memory of cases, specifically those with a similar problem description to the current case. In order to carry out this process, a cosine similarity-based algorithm is applied, allowing the retrieval of those cases which are at least 90% similar to the current case. The cases recovered are used to train the mixture of neural networks implemented in the reuse phase; the neural network with the sigmoidal function is trained with the retrieved cases that were an attack or not, whereas the neural network with hyperbolic function is trained with all the recovered cases (including the suspects). A preliminary analysis of correlations is required to determine the number of neurons of the input layer of the neuronal networks. Additionally, it is necessary to normalize the data (i.e., all data must be values in the interval [0,1]). The data used to train the mixture of networks must not be correlated. With the cases stored after deleting correlated cases, the inputs for training the mixture of networks are normalized. It is considered to be two neural networks. The result obtained using a mixture of the outputs of the networks provides a balanced response and avoids individual tendencies (always taking into account the weights that determine which of the two networks is more optimal). Figure 2 explains the four steps of CBR cycle, which incorporates a mixture of neural networks through an algorithm. This strategy of classification is carried out within the Anomaly CBR-BDI agent. This Anomaly CBR-BDI agent is located on a strategic level into the architecture.

```
Algorithm_Solution_CBR(new_case)  (CBR Algorithm)
Begin
  cases:=Retrieve(new_case)   (case retrieval function)
  assessment:=Reuse(cases[], new_case) (result of the classification)
  decision:=Revise(new_case, assessment) (revision of the classification)
  If decision then  (If decision is accepted)
     Retain(new_case, solution) (update of the memory base)
  End if
End
Algorithm_Retrieve(new_case)  (Retrieve Algorithm)
Begin
   SQL_Query:=Select cases from Tb_Cases Where
   If [User] and [Ip_adress] then   (If User and Ip Adrress is recognized)
     SQL_Query+="user=[user]and IP_Adress=[IP_adress] and Command_type =
     [Command_type] and Affected_table=[Content(Affected_table)]"
   Else
       If [User] then (If only one User is recognized)
         SQL_Query+="User=[User] and Command_type=[Command_type]and
         Affected_table=[Content(Affected_table)]"
         Else
           If [Ip_adress] then (If only one Ip Adress is recognized)
               SQL_Query+="IP_adress=[IP_adress] and Command_type=
               [Command_type] and Affected_table=[Content(Affected_table)]"
           Else  (If user and Ip Adress are not included in the query)
               SQL_Query+="Command_type=[Command_type] and
               Affected_table=[Content(Affected_table)]"
           End If
       End If
   End If
   cases[]:=executeQuery(SQL_Query) (recovering of the cases in database)
   cases[]:=fsimilirity_cosine(cases[]) (cosine similarity-based algorithm)
   cases[]:=fcorrelation(cases[]) (eliminating correlated cases)
End
Algorithm_Reuse(cases[], new_case) (Reuse Algorithm)
Begin
  blenew_case=false
  If description_new_case<>descripcion_previous_case then
     blnew_case=true  (If user or Ip_adress are different of previous case)
  End If
  If blnew_case then   (If is true then training neural network)
    Input:=Retrieve_Input(cases[]) (Retrieval of input)
    Output:=Retrieve_Output(cases[]) (Retrieval of ouput)
    (Training of the neural network)
    error_training:=Training_Neural_Network(Input, Output)
    if (error_training)=low then
       (Classification by mixture of neural network)
       assessment:=Classification_Neural_Network(new_case)
    Else
       Exception(Error_Code, Description) (Imposible to classify new case)
    End If
  Else
   (Classification by mixture of neural network)
    assessment:=Classification_Neural_Network(new_case)
  End If
End
Algorithm_Revise(new_case, assessment)  (Revise Algorithm)
Begin
   Boolean decision
   If new_case=complete and assessment=optimal then (Evaluation by an expert)
      decision:=true
   End If
End
Algorithm_Retain(new_case, solution) (Retain Algorithm)
Begin
   executeUpdate(new_case, solution) (Update case memory with new case)
End
```

Fig. 2. CBR Cycle Algorithm for classifying SQL query

Additionally, we mean to detect attacks, so if one only network with a sigmoidal activation function was used, then the result provided by the network would tend to be attack or not attack, and no suspects would be detected. On the other hand, if only one network with a hyperbolic tangent activation was used, then a potential problem could exist in which the majority of the results would be identified as suspect although they were clearly attack or not attack. The mixture provides a more efficient configuration of the networks, since the global result is determined by merging two filters. This way, if the two networks classify the user request as an attack, so too will the mixture; and if both agree that it is not an attack, the mixture will as well be. If there is not concurrence, the system uses the result of the network with the least error in the training process or classifies the user request as suspect. In the reuse phase the two networks are trained by a back-propagation algorithm for the same set of training patterns (in particular, these neural networks are named Multilayer Perceptron), using a sigmoidal activation function (which will take values in [0,1], where 0 = Illegal and 1 = legal) for a Multilayer Perceptron and a hyperbolic tangent activation function for the other Multilayer Perceptron (which take values in [-1,1], where -1 = Suspect, 0 = illegal and 1 = legal). The response of both networks is combined, obtaining the mixture of networks y^2; where the superscript indicates the number of mixed networks

$$y^2 = \frac{1}{\sum_{r=1}^{2} e^{-|1-r|}} \sum_{r=1}^{2} e^{-|1-r|} y^r \quad (1)$$

$$Error = \frac{1}{P} \sum_{i=1}^{P} \left| \frac{Forecast_P - Target_P}{Target_P} \right| \quad (2)$$

The number of neurons in the output layer for both Multilayer Perceptrons is 1, and is responsible for deciding whether or not there is an attack. The error of the training phase for each of the neural networks, can be quantified with formula (2), where P is the total number of training patterns.

5 Results and Conclusions

SQL injection attacks on databases suppose a serious threat against information systems. This paper has presented a distributed hierarchical multi-agent architecture incorporating a novel type of agent based on the CBR-BDI model [8] specially designed for detecting and blocking SQL injection attacks. This CBR-BDI agent handles a great adaptation and learning capacities using a CBR mechanism. In addition, it incorporates the prediction capabilities that characterize neural networks. As a result, an innovative and robust solution has been presented allowing a significant reduction of the error rate during the classification to attacks and a different way to tackle SQL injection attacks using a distributed and hierarchical approach. To check the validity of the proposed model many tests were done. These tests were executed on a memory of cases, specifically developed to generate malicious queries. In Table 3 it is possible to observe techniques for predicting attacks at the database layer and the errors associated with misclassifications. All the techniques presented in Table 3 have been applied under similar conditions to the same set of cases, taking into account the same problem common to all the methods. Note that the technique proposed in this paper provides the best results, with an error in only 0.5% of the cases.

As shown in Table 3, the Bayesian is the most accurate statistical method since it is based on the likelihood of the events observed. But it needs determining the initial

Table 3. Results obtained after testing different classification techniques

Forecasting Techniques	Successful (%)	Approximated Time (secs)
Anomaly Agent (CBR-BDI)	99.5	2
Back-Propagation Neural Networks	99.2	2
Bayesian Forecasting Method	98.2	11
Exponential Regression	97.8	9
Polynomial Regression	97.7	8
Linear Regression	97.6	5

parameters of the algorithm. Considering the errors obtained with the different methods, the Anomaly Agent and Bayesian methods provide the better results. Because of the non linear behaviour of the hackers, linear regression offers the worst results, followed by the polynomial and exponential regression. This can be explained by looking at hacker behaviour: as the hackers break security measures, the time for their attacks to obtain information decreases exponentially. The empirical results show that the best methods are those that involve the use of neural networks and, if it is considered a mixture of two neural networks, the predictions capabilities are remarkably improved. These methods are more accurate than statistical methods for detecting attacks to databases as the behaviour of the hacker is not linear, dynamic and chaotic.

Acknowledgements. This development has been partially supported by the Spanish Ministry of Science project TIN2006-14630-C03-03.

References

1. Halfond, W., Orso, A.: AMNESIA: Analysis and Monitoring for Neutralizing SQL-injection Attacks. In: 20th IEEE/ACM international Conference on Automated software engineering, pp. 174–183. ACM, New York (2005)
2. Kosuga, Y., Kono, K., Hanaoka, M., Hishiyama, M., Takahama, Y.: Sania: Syntactic and Semantic Analysis for Automated Testing against SQL Injection. In: 23rd Annual Computer Security Applications Conference, pp. 107–117. IEEE Computer Society, Los Alamitos (2007)
3. Valeur, F., Mutz, D., Vigna, G.: A Learning-Based Approach to the Detection of SQL Attacks. In: Conference on Detection of Intrusions and Malware and Vulnerability Assessment, Vienna, pp. 123–140 (2005)
4. Rietta, F.: Application layer intrusion detection for SQL injection. In: 44th annual Southeast regional conference, pp. 531–536. ACM, New York (2006)
5. Abraham, A., Jain, R., Thomas, J., Han, S.Y.: D-SCIDS: distributed soft computing intrusion detection system. Journal of Network and Computer Applications 30, 81–98 (2007)
6. Woolridge, M., Wooldridge, M.J.: Introduction to Multiagent Systems. John Wiley & Sons, Inc., New York (2002)
7. Aamodt, A., Plaza, E.: Case-based reasoning: foundational issues, methodological variations, and system approaches. AI Commun. 7, 39–59 (1994)

8. Laza, R., Pavon, R., Corchado, J.M.: A Reasoning Model for CBR_BDI Agents Using an Adaptable Fuzzy Inference System. In: Conejo, R., Urretavizcaya, M., Pérez-de-la-Cruz, J.-L. (eds.) CAEPIA/TTIA 2003. LNCS, vol. 3040, pp. 96–106. Springer, Heidelberg (2004)
9. Anley, C.: Advanced SQL Injection. In: SQL Server Applications (2002), http://www.nextgenss.com/papers/advancedsqlinjection.pdf
10. Christensen, A.S., Moller, A., Schwartzbach, M.I.: Precise Analysis of String Expressions. In: Cousot, R. (ed.) SAS 2003. LNCS, vol. 2694, pp. 1–18. Springer, Heidelberg (2003)
11. Su, Z., Wassermann, G.: The essence of command injection attacks in web applications. In: 33rd Annual Symposium on Principles of Programming Languages, pp. 372–382. ACM Press, New York (2006)
12. Huang, Y., Huang, S., Lin, T., Tsai, C.: Web application security assessment by fault injection and behavior monitoring. In: 12th international conference on World Wide Web, pp. 148–159. ACM, New York (2003)
13. Skaruz, J., Seredynski, F.: Recurrent neural networks towards detection of SQL attacks. In: 21th International Parallel and Distributed Processing Symposium, pp. 1–8. IEEE International, Los Alamitos (2007)
14. Ramasubramanian, P., Kannan, A.: Quickprop Neural Network Ensemble Forecasting a Database Intrusion Prediction System. In: 7th International Conference Artificial on Intelligence and Soft Computing, Neural Information Processing, vol. 5, pp. 847–852 (2004)
15. Corchado, J.M., Bajo, J., Abraham, A.: GerAmi: Improving Healthcare Delivery in Geriatric Residences. Intelligent Systems 23, 19–25 (2008)
16. Corchado, J.M., Bajo, J., de Paz, Y., Tapia, D.: Intelligent Environment for Monitoring Alzheimer Patients, Agent Technology for Health Care. Decision Support Systems 44(2), 382–396 (2008)
17. Corchado, J.M., Gonzalez-Bedia, M., De Paz, Y., Bajo, J., De Paz, J.F.: Replanning mechanism for deliberative agents in dynamic changing environments. Computational Intelligence 24(2), 77–107 (2008)

Trusted Computing: The Cornerstone in the Secure Migration Library for Agents[*]

Antonio Muñoz, Antonio Maña, and Daniel Serrano

University of Malaga
{amunoz,amg,serrano}@lcc.uma.es

Abstract. The agent paradigm can play an important role and can suit the needs of many applications in new emerging Ambient Intelligence scenarios. Unfortunately the lack of security is hindering the application of this technology in real world applications. The problem known as malicious hosts is considered the most difficult to solve in mobile agent. Therefore, we address this problem by means of a secure migration process using for that the Trusted Computing technology. In this paper we aim to study the benefits in the introduction of trusted computing technology can bring to the mobile agent technology.

1 Introduction

A well accepted definition of software agent is found in [12]: "An agent is an encapsulated computer system that is situated in some environment and that is capable of flexible, autonomous action in that environment in order to meet its design objectives". More important characteristics of an agent are that are highly agree upon to include: Autonomy (behaves independent according to the state it encapsulates), proactiveity (able to initiate without external order), reactivity (react in timely fashion to direct environment stimulus), situatedness (ability to settle in an environment that might contain other agents), directness (agents live in a society of other collaborative or competitive groups of agents), and the social ability to interact with other agents, possibly motivated by collaborative problem solving.

One of the most relevant features of agent technology is that an agent is supposed to live in a society of agents; multi-agent systems (MAS). A MAS is known as a system composed of several agents collectively capable of reaching goals that are difficult to achieve by an individual agent of monolithic system. These represent a natural way of decentralization, where there are autonomous agents working as peers, or in teams, with their own behaviour and control. Mobile agents are a specific form of mobile code and software agent paradigms. However, in contrast to the Remote evaluation and Code on demand paradigms,

---

[*] Work partially supported by E.U. through projects SERENITY (IST-027587) and OKKAM (IST- 215032) and DESEOS project funded by the Regional Government of andalusia.

mobile agents are active in that they may choose to migrate between computers at any time during their execution. This makes them a powerful tool for implementing distributed applications in a computer network. Agent migration consists on a mechanism to continue the execution of an agent on another location. This process includes the transport of agent code, execution state and data of the agent. In an agent system, the migration is initiated on behalf of the agent and not by the system. The main motivation for this migration is to move the computation to a data server or a communication partner in order to reduce network load by accessing a data server a communication partner by local communication. Then migration is performed from a source agency where agent is running to a destination agency, which is the next stage in agent execution. This migration can be performed by two different ways. First way is by moving, that is, the agent moves from the source agency to the destination one. And the second way is by cloning. In this case, the agent is copied to the destination agency. From this moment on, the two copies of the agent coexist executing in different places. In the remainder of this paper and at least stated explicitly we will use the term migration to refer to both agent cloning and agent moving. Software agents are a promising computing paradigm. It has been shown [1] that scientific community has devoted important efforts to this field. Indeed, many applications exist based on this technology. However, all this efforts are not useful in practice due to the lack of a secure environment for agent-based applications. Let us introduce an example to highlight the lack of security in mobile agent systems. This example consists of a mobile agent that plans to migrate and visit sites run by airlines searching for the cheaper flight from Barcelona to London. The customer dispatches the agent to the first airline, which is the first agency in the agent plan, where the agent queries the flight database in this agency. Thus the agent stores the results in its environment and migrates to the second airline where it again queries the database. This process is repeated for every airline in the agent plan. Finally the agent compares flight prices and migrates to the chosen agency to book the desired flight and returns to the customer with their results. The customer can expect that every airline provides true information on flight schedules and fares in an attempt to win her business, just as we assume nowadays that the reservation information the airlines provide over the telephone is accurate, although it is not always complete. Nevertheless, the airlines are in a competitive relation with each other and there are several kinds of attacks they may attempt. For instance, the second airline may be able to corrupt the flight information of the first airline raising the price of its competitor. Thus, the mobile agent cannot decide its best price flight since the agency has the ability to manipulate this decision. In [1] two different solutions are presented to provide a security framework for agents building. The first of these is a software solution based on the protected computing approach. The second one consists on a hardware based solution that takes advantage of the TPM technology. In this paper we use that theoretical and abstract solution to build a solution for the secure migration of software agents. The remainder of this paper is structured as follows. In section 2, we review the previous work in the agent protection. Section 3 deals the role of trusted computing in the agent protection. In section 4, we describe our solution a tangible approach for the agent protection. Finally we conclude with section 5.

2 Previous Work

The hardware-based protection infrastructure described in the present paper takes advantage of the recent advances in trusted hardware. The basic idea behind the concept of Trusted Computing is the creation of a chain of trust between all elements in the computing system, starting from the most basic ones. Consequently, platform boot processes are modified to allow the TPM to measure each of the components in the system and securely store the results of the measurements in Platform Configuration Registers (PCR) within the TPM. This mechanism is used to extend the root of trust to the different elements in the computing platform. Therefore, the chain of trust starts with the aforementioned TPM, which analyses whether the BIOS of the computer is to be trusted and, in that case, passes control to it. This process is repeated for the master boot record, the OS loader, the OS, the hardware devices and finally the applications. In a Trusted Computing scenario a trusted application runs exclusively on top of trusted and pre-approved supporting software and hardware. Additionally the TC technology provides mechanisms for the measurement (obtaining a cryptographic hash) of the configuration of remote platforms. If this configuration is altered or modified, a new hash value must be generated and sent to the requester in a certificate. These certificates attest the current state of the remote platform.

Several mechanisms for secure execution of agents have been proposed in the literature with the objective of securing the execution of agents. Most of these mechanisms are designed to provide some type of protection or any specific security property. In this section we will focus on solutions that are specifically tailored or especially well-suited for agent scenarios. More extensive review of the state of the art in general issues of software protection can be found in [6].Some protection mechanisms are oriented to the protection of the host against malicious agents. Among these, SandBoxing [7], proof-carrying code [8], and a variant of this technique, called proof-referencing code [9]. One of the most important problems of these techniques is the difficulty of identifying which operations (or sequences of them) can be permitted without compromising the local security policy. Other mechanisms are oriented toward protecting agents against malicious servers. Among them the concept of sanctuaries [10] was proposed. Several techniques can be applied to an agent in order to verify self-integrity and avoid that the code or the data of the agent is inadvertently manipulated. Anti-tamper techniques, such as encryption, checksumming, anti-debugging, anti-emulation and some others [11] [13] share the same goal, but they are also oriented toward the prevention of the analysis of the function that the agent implements. Additionally, some protection schemes are based on self-modifying code, and code obfuscation [14]. There are techniques that create a two-way protection. Some of these are based on the aforementioned protected computing approach. Finally, the use of trusted computing in agent systems has been put forward in [2, 3, 4]. In [2, 3] this technology is applied to achieve the user privacy in static agents. SMASH [4] is a system that uses the trusted computing technology to form a middleware instance. However, the use of these techniques for agent-based developers entails a hard challenge due to the lack expertise in security of this profile of developers. In this paper we present a solution for the malicious host problem transparent in its usage for agent-based systems developers.

3 The Role of Trusted Computing in the Agent Protection

We aforementioned that is essential to integrate strong security mechanisms in agent based systems to achieve a reasonable security level to support real world applications. For this reason, we present a hardware-based mechanism to provide security to agent systems. The TPM provides mechanisms, such as cryptographic algorithms, secure key storage and remote attestation that provides important tools to achieve a high level of security. The most relevant mechanism to build our solution is the remote attestation. This is the process by which a platform declares its current operating environment-software state as recorded in a dedicated set of TPM registers named platform configuration registers (PCRs).

Unfortunately, the TPM technology is very complex and so are the procedures and action sequences to use it. The access to the device and the use of TPM functionality is not an easy task, especially for average software developers. Therefore, we have developed a library that provides access to the TPM functionalities to software agents. Although the main advantage of this approach is that developers of agent systems do not need a security expertise, and they are able to use easily appropriate security mechanisms. Hardware-based solutions can be built on the basis of different devices. Among them we have selected the TPM for our solution because it provides features such as cryptography capabilities, secure storage, intimate relation to the platform hardware, remote attestation, etc. Remote attestation is perhaps the most important feature; it allows the computer to recognize any unauthorized/authorized changes to software. If certain requirements are met then the computer allows the system to continue its operations. This same technology that TPM utilizes can also be applied in agent-bases systems to scan machines for changes to their environments made by any entity before the system boots up and accesses the network. Additional reasons are the support of a wide range of industries and the availability of computers equipped with TPM In summary, this element is the cornerstone of our approach and the security of our system is focused on it. The use of TPM in these systems is as follows. Each agency takes measures of system parameters while booting to determine its security, such as BIOS, keys modules from Operating System, active processes and services in the system. Through these parameters an estimation of the secure state of the agency can be done. Values taken are stored in a secure way inside the trusted device, preventing unauthorized access and modification. Agencies have the ability to report previously stored configuration values to other agencies in order to prove their security. Our system takes advantage of the remote attestation procedure provided by the TPM in order to verify the security of the agencies. In this way agents can verify the security of the destination agency before migration.

4 A Library to Build Secure Agent Based Systems

This Section introduces the Secure Migration Library for Agents that is a platform for the agent migration on top of JADE. In order to provide a secure environment this library makes use of a TPM. Thus, the migrations are supported by a trusted platform. TPM4Java provides access to the TPM functions from the Java programming language, but TPM4Java was extended, as described in next sections. Among the main objectives of this library we highlight the following; (i) to provide a secure environment, which agents can be securely executed and migrated in is the most important target of this library. Although some features are desirable to meet such as (ii) easily to integrate the TPM technology with JADE, (iii) to provide mechanisms that are simple to use from the programmers point of view, (iv) easily adaptable with another security devices such as smartcards, (v) a flexible interface to allow the users to exploit the library to solve a wide range of problems, (vi) extendedable with new functionalities in the future. Despite of all these features, the most important function provided by the library is the secure migration mechanism. This mechanism is based on the remote attestation of the destination agency. Before, the migration actually takes place to guarantee that agents are always executed in a secure environment. Additionally we provide a secure environment for agent execution. A Secure environment means that a malicious agent is not able to modify the host agency. Our library was designed according to the following set of requirements. On the one hand, the functional requirement set is the provision of a mechanism for secure agent migration in JADE platform. The goal was to make possible for agents to extend their trusted computing base to the next destination agency before migrating to it. It is important to mention that each agency must provide local functionality to perform the secure migration mechanism, which requires the presence of the TPM for remote attestation. Similarly each agency must provide the functionality to allow to other agencies to take remote integrity measures to determine whether its configuration is secure. In order to support the previous processes, the library includes a protocol that allows two agencies to exchange information about their configurations. This exchange establishes the bases of the mutual trust between agencies. Concerning the non-functional requirements, the library must seamlessly integrate in the JADE platform, in such a way that its usage does not imply modifications in the platform. Additionally, the operation of the library must be transparent to the user. We decided to create an additional abstraction layer that provides generic services in order to facilitate the adaptation of the library to new devices. In this way the generic services keep unchanged. One final requirement has to support both types of migration (moving and cloning).

4.1 The Basis: Secure Migration Protocol

Along this section we present the protocol that provides the security to the migration. Firstly, we analyse the different attestation protocols as well as the secure migration protocols, studying their benefits, concluding with our own secure migration protocol. A first approach of a secure migration protocol is in [1]. This protocol provides some important ideas to take into account during the design process of the final protocol. Among them we highlight that the agent trusts in its platform to check the migration security, so that the necessity of the use of TPM to obtain and report configuration data by a secure way. Other protocol that provides interesting ideas to take into account when we developed the secure migration system is in [11]. The most relevant ideas provided are; an agent from the source agency requests to TPM the signed values from PCRs. The protocol shows how the agent obtains platform credentials. These credentials together with PCRs signed values allow to determine if destination configuration is secure. Finally, we analyse the protocol from [11]. More relevant ideas from this protocol are; the use of an AIK to sign the PCR values, the use of a Certification Authority that validates the AIK, and the use of configurations to compare received results from remote agency. We designed a new protocol based on the study of these three protocols, in such a way that we took advantage of the appeals provided by each of them. Our protocol has some common features, that is, the agency provides to the agent the capacity to migrate by a secure way. Also the agency uses a TPM that provides configuration values stored in PCRs. The TPM is used to sign the PCRs values using a specific AIK for the destination agency. In such a way that data receiver knows the identity of the remote TPM. A Certification Authority generates the needed credentials to correctly verify the AIK identity. Together with the signed PCRs values the agency provides AIK credentials in such a way that the requester can correctly verify that data comes from agency TPM. Following we define the 18 steps protocol, used to perform secure migration. Secure migration protocol is composed of the following stages: We can observe that the protocol fulfills the five main characteristics we mentioned previously. Hence, we have a clear idea of the different components of the system as well as the interaction between them to provide the security in the migration.

4.2 Verification of Secure Migration Protocol with AVISPA

Secure Migration protocol previously described is the basis of this research. Thus we want to build a robust solution, for this purpose a validation of this protocol is the next step in this research. Among different alternatives we selected a model checking tool called AVISPA. AVISPA is an automatic push-button formal validation tool for Internet security protocols, developed in a project sponsored by the European Union. It encompasses all security protocols in the first five OSI layers for

Algorithm 1. Secure migration protocol

1: Agent requests to source agency S migrate to destination agency D.
2: Source agency S sends to destination agency D a request query.
3: Destination agency D accepts this query and sends a nonce (random bits used to avoid replay attacks) as well as PCR values to know '(NONCEd,PCRId)'.
4: A TPM_Quote ('getConfig(NONCEs,PCRIs)') is done in the source agency. This consists on the source agency S requests to source TPM takes measures of the PCR together with the nonce all signed. The signature produced is 'SIG(PCRs(PCRId),NONCEd,AIKsd)'.
5: TPM returns requested data 'SIG(PCRs(PCRId),NONCEd,AIKsd)'.
6: Source agency S obtains AIK credentials from his credentials repository 'AIKsd'.
7: S sends to D PCR values requested together with the nonce all signed. S sends AIK credentials, these contains the public key related to the private key used to sign these data. Besides S sends a nonce and PCR values to know ('SIG(PCDd(PCRId),NONCEs),AIKds,PCRd(PCRIs),CREDds').
8: D validates the authenticity of the received key verifying the credentials by means of CA public key that generated those credentials ('verify(CREDsd,PCRs)').
9: Destination agency D verifies the signature of PCR values and nonce received using the AIK public key ('verify(SIG(PCRs,PCRId),NONCEd,AIKsd)').
10: Destination agency D verifies that PCR values ('check(PCRs,TrustSetd)') received belongs to the set of acceptable values and therefore source agency S configuration is secure.
11: A TPM_Qoute ('getConfig(NONCEs,PCRIs)') is done in destination Agency. Then destination agency $\overline{\text{D}}$ requests to destination TPM the PCR values S together with the nonce all signed, the signature is 'SIG(PCRd(PCRIs),NONCEs,AIKds'.
12: TPM returns requested signed data 'SIG(PCRd(PCRIs),NONCEs,AIKds'.
13: Destination agency D obtains AIK credentials from its credentials repository calling to 'getCredentials(AIKds)'.
14: D sends to source agency S PCR the values requested and nonce signed '(remoteAttestResults(SIG(PCDd(PCRId),NONCEs),AIKds,PCRd(PCRIs),CREDds))'. Besides D sends AIK credentials, these contain the public key related to the private key used to sign data.
15: Source agency S validates the authenticity of received key verifying the credentials by means of the CA public key that generated those credentials, by means of 'verify(CREDds,PCRs)'.
16: Source agency S verifies PCR signature values ('verify(SIG(PCDd(PCRIs),NONCEd),AIKds)') and the nonce received using the AIK public key.
17: Source agency S verifies that PCR values received ('check(PCDd,TrustSets)') belongs to the set of valid values and therefore destination agency D configuration is secure. Henceforth a mutual trustworthy exists between source and destination agencies.
18: Source agency S sends to agent the confirmation to migrate to D ('migrateto(Destination Agency)').

more than twenty security services and mechanisms. Furthermore this tool covers (that is verifiable by it) more than 85 of IETF security specifications. AVISPA library available on-line has in it verified with code about hundred problems derived from more than two dozen security protocols. AVISPA uses a High Level Protocol Specification Language (HLPSL) to feed a protocol in it; HLPSL is an extremely expressive and intuitive language to model a protocol for AVISPA. Its operational semantic is based on the work of Lamport on Temporal logic of Actions. Communication using HLPSL is always synchronous. Once a protocol is fed in AVISPA and modelled in HLPSL, it is translated into Intermediate Format (IF). IF is an intermediate step where re-write rules are applied in order to further process a given protocol by back-end analyzer tools. A protocol, written in IF, is executed over a finite number of iterations, or entirely if no loop is involved. Eventually, either an attack is found, or the protocol is considered safe over the given number of sessions. System behaviour in HLPSL is modelled as a 'state'. Each state has variables which are responsible for the state transitions; that is, when variables change, a state takes a new form. The communicating entities are called 'roles' which own variables. These variables can be local or global. Apart from initiator and receiver, environment and session of protocol execution are also roles in HLPSL. Roles can be basic or composed depending on if they are constituent of one agent or more. Each honest participant or principal has one role. It can be parallel, sequential or composite. All communication between roles and the intruder are synchronous. The communication channels are also represented by the variables carrying different properties of a particular environment. The language used in AVISPA is very expressive allowing great flexibility to express fine details. This makes it a bit more complex than Hermes to convert a protocol into HLPSL. Further, defining implementation environment of the protocol and user-defined intrusion model may increase the complexity. Results in AVISPA are detailed and explicitly given with reachable number of states. Therefore regarding result interpretation, AVISPA requires no expertise or skills in mathematics contrary to other tools like HERMES[15] where a great deal of experience is at least necessary to get meaningful conclusions.

```
SUMMARY
  SAFE
DETAILS
  BOUNDED_NUMBER_OF_SESSIONS
PROTOCOL
  /home/anto/avispa/avispa-1.1/testsuite/results/protocolo_2.txt.if
GOAL
  as_specified
BACKEND
  OFMC
COMMENTS
STATISTICS
  parseTime: 0.00s
  searchTime: 3.23s
  visitedNodes: 4 nodes
  depth: 200 plies
environment()
```

These results show that the summary of the protocol validation is safe. Also some statistics are shown among them depth line indicates 200 plies, but this process has been performed for 250, 300, 400, 500, 1000 and 2000 of depth values with similar results. Appendix A depicts the HLPSL code to verify the Secure Migration Protocol in AVISPA. For this purpose we check the authentication on pcra(pcrib), pcrb(pcria), aikab and aikba, due to these are exchanges values among agent A and B.

4.3 Design and Deployment of the Secure Migration Library for Agents

This section analyses the deployment and design of the Secure Migration Library for Agents, for this purpose we study the architecture of this in details as well as we show the components and their related functionalities. Main use case consists on a user that uses this library to develop a regular, but secure multi-agent system. Traditionally in these kind of systems, the user defines the set of agents that compound the system. Concretely Agent class extends the Agent class defined in JADE. Among the basic functionality of a JADE Agent we found the compatibility with inter-containers migration. Two main services are provided by our library that uses the AgentMobility Service to perform a secure inter-platform migration in the same platform, and SecureInterPlatformMobility service that uses the InterPlatformMobility service to perform the secure intra-platform migration. Previously we mentioned that JADE Agent class provides two "insecure" migration methods, then we created a new class that is extended by this class and redefines migration methods to perform them securely.

5 Conclusions and Future Work

Among the future work that we consider interesting to address is the development of a migration library that implements the anonymous direct attestation. In this paper we presented the library based on an attestation protocol which uses a Certification Authority this concerns some disadvantages. Some of these disadvantages are that is needed to generate a certificate for each key used for the attestation, this implies that many request are performed to the CA and this is a bottleneck of the system. Also we found that if verification authority and CA act together the security of the attestation protocol can be violated. The use of the anonymous direct attestation protocol provides some advantages. Certify issuer entity and verifier entity can not act together to violate the system security, then one unique entity can be verifier and issuer of certificates. Last but not least of the advantages is that the certificates only need to be issued once which solves the aforementioned bottleneck. These advantages show that anonymous attestation protocol is an interesting option for attestation.

Improve the keys management system of the library. Our library uses RSA keys for attestation protocol that must be loaded in TPM to be used. However the space for available keys in the TPM is very limited, them it must be carefully managed to avoid arisen space problems. Our library handle the keys in such a way that only one key is loaded in TPM simultaneously, then keys are loaded when will use and downloaded when used. This procedure is not of maximum efficiency due to constantly are the keys loading and downloading. Indeed there is the possibility that we download the same key that we will use in next step. A possible improvement in key management is establish a cache in such a way that several keys can be loaded simultaneously in TPM and that these will be downloaded when we need more available space for new keys. For this we need to develop a keys replace policy to determine which key delete from TPM to load a new key in the cache.

Another future work is to integrate new functionalities to secure migration services to manage concurrent requests. Secure migration services implemented in the library provide secure migration to a remote container. However they have capacity to handle one unique request simultaneously, then migration requests arriving while migration is performed are refused. This happens due to TPM management of the problem together with the aforementioned key management problem. A possible improvement of the library is provided to secure migration services with the capability to accept and handle several migration requests simultaneously. For this is necessary to develop the previous described capability of several keys simultaneously loaded in the TPM. This improvement implies several difficulties related to key management. There is a possibility that the service must attend several requests which use different keys, forcing to manage the keys in such a way that these are available when are needed. Also we can find the possibility in which any key can be replaced because all are being used, in this case the request might wait until its key is loaded.

References

1. Maña, A., Muñoz, A., Serrano, D.: Towards secure agent computing for ubiquitous computing and ambient intelligence. In: Indulska, J., Ma, J., Yang, L.T., Ungerer, T., Cao, J. (eds.) UIC 2007. LNCS, vol. 4611, pp. 1201–1212. Springer, Heidelberg (2007)
2. D'Anna, L., Matt, B., Reisse, A., Van Vleck, T., Schwab, S., LeBlanc, P.: Self-protecting mobile agents obfuscation report. Technical Report 03-015, Network Associates Laboratories (2003)
3. Pearson, S.: How trusted computers can enhance for privacy preserving mobile applications. In: Proceedings of the 1st International IEEE WoWMoM Workshop on Trust, Security and Privacy for Ubiquitous Computing, Taormina, Italy, pp. 609–613. IEEE Computer Society, Washington (2005)
4. Riordan, J., Schneier, B.: Environmental key generation towards clueless agents. In: Vigna, G. (ed.) Mobile Agents and Security. LNCS, vol. 1419, pp. 15–24. Springer, Heidelberg (1998)
5. JADE project official home page, http://jade.tilab.com
6. Hachez, G.: A Comparative Study of Software Protection Tools Suited for E-Commerce with Contributions to Software Watermarking and Smart Cards. PhD Thesis. Universite Catholique de Louvain (2003),
http://www.dice.ucl.ac.be/~hachez/thesis_gael_hachez.pdf
7. Gosling, J., Joy, B., Steele, G.: The Java Language Specification. Addison-Wesley, Reading (1996)
8. Necula, G.: Proof-Carrying Code. In: Proceedings of 24th Annual Symposium on Principles of Programming Languages (1997)
9. Gunter Carl, A., Peter, H., Scott, N.: Infrastructure for Proof-Referencing Code. In: Proceedings of Workshop on Foundations of Secure Mobile Code (March 1997)
10. Yee, B.S.: A Sanctuary for Mobile Agents. Secure Internet Programming (1999)
11. Trusted Computing Group: TCG Specifications (2005),
https://www.trustedcomputinggroup.org/specs/
12. Wooldrigde, M.: Agent-based Software Engineering. In: IEE Proceedings on Software Engineering, vol. 144(1), pp. 26–37 (February 1997)
13. Stern, J.P., Hachez, G., Koeune, F., Quisquater, J.J.: Robust Object Watermarking: Application to Code. In: Pfitzmann, A. (ed.) IH 1999. LNCS, vol. 1768, pp. 368–378. Springer, Heidelberg (1999),
http://www.dice.ucl.ac.be/crypto/publications/1999/codemark.pdf
14. Collberg, C., Thomborson, C.: Watermarking, Tamper-Proofing, and Obfuscation - Tools for Software Protection. University of Auckland Technical Report 170 (2000)
15. Hussain, M., Seret, D.: A Comparative study of Security Protocols Validation Tools: HERMES vs. AVISPA. In: Proceecings of International Conference on Advanced Communication Technology (ICACT) 2006, IEEE Communication Society, Los Alamitos (2006)

Negotiation of Network Security Policy by Means of Agents

Pablo Martin, Agustin Orfila, and Javier Carbo

Universidad Carlos III de Madrid, Avda. de la Universidad 30, 28911 Leganes, Spain
100029778@alumnos.uc3m.es, adiaz@inf.uc3m.es, jcarbo@inf.uc3m.es

Abstract. Nowadays many intranets are deployed without enforcing any network security policy and just relying on security technologies such as firewalls or antivirus. In addition, the number and type of network entities are no longer fixed. Typically, laptops, PDAs or mobile phones need to have access to network resources occasionally. Therefore, it is important to design flexible systems that allow an easy administration of connectivity without compromising security. This article shows how software agents may provide secure configurations to a computer network in a distributed, autonomous and dynamic manner. Thus, here we describe the system architecture of a prototype, the negotiation protocol it uses and how it works in a sample scenario.

1 Introduction

In the last decade the number of local networks, particularly small ones, has grown considerably motivated by wireless technology. Many of these networks are usually deployed in environments where users lack sufficient technical knowledge and, as a consequence, they are not being properly protected. Thus, security policies are often not defined and network firewalls and antivirus are the unique defensive measures. This situation motivates the creation of systems that assist in security administration. In this paper we propose an agent system that dynamically and autonomously agrees if network services are allowed, depending on the security requirements of the entities involved, the threat model for each scenario and the vulnerability level of the network elements. Thus, each agent imposes the security constraints for the entity it represents. Then, agents negotiate in order to fulfill their interconnection goals. Different global configurations can be agreed and a network administrator is able to know the network policy negotiated at any moment.

The idea of using software agents for this purpose is relatively new. Traditionally, a network security policy determines what is allowed and what is not in a network [3]. It is centrally defined and centrally or distributely enforced [5]. Although this approach is more straightforward and rigorous, it demands a management and technical effort to define it and enforce it. This effort is not usually done in many medium and small networks what leaves them unprotected. In addition, current networks have become highly dynamic and heterogeneous due to mobile devices, what makes a centralized configuration harder to manage. These facts promote the exploration of the agent paradigm to provide a flexible, intelligent, dynamic and autonomous solution that improves the security administration of these networks. Related work include higher-level approaches like the one on policy-based governance by multiagent systems by Udupi and Singh [7, 8]. The multiagent architecture they propose is proactive (supporting policy monitoring, governance and enactment) and focused on virtual organizations (VO) while ours is reactive and focused on local area networks. In addition, Krügel and Toth [6] developed *Sparta* a system that allows to detect security policy violations and network intrusions in a heterogeneous, network environment. In order to fulfill this goal they use mobile agents for the task of correlating events in a fully decentralized manner. *Sparta* focuses in the detection of security policy violations but not in the policy definition itself.

Next section describes our proposed agent system architecture. This is followed by the exposition of the negotiation protocol between agents. Then, a sample scenario is explained in order to illustrate system functionality. Finally, the article ends up with the main conclusions.

2 System Architecture

In our proposal, each computer within the network has its own agent acting on behalf of it. In addition, there is an agent for each user of the network. When a user launches the execution of an agent, he determines the desired level of security by choosing a predefined security profile. Then the user informs his own agent about the operation he wants to perform and the agent carries out this task starting the negotiation with the other agents involved in the required task. Depending on the success of the negotiation, the operation will be permitted or blocked[1]. A schema of the agent system design developed is showed in Figure 1.

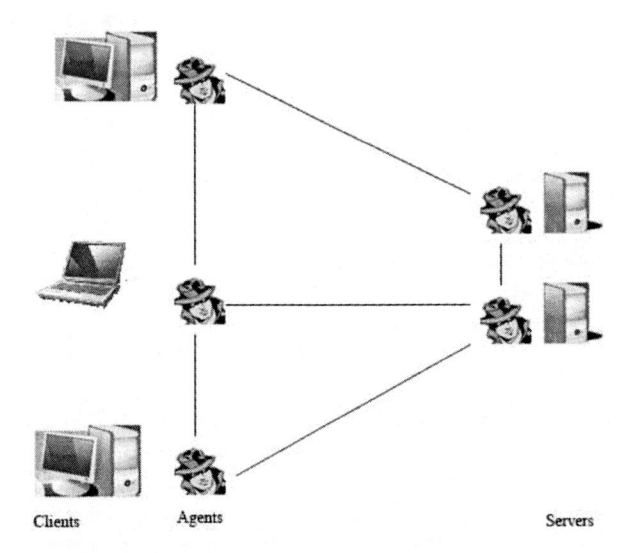

Fig. 1. Agent system architectural design. Every entity has an agent that negotiates on behalf of it.

[1] Security parameters may change during the negotiation process.

Table 1. Agents' roles and services

Agent Role	Service
Web server	Configures web server's security
	Negotiates security
Mail server	Configures mail server's security
	Negotiates security
Client	Configures client's security
	Negotiates security
Monitoring system	Communicates the alert level of the network
Firewall	Configures firewall's security
	Negotiates security
Vulnerability scanner	Provides vulnerability level reports from the requested agents
Logger	Monitors messages exchanged between agents

Roughly, the functionality of the system can be summarized as follows. The client agent negotiates to provide to the client the permission to perform the requested operation. Simultaneously, the rest of the agents involved in the operation impose restrictions due to their own security profile in order to ensure the security of the operation. This results in a complex communication scheme in which the agents are able to change their configurations to achieve their requested operations. In order to provide such functionality, our system comprises of several roles of agents according to the services that they provide in the network. Table 1 shows these roles and their corresponding services in our prototype.

There are configuration variables that may change during the negotiation process. Some of these are: the allowed number of incoming/outgoing connections, trusted certificate ownership, open ports, maximum simultaneous connections from an authorized IP, remote administration allowed, automatic updates, etc. A set of pre-configured profiles or security levels for agents can be defined according to these configuration variables. These profiles allow non-expert users to configure agent's security according to their needs just choosing one of these three levels: low, medium and high.

- Low level corresponds to an agent that does not impose any constraints. It does not enforce the principle of least privilege[2]. The principle establishing that users, servers and applications may have only the necessary privileges required to perform their tasks. A network with all agents configured with this security profile is equivalent to a network without the agents.
- Medium level does not enforce the principle of least privilege but it restricts some services to improve security.
- High level offers every security mechanism available. The agents play exclusively their assigned role. It enforces the principle of least privilege.

---

[2] The principle establishing that users, servers and applications may have only the necessary privileges required to perform their tasks [3].

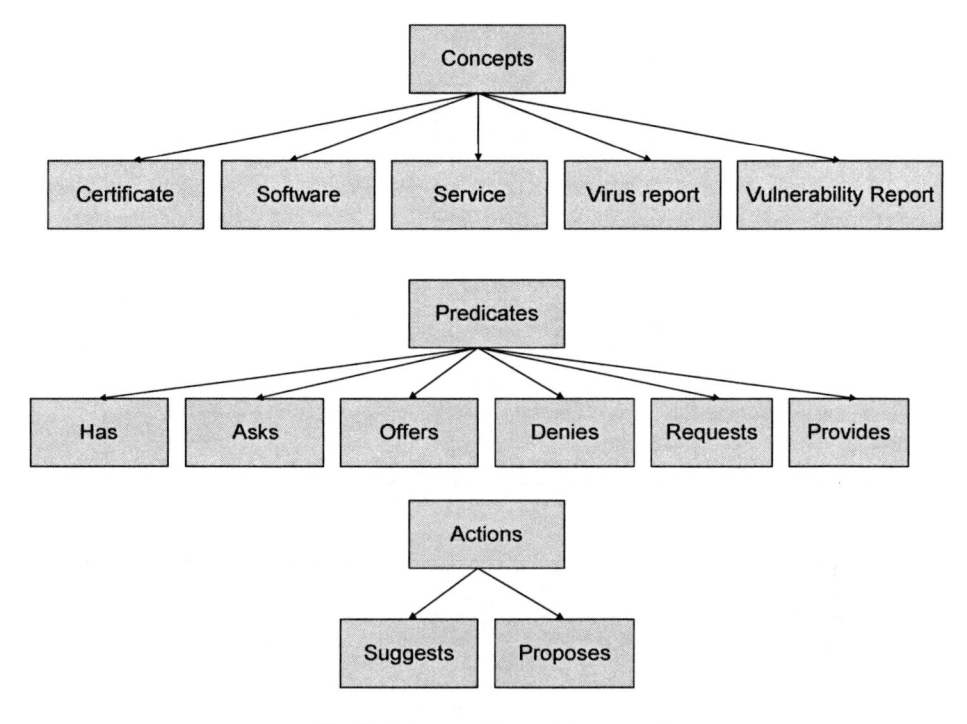

Fig. 2. Schema of the ontology used

These three levels must be present in every agent and determine the security configuration notwithstanding the fact that variables can also be changed individually for advanced users.

As shown in Table 1, our design corresponds to a purely communicative agent in accordance with the fact that agents have a perception of the environment that only comes from messages from other agents. The knowledge included in these messages is represented by an ontology which let agents understand their communicative intentions properly and act in consequence. The predicates/actions, concepts (elements), and relationships that form our ontology are shown in Figure 2.

3 Negotiation Protocol

Agents must negotiate their security configurations to provide the services requested by the user. This negotiation will be based on a message exchange with different contents expressed with terms, predicates and actions of the ontology. Since negotiation could terminate in decontrolled situations due to agents are not aware of the global state of the network, it is necessary to define a protocol that contains a set of authorized states in the network. Each service will have a set of valid configurations depending on the level of security assigned to the computer providing this service. A network is considered in a valid state if and only if all the services provided are performed with valid security configurations. During negotiation agents internally perform as a rule-based system.

They check each rule and ask the rest of agents for the information they need. These rules allow changes in configuration and, therefore, agents will use them to change their settings in order to reach their goals. Messages fulfill FIPA standard [4]. The different phases of negotiation are described as follows:

1. The beginning of negotiation. An agent representing a client requests any service or tries to perform an action.
2. Feasible conditions checking. At this stage, messages about agent configurations are exchanged. Server may change its configuration temporarily (just while an operation is performed) and can suggest changes in the client's configuration.
3. Acceptance/denial of the operation. If phase 2 reaches a stage on the network that becomes the operation feasible, the server communicates it to the client. Otherwise, the server denies the connection and gives the client some information about the causes and possible alternatives.
4. Proposals. If the operation was denied, the client proposes some changes in his configuration or asks the server to look for alternatives. If the server accepts the changes proposed in the first case, then we go back to step 2. Otherwise, the server informs the client of another reason, if any, for the denial of the operation so that the client can generate new proposals. If no proposal is possible the request is denied.

Currently, our prototype implements the negotiation protocol for four scenarios. It has been developed in JAVA and the agents use JADE platform [2] to work.

4 A Sample Scenario

In order to illustrate the operation of the system, this section exposes one of the scenarios developed by the prototype. First, an analysis of the threats for the scenario at issue and an examination of the measures that can mitigate these threats are provided. Second, the different combinations that make a specific operation possible is shown. Finally, a diagram showing how messages are exchanged between the agents in the negotiation process is depicted.

4.1 Browsing from an Internal Client

This scenario illustrates the situation where a typical windows client in the intranet wants to connect to a web server that is not in the intranet. For this scenario, only three of the agent roles described in Table 1 need to be considered: client, firewall and vulnerablility scanner. A simplified threat model for this operation is summarized in Table 2. As it is shown, the main threats are three. First, the client can download malware, such as virus, trojans or spyware, that infects the intranet. The main security mechanism to avoid this threat is to force the client to have an updated antivirus. Of couse it would be better not to allow the client to download software but this situation is too restrictive. Second, the client browser can have vulnerabilities that cause an infection just visiting certain web pages. In order to avoid this threat, the browser should be updated with the corresponding security patches. A vulnerability scanner (like nessus [1]) can scan the client in order to know if the browser is vulnerable or not. Third, technologies like

Table 2. Simplified threat model for the browsing scenario

Threat	Security mechanism
Download and execute malware	Update antivirus
Automatic exploitation of browser vulnerabilities	Update browser
Execution of dangerous programs through browser	Disallow ActiveX controls

ActiveX may execute dangerous programs. Restricting this kind of controls is a good preventive measure in certain cases.

Having in mind the preceeding analysis, we can elaborate a table with the security configuration combinations that make this operation possible depending on the desired level of security. This level is established by the administrator on the firewall through its corresponding agent. As can be seen in Table 3, an antivirus installed in the client is necessary for every possible combination. Depending on the level of security, the firewall agent would demand to have it updated more or less recently. In addition, browsing will be possible depending on the vulnerability level of the client browser and on the possiblity of restricting ActiveX execution.

Figure 3 shows a possible sequence of negociation steps for the case the level of security is set to high. First the client agent (CA) tries to connect to an external web page. The firewall agent (FA) does not have any rule for this operation so it begins checking the fourth configuration of Table 2 (because is the less restrictive regarding the security level). Thus, FA asks the CA if it has an updated antivirus. CA sends back the antivirus report that states it has not been updated in two days. As a consequence, FA blocks the connection and informs the client about the cause: having the antivirus out of date. Then, the CA proposes the FA to update it and the FA accepts it. The client downloads and install the update and the CA sends a hash of the update signed by the antivirus provider. At this moment, the FA proceed to verify the autenticity and integrity of the update. Once verified the FA asks the client to scan itself with the recently updated antivirus and ask him for a report. The client is not infected and the CA sends the report to FA. Accordingly, FA goes to the next step of the protocol and communicate to the CA the need of scanning the client for vulnerabilities. CA accepts and, as a

Table 3. Allowed configurations for the browsing scenario. Dash represents any value

Combination	Firewall	Client		Vulnerability Scanner
	Security level	Antivirus update	ActiveX allowed	Vulnerability level
1	Low	Last 24h	-	-
2	Medium or Low	Last 12h	-	Low or medium
3	Medium or Low	Last 12h	No	-
4	-	Last 8h	-	Low
5	-	Last 8h	No	Medium

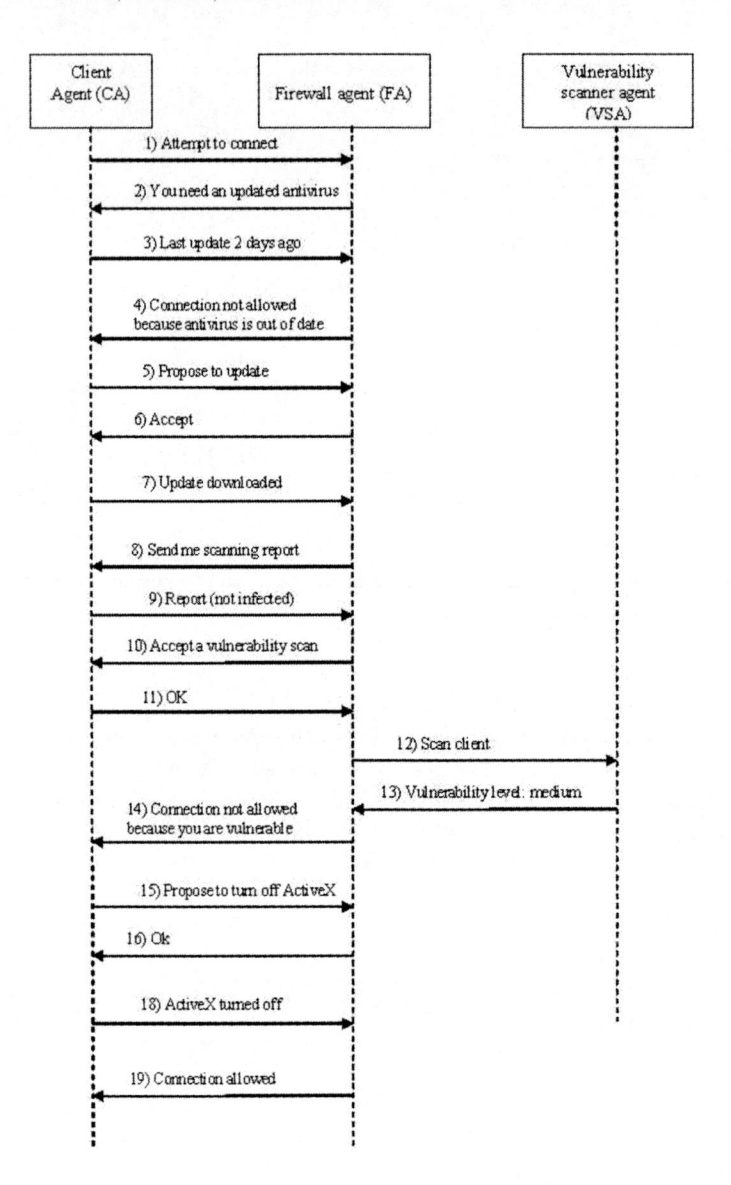

Fig. 3. Negotiation diagram for the browsing scenario when the security level is set to high

consequence, the FA asks the vulnerability scanner agent (VSA) to scan the client. The result is a medium vulnerability level and it is reported to FA. As a consequence, the FA blocks the connection (security configuration number 4 is not fulfilled) and informs the CA about the cause. Then, the CA proposes to desactivate ActiveX execution and the FA accepts. The CA proceeds and communicates it to the FA. Finally, security configuration number 5 fulfills and, consequently, the FA accepts the connection and notifies it to the CA. At this moment the FA aggregate a rule in the firewall to allow this connection. Nevertheless, it is not a permanent rule because the conditions can change (for instance, the antivirus can become out of date).

5 Conclusion

In this paper we have introduced the idea of negotiating the network security policy by means of software agents. This makes sense in those networks where no security policy has been defined in advance. Unfortunately this is a typical situation because administrators of small networks do not usually have the knowledge or the time to define and enforce one. Futhermore, the every day most deployed ad-hoc networks demand a flexible, distributed and autonomous admnistration. Our work focuses on developing an agent system that makes the process of security administration easier and dynamic. Thus, agents represent entities in the network and their goals are to protect the network while keeping an acceptable level of connectivity. The main problem faced was to offer a deterministic level of security that limits the intrinsic uncertainty of negotiation. However, the design of negotiation protocols for different scenarios according to the corresponding threat models has lead to deterministic solutions. The prototype we have implemented shows that a distributed conception of security based on agent paradigm is promising. Future work will involve the study of how agents can enforce what has been negotiated, the analysis of a threat model against the own agent system and a higher-level formalized architecture. In addition, more complex scenarios will be studied in detail.

References

1. Beale, J., Deraison, R., Meer, H., Temmingh, R., Walt, C.V.D.: Nessus Network Auditing. Syngress Publishing (2004)
2. Bellifemine, F., Caire, G., Poggi, A., Rimassa, G.: Jade: A software framework for developing multi-agent applications. lessons learned. Information and Software Technology 50(1-2), 10–21 (2008)
3. Bishop, M.A.: The Art and Science of Computer Security. Addison-Wesley Longman Publishing Co., Inc., Boston (2002)
4. FIPA: FIPA ACL Message Structure Specification. FIPA (2001), http://www.fipa.org/specs/fipa00061/
5. Ioannidis, S., Keromytis, A.D., Bellovin, S.M., Smith, J.M.: Implementing a distributed firewall. In: CCS 2000: Proceedings of the 7th ACM conference on Computer and communications security, pp. 190–199. ACM, New York (2000)
6. Krügel, C., Toth, T., Kirda, E.: Sparta, a mobile agent based instrusion detection system. In: Proceedings of the IFIP TC11 WG11.4 First Annual Working Conference on Network Security, pp. 187–200. Kluwer, B.V., Deventer (2001)
7. Udupi, Y.B., Singh, M.P.: Multiagent policy architecture for virtual business organizations. In: Proceedings of the IEEE International Conference on Services Computing, SCC 2006, pp. 44–51. IEEE Computer Society, Washington (2006)
8. Udupi, Y.B., Singh, M.P.: Governance of cross-organizational service agreements: A policy-based approach. In: Proceedings of the 2007 IEEE International Conference on Services Computing, SCC 2007, pp. 36–43. IEEE Computer Society, Salt Lake City (2007)

A Contingency Response Multi-agent System for Oil Spills

Aitor Mata, Dante I. Tapia, Angélica González, and Belén Pérez

Departamento Informática y Automática
Universidad de Salamanca
Plaza de la Merced s/n, 37008, Salamanca, Spain
University of Salamanca, Spain
{aitor,dantetapia,angelica,lancho}@usal.es

Abstract. This paper presents CROS, a contingency response multi-agent system for oil spills situations. The system makes use of a Case-Based Reasoning system which generates predictions to determine the probability of finding oil slicks in certain areas of the ocean. CBR uses past information to generate new solutions to the current problem. The system employs a distributed multi-agent architecture so that the main components of the system can be accessed remotely. Therefore, all functionalities can communicate in a distributed way, even from mobile devices. The core of the system is a group of deliberative agents acting as controllers and administrators for all functionalities. The system has been used to predict real oil spill situations. Results have demonstrated that the system can accurately predict the presence of oil slicks in determined zones. It has been demonstrated that using a distributed architecture can enhance the overall performance of the system.

Keywords: Oil Spill, Multi-Agent Systems, Case-Based Reasoning, Distributed Architectures.

1 Introduction

The response to minimize the environmental impact when an oil spill is produced must be precise, fast and coordinated. The use of contingency response systems can facilitate the planning and tasks assignation when organizing resources, especially when multiple people are involved.

This paper presents CROS, a contingency response multi-agent system for helping manage these situations. This system deploys a prediction model which makes use of intelligent agents and Case-Based Reasoning systems to determine the possibility of finding oil slicks in a certain area of the ocean. It also applies a distributed multi-agent architecture based on Service Oriented Architectures (SOA), modeling most of the system's functionalities as independent applications and services. These functionalities are invoked by deliberative agents acting as coordinators.

Agents and multi-agent systems have been successfully applied to several scenarios, such as education, culture, entertainment, medicine, robotics, etc. [1]. Agents have a set of characteristics, such as autonomy, reasoning, reactivity, social abilities, pro-activity, mobility, organization, etc. which allow them to cover several needs for developing contingency response systems [2].

Predicting the behavior of oceanic elements is a quite difficult task. In this case, the prediction is related with external elements (oil slicks) and this makes the prediction even more difficult. The open ocean is a highly complex system that may be modeled by measuring different variables and structuring them together. Some of these variables are essential to predict the behavior of oil slicks. It is necessary to know the previous positions of oil slicks in order to predict the future presence in a specific area. That knowledge is provided by the analysis of satellite images which reveal the precise position of the slicks.

The system presented in this paper generates as a solution a probability of finding oil slicks for different geographical areas after an oil spill. Predictions are created using a Case-Based Reasoning system. The cases used by the CBR system contain information about the oil slicks (size and number) and atmospheric data (wind, ocean currents, salinity, temperature, height and pressure). CROS combines artificial intelligence techniques in order to improve the efficiency of the CBR system, thus generating better results. CROS has been trained using historical data acquired during the Prestige oil spill at the Galician west coast in Spain, from November 2002 to April 2003. Most of the data used by CROS has been acquired from the ECCO (Estimating the Circulation and Climate of the Ocean) consortium [3]. Position and size of the slicks has been obtained by treating SAR (Synthetic Aperture Radar) satellite images [4]. The development of agents is an essential piece in the analysis of data from distributed sensors and gives those sensors the ability to work together and analyze complex situations.

Next, the oil spill problem is presented showing the difficulties and the possibilities of finding solutions to this problem. Afterwards, the main components of the system, including its architecture are described. Finally, the results and conclusions are presented.

2 Different Approaches to the Oil Spill Problem

It is very important to determine if an area will be contaminated or not after an oil spill. To do so, it is necessary to know how the slicks generated by the spill behave for concluding about the presence of contamination in a specific area. First, position, shape and size of the oil slicks must be identified. One of the most precise ways to acquire that information is by using satellite images. SAR (*Synthetic Aperture Radar*) images are the most commonly used to automatically detect this kind of slicks [5]. Satellite images show certain areas where it seems to be nothing (e.g. zones with no waves) as oil slicks. Figure 1 shows a SAR image on the left side, which displays a portion of the Galician west coast with black areas corresponding to oil slicks. On the right side of Figure 1 an interpretation of the SAR image after treating the data is shown. SAR images make it possible to distinguish between normal sea variability and oil slicks. It is also important to make a distinction between oil slicks and look-alikes. Oil slicks are quite similar to quiet sea areas, so it is not always easy to discriminate between them. This can lead to mistakes when trying to differentiate between a normal situation and an oil slick. This is a crucial aspect in this problem that can be automatically managed by computational tools [6]. Once the slicks are correctly identified, it is also crucial to know the atmospheric and maritime situation

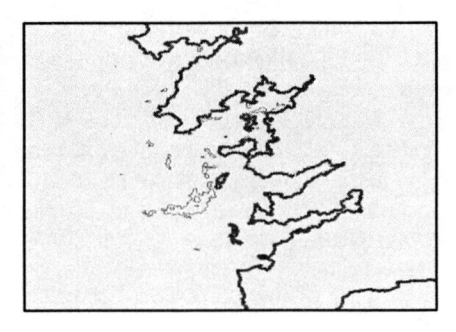

Fig. 1. Satellite image of an oil spill near the Galician west coast in Spain (left) and its interpretation by the CROS system (right)

that is affecting the zone at the moment that is being analyzed. Information collected from satellites is used to obtain the atmospheric data needed. That is how different variables such as temperature, sea height and salinity are measured in order to obtain a global model that can explain how slicks evolve.

There are different ways to analyze, evaluate and predict situations after an oil spill. One approach is simulation, where a model of a certain area is created introducing specific parameters (weather, currents and wind) and working along with a forecasting system. Using simulations it is easy to obtain a good solution for a certain area, but it is quite difficult to generalize in order to solve the same problem in related areas or new zones. Nevertheless arriving at these kinds of solutions requires a great data mining effort. Different techniques have been used to achieve this objective, from fuzzy logic to negotiation with multi-agent systems. One of these techniques is Case-Based Reasoning which is described in the next section.

3 CROS: A Contingency Response Multi-agent System for Oil Spill

CBR has already been used to solve maritime problems in which different oceanic variables were involved [7]. In CROS, the data collected from different satellites is processed and structured as cases. Table 1 shows the main variables that defines a case. Cases are the key to obtain solutions to future problems through a CBR system. The functionalities of CROS can be accessed using different interfaces executed on PCs or PDAs (Personal Digital Assistant). Users can interact with the system by introducing data, requesting a prediction or revising a solution generated (i.e. prediction). The interface agents communicate with the services through the agents' platform and vice versa.

The interface agents perform all the different functionalities which users can make use for interacting with CROS. The different phases of the CBR system have been modeled as services, so each phase can be requested independently. For example, one user may only introduce information in the system (e.g. a new case), while another user could request a new prediction.

Table 1. Variables that define a case

Variable	Definition	Unit
Longitude	Geographical longitude	Degree
Latitude	Geographical latitude	Degree
Date	Day, month and year of the analysis	dd/mm/yyyy
Sea Height	Height of the waves in open sea	m
Bottom pressure	Atmospheric pressure in the open sea	Newton/m^2
Salinity	Sea salinity	ppt (parts per thousand)
Temperature	Celsius temperature in the area	°C
Area of the slicks	Surface covered by the slicks present in the analyzed area	Km^2
Meridional Wind	Meridional direction of the wind	m/s
Zonal Wind	Zonal direction of the wind	m/s
Wind Strength	Wind strength	m/s
Meridional Current	Meridional direction of the ocean current	m/s
Zonal Current	Zonal direction of the ocean current	m/s
Current Strength	Ocean current strength	m/s

All information is stored in the case base and CROS is ready to predict future situations. A problem situation must be introduced in the system for generating a prediction. Then, the most similar cases to the current situation are retrieved from the case base. Once a collection of cases are chosen from the case base, they must be used for generating a new solution to the current problem. Growing Radial Basis Functions Networks [8] are used in CROS for combining the chosen cases in order to obtain the new solution.

CROS determines the probability of finding oil slicks in a certain area. CROS divides the area to be analyzed in squares of approximately half a degree side for generating a new prediction. Then, the system determines the amount of slicks in each square. The squares are colored with different gradation depending on the quantity of oil slicks calculated.

Figure 2 shows the structure of CROS. There are four basic blocks in CROS: Applications, Services, Agent Platform and Communication Protocol. These blocks provide all the system functionalities:

Applications. These represent all the programs that users can use to exploit the system functionalities. Applications are dynamic, reacting differently according to the particular situations and the services invoked. They can be executed locally or remotely, even on mobile devices with limited processing capabilities, because computing tasks are largely delegated to the agents and services.

Services. These represent the activities that the architecture offers. They are the bulk of the functionalities of the system at the processing, delivery and information acquisition levels. Services are designed to be invoked locally or remotely. Services can be organized as local services, web services, GRID services, or even as individual stand alone services. CROS has a flexible and scalable directory of services, so they can be invoked, modified, added, or eliminated dynamically and on demand. It is absolutely necessary that all services follow a communication protocol to interact with the rest of the components.

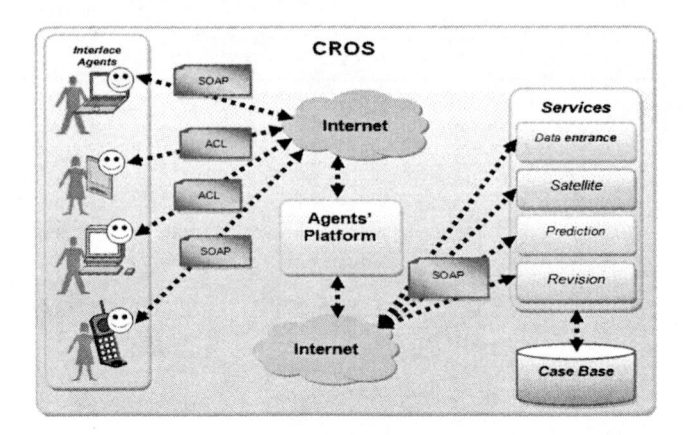

Fig. 2. CROS structure

Agent Platform. This is the core of the system, integrating a set of agents, each one with special characteristics and behavior. An important feature in this architecture is that the agents act as controllers and administrators for all applications and services, managing the adequate functioning of the system, from services, applications, communication and performance to reasoning and decision-making. In CROS, services are managed and coordinated by deliberative BDI agents. The agents modify their behavior according to the users' preferences, the knowledge acquired from previous interactions, as well as the choices available to respond to a given situation.

Communication Protocol. This allows applications and services to communicate directly with the Agents Platform. This protocol is based on SOAP specification to capture all messages between the platform and the services and applications [9]. Services and applications communicate with the *Agents Platform* via SOAP messages. A response is sent back to the specific service or application that made the request. All external communications follow the same protocol, while the communication among agents in the platform follows the FIPA Agent Communication Language (ACL) specification.

Agents, applications and services in CROS can communicate in a distributed way, even from mobile devices. This makes it possible to use resources no matter its location. It also allows the starting or stopping of agents, applications, services or devices separately, without affecting the rest of resources, so the system has an elevated adaptability and capacity for error recovery. Users can access to CROS functionalities through distributed applications which run on different types of devices and interfaces (e.g. computers, PDA).

Interface Agents are a special kind of agents in CROS designed to be embedded in users' applications. These agents are simple enough to allow them to be executed on mobile devices, such as cell phones or PDAs because all high demand processes are delegated to services. CROS defines three different *Interface Agents*:

CROS also defines three different services which perform all tasks that the users may demand from the system. All requests and responses are handled by the agents. The requests are analyzed and the specified services are invoked either locally or remotely. Services process the requests and execute the specified tasks. Then, services send back a response with the result of the specific task. In this way, the agents act as interpreters between applications and services in CROS. Next, CBR system used in CROS is explained.

3.1 Data Input Service

When data about an oil slick is introduced, CROS must complete the information about the area including atmospheric and oceanic information: temperature, salinity, bottom pressure, sea height. CROS uses Fast Iterative Kernel PCA (FIKPCA) which is an evolution of PCA [10]. This technique reduces the number of variables in a set by eliminating those that are linearly dependent, and it is quite faster than the traditional PCA. To improve the convergence of the Kernel Hebbian Algorithm used by Kernel PCA, FIK-PCA set η_t proportional to the reciprocal of the estimated values. Let $\lambda_t \in \mathfrak{R}^r_+$ denote the vector of values associated with the current estimate of the first r eigenvectors. The new KHA algorithm sets de i^{th} component of η_t to the files.

$$[\eta_t]_i = \frac{1}{[\lambda_t]_i} \frac{\tau}{t+\tau} \eta_0 , \qquad (1)$$

When introducing the data into the case base, Growing Cell Structures (GCS) [11] are used. GCS can create a model from a situation organizing the different cases by their similarity. If a 2D representation is chosen to explain this technique, the most similar cells (i.e. cases) are near one of the other. If there is a relationship between the cells, they are grouped together, and this grouping characteristic helps the CBR system to recover the similar cases in the next phase. When a new cell is introduced in the structure, the closest cells move towards the new one, changing the overall structure of the system. The weights of the winning cell ω_c, and its neighbours ω_n, are changed. The terms ε_c and ε_n represent the learning rates for the winner and its neighbors, respectively. x represents the value of the input vector.

$$\omega_c(t+1) = \omega_c(t) + \varepsilon_c(x - \omega_c) \qquad (2)$$

$$\omega_n(t+1) = \omega_n(t) + \varepsilon_n(x - \omega_n) \qquad (3)$$

Once the case base has stored the historical data, and the GCS has learned from the original distribution of the variables, the system is ready to receive a new problem. When a new problem comes to the system, GCS are used once again. The stored GCS behaves as if the new problem would be stored in the structure and finds the most similar cells (cases in the CBR system) to the problem introduced in the system. In this case, the GCS does not change its structure because it has being used to obtain the most similar cases to the introduced problem. Only in the retain phase the GCS changes again, introducing the proposed solution if it is correct.

3.2 Prediction Generation Service

When a prediction is requested by a user, the system starts recovering from the case base the most similar cases to the problem proposed. Then, it creates a prediction using artificial neural networks. Once the most similar cases are recovered from the case base, they are used to generate the solution. Growing RBF networks [12] are used to obtain the predicted future values corresponding to the proposed problem. This adaptation of the RBF networks allows the system to grow during training gradually increasing the number of elements (prototypes) which play the role of the centers of the radial basis functions. The creation of the Growing RBF must be made automatically which implies an adaptation of the original GRBF system. The error for every pattern is defined by (4).

$$e_i = {1}/{p^*} \sum_{k=1}^{p} ||t_{ik} - y_{ik}||, \qquad (4)$$

Where t_{ik} is the desired value of the k_{th} output unit of the i_{th} training pattern, y_{ik} the actual values of the k_{th} output unit of the i_{th} training pattern.

Once the GRBF network is created, it is used to generate the solution to the proposed problem. The solution proposed is the output of the GRBF network created with the retrieved cases. The input to the GRBF network, in order to generate the solution, is the data related with the problem to be solved, the values of the variables stored in the case base.

3.3 Revision Service

After generating a prediction, the system needs to validate its correction. CROS can also query an expert user to confirm the automatic revision previously done. The system also provides an automatic method of revision that must be also checked by an expert user which confirms the automatic revision.

Explanations are a recent revision methodology used to check the correction of the solutions proposed by CBR systems [13]. Explanations are a kind of justification of the solution generated by the system. To obtain a justification to the given solution, the cases selected from the case base are used again. As explained before, a relationship between a case and its future situation can be established. If it is considered the two situations defined by a case and the future situation of that case as two vectors, a distance between them can be defined, calculating the evolution of the situation in the considered conditions. That distance is calculated for all the cases retrieved from the case base as similar to the problem to be solved. If the distance between the proposed problem and the solution given is not greater than the average distances obtained from the selected cases, then the solution is a good one, according to the structure of the case base. If the proposed prediction is accepted, it is considered as a good solution to the problem and can be stored in the case base in order to solve new problems. It will have the same category as the historical data previously stored in the system.

4 Preliminary Results

CROS uses different artificial intelligence techniques to cover and solve all the phases of the CBR cycle. Fast Iterative Kernel Principal Component Analysis is used to reduce the number of variables stored in the system, getting about a 60% of reduction in the size of the case base. This adaptation of the PCA also implies a faster recovery of cases from the case base (more than 7% faster than storing the original variables).

The predicted situation was contrasted with the actual future situation. The future situation was known, as long as historical data was used to develop the system and also to test the correction of it. The proposed solution was, in most of the variables, close to 90% of accuracy. For every problem defined by an area and its variables, the system offers 9 solutions (i.e. the same area with its proposed variables and the eight closest neighbors). This way of prediction is used in order to clearly observe the direction of the slicks which can be useful in order to determine the coastal areas that will be affected by the slicks generated after an oil spill.

Table 2. Percentage of good predictions obtained with different techniques

Number of cases	RBF	CBR	RBF + CBR	CROS
100	45 %	39 %	42 %	43 %
500	48 %	43 %	46 %	46 %
1000	51 %	47 %	58 %	64 %
2000	56 %	55 %	65 %	72 %
3000	59 %	58 %	68 %	81 %
4000	60 %	63 %	69 %	84 %
5000	63 %	64 %	72 %	87 %

Table 2 shows a summary of the results obtained after comparing different techniques with the results obtained using CROS. The table shows the evolution of the results along with the increase of the number of cases stored in the case base. All the techniques analyzed improve its results while increasing the number of cases stored. Having more cases in the case base, makes easier to find similar cases to the proposed problem and then, the solution can be more accurate. The "RBF" column represents a simple Radial Basis Function Network that is trained with all the data available. The network gives an output that is considered a solution to the problem. The "CBR" column represents a pure CBR system, with no other techniques included; the cases are stored in the case base and recovered considering the Euclidean distance. The most similar cases are selected and after applying a weighted mean depending on the similarity of the selected cases with the inserted problem, a solution s proposed. The "RBF + CBR" column corresponds to the possibility of using a RBF system combined with CBR. The recovery from the CBR is done by the Manhattan distance and the RBF network works in the reuse phase, adapting the selected cases to obtain the new solution. The results of the "RBF+CBR" column are, normally, better than those of the "CBR", mainly because of the elimination of useless data to generate the solution. Finally, the "CROS" column shows the results obtained by CROS, obtaining better results that the three previous analyzed solutions.

5 Conclusions and Future Work

CROS is a new solution for predicting the presence of oil slicks in oceanic areas after an oil spill. This system presents a distributed multi-agent architecture which allows the interaction of multiple users at the same time. Distributing resources also allows users to interact with the system in different ways depending on their specific needs for each situation (e.g. introducing data or requesting a prediction). This architecture becomes an improvement with previous tools where the information must be centralized and where local interfaces where used. With the vision introduced by CROS, all the different people that may interact with a contingency response system can collaborate in a distributed way, being physically located in different places but interchanging information in a collaborative mode.

CROS makes use of a Case-Based Reasoning system for creating new solutions and predictions using past solutions given to past problems. It has been demonstrated that the CBR system generates consistent results. The structure of the CBR system has been divided into services in order to optimize the overall performance of CROS.

Generalization must be done in order to improve the system. Applying the methodology explained before to diverse geographical areas will make the results even better, being able to generate good solutions in more different situations. The current system has been mainly developed using data from the accident of the Prestige in the north-west coast of Spain. With that information, CROS has been able to generate solutions to new situations, based on the available cases. If the amount and variety of cases stored in the case base is increased, the quality of the results will also be boosted.

Although the performed tests have provided us very useful data, it is necessary to continue developing and enhancing CROS. The number of possible interfaces can be augmented, including independent sensors that may send information to the system in real-time. The data received by the system must be analyzed in order to detect new spills and to generate fast and accurate solutions to existing problems without the direct intervention of the users. Then, the system will not only be a contingency response but also a kind of supervising system especially in dangerous geographical areas.

References

[1] Corchado, J.M., Bajo, J., De Paz, Y., Tapia, D.I.: Intelligent Environment for Monitoring Alzheimer Patients, Agent Technology for Health Care, Decision Support Systems (in press, 2008)
[2] Yang, J., Luo, Z.: Coalition formation mechanism in multi-agent systems based on genetic algorithms. Applied Soft Computing Journal 7(2), 561–568 (2007)
[3] Menemenlis, D., Hill, C., Adcroft, A., Campin, J.M., et al.: NASA Supercomputer Improves Prospects for Ocean Climate Research. EOS Transactions 86(9), 89–95 (2005)
[4] Palenzuela, J.M.T., Vilas, L.G., Cuadrado, M.S.: Use of ASAR images to study the evolution of the Prestige oil spill off the Galician coast. International Journal of Remote Sensing 27(10), 1931–1950 (2006)

[5] Solberg, A.H.S., Storvik, G., Solberg, R., Volden, E.: Automatic detection of oil spills in ERS SAR images. IEEE Transactions on Geoscience and Remote Sensing 37(4), 1916–1924 (1999)
[6] Ross, B.J., Gualtieri, A.G., Fueten, F., Budkewitsch, P., et al.: Hyperspectral image analysis using genetic programming. Applied Soft Computing 5(2), 147–156 (2005)
[7] Corchado, J.M., Fdez-Riverola, F.: FSfRT: Forecasting System for Red Tides. Applied Intelligence 21, 251–264 (2004)
[8] Karayiannis, N.B., Mi, G.W.: Growing radial basis neural networks: merging supervised andunsupervised learning with network growth techniques. IEEE Transactions on Neural Networks 8(6), 1492–1506 (1997)
[9] Cerami, E.: Web Services Essentials Distributed Applications with XML-RPC, SOAP, UDDI & WSDL. O'Reilly & Associates, Inc., Sebastopol (2002)
[10] Gunter, S., Schraudolph, N.N., Vishwanathan, S.V.N.: Fast Iterative Kernel Principal Component Analysis. Journal of Machine Learning Research 8, 1893–1918 (2007)
[11] Fritzke, B.: Growing cell structures—a self-organizing network for unsupervised and supervised learning. Neural Networks 7(9), 1441–1460 (1994)
[12] Ros, F., Pintore, M., Chrétien, J.R.: Automatic design of growing radial basis function neural networks based on neighboorhood concepts. Chemometrics and Intelligent Laboratory Systems 87(2), 231–240 (2007)
[13] Plaza, E., Armengol, E., Ontañón, S.: The Explanatory Power of Symbolic Similarity in Case-Based Reasoning. Artificial Intelligence Review 24(2), 145–161 (2005)

V-MAS: A Video Conference Multiagent System

Alma Gómez-Rodríguez, Juan C. González-Moreno, Loxo Lueiro-Astray,
and Rubén Romero-González

Dpto. de Informática, University of Vigo
Ed. Politécnico, Campus As Lagoas, Ourense, 32004, Spain
{alma,jcmoreno}@uvigo.es, {loxomp,rrglez}@gmail.com
http://gwai.ei.uvigo.es/

Abstract. In this work an original case study corresponding to a new multiagent system implementation is presented. The built system is a distributed videoconferencing software which is integrated in an e-learning environment. The system provides the user with specific functionalities including storage of data and multiple profile management. It has been developed following a well know multiagent methodology: INGENIAS and implemented using the Java Multimedia Framework. The use of the system in a university domain has yield to promising results. Moreover, the software has been designed in an open way resulting in a portable system, which can be used in other domains different form e-learning such as e-Health-care or e-Business.

Keywords: Video conference, Multiagent Systems, INGENIAS Methodology.

1 Introduction

Thanks to the facilities provided by Internet, there is an important boom of the e-learning environments. The e-learning frameworks provide a method of apprenticeship without time or space constraints. Students can learn independently in their own place, but having also the possibility of working and collaborating with people from other places.

FAITIC [8] is a web portal which offers a range of services related to e-learning. With this portal, the University of Vigo tries to offer new possibilities to undertake a continued learning process using information technologies.

Trying to improve the functionality of the framework described previously, a new videoconference system with support for private conversations was required. Due to the special characteristics of the environment, it was decided to implement the system using an agent philosophy. As it is know, the intrinsic characteristics of agents (autonomy, reactivity, proactivity, etc.) provide a good approach in the solution of distributed complex problems [12]. This was very useful in the system required, so it was designed using a methodology for Multiagent Systems (MAS) [15], in particular INGENIAS, and implemented incorporating the usual characteristics of agents.

It can be thought that software agents are not strictly for use in this type of platforms. Other well-known solutions, such as Isabel, Skype or MSN are based on a heavy client-server architecture whose management is done by central nodes. The use of agents in this domain provides multi-processing, autonomy and decentralization and distributes the system workload among different nodes.

Other works have addressed the use of MAS in multimedia domains [4,7]. In particular, the first paper presents the development of an innovating appliance that incorporates interactive services of leisure and information, offered through a high-quality user interface integrated in an agent-based network. The paper [7] is focused in the proposal of a new approach based on intelligent agents for the initiation, control and adaptation of a multimedia session. Taking into account the advantages of the approaches based on MAS for this domain, the main aim of this paper is describing the development of the MAS system constructed for providing a videoconference application integrated in FAITIC.

The structure of the remaining of the paper follows. Section 2 does a more detailed description of the system to construct, while section 3 justifies the utility of constructing the system as a MAS and introduces the system architecture. Next, section 4 details the results of system design and implementation, including also a brief explanation of how the system works. Finally, section 5 introduces the conclusions and future work.

2 Problem Description

As it has been introduced in the previous section, the department responsible of Quality Assurance and Innovation in the University of Vigo is using a tool, called FAITIC for e-learning. This tool has been developed by the Area of Information and Communication Technology, pertaining to the University and offers an e-learning service using Internet which complements the habitual to-face teaching. Following this philosophy, a software called TEMA [1] based on the Claroline CMS (Content Management System) [3] has been developed. This program offers a centralized service for the students and also a management system for teachers.

This software has been evaluated, following quality assurance criteria, and several important shortcomings have been detected. In particular, one of the tools integrated in TEMA is the videoconferencing, which has the problem that there is no way of having private conversations between users; that is, all the conversations are visible to everybody. It is though that allowing users to talk in private conversations will improve the comprehension of the topic discussed and will facilitate the monitoring and exchange of ideas between the two of them. Moreover, as the number of users increases, the number of participants wishing to share knowledge with one particular user in private also increases.

Trying to overcome the problem previously introduced, the videoconferencing system available in TEMA has been substituted by a new one with the following features:

- Establishing conferences between users privately. This underscores the problem addressed in the previous paragraph, referring to the absence of private conversations.
- Integration into the TEMA platform as an additional module. This allows the module to have access to the services it provides and to interact with users in the

existing system. This requirement is essential if the current videoconferencing system is to be replaced.
- Transmission Audio/Video in real time (TAVTR), the transmission and reception of signals of video and audio must be done in real time for user satisfaction.
- Having an intuitive and adaptive interface. The program must be provided with a graphical interface that allows the end user to communicate through the system in a simple and attractive way.
- Portability to different platforms. It is very important to ensure the proper functioning of the video conference regardless of the hardware platform involved in the communication (PDAs, mobile phone, Web browser, etc.).

3 The Multiagent System

The TEMA platform is in continuous expansion. That means that developing software for this platform involves using software techniques which allow high heterogeneity, and scalability, in the growing of data or in the number of problems to address. Attending these issues, it has been decided to develop the system described in previous section using a Multiagent System (MAS).

3.1 System Architecture

Taking into account the system objectives introduced in the previous section, the following architectural requirements and system agents have been established:

1. Communication subsystem: *Transmitter*. It provides the necessary logic to initiate a communication. An agent or a set of agents will provide this functionality to the MAS.
2. Communication subsystem: *Receiver*. This subsystem provides the necessary complement to establish a communication between two entities. This functionality will be provided by one or more agents.
3. Monitoring subsystem. During the establishment and maintenance of communication, some help may be needed for maintaining the means of transmission in its proper functioning.
4. Control subsystem. Once the system is operating, it will be controlled by a subsystem developed for this purpose. This functionality will be implemented by one or more agents responsible for the overall control of the MAS.
5. Storage of data transmitted. The MAS will have the possibility of storing data of the video and audio signals.
6. Viewer of stored Data. It will allow displaying the stored data using a player.
7. Multiple profiles manager. It allows the user to select among several possible profiles for starting the communication.

In addition, the user interface must be intuitive and adaptable. This means that the GUI application will be developed trying to provide the user with a fast and intuitive way to perform the tasks. This will imply to introduce one or more Interface Agents in the MAS. Once the functional requirements have been presented, an initial diagram of MAS architecture can be shown; this is the purpose of Figure 1.

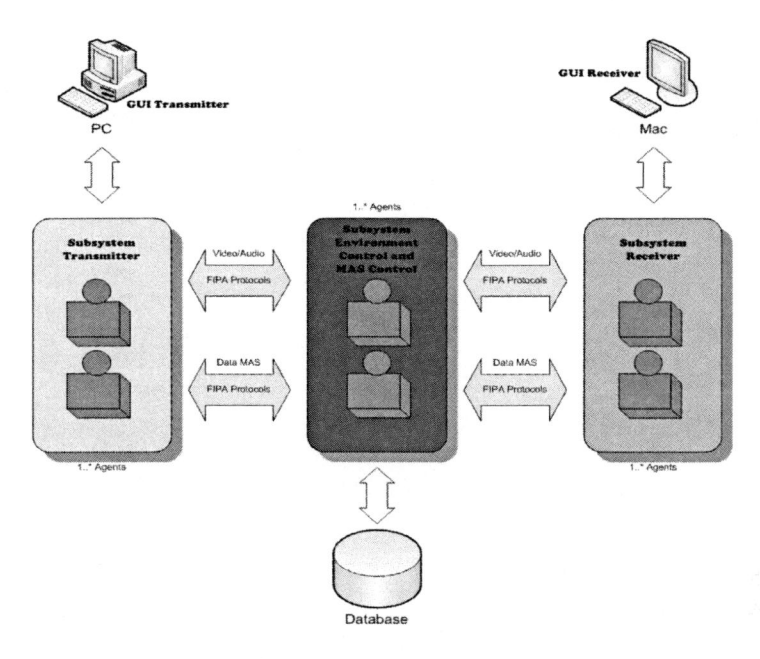

Fig. 1. System architecture

In addition, there are other non functional requirements taken into account in the proposed architecture. These are:

- Response time: The application must be capable of performing the transmission of video and audio in real time, that is, ensuring that there are no delays in transmission and that the data will be received without cuts.
- Memory: It is intended that the application consumes as little memory as possible to ensure that the machine will not be slowed down and allow the execution of other applications in parallel.
- Robustness: It is expected that the application is able to react to invalid user data. For instance, the user should not be allowed to interact with the application until he has been properly authenticated.
- Fault Tolerance: It is expected that the application is able to detect any errors that may arise during implementation and act accordingly to prevent the execution to abort. The application notifies the user through an error message and tries to retrieve from it.
- Practical value: The implementation will improve what currently exists, introducing the capacity to conduct private conversations between users.
- Useful: It is intended that the application has a user friendly interface, allowing a rapid learning and a bigger satisfaction of user.

4 Development of the System

The system described in previous sections has been developed following a specific MAS methodology: INGENIAS. INGENIAS covers analysis and design of MAS, and it is intended for general use, with no restrictions on application domain [17]. It is based on UML diagrams and Rational Unified Process (RUP) [13, 6], trying in this way to facilitate its use and apprenticeship. New models are added and the UML ones are enriched to introduce agent and organizational concepts. Nowadays, the metamodels proposed cover the whole life cycle and capture different views of the system. In the latest years, INGENIAS metamodels have demonstrated their capability and maturity, as supporting specification for the development of Multiagent Systems [16,17,18].

Recently the INGENIAS Agent Framework (IAF) [9] has been proposed taking into account the experience in application of INGENIAS methodology during several years enabling a full model driven development. This means that, following the guidelines of the IAF, an experienced developer can focus most of its effort in specifying the system, converting a great deal of the implementation in a matter of transforming automatically the specification into code. A MAS, in IAF, is constructed over the JADE platform [5]. It can be distributed along one or several containers in one or many computers. To enable this feature, the IAF has means of declaring different deployment configurations.

All these reasons make INGENIAS a very suitable methodology for a system like the one addressed in this paper.

4.1 System Design

Following INGENIAS methodology, the first diagram constructed was the Environment Model, which is shown in Figure 2.

In the environment model (Figure 2), all the three subsystems that make up the core of Videoconferencing MAS can be seen. Each subsystem is identified with an agent in the model, so there are three agents: *TransmitterAgent*, *ReceptorAgent* and *SystemAgent*. The collaboration of three agents achieved the main goal, that is,

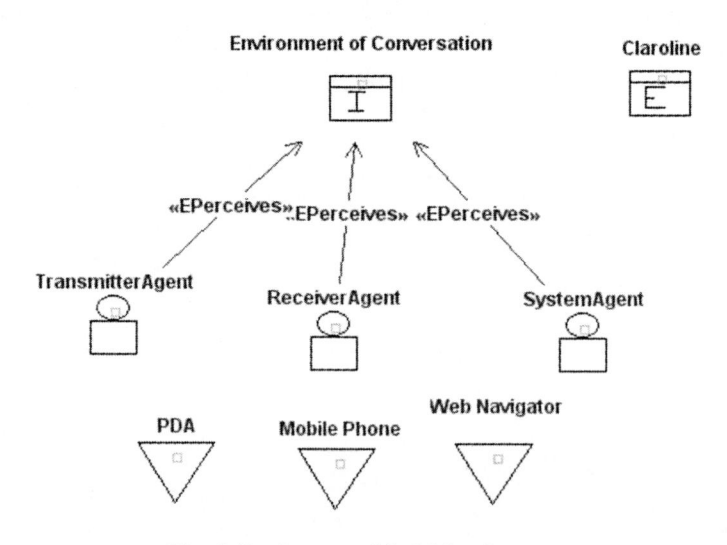

Fig. 2. Environment Model for the system

establishes an environment that allows a videoconversation which meets the system requirements. In the environment model, the various resources from which a user can access the system (PDA, mobile phone, browser) are also shown.

In the model of the figure, several important components of the system are shown. In first place, the Environment of Conversation is in charge of opening a private connection (staring a private communication channel between sender and receiver to carry out the conversation), sending messages exchanged during the life of the conversation or closing the connection, that is, when the sender or receiver ask for the completion or abandonment of the private conversation that they were keeping.

The *TransmitterAgent*, included in the model, is responsible for making the request to start a private conversation indicating the receiver with whom it wants to communicate. In addition, it asks the system for the finishing of the private conversation that it was keeping.

The *ReceptorAgent* accepts the creation of a connection to the sender if it wants to have a private conversation with him. It also sends send the system its response to the request for the initiation of a private conversation sent by the sender and asks at any moment for the abandonment of a private conversation with a particular sender.

The *SystemAgent* manages the private conversations. That means that it is the one which receives the request for initiation of private conversation, sends the request to the receiver, starts the communication link between the sender and receiver and controls the flow of conversation. When required, it will be also responsible of closing the communication link between sender and receiver. This closure may be caused by an abandonment of the conversation by the receiver or by the end of the conversation by the transmitter.

Finally, Claroline platform is also included in the Environmental model, trying to show how it was used to access the database and retrieve information from available users. From this list, the receiver selected for maintaining the private conversation is chosen.

The Roles/Task Model, introduced in Figure 3, shows the roles that accomplish the agents (from the Environment Model) to fulfill the requirement of the MAS. The descriptions associated with each role have been intentionally omitted, because there

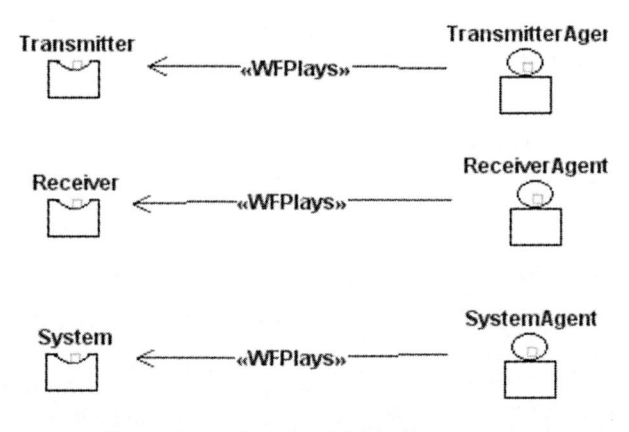

Fig. 3. Roles/Task Model for thesystem

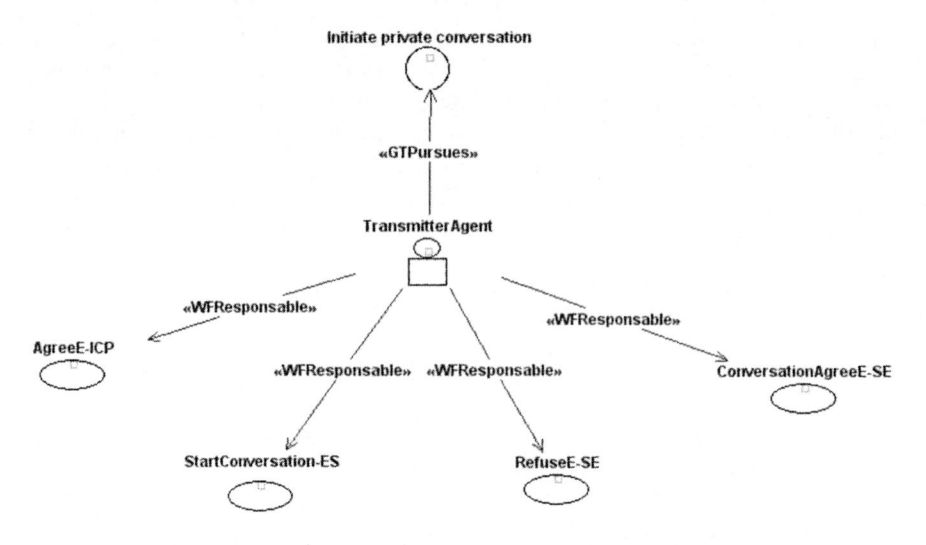

Fig. 4. Agent/Task Model for the Transmitter Agent

is a direct correspondence with their repective agents. Thus, the features described by the role System will be analogous to the ones of its related agent: the *SystemAgent*.

Other important issue in the system design is to indicate the tasks accomplished by each agent. The agents and tasks model, introduced in Figure 4, reflects the different tasks to be performed by a particular agent; in this case the *TransmitterAgent*, in order to achieve the purpose it was designed form, that is, to initiate a private conversation.

In the Figure 4, several important issues in the behavior of Transmitter Agent are shown. The global goal of the agent is to *Initiate a Private Conversation*, when a user (transmitter) wants to start a private conversation with other of the users (receiver). The conversation will be accepted or rejected depending on the answer of the receiver or in the system state known by *SystemAgent*. In the Figure, the agent in charge of fulfilling the previous goal, which is Transmitter Agent, is also shown. It performs the tasks: Agree-ICP, StartConversation, RefuseE-SE, ConversationAgreeE-S. *Agree-ICP* task reflects the fact occurring when a petition for connection to a receiver is answered affirmatively. The task *StartConversation* indicates that the agent takes the user's request to start a conversation with another agent of the MAS. In this case the agent is the start point of the petition of connection with the receiver. *RefuseE-SE* is the task produced when the agent receives notification about the negation of receiver for having a conversation. Finally, *ConversationAgreeE-SE* is produced when the agent is notified that its request has been accepted by the recipient.

The system has been completely designed using INGENIAS models, but, because of lack of space, only some of the most significant ones have been introduced in this section.

4.2 Implementation

Although INGENIAS methodology provides important facilities for automatic implementation of the system, this utility has been used only partially. Once the code has been generated through the IAF taking into account the results of previous phases, the Java Multimedia Framework (JMF) [10] has being used to support the treatment of audio and video. This must be done to fulfil user requirements which indicate that the system must be developed using Java SLK [11] and JMF.

4.3 Resulting Tool

At the moment, V-MAS is under beta-test for a set of users, which provide a feed back of reasonable satisfaction. The system has a screen to access, through login and password, to the videoconferencing MAS. As the software is integrated with TEMA, it must access the database integrated in Claroline CMS to verify user and permissions.

5 Conclusions and Future Work

This paper shows how to solve an underlying problem in a system through MAS. The new system has been integrated with a legacy one using the cooperation and coordination among agents. Moreover, the use of MAS specific methodologies, such as IN-GENIAS, allows users to develop this kind of systems in a driven, rapid and robust manner. Although one could have thought of developing the system in a simplest way, using for instance traditional object-oriented approach, creating the system in this latest manner would cause a negative impact on scalability, coupling of subsystems or problem solving methods.

The final adopted solution is based on agents as they allow accomplishing the main system requirements and covers the possibility of further improvements on the videoconference, scalability or portability of the system.

From a functional point of view, in the future improvements in the system attending new approaches in the structure of the business logic will be taken into account. In the future, an agent will be incorporated to the system to adapt the data transmission to available means.

Continuing this line of development, the need of integrating a subsystem capable of managing user groups has been identified. The subsystem will detected the theme about which connected users are discussing and will give them the possibility of creating a discussion group for all those implied in the conversation.

Taking into account the power of V-MAS and trying to increase its dissemination, authors have the intention of incorporating the system to various platforms such as CMS Joomla, Drupal, Xoops, and so on. Moreover, its integration into new platforms will continue because the tool is adhered to the e-Health platform [14] and Sie-Health [18]. In this latest case, V-MAS will perform all the functions that that platform of electronic health-care needs.

Acknowledgments. This work has been supported by the project Methods and tools for agent-based modeling supported by Spanish Council for Science and Technology with grant TIN2005-08501-C03-03.

References

1. Software TEMA, http://faitic.uvigo.es/index.php
2. Adamson, C.: Quick Time for Java: A Developer's Notebook. O'Reilly Media, Inc., Belgium (2005)
3. Tihon, A.: Claroline: exportación de recorridos pedagógicos en formato scorm (2004-2005), http://www.claroline.net/dlarea/memoire_amand_tihon_2005.pdf
4. Ceccaroni, L., Verdaguer, X.: Agent-oriented, multimedia, interactive services in home automation. In: Proceedings of the Second European Workshop on Multi-Agent Systems (EUMAS 2004), Barcelona, Spain (2004)
5. Bellifemine, F.L., Caire, G., Greenwood, D.: Developing MultiAgent Systems with JADE. Wiley, Chichester (2007)
6. Booch, G., Rumbaugh, J., Jacobson, I.: Unified Modeling Language User Guide, 2nd edn. The Addison-Wesley Object Technology Series. Addison-Wesley Professional, Reading (2005)
7. Botía, J.A., Gómez-Skarmeta, A.F., Ruiz, P.: Multimedia session capabilities negotiation using software agents. In: Intelligent Agent Technology (IAT 2004) (2004)
8. http://faitic.uvigo.es/
9. Gómez-Sanz, J.: Ingenias Agent Framework. Development Guide V.1.0. Technical report, Universidad Complutense de Madrid
10. Gordon, R., Talley, S.: Essential JMF: Developer's Java Media Players. Prentice Hall PTR, Upper Saddle River (1999)
11. Gosling, J., Joy, B., Steele, G., Bracha, G.: The Java Language Specification, 2nd edn. Addison-Wesley, Boston (2000)
12. Jennings, N.R.: Agent-based computing: Promise and perils. In: Thomas, D. (ed.) Proceedings of the 16th International Joint Conference on Artificial Intelligence (IJCAI 1999), vol. 2, pp. 1429–1436. Morgan Kaufmann Publishers, San Francisco (1999)
13. Kruchten, P.: The Rational Unified Process: An Introduction, 2nd edn. Addison-Wesley Professional, Reading (2000)
14. Maheu, P., Whitten, M.: E-Health, Telehealth and Telemedicine. Jossey-Bass,
15. Mas, A.: Agentes Software y Sistemas Multiagente. Prentice-Hall, Englewood Cliffs (2005)
16. Pavón, J., Gómez-Sanz, J.: Agent oriented software engineering with INGENIAS. In: Mařík, V., Müller, J.P., Pěchouček, M. (eds.) CEEMAS 2003. LNCS, vol. 2691, pp. 394–403. Springer, Heidelberg (2003)
17. Pavón, J., Gómez-Sanz, J., Fuentes, R.: Model driven development of multi-agent systems. In: Rensink, A., Warmer, J. (eds.) ECMDA-FA 2006. LNCS, vol. 4066, pp. 284–298. Springer, Heidelberg (2006)
18. Gómez-Sanz, J.J., Pavón, J.: Implementing multi-agent systems organizations with INGENIAS. In: Bordini, R.H., Dastani, M., Dix, J., El Fallah Seghrouchni, A. (eds.) PROMAS 2005. LNCS, vol. 3862, pp. 236–251. Springer, Heidelberg (2006)
19. Loxo Lueiro Astray José Ramón Marra Garachana César Parguiñas Portas Castor Sánchez Chao Francisco Javier Rodríguez Martínez Juan Carlos González Moreno Rubén Romero González, Julián Martínez Meira. Sie-health, information system of e-health (2008), http://www.siehealth.org

Online Scheduling in Multi-project Environments: A Multi-agent Approach

José Alberto Arauzo[1], José Manuel Galán[2], Javier Pajares[1], and Adolfo López-Paredes[1]

[1] Social Systems Engineering Centre (INSISOC), University of Valladolid
{arauzo,pajares,adolfo}@insisoc.org
[2] INSISOC, University of Burgos
jmgalan@ubu.es

Abstract. In this paper, we propose a multi-agent system and an auction inspired mechanism for online scheduling in multi-project environments. Agents are resources and projects. Projects demand resources for fulfilling their scheduled planned work, whereas resources offer their capabilities and workforce. An auction inspired mechanism is used to allocate resources to projects; and the price of resources emerges and changes over time depending on supply and demand levels in each time slot. By means of this multi-agent system, we are able to overcome most of the problems faced in multi-project scheduling as changes in resources capabilities, allocation flexibility, changes in project strategic importance, etc.

Keywords: multi-agent systems, multi-project environments, auction based allocation resources, project scheduling and control.

1 Introduction

In practice, firms work in multi-project environments, where several projects are running at the same time and share the same limited resources. Every project needs scheduled activities to be done, and some projects have higher priority than others because of its contribution to financial returns or to strategic purposes. Therefore, portfolio and program managers should coordinate this complex system, deciding which resources will be allocated to what project and when.

Moreover, initial scheduled must be revisited because of strategy changes (depending, for instance, on external economic data), uncertainty or project overruns. The traditional scheduling and control systems based on hierarchical and centralized architectures, are not flexible enough to adapt themselves to the dynamism and complexity needed multi-project environments.

For this reason successive proposals to improve the scheduling and control in a multi-project environment are appearing continually. The recent paradigm of Multi-agent Systems, which offers new techniques to face complex unsolved problems, can help to find promising solutions. As a matter of fact, it is currently a very active field of investigation.

In a multi-project environment, managers have to prioritise and select the projects that will be added to the portfolio taking into account their expected profits, returns and their alignment with the firm strategy.

Priority ranking changes over time because of the addition of new interesting projects, changes in corporate strategy or simply, because of feedback information about individual project overruns affecting their expected returns. As priority changes, resources have to be reallocated to meet the requirements of the most interesting projects. At the same time, individual projects belonging to the portfolio have disruptions, delays and over-costs in their activities.

Overruns and priority changes take place in parallel, and as a consequence, conflicts between projects emerge, as individual projects compete for the same scarce resources. Multi-project management requires the development of *dynamic* resource constrained project scheduling systems.

Unfortunately, the research in this area has not converged to one solution or scheduling rule robust enough to hold in the general case [1]. Hans et al [2] review existing literature in hierarchical approaches and propose a generic project planning and control framework for helping management to choose between planning methods, depending on organisational issues. Kao et al [3] suggest an event-driven approach to develop a trade-off decision framework for scheduling and re-scheduling. Cohen et al. [4] analyse the performance of Goldratt´s critical chain in multi-project environments.

In this paper, we propose a novel approach for online dynamic scheduling in multi-project environments. We use a multi-agent system where the agents are the resources and the projects; the price of the work done emerges from an auction.

Projects have scheduled work to be done by different resources. Resources are endowed with some capabilities (knowledge, work force, etc.) that are needed to do the work. Projects demand resources over time and resources offer their capabilities and time availability. There is an auction process, and the price of resource-time slots emerges endogenously as a result of supply and demand.

In our research, we are mainly concerned with helping project portfolio managers to take decisions about resources and about portfolio composition.

Under this powerful framework, we are able to analyse some hard problems in portfolio project management: under and over usage of resources; key resources and resources that should be added to the firm; the role of flexibility in human resources, that is, the role of people exhibiting several capabilities at the same time; or the addition or deletion of individual projects; priorities in project execution, etc.

The rest of this paper is organised as follows. First, we will discuss the role of agent based methodology in project scheduling. Then, in section 3, we will formally define the problem we want to face with, that is, the scheduling problem with project rejection. In section 4, we will explain the main features of our multi-agent approach to portfolio project management and we will see some of preliminary results in section 5. We will end with the main conclusions of our work.

2 Multi-agent Systems and Multi-project Environments

Projects are characterized by complexity (they include many components and dependencies), uncertainty (about the availability of resources, task durations), dynamic behaviour (changes in the scope of the project, adding or removing unexpected tasks, re-scheduling processes) and are inherently distributed (each task may be completed by different resources or in different geographical locations).

In the case of a multi-project environment, each one of these features is severely intensified. More projects mean more complexity, more uncertainty sources, more complex dynamics, and problems more dispersed since often not only the tasks are distributed but the management of each project too.

Multi-agent systems may be particularly appropriate to deal with the complexity of multi-project environments, as agents can be abstracted as task, as resources, as project managers, etc. As Jennings and Wooldridge point out, multi-agent systems are suitable for problems having the following properties: *complexity, openness*, with *dynamical and unknown environments changing over time* (uncertainty) and *ubiquity* [5] and [6].

The decentralized approach in project management has been used since the last decade [7], but it is in the last years when market based approaches [8] are receiving a growing interest. Recently, Lee, Kumara and Chatterjee [9] [10] have proposed a multi-agent based dynamic resource scheduling for distributed multiple projects using market mechanisms. Following the same research line Confessore et al [11] propose an iterative combinatorial auction mechanism as coordination mechanism to resolve the same problem.

Other examples of agent based approaches in project management field can be found in the work of Kim and colleagues [12], [13], [14] in Wu and Kotak [15] or in Cabac [16]. As underlined in the introduction, our work makes use of a market metaphor and an auction mechanism to help project managers to take portfolio decisions about resources and about portfolio composition.

3 The Scheduling Problem with Project Rejection

We define a multi-project scheduling problem with opportunity of project rejection as follows:

At any instant t there are I projects in the system, each one denoted by i. Each project is characterized by a value V_i, that can be interpreted as the revenue obtained for the project, a weight w_i representing the strategic importance given to the specific project, a desirable delivery date D_i, a limit delivery date D_i^* that cannot be exceeded, an arrival date of the project to the system, B_i, and a limit answer date R_i that represents the latest date to decide to reject the project.

The system is considered dynamic: while some projects are being developed other projects can be included or rejected in real time.

Each project i consists of J_i activities, each one denoted by ij where $i \in \{1, 2,..., I\}$ and $j \in \{1, 2,..., J_i\}$. They have associated a workload d_{ij}. A set of M resources is given. Each resource $m \in \{1, 2, 3...M\}$ can just be assigned simultaneously to one activity.

A set Γ of K competences $\Gamma=\{k_1, k_2, ... k_K\}$ are necessary to complete the projects. Each resource is endowed with a given cost rate per unit of time C_m and a subset $\Gamma_m=\{k_{1m}, k_{2m}, ... k_{fm}\}$ of competences, each k_{lm} in a certain grade or ability to perform an activity e_{lm}, where $e_{lm} \in (0,1]$. Thus, each resource is defined by the set $\{C_m, H_m\}$ where $H_m=\{(k_l, e_{lm}),...,(k_f, e_{fm})\}$ with $k_l,..., k_f \in \Gamma$.

Every activity is associated with a competence c_{ij}. Any activity ij with a given c_{ij} can be performed by a resource m if $c_{ij} \in \Gamma_m$. The duration of the activity ij depends on the resource assigned to perform it according to d_{ij}/e_{ijm}, where e_{ijm} is the ability of the resource m in the competence c_{ij}.

In this first simplified model we assume that the activities of any project should be performed sequentially in the order defined by j and only one resource can be assigned to an activity. We also assume that once some resource has begun a task, the activity cannot be interrupted; the resource needs to finish it to be assigned to any other activity.

We include explicitly the option of reassigning resources in real time when a new project arrives to the system. The activities can be assigned to any resource that has the specific competences to perform it.

The overall efficiency of the system will be evaluated by the average benefit obtained in a certain time interval T according to:

$$Efficiency = \frac{B_T}{T} = \frac{\sum_i (V_i - Cost(i))}{T} \qquad (1)$$

Where i are each one of the projects finished in T and $Cost(i)$ is the cost to complete the i project. This cost has two components, the direct resource cost and the delay cost:

$$Cost(i) = \sum_j C_{\overline{m}(j)} \cdot \frac{d_{ij}}{e_{ij\overline{m}}} + w_i \cdot (D_i - F_i)^2 \qquad (2)$$

The first addend corresponds to the direct resource cost to finish each activity j. $\overline{m}(j)$ denotes the resource selected to comply with activity j. The second addend is the delay cost associated to the project, where F_i is the real delivery date.

The problem considers the decision to reject projects. This could happen in any of the following cases:

- The revenue obtained from the project does not compensate the costs.
- Our scheduling exceeds the D_i^* of the project.
- The impact on the scheduling of the rest of the projects is not acceptable. This may happen for two causes. First, if the new project obliges to delay a committed project beyond D_i^*, it will be rejected. If not, but the inclusion of the new project increases the delay costs of the other projects more than the direct benefit obtained for the project, it will also be rejected.

4 A Multi-agent System for Online Multi-project Scheduling

Traditionally, scheduling problems have been solved offline and simultaneously by a centralized decision-maker that use a global optimisation model. Instead of previous approach, we propose in our systems several decision-makers modelled as agents. We consider two kinds of agents: project agents and resource agents. At any moment, the system has as many project agents as projects are ordered. Each one represents a determined project characterized by its tasks, precedence relationships, due date, value, local programs and their execution state. Their goal is to look for contracts with resources that can perform the required activities and hence completing successfully the project. The system includes as many resource agents as resources are considered. They are defined by their competences, availability, local programs, cost rate, and pending task queue.

Every agent in the system can communicate with others by sending messages. The interaction mechanism is ruled by means of a combinatorial auction where a theoretical basis is provided for structuring message sequencing, bid evaluation, and price updating. This auction based multi-project scheduling approach is founded on Lagrangian Relaxation [17], [18]), a decomposition technique for mathematical programming problems.

In our distributed multi-project system, the decision-making is decentralized since each project creates its own schedule (local schedule). The auction mechanism ensures that local schedules are nearly compatible (several projects don't use the same resource at the same moment) and globally efficient.

Each unit of time (time slot) on each resource is modelled as a 'good' that can be sold in an auction where each resource acts as a seller. The number of sellers is equal to the number of resources in the system and each resource proposes a price for the time slots from the current time to the end of the scheduling horizon. The scheduling horizon changes dynamically by coinciding with the latest time slot that some project has asked at any moment.

Each project agent is a 'bidder' that participates in auctions by asking the resource agents for the set time slots that it requires to execute its pending task at the current time. It will try to find its time slots through the resource set while incurring the minimum possible cost. This cost has two components, the sum of the price of the selected time slots and the delay cost (expression 3).

$$TC_i = \sum_{mt \in Z_i} p_{mt} + w_i \cdot (D_i - F_i)^2 \qquad (3)$$

Where: TC_i is the total cost of project, p_{mt} is the price of the time slot (t) of the resource (m), and Z_i the set of selected time slots for project (i)

To minimize the cost, the project agents use a dynamical programming algorithm where all possible combination of time slots and resources are considered [19]. In their decision, they take into account that only those resources endowed with the necessary competences can carry out a certain activity. Moreover, the number of time slots necessary to complete the task are determined the ability of the resource in the competence. Each project agent will regard as scheduling horizon the time slot which goes from the current time to the limit delivery date (D_i^*). If some project agent can't find a set of time slots in such a manner that it allows to schedule tasks before D_i^*, with a smaller cost than its value (V_i), then it will not ask for any set of time slots. This implies that the project is unprofitable at the correspondent round of bidding and must be rejected.

Each project agent determines the price charged for the time slots with the purpose of reducing resource conflicts and maximizing their revenue. In order to get this goal a subgradient optimization algorithm is used to adjust prices at each round of bidding. By means of this algorithm the resource agents increase the price of the time slots where there is conflict (more than a project have asked for this time slot) and reduce the price of the time slots that have not been demanded. The process of price adjustment and bid calculation continues indefinitely. At each round of bidding the resource conflicts will be lower and lower.

At the first round of bidding, the time slots prices for the resource (m) are equal to the resource cost rate (C_m). At the rest of bidding round, the prices will be updated by means of the expression 4.

$$p_{mt}^{n+1} = \max\{C_m, p_{mt}^n + \alpha^n \cdot g_{mt}^n\} \qquad (4)$$

Where:

p_{mt}^{n+1} : price of the time slot (t) of resource (m) at the round (n+1).

p_{mt}^n : price of the time slot (t) of resource (m) at the round (n).

α^n : step at the round (n). It decreases when (n) increases.

$g_{mt}^n = a_{mt}^n - 1$: subgradient, where a_{mt}^n is the demand of slot (t) of resource (m).

By means of the described auction mechanism, project agents build compatible and globally efficient local schedules for their pending activities. Moreover, at the same time agents interact through a complementary process to make firm agreements. These agreements determine fixed programs for earliest scheduled tasks. When these agreements are obtained, project agents will never consider the tasks included as firm contracts as pending.

5 Case Study: The Role of Resource Capabilities

In this section, we show the results of some preliminary simulations, focusing on the role of resource capabilities in multi-project environments. In the final version of the paper, we will extend the results, and we will show a wider set of applications and its implications on managerial decisions.

We consider three different resources (R1, R2 and R3), endowed with the competences C1, C2 and C3 respectively. In table 1, we show a portfolio of six projects, and the tasks needed to complete the project. Each task is defined by means of the pertaining competence and expected time to be completed. The arrival date is the date when

Table 1. Dynamic portfolio of projects

Project	Tasks			Arrival Date	Starting Date	DD 1	DD 2	Value
	Task 1	Task 2	Task 3					
P1	C1 50	C2 10	C3 20	-30	0	120	180	2500
P2	C3 10	C1 60		-30	0	180	240	3500
P3	C2 15	C1 50		-30	0	120	180	7000
P4	C3 20	C1 45	C2 10	30	120	240	270	4000
P5	C2 15	C3 10	C1 60	30	120	240	270	3000
P6	C3 10	C2 20	C1 50	-30	0	120	180	5000

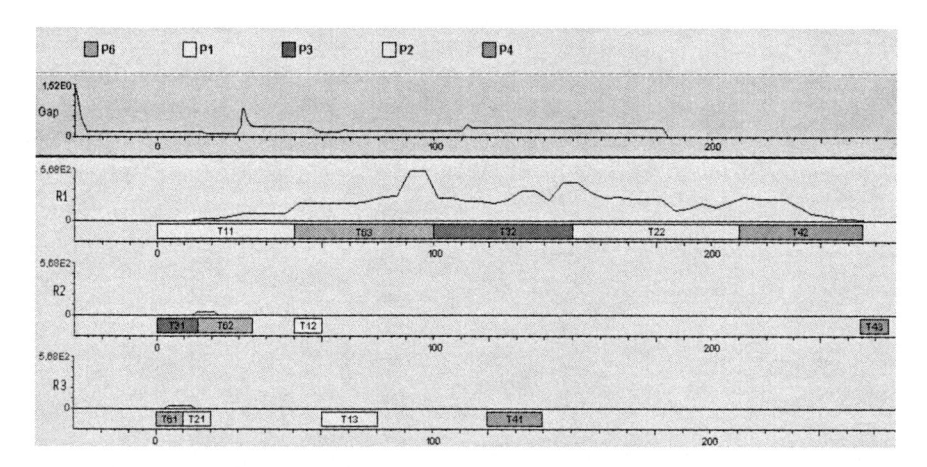

Fig. 1. Tasks performed by resources. Tij denotes Task j of project Pi

the project is included in the system. Projects can start-up in the starting date; otherwise, they should have been rejected before this date. Due Date 1 (DD1) is the most desirable duration whereas Due Date 2 (DD2) is the maximum allowed. All the projects have a weight of 1.

In figure 1, we show the evolution of the tasks performed by each resource and the prices of the time slots. We also show the evolution of the Duality Gap (upper side of figure 1). The prices of time slots are the solution of the dual problem and the duality gap is a measure of the difference between the primal and dual objective function, so it gives a measure of the quality of the solution. The smaller the duality gap, the more representative the prices, and the better the solution will be.

When the first projects are included in the system, the duality gap is high. But the, the price formation mechanism makes the prices to stabilise, and the gap becomes smaller.

As new projects are required in time 30, the gap increases because the prices of the resources are not stable at all. After some time, the prices changes to adapt themselves to the new system conditions and the gap decreases again. Finally, the value of the objective function is 19402.

We have to remark that project P5 has been rejected although it has high value, but its value was not available at time 0, when projects P1, P2, P3 and P6 were) waiting to star-up.

The simulation not only gives us the dynamic schedule and the refused projects, but the value of each resource as well. For instance, in figure 1, the prices of resource R1 are very high during all time slots. This means that the resource competence is very valuable (bottle neck), so if the firm is going to be engaged in similar project in the nearby future, it would be useful to include more resources with the same competences. In the other side, prices of resources R2 and R3 are small, although they are working in different tasks during the simulation.

So we should inquiry about the possibility of enhancing the range of capabilities of resources R2 and/or R3; for instance, in the case of human resources, this can be done by means of training.

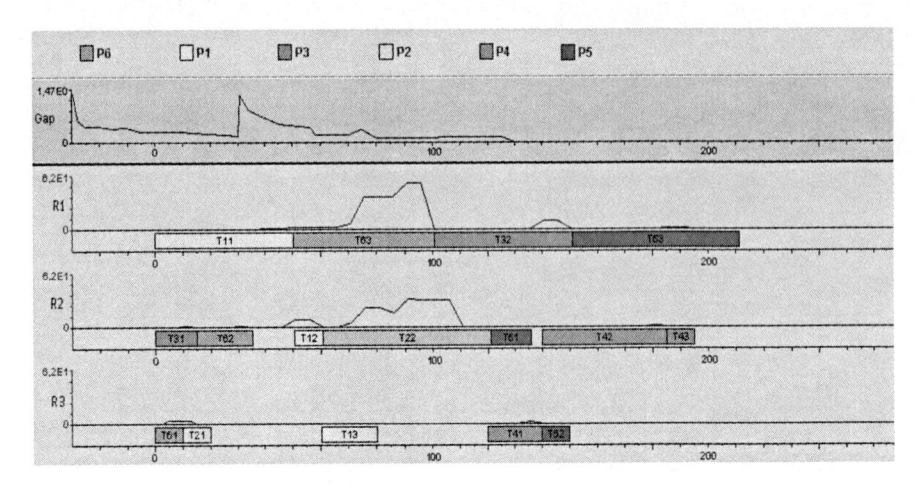

Fig. 2. Tasks performed by resources (competences of resource R2 increased)

In figure 2, we show the evolution of the system when the resource R2 is also endowed with the competence C1. Comparing with the previous case, now the price range is lower for resource R1 and bigger for R2. Some tasks exclusively preformed by R1, are now made by R2. This shows the system is capable to use the flexibility of resource R2 to improve the global performance. Now, the project P5 is accepted and executed, and the objective function has been increased from 19402 to 24039.

6 Conclusions

The most widely accepted idea, which is the one we have followed in our study, consists in building an agent society. In this society the physical agents that represent the real resources interact with other software agents that represent the existing projects. These agents negotiate according to certain schemes, and the result of this interaction will be an acceptable performance of the system.

According to these ideas we have developed a small prototype and we intend to verify the adaptation of these techniques for our problem. We proposed a multi-agent system with two types of agents: projects and resources, and we propose an auction mechanism to distribute scheduling and control activities in the set of agents. By means of this auction mechanism, time slots of resources are valued and these values are used to make scheduling and control decision.

The proposed case study show some of the capabilities of our system to deal with some of the decisions manager need to take within multi-project environments. The system allocate dynamically resources to projects, and decides what projects to accept or rejects taking into account project value, profitability and (feedback) operational information. We also show how it is possible to discover which resources are the most valuable, so they should be added to the firm.

Preliminary results show efficient performance, but there are still many matters to investigate. Future works will be devoted to test the proposed approach on more study cases or even on real cases. We can add complexity to the structure of the projects, and we must improve same aspects of the auction mechanism such as convergence and stability.

References

1. Anavi-Isakov, S., Golany, B.: Managing multi-project environments through constant work-in-process. International Journal of Project Management 21, 9–18 (2003)
2. Hans, E.W., Herroelen, W., Leus, R., Wullink, G.: A hierarchical approach to multi-project planning under uncertainty. Omega 35, 563–577 (2007)
3. Kao, H.P., Wang, B., Dong, J., Ku, K.C.: An event-driven approach with makespan/cost tradeoff analysis for project portfolio scheduling. Computers in Industry 57, 379–397 (2005)
4. Cohen, I., Mandelbaum, A., Shtub, A.: Multi-project scheduling and control: a process-based compartative study of the critical chain methodology and some arternatives. Project Management Journal 35(2), 39–50 (2004)
5. Jennings, N.R., Wooldridge, M.J.: Applying agent technology. Applied Artificial Intelligence 9, 357–369 (1995)
6. Wooldridge, M.J.: An Introduction to Multi-agent Systems. John Wiley & Sons Ltd., New York (2002)
7. Yan, Y., Kuphal, T., Bode, J.: Application of Multi-Agent Systems in Project Management. In: Working Notes of the Agent-Based Manufacturing Workshop, Minneapolis, MN, pp. 160–170 (1998)
8. Clearwater, S.: Market-Based Control: A Paradigm for Distributed Resource Allocation. World Scientific, Singapore (1996)
9. Kumara, S.R.T., Lee, Y.H., Chatterjee, K.: Distributed multi-project resource control: A market-based approach. CIRP Annals - Manufacturing Technology 51, 367–370 (2002)
10. Lee, Y.H., Kumara, S.R.T., Chatterjee, K.: Multi-agent based dynamic resource scheduling for distributed multiple projects using a market mechanism. Journal of Intelligent Manufacturing 14, 471–484 (2003)
11. Confessore, G., Giordani, S., Rismondo, S.: A market-based multi-agent system model for decentralized multi-project scheduling. Annals of Operations Research 150, 115–135 (2007)
12. Kim, K., Paulson, J., Levitt, R.E., Fischer, M.A., Petrie, J.: Distributed coordination of project schedule changes using agent-based compensatory negotiation methodology. Artificial Intelligence for Engineering Design, Analysis and Manufacturing: AIEDAM 17, 115–131 (2003)
13. Kim, K., Paulson, J.: Multi-agent distributed coordination of project schedule changes. Computer-Aided Civil and Infrastructure Engineering 18, 412–425 (2003b)
14. Kim, K., Paulson, J.: Agent-based compensatory negotiation methodology to facilitate distributed coordination of project schedule changes. Journal of Computing in Civil Engineering 17, 10–18 (2003a)
15. Wu, S., Kotak, D.: Agent-based collaborative project management system for distributed manufacturing. In: Proceedings of the IEEE International Conference on Systems, Man and Cybernetics, pp. 1223–1228 (2003)
16. Cabac, L.: Multi-agent system: A guiding metaphor for the organization of software development projects. In: Petta, P., Müller, J.P., Klusch, M., Georgeff, M. (eds.) MATES 2007. LNCS (LNAI), vol. 4687, pp. 1–12. Springer, Heidelberg (2007)
17. Luh, P.B., Hoitomt, D.J.: Scheduling of Manufacturing Systems Using the Lagrangian Relaxation Technique. In: IFAC Work Shop on Discrete Event System Theory and Applications in Manufacturing and Social Phenomena, Shenyang, China (1991)
18. Zhao, P.B., Luh, J.: Surrogate Gradient Algorithm for Lagrangian Relaxation. Journal of Optimization Theory and Applications 100(3), 699–712 (1999)
19. Wang, J., Luh, P.B., Zhao, X., Wang, J.: An Optimization-Based Algorithm for Job Shop Scheduling. Sadhana, a Journal of Indian Academy of Sciences 22, Part 2, 241–256 (1997)

Experiencing Self-adaptive MAS for Real-Time Decision Support Systems

Jean-Pierre Georgé[1], Sylvain Peyruqueou[2], Christine Régis[1], and Pierre Glize[1]

[1] IRIT - Institut de Recherche en Informatique de Toulouse
118 route de Narbonne, 31062 Toulouse Cedex 4, France
{george,regis,glize}@irit.fr
[2] UPETEC – Emergence Technologies for Unsolved Problems
Parc Technologique du Canal - 10 avenue de l'Europe
31520 Ramonville Saint Agne, France
sylvain.peyruqueou@upetec.fr

Abstract. Hydrological phenomena are often very dynamic and depend on numerous criteria. The STAFF software is an adaptive model for flood forecast based on self-organizing multi-agent systems. It is operational since 2002 in the Midi-Pyrenees region in France. The aim of this paper is to show the relevance of our approach to model complex natural systems by focusing on the results, architecture and self-organization mechanisms of a real world application. The main idea is to let the artificial system self-adapt towards the adequate model by confronting it to real data, thus ensuring that the resulting model represents reality. Moreover, since the MAS is *constantly adapting*, we obtain a dynamic and autonomous system that can take into account any future dynamics (strong perturbations, sensor breakdowns...) and able to provide decision-makers with usable information *anytime*.

Keywords: Cooperative agents, self-organisation, emergence, adaptation, flood forecast.

1 Introduction

Floods are sometimes violent phenomena, particularly in the Pyrénées mountains [3] and designing decision support systems for flood forecast is a critical but complex problem. Their forecast requires specialised and experiencing human experts or computer software able to relevantly predict their real-time evolution. This need is imposed by legal responsibility devoted to the DIREN (*Direction Régionale de l'Environnement*, a public organism of the French environment ministry) to inform the prefect and other public structures on risks due to increasing water level.

More than one hundred sensors (see section 2.1) are distributed on the Garonne basin in order to perceive in real-time the evolution of the water level of the different rivers. For twenty years several physico-hydrological models are used in the DIREN as decision support systems in order to fulfil its mission. Unfortunately these models have some serious insufficiencies addressed by the work described in this paper.

Hydrology is a rich scientific research domain but many works focus on real-time physical and spatial modelling where terrain data is a main input (like in [8]). On the contrary, we aim at achieving a predictive functional model, abstracted from the underlying hydro-physical data.

In this work new models are created based on self-organizing multi-agent systems. The generic MAS model is called STAFF[1] for Software Tool for Adaptive Flood Forecast. Regarding the novelty of this technology, this work corresponds to a concrete technology transfer. Three actors having diverse objectives (applicative, technological and theoretical) were involved in this transfer process:

1. **The end user.** The DIREN service has been using a lot of mathematical models for thirty years.
2. **The laboratory.** The SMAC team in IRIT proposes a theory for designing adaptive complex systems based on cooperative self-organization of their micro-level components [6]
3. **The SME.** Because software systems are more and more complex, IT service providing companies need to acquire new technologies (methods, languages, middleware,...).

In the following, to emphasise the strengths of the STAFF flood forecast system, we first present the experimental results, in particular in difficult situations like missing or noisy data. We also discuss the important characteristics of the system such as ease of deployment and robustness. In the second part, we then show how the two-level multi-agent system which is detailed offers an adaptation process taking into account this difficult dynamic environment thanks to the cooperative behaviour of the agents and to their self-organization.

2 STAFF Performance on Real Data

Since 2002, STAFF is running under the software platform *Sophie* [4]. The results presented below are extracted from real behaviours given in this platform. For all the graphics, x-axis represents time over ten days (240 hours), whereas y-axis gives the water level (metres) on the selected river. The curve in bold in each figure is the real river level evolution.

2.1 Forecast with Missing Data

The figure 1 compares STAFF with a physico-hydrological model specifically tuned for this river. Unfortunately this last model is unavailable during 36 hours (its curve is not always present) because of the main data required for computing the mathematical function are missing due to the breakdown of the associated sensors.

During the same period, STAFF (the third curve) is always available because it computes an affine function where unavailable dates are forced to 0, leading to erroneous results. But STAFF is a real-time learning software allowing to compensate data by other ones. The result is certainly sub-optimal but the starting flood is sufficiently well predicted. When the sensors are repaired, a new learning process occurs allowing the reintegration of the corresponding data in the computation.

---

[1] The authors thank Thomas Sontheimer, of Artal technologies company, who was involved in the design of Staff.

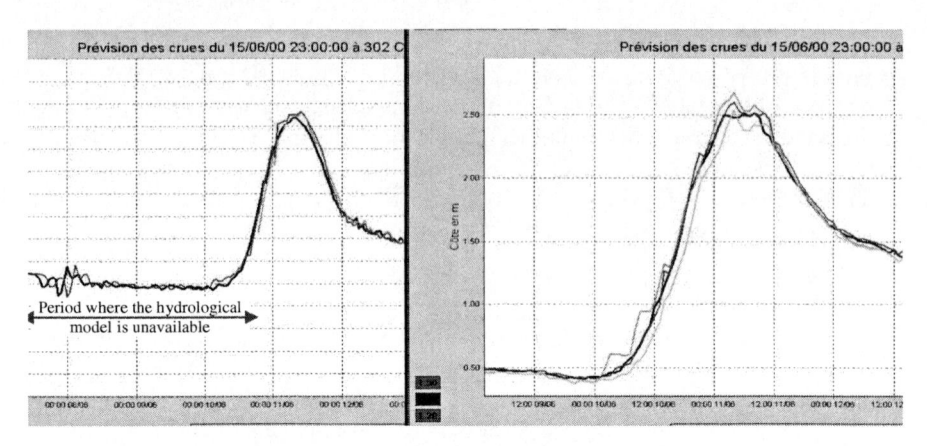

Fig. 1. Missing Data

Fig. 2. Intermediary Station

2.2 Flood Forecast Given for an Intermediary Station

In the figure 2, the STAFF model is used three hours ahead. The real height is in bold. STAFF's curve is situated between the two standard models curves and so fits the best with the real situation.

From evidence, STAFF is closed to the real river dynamic, whereas the two classical models, specifically tuned- are not so well. This is the general observation which could be done on the dozen of places where classical models are installed in Midi-Pyrénées.

2.3 Forecast with Noisy Data

In figure 3, the two noisy curves indicate the evolution levels of the two important upstream rivers (discontinuous curve for STAFF, real height in bold). We can observe two distinct phenomena on these stations:

- The station with the highest curve has three important periods of missing data (when the flood starts, at the middle and on the maximum).
- The station with the lowest curve is very noisy due to non natural evolution. In fact, its location on the river is near a dam constructed to produce electricity.

The STAFF model (discontinuous curve) gives not very precise results, reflecting the noisy and absent data observed on the upstream stations. Nevertheless STAFF gives some idea about the river fluctuation useful for the expert decision. During the past years some hydrological models were tested and given up on this station because they gave no relevant result.

2.4 System Analysis

Concrete results obtained by the self-modelling software by its 22 instances installed in the DIREN service are a good indicator of its real adaptive capability. It possess also some other interesting characteristics:

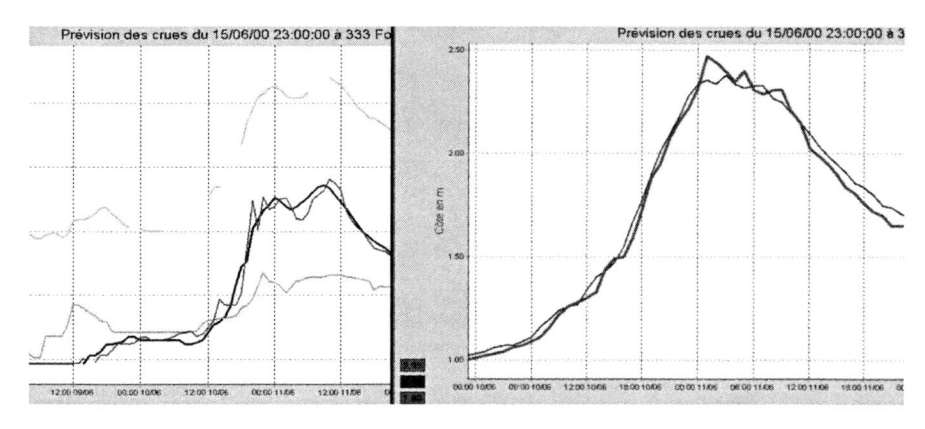

Fig. 3. Upstream Noisy Stations   **Fig. 4.** Upstream Station Result

- Deployment facility. The first phase of self-organization requires an initial collection of historical data in order to obtain a relevant behavior in a given location by automatic training. This phase requires at the most a day work comparing to many months for usual model calibration.
- If historical data are not available, STAFF can be directly deployed because it learns directly on real river level evolution. This starting from scratch leads to a non relevant forecast, but its learning real-time capability allows a sensible improvement during the first hours.
- Robustness. The sensor network is subject to failure such as sensors breakdown or communication problems. This is very problematic for a usual model when one of its central data is unavailable. In this situation, the STAFF adaptive capability allows to recompute a new sub-optimal model. Even it is not totally accurate STAFF gives responses in a very wide spectrum of configurations.
- Evolution. A river bank could change after a severe flood, implying the obsolescence of a usual model. On the opposite, the STAFF model calibration could be let automatically on real-time or explicitly improved by new historical data.

3 The Self-organizing Agents

The aim of STAFF was to be able to compute a flood forecast at any point in the basin without any prior information (either physical or hydrological). It only utilizes the data gauges in the basin and the real current river level at the point for which STAFF must provide a flood forecast. It must give such forecasts in real time.

3.1 Theoretical and Methodological Background

Based on the well-discussed advantages of cooperation [1], we have developed the AMAS (Adaptive MultiAgent Systems) theory which is based upon the following theorem describing the relation between cooperation in a system and the resulting functional adequacy[2].

Theorem. *For any functionally adequate system, there is at least a cooperative internal medium system that fulfils an equivalent function in the same environment.*

Definition. *A cooperative internal medium system is a system where no Non-Cooperative Situations exist.*

Definition. *An agent is in a Non-Cooperative Situation (NCS) when: (1) a perceived signal coming from the environment is not understood or is ambiguous; (2) perceived information does not produce any activity of the agent; (3) the conclusions are not useful to others.*

Our definition of cooperation is based on three local meta-rules the designer has to instantiate according to the problem to solve:

Metarule1 (c_{per}): *Every signal perceived by an agent must be understood without ambiguity.*

Metarule2 (c_{dec}): *Information coming from its perceptions has to be useful to its reasoning.*

Metarule3 (c_{act}): *This reasoning must lead the agent to make actions which have to be useful for other agents and the environment.*

The theorem of functional adequacy means that we only have to use (and hence understand) a subset of particular systems (those with cooperative internal mediums) in order to obtain a functionally adequate system in a given environment [7]. The designer provides the agents with local criterion to discern between cooperative and NCSs. The cooperative attitude between agents constitutes the engine of self-organization. The agents have to try to choose the more cooperative action when they can and also when NCSs occur to detect them and to remove them. Depending on the real-time interactions the multi-agent system has with its environment, the organization between its agents emerges and constitutes an answer to the difficulties of complex systems modelling (indeed, there is no global control of the system) [5].

3.2 The STAFF Architecture

The upper level is a MAS (figure 5) which computes a forecast for the water level continuously. Each agent in this level (called "hourly agents") predicts the changes occurring during the next hour (for example between t+3 and t+4). Each hourly agent is a MAS itself, composed of agents on the second level. An agent in this second level is associated with a sensor and is trying to find the influence of its measurement during that time period for the hourly agent. We use these notations:

---

[2] "Functional" refers to the "function" the system is producing, in a broad meaning, i.e. what the system is doing, what an observer would qualify as the behaviour of a system. And "adequate" simply means that the system is doing the "right" thing, judged by an observer or the environment. So "functional adequacy" can be seen as "having the appropriate behaviour for the task".

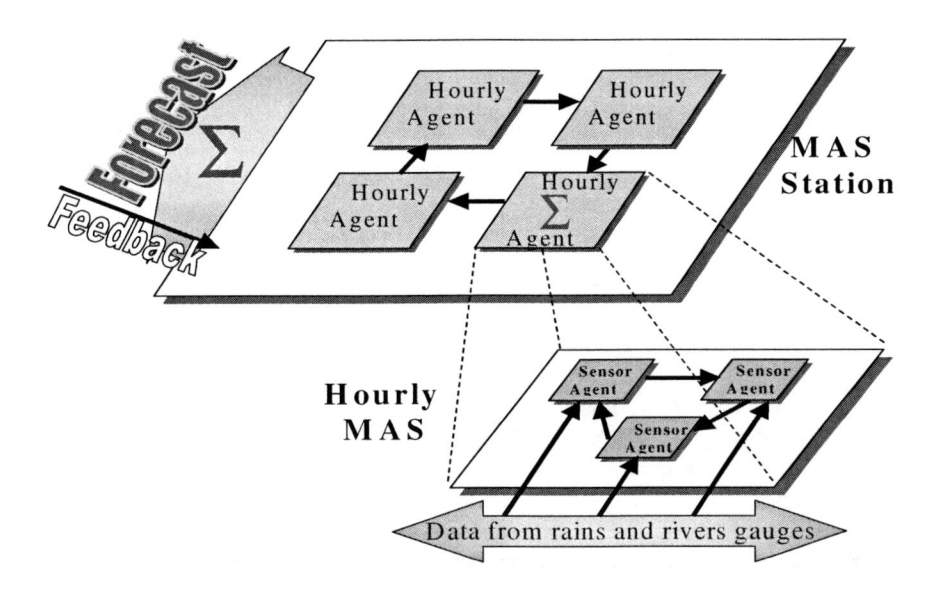

Fig. 5. The two MASs Levels of STAFF

- Si is the raw value provided by the sensor agent associated with the sensor i (which should correspond indifferently to a rain gauge or a river gauge without any additional information on its location).
- ωi is the weight which will be applied to the entry value i. This is an integer varying between 0 and 2000. A value less than zero means that this sensor is currently irrelevant to explain the output (the hourly forecast).
- Δω is the minimal value of the increment applied to the weight ωi. This value is 1 in the current system.
- The 'forecast' provided by the hourly agent is Fk (from time t+k-1 to time t+k) calculated as the weighted sum of all the entries greater than zero. Fk = Σ (ωi*Si), ∀i/ωi>0. This simple affine function is the change of the river level.
- The global forecast is the result given by the MAS station. It is the sum of the Fk (for a forecast for time t four hours in the future, it is the sum of the changes during the four hours preceding t).
- The feedback to each station is the actual change in the river level between time t-1 and t.

3.3 Self-organisation in the MAS

We encompassed each sensor with an agent that is in charge of determining its influence as a weight. Each entry (typically one thousand of them) comes from sensors in the Garonne hydrological basin. At a station, all the sensors intervene as a pondered sum of their weights in order to compute the change in water level over one hour. Thanks to this systematic adjustment, the modelling is adaptive and the relation non-linear over time in spite of using a simple balanced sum, since the weights will change over that time.

The aim of a sensor agent is to adjust the extent to which the measure it is associated with affects the hourly forecast. The typical non-cooperative situation for a sensor agent consists in a bad evaluation of its influence. This case appears every time the hourly agent it is working for has to readjust its forecast. Non-cooperative situations depend on the notions of correlation and influence. Correlation indicates whether a measure has to be used to compute the forecast and if so the level of influence. There are three types of non-cooperative situations a sensor agent may face:

1. The entry value of an agent is not correlated with the feedback (the station evolution). This is a non-cooperative situation of uselessness because the agent cannot, at this moment, explain the output. In this case the weight must be diminished: $\omega_i = \omega_i - \Delta\omega$.
2. The entry value of an agent is correlated with the feedback, but ω_i is currently negative (this entry was in the past mainly uncorrelated). This agent could be useful and the weight value must be increased: $\omega_i = \omega_i + \Delta\omega$.
3. The entry is correlated but the forecast given previously (i.e. the forecast for the current time t, for which we know now the feedback value) is erroneous. The wrong influence of the agent must be modified in order to decrease the forecast error. Since in this situation many agents will act in the same way the adjustment, $\omega_i = \omega_i + \text{Sign}(\text{Feedback} - \text{Forecast}) * \Delta\omega$, is concurrent.

In order to take into account the very latest sensor information, the forecast delay for an hourly agent has to be the shortest possible. The inferior limit for the forecast delay is equal to the range period of the sensors which is one hour for the Garonne basin.

Thus we have built a multi-agent system made of several forecasting agents (the hourly agents), each having a forecast range of one hour. The number of these agents in the MAS at the station depends on the number of hours in the forecast delay. Each of these agents computes its forecast for its own period of one hour. The sum of the hourly agents gives the final forecast associated with the station.

The presence of several hourly agents in the same system may lead to conflict situations between their respective results. STAFF is programmed to take into consideration only the most critical of them leading to the following treatment:

1. The first forecasting agent F1 compares its previous forecast with the last station measure (Feedback). A difference is a conflict between the forecast value and the real value, which implies an adjustment inside the corresponding hourly agents.
2. F2 (giving the change forecast between 1 and 2 hours in the future) compares its previous forecast (the one it gave 1 hour in the past) with the new forecast of F1 (so for the same time in fact). The difference is interpreted as a conflict that implies an adjustment inside the corresponding hourly MAS. This is repeated with the other forecast agents.

4 Conclusion

Since 2002, STAFF has been operational on the *Sophie* software platform in France. STAFF's results indicate its real adaptation capacity based on a process of self-organization distributed among agents on a two-level architecture. We have shown in part 3 that each level is necessary to tend at best towards the functional adequacy. The global function is really emergent compared with the agents' activity which realizes very simple treatments and compared with their self-organization process conducted by adjustment rules which do not depend on the expected global function.

This real application gives a positive experimental feedback on the relevance of the underlying AMAS theory. STAFF and the other applications using the AMAS theory are example of collaborative emergence, in which the goal is not to obtain a given end state but a never ending adaptation process because the systems are plunged into a dynamic environment. We think that it is an important class of MAS simulation hard problems and having specific characteristics (dynamic environment, non termination, huge space search). As a long term theoretical perspective we work on searching for formal demonstration of convergence, robustness and efficiency of this class of systems. From an engineering point of view, we currently enrich the ADELFE methodology by adding more precise guidelines and tools.

References

[1] Axelrod, R.: The Evolution of Cooperation. Basic Books, New York (1984)
[2] Bernon, C., Gleizes, M.-P., Peyruqueou, S., Picard, G.: ADELFE: A methodology for adaptive multi-agent systems engineering. In: Petta, P., Tolksdorf, R., Zambonelli, F. (eds.) ESAW 2002. LNCS (LNAI), vol. 2577, pp. 156–169. Springer, Heidelberg (2003)
[3] Dupouyet, J.P., Vidal, J.J.: La modernisation de l'annonce et de la prévision des crues dans le bassin de la Garonne. SHMA – Direction Régionale de l'Environnement Midi-Pyrénées (1991)
[4] Georgé, J.-P., Gleizes, M.-P., Glize, P., Régis, C.: Real-time Simulation for Flood Forecast: an Adaptive Multi-Agent System STAFF. In: Proceedings of the AISB 2003 symposium on Adaptive Agents and Multi-Agent Systems, University of Wales, Aberystwyth, April 7-11 (2003)
[5] Georgé, J.-P., Edmonds, B., Glize, P.: Making Self-Organising Adaptive Multiagent Systems Work. In: Bergenti, F., Gleizes, M.-P., Zombonelli, F. (eds.) Methodologies and Software Engineering for Agent Systems, pp. 319–338. Kluwer, Dordrecht (2004)
[6] Gleizes, M.P., Camps, V., Glize, P.: A theory of emergent computation based on cooperative self-organization for adaptive artificial systems. In: Fourth European Congress on Systemic (1999), http://www.irit.fr/SMAC
[7] Gleizes, M.P., Camps, V., Georgé, J.-P., Capera, D.: Engineering systems which generate emergent functionalities. In: Weyns, D., Brueckner, S.A., Demazeau, Y. (eds.) EEMMAS 2007. LNCS, vol. 5049, pp. 58–75. Springer, Heidelberg (2008)
[8] Servat, D., Perrier, E., Treuil, J.-P., Drogoul, A.: When agents emerge from agents: Introducing multi-scale viewpoints in multi-agent simulations. In: Sichman, J.S., Conte, R., Gilbert, N. (eds.) MABS 1998. LNCS (LNAI), vol. 1534, pp. 183–198. Springer, Heidelberg (1998)

Induced Cultural Globalization by an External Vector Field in an Enhanced Axelrod Model

Arezky H. Rodríguez[1], M. del Castillo-Mussot[2], and G.J. Vázquez[3]

[1] Academia de Matemáticas, Universidad Autónoma de la Ciudad de México, Mexico City, Mexico
arezky@gmail.com
[2] Departamento de Estado Sólido, Instituto de Física, Universidad Nacional Autónoma de México (UNAM), Apdo. Postal 20-364, San Ángel 01000, México D.F., Mexico
11dlcstll@yahoo.com
[3] Departamento de Estado Sólido, Instituto de Física, Universidad Nacional Autónoma de México (UNAM), Apdo. Postal 20-364, San Ángel 01000, México D.F., Mexico
jorge@fisica.unam.mx

Abstract. A new model is proposed, in the context of Axelrod's model for the study of cultural dissemination, to include and external vector field (VF) which describes the effects of mass media on social systems. The VF acts over the whole system and it is characterized by two parameters: a non-null overlap with each agent in the society and a confidence value of its information. Beyond a threshold value of the confidence there is induced monocultural globalization of the system lined up with the VF. Below this value, the multicultural states are unstable and certain homogenization of the system is obtained in opposite line up according to that we have called *negative publicity* effect. Three regimes of behavior for the spread process of the VF information as a function of time are reported.

1 Introduction

Agent-Based Models (ABMs) [5, 11] are computer simulations of the local interactions of the members of a population which could be plants and animals in ecosystems [15], vehicles in traffic, people in society [1], etc. Locally interactions at lower-level give rise to the spontaneously emergence of higher-level organizations whose properties are not possessed by the individuals neither directly determined by them. Complex and non-linear phenomena have attracted the attention of the scientific community to study the interplay between the lower and higher levels of organizations [20]. These models typically consist of an environment or framework in which the interactions occur among some number of individuals defined in terms of their behaviors (procedural rules) allowing the tracking of the characteristics of each individual through time.

There are lots of applications to model different aspects of dynamics in society [7, 16, 26]. Specifically, the Axelrod model [1, 2, 8] is an ABM designed to investigate the dissemination of culture among interacting agents in a society. In this model, society is represented by a lattice composed by a 2-dimensional array of vectors (agents) with a number of entries called "features". The definition of its culture is given by the set of traits an agent has in its features. The Axelrod's model has been exhaustively implemented to study a great variety of problems: the nonequilibrium phase transition between monocultural and multicultural states [18], the cultural drift driven by noise [19, 22], nominal and metric features [10, 17], propaganda [6], time evolution dynamics [25], the resistance of a society to the spread of a foreign cultural traits [4], finite size effects [24], the impact of the evolution of the network structure with cultural interaction [9] among others.

Some works have been done including in the Axelrod model an extra agent acting as a vector field (VF) over the whole society with the purpose of simulating a mass media effects [12, 13, 14, 23]. In all of them, the interaction between the external VF and agents is similar to that between an agent and its neighbors: they interact only if they have at least one common trait in their corresponding features. In this formalism, the inclusion of an external field, which does not change its values on time, introduces an asymmetry on the lattice, which can now be described as composed by two groups of agents: group A where agents have trait(s) in common with the VF and group B whose do not. All the interactions can now be classified as follows: agents from group A with VF (VF-A), agents from group B with VF (VF-B), between agents from group A (A-A), between agents from group B (B-B) and finally between agents from group A and B (A-B). The VF-B interaction is a null interaction because agents from group B do not share traits with the VF. In this way, the only opportunity of agent B to acquire one VF trait is through a diffusion mechanism with the combined interactions VF-A plus A-B. In those models it is also included the strength of the VF as a probability P of interactions VF-A and VF-B while the probability of interactions A-A, B-B and A-B are given then by $1 - P$. Therefore the diffusion mechanism has very low probability as P increases and agents from group B are set apart from the VF information. Furthermore, the mechanism VF-B can be very active (time consuming) but with null effects, and the internal relaxing mechanism B-B is not able to drive agents to the final state in an efficient way. Thus, the final absorbing states obtained in those previous models consisting in multicultural states when the VF strength is increased is then not surprising [12, 13, 14, 23].

Our current interest is to develop a new model for the inclusion of an extra agent acting as a vector field (VF) over the *whole* society to overcome the difficulties achieved by the previous models described above. As mentioned in Ref. [23], the media information is socially processed through personal networks. Then, models will be more realistic is they allow a strong interaction with the VF without loss of interchanges between agents on the lattice. It is also important to say that mass media designs its publicity in a clever way. As mentioned by the anthropologist Gregory Bateson: to produce a change it is necessary to be different but, at the same time, it is necessary to be "close enough" to be taken into account [3]. When acting over the society, mass media always try to have something in common with the people chosen as target of publicity or propaganda. It is designed to offers attractive materials for the whole society: news, sports, soap operas, movies, cartoons, music, arts, etc.

Our goal here is to develop a model to include this effect considering an additional non-zero probability of all agents to copy a trait from the VF, even if they do not share any trait of their features. Section 2 is devoted to that purpose. In section 3 it is exposed some numerical calculations and finally some conclusions are outlined in section 4.

2 The Model

The system consists of L^2 agents as the sites of a square lattice. The state of an agent i is defined as a vector of F *nominal* components called features given by $\sigma_i = (\sigma_{i_1},...,\sigma_{i_f},...,\sigma_{i_F})$ which characterize the nominal F-dimensional culture of the corresponding agent. This way, each agent has four nearest neighbors but as the fifth it is introduced the VF $\mathcal{M}$ with nominal features $\sigma_{\mathcal{M}} = (\sigma_{\mathcal{M}_1},...,\sigma_{\mathcal{M}_f},...,\sigma_{\mathcal{M}_F})$. The VF intents to simulate and external mass media or publicity which acts over the whole society. Then, each agent can interact with five agents: its four nearest neighbors and the VF $\mathcal{M}$, all with equal probability $1/5$. Additionally, each feature σ_{i_f} and $\sigma_{\mathcal{M}_f}$ can take any of the values in the set $\{0,1,...,q-1\}$ which are the corresponding cultural traits of an agent i or the VF $\mathcal{M}$. Initially, the values of the vectors σ_i and $\sigma_{\mathcal{M}}$ are randomly and independently set with one of the q^F state vectors with uniform probability.

The interaction between different agents is possible only when the two vector have an overlap $0 < l < 1$ where the overlap between two agent i and j is the number of shared traits and it is given by $l(i,j) = \sum_{f=1}^{F} \delta_{\sigma_{i_f},\sigma_{j_f}}$. Here δ is the Kronecker symbol. The probability, which we call here *nominal* probability, of the interaction between two agents is given by $p(i,j) = l(i,j)/F$. In general, the situation $p(i,j) = 0$ is possible when the overlap between two agents is zero, but the case where the probability between an agent and the VF is zero is not an acceptable situation for a publicity (or mass media) which intentionally designs its interaction in such a way that always there are features which have traits in common with the agent subject of the influence to guarantee that the connection is active.

In order to include this important effect in our model, we included some *effective* features that the VF always shares with each agent when interacting, besides nominal features, with the purpose to simulate phenomenologically in a simple way the almost omnipresent force of today's publicity or propaganda in mass media that offers something for all tastes and ages (magazines, radio programs, TV series, etc). Note that the effective features are related with the dynamics between the agents and the VF while the nominal features are related with the dynamics between agents inside the lattice. The specific nature in real society of the effective features is not of importance here. It will be different for different agents, but the intention is to take into account the specific design of the publicity that mass media does to attract everyone. For $\varepsilon/F < 1$ where ε is defined as the "effective feature", the VF and the agent share more nominal than effective features and the VF can be considered as a "perturbation" to the internal interaction between different agents in the society. In our model, the parameter ε not only takes natural values, but also fractional values, as it will be seen later.

Therefore, the probability of interaction between the external vector and an agent, which we call here *extended* probability, is written as

$$p(i,\mathcal{M}) = \frac{l(i,\mathcal{M}) + \varepsilon}{F + \varepsilon} = \frac{l(i,\mathcal{M})/F + \varepsilon/F}{1 + \varepsilon/F} \qquad (1)$$

where $l(i,\mathcal{M})$ is the overlap of the nominal features between agent i and the VF. The probability is zero only when there are not effective features ($\varepsilon = 0$) and the overlap between the agent i and the VF is zero. In contrast, in all the other cases the probability

Induced Cultural Globalization by an External Vector Field   313

Case A:

$p' = \mathcal{C}$

$$\begin{pmatrix} 1 \\ 2 \\ 3 \end{pmatrix} \begin{pmatrix} 0 \\ 0 \\ 0 \end{pmatrix}$$

$i \quad \mathcal{M}$

$$p(i, \mathcal{M}) = \frac{\epsilon/F}{1 + \epsilon/F}$$

Case B:

$p' = \mathcal{C}$

$$\begin{pmatrix} 1 \\ 2 \\ 3 \end{pmatrix} \begin{pmatrix} 0 \\ 2 \\ 1 \end{pmatrix}$$

$i \quad j$

$p(i, j) = 1/3$

Case C:

$p' = 1 - \mathcal{C}$

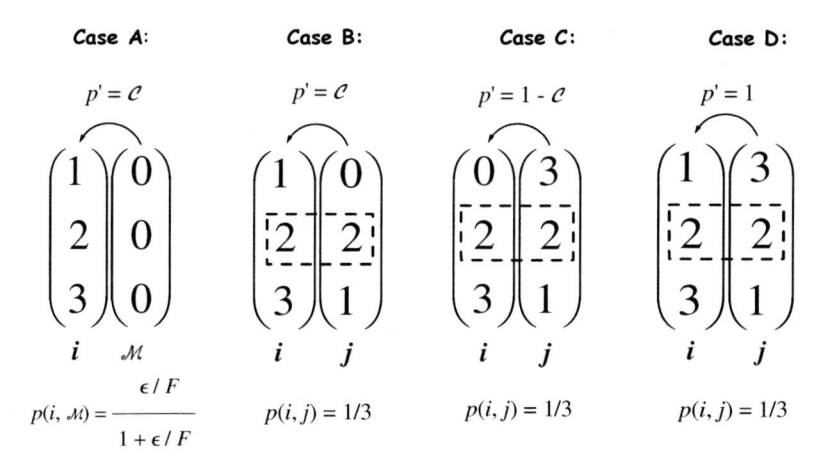

$$\begin{pmatrix} 0 \\ 2 \\ 3 \end{pmatrix} \begin{pmatrix} 3 \\ 2 \\ 1 \end{pmatrix}$$

$i \quad j$

$p(i, j) = 1/3$

Case D:

$p' = 1$

$$\begin{pmatrix} 1 \\ 2 \\ 3 \end{pmatrix} \begin{pmatrix} 3 \\ 2 \\ 1 \end{pmatrix}$$

$i \quad j$

$p(i, j) = 1/3$

Fig. 1. Four possible cases of interaction for a system with $F = 3$ features. Shared features are indicated inside a dashed rectangle. The probability of interaction $p(i, j)$ (or $p(i, \mathcal{M})$ in Case A) is indicating below. In each case the trait in σ_{i_1} will be deleted by coping trait σ_{j_1} (or by trait $\sigma_{\mathcal{M}_1}$ in Case A). The probability of coping/deleting trait 1 is given by a) $p' = \mathcal{C}$, b) $p' = \mathcal{C}$, c) $p' = 1 - \mathcal{C}$ and c) $p' = 1$.

$p(i, \mathcal{M})$ is always different from zero. For $\varepsilon = 0$ the values for the case with no effective features are recovered. As expected, the probability is lager for larger values of the effective features ε for a given value of $l(i, \mathcal{M})$ and also increases for larger values of the overlap $l(i, \mathcal{M})$ at a given number of ε. Finally, when the agent i and the VF share all the nominal features ($l(i, \mathcal{M}) = 1$) the probability is always one for any value of ε.

In Ref. [21] the authors have used an expression similar to Eq. (1), but with $\varepsilon = 1$ and, as in Ref. [13], they have modelled the strength of the VF as a probability of the interaction between agents and the VF. Increasing values of this probability implies a decreasing value of the probability for agents interacting between each other and then the corresponding diffusion of traits values between agents can be stopped which is not a realistic or desirable effect. They have found that only monocultural states are obtained and only by introducing a noise rate it is possible to drive the system to a multicultural final state.

In our case, we are not interested to include the effects of random perturbation effects, but we instead introduce another parameter related with the confidence of the information belonging to the VF. As "confidence" we understand here the credibility granted by agents to the information possessed by the VF. It is included as an extra probability $\mathcal{C}$ for agent i to copy an entry directly from the VF or an entry from another agent j that belongs to the VF either. It is also included as an extra probability $1 - \mathcal{C}$ when agent i, when coping a trait, deletes and information the VF possesses in the same feature.

To clarify this important concept we show in Fig. 1 four situations of interaction which resume all the possible cases. In Case A it is described an interaction between agent i and the VF which has been set to (0,0,0) without lost of generality. None of the nominal features are shared and then the probability of interaction $p(i, \mathcal{M})$ depends only from the value of the effective features ε according to the expression in the figure and is always larger than zero. In the practical case when agent i copies, for example the first entry, then the VF will be copied with probability $p' = \mathscr{C}$ which characterize the confidence of the information possesses by the VF. In the next cases, the interaction occurs between agents i and j which only share one trait of three posibles. The probability of interaction is then given by $p(i,j) = 1/3$ in all these cases. In Case B the nominal feature that agent i selects to copy from agent j coincides with the value the VF has in the same feature. Then, as in Case A, the corresponding trait is copied with probability $p' = \mathscr{C}$. In Case C, when coping, agent i will delete its trait which is equal to that possessed by the VF. Then, it is deleted with probability $p' = 1 - \mathscr{C}$. Finally, in Case D the traits copied and deleted are not related with the VF and then they are copied with probability $p' = 1$. Note that according with these rules of interaction, when the confidence of traits possessed by the VF are $\mathscr{C} = 1$, these traits are always copied with probability $p' = 1$ and never deleted. Otherwise, if the confidence of the traits possessed by the VF is $\mathscr{C} = 0$, this traits are never copied ($p' = 0$) and are always deleted with probability $p' = 1$. Then, starting from the initial condition described above, the system evolves by iterating the following steps:

(1) Select at random an agent i on the lattice, which is the active element.
(2) Select at random, with equal probability, an agent of interaction. It could be one of the four nearest neighbors or the VF.
(3) Calculate the overlap $l(i,s)$ where $s = j$ for the neighbor or $s = \mathscr{M}$ for the VF. If $s = \mathscr{M}$, the agent i and the VF interact with the extended probability $p(i,\mathscr{M})$. If $s = j$ and $0 < l(i,j) < F$, agents i and j interact with the nominal probability $p(i,j)$.
(4) In case of interaction between agent i and agent s, choose a position trait h at random such that $\sigma_{i_h} \neq \sigma_{j_h}$ (or $\sigma_{i_h} \neq \sigma_{\mathscr{M}_h}$) and then set $\sigma_{i_h} = \sigma_{j_h}$ (or $\sigma_{i_h} = \sigma_{\mathscr{M}_h}$) according to:
   (4.1) if $s = \mathscr{M}$ then set $\sigma_{i_h} = \sigma_{\mathscr{M}_h}$ with probability $p' = \mathscr{C}$,
   (4.2) if $s = j$ and $\sigma_{j_h} = \sigma_{\mathscr{M}_h}$ then set $\sigma_{i_h} = \sigma_{j_h}$ with probability $p' = \mathscr{C}$,
   (4.3) if $s = j$ and $\sigma_{i_h} = \sigma_{\mathscr{M}_h}$ then set $\sigma_{i_h} = \sigma_{j_h}$ with probability $p' = 1 - \mathscr{C}$.
   (4.4) if $s = j$ and both $\sigma_{j_h} \neq \sigma_{\mathscr{M}_h}$ and $\sigma_{i_h} \neq \sigma_{\mathscr{M}_h}$, then set $\sigma_{i_h} = \sigma_{j_h}$ with probability $p' = 1$.

Before studying the effects of the VF in our model let us review the original Axelrod's model. The computational dynamics of this model ends when the system reaches an absorbing state characterized by either $l(i,j) = 0$ or $l(i,j) = F$ for all pairs of closed neighbors (i,j). A class of absorbing state, given by q^F different configurations is called the "monocultural" state which corresponds to the case where $l(i,j) = F$ for all pairs of closed neighbors. In this case, all agents in the network share the same trait at each feature ($\sigma_{i_h} = \sigma_{j_h}$ for all (i,j)) and the dynamics ends. Another class of absorbing state is called "multicultural" state and consist of at least two (or more) homogeneous domains which agents have cultural traits completely different. This way, two agents belonging to two different domains have zero overlap. The multicultural state is reached when each agent in the lattice has full of null overlap with all its neighbors. In these cases, a domain is given by a set of contiguous sites with identical state vector.

Previous works have shown that the system reaches monocultural or multicultural states in dependence of lower ($q < q_c$) or higher values ($q > q_c$) of the cultural diversity

q [8]. To characterize the transition it has been considered two different order parameters: the average fraction of different cultural domains or the average number of agents in the biggest domain $< S_{max} >$ normalized to the number of lattice elements. In our case, it will be shown how our model produces a complex pattern of social behavior with affinities and repulsions to the influence of the VF in dependence of the effective features ε and the confidence $\mathscr{C}$ of the information.

3 Numerical Results

Numerical simulations have been carried out in lattices with $L^2 = 30 \times 30$ agents and $F = 4$ features each. Different absorbing states have been found and we report the average realization number of agent in the largest domain over 50 different initial conditions.

Figure 2 shows the calculations of $< S_{max} >$ as a function of q at the absorbing state. Each panel shows the result for a certain value of the ratio ε/F and different values of the confidence $\mathscr{C}$ in increasing order. The result using the Axelrod model without VF is included for comparison with full square dots. In this case, the system reaches a monocultural state at $q < q_c \approx 18$ and a multicultural state at $q > q_c$. In panel a) we set $\varepsilon = 1.0$. It can be seen that when the value of the confidence $\mathscr{C}$ is small ($\mathscr{C} = 0.05$), the results are very close to those of Axelrod model. The monocultural states remain unchanged at $q < q_c$ but the multicultural state is less "robust" and higher values of $< S_{max} >$ are obtained. For increasing values of $\mathscr{C}$, higher values of $< S_{max} >$ for $q > q_c$ are obtained, as seen with $\mathscr{C} = 0.10$ (hence the number of different domains in the multicultural state are smaller) and finally, at $\mathscr{C} = 0.15$, the multicultural states vanish and the system remains in monocultural states for all values of q. Then, it can be concluded that the increasing value of the confidence induces an homogenization of the cultural information that the system has, even at those values of q where the system

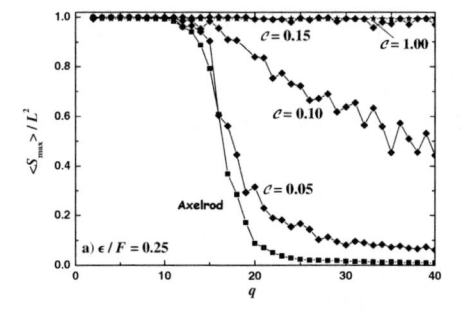

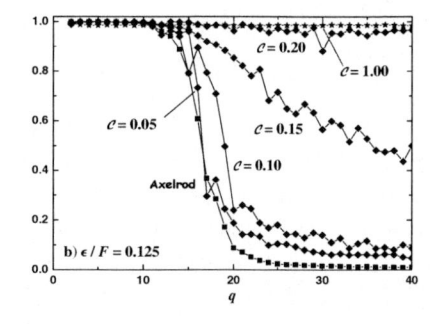

Fig. 2. Calculation of the normalized average number of agent in the greatest domain at an absorbing state as a function of q averaged over 50 realizations. The values of the ratio ε/F are a) 0.25 and b) 0.125. At each panels, different value of the confidence $\mathscr{C}$ is taken into account in increasing order. The case of the Axelrod's model without VF is included in each panel with full square dots. Full rhombus indicate that the corresponding absorbing states do not share the information possessed by the external field, while full stars indicate full coincidence of the absorbing state with the VF.

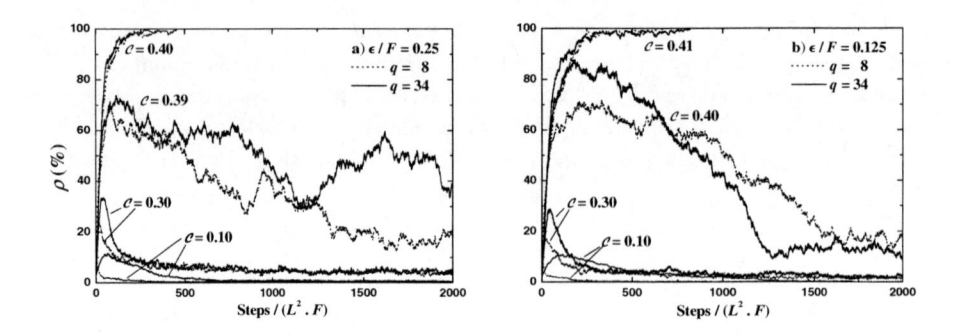

Fig. 3. Percent of the VF information in the lattice as a function of time for different values of the ratio ε/F and the confidence $\mathscr{C}$. In straight lines is used $q = 34$ while in dotted lines it is used $q = 8$.

reaches multicultural states when there is no VF. Then, multicultural states are unstable for increasing values of the confidence $\mathscr{C}$ at this value of the effective trait ε. In panel b) we set $\varepsilon = 0.5$. It can be seen that multicultural states at $q > q_c$ are again obtained for low values of $\mathscr{C}$ but now higher values are needed for the confidence to produce a cultural homogenization ($< S_{max} \approx 1 >$). This can be seen when comparing the results for $\mathscr{C} = 0.15$ in both panels.

Nevertheless, care has to be taken when analyzing the information of the induced monoculture at $q > q_c$ when the VF is present (as the case $\mathscr{C} = 0.15$ in Fig. 2 a) and $\mathscr{C} = 0.20$ in Fig. 2 b)). It is interesting to know whether the greatest domain in the absorbing state characterized by $< S_{max} >$ has, or not, the information possesses by the VF in the corresponding nominal features. This information is indicated in Fig. 2, where all the full dots show absorbing states where the corresponding greatest domain does not possess the information of the VF in any of its features. That is, the overlap between VF and the cultural state of the largest domain is zero. Only at those absorbing states indicated by full stars the corresponding biggest domain fully shares the information at nominal features of the VF. Then, it is obtained the interesting result that at low confidence values $\mathscr{C}$, the culture homogenization induced by the VF results in a negation or cancellation by the agents of the lattice of the information possesses by an external media. Here we call this phenomenon *negative publicity* effect, and it represents the process occurring in society when a group or different groups of people gather together, physically or intellectually, against an external action they consider misconceived.

In order to study in more detail how the information of the VF spreads (or not) into the society, we have defined the parameter ρ which gives the percent of the total amount of the information that agents share in their nominal features with the VF as a function of time. It is given by

$$\rho = 100 \times \frac{1}{L^2 F} \sum_{i=1}^{L^2} \sum_{f=1}^{F} \delta_{\sigma_{i_f}, \sigma_{\mathscr{M}_f}} \qquad (2)$$

and it is shown in Fig. 3. The calculation has been done for different values of the ratio ε/F and different values of the confidence $\mathscr{C}$ taking in consideration only one initial state. In straight lines it is considered $q = 34$ while in dotted lines $q = 8$. When the dynamics starts from random initial conditions, the information possessed by the VF is already present in the lattice and are shared by some agents. This anisotropy gives rise to the strong increase of its percent at the beginning. Nevertheless, if the confidence value is low, the information is avoided by agents when traits are copied and it percent decreases with time after some maximum is achieved. This can be seen for $\mathscr{C}$ equal or below 0.39 and 0.40 in Fig. 3 a) and b) respectively. It is necessary a higher value of $\mathscr{C}$ for drive the system to a monocultural state with all agents aligned with the VF information. On the other hand, if the confidence value is high enough, the percentage of the VF information increases continuously until the absorbing state is reached. This can be seen for $\mathscr{C} = 0.40$ and 0.41 in Fig. 3 a) and b). At the same time, there is a sharp discontinuity between the regions of confidence values where the system remains monoculture and multiculture, as can be seen between $\mathscr{C} = 0.39$ and 0.40 in Fig. 3 a) and between $\mathscr{C} = 0.40$ and 0.41 in Fig. 3 b). For higher values of $\mathscr{C}$ the system reaches the monocultural state in only few time steps, while for intermediate values of $\mathscr{C}$ the value of ρ changes slowly, at least until $2000L^2F$ time steps where the simulation was artificially aborted. At those values of $\mathscr{C}$ the system seems to be in a quasi-stationary state and no absorbing state was found in these regions. For enough low values of $\mathscr{C}$, as those shown in Fig. 2, the system reaches the multicultural state and there is almost no VF information on the lattice, as seen for $\mathscr{C} = 0.10$ in Fig. 3 in all panels.

4 Conclusions

It is developed here a new model for the inclusion of an external vector field in the Axelrod model at zero temperature to describe the effects of the mass media on a social system. The clever design of publicity which allows the mass media to have influence over the whole society was included as a non-zero *extended* probability of traits being copied. This important effect is related with a parameter ε which can be interpreted as an extra effective feature or features the mass media could have with all agents in society, beyond the nominal features. This effective feature(s) is(are) used by the VF to reinforce the frequency of certain information already present on the society or to introduce a new one. It is also included in the model a *confidence* value of the information possessed by the VF. It is modelled as a probability of copying/deleting VF information which represents the criteria a person (or a group of persons) has (have) about what the mass media is proposing to the society. Three different regimes were found. First, for very low values of this confidence, the dynamics recovers the Axelrod model with no external field, but an increasing value produces an homogenization on the society which would be multicultural without the external influence. This cultural homogenization is lined up against the acting influence of the VF with a zero overlap with the VF information. We have called here *negative publicity* to this effect. It simulates the behavior of people in society who gathers together against the external information they estimate wrong or incorrect. Second, for large values of the confidence, the system reaches a quasi-stationary state with only slow changes of the amount of VF information in the society and no absorbing state was found for the amount of time steps tested. Finally, in the third case, higher enough values of the confidence produce a completed homogenized lattice lined up with the VF information, a situation that could be the purpose of mass media, politic party, etc. These results are qualitatively in agreement with the intuitive idea that with enough bombardment of the mass media information, that is accepted as valid and trusted by people, there will be a strong induced culture homogenization in the society. In other words, when people assume this information as personal the cultural differences tend to disappear. It is important to mention that the confidence parameter $\mathscr{C}$ can be experimental measured and, therefore, it could advance the results expected on a society when designing publicity.

Acknowledgements

The authors thank UNAM and CONACyT for partial financial support through Grants IN-114208 and 45835-F respectively. AHR also thanks psychologist Gezabel Guzmán for helpful suggestions and enlightening discussions.

References

1. Axelrod, R.: J. Conflict Resolut. 41, 203 (1997)
2. Axelrod, R.: The Complexity of Cooperation. Princeton, New Jersey (1997)
3. Bateson, G.: Mind and Nature: A Necessary Unity. Bantam Books, Toronto (2002)
4. Boccara, N.: arXiv:nlin/0611035v1 (2006)
5. Bonabeau, E.: PNAS 99, 7280 (2002)
6. Carletti, T., Fanelli, D., Grolli, S., Guarino, A.: Europh. Lett. 74, 222–228 (2006)
7. Castellano, C., Fortunato, S., Loreto, V.: arXiv:physics/0710.3256
8. Castellano, C., Marsili, M., Vespignani, A.: Phys. Rev. Lett. 85, 3536 (2000)
9. Centola, D., González-Avella, J.C., Eguíluz, V.M., Miguel, M.S.: J. Conflict Resolut. 51, 2007 (2007)
10. Flache, A., Macy, M.W.: arXiv:physics/0604201v1 (2006)
11. Goldstone, R.L., Janssen, M.A.: TRENDS in Cognitive Sciences 9, 424 (2005)
12. González-Avella, J.C., Cosenza, M.G., Klemm, K., Eguíluz, V.M., Miguel, M.S.: J. Artif. Soc. Soc. Simul. 10, 9 (2007)
13. González-Avella, J.C., Cosenza, M.G., Tucci, K.: Phys. Rev. E 72, 065,102(R) (2005)
14. González-Avella, J.C., Eguíluz, V.M., Cosenza, M.G., Klemm, K., Herrera, J.L., Miguel, M.S.: Phys. Rev. E 73, 046, 119 (2006)
15. Grimm, V., Railsback, S.F.: Individual-based Modeling and Ecology. University Press, Princeton (2005)
16. Helbing, D.: Quantitative Sociodynamics. Kluwer Academic Publishers, Dordrecht (1995)
17. Jacobmeier, D.: Int. Journ. Modern Phys. C 4, 633 (2005)
18. Klemm, K., Eguíluz, V.M., Toral, R., Miguel, M.S.: Phys. Rev. E 67, 026, 120 (2003)
19. Klemm, K., Eguíluz, V.M., Toral, R., Miguel, M.S.: Phys. Rev. E 67, 045, 101(R) (2003)
20. Laughlin, R.: A Different Universe: Reinventing Physics from the Bottom Down. Basic Books, New York (2006)
21. Mazzitello, K.I., Candia, J., Dossetti, V.: Int. J. Mod. Phys. C 18, 1475 (2007)
22. Sanctis, L.D., Galla, T.: arXiv:0707.3428v1 (2007)
23. Shibanai, Y., Yasuno, S., Ishiguro, I.: J. Conflict Resolution 45, 80 (2001)
24. Toral, R., Tessone, C.J.: arXiv:physics/0607252v1 (2006)
25. Vázquez, F., Redner, S.: EPL 78, 18, 002 (2007)
26. Weidlich, W.: Sociodynamics. Dover Publications, New York (2000)

Towards the Implementation of a Normative Reasoning Process

Natalia Criado, Vicente Julian, and Estefania Argente

DSIC Universidad Politécnica de Valencia
Camino de Vera s/n, 46022 Valencia, Spain
{ncriado,vinglada,eargente}@dsic.upv.es

Abstract. Multi-agent systems have been employed for modeling dynamical and complex systems. In order to control these agent societies, the normative theory has arisen inside multi-agent system area as a coordination mechanism, being a key element in open systems. In this paper, a new normative reasoning process is presented. It allows agents to consider the existence of a dynamical normative context that regulates their behaviors.

Keywords: Multi-agent Systems, Norm Implementation, Normative Reasoning.

1 Introduction

A normative multi-agent system (MAS) is organized through mechanisms for representing, communicating, distributing, detecting, creating, modifying and imposing norms; as well as mechanisms for deliberating on norms and detecting norm fulfillment and violation [1]. Thus, an autonomous normative agent is able to include social norms in their decisions and able to react to norm violations.

This paper proposes a normative reasoning process that allows agents to consider the existence of a dynamical normative context that regulates their behaviors. Open systems, in which heterogeneous and autonomous agents work together, make use of norms as a coordination mechanism. More concretely, this work is focused on allowing an external agent, which has been developed independently of the MAS, to provide its functionality inside the system. For ensuring social order, all agents are informed about the norms that regulate this society. Thus, the external agent should take into consideration the norms that define constraints on its functionality, i.e. who is authorized to make use of it, when and how this functionality can be provided.

With the purpose of giving support to external agents to respect norms, an implementation of norms is presented in this paper. It allows agents to consider the existence of social norms that regulate their behaviors. Our normative reasoning process enables agents to take their decisions according to the dynamical normative context. The rest of the paper is structured as follows. Section 2 briefly describes a state of the art on Normative MAS. Section 3 contains a description of an open architecture for virtual organizations. A detailed description of the reasoning process implementation is contained in section 4. Finally, section 5 contains conclusions and future works.

2 State of the Art

The need for coordinating individual behaviors to achieve the desired social behavior of a MAS has been an important research area for a long time. With this aim, several works have focused on the integration of the normative theory inside the MAS field. Initial works such as [2] [3], which focused on the incorporation of normative concepts inside agents, tried to employ results of previous proposals on legal theory [4]. They have a theoretical point of view, describing agent's obligations and prohibitions as logical propositions by means of deontic logics [5]. More recently, proposals concerning the employment of norms inside MAS have evolved to a more practical conception of norms [6]. These works are aimed at overlapping the gap between the theoretical specification of norms, based on deontic logics, and the implementation of executable norms inside real MAS applications.

The employment of norms as an infrastructure for achieving coordination and cooperation among heterogeneous agents implies the need for the design of reasoning mechanisms, in order to allow agents to consider the existence of norms in their reasoning process. All this process has been generically named as *normative learning* [7]. It covers two different sub processes: the norm emergence and the norm acceptance processes.

The norm emergence problem concerns the study of mechanisms that allow agents to reach an agreement on the definition of a norm that regulates its society [8, 9]. The norm acceptance problem covers the analysis of the motivations for norm acceptance [10], as well as the investigation of the influence of norms into the agent's behavior. Regarding this second research field, in [11] a modification of the BDI architecture is proposed. It allows the development of agents that deliberate on their obligations and prohibitions. With the same aim, in [12] a general normative agent architecture is illustrated. One of the main drawbacks of these two works is that they have an agent-centric point of view, since they do not conceive the agent social environment. In addition, they propose a theoretical formalism which cannot be easily implemented in a computational model.

The objective of the present work is to allow the regulation of open and dynamical societies by means of the definition of norms. More concretely, our approach deals with the problem of developing norm-regulated virtual organizations, providing a framework for the cooperation of heterogeneous agents. Following this same idea, in [13] it is presented a model of social organizations in which permissions, obligations and commitments can dynamically emerge as a consequence of agents' interactions. This approach offers a theoretical model for dynamic agent societies, but it does not provide a framework for modeling practical applications.

Regarding works on the implementation of norms inside agent societies, in [14, 15] an implementation of norms for Electronic Institutions (EI) is proposed. This approach conceives norms as a static specification of the desired behavior. Thus, it does not allow agents to reason on norms that might dynamically change for adapting the institution to the environmental changes. For overlapping this drawback, a proposal of a dynamical norm implementation is shown in [16]. An EI provides a framework for heterogeneous agent cooperation. However, it is not an open environment in its broadest sense, as agents participate inside the institution through the infrastructure provided by the EI. Thus, the behavior of external agents is completely controlled by the institution. In addition, an external agent must be developed according to the concrete protocol specified by the EI. Otherwise, the external agent will not be able to perform the desired tasks inside the institution. EIs do not provide any mechanism to external agents for obtaining information about the process for providing or requesting functionalities inside a specific EI.

Current section illustrates an overview of works on normative MAS. As argued before, these proposals have some drawbacks to giving support to open virtual organizations. An open architecture for overlapping these difficulties is presented in the next section.

3 THOMAS: An Architecture for Open Virtual Organizations

With the aim of giving support for developing open MAS in its broadest sense, an open abstract architecture known as THOMAS has been proposed[1]. It has the goal of integrating both MAS and service-oriented computing technologies as the foundation of virtual organizations. Therefore, agents provided functionality, as well as the functionality offered by the THOMAS framework are described and published employing Web Services standards. More concretely, an external agent interested on being a THOMAS member, is enabled to obtain the specific knowledge concerning services for becoming a member. Similarly, this external agent will be allowed to provide its functionality inside the THOMAS architecture. Thus, an agent does not have to be designed according to a specific virtual organization, since all services needed for providing and requesting functionalities are published and described by means of service standards. Therefore, external entities may become organization members and offer their own implementation of the required services.

A control mechanism is needed to regulate the system behavior and to allow the achievement of the desired global goal. More concretely, this control is performed by means of a normative system. Norms are thus an essential tool for ensuring social order inside open societies, in which heterogeneous and self-interested agents work together. Therefore, the THOMAS architecture permits the definition of norms for controlling who can provide a service, when and under

---

[1] http://www.fipa.org/docs/THOMASarchitecture.pdf

which circumstances. These norms are stored by the *Organizational Manager Service*, which is an internal component of THOMAS in charge of managing agent organizations. As a consequence, when an agent becomes a THOMAS member, it is informed about its addressed norms. Then, this external agent is capable of considering the constraints over the services that it can provide and/or request.

In the present paper, a mechanism for considering norms inside the agent's decision making process is proposed. More concretely, it is a normative reasoning process aimed at giving support to external agents, which are informed about their expected behavior, defined by means of norms that regulate service accesses. Following, the usefulness of the developed normative reasoning process is described.

3.1 The Normative Reasoning Process

As commented before, this work proposes a mechanism for permitting agents to consider norms in their decision making process. Therefore, any agent informed by the organization about its addressed norms would be able to employ the developed normative reasoning mechanism in order to establish:

- The set of services that it is allowed to register, i.e. which service implementations it is authorized to provide.
- The set of services that can be requested to other organization members, i.e. the services that the agent is allowed to utilize.
- The conditions for the provision and request of services, i.e. when a specific service can be provided or requested, which agents can make use of it, etc.

Currently, this normative reasoning process has been implemented as an agent behaviour. However, our aim is to integrate this functionality as a THOMAS service which can be requested by "external agents" in order to know its permitted actions in a specific moment. Therefore, This service will allow non norm-aware agents to behave correctly without needing to include a normative reasoning mechanism inside them.

As an example, an external agent that acts as a service *provider* inside the organization is employed in order to illustrate the proposed normative reasoning process. More specifically, the *provider* agent offers an implementation of some of the specific services registered in the system. Once it is registered as a service *provider* in THOMAS, it will be informed about its addressed norms. Therefore, it will employ the developed normative process for controlling those norms which are related to the request and provision of its own offered services.

The following section contains a detailed description of the normative reasoning process. The process for incorporating norms to agent's beliefs does not depend on the case study. The only aspect of the norm implementation that is dependent on the example is the interpretation of norms made by the *provider*, i.e. how it considers norms in its decision making process.

4 Implementation of the Normative Reasoning Process

In this section, the phases that cover the process of translating a formal norm into a set of executable rules are described. The developed implementation allows agents to guide their behavior in accordance with a dynamic normative context, in which norms might be created or deleted for adapting an agent organization to the environmental changes.

The normative reasoning process has been implemented by means of a rule-based system in Jess[2]. Figure 1 shows a schematic view of this process. More concretely, a formal norm expressed by means of a deontic language is transformed into an instance of the norm template. This instance is asserted as a fact into the Jess reasoning process. The rule-based system has rules for detecting the activation and fulfillment of norms. Therefore, when the activation conditions of the norm occur, then the norm instance changes into an active norm. Thereby, the *provider* agent is capable of taking into account its addressed norms each time it receives a service request.

The dynamical feature of norms that regulate the behavior of agents causes the need for a function for registering a new norm as a belief. Similarly, a function for deleting or modifying a previous norm is also needed. The process performed by this function is:

- (i) *Normative Analysis*. The formal specification of a norm is translated into an operational norm computable by a rule-based system.
- (ii) *Norm Activation Detection*. A new fact corresponding to the new norm is asserted into the belief base.

Following, these two processes are detailed together with a description of the function that allows a *provider* agent to take into account norms on its decision process.

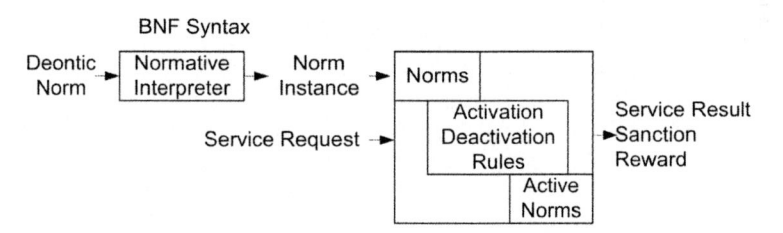

Fig. 1. Overview of the Automatic Normative Reasoning process

4.1 Normative Analysis

Once the provider is registered as a member of the organization, it is informed about its addressed norms. These norms define its expected behavior in terms of constraints on the provision of its services, such as when it can provide a service, on which circumstances, etc. These constraints are expressed by means

---

[2] http://herzberg.ca.sandia.gov/

of a normative formal language, which is described in [17]. Then, the agent must translate the formal definition of norms into a normative fact, in order to be computable by its rule-based system.

The normative analysis phase has two different purposes: on the one hand, it analyzes whether the new formal norm belongs to the subset of norms that can be supervised by the *provider* agent; on the other hand, it translates the formal expression of a norm into a norm fact in order to be automatically controlled by the rule-based system. This process is performed by means of an interpreter that has been automatically built employing the JavaCC[3] tool.

The normative language employed for norm representation is a general-purpose language for service access control [17]. As previously mentioned, the *provider* agent is not capable of controlling any norm expressed through the proposed normative language, since it needs extra capabilities or information in order to take a norm into consideration. In this sense, the set of verifiable norms is a subset of the norms defined for regulating its services. Therefore, controllable norms are *regulative* norms that define ideal behavior in terms of obligations, prohibitions and permissions concerning access to the services offered by the *provider* agent.

Our approach defines norms as a deontic control affecting to an entity on the request of a specific service. The state condition for the activation of the norm is defined in terms of boolean conditions that can be expressed on some variables, identifiers or service results. On the other hand, a temporal condition for the activation/deactivation of the norm can be expressed as a deadline, an action or a service result. Sanctions and rewards can be also defined to persuade agents to behave correctly. As an example, Norm (1a) is formally expressed using the normative language designed in [17]. In addition, its translation into normative facts is shown in (1b). More concretely, Norm (1a) allows an agent to request the *InformationSearch* service after it has registered as a client using the *ClientRegister* service.

$$PERMITTED\ ?agent:?role\ REQUEST\ InformationSearch$$
$$AFTER\ ([?agent]REQUEST\ ClientRegister \qquad (1a)$$
$$CONTENT(agentid\ ?agent))=Provided$$
$$\updownarrow$$
$$(norm\ (deonticConcept\ permitted)\ (entity\ agent\ ?agent\ role\ ?role)$$
$$(serviceID\ InformationSearch) \qquad (1b)$$
$$(after\ "Request(?agent\ ClientRegister\ agented\ ?agent\ provided)")$$

4.2 Norm Activation Detection

The previous phase allows the *provider* agent to translate formal specifications of norms into normative facts computable by the Jess system. As previously argued, norms are not active at all times. In fact, the activation of norms might be defined in terms of state and temporal conditions. As a consequence, a set of rules for detecting the activation and deactivation of norms are needed. Therefore, an

---
[3] https://javacc.dev.java.net/

agent should only take into consideration the norms that are active in each moment.

The norm implementation implies the detection of temporal situations and conditions. It is based on the implementation of norms proposed in [14, 15]. The way in which the *provider* agent interprets norms depends on the norm deontic type (obligation, permission or prohibition). Thus, norms are classified into: *service access norms*, which define permissions and prohibitions on the use of services; and *obligation norms*, which specify an obligation control over a service request. As a consequence, the defined rules for norm implementation depend on the concrete norm type. Following, details about the implementation of each type of norm are commented:

- *Service Access Norms*: this kind of norms defines prohibitions or permissions on the use of the provided services. Our implementation of norms defines permissive norms as exceptions to more general prohibitions. The management of a new service access norm implies the addition of three new rules to the rule-based system: one for detecting the norm activation and two rules for deactivating the norm in case of detecting the end event. Table 1 contains the portion of source code that corresponds to the addition of a new norm. Rule (i) detects the occurrence of the start event together with the satisfaction of the condition and it activates the norm. Rule (ii) is in charge of deactivating the norm if the activation condition is not true. Finally, the last Rule (iii) retracts the norm from the set of active norms if the end event is detected.
- *Obligation Norms*: these norms are that ones that oblige *client* agents to request a specific service. Thus, the *provider* agent does not implement these norms directly, since it cannot force other agents to make a service request. However, it is capable of detecting that an obligation has not been fulfilled and then it needs to carry out its related sanction. Otherwise, it will perform actions defined as a reward. More specifically, when the *provider* agent detects the activation of an obligation it asserts an expectation into the rule-based process. If the action is performed it retracts the expectation from the fact base and carries out the reward. On the contrary, if the end event is detected and the action has no been performed yet the *provider* agent carries out the sanction.

As previously mentioned, a formal specification of a norm is asserted as a fact into a rule-based system in order to enable the automatic detection of its activation and deactivation. The following section describes how the decision making process of a *provider* agent takes into account these active norms.

4.3 Norm Reasoning

This section illustrates the way in which the *provider* agent interprets norms, i.e. how it deliberates on the set of active norms during its decision making process. This is a domain dependent aspect of our norm implementation, since the norm semantics, i.e. the concrete sense of obligations, prohibitions and permissions, depends on the concrete application domain.

Table 1. Function that manages the creation of a new service access norm

```
(defrule newNorm
?f<-( norm (normid ?normid) (deonticConcept ?deon) (if ?condition)
  (before ?before) (after ?after)...)
(test(or (eq ?deon forbidden)(eq ?deon permitted)))
=>
;i) Activation Rule
(build (str-cat "(defrule activateNorm "?normid" "?condition" "?after"
=>
(assert(activenorm(normid "?normid") (deonticConcept " ?deon ") )) )"))
;ii) Conditional Deactivation Rule
(bind ?ruledelIF (str-cat "(defrule deactivateNormIF"?normid"
?f<-(activenorm(normid "?normid") (deonticConcept "?deon") )
(not "?condition")
=>
(retract ?f))")) "
;iii) Temporal Deactivation Rule
(build (str-cat "(defrule deactivateNormBEFORE"?normid"
?f<-(activenorm(normid "?normid") (deonticConcept "?deon") ) "?before"
=>
(retract ?f))")))
```

As explained before, norms are classified into service access norms and obligation norms. The *provider* agent controls the fulfillment and violation of obligations, and it is in charge of performing sanctions and rewards. On the other hand, when the *provider* detects activation and deactivation of service access norms, it should update its beliefs corresponding to active norms in order to deliberate on this kind of norms each time it provides a service. In this sense, whenever the *provider* agent

Table 2. Source code of the function that checks whether an action is permitted

```
(deffunction isAllowed (?agent ?service $?info)
;i)Checks agent addressed permissive norms
(if(< 0 (countAgentActiveNorms ?agent permitted ?service))
then(return TRUE))
;ii)Checks agent addressed prohibition norms
(if(< 0 (countAgentActiveNorms ?agent forbidden ?service))
then(return FALSE))
;iii)Checks role addressed permissive norms
(if(< 0 (countRoleActiveNorms ?agent permitted ?service))
then(return TRUE))
;iv)Checks role addressed prohibition norms
(if(< 0 (countRoleActiveNorms ?agent forbidden ?service))
then(return FALSE))
;v)None norm is found
(return TRUE))
```

receives a new service request from any *client* agent, then the *provider* should revise the set of active norms in order to decide whether provide the requested service.

The determination of the allowed actions to a *client* agent that requests a certain service offered by the *provider* agent is made by employing a criterion known as *lex specialis* [18]. This criterion for norm precedence claims that in case of norm conflict, the most specific norm precedes the rest of norms. Therefore, the normative analysis begins with checking permissive norms addressed to the agent, as seen in Table 2 (i). If there is not any norm, prohibition norms are taken into consideration (ii), since permissive norms are defined as exceptions to a more general prohibition norm. In the last case, permissive and prohibition norms addressed to the roles played by the *client* agent are checked (iii and iv). If none norm is found, the service is defined as permitted by default (v).

5 Conclusions and Future Works

A normative MAS defines norms as a mechanism for coordinating and controlling agent behaviors. Therefore, it disposes of several mechanisms for creating, deleting, enforcing and diseasing norms. In this kind of systems, agents must be capable of reasoning about their addressed norms in order to consider these restrictions in their decision making processes.

In this paper, a new normative reasoning process has been presented. It extends previous works since it allows the automatic implementation of norms in order to allow the dynamical change of the normative context. This implementation has been employed for controlling access to services provided by the THOMAS architecture. This open abstract architecture needs norms for controlling access to services in order to manage the organization dynamics.

The definition of complex conditions and deadlines as well as the implementation of complex norms are the future extensions to this work. More specifically, the management of complex norms implies different tasks such as delegation of norm control, definition of entities specialized on norm supervision, etc. On the other hand, the detection and solution of normative conflicts and inconsistencies remains as an open issue that should be considered in the next versions of the normative reasoning process.

Acknowledgements. This work is supported by TIN2005-03395 and TIN2006-14630-C03-01 projects of the Spanish government, FEDER funds and CONSOLIDER-INGENIO 2010 under grant CSD2007-00022, FPU grant AP-2007-01256 awarded to N.Criado.

References

1. Boella, G., van der Torre, L., Verhagen, H.: Introduction to the special issue on normative multiagent systems. Auton. Agents Multi-Agent Syst. 17, 1–10 (2008)
2. Boman, M.: Norms in artificial decision making. Art. Int. and Law 7(1), 17–35 (1999)

3. van der Torre, L., Tan, Y.-H.: Diagnosis and decision making in normative reasoning. Artificial Intelligence and Law 7(1), 51–67 (1999)
4. Conte, R., Falcone, R., Sartor, G.: Introduction: Agents and Norms: How to Fill the Gap? Artificial Intelligence and Law 7(1), 1–15 (1999)
5. Dignum, F.: Autonomous Agents with Norms. Artif. Intell. Law 7, 69–79 (1999)
6. Boella, G., van der Torre, L., Verhagen, H.: Introduction to normative multiagent systems. In: Normative Multi-agent Systems, Dagstuhl Seminar Proceedings (2007)
7. Verhagen, H.: Norm Autonomous Agents. PhD thesis, Stockholm University (2000)
8. Walker, A., Wooldridge, M.: Understanding the emergence of conventions in multi agent systems. In: ICMAS 1995, pp. 384–390 (1995)
9. Savarimuthu, B.T.R., Cranefield, S., Purvis, M.: Role model based mechanism for norm emergence in artificial agent societies. In: Sichman, J.S., Padget, J., Ossowski, S., Noriega, P. (eds.) COIN 2007. LNCS, vol. 4870, pp. 203–217. Springer, Heidelberg (2008)
10. López, F., Luck, M., d'Inverno, M.: Constraining autonomy through norms. In: AAMAS, pp. 674–681 (2002)
11. Dignum, F., Morley, D., Sonenberg, L., Cavedon, L.: Towards socially sophisticated BDI agents. In: ICMAS, pp. 111–118 (2000)
12. Castelfranchi, C., Dignum, F., Jonker, C.M., Treur, J.: Deliberative normative agents: Principles and architecture. In: Jennings, N.R. (ed.) ATAL 1999. LNCS, vol. 1757, pp. 364–378. Springer, Heidelberg (2000)
13. Castelfranchi, C.: Formalising the informal? Dynamic social order, bottom-up social control, and spontaneous normative relations. J. Applied Logic 1, 47–92 (2003)
14. García-Camino, A., Noriega, P., Rodríguez-Aguilar, J.A.: Implementing norms in electronic institutions. In: EUMAS, pp. 482–483 (2005)
15. da Silva, V.T.: Implementing norms that govern non-dialogical actions. In: Sichman, J.S., Padget, J., Ossowski, S., Noriega, P. (eds.) COIN 2007. LNCS, vol. 4870, pp. 232–244. Springer, Heidelberg (2008)
16. da Silva, V.T.: From the specification to the implementation of norms: an automatic approach to generate rules from norms to govern the behavior of agents. Auton. Agents Multi-Agent Syst. 17(1), 113–155 (2008)
17. Argente, E., Criado, N., Julián, V., Botti, V.: Designing norms for Virtual organizations. In: CCIA, pp. 16–24 (2008)
18. Boella, G., van der Torre, L.: Permissions and obligations in hierarchical normative systems. In: ICAIL, pp. 109–118 (2003)

Negotiation Exploiting Reasoning by Projections

Toni Mancini

Dipartimento di Informatica, Università di Roma "La Sapienza", Italy
tmancini@di.uniroma1.it

Abstract. We present a framework that allows two self-motivated distributed agents to perform, in an efficient way, negotiations aiming at achieving mutually satisfactory agreements, when privacy of information is an issue, and no central authority could be used. In particular, each agent has her own constraints to satisfy, as well as her own utility function. Such issues are kept private, and cannot be disclosed to the counterpart. Negotiation is hence carried out by exchanging proposals and by performing sophisticated forms of reasoning on the remote agent's offers, by trying to infer some characteristics of the counterpart, in order to achieve efficient process convergence.

1 Introduction

Automated negotiation among rational agents is an important topic in Distributed Artificial Intelligence, being necessary in several domains, e.g., distributed resource allocation [3], distributed scheduling [11], e-business [9] and, in general, in applications in which: *(i)* no single agent can achieve her own goals without interaction with the others (or she is expected to achieve more utility with interaction), and *(ii)* constraints of various kinds (e.g., security or privacy) forbid the parties to communicate their desiderata to others (the counterpart or a trusted authority), hence traditional centralized approaches, like Mathematical or Constraint Programming cannot be used. In the literature, various protocols and algorithms for negotiation have been proposed (cf., e.g., [13, 9, 12, 8, 10]), all of which can be classified according to several factors, like the negotiation objects, the agents' decision making models, the degree of cooperation and the level of privacy about each agents' constraints and preferences, or the communication and computation costs.

In this paper, we focus on negotiation between two self-motivated, competitive agents, which aim to find a mutually satisfactory agreement without disclosing their private information (constraints and preferences). Differently from much work existing in the literature, *we explicitly deal with the fact that, in many scenarios, agents face with unknown counterparts, for which no strong assumptions (even probabilistic ones) can be made*: understanding their characteristics and behavior is part of the job of the agent. One of the major prices of this setting is that even the termination of the negotiation process is not guaranteed. To this end, we focus on the local agent only (the only one upon which we have control), which aims at *maximizing* her own private utility function (*competitiveness*) while *guaranteeing termination and efficiency* of the negotiation process. In order to do so, she behaves in a sort of *non-obstructionist way*, and acts very carefully in order to avoid the process to go thrashing and last indefinitely, relying on initial expectations about the reasoning capabilities of the counterpart, but being ready to revise them, by means of a continuous monitoring of her actions, in case a contradicting behavior is observed.

The starting point of our research is the (cooperative) framework described in [1], in which the task of finding a mutually satisfactory agreement is particularly efficient. In such a framework, in which privacy of information is a major concern, negotiation proceeds by exchanging *proposals*. The other agent can either accept the deal or propose a counter-offer. An agent is also able to *reason* on the other agent's proposals. This form of reasoning is called *reasoning by projections*.

In this paper we extend the framework above in the following directions (Sec. 2): *(i)* we provide the agent with the ability to *dynamically estimate and reason on the quality of reasoning made by the counterpart*, and to consequently adapt her behavior *exploiting different heuristics* suitable for the different cases; *(ii)* we provide the agent with the possibility of *defining her own* (private) *utility function* to maximize. Given that the presence of utility functions could make the agent interested in refusing acceptable deals in order to pursue better ones, guarantee of termination can easily be lost. We equip the agent with capabilities that allow her to *still guarantee convergence*, by reasoning on the impact that her refusals have on the counterpart potential reasoning. A *full functional implementation* is briefly described and experimentally evaluated (Sec. 3, details in Appendix[1] due to space reasons).

2 The Framework

When dealing with negotiation, there are several aspects that must be taken into account. We address the reader to [1] for a discussion on the most important ones. Here, we only describe the assumptions behind our theoretical framework.

Negotiation framework. When two agents start a negotiation, they already agree on the relevant *variables* (or issues). Following [1], we assume that for each negotiation there is a finite list of *real variables* that are involved. Hence, negotiation spaces can be regarded as multi-dimensional real vector spaces. Agents have their own private *feasibility regions* (R_{loc} and R_{rem}). This setting is different from what is often assumed in the literature, since agents don't even know (or probabilistically estimate) possible counterpart's *types*, *bounds* or *most preferred values* for the variables (in other words, it is not a *split-the-pie* game [5] although with incomplete information [10]). Negotiation proceeds with agents alternatively exchanging *proposals* (aka *deals*), as single points in the space of variable assignments. The counterpart can either accept the deal or decline it, by sending a *counter-offer*.

Characteristics of the local agent. We assume that R_{loc} is *convex* (i.e., all points between two acceptable points are acceptable as well), and more particularly *limited* and defined by means of *linear constraints*, hence it is a *bounded polyhedron*. We also assume that R_{loc} is *stable* during time. Convexity and stability, which play a key role in our approach, are very common in many scenarios of practical utility (cf., e.g., [1] and Appendix A.2 for an example). Admissible proposals are in general not equally worth for the agent, that may have her own *utility function*, preferring some solutions to other ones. We assume that the utility function, to be *maximized* by the local agent, is *linear*.

---

[1] Available online at http://www.dis.uniroma1.it/~tmancini

Reservation utility can be, wlog, embedded in R_{loc} by additional constraints: hence, any agreement in R_{loc} would be better than failure.

The agent has two *goals*: *(i)* to pursue deals that give her high utility, and *(ii)* to guarantee termination of the negotiation, i.e., either to find a point (agreement) in $R_{loc} \cap R_{rem}$, or to logically prove that none exists. Of course, such goals cannot be formally guaranteed, since they depend on the unknown characteristics and behavior of the remote agent. In other words, local agent follows a *heuristic approach*.

The agent is *logically omniscient*, i.e., she is able to compute all logical consequences of the information she has. Moreover, she is *non-obstructionist*, in the sense that she exploits all her knowledge and reasoning capabilities on the behavior of the counterpart in order to keep the negotiation efficient: *(i)* she never proposes a deal that she believes it will be rejected, and *(ii)* never refuses an acceptable offer, if she believes that no better agreement (for her) can be reached. However, given that, in our scenario, the local agent has no information about constraints, goals, preferences, and strategy of the other, this form of collaboration turns out to be much weaker than that which can be exploited in other frameworks, and desirable properties like agreements' Pareto optimality or maximum social welfare cannot be formally guaranteed (but can be reached by a suitable post-negotiation cooperative step, starting from an already mutually satisfiable agreement).

Beliefs about the remote agent. Even if agent has no knowedge about the counterpart, she makes some *initial* (and favorable) assumptions on her, ready to be revised in case an unexpected behavior is observed. In particular she initially assumes that the counterpart: *(i)* Knows that R_{loc} is convex and stable; *(ii)* Has *at least* her own reasoning capabilities (hence is able to reason on the local agent's own behavior, avoiding obstructionism, cf. later), and *(iii)* In turn, believes the same about her (hence, e.g., knows that local agent assumes R_{rem} convex). Local agent maintains and continuously revises a *knowledge base* with the following information: *(i)* All proposals made ($\mathbb{P}$) and offers received ($\mathbb{O}$); *(ii)* A Boolean flag (X) indicating whether R_{rem} is believed conveX; *(iii)* A Boolean flag (C) indicating whether the counterpart is believed non-obstructionist (partially Collaborative). Such flags are initially both set to *true*, but will be revised in case of unexpected remote behaviors.

Protocol rules. These are formal rules supposed to always be respected by the agents. In particular, cf. [1], "No cheat" seems to be fundamental in order to guarantee convergence. There (fully collaborative framework and no utility functions), agents cannot propose deals they are not willing to accept, and agree on the first offer that belongs to their feasibility region. Here, given that the presence of utility function makes the local agent interested in rejecting acceptable offers in order to pursue better deals, and given that local agent has no strong guarantees about the counterpart, we relax the "No cheating" rule as follows: *(i)* Local agent never proposes deals she is not willing to accept, and *(ii)* never behaves in ways that would prove herself to be obstructionist or not convex to a counterpart with the same reasoning capabilities. However, given that the actual characteristics of the remote agent are unknown, local agent cannot enforce symmetric protocol rules to her, and is ready to tune her behavior when the counterpart proves to be less cooperative.

Negotiation with no utility function. In order to describe the form of reasoning used by the local agent in the simplest case, when she has no utility function and believes the

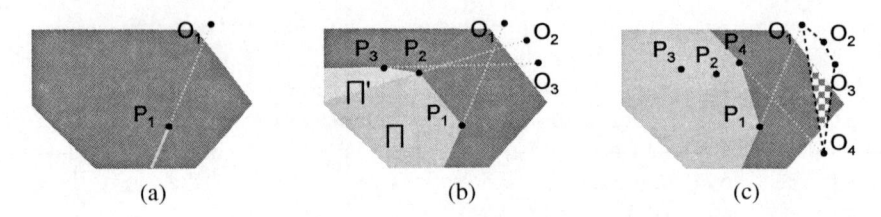

Fig. 1. (a) Agent's feasibility region along with the first proposal. (b) Regions that can be excluded from proposals after the second (Π) and third (Π') iteration (light areas). (c) 4th iteration: local agent proposes P_4, and receives counter-offer O_4. $R_{\text{loc}} \cap conv(\mathbb{O})$ (textured) now is non-empty: any point there should be good for both agents. Agent may successfully resolve by proposing P_5 there.

same about the counterpart, we show an example concerning a negotiation that involves just two real variables (only to graphically show the ongoing process). R_{loc} is shown in Fig. 1(a); all points are equally worth. Fig. 1(a) also shows the first deal P_1 proposed by the agent, which is not accepted by the counterpart, that in turn offers O_1. Information stored in KB allows the agent to infer new constraints, which add on those of R_{loc}. An obvious example is $\mathbb{P} \wedge \mathbb{O} \vdash (P_1 \notin R_{\text{rem}})$, since the remote agent, offering O_1, implicitly rejects P_1. Interestingly, the local agent, relying on the current assumptions she has about the remote party (initially believed convex), is able to infer that all the points lying on the same line joining P_1 and O_1, below P_1 (highlighted in solid light gray) do not belong to R_{rem}. The reasoning is as follows: *(i)* in case there would be a point $S \in R_{\text{rem}}$ there, by convexity of R_{rem} all points in the segment $\overline{SO_1}$ would belong to R_{rem}; *(ii)* however, since also P_1 lies on $\overline{SO_1}$, a contradiction arises.

As for the next iteration, the local agent proposes P_2, which is again refused by the counterpart, which in turn offers O_2 (cf. Fig. 1(b)). The reasoning that agent can perform at this stage is much more interesting (cf. [1]), and exploits also the initial assumption of non-obstructionism ($C = true$) of the counterpart:

1. By convexity of R_{rem}, all points in $\overline{O_1 O_2} \in R_{\text{rem}}$. Hence, in case $\overline{O_1 O_2} \cap R_{\text{loc}} \neq \emptyset$, she would expect that any point in such set will be acceptable for the counterpart, and would terminate the negotiation process successfully by choosing one of them. So, let us assume that the intersection is empty. The agent may perform additional reasoning in order to exclude, from the set of all possible new proposals, the whole area Π highlighted in medium gray:

    - Since the counterpart offered O_2 after having received P_2, the agent knows that $R_{\text{rem}} \cap \overline{P_1 P_2} = \emptyset$, since also the remote one is assumed to be collaborative ($C = true$), hence should assume convexity of R_{loc} (in absence of a remote utility function, a non-obstructionist remote agent would accept the first acceptable deal, simply because no better agreements are possible);
    - Let us assume that there is a point $O \in R_{\text{rem}} \cap \Pi$; from the last assertion and the convexity hypothesis of R_{rem}, R_{rem} must contain all points of the triangle $\widehat{O_1 O O_2}$; now, since such a triangle contains at least one point of the line segment $\overline{P_1 P_2}$, a contradiction arises.

Such form of reasoning can be performed at *any* iteration, in order to reduce the so called *active region*, i.e., the set of all candidate points in R_{loc} for the new proposal. In particular, at any step, when the agent has proposed points $\mathbb{P}$ and received offers $\mathbb{O}$, the choice of next proposal P is made by considering, in order, two alternatives:

(Resolve) If $conv(\mathbb{O}) \cap R_{\text{loc}} \neq \emptyset$, where $conv(\mathbb{O})$ is the convex hull of the received offers, then (under current assumptions) any point in such intersection will be acceptable by both agents. The agent will choose P among them, expecting the successful termination of the process.

(Propose) Otherwise, P is chosen in $R_{\text{loc}} - \Pi(conv(\mathbb{O}), conv(\mathbb{P}))$, where, $\Pi(R, R')$ is the *projection* of region R over R' (defined as the set of all points Q' for which there exists $Q \in R$ such that $\overline{QQ'}$ intersects R'). In case also such set is empty, current assumptions imply that $R_{\text{loc}} \cap R_{\text{rem}} = \emptyset$, hence no agreements exist.

Formal definitions of convex hull and projection are given in Appendix A.1. Fig. 1(b) shows the scenario resulting after points P_3 and O_3 have been exchanged (the whole area Π' colored in light and medium gray can be safely excluded), while Fig. 1(c) shows the situation after the next step, where point O_4 has been received: since now the intersection of R_{loc} with the convex hull of the received offers is non-empty, any point in the textured area is (by current assumptions) a guaranteed good deal.

Such form of reasoning allows to exclude large areas from R_{loc}. In [1], the agent is supposed to propose *vertices*, to guarantee convergence. In fact, assuming a collaborative remote party, if all vertices have been rejected, then no possible agreement can be found (the whole R_{loc} –convex– is excluded from the active region). This can be seen in Fig. 1: if $P_1, \ldots, P_4$ were vertices, the active (dark) region would be much smaller, and convergence faster. However, convergence does not ensure efficiency of termination: in the worst case, a number of proposals proportional to the number of vertices, hence easily exponential in that of the variables is needed to terminate [1].

The question that arises is then the following: *which is the best vertex to propose, in case the alternative* (resolve) *is not feasible?* According to the *fail-first* principle, we equip the local agent with the following heuristic for next vertex proposal:

Best vertex. *The new proposal is the vertex (of R_{loc} and active) which, in case of rejection, will make the highest number of vertices excluded by the active region of the next step.*

This heuristic (that of course cannot take into account the effects, on the exclusion of other vertices, of the counter-offer that will be received upon rejection of the proposal being computed), clearly tends at keeping the negotiation efficient (cf. also Sec. 3), maximally reducing the number of remaining vertices to propose.

Revision of belief about remote agent. Assume we are in the situation of Fig. 1(c), i.e., alternative *(resolve)* is feasible ($conv(\mathbb{O}) \cap R_{\text{loc}} \neq \emptyset$). The agent will propose an arbitrary deal in such intersection, e.g., point P_5 of Fig. 2(a), expecting acceptance. *What if the remote agent refuses this deal?* Unsurprisingly, local agent infers that R_{rem} is not convex, and revises KB assuming $\neg X$. Consequently, local active region becomes larger, cf. Fig. 2(a): only points in $conv(\mathbb{P})$ can be excluded (the collaborative counterpart would already offered them if possible, because they would be assumed in R_{loc}).

As a second example, assume that, at a given point, e.g., from Fig. 1(b), *remote agent offers a deal belonging to the projection she is supposed to compute,* i.e., $\Pi(conv(\mathbb{P}), conv(\mathbb{O}))$, e.g., offer O_4 in the textured area of Fig. 2(b). Local agent would immediately notice that her counterpart is obstructionist. The reasoning mimics the one that a collaborative counterpart would perform: *(i)* A collaborative remote agent would believe that the local one has a convex region, because she never refused a point O_i in $conv(\{P_1, \ldots, P_{i-1}\})$, and is collaborative, because she never behaved in contradiction with such hypothesis. *(ii)* Hence, she would infer that, if O_4 was acceptable, so would be all points in $conv(\{O_4, P_1, P_2, P_3\})$. *(iii)* Since by construction, such a region has non-empty intersection with $conv(\{O_1, O_2, O_3\})$, the remote agent would expect that the collaborative local one should have already proposed a deal in such intersection, which is not the case. *(iv)* Hence, the remote agent cannot be collaborative (since she doesn't properly assume convexity and collaborativeness of the local agent that have not been violated so far). The local agent thus revises her KB, inferring $\neg C$. Also in this case the active region becomes different, cf. Fig. 2(b). Two more scenarios may arise that let the agent discover violations, by the counterpart, of the convexity or collaborativeness assumptions, one of which is shown in Fig. 2(c). Due to space reasons, they are discussed in Appendix A.3.

In case the counterpart is believed to be either not convex or not collaborative, Best vertex heuristic loses its main advantages: if the remote agent is not convex, the local one cannot exclude any projection, but only the convex hull of her own proposals ($conv(\mathbb{P})$, cf. Fig 2(a&c), light areas). If not collaborative, proposing vertices does not even guarantee termination (although, if still believed convex, the agent could exclude points in $\bigcup_{P \in \mathbb{P}} \Pi(conv(\mathbb{O}), P)$, cf. Fig. 2(b)). To this end, focusing on the case of a non-convex but collaborative remote party, in principle proposing either vertex is equally worth. However, by following the intuitive approach of trying to meet as much as possible the counterpart's last offer we devise a second heuristic:

Closest vertex. *The new proposal is the vertex (of R_{loc} and active) closest to the last offer.*

Since the remote agent is believed collaborative, when all vertices of R_{loc} (bounded polyhedron) have been proposed, the active local region becomes empty. In case of a non-collaborative remote agent, no guarantee of termination is possible. Hence, the local agent could either *(i)* abort the negotiation; *(ii)* propose random points in her active region (**Any point** heuristic) or, as suggested in [4], *(iii)* approach the opponent last offers proposing internal points of R_{loc}. This last heuristic (which mimics the approach typical of local search) is claimed to work well in several cases.

It is worth noting that, every time the local active region becomes empty, the agent has a proof that, if her current assumptions are valid, then no agreement actually exists. But of course, it might be the case they are wrong, but have never been violated by the counterpart so far. As above, the local agent may in principle decide either *(i)* to abort the negotiation (trusting in her assumptions), or *(ii)* to deliberately relax them in order to have additional deals to propose. However, in the latter case she may herself appear obstructionist (if her assumptions were indeed valid). Also, switching to $\neg C$ would make any guarantee of convergence lost.

Negotiation in presence of utility function. Heuristics above might behave very bad in case an utility function does exist (e.g., it would be not hard to figure out situations in which the Best vertex is the point with lowest utility). However, facing with utility

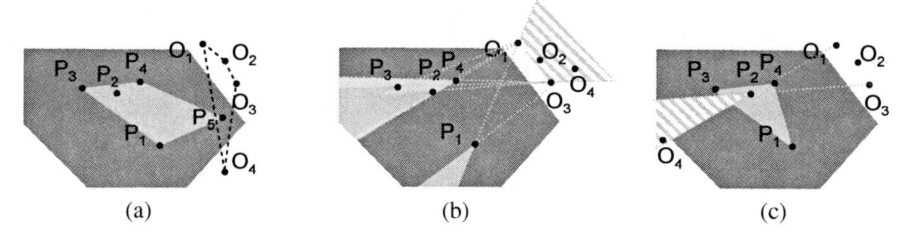

Fig. 2. (a) 5th iteration: Remote agent refuses $P_5 \in R_{loc} \cap conv(\mathbb{O})$, making a counter-offer (not shown). The local agent infers $\neg X$. (b) 4th iteration: Remote agent offers O_4 in $\Pi(conv(\mathbb{P}), conv(\{O_1, O_2, O_3\}))$ (textured area, unlimited). Local agent infers $\neg C$. (c) 4th iteration: Remote agent offers O_4 in $\Pi(conv(\mathbb{O}), P))$ for some $P \in \mathbb{P}$ (textured area, with $P = P_2$, cf. Appendix A.3). Local agent infers $\neg X$.

functions is problematic in general, since the guarantee of convergence can be easily lost on the way, because agents may have interest in rejecting acceptable offers, hoping to increase their utility (in the following, we call these actions *lies*, because agents are deliberatively making the negotiation longer than that it could be). In order to still guarantee termination, lies should have no negative impact in the effectiveness of the counterpart's potential reasoning. In particular, we assume that local agent always acts in order to *appear* (rather than to be, given the presence of an utility function) convex and non-obstructionist: she will never refuse a deal if such action would give the counterpart enough information to infer she is not convex. On the other hand, as for other offers and her own proposals, she will follow more careful strategies, deeply exploiting, by additional reasoning, what privacy of information guarantees, i.e., the potential discrepancy between her actual behavior and that *perceived* by the counterpart.

In particular, when rejecting an acceptable offer O may have an impact on process convergence? The answer is pretty simple, if we consider what a collaborative counterpart could infer. Rejecting $O \in conv(\mathbb{P})$ would give to remote agent evidence that the local one is *not* convex (cf. Fig. 2(a), reversing the roles). Hence, such offers *cannot be rejected*, in order to maintain guarantee of convergence.[2] In the other cases ($O \in R_{loc} - conv(\mathbb{P})$), acceptable offers may safely be rejected (we call these: *safe* lies). However, in such situations, other constraints must be taken into account by local agent when making subsequent proposals in order to "lying with impunity" and still appear convex and collaborative: *(i)* The agent must never propose a point P in $\bigcup_{O \in \mathbb{O}} \Pi(conv(\mathbb{P}), O)$, because this would show to counterpart that (rejected) $O \in conv(\mathbb{P} \cup \{P\})$, hence R_{loc} is not convex; *(ii)* In case the counterpart can be believed as convex, the agent must never propose a point P in $\Pi(conv(\mathbb{P}), conv(\mathbb{O})) - (conv(\mathbb{O}) - conv(\mathbb{O} - \{O_{last}\}))$, with O_{last} being the last received offer (proof in Appendix A.4).

In order to allow the agent to autonomously decide whether to lie or not, an *acceptance policy* can be defined for her. Following (and slightly generalizing) the main approach [5] of accepting an offer if *better* than the counter-proposal the agent would make next, we define an acceptance policy as a real $\xi \in [0..1]$: offer $O \in R_{loc}$ would be accepted iff $u_{loc}(O) \geq u_{loc}(P_{next}) - span \cdot \xi$, where *span* is the absolute difference of the extreme values of $u_{loc}()$ in R_{loc}, and P_{next} is the point that would be chosen by

---

[2] This is not a big deal, since, by construction, $u_{loc}(O) \geq u_{loc}(P)$ for some past proposal $P \in \mathbb{P}$.

following alternative *(propose)*. Thus, acceptance policies vary between two extremes: *(i)* Any acceptable offer is accepted ($\xi = 1$ –this neutralizes the need for lies and minimizes the negotiation length); and *(ii)* An acceptable offer is accepted iff its utility is higher than that of the counter-proposal P_{next} that would be chosen next, upon rejection ($\xi = 0$).

As for the overall strategy in presence of a linear utility function $u_{loc}()$ to maximize, the agent maintains two distinct feasibility regions, R_{loc} (the real one), and $R_{loc}^k = R_{loc} \cap \{P : u_{loc}(P) \geq k\}$ (with k varying during the process). A linear $u_{loc}()$ guarantees that also R_{loc}^k is a bounded polyhedron. The agent will use $R_{loc} \cap \{P : u_{loc}(P) \geq u_{loc}(P_{next}) - span \cdot \xi\}$ to check whether an offer is acceptable or for resolving a negotiation (according to the acceptance policy), and, as for alternative *(propose)*, she will search for deals in R_{loc}^k only.

1. At the beginning, k is fixed to the highest utility value: $k = \max(u_{loc}(P) : P \in R_{loc})$.
2. The local agent proceeds as in the case with no utility function (using R_{loc} and the acceptance policy as for alternative *(resolve)* –as long as $X = true$– and R_{loc}^k as for *(propose)*, taking also care of her lies), until one of two conditions holds: *(i)* the active region is empty; or *(ii)* the counterpart proves to be not collaborative.
3. In case *(i)*, the agent has a proof that no agreements (modulo safe lies on remote side) can be found in R_{loc}^k. Hence, she will perform a *utility concession* to the counterpart, by lowering k by a given $\delta > 0$, hence considering $R_{loc}^{k-\delta} \supseteq R_{loc}^k$. This will produce (in case of $\supset$) new vertices to propose. Negotiation aborts when $R_{loc}^k = R_{loc}^{k-\delta} = R_{loc}$, meaning that no agreements exist at all (modulo lies at both sides, and except for wrong assumptions, dealt with by unilateral relaxation).
4. As with no utility function, in case *(ii)*, any guarantee of termination is lost (the agent cannot count on the capabilities of the remote one). Also here, she can in principle proceed in different ways, but needs to have a mechanism for conceding in terms of utility even if the current active region is non-empty (to avoid to remain stuck forever with unfruitful proposals). As an example, she can decrease k every a fixed number of steps, or immediately reduce herself to the case with no utility function. The efficacy of either option is expected to strongly depend on the application domain, hence it will not be considered here.

From point 2 it can be observed that, in order to lie, it is not necessary to reject an acceptable offer. Consider the situation in Fig. 1(c). The local agent believes, by the convexity assumptions about the counterpart, that all points in $conv(\mathbb{O}) \cap R_{loc} \neq \emptyset$ are acceptable deals. To behave collaboratively, she should propose there. However, she can safely lie, proposing in a different region, as long as the counterpart has no evidence of this non-collaborative behavior.

The overall strategy causes a quantization of the utility function: for a given k, all points with utility between k and $k + \delta$ (δ being the *quantization value*) are considered equally worth. This is important to preserve convergence (the process reduces to a repeated but finitely long application of the basic strategy), and results in δ that should be given explicitly by the user, since, as ξ, it intrinsically depends on the application domain: the lower are δ and ξ, the longer could be the process; moreover, the lower is ξ, the higher is the risk that the process fails even if agreements exist, because all the ones found have been rejected by lies. An important property of this strategy is that *all the reasoning made in the previous steps* (in terms of reduction of the active region) *remains*

(a) (b) (c)

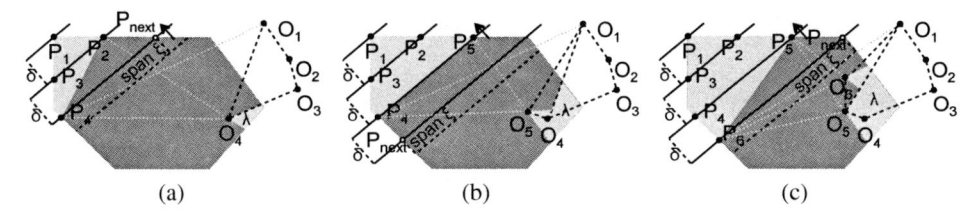

Fig. 3. Negotiation in which the agent has a linear utility function to maximize (black solid line, utility growing North-West), exploits Best vertex heuristic, and has an acceptance policy $\xi < 1$ (never accepting deals below the dashed line). (a) 4th step, she lowers her utility threshold to $U_{max} - 2\delta$, having a proof that no agreement can be found in $R_{loc}^{U_{max}-\delta}$ –previous steps not shown. The best vertex is P_4, which is rejected by the counterpart that in turn offers $O_4 \in R_{loc}$. Local agent may safely refuse it, as long as she still appears convex and collaborative, not proposing in $\lambda = \bigcup_{O \in \mathbb{O}} \Pi(conv(\mathbb{P}), O) \cup (\Pi(conv(\mathbb{P}), conv(\mathbb{O})) - (conv(\mathbb{O}) - conv(\mathbb{O} - \{O_{last}\})))$, O_{last} being the last offer). (b)&(c) 5th and 6th steps, with additional lies.

valid: a proposal unacceptable for the counterpart will remain such when lowering k (as long as belief about the counterpart doesn't change). This preserves efficiency. Fig. 3 shows a fragment of a negotiation process where the local agent has an utility function to maximize, proposes vertices (as required to guarantee termination), and exploits safe lies.

3 Experiments and Perspectives

In order to assess the effectiveness of this approach, we built a Java P2P application fully implementing the present framework, and developed a random generator to produce a great number of negotiation scenarios (consisting of pairs of random polyhedra with controllable probability of intersection, and random utility functions). Experiments show (detailed results in Appendix A.6 and summarized in Table 1) that iAgree is able to deal with quite complex negotiation instances (e.g., 3 variables, 45 vertices per side) in reasonable time, that Best vertex plays a role in efficiency (saving up to 18% in terms of nb. of rounds wrt Closest vertex), and that moderately loose acceptance policies ($\xi \sim 0.4$) may lead to both high success ratios (existing agreements are actually found

Table 1. Some results for 3 vars (#vertices in 16..45, ~35 on average). 200 random instances solved for each combination of ξ_1, ξ_2 in $\{0, .2, .4, .6, .8, 1\}$. Success ratio = #agr. found/#sat inst. Agr. quality = $(u_{loc}(agr.) - min_{P \in R_{loc} \cap R_{rem}} u_{loc}(P))/(max_{P \in R_{loc} \cap R_{rem}} u_{loc}(P) - min_{P \in R_{loc} \cap R_{rem}} u_{loc}(P))$.

Average results over 200 instances	$\xi_{loc}=\xi_{rem}=0$	$\xi_{loc}=\xi_{rem}=.4$	$\xi_{loc}=\xi_{rem}=1$	$\xi_{loc}=0,\xi_{rem}=1$	$\xi_{loc}=1,\xi_{rem}=0$
Number of rounds	25	27	34	29	28
Reasoning time (local agent, sec)	27	30	35	30	30
Success ratio	~60%	~95%	100%	~75%	~85%
Agreements' quality (local agent)	~65%	~75%	~65%	~83%	~48%
Rounds saved by BestVtx wrt ClsVtx	0%	6%	7%	18%	4%

in $\sim 95\%$ of the cases) and agreement quality (close to optimum of $\sim 75\%$ in terms of utility).

The effectiveness of the approach let many important directions for future work emerge: *(i)* investigating the possibility of handling variables on discrete domains (potentially exploiting the current form of reasoning to perform suitable relaxations) and non-linear constraints and utility functions (cf., e.g., [7]); *(ii)* devising new strategies and heuristics; *(iii)* dealing with deadlines [6] (e.g., by a dynamic handling of δ and ξ); *(iv)* dealing with negotiations among more than two agents, extending the reasoning capabilities of the agent to cope with multiple counterparts; *(v)* dealing with non-convex regions, e.g., *union* of polyhedra, also exploiting local search and hybrid techniques [4, 7].

Acknowledgments. The author thanks co-authors of the preliminary version of iAgree [2]: Marco Cadoli (who introduced, already in [1], projections as a mean to perform reasoning in cooperative negotiations) and Guido Chella (who implemented the first version of the system during his Bachelor thesis).

References

1. Cadoli, M.: Proposal-based negotiation in convex regions. In: Klusch, M., Omicini, A., Ossowski, S., Laamanen, H. (eds.) CIA 2003. LNCS, vol. 2782, pp. 93–108. Springer, Heidelberg (2003)
2. Cadoli, M., Chella, G., Mancini, T.: iAgree: a system for proposal-based negotiation among intelligent agents. In: Proc. of ECAI 2006. IOS Press, Amsterdam (2006) System demonstration
3. Conry, S.E., Kuwabara, K., Lesser, V.R., Meyer, R.A.: Multistage negotiation for distributed constraint satisfaction. IEEE Trans. on Systems, Man and Cybernetics 21(6), 462–477 (1991)
4. Costantini, S., Tocchio, A., Tsintza, P.: Experimental evaluation of a heuristic approach for P2P negotiation. In: Proc. of RCRA 2007 (2007)
5. Faratin, P., Sierra, C., Jennings, N.R.: Using similarity criteria to make issue trade-offs in automated negotiations. Artif. Intell. 142(2), 205–237 (2002)
6. Fatima, S.S., Wooldridge, M., Jennings, N.R.: Multi-issue negotiation with deadlines. J. of Artif. Intell. Research 27, 381–417 (2006)
7. Ito, T., Hattori, H., Klein, M.: Multi-issue negotiation protocol for agents: Exploring nonlinear utility spaces. In: Proc. of IJCAI 2007, pp. 1347–1352. Morgan Kaufmann, San Francisco (2007)
8. Jennings, N., Faratin, P., Lomuscio, A., Parsons, S., Wooldridge, M., Sierra, C.: Automated negotiation, prospects, methods and challenges. Group Dec. and Negot. 10, 199–215 (2001)
9. Jennings, N.R., Norman, T.J., Faratin, P., O'Brien, P., Odgers, B.: Autonomous agents for business process management. Applied Artif. Intell. 14(2), 145–189 (2000)
10. Lin, R., Kraus, S., Wilkenfeld, J., Barry, J.: Negotiating with bounded rational agents in environments with incomplete information using an automated agent. Artif. Intell. 172(6–7), 823–851 (2008)
11. Sycara, K.P., Roth, S., Sadeh, N., Fox, M.: Distributed constrained heuristic search. IEEE Trans. on Systems, Man and Cybernetics 21(6), 446–461 (1991)
12. Yokoo, M., Katsutoshi, H.: Algorithms for distributed constraint satisfaction: A review. Autonomous Agents and Multi-Agents Systems 3(2), 185–207 (2000)
13. Zlotkin, G., Rosenschein, J.S.: Mechanisms for automated negotiation in state oriented domains. J. of Artif. Intell. Research 5, 163–238 (1996)

A JADE-Based Framework for Developing Evolutionary Multi-Agent Systems

Bertha Guijarro-Berdiñas, Amparo Alonso-Betanzos, Silvia López-López, Santiago Fernández-Lorenzo, and David Alonso-Ríos

University of A Coruña, Campus de Elviña, 15071, A Coruña, Spain
cibertha@udc.es, ciamparo@udc.es, silvial19@hotmail.com,
sfernandezl@udc.es, dalonso@udc.es

Abstract. Evolutionary agents are flexible, agile, capable of learning, and appropriate for problems with changing conditions or where the correct solution cannot be known in advance. Evolutionary Multi-Agent systems, therefore, consist of populations of agents that learn through interactions with the environment and with other agents and which are periodically subject to evolutionary processes. In this paper we present a JADE-based programming framework for creating evolutionary multi-agent systems with the aim of providing all the necessary infrastructure for developing multi-agent systems of this type. Through its graphical interface, the framework allows to easily configure the parameters of the multi-agent system, to hold complete control over its execution, and to collect performance data. This way the development of an evolutionary MAS is simplified and only little pieces of code have to be written in order to apply the framework to a particular problem. Along this paper, the features of the framework are described and its capabilities and usage are illustrated through its application to the tic-tac-toe problem.

1 Introduction

There are currently several programming frameworks to facilitate the development of multi-agent systems (MAS). However, these frameworks are mainly limited to traditional deliberative and reactive agent architectures. These architectures have some disadvantages that could be addressed by using evolutionary agent architectures, but at present there does not exist a framework of this type.

The architecture of an agent specifies its internal structure and determines the mechanisms that the agent uses for reacting to the environment, communicating with other agents, learning, etc. Classically, there are three agent architectures [10], depending on the type of reasoning: deliberative, reactive and hybrid. However, all three present some drawbacks, such as the high need for computational resources, the relatively slow reactions to the environment (deliberative), practically impossible long-term reasoning and difficulties in learning from experience (reactive), and lack of conceptual and semantic clarity, and problems of anticipated interactions and control between vertical and horizontal layers in the hybrid models.

A possible interesting alternative is the design of evolutionary agents. An evolutionary agent is an agent that is subject to evolutionary processes through computational techniques. These processes consist in applying methods like natural selection, crossover, mutation, and so forth, to a population, thus retaining the best genes and propagating them to the subsequent generations. As a consequence, the agents are able to improve their behavior over time and to become better adapted to changes in the environment. Evolution can also be combined with learning algorithms to obtain significantly better results than using both separately, as demonstrated in [3]. A summary of the interactions between evolution and learning can be found in [6].

The literature describes different approaches for introducing evolutionary computation into intelligent agents. One of them is to evolve the set of production rules of a single agent. This approach is recommended for problems in which the actions of different agents in the same environment can be incompatible, or when it is unfeasible to simulate distinct parallel environments. An example is dynamic rule prioritization [8]. A different approach is to regularly evolve a whole population of agents, and examples include the FuzzyEvoAgents [1], the Amalthaea system [7], and the InfoSpider system [5]. This second approach is suitable for problems in which the environment can be simulated computationally and where several agents operate concurrently in the same environment. In summary, evolutionary agents are flexible, agile, capable of learning, and appropriate for problems with changing conditions or where the correct solution cannot be known in advance.

The aim of this paper is to present a programming framework for developing evolutionary multi-agent systems. In order to implement evolutionary computation, the framework follows the second approach mentioned above, that is, the systems are composed of agents that, besides learning individually, are also evolved at regular intervals. The framework will facilitate the development of MAS by implementing all the general structures needed, and leaving the users the programming of only the specific characteristics of the problem under study.

2 Characteristics of the Evolutionary MAS to Be Developed under Our Framework

Our multi-agent systems consist of agents that are, on the one hand, constantly learning through interacting with the environment and with other agents, and, on the other hand, periodically subject to automatic population-level evolutionary processes.

In its interaction with the environment, an individual agent makes decisions based on a set of rules that indicate which action it should take when faced with a particular state of the environment. An agent can initially possess an empty rule base or it can start with a set of predetermined rules. When an agent receives a stimulus from the environment, it can choose between exploiting one rule in its rule base (if applicable), or exploring (i.e., creating a new rule). At regular intervals, the agent can check its progress by means of a reinforcement scheme. Reinforcement consists in rewarding or penalizing the agent depending on whether its past actions have helped it in getting closer to its goals. In our approach, positive reinforcement consists in increasing the weights of the most recent rules (those used since the last time reinforcement was applied) if they led to success, whereas negative reinforcement consists in decreasing them. This reinforcement scheme is what enables the agent to learn at the most basic level; that is, to memorize which rules are most appropriate given a specific state of the environment. Another feature of our approach is that it allows for implementing agent imitation. Imitation is a mechanism where agents copy the characteristics or strategies of other successful agents (this is widely used in areas where competition is essential, such as e-commerce [9]).

In addition to all this, an evolutionary algorithm is periodically executed with the aim of propagating the best rules to the next generations. A fitness scheme is used to determine which agents should survive, which ones should be selected for reproduction, and which ones ought to disappear from the population.

3 Requirements and Functionalities of the Framework

In order to be able to create evolutionary multi-agent systems, to manage their execution, and to check their performance, a generic programming framework was developed. The main requirement for our work is to provide a ready-made infrastructure for developing evolutionary multi-agent systems, flexible enough so as to allow to adapt the MAS to multiple problems, while requiring from the user only the writing of little pieces of code, describing the specific problem at hand.

Regarding the evolutionary aspects, these requirements also need to be met:

- To make the evolutionary MAS adaptable, and capable of learning.
- To allow the possibility of monitoring the progress of the different populations of agents by collecting data, obtaining statistical measurements and drawing graphs.
- To make the resulting MAS flexible and reliable. The user should be able to:
  - Modify all the parameters of the MAS at different levels (global system, agents, populations, etc.).
  - Control the execution of the MAS (i.e., to start, pause, and restart the system).
  - Manipulate specific agents and populations (e.g., to kill or copy agents).
  - Save and retrieve the state of the system.

In addition to these specific requirements, the existing standards and available tools in this area were studied with the aim of making our framework conforming to standards. In multi-agent systems, agents need to communicate among themselves to achieve their ultimate goals. In order to communicate, several standards have been developed. Nowadays, the standard is FIPA-ACL [2]. Therefore, inter-agent communication will be performed through message passing.

4 Design of the Framework

Figure 1 shows a simplified class diagram of the framework. The main classes allow to instantiate and manage all the essential elements of the system, that is, agents, populations of agents, environments, actions on the environment, agent rules, and evolutionary algorithms. These main classes are:

- System: This class controls the functioning of the global system, which includes managing the agents and their environments, and checking if the end conditions of the system are met (e.g., specific goals or a given interval of time).

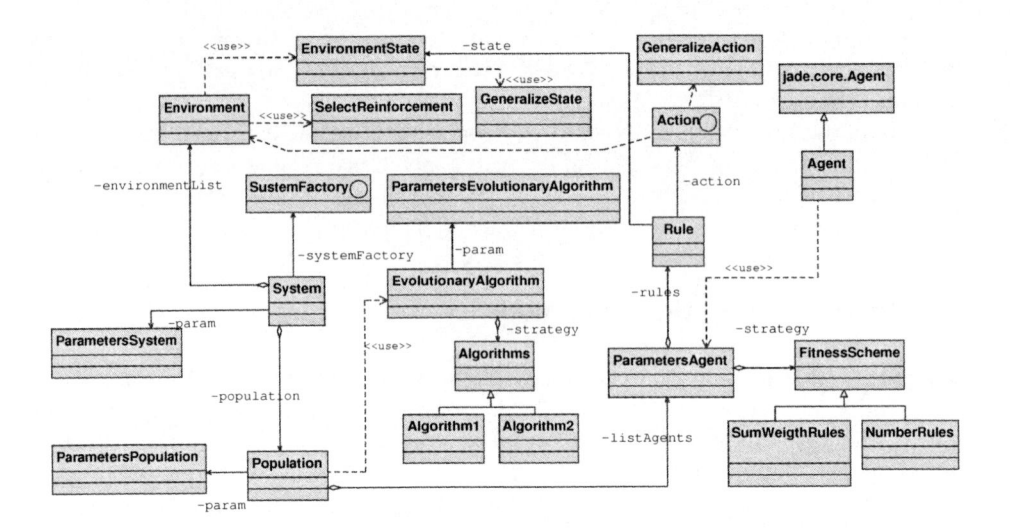

Fig. 1. Class diagram of the framework

- Population: A grouping of agents. In our design, the global MAS is composed of distinct populations capable of possessing several specific attributes that distinguish them from the others. Populations evolve separately. That is, all the agents of a population, and only them, are subject to the same evolutionary algorithm.
- Agent: This class models an individual intelligent agent. As with populations, agents are not identical: in addition to possessing an individual rule base, an agent has several other defining characteristics, as it will be described later.
- Environment: This class models the environment in which a given set of agents are situated. The precise number of agents of an environment depends on the characteristics of the problem being addressed (it could be anything from one agent to the entire population). Likewise, the exact characteristics of the environment are entirely dependent upon the problem in question.
- EnvironmentState: A specific configuration of the environment. Changes in the state of the environment are the stimuli to which the agents react.
- Action: An operation on the environment, performed by an agent. Again, the specific implementation of an action depends on the problem being solved.
- Rule: An individual rule in the rule base of an agent. This rule consists of situation-action pairs representing the antecedent and the consequent of the rule, respectively. As mentioned, a situation is a particular state of the environment, and an action is an operation on the environment.
- Evolutionary algorithm: The genetic algorithms applied to the populations.

The framework, thus, provides the basic structure for an evolutionary multi-agent system. As described above, some aspects of the MAS are entirely dependent upon the problem being addressed, namely, *environment, environment state*, and *action*. Nevertheless, the framework provides generic classes (superclasses or interfaces) for them. This way, all that is necessary in order to apply the framework to a particular problem is to write classes for these three elements. The rest of the elements of the MAS are instantiated from the classes of the framework itself.

Finally, agents can be programmed using specific programming paradigms or–as is more frequent in recent times–using a generic language. In the latter case, Java is among the most popular at present, mainly because of its multi-platform capabilities. JADE (Java Agent DEvelopment Framework) [4] is a powerful and widely-used software framework fully implemented in the Java language and aimed at the development of multi-agent systems [10]. JADE provides many interesting features, such as a generic agent architecture, concurrent agent execution, standard mechanisms for manipulating agents, ACL-FIPA communication, and a directory service to accelerate agent search. Therefore, we have chosen JADE as the implementation tool for our framework in order to take advantage of these features.

4.1 Parameters of the Multi-Agent System

The parameters of the MAS allow to define its characteristics and modify its functioning. They are classified into four categories: the *system* in general, the *populations*, the *agents*, and the *genetic algorithms*. Some attributes are present at different levels of granularity. For example, agent-level parameters can be defined at a population level, providing thus a default value for all the agents. It should be noted that higher-level parameters place restrictions on lower-level parameters–the parameters cannot contradict each other. Likewise, the parameters for all populations can be configured at once at the system level, but it is also possible to choose different parameters for specific populations. The parameters are:

- System-level parameters:
  - Number of populations.
  - Execution mode: The execution of the system can be carried out by turns (in every cycle, each agent has one turn to perform one action on the environment, and must wait until the other agents in the environment have ended their turn) or concurrently (all agents operate simultaneously). Board games are an example of the former, whereas the latter is typical of real-time situations.
  - End conditions, which are divided into: a) completing a maximum amount of time or a maximum number of cycles, b) reaching a desired fitness level, and c) learning a specific number of rules.
  - Evolutionary step: The amount of time or the number of cycles that pass between each execution of the genetic algorithm.
  - Fitness scheme: This determines how the fitness of the agents is measured. The user can choose between different fitness schemes or even define a new one adapted to the characteristics of the problem being addressed. For example, using problem-specific heuristics. By default, the system uses a standard, generic scheme: the sum of the weights of the rules of the agent [6].
- Population-level parameters:
  - Number of agents.
  - End conditions, analogous to the end conditions of the system but including also the case of reaching an evolutionary dead end (i.e., when the fitness fails

to improve after a given amount of time or cycles). This is determined by two parameters: a) minimum fitness increment, and b) time or number of cycles.
- Agent-level parameters:
  - Maximum number of rules in the rule base. When the rule base reaches its maximum size, the worst rules are pruned.
  - Exploration rate: When an agent has to perform an action on the environment, this parameter determines the probability of experimenting with creating a new rule instead of using an already known one. The exploration rate can be a fixed value or it can vary dynamically on the basis of the number of rules.
  - Rule selection strategy: This parameter dictates which rule is chosen during exploitation. The strategy can be either deterministic (i.e., the rule with the highest weight is always selected) or probabilistic (the selection is based on probabilities, which in turn are based on the weights).
  - Imitation rate: The probability of deciding to imitate the characteristics of other, better performing agents.
  - Imitation strategy: This determines which agent characteristics are copied. An agent can imitate both the exploration rate and the rule selection strategy of other agents, and it is also possible to copy their rules.
- Genetic algorithm parameters:
  - Type of algorithm: The users can select a predefined algorithm or provide one of their own. The default algorithm is a one-point crossover.
  - Reproduction rate: When the genetic algorithm is executed, this parameter indicates the percentage of agents that will be chosen for reproduction (i.e., those that will serve as progenitors of new agents).
  - Elitism rate: Percentage of agents that automatically pass, with their rule base unaltered, to the next generation.
  - Mutation rate: The probability that an agent will undergo a mutation, thus adding randomness to the evolutionary process.

5 The System at Work

In this section, the functioning of the framework will be illustrated by its application to the tic-tac-toe problem. As shown in Figure 2, the framework classes provide the basic infrastructure and it is only necessary to write a few classes to define those elements specific to the problem (i.e., board, piece, board state, move, and, in this case, functions for generalizing analogous states and analogous moves). Moreover, the framework offers templates for additional elements such as fitness schemes. The framework has a graphical interface (see Figure 3) that allows the user to control the system. The top of the window contains menus for opening and saving MAS files and for viewing help pages. In order to run the MAS, first of all, the user must determine its parameters. The main area of the graphical interface offers several tabbed panes for configuring the parameters for the system, the populations, and the agents. The bottom of the window provides buttons for starting, pausing, and restarting the MAS. In order to monitor the progress of the agents in real time, the user can click the "Running View" tab to view metrics (see Figure 4) and graphs on the performance of the agents. While the system

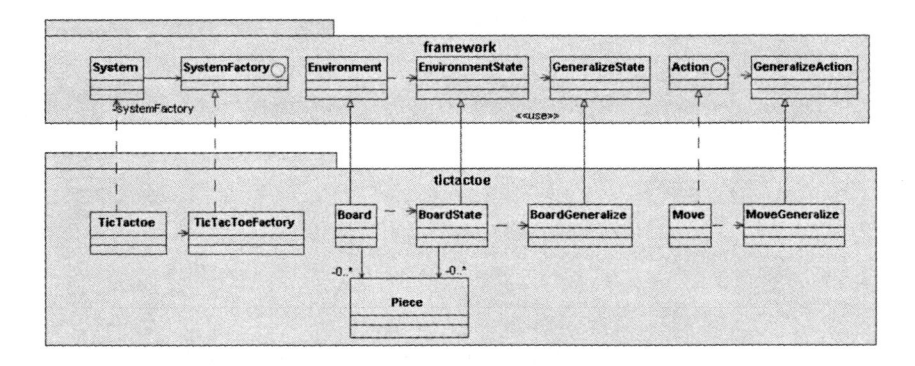

Fig. 2. Example of the application of the framework. An MAS for solving the tic-tac-toe problem.

Fig. 3. Graphical interface of the framework, containing menus for file management and tabbed panes for configuring the parameters and viewing the execution of the MAS

is running, the user can also modify the parameters and manipulate the agents (e.g., to kill them or to inject rules).

Several experiments were carried out to test the framework. We used systems of the same type, consisting of one population of ten evolutionary agents playing tic-tac-toe in pairs. The evolutionary algorithm was set up to be executed every two cycles, and the exploration rate was configured to vary dynamically. The graph in Figure 5 depicts the evolution of one of the experiments over 1700 cycles, showing the average fitness,

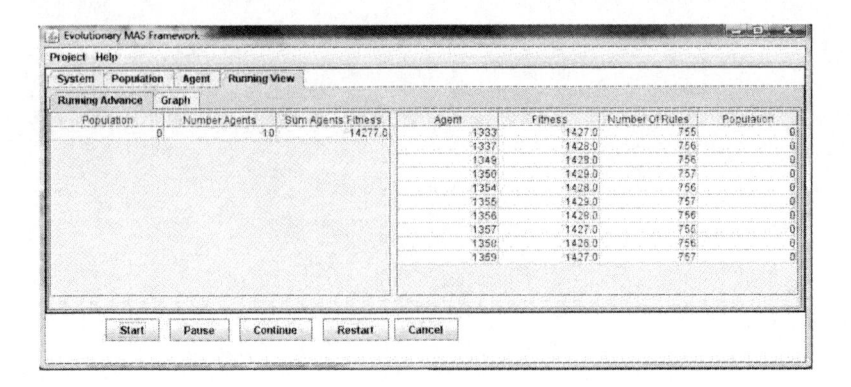

Fig. 4. Real-time metrics on the performance of the tic-tac-toe MAS

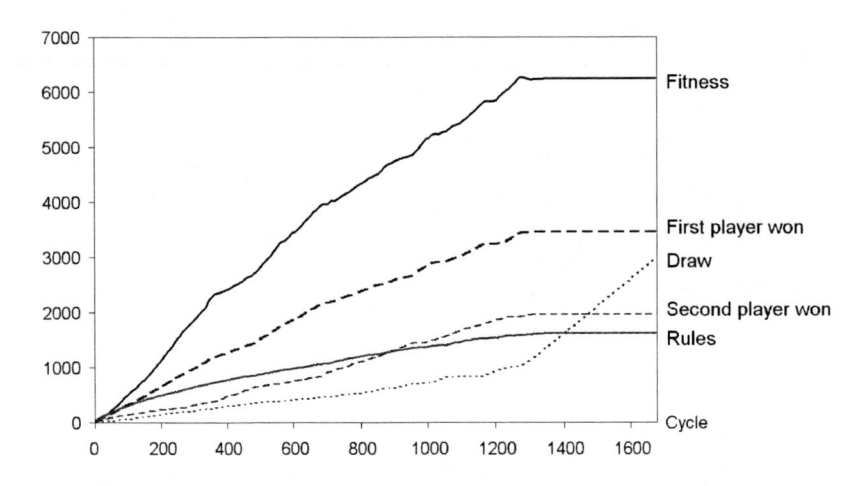

Fig. 5. Graph on the performance of the tic-tac-toe MAS, showing the average fitness of the population, the total number of games won by the agents that played first, the games that ended in a draw, the games won by the agents that played second, and the average number of rules

the average number of rules, the total number of games won up to that point by the agents that played first, the total number won by the agents that played second, and the total number of games that ended in a draw. It can be observed that the fitness and the number of rules increase progressively, until both reach a plateau at around 1300 cycles. We can also see that, up to that point, the total number of games won by either player also increased progressively. It is at this point that the playing skill of the agents reaches its peak, as can be discerned from the fact that the total number of draws increases exponentially (in tic-tac-toe, if both agents play flawlessly, the game always ends in a draw). In other words, emergent behavior appears.

6 Conclusions

In this work, a framework for the development of evolutionary multi-agent systems has been presented. It is based on JADE, an open-source platform for developing multi-agent systems, FIPA-compliant, and fully implemented in Java. The MAS constructed with the framework consist of populations of intelligent agents that are capable of learning at three levels. Firstly, agents learn individually from their interaction with the environment. A reinforcement scheme indicates whether their actions have been correct or not. Secondly, agents can decide to imitate the characteristics of other agents. Thirdly, an evolutionary algorithm is regularly executed in order to share the best genes among a population.

This approach is thus particularly appropriate for problems with the following characteristics: environmental conditions change unpredictably; the correct way of solving the problem is not known in advance; it is not advisable or feasible to have complicated agent mechanisms.

The programming framework provides all the necessary infrastructure for developing multi-agent systems based on our approach. That is, the framework offers generic classes for creating agents, populations, environments, evolutionary algorithms, and so forth. In order to apply the framework to a specific problem, it is only necessary to write code for those elements that describe it (namely, the characteristics that define the environment, the possible states of the environment, the actions on the environment, and, if applicable, functions for generalizing analogous states and analogous actions). The rest of the elements are supplied by the framework. The framework also provides a graphical interface that allows to configure the parameters of the MAS at different levels (global system, agents, populations, etc.), to have complete control over its execution, and to collect information on its performance.

Whereas the literature on evolutionary agents is largely focused on ad hoc systems, our framework is intended to be general, so instead of limiting ourselves to a particular architecture from the literature we have opted for flexibility. The framework gives the users the possibility of configuring many parameters and using optional features such as imitation.

Regarding future work, and given that the framework is based on JADE, an interesting extension would be to allow the manipulation of pre-existing JADE MAS.

Acknowledgements. This work has been funded in part by the Xunta de Galicia under project PGIDT06PXIB105205PR.

References

1. Di Nola, A., Gisolfi, A., Loia, V., Sessa, S.: Emerging Behaviors in Fuzzy Evolutionary Agents. In: 7th European Congress on Intell. Tech. and Soft Comput., EUFIT 1999 (1999)
2. FIPA, Foundation for Intelligent Physical Agents (last accessed 30/10/2008) (2008), http://www.fipa.org
3. Hinton, G.E., Nowlan, S.J.: How Learning Can Guide Evolution. Complex Systems 1, 495–502 (1987)

4. JADE, Java Agent DEvelopment Framework (last accessed 30/10/2008) (2008), http://jade.tilab.com
5. Menczer, F., Monge, A.E.: Scalable Web Search by Adaptive Online Agents: An InfoSpiders Case Study. In: Klusch, M. (ed.) Intelligent Information Agents: Agent-Based Information Discovery and Management on the Internet, pp. 323–347 (1999)
6. Mitchell, M., Forrest, S.: Genetic Algorithms and Artificial Life. Artificial Life 1, 267–289 (1994)
7. Moukas, A.: Amalthaea: Information Discovery and Filtering Using Multiagent Evolving Ecosystem. Applied Artificial Intelligence 11(5), 437–457 (1997)
8. Nonas, E., Poulovassilis, A.: Optimisation of active rule agents using a genetic algorithm approach. In: Quirchmayr, G., Bench-Capon, T.J.M., Schweighofer, E. (eds.) DEXA 1998. LNCS, vol. 1460, pp. 332–341. Springer, Heidelberg (1998)
9. Wan, Y.: A New Paradigm for Developing intelligent Decision-Making Support Systems (i-DMSS): A Case Study on the Development of Comparison-Shopping Agents. In: Gupta, J.N.D., Forgionne, G.A., Mora, M. (eds.) Intelligent Decision-making Support Systems Foundations, Applications and Challenges, pp. 147–165 (2006)
10. Wooldridge, M.: Introduction to MultiAgent Systems. John Wiley and Sons, New York (2002)

A Multi-agent Approach for Web Adaptation

A. Jorge Morais

Universidade Aberta (Portuguese Open University), Lecturer
Faculty of Engineering of the Univeristy of Porto, PhD Student
Laboratory of Artificial Intelligence and Data Analysis (LIAAD – INESC Porto L. A.),
PhD Student Researcher

Abstract. Web growth has brought several problems to users. The large amount of information that exists nowadays in some particular Websites turns the task of finding useful information very difficult. Knowing users' visiting pattern is crucial to owners, so that they may transform or customize the Website. This problem originated the concept known as Adaptive Website: a Website that adapts itself for the purpose of improving the user's experience. This paper describes a proposal for a doctoral thesis. The main goal of this work is to follow a multi-agent approach for Web adaptation. The idea is that all knowledge administration about the Website and its users, and the use of that knowledge to adapt the site to fulfil user's needs, are made by an autonomous intelligent agent society in a negotiation environment. The complexity of the problem and the inherently distributed nature of the Web, which is an open, heterogeneous and decentralized network, are reasons that justify the multi-agent approach. It is expected that this approach enables real-time Web adaptation with a good level of benefit to the users.

1 Introduction

The World Wide Web has experienced a quick growth during the past decade. Nowadays, having a Website is almost mandatory for organizations, in order to easily deliver information to the general audience. With the advent of e-commerce (electronic commerce), organizations became more concerned with the problem of organizing all the information efficiently, so that it may be easy to find every product a user is searching or the organization wants clients to purchase, or to present useful information to the public.

Let us consider, for instance, that a client is searching a laptop with some particular characteristics (processor, memory, hard disk, screen size, etc.). The Website structure may be organized by different suppliers, which means clients will have to search inside each one in order to compare models. Suppose that this is a standard procedure of most of the users. Discovering such a pattern might lead to considering a solution like suggesting similar laptops from other suppliers in the same page. Another solution might be reorganizing the Website structure to organize laptops by characteristics instead of suppliers.

The problem of dealing with large sets of data was already the motivation for the area of Data Mining and Knowledge Discovery [1]. The idea was to take advantage of the large quantity of data from previous transactions that were kept in organizations, finding useful information that is not easily reachable. This led to the use and development of a set of algorithms that automatically extracted and discovered important patterns, and represented them in an understandable way. Considering the large number of pages in the Web, it became natural the application of Data Mining and Knowledge Discovery to the Web scope, which resulted in the new area of Web Mining [2][3]. Web Mining can be defined as the application of data mining algorithms to extract and discover useful information (documents and services) from the Web. The results may enable the Website owner to improve its structure by reorganizing the Website or providing users navigation assistance.

There are three main research sub areas of Web Mining. Web Content Mining is focused on the content of Web pages, mainly the search of documents satisfying a given criteria, and summarization. Web Structure Mining takes into account the structure of the Website, its pages, and hyperlinks between them, enabling categorization of Web pages and finding similarities and relationships between Websites. Web Usage Mining tries to extract and discover useful information from previous visits to the Website, and is useful for Web adaptation and market-based analysis.

The problem of Web adaptation is not new. Some work has already been made in this particular area in the last years. One of the current solutions that are being proposed for this problem is using autonomous agents. Multi-Agent Systems [4] is a research area that has been in great development over the last decade, and has some particular characteristics that fit in this problem. In fact, it was already proposed to use a multi-agent approach, because of its flexibility and its capability of dynamic adaptation to the Web applications needs [5]. Moreover, Multi-Agent Systems are already used for automatic retrieval and update of information in Websites [6]. An implementation of a recommender system using this approach was already proposed in [7].

This paper is organized as follow. We start by presenting previous approaches and applications in the area of recommender systems and multi-agent systems. Then we present an overview of the project, showing its scope and previous work. Finally, we describe some conclusions and future work.

2 Previous Approaches and Applications

A global vision on adaptive Web sites based on user interaction analysis is given in [8]. In fact, only less ambitious approaches were proposed, such as reorganization of the Website [9], use of recommendations in the pages [10], automatic categorization of user actions [11], or seek of relevant Web sequence paths using Markov models [12].

Recommendation systems include the combination of clustering with nearest neighbour algorithms [13], Markov chains and clustering [14], association rules [15], and collaborative and content-based filtering [16].

Web dynamics has been controlled, for instance, by efficient incremental discovery of sequential Web usage patterns [17], and on-line discovery of association rules [18].

Data-driven categorization of a Website usability may be done by typical usage patterns visualization [11] or with objective metrics [19].

Some platforms, like WebWatcher, use previous users' knowledge to recommend links [20]. AVANTI implements an adaptive presentation based on a model constructed from user actions [21]. WUM infers a tree structure from log records enabling experts to find patterns with predefined characteristics [22]. In [23] it was proposed an integrated tool (HDM) to discover access patterns and association rules from log records in order to automatically modify hypertext organization.

In [5] it was proposed the use of a multi-agent platform for personalization of Web-based systems, given the flexibility of this approach and its dynamic adaptation to Website needs. Multi-agent approaches for developing complex systems, like Web adaptation, were defended in [24]. Intelligent agents may also be an important contribution for autonomic computing [25]. Such systems main characteristics are being complex systems with self-administration, self-validation, self-adjustment and self-correction. Web adaptation systems should also have these characteristics, because Website environment dynamics requires either a high degree of system automation or high allocation of human resources. Another important usage of multi-agent systems in this issue is the automatic collection and actualization of information in Websites [6].

There is already an implementation of a recommendation system [7]. In this work the author implements a market-based recommender system that is stated to be Pareto efficient, to maximize social welfare and all agents are individually rational. The author distinguishes between Internal Quality (INQ) and User Perceived Quality (UPQ). INQ of a specific recommendation is the sum of the weighted evaluation scores made of different techniques on different properties of a recommendation while UPQ is introduced by users. Agents are unaware of UPQ, and they must infer from the reward mechanism. A generalized first-price sealed-bid auction protocol is used to minimize the time for running the auction and the amount of communication generated.

3 Project Overview

This project is a part of an undergoing work in the area of Adaptive Websites. We will now describe its scope, the preliminary work and compare it with related work in the same area.

3.1 Scope of the Project

The work to develop during this thesis is to be implemented in a web adaption platform [26] that was developed during the site-o-matic project[1], and is already implemented and running (except the part described in this paper). The architecture of the platform is presented in figure 1.

As the client opens the Website page in a browser, the information about the current page content and the actions performed are recorded in the data warehouse. The client browser also performs an automatic adaption request to broker. The broker acts as an intermediary between the client and the adapters, requesting hyperlinks to be shown in the browser. This activity is recorded in a logger, which also feeds the data warehouse. Finally the broker sends suggested hyperlinks to the client browser. The process will be repeated whenever the user chooses another hyperlink or performs some particular action in the browser (e.g., selecting a portion of the text with the mouse).

---

[1] Supported by the POSC/EIA/58367/2004/Site-o-Matic Project (Fundação Ciência eTecnologia), FEDER e Programa de Financiamento Plurianual de Unidades de I & D.

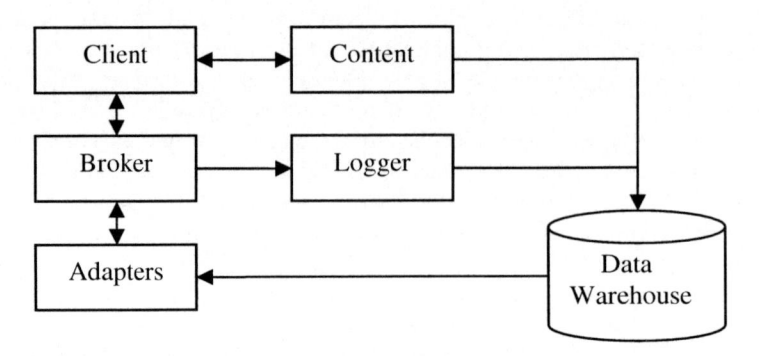

```
        ┌─────────┐      ┌─────────┐
        │ Client  │◄────►│ Content │───────────┐
        └─────────┘      └─────────┘           │
             ▲                                 │
             ▼                                 │
        ┌─────────┐      ┌─────────┐           │
        │ Broker  │─────►│ Logger  │           │
        └─────────┘      └─────────┘           │
             ▲                                 ▼
             ▼                            ┌─────────┐
        ┌─────────┐                       │  Data   │
        │Adapters │◄──────────────────────│Warehouse│
        └─────────┘                       └─────────┘
```

Fig. 1. Web adaption platform

Our propose is that the construction and management of the Knowledge Base with information about the Website and its users, and the use of that knowledge to perform the adaptation according to users behaviour, should be done by a multi-agent system with negotiation. Each agent goal is to optimize the quality of the Website user experience by choosing adequate contents and to represent them in an appropriate manner (for instance, with an appropriate format that calls the attention of the user). They must collect and produce knowledge about users, eventually specializing themselves in a given segment of users or contents. Depending on the recommender system algorithm, agents will try to adapt a particular client using its or all users (or part of them) past behaviour.

Therefore, agents will have to compete for the limited space in the browser and the user's attention. They also have to compete for modelling each user. This attribution may be made randomly, centralized, or as result of a negotiation or an auction between agents. Each agent will have its virtual financial resources (utilities) that rise when good decisions are made and decrease with the acquisition of user modelling and knowledge acquisition.

In the scope of the architecture of the platform in figure 1, the broker will be modified in order to perform an auction and manage the utilities received from and paid to the agents. The adapters will have to be given the additional capability of performing intelligent bids in order to maximize their revenues. When a request is made to the broker, all adapter agents will make a single bid, and the broker choose the best offer, informing the winner and the losers (with the value of the winning proposal, so that they can use this information in future bids). The agent that wins the auction will be given the task of adapting the Website for the respective client until the end of the session and will be rewarded according to the success of its recommendation.

3.2 Preliminary Work

A small project was made in the scope of the course Multi-Agent Systems, in the Doctoral Programme in Informatics Engineering at Faculty of Engineering of the University of Porto, and consisted in a simple recommendation system, developed using the JADE platform [27].

The system used an applet in the client-side and a multi-agent system in the server-side. After a request from the user browser, three agents with different recommendation algorithms competed for the recommendation of hyperlinks that were sent to the applet in the client browser, offering an amount of its virtual financial resources. If users followed the hyperlinks, the agent was rewarded according to the number of pages and the order in which the respective hyperlinks were presented.

The project was important for studying and using the platform, and also showed that the approach was pertinent. However, some significant amount of time was usually spent while loading the applet. For this reason another concept will be used: AJAX [28]. It is not yet possible to estimate the time that will take to make an adaption, but since AJAX acts asynchronously and the adaption is processed in the server side, it is expected that it takes a very small amount of time.

3.3 Comparison with Related Work

This proposal approach is different in many aspects from the recommender system proposed in [7], although some knowledge about that work may be useful (for instance, some tests that were already made about the number of hyperlinks that should be used in recommendation, and some parameters that were already tuned). The cited work does not take into account the evolution of the market, which is not suitable for a daily updated Website. For instance, in an online news journal a user might classify a Web page as very interesting today but tomorrow it could loose all of its interest. But next week, if the issue of that Web page was again on the top of the news, it would be interesting again.

There are also other aspects that our work can improve such as taking into account client behaviour during each session and obtaining user's quality score implicitly rather than expecting their explicit scores.

4 Conclusions and Future Work

The current work is in an early stage. Rather than final conclusions we will present the expected contribution of the thesis and the respective work plan.

4.1 Expected Contribution of the Thesis

The main contribution of this thesis is the use of a market-based approach, using autonomous agents. Multi-Agent Systems [4] are a research area that has been in great development over the last decade, and have some particular characteristics that fit in this problem. In fact, in [5] it was proposed the use of a multi-agent approach, because of its flexibility and its capability of dynamic adaptation to the Web applications needs. Moreover, Multi-Agent Systems are already used for automatic retrieval and update of information in Websites [6].

It is our belief that design, construction, validation and management of a dynamic Website, capable of adapting itself automatically to the user needs, following high-level specifications, are tasks sufficiently complex to justify this approach. It is even more pertinent if we consider large Websites, where manual organization or definition of explicit rules is not easy, making the task of improving user experience harder.

4.2 Work Plan

After defining the Web adaptation strategy, using Web mining algorithms and Multi-Agent Systems, and also the evaluation strategy for measuring its success, the primary goals are the design of the knowledge base and the definition of agents, possibly with some pilot implementations.

The platform has already some recommender systems implemented, using algorithms such as collaborative filtering, association rules, etc. The agents to be implemented will use these algorithms to perform adaption. The bidding strategy will be important for the success of the approach: if the market fails, an agent may become in a situation of monopoly, which means that the approach becomes useless. This means that all agents must have intelligent bid strategies, in order to maintain the equilibrium of the market.

Afterwards, a preliminary evaluation that may lead to important changes in the previous design phase will be held, resulting in the final specification for the implementation phase, with the integration of the Knowledge Base with log analysis problems and the implementation of the agents. Finally, the system will be evaluated, with eventual refinement of methods.

References

1. Fayyad, U.M., Piatetsky-Shapiro, G., Smyth, P., Uthurusamy, R. (eds.): Advances in Knowledge Discovery and Data Mining. AAAI/MIT Press, Menlo Park (1996)
2. Etzioni, O.: The World Wide Web: Quagmire or gold mine? Communications of the ACM 39(11), 65–68 (1996)
3. Cooley, R., Mobasher, B., Srivastava, J.: Web mining: Information and patterns discovery on the world wide Web. In: Proceedings of the ninth IEEE International Conference on Tools with Artificial Intelligence, Newport Beach, California, pp. 558–567 (1997)
4. Wooldridge, M.: An Introduction to MultiAgent Systems. John Wiley & Sons, Chichester (2002)
5. Ardissono, L., Goy, A., Petrone, G., Segnan, M.: A multi-agent infrastructure for developing personalized web-based systems. ACM Trans. Inter. Tech. 5(1), 47–69 (2005)
6. Albayrak, S., Wollny, S., Varone, N., Lommatzsch, A., Milosevic, D.: Agent technology for personalized information filtering: the pia-system. In: SAC 2005: Proceedings of the 2005 ACM symposium on Applied computing, pp. 54–59. ACM Press, New York (2005)
7. Wei, Y.Z.: A Market-Based Approach to Recommendation Systems, PhD thesis, University of Southampton (2005)
8. Perkowitz, M., Etzioni, O.: Towards adaptive web sites: Conceptual framework and case study. Artificial Intelligence 118(2000), 245–275 (2000)
9. Ishikawa, H., Ohta, M., Yokoyama, S., Nakayama, J., Katayama, K.: Web usage mining approaches to page recommendation and restructuring. International Journal of Intelligent Systems in Accounting, Finance & Management 11(3), 137–148 (2002)
10. El-Ramly, M., Stroulia, E.: Analysis of Web-usage behavior for focused Web sites: a case study. Journal of Software Maintenance and Evolution: Research and Practice 16(1-2), 129–150 (2004)
11. Berendt, B.: Using Site Semantics to Analyze, Visualize, and Support Navigation. In: Data Mining and Knowledge Discovery, vol. 6(1), pp. 37–59 (2002)

12. Borges, J.L.: A Data Mining Model to Capture User Web Navigation Patterns, PhD thesis, University College London, University of London (2000)
13. Mobasher, B., Dai, H., Luo, T., Nakagawa, M.: Discovery and Evaluation of Aggregate Usage Profiles for Web Personalization. In: Data Mining and Knowledge Discovery, vol. 6(1), pp. 61–82. Kluwer Publishing, Dordrecht (2002)
14. Cadez, I., Heckerman, D., Meek, C., Smyth, P., White, S.: Model-Based Clustering and Visualization of Navigation Patterns on a Web Site. In: Data Mining and Knowledge Discovery, vol. 7(4), pp. 399–424 (2003)
15. Jorge, A., Alves, M.A., Grobelnik, M., Mladenic, D., Petrak, J.: Web Site Access Analysis for A National Statistical Agency. In: Mladenic, D., Lavrac, N., Bohanec, M., Moyle, S. (eds.) Data Mining And Decision Support: Integration And Collaboration. Kluwer Academic Publishers, Dordrecht (2003)
16. Basilico, J., Hofmann, T.: Unifying collaborative and content-based filtering. In: Proceedings of Twenty-first International Conference on Machine Learning, ICML 2000. ACM Press, New York (2004)
17. Masseglia, F., Teisseire, M., Poncelet, P.: HDM: A client/server/engine architecture for real time web usage mining. In: Knowledge and Information Systems (KAIS), vol. 5(4), pp. 439–465 (2003)
18. Lin, W., Alvarez, S.A., Ruiz, C.: Efficient Adaptive-Support Association Rule Mining for Recommender Systems. In: Data Mining and Knowledge Discovery, vol. 6, pp. 83–105 (2002)
19. Spiliopoulou, M., Pohle, C.: Data mining for measuring and improving the success of web sites. In: Kohavi, R., Provost, F. (eds.) Journal of Data Mining and Knowledge Discovery, Special Issue on E-commerce, vol. 5(1-2), pp. 85–114. Kluwer Academic Publishers, Dordrecht (2001)
20. Armstrong, R., Freitag, D., Joachims, T., Mitchell, T.: WebWatcher: A learning apprentice for the world wide web. In: Proceedings of the AAAI Spring Symposium on Information Gathering from Heterogeneous, Distributed Environments, California, pp. 6–12 (1995)
21. Fink, J., Kobsa, A., Nill, A.: User-oriented adaptivity and adaptability in the AVANTI project. In: Designing for the Web: Empirical Studies, Microsoft Usability Group, Redmond, Washington (1996)
22. Spiliopoulou, M., Faulstich, L.C.: WUM: a tool for web utilization analysis. In: Proceedings of the International Workshop on the Web and Databases, Valencia, Spain, pp. 184–203 (1998)
23. Masseglia, F., Teisseire, M., Poncelet, P.: Real Time Web Usage Mining: a Heuristic Based Distributed Miner. In: Second International Conference on Web Information Systems Engineering (WISE 2001), vol. 1, p. 0288 (2001)
24. Jennings, N.R.: An agent-based approach for building complex software systems. Communications of the ACM 44(4), 35–41 (2001)
25. Kephart, J.O.: Research challenges of autonomic computing. In: ICSE 2005: Proceedings of the 27th International Conference on Software Engineering, pp. 15–22. ACM Press, New York (2005)
26. Domingues, M.A., Jorge, A.M., Soares, C., Leal, J.P., Machado, P.: A data warehouse for web intelligence. In: Neves, J., Santos, M.F., Machado, J.M. (eds.) EPIA 2007. LNCS, vol. 4874, pp. 487–499. Springer, Heidelberg (2007)
27. JADE (Java Agent DEvelopment Framework) (Website: access date: 01/11/2008), http://jade.tilab.com
28. Asynchronous Javascript And XML (AJAX), Mozilla Developer Center (access date: 01/11/2008), http://developer.mozilla.org/en/docs/ajax

A Multi-tiered Approach to Context and Information Sharing in Intelligent Agent Communities

Russell Brasser[1] and Csaba Egyhazy[2]

[1] Integrity Applications Inc., Chantilly, Virginia, USA
rbrasser@vt.edu
[2] Computer Science, Virginia Tech, Falls Church, Virginia, USA
cegyhazy@vt.edu

Abstract. This paper presents an expanded role for context in agent communities. We propose a multi-tiered approach in which the contextual information is segregated and encapsulated in three different contextual levels, namely Local, Focal, and Zonal. Each level contains specific kinds of contextual information, thereby pruning the search space and making the process of reasoning about context more manageable. These levels can be thought of as super-contexts within which each of the subordinate contexts become potentially valid. Such a multi-tiered approach helps organize, manage, select, and maintain the large amount of contextual information required by future multi-agent systems.

Keywords: Agent community, Context, Intelligent agents, Multi-agent system.

1 Introduction

For a software agent to be worthy of the label intelligent it must obviously behave appropriately in a given set of circumstances [1, 2]. Additionally, the decisions it makes should account for, and be guided by the situation at hand. In other words, an intelligent agent must consider the context in which it is operating. Furthermore, explicit representation of context allows agents to explicitly reason about context, not just within it. When expanding from a single intelligent agent to a multi-agent system, our understanding of the role of context must also expand. As Turner argues in [3], not only do each of the individual agents need to behave appropriately for their given context, but the entire system as a "corporate entity" must behave in a context appropriate manner as well. To accomplish this contextual information may need to be communicated among the individual agents. This is because an agent must consider the contexts of its peers so that they can all act in concert. Additionally, agents must include in their reasoning a context that takes into account their membership as part of a group, and the social structure of that group. In fact, contextual information is needed in the very development of that social structure. Context is needed to inform the communications protocols used, the network topology, and how the agents organize. Here the role of context can be seen as not only mediating the behavior of individual agents, but also shaping the social structure of the entire system.

1.1 Context in Agent Communities

Turner's work in [3] is focused on relatively homogenous networks of cooperative Autonomous Underwater Vehicles and sensor systems. As such it addresses only how agents in a multi-agent system can share their contextual information, and how to use contextual information to facilitate information sharing in general. It does not address the broader scope of the agent community. As defined in [4] an agent community is a stable, adaptive group of self-interested agents that share common resources and must coordinate their efforts to effectively develop, utilize and nurture group resources and organization. This definition falls a bit short, however, in that it does not account for the important trend toward pervasive computing. As pointed out in [5] the inherent openness of pervasive computing applications means that they may contain a heterogeneous mix of agents.

So let us expand our definition of an agent community to be a dynamic, adaptive, heterogeneous group of self-interested agents that share common resources and must coordinate their efforts to effectively develop, utilize, and nurture group resources and organization. For the multi-agent system it was assumed that all agents in the systems were designed to operate together. No such assumptions can be made about the agents in an agent community. Clearly the fact that these agents will operate together on some sort of platform means that there will be some commonality. Yet, in order to retain the open nature of the pervasive computing concept, that required level of commonality must be kept to a minimum. This will require agents to be able to change the ways in which they communicate based on which other agents they are interacting with. Furthermore, since agents may move freely from one community to another, they will also need a way to understand the context of which community they are currently operating in. Also, due to the variety of agents and their various intentions, any number of social constructs may arise within the community. For example, some agents may organize into cliques. A clique is a subset of the agents in a system such that all of the agents within the clique are highly cooperative with each other but less cooperative with agents who are not members of the clique. An example of a clique might be a group of agents who all cooperate to act as a service provider for the rest of the community. The self-interested nature of the members of the community means that agents may need a context dependant way to model such concepts as trust. Although two agents may be interacting cooperatively at one point in time, their individual goals may be competitive. It would be unwise in such a situation for one agent to freely share its information with the other since sharing that information could cause the agent to lose a competitive advantage. At the same time, a blanket policy of secrecy might limit or even block an agent's ability to participate in a cooperative relationship essential to achieving its goals. Instead, agents will need to have some understanding of the value of the information they hold, and what the risks and rewards are of sharing that information. Members of an agent community must manage multiple internal and external contexts at all times. In the next section we propose a simple multi-tiered approach to help manage, select, and maintain these enormous amounts of contextual information.

2 A Multi-tiered Approach

Agents operating in an agent community must manage an enormous amount of contextual information. They need to maintain and protect their individual contexts and data while interacting with a wide variety of other agents. In order to accomplish this, we propose using a multi-tiered approach in which the contextual information is segregated and encapsulated in three different contextual levels, the *Local, Focal,* and *Zonal*. Each of the contextual levels is described in detail next.

2.1 Zonal Level

The Zonal Level context represents the context of the community as a whole. All agents in the community must share this context, even if their internal representations differ. Also, the Zonal Level context is the only context that all agents are required to share. As such, this context defines both the necessary and sufficient requirements for membership in the agent community.

The Zonal Level context is comprised of three features: Common Knowledge, Required Behaviors, and Required Protocols. Common Knowledge is the body of knowledge that all agents in the community need in order to participate. This may include how to access available services, systems messages, or any other information that is deemed necessary. Required Behaviors is the set of most basic behavioral standards that all agents in the community must respect. In essence this feature defines the "societal rules" that govern the behavior of every member of the community. Required Protocols is essentially the definition of the most basic language that members of the community can use to communicate with one another.

Since every agent in the community shares the Zonal Level context, it forms the most basic foundation for all agent interactions. In order for agents in a multi-agent system to be social and interact, there needs to be some ground rules on which to form such interactions. Such rules are often built intrinsically into the design of the system. In relatively simple systems, these intrinsic rules may be sufficient to define the entire range of member interactions. However agent communities present such a vast range of possible interactions among a heterogeneous membership, that it is nearly impossible to design the details of all such relation-ships into the system a priori. What is needed instead is a set of foundational requirements from which complexity can be synthesized. That is what the Zonal Level context provides.

In order to produce a richly functioning agent community, it is necessary for the designer to take care in selecting the elements of the Zonal Level context. These should be minimal enough to be easily supported by a wide variety of agents, but must also be sufficient to support productive interaction between agents. It is by exercising the knowledge, protocols, and behaviors of the Zonal Level context that unfamiliar agents can begin to interact and learn about each other. Through these interactions, agents can discover common higher-level protocols and develop more complex relationships. The details of these relationships are what are stored in the Focal Level contexts.

2.2 Focal Level

The Focal Level consists of the contexts of individual agents relationships with other agents in the community. These contexts are built up dynamically as agents in the community interact. The exact contents of Focal Level contexts are specific to each agent, but must include such information as the nature of the relationship, agreed upon protocols, the interaction history, appropriate behaviors, and trust metrics. The nature of the relationship is important because it informs nearly every other aspect of the interactions. For instance it helps to determine which aspects of the interaction history are important, and which behaviors are appropriate or expected. Obviously, it also must play a role in determining the level of trust between the participants. Arguably, the nature of the relationship could be encoded intrinsically in the other aspects of the relationship. But storing it explicitly makes it available as a parameter to algorithms that may be common to many different kinds of relationships. Interaction history is the collection of data representing the important aspects of past interactions pertaining to this relationship. The exact pieces of data stored may vary greatly from agent to agent and from one relation-ship to the next. In some cases there may be no need to maintain any history at all, while others may require detailed historical context.

The agreed upon protocols are those protocols that were established as a result of the initial Zonal Level facilitated interactions. These may consist of anything from low-level communications protocols to high-level negotiation protocols. These are agreed upon through the course of the relationship, and they define the mechanism by which the agents interact. Of course each of the agents involved is be able to support each of the agreed upon protocols, otherwise they would not have been agreed to. The appropriate behaviors are the relationship specific behavioral algorithms that are appropriate for this relationship. Clearly these are highly dependant on the nature of the relationship. The behaviors that are appropriate when interacting collaboratively with another agent may not be appropriate at all in a competitive or service oriented relationship. The final piece of the Focal Level context is the trust metric. This is the collection of information and algorithms that define the degree of trust that an agent applies to the entity on the other side of the relationship. In other words the trust metric is what an agent uses to evaluate the risk that the other party may deliberately interfere with the agents ability to achieve its goals. This metric is critical to dealing with the major difference between agent communities and cooperative multi-agent systems. Specifically, it is the trust metric that allows agents to cope with the fact that some agents with whom they need to cooperate may also be in competition with them. By employing this metric an agent can evaluate how likely another party is to uphold its end of an agreement, or how risky it might be to share important information with that entity.

The trust metrics define the level of trust assigned to agent B from agent A, and is unlikely to have an appropriate meaning for agent B assigning trust to agent A. Still, each Focal Level context describes a relationship. That is to say that it de-scribes how two parties interact. So while it is not shared, an agents Focal Level context must be compatible with the corresponding context of the other party to the relationship. By "compatible" we mean that the Focal Level context for each party to a given relationship must represent the other parties to the relationship, and define the interface by which all parties to the relationship can interact; and to which all parties to the relationship have agreed. The first item simply indicates that the Focal Level context of parties A and B are not compatible if A's context is not for a relationship with B or if B's context is not for a relationship with A.. The second item is basically referring to the agreed upon protocols feature of the con-text. In particular it means that the context for each party must contain the same agreed upon protocols, and that each party must be capable of using the specific protocols. If the Zonal Level context provides the necessary features to begin an interaction, and the relationship and its Focal Level context are built up from the Zonal Level context through negotiation, then the corresponding contexts will naturally be compatible. Additionally, if the Focal Level contexts are malformed or become incompatible in some other way, the corresponding relation will become either non-functioning or non-productive. By non-functioning we mean that the parties will no longer be able to communicate. By non-productive we mean that incorrect or misunderstood communication will lead to the relationship failing to serve the needs of one or both of the parties. In either case, the relationship can be terminated and either abandoned or rebuilt from the Zonal Level.

2.3 Local Level

The Local Level is the most similar to contextual representations in other agent systems. This level is the agent's internal storage of context, BDI, knowledge, and so on. These are the internal features that guide the actions of each individual agent in pursuit of their goals. Along the way, there may be occasion for agents to exchange some of this information. In a cooperative multi-agent system this would not present a problem. However, in an agent community, as has been discussed many times before, there is no guarantee that another agent will be cooperative. Even if an agent is cooperating now, that agent may prove to be a competitor later on. With this uncertainty, an agent must be selective about what information it shares and with whom. If it is not selective enough, an agent may find that it has revealed information that causes it to be blocked from achieving its goals. On the other hand, if an agent is too restrictive about sharing information it may not be able to form the cooperative relationships it needs in order to advance its intents. This is where the Focal Level trust metrics come into play.

The purpose of the trust metric is to inform a risk verses reward evaluation of sharing a piece of information with another party. The trust metric uses information gained through a history of interactions to evaluate how likely it is that another entity can be trusted. In order to evaluate the potential risks or rewards of sharing information, the agent must have understood the value, sensitivity, and importance of every piece of data. The value of the data is an inverse valuation of the availability of the information. Readily available information is not very valuable, and rare information is more valuable. The sensitivity of information is a measure of how likely it is that it could be used to prevent the agent from achieving its goals. The importance of a piece of data refers to how important it is to the agent's ability to accomplish the goal or goals for which it is needed. This metric is primarily used in evaluating the reward of gaining a given piece of information against the cost, for example, of sharing another piece of information.

In addition to these data valuations, an agent must also be able to rank the importance of its goals. This is important to both the risk and reward calculations. For instance sharing risky information (high value and high sensitivity) that relates to a low ranked goal may be worth the reward of receiving high value, high importance information that is necessary for a highly ranked goal.

Certainly, each agent may have its own formulas for calculating risks and rewards, but we offer one here by way of example:

For a piece of data D relating to goal G:

Let $0 \leq Dv \leq 1$ be the value of D
Let $0 \leq Ds \leq 1$ be the sensitivity of D
Let $0 \leq Di \leq 1$ be the importance of D to G
Let $0 < Gr \leq 1$ be the rank of G

Let $0 \leq Rt < 1$ be the result of evaluating the trust metrics for relationship R

$$RISK = \frac{Dv \cdot Ds}{(1-Rt) \cdot Gr} \qquad REWARD = Dv \cdot Di \cdot Gr \cdot Rt \qquad (1)$$

In order for an agent to perform any of these evaluations, it must have these valuations available. This is where the Local Level differs from other contextual representations. In the Local Level any piece of information that may be shared with another entity is tagged with its value and sensitivity. In addition every goal is ranked in terms of importance. Finally, whenever a piece of information is identified as being needed to help accomplish a goal a context is established that links the information to the goal along with its importance to that goal. If a piece of in-formation is identified as being needed for a goal but is not yet known, it is still tagged with all of its valuations. This is done so that the agent can properly calculate the reward involved with gaining that knowledge.

3 Benefits

This multi-tiered approach to context organization aids in context selection. By limiting the search space for certain frequently used contexts, it allows the agent to devote more resources to reasoning and interacting rather than searching and matching potential contexts. In particular, the vast majority of what occurs in an agent community is agent interactions, and this approach focuses on separating the contexts specific to relationships and interactions from the contexts used for general reasoning. At the most basic level, the Zonal Level context provides a constant base context that serves as a fallback position for all agent interactions. Since the Zonal Level context applies to the entire community it can be pre-selected, and always in force. As more sophisticated relationships develop, the Focal Level con-texts will override some aspects of the Zonal Level context, but the Zonal Level context will provide the default protocols and behavior, should the Focal Level context fail. Additionally, since the Focal Level contexts are specific to agent interactions and relationships, context selection can be simplified by focusing on those features that identify the relationship, and disregarding unnecessary pieces of information. Since that extra information might be relevant to a reasoning con-text, it would have to be considered if the relationships were not explicitly called out in the Focal Level. If the relationship and reasoning contexts where intermingled, the complexity and runtime of searching for a proper context within which to interact would be detrimentally impacted.

By separating and categorizing different contextual aspects into multiple tiers, our approach will assist agent socialization in several ways. First it divides the contexts into the appropriate contextual levels. Next, the contexts in the Focal Level are subdivided by the parties involved in the relationship. This organization will make the process of searching for contexts more efficient by partitioning the search space. Additionally, information in the local level is organized into con-texts which associate it with the goals it is immediately applicable to. This can be used to aid information retrieval by allowing the data associated with the goals at hand to be readied for rapid access.

Another feature of our approach is the inclusion of the trust metric in the Focal Level relationship context. This allows the understanding and evaluation of risks to be evaluated in the context of the nature of the communication. To offer a real world example, suppose that you have a neighbor who is also your lawyer. As your lawyer, when this neighbor is acting in the context of being your legal counsel the neighbor must follow strict rules of confidentiality. However, when you are talking to your neighbor outside of the attorney-client context, no such expectation of strict confidentiality applies. Similarly, in an agent community where members may take on different roles at different times, agents should at times evaluate trust in the context of the nature of the relationship being engaged at that time. The inclusion of the trust metric in the Focal Level context, which is also dependant on the nature of the relationship, provides the agents with a means to handle such nuances. Additionally, the existence of the Zonal Level context gives the agents a starting point for dealing with the diverse membership of an agent community. By providing a minimum requirement for community membership, the Zonal Level context ensures that a minimum level of communication is possible with all members of the community. This provides a foundation upon which agents can discover or negotiate other protocols to support more sophisticated interactions. This model of Zonal Level minimum requirements and building up Focal Level con-texts for individual relationships supports a rich variety of interactions among heterogeneous membership.

Finally, the Zonal Level context plays another role in handling the dynamics of an agent community. Since membership in an agent community is dynamic, agent may join or leave the community at any time. If the details of the Zonal Level con-text were built intrinsically into the design of an agent, at best it would only be able to join other communities with the same set of basic behaviors and protocols. However, by encapsulating these required community features in an explicitly modeled context, agents can move freely from one community to another by changing the basic Zonal Level context being used.

A potential drawback to our proposed approach is the storage requirements for the large amount of contextual information, and the additional processing needed to reason about those contexts. Indeed these aspects certainly limit the applicability of this approach to mobile devices. However, these are certainly not onerous requirements for modern computers, and the pace of advancement in mobile computing power and memory will surely render this argument mute in short order.

Ultimately though, the impact of this will need to be tested through implementation. Conceptually, we believe that the advantages in agent sociability and community dynamics will eventually outweigh the costs. Our approach suggests a more dynamic and social sort of intelligent agent. One that is able to move freely between dynamic communities of agents. One that is able to evolve complex communications from simple ones. One that is able to participate in multi-faceted relationships, and balance both cooperative and competitive interactions. Ultimately this could lead to agents that behave much more like us.

4 Next Steps

We have identified two next steps for this research. The first is to define the trust and data valuation algorithms. While these may ultimately vary from one implementation to another, an appropriate set of example algorithms should be identified. These aspects are critical to the handling of multi-faceted relationships, and therefore cannot be neglected. There has been a great deal of work done on the subject of trust within multi-agent systems. This work should be considered and would form the foundation, if not the direct source, of the trust metric algorithm. The second is to design a prototype implementation. Such an implementation will need to include a variety of agents and an agent community for them to participate in. This prototype will help us demonstrate the practicality of the approach, and allow the costs to be evaluated against the benefits.

References

1. Turner, R.M.: Context-Mediated Behavior for Intelligent Agents. International Journal of Human–Computer Studies 48(3), 307–330 (1998)
2. Edmonds, B.: Learning and Exploiting Context in Agents. In: AAMAS. ACM, Bologna (2002)
3. Turner, R.M.: Exploiting Contextual Knowledge in Multiagent Systems. University of Maine Computer Science Technical Report, 2005. 01 (2005)
4. Sen, S., Saha, S., Airiau, S., Candale, T., Banerjee, D., Chakraborty, D., Mukherjee, P., Gursel, A.: Robust Agent Communities. In: Gorodetsky, V., Zhang, C., Skormin, V.A., Cao, L. (eds.) AIS-ADM 2007. LNCS, vol. 4476, pp. 28–45. Springer, Heidelberg (2007)
5. Bucur, O., Beaune, P., Boissier, O.: What is context and how can an agent learn to find and use it when making decisions? In: Pěchouček, M., Petta, P., Varga, L.Z. (eds.) CEEMAS 2005. LNCS, vol. 3690, pp. 112–121. Springer, Heidelberg (2005)
6. Klein, F., Giese, H.: Analysis and Design of Physical and Social Contexts in Multi-Agent Sys-tems using UML. In: ACM SIGSOFT Software Engineering Notes, vol. 30(4). ACM, New York (2005)

A Multiagent Distributed Design System

Ewa Grabska, Barbara Strug, and Grażyna Ślusarczyk

Faculty of Physics, Astronomy and Applied Computer Science,
Jagiellonian University, Reymonta 4, Cracow, Poland
uigrabsk@cyf-kr.edu.pl, barbara.strug@uj.edu.pl, gslusarc@uj.edu.pl

Abstract. This paper deals with a computer-aided distributed design system based on hypergraph grammars and cooperating agents. Hypergraph grammars are used to generate structures of designs in the form of specific graphs allowing for expressing multiargument relations among structure components. Agents perform actions defined by hypergraph grammars. The communication between the system of hypergraph grammars and agents is realized through a set of variables assigned to grammars and activating intelligent agents which simultaneously solve different design subtasks. During the design process solutions generated by agents are evaluated and on this basis the grammar system is modified. The approach is illustrated on the example of designing hotel grounds.

1 Introduction

A distributed model of computing is more and more popular, especially with rapid development of the Internet and availability of distributed development platforms. Such a model seems to be very useful in the domain of computer-aided design. Many design problems can be divided into a number of tasks, which can be solved by different cooperating agents contributing their capabilities to a common object being designed. Yet no attempt to find a formal model of such a cooperative distributed design has been proposed so far.

This paper deals with a linguistic approach [11] to distributed parallel design [2, 8] which is combined with a multiagent system [14, 12]. The formal linguistic model of design object structures is based on hypergraphs [3, 7, 9], i.e., graphs which enable us to express multi-argument relations of different types among components of a design object. Semantic design knowledge is expressed by hypergraph attributes which are assigned to hypergraph elements. Hypergraphs are generated by agents with the use of grammars which consist of hypergraph rules. Agents select rules to be applied according to the values of hypergraph attributes.

Usually a complex design problem is divided into a number of subproblems. Therefore in [4, 5] we have proposed a distributed design system consisting of several cooperating hypergraph grammars. In this paper we propose to combine the system of grammars with agents which simultaneously perform different design subtasks. Actions of each agent are defined by one grammar of the system. Agents of the same type, which solve subproblems of the same kind (i.e. designing spaces of the same kind for example kitchens, bathrooms, rooms), can work according to the same system grammar. The presented approach differs from the one described in [10], where agents perform inference on a system whose structure is represented using attributed graphs, as in our method agents act according to their grammars and simultaneously change the hypergraph representing the structure of the solution.

Solving the whole design problem requires the communication between the system of hypergraph grammars and agents. In [5] the communication of the system of grammars has been realized by a special set of variables. Now, each of these variables will be assigned to one grammar and will allow for activation of an agent defined by this grammar. Agents realizing subtasks described by hypergraph grammars work in parallel operating on the common hypergraph. In each design step an agent either performs a rewriting using a grammar rule or requests cooperation of other agents specified by variables occurring in the hypergraph.

The presented system of hypergraph grammars combined with agents generates a set of hypergraphs representing design task solutions compatible with the given design criteria. During the design process solutions generated by agents are evaluated and on this basis the grammar system is modified. These modifications result in new choices of rules available to agents. In this way the agents learn and adapt themselves to changing requirements of the design task.

The paper is organized in the following way. In section 2 the stages of the design process are outlined, section 3 deals with hypergraphs and hypergraph grammars used in our approach. In section 4 a distributed design system consisting of hypergraph grammars combined with cooperating agents is presented. The proposed approach is illustrated on the example of designing hotel grounds.

2 Design Process

The design process considered in this paper consists of different stages depicted in Fig. 1. In the generative part of the process, where hypergraphs representing design solutions are obtained, a distributed approach is used.

The process is started by the designer who formulates a set of design requirements on the basis of a design task (for instance after a discussion with a customer). These requirements together with legal norms and standards constitute the design criteria for a given project. The criteria are then matched against the existing designs from the design database (the "experience" part of a design process). If one or more designs satisfying the designer requirements are found they are presented to the designer.

It is rather rare that the user wants one of the "off-the-shelf" designs. But frequently the designer wants something similar to one of the presented samples or a combination of several of them. If a modification is desired a description of a selected design is copied from the database. This description should be

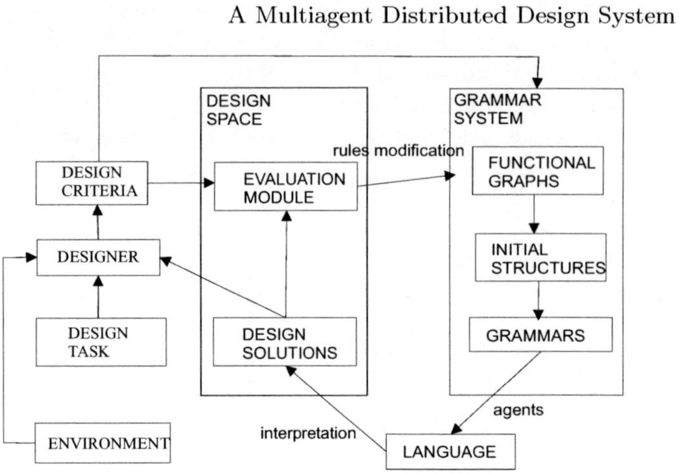

A Multiagent Distributed Design System

Fig. 1. A model of the design process

flexible enough to allow for easy modification. In a design model based on formal grammars the description of generation steps consists of a set of productions that were used to produce the design representation. Then the process can be "rewound" as much as needed and then different grammar productions can be used to satisfy user requirements.

If there is no design compatible with user requirements (or database is empty) the process starts from scratch. A functional graph is specified by the designer and converted into several initial structures. Then one of the initial structures is selected and a design generation process starts with the use of agents. Each type of agent is responsible for one design subtask and works on an independent part of the common graph representing a design. Each one of them is associated with a single grammar consisting of rules for designing a particular part of an object. Agents are responsible for selecting rules to be applied at the successive steps of generation. If some parts of a graph being generated should be developed by the same grammar, several agents can be associated with the same grammar.

The resulting graph structures are passed to the evaluation module which analyses them and evaluates their quality in respect to the design criteria. On the basis of the evaluation results the productions of grammars are modified and these modifications are passed to agents. The final graphs representing design can be passed to interpretation module and then visualized. Their descriptions are added to the design database.

3 Hypergraphs and Hypergraph Grammars

As in this paper design structures are represented by means of attributed hypergraphs generated by grammars, this section introduces the main concepts and

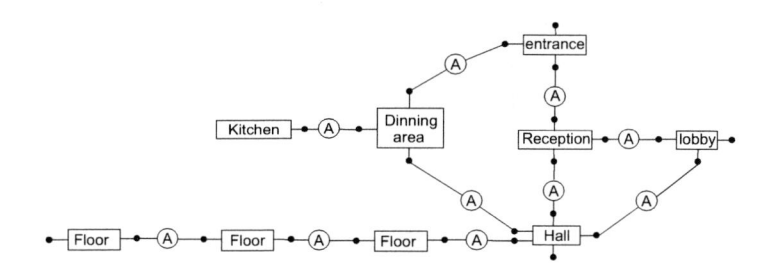

Fig. 2. A hypergraph representing a layout of a three-storey hotel

notations related to hypergraphs and hypergraph grammars. Our approach is based on a formal model of hypergraphs presented in [9] and extended in [13].

A hypergraph consists of hyperedges and nodes. Hyperedges of the hypergraph are labelled by names of the corresponding components or relations. Hyperedges representing components are non-directed. Hyperedges corresponding to asymmetric relations are directed. Hypergraph nodes express potential connections between hyperedges. To each hyperedge a sequence of source and target nodes is assigned. The number of all nodes assigned to a hyperedge specifies its type. As a hypergraph can be seen as a hyperedge on the higher level of detail, the external nodes which determine the type of a hypergraph are also specified.

To represent features of components and relations between them, attributing of hyperedges and nodes is used. Values assigned to attributes specify the design requirements which should be met by solutions.

In Fig. 2 a hypergraph representing a layout of a three-storey hotel is depicted. Hyperedges representing components are drawn as rectangles and labelled with names of elements they represent. The layout consists of 9 components, a reception, dining area, kitchen, lobby, entrance, hall and three floors. Relational hyperedges depicted as circles labelled A denote the accessibility between areas.

Let $[i]$ denote the interval $\{1\ldots\}$ for $i \geq 0$ (with $[0] = \emptyset$). Let $\Sigma_E = \Sigma_C \cup \Sigma_R$, where $\Sigma_C \cap \Sigma_R = \emptyset$, and Σ_V be fixed alphabets of hyperedge and node labels, respectively. Let A_V and A_E denote sets of node and hyperedge attributes, respectively.

Definition 1. *An attributed hypergraph over $\Sigma = \Sigma_E \cup \Sigma_V$ is a system $G = (E_G, V_G, s_G, t_G, lb_G, ext_G, att_V, att_E)$, where:*

1. $E_G = E_C \cup E_R$, where $E_C \cap E_R = \emptyset$, is a finite set of hyperedges, where elements of E_C represent object parts, while elements of E_R represent relations,
2. V_G is a finite set of nodes,
3. $s_G : E_G \to V_G^*$ and $t_G : E_G \to V_G^*$ are two mappings assigning to hyperedges sequences of source and target nodes, respectively, in such a way that $\bigvee e \in E_C \; s_G(e) = t_G(e)$,
4. $lb_G = (lb_V, lb_E)$, where:
   - $lb_V : V_G \to \Sigma_V$ is a node labelling function,

- $lb_E : E_G \to \Sigma_E$ is a hyperedge labelling function, such that $\bigvee e \in E_C$ $lb_E(e) \in \Sigma_C$ and $\bigvee e \in E_R$ $lb_E(e) \in \Sigma_R$,
5. $ext_G : [n] \to V_G$ is a mapping specifying a sequence of hypergraph external nodes,
6. $att_V : V_G \to P(A_V)$ and $att_E : E_G \to P(A_E)$ are two functions assigning sets of attributes to nodes and hyperedges, respectively.

Hypergraphs representing structures of design solutions are generated by hypergraph grammars. Let us denote by L_H a set of attributed hypergraphs and define $\pi : L_H \to \{TRUE, FALSE\}$ as a design predicate.

A hypergraph grammar is composed of a set of hypergraph edges with terminal and non-terminal labels, a set of hypergraph nodes, a set of productions and an axiom being its initial hypergraph.

Each grammar production is of the form $p = (l, r, \pi)$, where

- l contains only one attributed hyperedge,
- r is an attributed hypergraph with the same number of external nodes as the number of external nodes of l,
- π is a design predicate determining the production applicability.

Three productions of an example hypergraph grammar generating layouts of a hotel are presented in Fig. 3. The set of attributes assigned to the left side of the first production contains, among others, the area, a style and the number of floors. This production can be applied when the value of attribute area of the hyperedge labelled by H is greater than c_1, where $c_1 \in C$ and C is a set of constants for the design problem. The design predicate for p_3 is true when the number of floors defined by the appropriate attribute of the left side of the production is larger than 1.

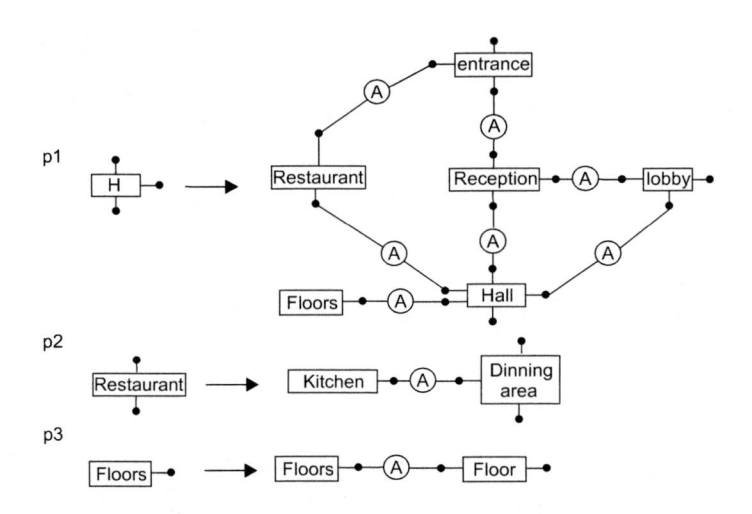

Fig. 3. Selected productions of a hypergraph grammar generating layouts of a hotel

The application of the production p to a hypergraph H consists in substituting r for a hyperedge isomorphic with l and replacing external nodes of the hyperedge being removed by the corresponding external nodes of r. To apply a given production an agent checks conditions imposed on values of hypergraph attributes expressed by π. In this way an agent can select the grammar productions leading to the design having required properties or style. During the application of the production the values of the attributes of r are established on the basis of the values of attributes of the hypergraph replaced by l.

Application of the productions presented in Fig.3 leads to the hypergraph shown in Fig.2. After the application of the production p_1 the value of the attribute *number of floors* is transferred from the left hand side of the production to the same attribute assigned to the hyperedge labelled Floors on the right hand side. During the application of the production p_3 the value of the attribute *number of floors* is decreased by 1 while being transferred from the hyperedge of the left side of the production to the hyperedge labelled Floors on the right side.

4 Hypergraph Grammar System with Agents

This section presents a distributed design system consisting of several hypergraph grammars defining the behaviour of cooperating agents. The communication between system grammars and the agents is realized by a special set of variables.

Let $Y = \{y_1 \ldots y_n\}$ be a set of variables, where each y_i invokes one agent and takes the value of an attributed hypergraph being an axiom of a hypergraph grammar from which the agent using this grammar starts generation of a subtask solution. It should be noted that the number of agents defined in a system can be greater than the number of grammars, as agents, which solve similar subproblems, can work according to the same system grammar. Agents which use the same system grammar are called agents of the same type.

In our functional-structural approach at the first step of conceptual design the designer specifies functional requirements and constraints in the form of a functional graph. On the basis of this graph the initial design structures in the form of attributed hypergraphs containing component hyperedges labelled only by elements of Y are created. These elements specify the grammars which will be used by agents working simultaneously at the beginning of the design process. Converting a functional graph into the initial structures is performed with the use of a functional structures editor implemented in Java [1, 6].

Definition 2. *A hypergraph grammar system with agents is a tuple* $HGSA = (Y, S, G_1, \ldots, G_n, A_1, \ldots A_m)$, *where*

1. *Y is a set of variables and $\#Y = n$, where n is the number of hypergraph grammars in HGSA,*
2. *S is a finite set of initial design structures in the form of attributed hypergraphs composed of nodes and hyperedges labelled only by elements of Y,*
3. *$G_1, \ldots, G_n$ are hypergraph grammars,*
4. *$A_1, \ldots, A_m$, where $m \geq n$, are system agents.*

Hypergraphs, being design structures, contain hyperedges labelled by elements of Y. In the design process each $y_i \in Y$ takes the value of an axiom of a hypergraph grammar G_i and activates agent A_j. Each new variable of Y which occurs in a hypergraph generated by agent A_j causes one other agent to be activated. Thus, agents not only simultaneously develop different parts of a derived hypergraph but also delegate the derivation of some parts to other agents.

Definition 3. *Let $h', h'' \in L_H$. A derivation process in a hypergraph grammar system with agents $HGSA = (Y, S, G_1, \ldots, G_n, A_1, \ldots A_m)$ is composed of direct derivations of the two following forms:*

1. *h'' is directly derivable from $h' (h' \Rightarrow_1 h'')$ iff*
   - *there exists a hyperedge h in h' labelled by $y_i \in Y$ and isomorphic with the axiom x_i of G_i and*
   - *$\pi(h') = TRUE$, and*
   - *h'' is isomorphic with the result of replacing h in h' by g, which is an attributed hypergraph generated by agent A_j starting from x_i, and substituting external nodes of h by corresponding external nodes of g.*
2. *h'' is directly derivable from $h' (h' \Rightarrow_2 h'')$ iff*
   - *there exists a production $p = (l, r, \pi)$ of G_i and*
   - *$\pi(h') = TRUE$, and*
   - *h is a hyperedge of h' isomorphic with l and h'' is isomorphic with the result of replacing h in h' by r and substituting external nodes of h by the corresponding external nodes of r.*

The generation of a hypergraph representing a solution starts from an initial design structure and is continued by successively activated agents. The derivation process takes place in parallel. At first all hyperedges of the chosen initial design structure $h_0 \in S$ labelled by variables of Y activate agents which start generation of subhypergraphs from axioms of the corresponding grammars. Then, in each time unit each agent either performs a rewriting step using a rule of its grammar or activates agents for all hyperedges of the derived subgraph labelled by elements of Y. The invoked agents start their generations from the axioms of the corresponding grammars. When a hypergraph labelled only by terminal labels is generated an agent replaces it for the hyperedge labelled by the variable which activated this agent.

The language generated by a hypergraph grammar system with agents is a set of hypergraphs, where all hyperedges have terminal labels, and generated by different agents starting from one of the initial design structures.

Definition 4. *Let $HGSA = (Y, S, G_1, \ldots, G_n, A_1, \ldots A_m)$ be a hypergraph grammar system with agents. The language generated by $HGSA$ is a set $L(HGSA) = \{h \in L_H | h_0 \Rightarrow_1^+ h_i \{\Rightarrow_1, \Rightarrow_2\}^* h\}$, where $h_0 \in S$ and all hyperedges of h have terminal labels.*

Our hypergraph grammar system used to design hotel grounds is composed of the following grammars: H generating different layouts of hotel interiors,

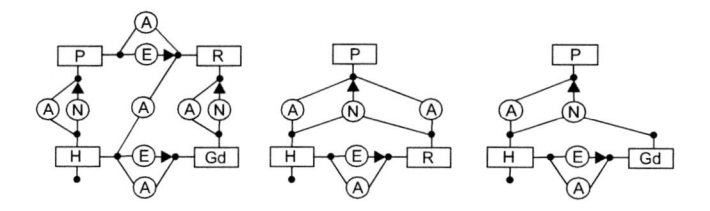

Fig. 4. Three examples of initial design structures

Floor generating hotel floor-layouts, *Gd* generating different garden arrangements, *R* generating arrangements of recreation areas and *P* generating arrangements of a parking space. Therefore the set of class variables Y contains labels $H, Floor, Gd, R, P$.

Selected three initial design structures of the set S are presented in Fig. 4. The first one corresponds to a hotel with a parking space on the north side (denoted by a directed hyperedge labelled N) accessible from a hotel (denoted by an undirected hyperedge labelled A, a garden on the east side (hyperedge label E) accessible from a hotel, and a recreation area located on the east of a parking space, on the north of a garden and accessible from a hotel, garden and parking space. The second and third initial design structures correspond to the hotel grounds without a garden and without a recreation area, respectively.

After choosing one of the initial design structures the derivation process starts. The labels H, Gd and R in the hypergraphs of S denote that three agents will be activated, each one using a hypergraph grammar of the corresponding type. Selected productions of the hypergraph grammar H generating hotel layouts are shown in Fig. 3. When an agent using grammar H obtains a hypergraph shown in Fig. 2 it will activate three new agents generating floor-layouts using the grammar *Floor*, as indicated by three nodes labelled *Floor* in this hypergraph.

The elements of the language generated by the proposed system of hypergraph grammars combined with agents are interpreted in the form of designs and evaluated by the designer. On the basis of this evaluation the grammar system can

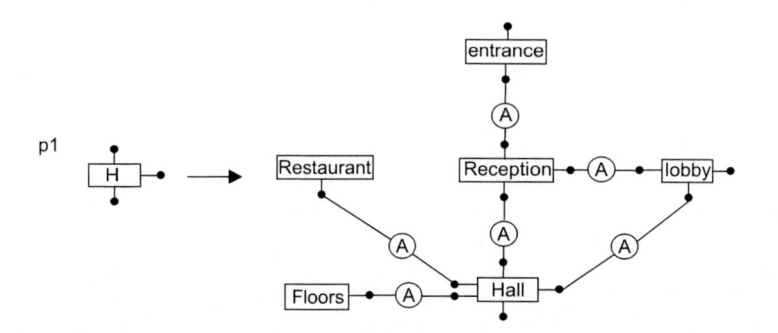

Fig. 5. A modified rule of a hypergraph grammar generating hotel layouts

be modified. For example, if the designer is not satisfied with a hotel layout (Fig. 2), where the restaurant is directly accessible from the main entrance of the hotel, the grammar H will be modified. Its first rule after the modification is shown in Fig. 5.

5 Conclusions

This paper presents a new approach to a distributed design system with the use of a system of hypergraph grammars combined with agents. Actions of each agent are defined by one grammar of the system. Cooperating agents simultaneously perform different design subtasks. Communication between agents is realized by a set of variables assigned to the grammars of the system. On the basis of the evaluation of the solutions generated by agents the grammars of the system can be modified. As a result the agents can adapt to changing requirements of the design task by using new grammar rules.

In the classical approach to design four types of design actions are distinguished, namely perceptual, physical, functional and conceptual. In the presented approach our agents are responsible only for selecting productions to be applied and using them to modify hypergraph structures. Thus only physical and functional design actions are simulated. In future research the remaining types of design actions will be simulated by agents. Such a simulation can provide a designer with a visual transformation of design specifications into a final design description. The agent's perceptual actions will be supported by hypergraph mining methods.

Moreover we are going to study the influence of the styles of particular design objects generated by appropriate hypergraph grammars on the impression of the whole design solution.

References

1. Borkowski, A., Grabska, E., Hliniak, G.: Function-structure computer aided. Machine Graphics and Vision 8(3), 367–381 (1999)
2. Csuhaj-Varju, E., Dassow, J., Kelemen, J., Paun, G.: Grammar systems. A grammatical approach to distribution and cooperation. Topics in Computer Mathematics, vol. 8. Gordon and Breach Science Publishers, Yverdon (1994)
3. Grabska, E., Palacz, W.: Hierarchical graphs in creative design. MG&V 9(1/2), 115–123 (2000)
4. Grabska, E., Strug, B.: Applying cooperating distributed graph grammars in computer aided design. In: Wyrzykowski, R., Dongarra, J., Meyer, N., Waśniewski, J. (eds.) PPAM 2005. LNCS, vol. 3911, pp. 567–574. Springer, Heidelberg (2006)
5. Grabska, E., Strug, B., Ślusarczyk, G.: A Graph Grammar Based Model for Distributed Design. In: International Conference on Artificial Intelligence and Soft Computing, ICAISC 2006, pp. 440–445 (2006)
6. Grabska, E., Ślusarczyk, G., Papiernik, K.: Interpretation of objects represented by hierarchical graphs. In: Proceedings International Conference Computer Recognition Systems, KOSYR 2003, pp. 287–293 (2003)

7. Grabska, E., Grzesiak-Kopeć, K., Lembas, J., Łachwa, A., Ślusarczyk, G.: Hypergraphs in Diagrammatic Design. In: Computational Imaging and Vision, pp. 111–117. Springer, Heidelberg (2006)
8. Kelemen, J.: Syntactical models of cooperating/distributed problem solving. Journal of Experimental and Theoretical AI 3(1), 1–10 (1991)
9. Minas, M.: Concepts and Realization of a Diagram Editor Generator Based on Hypergraph Transformation. Science of Computer Programming 44, 157–180 (2002)
10. Provan, G.: Multi-Level Modeling and Distributed Agent-Based Inference: the Role of System Structure. In: 11th IEEE International Conference on Engineering of Complex Computer Systems (ICECCS 2006), pp. 217–226 (2006)
11. Rozenberg, G.: Handbook of Graph Grammars and Computing by Graph. Transformations, Applications, Languages and Tools, vol. 2. World Scientific, London (1999)
12. Simeoni, M., Staniszkis, M.: Cooperating graph grammar systems. In: Paun, G., Salomaa, A. (eds.) Grammatical models of multi-agent systems, pp. 193–217. Gordon and Breach, Amsterdam (1999)
13. Ślusarczyk, G.: A grammar-based multiagent system in dynamic design. Artif. Intell. Eng. Des. Anal. Manuf. 22(2), 129–145 (2008)
14. Wooldridge, M.J.: Intelligent agents. In: Weiss, G. (ed.) Multiagent Systems: A Modern Approach to Distributed Artificial Intelligence, pp. 27–77. MIT Press, Cambridge (1999)

A Realistic Approach to Solve the Nash Welfare

A. Nongaillard[1,2], P. Mathieu[1], and B. Jaumard[3]

[1] LIFL, University of Lille, Villeneuve d'Ascq, France
   forename.name@lifl.fr
[2] CSE
[3] CIISE, Concordia University, Montreal, Canada
   bjaumard@ciise.concordia.ca

Abstract. The multi-agent resource allocation problem is the negotiation of a set of resources among a population of agents, in order to maximize a social welfare function. The purpose of this study is the definition of the agent behavior which leads, if possible, to an optimal resource allocation at the end of the negotiation process as an emergent phenomenon. This process can be based on any kind of contact networks. Our study focuses on a specific notion: the Nash product, which has not the drawbacks of the other widely used notions. However, centralized approaches cannot handle large instances, since the social function is not linear. After a study of different bilateral transaction types, we underline the most efficient negotiation policy in order to solve the multi-agent resource allocation problem with the Nash product and provide an adaptive, scalable and anytime algorithm.

1 Introduction

The multi-agent resource allocation problem has been studied for a long time, either within a centralized framework or a distributed one. In studies with a centralized approach, agents report their preferences on the resources to a specific agent, e.g., an auctioneer, who then determines the final resource allocation. Within this context, authors [17] have suggested different transaction models for given types of auctions. At the opposite, in studies based on a distributed framework, an initial resource allocation evolves by means of local negotiations among agents. The convergence of such a negotiation process can be viewed as an emergent phenomenon, due to local negotiations among agents. The advantages are adaptability and dynamicity of the system, while keeping privacy for all users. However, assumptions have been implicitly made. Indeed, an agent is able to communicate and negotiate with all other agents. This assumption is not always plausible as soon as real world applications are considered.

The evaluation of a resource allocation is usually made by means of notions from the social welfare theory [2]. The most widely used notions such as the utilitarian welfare or the egalitarian welfare may have some undesirable effects on the resource allocation which is obtained. In this work, we choose to focus on the Nash product, which has not these drawbacks (see Section 2.3). Considering this welfare, an approach based on a centralized framework is not efficient. The determination of the optimal resource allocation is a complex and time-consuming problem which can only be solved in a very large amout of time.

In this study, our purpose is the definition of the best agent behavior, in order to ensure the convergence of the negotiation process towards a socially optimal resource allocation, or when the need arises, towards a socially close allocation. We provide a scalable and anytime algorithm which can be based on any kind of contact network, when the Nash product is considered as social welfare measure. The solution which is provided in this paper is adaptive, new agents can be added with new resources during the negotiation process, while with a centralized approach, such a thing is not possible without restarting the whole solving process with the new data.

After a presentation of related studies in Section 1.1, we define basic notions in Section 2. Section 3 presents a centralized approach, and Section 4 presents our distributed approach. Section 5 presents the experiment protocol, the evaluation criteria, and an analysis of the results.

1.1 Related Works

Lots of studies focus on mathematical properties of the multi-agent resource allocation problem. In [16], the author handles the properties of the allowed transactions and establishes a classification of the basic transactions along with theorems on the existence or the non-existence of a transaction sequence leading from any initial resource allocation to an optimal one. However, no process are provided to reach an optimal resource allocation. Along the same lines, mathematical properties on some classes of utility functions and payment functions are studied in [7] in order to design negotiation processes, which terminate after a finite number of iterations. [8] presents the impact of the acceptability criterion, the utility function and the transaction properties on the society welfare, without regard for the agent behavior which leads the negotiation process to such a socially optimal resource allocation. In other studies, authors define criterion which favors equitable deals [9] and others study the envy-freeness in the resource allocation process [4, 6]. In [14, 15], agent behaviors are studied, but only in the case of the utilitarian welfare, for which an obvious centralized solution exists. The notion of neighborhood is seldom considered, whereas it is one of the most important points for real world applications.

2 Multi-agent Resource Allocation Problem

2.1 Definitions and Notations

The multi-agent resource allocation problem is based on a population $\mathscr{P} = \{a_1,\ldots,a_n\}$ of agents, and on a set $\mathscr{R} = \{r_1,\ldots,r_m\}$ of available resources, which are assumed indivisible and static.

This set of resources $\mathscr{R}$ is initially distributed over the population of agents $\mathscr{P}$. Each agent a owns a bundle of resources, $\mathscr{R}_a$. A resource allocation A is a partitioning of the resources in $\mathscr{R}$ among the agents of $\mathscr{P}$, $A = \{\mathscr{R}_1,\ldots,\mathscr{R}_n\}$. $\mathscr{A}$ is the set of all the possible allocations. The preferences of the agents are expressed by means of 1-additive utility function [5, 12]. $u_a : \mathscr{R}^{m_a} \to \mathbb{R}$ and $u'_a : \mathscr{R} \to \mathbb{R}$ with the following relationship: $u_a(\mathscr{R}_a) = \sum_{r \in \mathscr{R}_a} u'_a(r)$. Even if their mathematical definitions are different, since they are used in the same purpose, u_a will be used equally in order to simplify the notations.

2.2 Contact Network

The relationships among the agents can be represented by means of a graph: the contact network. A link between two agents means that they are able to communicate between them. Most of the studies rely on the hypothesis of a complete contact network. Any agent is able to negotiate with all other agent in the population. Such a hypothesis has a strong impact on the negotiation process, and it is not realistic as soon as real world applications are considered. For instance, in the case of social networks, a person only knows a subset of the overall set of actors in the network. Thus, the neighborhood of an agent a, denoted by $\mathcal{N}_a$, define the set of agents with who he is able to talk. In this study, we consider that the contact network can be any connected graph, ranging from complete graphs to small-world graph [1], including structured graphs like rings, trees, or grids.

2.3 Social Welfare

In order to evaluate a resource allocation, notions which come from the social welfare theory are considered [2, 13]. Several notions exist, and each has advantages and drawbacks.

Definition 1. *The **utilitarian welfare** of a resource allocation A, denoted by $sw_u(A)$, corresponds to the summation of the agent utilities:* $sw_u(A) = \sum_{a \in \mathcal{P}} u_a(\mathcal{R}_a)$.

Definition 2. *The **egalitarian welfare** of a resource allocation A, denoted by $sw_e(A)$, corresponds to the utility of the poorest agent:* $sw_e(A) = \min_{a \in \mathcal{P}} u_a(\mathcal{R}_a)$.

Definition 3. *The **Nash product** of a resource allocation A, denoted by $sw_N(A)$, corresponds to the product of the agent utilities:* $sw_N(A) = \prod_{a \in \mathcal{P}} u_a(\mathcal{R}_a)$.

In order to illustrate the difference among these notions, let us consider a population of 3 agents, $\mathcal{P} = \{a_1, a_2, a_3\}$, and a set of 6 available resources, $\mathcal{R} = \{r_1, r_2, r_3, r_4, r_5, r_6\}$. Their preferences are expressed by means of a utility function, as described in Table 1. Optimal social values are gathered in Table 2 with a corresponding resource allocation. The utilitarian welfare considers the welfare of the whole agent community, without concern about the individual welfare, and then can lead to resource allocations where one agent, a_3, does not get any resource. Some agents can be neglected, especially if, for each resource, there exists another agent who estimates more this resource.

Table 1. Agent preferences

Agents	r_1	r_2	r_3	r_4	r_5	r_6
a_1	10	7	10	9	2	1
a_2	6	10	3	4	8	6
a_3	1	2	1	2	1	3

Table 2. Optimal social values

Social welfare	Value	Resource allocation
sw_u	53	[$\{r_1, r_3, r_4\}\{r_2, r_5, r_6\}\{\}$]
sw_e	6	[$\{r_1\}\{r_5\}\{r_2, r_3, r_4, r_6\}$]
sw_N	1800	[$\{r_1, r_3\}\{r_2, r_5\}\{r_4, r_6\}$]

At the opposite, the egalitarian welfare considers only the individual welfare, and then leads to resource allocations where every agent owns at least one resource. No agent is neglected, but an agent with low preferences, like a_3, drains the resources, and the resulting allocation may be very unbalanced. In between, the Nash product is a compromise which leads to more balanced resource allocations, avoiding such a draining phenomenon, and where no agent is neglected. This notion can only be used when utility values are positive. Moreover, small variations in an allocation lead to very large variations of the welfare: for instance, a simple exchange of r_1 with r_6 leads to a decrease of the social value from 1800 to 594.

3 A Centralized Approach

Of course, this resource allocation process can be solved using a centralized framework. The optimal social value, and a corresponding resource allocation, can be determined by means of the following mathematical model. The boolean variables x_{ra} represent the ownership of a resource r by an agent a, with $r \in \mathcal{R}, a \in \mathcal{P}$. Then, the optimal value of the Nash product can be found by solving this equation system:

$$sw_N^\star = \begin{cases} \max \prod_{a \in \mathcal{P}} \sum_{r \in \mathcal{R}} u_a(r) x_{ra} \\ \text{subject to:} \sum_{a \in \mathcal{P}} x_{ra} = 1 \quad r \in \mathcal{R} \\ \phantom{\text{subject to:}} x_{ra} \in \{0,1\} \quad r \in \mathcal{R}, a \in \mathcal{P}. \end{cases}$$

Such a system cannot be handled in a classic way since the objective function is not linear. However, an estimation could be made with the following method. First, a *Lagrangian* relaxation is used [10]. This method can solve non linear equation system if the objective function is convex. However, it is not the case when the Nash product is considered, and a multi-start algorithm has to be combined with the relaxation using a sampling small enough for initial solutions. Starting from multiple initial solutions may avoid local optima when non convex function are considered [11].

Moreover, since resources are not divisible, an integer solution has still to be found. Indeed, the relaxation changes the discrete value set $\{0,1\}$ into a continuous value set $[0,1]$. In order to obtain an integer solution, a *branch-and-bound* algorithm is then used, guided by the values that have been provided by the relaxed solution.

Such a method cannot certify the optimality of the found solution. This centralized approach is not really scalable, consequently of the nonlinearity of the objective, and of the exponential solution space. A resource allocation problem with n agents and m resources leads to a solution space of size n^m ($n \ll m$). Moreover, the implicit assumption of a complete contact network has been made with such a model. Large sets of constraints have to be added to this model in order to prohibit exchanges between agents who are not related.

Since such a method is not scalable, we developed two scalable heuristics in order to determine an estimation of the optimal social value. The first heuristic is focused on the resource value. The first step of this algorithm is to allocate each resource to the agent who estimates it the most. The second step is to be sure that all agents own at least one resource, otherwise, it looks for picking up the resource maximizing the social value to an agent who has at least two resources. The second heuristic is focused on the resource distribution uniformity, by allocating successively the best remaining resource to each agent.

4 A Distributed Approach

Our proposition is then to use a distributed approach. The purpose of such an approach is to define the agent behavior which leads to a socially optimal resource allocation, as an emergent phenomenon at the end of the negotiation process. At the opposite of centralized approach, our distributed approach can be based on any kind of communication network, as discussed in Section 2.2. The question is: "which behavior must we give to the agent in order to obtain a good Nash welfare as emergent phenomenon ?"

4.1 Acceptability Criteria

Such criteria have a strong impact on negotiation processes. Indeed, if an agent can accept any kind of deal, then the negotiation process will not be able to stop. Even if the resource allocation process reaches an optimal state, the agents will continue to negotiate among them and leave the optimum. Moreover, there is not guarantee that such a process reaches one time a suitable resource allocation. The acceptability criteria help the agent to determine whether a transaction is profitable or not. An agent has to based his decision on an acceptability criteria, with respect to the agent behavior. Such criteria restrict a lot the set of possible transactions among the agents. A negotiation process ends when no agent in the population is able to find an acceptable deal.

Let two agents, a and a', illustrate the considered criteria. The agent a initiates a transaction $\delta(A, A')$ with an agent a': the initial resource allocation $A = [\ldots, \mathscr{R}_i, \ldots, \mathscr{R}_j, \ldots]$ evolves towards a new one A'.

Definition 4. *A* ***rational agent*** *is an agent who only accepts transactions that increase his utility. If the agent a is rational, he accepts a transaction only if:* $u_a(\mathscr{R}'_a) > u_a(\mathscr{R}_a)$.

The rationality criterion is the most widely used in the literature, especially in the case of non cooperative and selfish agents.

Definition 5. *A* ***rational transaction*** *is a transaction in which all involved agents are rational. If a transaction is rational, involved agents accept it if:* $u_a(\mathscr{R}'_a) > u_a(\mathscr{R}_a)$ *and* $u_{a'}(\mathscr{R}'_{a'}) > u_{a'}(\mathscr{R}_{a'})$.

However, this criterion restricts a lot the set of possible transactions, and may lead the negotiation process to a sub-optimal resource allocation.

Another criterion that ensures the end of the negotiation process after a finite number of transactions is the sociality. This criterion is based on a local evaluation of the social welfare evolution.

Definition 6. *A* ***social agent*** *is an agent who can only accept transactions that increase the considered social welfare function of the multi-agent system.*

Definition 7. A *social transaction* is a transaction which causes an increase of the considered social welfare function. Such a transaction can only be accepted by the involved agents if: $sw(A') > sw(A)$, $A, A' \in \mathcal{A}$ such that $A \xrightarrow{\delta} A'$.

The determination of the social value associated to a resource allocation needs global information: Indeed, it is essential to have the value of the utility of each agent. However, it is possible to determine the variation of this social value with local information. From the agent point of view, the acceptability can be determine from information given by involved agents. It is then not necessary to determine its value.

$$sw_N(A') > sw_N(A) \iff u_a(\mathcal{R}'_a) * u_{a'}(\mathcal{R}'_{a'}) > u_a(\mathcal{R}_a) * u_{a'}(\mathcal{R}_{a'}).$$

where $\mathcal{R}_a$ and $\mathcal{R}'_a$ is the bundle of a before and after the deal. Since a finite number of agents are involved in a transaction, two in the case of bilateral transactions, only their resource bundle change. Then the utility of the agents that are not involved in this transaction can be considered as a constant value.

4.2 Transaction Kinds

Our study is restricted to bilateral transactions, i.e., transactions involving simultaneously two agents. Indeed, multilateral transactions are too much complex and time-consuming to be optimally determined, especially when the Nash product is considered. Moreover, our aim is to define the simplest agent behavior in order to favor the scalability of the algorithm. Three kinds of bilateral transactions can be distinguished. Others are combinations of these basic transactions. In each case, the transaction is initiated by a, in which is involved one of his neighbor a'. They own respectively m_a and $m_{a'}$ resources in their bundle.

First, the *gift* transaction. The initiator a can only give one resource to a'. Only m_a gifts are possible. The gift transaction cannot be rational for the initiator and is always rational for the agent participant (since utilities are positive).

Then, the *swap* transaction. Each agent provides a unique resource. This deal is symmetric: the number of resources per agent cannot vary. Hence, an optimal solution can be reached only if the initial allocation has the same resource distribution as one of the optimal allocation. $m_a \times m_{a'}$ swaps are possible.

Finally, the *cluster-swap* (CS). Each agent can involve a subset of their resources. At the opposite of the swap, it can be asymmetric. The cluster-swap contains the gift and the swap transactions. $2^{m_a+m_{a'}}$ are possible

In the experiments of Section 5, besides "pure" negotiation policies which use only one transaction kind, a "mixed" policy is defined: the swap+gift policy (S+G) in which the initiator tries first to find an acceptable swap, and a gift if the need arises. Agents use these policies according a specific behavior, which is defined in the next Section.

4.3 Agent Behavior

A negotiation can be managed in many different ways. Indeed, during a negotiation, if the participant rejects the offer, three alternatives arise: (i) the initiator gives up and stops the negotiation, (ii) he selects another neighbor, or (iii) he changes the offered

Algorithm 1. Behavior of the initiator a

Sorts his resource bundle $\mathcal{R}_a$;
forall $r \in \mathcal{R}_a$ **do**
　　forall $a' \in \mathcal{N}_a$ **do**
　　　　if δ_a *is acceptable* **then**
　　　　　　Performs the transaction δ_a ;
　　　　　　Ends the negotiation ;
　　　　end
　　end
end

resource. Based on these, various behaviors have been designed, implemented and evaluated. Nevertheless, we always assumed that each agent tries to give first his resource associated with the lowest utility. Only the behavior which leads to the best results is presented in the sequel.

This agent behavior is flexible and volatile, which means that the initiator can change either the selected neighbor or the offered resource. Such a behavior is "complete", meaning that according to kind of allowed transactions, if an acceptable transaction exists in the neighborhood, it will be identified. This completeness, which leads to greater results, has a cost. Costless behaviors can be designed, according to the application and its quality requirements.

5 Experiments

5.1 Experimental Setup and Evaluation

During the experiments, various contact networks have been generated, some complete and some Erdos-Renyi networks [3]. The mean connectivity degree of such networks is $\frac{n}{4}$, which means that an agent can talk at most to 25% of the population. The resources are initially distributed randomly. The preferences are also generated randomly with values in 1..100. During the negotiation processes, the speech turn is uniformly distributed over the agents. For the different population's sizes, the different allocation kinds, and the different networks, 100 instances are run each time.

The evaluation of negotiation processes is not an obvious issue. It is always possible to find a metric which makes a process the best. Various metrics can be considered like the number of performed deals, the number of exchanged resources, the number of speech turns or the number of attempted transactions. The relative standard deviation among the social value of emergent allocations are also considered: a large value means that the considered negotiation process is very sensitive to the initial resource allocation, and thus the quality of the emergent allocation quite varies. Finally, a comparison with the estimation that is obtained with centralized heuristics is made over 100000 instances. A ratio is computed to determine the gap between them: the distributed social value over the centralized social value.

Table 3. Computation time

Mean number of resources per agent	Number of agents 5	25	50
5	50ms	250ms	600ms
25	150ms	45s	4min
50	6s	5min	25min

Table 4. Number of performed deals

Mean number of resources per agent	Number of agents 5	25	50
5	35	350	900
25	150	1800	4600
50	400	4000	9000

5.2 Result Analysis

First, results related to complete networks. The social swap+gift policy is more robust than others, with a relative standard deviation of 3.03% among the social values for instances with 50 agents and 300 resources. This policy is less sensitive to the initial allocation. At the opposite, the swap policy is not reliable, because strongly sensitive to the initial allocation, with a social deviation of 114.09%. The swap policy is not enough flexible to avoid a local optimum. The swap+gift policy leads to socially greater allocation than than ones reached by other policies, but is more expensive in terms of attempted transactions(until 6 times). The cluster-swap policy is the most expensive in time and in attempted transactions, the social deviation is large enough to not compensate the additional costs, and thus is not interesting. Computation times, shown in Table 3, and the corresponding number of performed deals, shown in Table 4, are obtained for the swap+gift policy on complete contact networks. Even an instance involving 50 agents and 2500 resources remains scalable in a reasonable time.

Experiments on Erdos-Renyi networks bring about similar conclusions. The social swap+gift policy still obtains best results, with the greatest social value, coupled with the lowest deviation ($\simeq 17\%$). The value of this deviation depends strongly of the network topology. Greater is the mean connectivity, lower will be the relative standard deviation among the social values.

Finally, complex transactions such as the cluster-swaps, lead to a higher number of attempted transactions, a larger computation time, but a lower number of speech turns. At the opposite, simpler transactions such as gifts, lead to short negotiations in time, but more speech turn are required before the end of the processes. However, when the Nash product is considered, the gift policy or the swap policy are not enough flexible to leave local optima and then leads to weaker allocations. The swap+gift policy is a good compromise between scalability and complexity.

Since the swap+gift policy leads to the greatest social value, it has been compared to the value obtained by the centralized heuristics, defined in Section 3. First, the comparison to the heuristic which focused on the resource value. The distributed negotiation process leads to better results on 99.97% of the instances, with a social improvement of 140.86%, whereas when it leads to worst results, the gap is only of 1.13%. During our experiments, the second heuristic, which is focused on the resource distribution, never reaches a better allocation than our distributed negotiation process. The gap between the social value are huge, more than 10000%.

Thus, the social swap+gift policy is a flexible policy which leads to socially efficient allocations as well on complete networks as on Erdos-Renyi networks. Negotiation processes end in scalable time, for a small additional costs in terms of attempted transactions.

6 Conclusion

A centralized approach makes the solution of the resource allocation problem very complex and time-consuming, as soon as the Nash product is considered as the social welfare function. In this study, we have designed a negotiation process among agents which leads to the emergence of a suitable resource allocation, by means of local negotiations. This solving method is scalable, robust in terms of solution quality, and adaptable. At the opposite of a centralized approach, ours can be based on any type of contact network, and the addition of new agents (or new resources) is possible and does not need to restart of the negotiation process. Moreover, it is also an "anytime" algorithm: The quality of the solution increases as transactions go along, and the solving process can be interrupted anytime.

References

1. Albert, R., Barabási, A.: Statistical mechanics of complex networks. Reviews of Modern Physics 74(1), 47–97 (2002)
2. Arrow, K., Sen, A., Suzumura, K.: Handbook of Social Choice and Welfare. Elsevier, Amsterdam (2002)
3. Bollobás, B.: Random Graphs. Cambridge University Press, Cambridge (2001)
4. Bouveret, S., Lang, J.: Efficiency and envy-freeness in fair division of indivisible goods: logical representation and complexity. In: IJCAI 2005, pp. 935–940 (2005)
5. Chevaleyre, Y., Endriss, U., Estivie, S., Maudet, N.: Multiagent resource allocation with k-additive utility functions. In: Proc. DIMACS-LAMSADE Workshop on Computer Science and Decision Theory. Annales du LAMSADE, vol. 3, pp. 83–100 (2004)
6. Chevaleyre, Y., Endriss, U., Estivie, S., Maudet, N.: Reaching Envy-free States in Distributed Negotiation Settings. In: IJCAI 2007, pp. 1239–1244 (2007)
7. Chevaleyre, Y., Endriss, U., Lang, J., Maudet, N.: Negotiating over Small Bundles of Resources. In: AAMAS 2005, pp. 296–302 (2005)
8. Endriss, U., Maudet, N., Sadri, F., Toni, F.: Negotiating socially optimal allocations of resources. Journal of Artificial Intelligence Research 25, 315–348 (2006)
9. Estivie, S., Chevaleyre, Y., Endriss, U., Maudet, N.: How Equitable is Rational Negotiation? In: AAMAS 2006, pp. 866–873 (2006)
10. Fisher, M.: The Lagrangian Relaxation Method for Solving Integer Programming Problems. Management Science 50(13), 1861–1871 (2004)
11. Hickernell, F., Yuan, Y.: A Simple Multistart Algorithm for Global Optimization. OR Transactions 1(2) (1997)
12. Miranda, P., Grabisch, M., Gil, P.: Axiomatic structure of k-additive capacities. Mathematical Social Sciences 49, 153–178 (2005)
13. Moulin, H.: Choosing from a Tournament. Social Choice and Welfare 3(4), 271–291 (1986)
14. Nongaillard, A., Mathieu, P., Jaumard, B.: A Multi-Agent Resource Negotiation for the Utilitarian Welfare. In: ESAW 2008 (2008)
15. Nongaillard, A., Mathieu, P., Jaumard, B.: La négociation du bien-être social utilitaire. In: JFSMA 2008, EU, France, Brest, pp. 55–64 (2008)
16. Sandholm, T.: Contract Types for Satisficing Task Allocation: I Theoretical Results. In: AAAI Spring Symposium: Satisficing Models, vol. 99, pp. 68–75 (1998)
17. Sandholm, T.: Algorithm for optimal winner determination in combinatorial auctions. Artificial Intelligence 135(1-2), 1–54 (2002)

A Study of Bio-inspired Communication Scheme in Swarm Robotics

P.N. Stamatis[1], I.D. Zaharakis[1], and A.D. Kameas[1,2]

[1] Computer Technology Institute, Designing Ambient Information Systems (DAISy) group,
  N. Kazantzaki str, University Campus, 26500 Patras, Hellas
[2] Hellenic Open University, School of Sciences and Technology,
  23 Sahtouri str, 26222 Patras, Hellas
  {stamatis,jzaharak,kameas}@cti.gr

Abstract. In this study two bio-inspired communication schemes (stigmergy and hormone based communication) are evaluated according to the contribution of each to the self organisation of a multi agent robotic system. The evaluation is carried on by the simulation of a test case scenario and it is based on measurable criteria that only depend on triggered agent actions. The results present the different situations where each of the communication schemes performs better in terms of task accomplishment.

1 Introduction

In current Artificial Intelligence (AI) research, there is an increased interest on how to engineer autonomous entities (hereafter called agents) with limited capabilities in both peripherals (sensors, actuators) and computational resources (processors, memory, communication, etc), yet simultaneously, exhibiting robustness and behavioural agility. Swarm Intelligence field contributes to this effort by focusing on the emergent collective intelligence of (unsophisticated) agents which interact locally with their environment [2, 9]. These unsophisticated agents are referred in the literature as simple reflex agents [12] or purely reactive agents [16], thought the latter are more hardware oriented.

This study concerns the application of a multi agent robotic system approach into a real environment, attributed by inaccessibility, dynamicity, non-determinism and continuity [12]. In addition to the challenges of the environment, robotic individuals may have limited sensing, processing and actuation power due to known technological issues like energy storage cost and miniaturization constraints. Taking into account these limitations, our engineering perspective is based on swarm intelligence and emergent collective intelligence in both computational and behavioural metaphors.

Our aim is to study and evaluate how an unsophisticated bio-inspired communication scheme contributes to purely reactive robotic agent societies and our methodology to achieve this aim is by simulating a real-life problem. The contribution will be evaluated through comparison of the numerical representation of the system's self-aggression property for identical test case scenarios, using no communication, stigmergic and hormone communication. The self-aggression measure is a commonly used measure of evaluation in swarm systems [13].

The structure of paper is as follows: in section 2, we discuss the background and motivation of this study. In section 3, we present the target domain of research and we focus on a particular test case scenario. Section 4 describes the engineering approach for the particular application including a description of the communication algorithm used by both schemes sections. Finally, sections 5 and 6 describe the simulation results analysis and conclusions respectively.

2 Background

Group formation and group gathering are very common examples of self organization forms. Studies in [1, 5, 6] focus on gathering distributed algorithms for autonomous mobile agents taking under consideration the limited visibility, lacking of memory and explicit communication capabilities for each individual. Proving the point convergence for agents initially placed in a compass, the distributed algorithm in [1] seems to be quite robust against sensor and control errors. In the sequel of this section we describe the role of the indirect communication in Swarm Intelligence applications.

The term "stigmergy" was coined by Grasse [7] to describe a form of indirect communication in societies of insects that is performed implicitly by altering the state of the environment in a way that will affect the behaviours of others for whom the environment is a stimulus. In Computer Science, "stigmergic behaviour" is commonly used to describe a broad class of agent collaboration mechanisms that rely on information exchange through a shared environment. Tasks like spatial sorting [8] path exploration [9], foraging [11] and task allocation [10] are common applications of multi agent systems using stigmergic-like communication. In [8], physical robots (U-Bots) where used in several collective robot experiments in order to examine the possible role of stigmergy and self-organization in the simple task of sorting two types of objects, concluding that stigmergy mediates self-organization.

Another bio-inspired solution to indirect communication is hormone-inspired communication, that is a communication scheme that is inspired by the way cells living in different places of the body communicate using hormones: chemical messengers originating from a cell and received from another cell inside an organism's physical space. Cells usually communicate by releasing chemical messengers targeted for cells that may not be immediately adjacent. Although some messengers travel only short distances, hormones are used by both animals and plants for signalling at greater distances (inside the living organism). In animals, hormones emitted by a cell can travel long distances through blood vessels and reach target cells [4]. This messenger based communication theory can be used in a communication analogy using digital signalling. The Digital hormone conceptualization is already used in the SI bibliography, such as the DHM model for the control of robot swarming in a distributed manner [15]. As a consequence, digital signals are modelled as digital hormones (in respect to chemical hormones) based the following assumptions:

- Hormones are intended to find target cells using the physiology of the body (e.g., blood vessels in animals). Digital hormones use the communication infrastructure (e.g., open or fixed network) to spread;
- Hormone communication is not based on hormone persistence to increase the possibility of reaching a target cell. Hormones not reaching the target cells are

absorbed by the organism. In an analogy, digital signals do not persist; signals received by nobody are ignored by the system.
- A single messenger is enough to trigger a cell but in any case, the messenger is attached to the target cell. Therefore, a number of hormone messengers are released so as to trigger numbers of reached target cells. In contradiction, a single digital hormone signal can trigger every target receiver according to the transmission range. The transmission range is presumed analogous to the number of messengers released so as to trigger target cells.
- Hormones do not carry information. Different hormone varieties are used to trigger different cell operation. Thus, digital signals data are assumed that do not contain any contextual information or symbolic meaning. Signal data can only be used as identifiers of different digital hormone variations.

3 Target Area and a Scenario of Application

Using a particular test case scenario, we present a simulation of a society of simple, homogeneous, resource-limited agents incapable of dealing with a fault as individuals. The simulation environment considered in this study is originally described as a case scenario of the SOCIAL project [14]. The particular environment is considered to be a secondary oil production facility tank, suffering the scale formation problem. This problem varies in severity depending on the composition of the make-up water used in water flooding. Scale deposits consisting either of Calcium carbonate or Calcium and Barium sulphate cause clogging in the pipes and damage to the pumping systems because of the formation of tenacious scale deposits.

Scale formation is distributed while the difficulty and the rough conditions of the environment (narrowness, high temperatures etc) make a multi agent approach typically well-fitted to solve this technical problem. Specifically, the test case scenario involves small-sized robotic agents placement and circulation inside one compartment of the fluidic environment continuously sensing for "fault traces". The "fault trace" is an environmental stimulus that signals the presence of scale formation, corresponding to an altered chemical/physical property e.g., pH/conductance. The trace is detectable using particular sensors mounted on the agents – pH sensors in our case – that record signal values depending on their spatial distribution with respect to the damage. Agents can deal with the fault (e.g., by releasing a suitable chemical substance) or – in the case where communication is used – by emitting a "fault detection" signal. When an agent perceives a "fault trace" or a "fault detection" signal through its relative sensors/antenna, it starts converging with other nearby agents that might have perceived the environmental variation. If the agent moves closer to the tube and this "fault trace" becomes intense enough, the agent starts trailing towards the tube. When eventually the agent reaches the tube, in an area considered as fault area, it releases a quantity of its load suitable for dealing with the fault.

4 Engineering Approach

In this study, the agent is modelled according to the subsumption architecture as a hierarchical structure organizing a particular set of primitive behaviours. A primitive behaviour may be thought as an individual building block which continually maps perceptual input (direct data from sensors) to an actuation output, according to [3]. Each agent is the bearer of a set of basic behaviours, yet it may exhibit a broad set of emergent or even composed behaviours. The subsumption architecture is found to be a well-fitting approach for the implementation of the mobile multi agent system due to very important advantages like reactiveness, simplicity, economy, computational tractability, robustness against failure and elegance [16].

Each agent bears seven different primitive behaviours (*Wander, Disperse, Aggregate, Move to target area, Avoid Kin, Avoid Obstacles* and *Repair*), hierarchically structured according to the subsumption scheme, as presented in Fig. 1. The more critical behaviours for the survival or the task accomplishment of the agent are placed higher in the scheme. The dotted lines represent the hierarchy between behaviours according to the communication mechanism, assuming that communication is used. In that case, not all but *Repair, Move to Target Area* and *Aggregate* behaviours can use the communication module in parallel to the other actuators. More specifically, there are two suppression control mechanisms that apply in parallel using the same hierarchical tree.

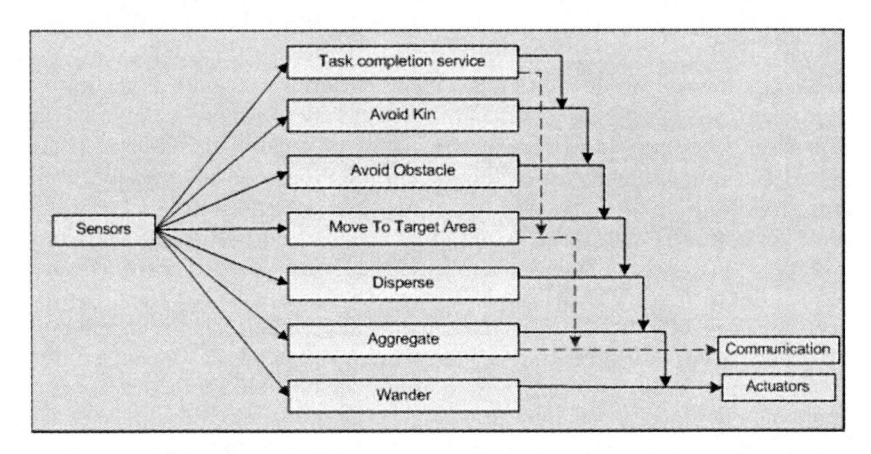

Fig. 1. The hierarchical schema of the seven basic behaviours. Although behaviours run in parallel, only the higher in importance (top) triggered behaviour will take the control of the actuators.

The communication algorithm used by both stigmergic and digital hormone schemes is described in this section. Both schemes use a simple "warm-or-cold" algorithm by comparing the most recent communication input value according to a previous communication input value. In the case of stigmergic communication, pheromone concentration is used as the "message" indicator: if the pheromone concentration increases as the agent moves, *Aggregate* maintains the direction of the agent. If not, then the behaviour changes the direction stochastically. In the case of digital communication, a predefined signal series can be used as "message" indicator: if the agent perceives the same or another "higher" message according to the predefined message sequence, maintains the direction of move. In different case, the direction changes randomly.

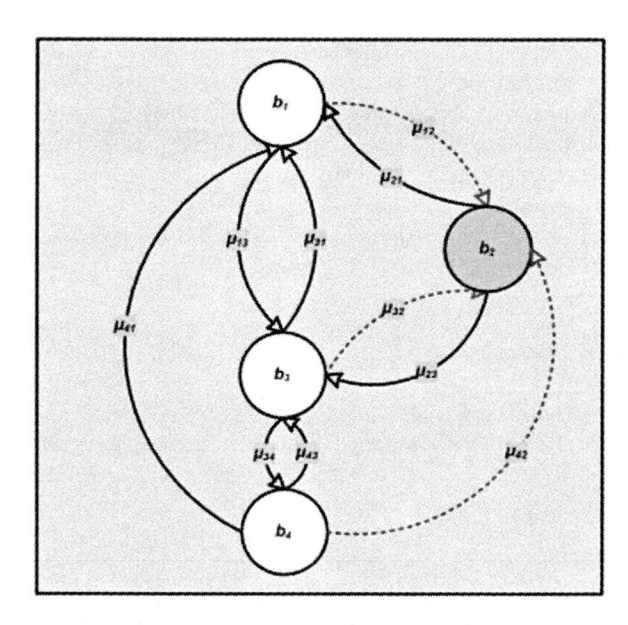

Fig. 2. Transitions between different composed behaviours in macroscopic view

During a time interval, a sequence of competitive behaviours can take the control of an individual's actuators and/or communication module; none the less, the pattern of individual basic behaviour triggering sequence can be categorized macroscopically by four states of exhibited composed behaviours. In detail:

a) *Searching* behaviour (b_1): this behaviour involves the *Wandering, Avoid Kin & Obstacle* and *Disperse* behaviours in random sequence. In a macroscopic view, agents in this state are found to circulate stochastically.
b) *Direct fault detection* (b_2): this composed behaviour involves the *Move to Target Area, Aggregate* and *Avoid Kin & Obstacle*; *Aggregate* is triggered directly by the target area stimulation tracing. Macroscopically, agents inside the area of the fault perturbation coil into more dense groups. These groups have the propensity to approach the surface in which the fault is located.
c) *Indirect fault detection* behaviour (b_3): involves *Aggregate* (triggered indirectly by receiving a communication signal), *Avoid Kin & Obstacle*. This state is observed with agents outside but near the fault stimulation area, tending to coil into groups
d) *Task completion* (b_4): this behaviour involves the *Repair* primitive behaviour. This behaviour is observed by dense formations of immobilized agents very close to the fault.

In a particular time interval, each agent can exhibit one of these four composed behaviours that maintain the competitive association of the basic behaviours. In Fig. 2, a graphical representation of the transition between composed behaviours is presented. The connection curves (μ_{xy}) represent the composed behaviour transition in a single individual. The dotted transitions are associated with the (external) communication between behaviours of different agents. These particular transitions are established only when a communication action is enabled. Transitions between behaviours occur a) when sensing input satisfies the threshold triggering conditions of a behaviour (μ_{13}, μ_{23}, μ_{34}), b) when lack of sensing input disable a behaviour (μ_{31}, μ_{43}, μ_{41} and μ_{21}) and c) when a behaviour of one or more individuals affects the behaviour of another agent ($\mu_{12} \mu_{32} \mu_{42}$). It is clear that communication (represented by the $\mu_{12} \mu_{32} \mu_{42}$ curves, directed to b_2) is intended to alter behaviour of the agent to *Indirect fault detection*.

This composed behaviour transition scheme is used for the evaluation of the simulation series discussed in the simulation section.

5 Simulation and Evaluation

For the purpose of the study, the SimAgent simulator was used. SimAgent is a java-based 3D simulator built to experiment with the effect of different basic behaviours, agent parameters of operation and sensing and finally different basic behaviour interaction architectures. The simulator considers a closed environment in which agents are placed, similar with the test case scenario described in section 30. Agents are represented with conic objects, each of operating independently. When fault areas occur (in predefined locations or by manual interference - mouse click on the pipe surface), the agents are meant to rally around this area and deal with the fault. A screenshot of the simulation is presented by Fig. 3.

Three series of simulations run using no communication, quantitative stigmergic communication (gradient trail construction using a single pheromone type) and digital hormone communication (using 3 different signal variations for the communication algorithm). Each series was done for 64 and 200 agents respectively. The small number of agents (64) is actually the least number of agents needed to cover (by cumulating each individual sensing range in an ideal agent distribution) the closed environment. The large number of agents (200) is the maximum limit of agents for the simulator correct operation. Several simulation parameters like basic behaviour triggering

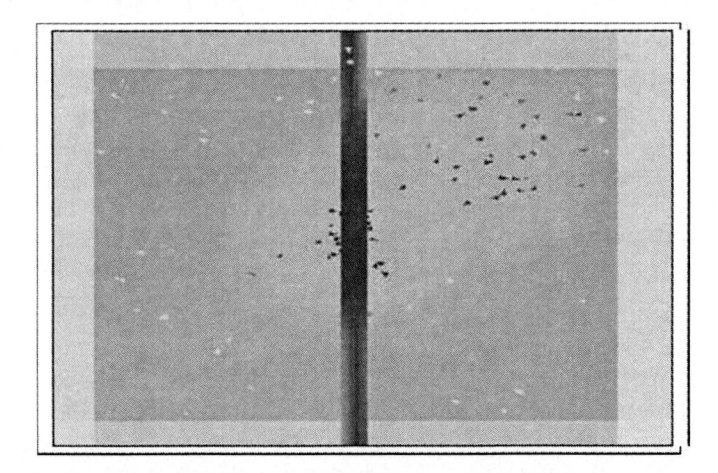

Fig. 3. A simulation screenshot, showing a group of agents to rally around an area of interest inter alia

thresholds and hierarchy were fixed for all simulation series. The purpose of the simulation is to experiment with the two different communication schemes and not other factors of the emerging system self organization. The major evaluation criterion is the overall agent *Direct fault detection* and *Task completion* emergence around the area of interest. For that purpose, each simulation series repeated for 20 times. Then, the results are averaged so as to present a median performance.

The combining results of using both communication schemes in comparison to the scheme without communication for 64, 120 and 200 agents are depicted by Figures 4, 5 and 6 respectively. In each plot two different trend lines of $f(\Delta\text{agents})$ are presented for each communication scheme. Δagents is the difference of the number of agents exhibiting b_2 or b_4 behaviour using communication (simulation using stigmergy or digital hormones respectively) minus the number of agents that exhibiting b_2 or b_4 without using communication at all. Concluding, it is found that in small numbers (64), not only digital communication results better than stigmergic communication but the latter delays remarkably to achieve positive profit, regarding the no communication approach.

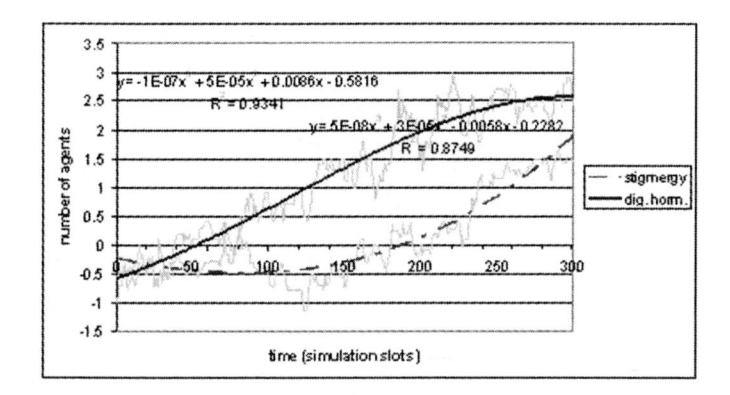

Fig. 4. Plot representing the relevant $f(\Delta\text{agents})$ over simulation time for stigmergic and digital hormone communication. The simulation series included a number of 64 homogenous agents.

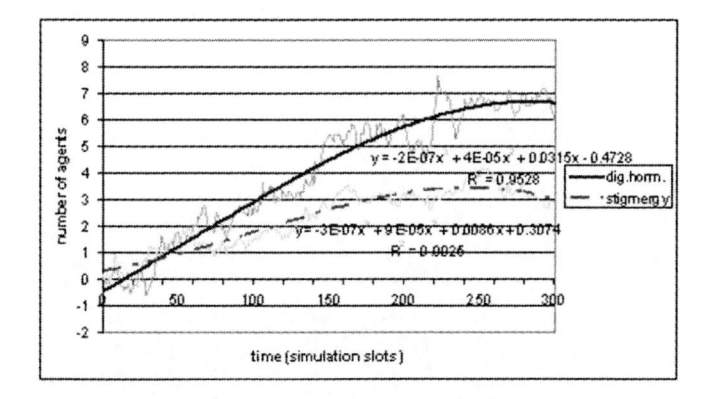

Fig. 5. Plot representing the relevant $f(\Delta\text{agents})$ over simulation time for stigmergic and digital hormone communication. The simulation series included a number of 120 agents.

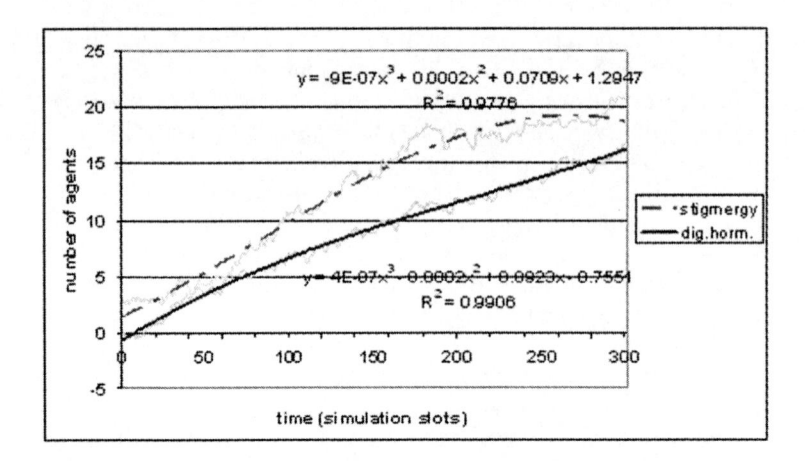

Fig. 6. Plot representing the relevant $f(\Delta \text{agents})$ over simulation time for stigmergic and digital hormone communication. The simulation series included a number of 200 agents.

According to the simulation results for using 120 agents, digital hormones present superior performance instead of stigmergic communication. On the other hand, the stigmergic trend line is positive in most of simulation time, presenting clear evidence of increasing performance with the increase of the agent group size. In large numbers (200) this situation is reverted.

Hormones were used not only to invoke the aggregation behaviour of each individual but to give implicit feedback to the communication algorithm. While, in respect of each pheromone area of effect, digital signals are spread out the extended area, it is the lack of persistence (in respect of the pheromones) that leads to poorer performance when large groups of agents are used. This problem can be engineered by using several techniques e.g., beacon so as to make a signal more persistent by means of continuing transmission by the transmitter; also, digital hormones are not fitted well for gradient fields (orientation as implicit information), especially when the agent number of the system is large. Also, mobility of individuals results in a dynamic and unstable gradient field around the area, regarding the relative position of receivers with the transmitters.

Digital hormone range is a feature that must be taken under consideration when triggering behaviours of distant individuals. Although a digital hormone can spread over a large area covering the sensing (and actuation) capabilities of the individuals, increasing the signal range will not necessarily improve the agent effectiveness. Even if the proper behaviour is triggered by a distant signal, myopic agents are unlikely to react effectively, because the information perceived in their local vicinity may not be enough or proper. For example, in Fig. 3 a group of agents are drifted away from the fault because of the "bad use" of aggregation behaviour. This situation occurs because the agents aggregate towards other individuals that are not aware of the fault and just passing by the fault aware agents. Thus, a trade-off between signal strength and agent sensing/actuation capabilities should be found. Nonetheless, digital hormones can be used for the gradient field generation only if physical pheromone communication is a non affordable or impractical solution.

6 Summary

This paper presents the outcome of an early study of applying digital pheromones so as to create gradient fields. The outcome is satisfactory, meaning that this communication scheme is found to fit (better than using no communication) to the particular problem domain. More work has to be done including exhaustive parameter experimentation and communication algorithm modifications in order to achieve a more fitted solution to the particular problem.

Refferences

[1] Ando, H., Oasa, Y., Suzuki, I., Yamashita, M.: A Distributed Memoryless Point Convergence Algorithm for Mobile Robots with Limited Visibility. IEEE Trans. on Robotics and Autom. 15(5), 818–828 (1999)
[2] Bonabeau, E., Dorigo, M., Theraulaz, G.: Swarm Intelligence: From Natural to Artificial Systems. Oxford University Press, Oxford (1999)
[3] Brooks, R.: Intelligence without Reason. In: Proceedings of 12th Int. Joint Conf. on Artificial Intelligence, Sydney, Australia, pp. 569–595 (1991)
[4] Campbell, N., Reece, J.: Biology, 7th edn. The Benjamin Cummings Publishing Company Inc. (2005)
[5] Flocchini, P., Prencipe, G., Santoro, N., Widmayer, P.: Gathering of asynchronous oblivious robots with limited visibility. In: Proc. 18th Annual Symposium on Theoretical Aspects of Computer Science (2001)
[6] Gordon, N., Wagner, I., Bruckstein, A.: Gathering multiple robotic a(ge)nts with limited sensing capabilities. In: Dorigo, M., Birattari, M., Blum, C., Gambardella, L.M., Mondada, F., Stützle, T. (eds.) ANTS 2004. LNCS, vol. 3172, pp. 142–153. Springer, Heidelberg (2004)
[7] Grasse, P.: La reconstruction du nid et les coordinations inter-individuelles chez bellicositermes natalensis et cubitermes. sp. la theorie de la stigmergie: essai d'interpretation du comportement des termites constructeurs. Insectes Soc. 61, 41–81 (1959)
[8] Holland, O., Melhuish, C.: Stigmergy, self-organization, and sorting in collective robotics. Artificial Life 5, 2 (1999)
[9] Kennedy, J., Eberhart, R.: Swarm Intelligence. Morgan Kaufmann Publishers, San Francisco (2001)
[10] Kube, C.R., Zhang, H.: Collective robotics: From social insects to robots. Adaptive Behavior 2(2), 189–219 (1993)
[11] Mataric, M.: Designing and Understanding Adaptive Group Behavior. Adaptive Behavior 4(1), 51–80 (1995)
[12] Russell, S., Norvig, P.: Artificial Intelligence: A Modern Approach, 2nd edn. Prentice Hall, Englewood Cliffs (2003)
[13] Rybski, P., Larson, A., Veeraraghavan, H., LaPoint, M., Gini, M.: Performance evaluation of a multi-robot search & retrieval system: Experiences with MinDART. Technical Report 03-011, Department of Computer Science and Engineering, University of Minnesota, MN, USA (2003)
[14] SOCIAL project, IST-2001-38911, http://www.socialspike.net/
[15] Shen, W.-M., Will, P., Galstyan, A., Chuong, C.-M.: Hormone-Inspired Self-Organization and Distributed Control of Robotic Swarms. Autonomous Robots 17(1), 93–105 (2004)
[16] Wooldridge, M.: Intelligent Agents. In: Weiss, G. (ed.) Multiagent Systems: A Modern Approach to Distributed Artificial Intelligence, pp. 27–77. MIT Press, Cambridge (1999)

Artificial Intelligence for Picking Up Recycling Bins: A Practical Application

Maria Luisa Pérez-Delgado and Juan C. Matos-Franco

Departamento Informática y Automática
Universidad de Salamanca
Escuela Politécnica Superior de Zamora, AV. Requejo, 33, 49022, Zamora, Spain
mlperez@usal.es, jcmatos@usal.es

Abstract. In this paper we address the problem of waste recycling in the province of Zamora. The recycling process requires the collection of the waste in an efficient way. This leads to the mathematical problem of vehicle routing, or some particular case of it, to define the optimal paths to collect the waste. This paper presents the application of artificial ants to plan the path to empty the recycling bins of a town called Toro, in the province of Zamora.

Keywords: Recycling, Mixed Rural Postman Problem, Artificial Ants.

1 Introduction

As defined by the World Bank, the goal of urban solid waste management is to collect, treat and dispose of solid waste generated by all urban population groups in an environmentally and socially satisfactory manner, using the most economical means available [1].

Increasingly, municipalities address urban environment issues related to solid waste management. Therefore, the management takes into consideration the collection, processing and recycling of waste.

Recycling means the conversion of a waste to form a new product. In order to make possible the recycling, waste separation is required. Separation can take place either before collection, on the waste generator side, or following the collection, in a centralized sorting plant.

Waste removal systems comprise specially designed collection containers and vehicles. For the collection of municipal solid waste, the system called "change the bin" is usually applied. This system employs standardized containers that are lifted up and emptied automatically into collection vehicles. The organization of waste collections allows performing a separate or combined collection of the different waste containers. Different waste containers can be collected in separate tours, or together in a single collection tour. It depends on the form and size of the containers, and how many containers are to be collected. The planning of collection tours is usually undertaken manually by experienced operators [2].

In this paper we propose the application of ant-based algorithms to plan the collection tours. The solution has been applied to the collection of paper bins and soft packages bins of Toro, a Spanish town situated in the province of Zamora.

The remainder of the paper is organized as follows. Section 2 describes the recycling system in the province of Zamora. It also describes the particular case of Toro. Section 3 introduces the ant-algorithm applied to solve the problem. Section 4 describes the solution applied to the problem. Finally, section 5 includes the conclusions of the work.

2   The Recycling of Waste in the Province of Zamora

Recently, the province of Zamora has implemented an integral system for management and treatment of waste. A project co-financed by the Cohesion Fund of the European Union, allowed designing the selective urban waste collection in the province. It allowed buying 250 glass bins, 580 paper bins, 1000 soft packages bins (Figure 1), and 3 trucks (Figure 2) to perform the selective collection of waste, [3].

The Regulatory Board for Urban Solid Waste Management of the province of Zamora is constituted by the Deputation of Zamora, 15 associations of the province, 10 municipalities not associated, and the Councils of Zamora, Benavente and Toro. The aim of the Board is to store, treat and eliminate the waste. It has 6 transference plants, situated in Palacios de Sanabria; Castrogonzalo; Villafáfila; San Vitero; Toro and Bermillo de Sayago; and a Centre for Waste Treatment (CWT) (Figure 3).

The CWT is situated in the city of Zamora. It has several rooms to classify and select the waste. It can process a maximum of 100.000 T per year. This centre is operative since 2001. From 2001 to 2007 it has processed 459.000 T of urban waste, [3]. The waste proceeding from different areas of the province are sent to the associated transference plant. The waste stored in the transference plants are finally sent to the CWT.

Fig. 1. Recycling bins for soft packages (yellow), paper (blue), and glass (green)

Fig. 2. Waste collection trucks

Fig. 3. Logistic Centre for waste management

The province has a service of selective pick up of paper and packages, which is performed as a whole. Two trucks are used for this purpose.

The amount of glass accumulated in each glass bin depends on its location and the time of the year. This situation makes difficult the establishment of paths for pick-up glass bins. Therefore, when a bin is full, the Council must phone to the service Ecovidrio to demand the pick-up.

2.1 Paper and Soft Packages Bins Collection in Toro

Toro is a town with over 10.000 inhabitants. It is situated in the East of Zamora (Spain). Figure 4 shows a map of Toro.

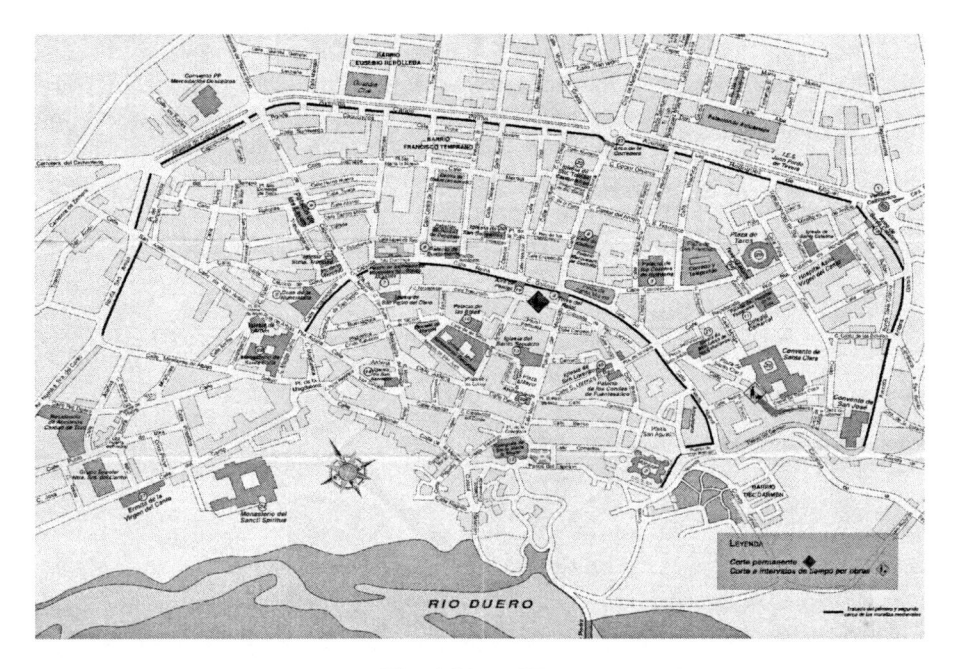

Fig. 4. Map of Toro

The town has 92 bins for recycling purposes: 22 paper bins, 36 soft packages bins and 34 glass bins. The glass bins are picked up when it is required by the Council. The bins for paper and soft packages are picked up once a week. There is a two-compartment truck for paper and packages collection. Another truck picks-up the glass bins. The trucks pick-up the bins and they put their content in the transference plant situated in the town. Periodically, the waste are transported from this transference plant to the CWT.

Our objective is to design a path to pick-up the glass bins and soft packages bins, being the total length of the path as small as possible.

3 The Ant Colony System Algorithm

Ant-based algorithms define a metaheuristic inspired by analogies to the foraging behaviour of biological ant colonies, [4]. It has been successfully applied to several NP-hard problems, [5].

The Ant Colony System (ACS) algorithm is an ant-based algorithm proposed by Dorigo and Gambardella, [6]. It was first applied to solve the well-known Travelling Salesman Problem (TSP). Given a set of n cities joined by weighted connections, the aim of the TSP is to find a closed path of minimum cost passing through all the cities once and only once, [7].

To solve a problem with n cities, we consider a set of m ants that cooperate in the search for a solution. We define a pheromone matrix, τ, that associates a pheromone trail τ(i, j), with each connection (i, j) of the TSP problem. Before starting the search for a solution, the pheromone of all connections is set to the value $\tau 0 = (n L)^{-1}$, L being the length of a nearest-neighbour solution to the problem. The pheromone allows ants to communicate among them, contributing in this way to the solution of the problem.

Each ant generates a complete tour starting from a randomly-selected city, and selecting the next city of its tour by applying a transition rule called pseudo-random proportional rule. If ant h is placed on city i, it chooses to move to city j according to the probability distribution (1).

$$j = \begin{cases} \arg \max_{l \in N_h^i} \{\tau_{il} \eta_{il}^\beta\} & \text{if } q \leq q_0 \\ J & \text{otherwise} \end{cases} \quad (1)$$

where $q0 \in [0, 1]$ is a parameter; q is a random value uniformly distributed in the interval $[0, 1]$; $\beta > 0$ is a parameter that determines the importance of the pheromone; N_i^k is the feasible neighbourhood for ant k when it is placed on city i (the set of cities accessible from city i and not yet visited by the ant); η_{ij} is the visibility of the connection (i,j) (for the TSP, the visibility of a connection is the inverse of the cost associated with such a connection); and J is a random variable selected according to the probability distribution (2).

$$p_{ij}^h = \begin{cases} \dfrac{\tau_{ij} \eta_{ij}^\beta}{\sum_{l \in N_i^h} \tau_{il} \eta_{il}^\beta} & \text{if } j \in N_i^h \\ 0 & \text{otherwise} \end{cases} \quad (2)$$

When each ant h has built its solution, S_h, it performs a local update of the pheromone, by applying the expression (3).

$$\tau_{ij} = (1 - \rho_L) \tau_{ij} + \rho_L \tau_0 \quad \forall (i, j) \in S_h \quad (3)$$

where $\rho_L \in (0, 1)$ is a value which determines the evaporation factor of the pheromone.

The pheromone is updated globally according to the expression (4), when all the ants have found a solution for the problem. This update is only performed by the ant hg, which constructed the best tour from the beginning of the algorithm, S_{hg}.

$$\tau_{ij} = (1 - \rho) \tau_{ij} + \rho \frac{1}{LS_{hg}} \quad \forall (i, j) \in S_{hg} \quad (4)$$

where $\rho \in (0, 1)$ is the global evaporation rate of the pheromone, and LS_{hg} is the length of the solution found by the ant hg.

The algorithm usually applies 2-OPT exchange to improve the path found by the best ant. This exchange is applied before local pheromone updating is performed.

The process is repeated until the solution converges, or until the predetermined maximum number of iterations for the algorithm has been performed.

4 The Solution to the Waste Pick-Up Problem

4.1 Mathematical Modeling of the Problem

To solve the waste pick-up problem, a set of roads and streets must be traversed, in order to pick up the bins situated along them. We must take into account that some roads and streets may not be traversed in both senses. The objective is to pick up all the bins traversing a path as sort as possible.

The urban solid waste collection problem can be classified as either a Traveling Salesman Problem [7] or a Vehicle Routing Problem [8]. Both problems are NP-hard [9]. In the first case, the problem is considered as a node routing problem, whereas in the second case it is considered as an arc routing problem. Ant-based algorithms have been applied to solve both problems, [6], [5], [10]

In [11]-[12] artificial ants are applied to solve the solid waste collection problem, considered as a node routing problem. The bins define the nodes of the graph, and the connections between any pair of bins define the arcs and edges of the graph.

In this paper we also apply ants, but we consider the problem as an arc routing problem, as proposed in [13]. To solve the problem, we have defined the graph associated to Toro (Figure 5). It is a mixed graph with 261 nodes, 672 edges and 83 arcs. The nodes are associated with street junctions. The edges and the arcs are associated with street sections between street junctions. Arcs represent single way streets, whereas edges correspond to two-way streets.

The problem can be modeled as a Mixed Rural Postman Problem (MRPP). Let $G=(V, E \cup A)$ be a weighted graph, where V is the set of points in the graph and $E \cup A$ is the set of connections, which have a nonnegative cost associated with them. E is the set of edges and A is the set of arcs. Let $F \subseteq E \cup A$, $F \neq \emptyset$, $F = E' \cup A'$, be a set of required connections, where E' is the set of required edges and A' is the set of required arcs. The aim of the MRPP is to find a closed tour of minimum length of G containing at least once each connection in F, [14]. Lenstra proved that the problem is NP-hard, [15].

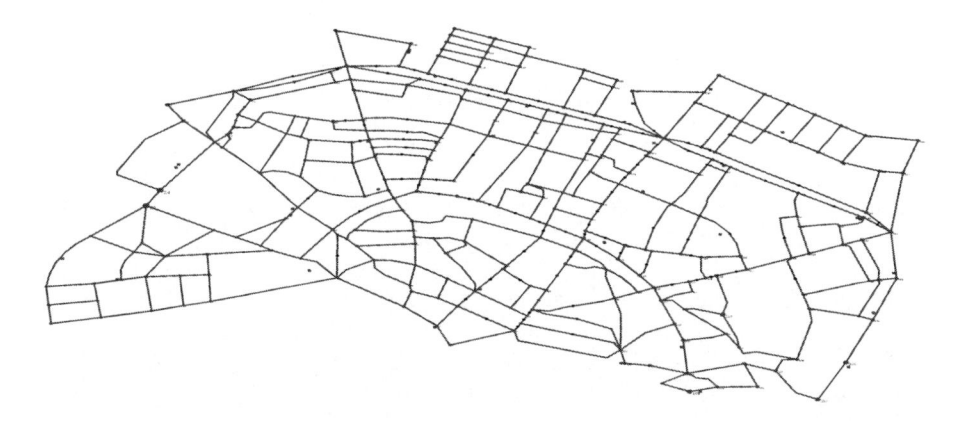

Fig. 5. Graph associated to the problem

In our problem, required links are associated with connections where there are collection containers. There are 34 required connections, with a total length equal to 2968,704 m.

The bins are situated on the sidewalks next to the roads. When a bin is located in a two-way street, such a street can be traversed in both directions, but it must only be traversed in one of the two directions. Therefore, the edge of the graph is replaced by two arcs with opposite directions (a required arc and a not required one).

4.2 Solution by Applying the Ant Colony System Algorithm

In a recent paper the ACS algorithm was applied to solve the Undirected Rural Postman Problem (URPP), [16]. To solve this problem, it was first transformed into an undirected TSP. The TSP was solved by applying the ACS algorithm. Finally, the TSP solution was transformed into a URPP solution.

In this paper, we apply the same steps to solve the problem. The transformation applied in [16] generated an undirected TSP. Nevertheless, in this case we must transform the MRPP into a directed TSP (DTSP).

The first step of the transformation consists of replacing each edge $e_{ij} \in E$ by two arcs, a_{ij} and a_{ji}, with the same costs as e_{ij}, (d_{ij}). These arcs are included in A, to yield an extended arc set, AX. Moreover, if $e_{ij} \in E'$, the arcs a_{ij} and a_{ji} are also included in A' to yield AX'. At the end of this step, the transformed graph GX=(V, AX) is directed.

Let $G_F = (V_F, F)$ be the subgraph of GX induced by F. To each node $i \in V_F$ we associate a set $S_i = \{s_i^j \mid j \in N(i)\}$, where N(i) represents the set of neighbours of node i in G_F.

We construct the complete weighted graph $G' = (V', E', c')$, where the set of nodes is:

$$V' = \bigcup_{i \in V_F} S_i \quad (5)$$

and the cost of the connections of the graph is calculated by applying expression 6:

$$c'(s_i^h, s_i^k) = 0 \quad \forall i \in V_F ; h, k \in N(i), h \neq k$$

$$c'(s_i^h, s_j^k) = \begin{cases} -M & \text{if } i = k, j = h \\ d(i, j) & \text{otherwise} \end{cases} \quad \begin{array}{l} \forall i, j \in V_F, i \neq j, \\ h \in N(i), k \in N(j) \end{array} \quad (6)$$

where d(i, j) represents the length of a shortest path between nodes i and j in GX, while M is a big value, that we will take as the sum of the costs of all connections in the graph.

When defining the DTSP graph, we apply the Floyd algorithm to determine the shortest paths among all the pairs of nodes of the graph GX. We store the cost of such paths as well as the information that allows reconstructing them. This information will be necessary in the last phase of our solution method.

When the DTSP is solved by applying ACS, a DTSP path is obtained. This path must be transformed into a MRPP path. Let $s_a^i, s_b^j, \ldots s_N^x$ be the sequence of n stops of the DTSP solution. Each node s_i^j in the DTSP path represents a required connection from i to j in the MRPP graph. Therefore, we must include in the MRPP path the arcs associated to the successive stops of the DTSP path. Then, we take pairs of consecutive DTSP stops, to determine a new stop or stops of the MRPP path. The first time stops s_a^i and s_b^j are considered, whereas the last time stops s_N^x and s_a^i are considered. Let s_i^h, s_j^k, be two consecutive stops of the DTSP path. If h≠j, we add to the MRPP path the stops of the shortest path from node h to node j in GX. The shortest path may be a single connection or a sequence of connections.

4.3 Computational Experiments

The algorithm has been coded using C language. The tests have been performed on a personal computer with a 1.5GHz Intel Centrino processor and 512M RAM, and running under the Linux operating system.

In the experiments, the following values were considered for the parameters: m = 10, $\alpha=1$, $\beta=2$, $\rho=\rho_L=\{0.1, 0.2, 0.01, 0.02\}$, $q_0=\{0.9, 0.98\}$, $\tau_0=1/(\rho\ L_{nn})$, as proposed in [5]; where L_{nn} is the length of a nearest-neighbour solution. Fifty independent runs were performed for each combination of values. Experimental results are not reported here. In Table 1 we only compare the best route obtained by the ants with the one used in the town. We observe that the cost of the solution reduces considerably. The average value of the nearest-neighbour solution considered to calculate τ_0 is 14996.58. The cost of this solution is smaller than the one applied in the town, but it is bigger than the best solution found by the ants.

The average time to reach a solution is smaller than 60 seconds.

Table 1. Cost of ants-solution versus practical solution

Ants	Practical
10533	15073.6922

5 Conclusion

This paper shows the application of the ACS algorithm to a real world problem. This problem can be modelled as a MRPP. A transformation of the MRPP into a DTSP allows us to apply the ACS algorithm for the TSP directly to our problem.

As previously mentioned, the definition of the paths to pick-up waste bins is usually based on the experience and intuition of a person. Nevertheless, we can observe that the application of artificial ants allows defining better paths.

The solution procedure applied to Toro can be applied to other towns. It can also be applied to pick up the containers situated on the roads connecting cities which only have one or two recycling bins.

Acknowledgements. This work has been partially financed by the Samuel Solórzano Barruso Memorial Foundation, of the University of Salamanca.

We thank the Regulatory Board for Urban Solid Waste Management of the province of Zamora for sending us the data used in our experiments.

References

1. World Bank home page, http://web.worldbank.org
2. Steiner, M., Wiegel, U.: The book of rubbish... A guide to the basics of waste management. Report: Waste training project
3. Plan para la gestión y tratamiento integral de residuos de la provincia de Zamora. Memoria Divulgativa. Diputación de Zamora
4. Deneubourg, J.L., Aron, S., Goss, S., Pasteels, J.M.: The self-organizing exploratory pattern of the argentine ant. J. Insect Behav. 3, 159–168 (1990)
5. Dorigo, M., Stützle, T.: Ant Colony Optimization. MIT Press, Cambridge (2004)
6. Dorigo, M., Gambardella, L.M.: Ant Colony System: A cooperative learning approach to the Travelling Salesman Problem. IEEE Trans. Evol. Comput. 1(1), 53–66 (1997)
7. Reinelt, G.: The Traveling Salesman. LNCS, vol. 840. Springer, Heidelberg (1994)
8. Toth, P., Vigo, D.: The vehicle routing problem. SIAM monographs on discrete mathematics and applications. Society for Industrial & Applied Mathematics, Philadelphia (2001)
9. Garey, M.R., Johnson, D.S.: Computers and intractability: a guide to the theory of NP completeness. W.H.Freeman & Co., New York (1979)
10. Bullnheimer, B., Hartl, R., Strauss, C.: Applying the Ant System to the vehicle routing problem. In: Voss, S., Martello, S., Osman, I., Roucairol, C. (eds.) Meta-heuristics: Advances and Trends in Local Search Paradigms for Optimization, pp. 285–296. Kluwer Academic Publisher, Dordrecht (1999)
11. Karadimas, N., Kouzas, G., Anagnostopoulos, I., Loumos, V.: Urban solid waste collection and routing: The ant colony strategic approach. International Journal of Simulation: Systems, Science, Technology 6, 45–53 (2005)
12. Karadimas, N., Papatzelou, K., Loumos, V.: Optimal solid waste collection routes identified by the ant colony system algorithm. Waste Management & Research 25, 139–147 (2007)
13. Beltrani, E., Bodin, L.: Networks and vehicle routing for municipal waste collection. J. Networks 4(1), 6–94 (1974)
14. Orloff, C.S.: A fundamental problem in vehicle routing. Networks 4, 35–64 (1974)
15. Lenstra, J.K., Rinnooy-Kan, A.H.G.: On the general routing problem. Networks 6(3), 273–280 (1976)
16. Pérez-Delgado, M.L.: A Solution to the Rural Postman Problem Based on Artificial Ant Colonies. In: Borrajo, D., Castillo, L., Corchado, J.M. (eds.) CAEPIA 2007. LNCS (LNAI), vol. 4788, pp. 220–228. Springer, Heidelberg (2007)

An Access Control Scheme for Multi-agent Systems over Multi-Domain Environments

C. Martínez-García[1], G. Navarro-Arribas[2], J. Borrell[1], and A. Martín-Campillo[1]

[1] Dept. of Information and Communication Engineering, Universitat Autònoma de Barcelona
{cmartinez,amartin,jborrell}@deic.uab.cat
[2] IIIA, Artificial Intelligence Research Institute, CSIC, Spanish National Research Council
guille@iiia.csic.es

Abstract. Multi-agent systems and mobile agents are enabling the deployment of applications in multi-domain environments. In these scenarios, different domains interact toward the same goal through resource sharing. As a result, there is the need to control the actions that an agent can perform in a foreign domain, with the only information of where it comes from and which roles does it hold in its own domain. However, this information will not be directly understandable as domains may not share the same role definitions.

MedIGS is a multi-agent middleware for the medical data sharing between hospitals which take part of a multi-domain environment. In this paper, a distributed access control for MedIGS is presented. Based on attribute conversion, this authorization scheme proposes a solution with a minimum impact in the local access control systems of the hospitals.

1 Introduction

Multi-agent technology and more precisely mobile agent systems have evolved as a powerful technology for distributed systems. Moreover, mobile agents, which have the ability to move between several locations over a computer network, are providing means to enable the interoperability of heterogeneous systems. This evolution has brought new challenges for security and access control, as there is the need of providing a global authorization scheme without the loss of autonomy and security of the systems or domains [6].

A clear example can be seen in the MedIGS (Medical Information Gathering System) [17] (see Section 2) application, which has been initially deployed in Hospital São João, Portugal's second largest hospital. The aim of MedIGS is to provide support for Virtual Electronic Patient Records (VEPR), which is the collection, distributed in several hospitals, of all the medical documents referring to a given patient. To do so it relies on mobile agent platforms, which interconnect several electronic health systems from different hospitals and/or departments. It allows a doctor to obtain information of a given patient by gathering all the data of such patient, which are disseminated over all the locations. For example, it may get a radiography from the radiology department of some hospital, get a result of a blood test from another hospital, and so on. The key point of MedIGS is that all the institutions involved are heterogeneous. They have different computer systems, use different formats to represent information, and may use different communication protocols. Mobile agents are used as an interoperability mechanism coping with this heterogeneity by searching information in situ in each institution.

Although MedIGS provides an appropriate solution to the problem of gathering information to draw up the VEPR, it still presents an important interoperability problem related to security and access control policies. Each institution has its own security policies and mechanisms, which normally are not compatible. They use different technology and different security models. So the MedIGS application had to develop a generic model based on Role-based Access Control (RBAC) [4] and adapt it to each different institution. It is clear that this security solution is not scalable and these kind of applications require a different approach to the interoperability of this security systems. The problem can be simplified considering the institutions as having a collection of subjects which have some security attributes assigned to them by the institution authority. Policies specify the attributes that a subject needs to hold in order to perform a given action on a given resource. This generic concept is broad enough to cope with most of the existing access control models, from RBAC to traditional file system-based mechanisms.

In this paper we present a model for allowing the interoperability of different security systems in a multi-agent environment (Section 3). The main advantage is that it avoids having to provide a generic access control policy for all the institutions. On the contrary, it allows the use of the local security policies in each platform to decide whether an action can be granted when it is requested by an agent from another institution, and thus with different security credentials. To do that, we provide an access control mechanism, and its integration analysis, (Section 4) based on converting attributes from one institution to another, so when an agent from one institution goes to another institution, its attributes can be converted to native attributes in the destination domain and this way use the local policies of this domain.

2 Medical Information Gathering System (MedIGS)

MedIGS is a multi-agent-system middleware with the purpose of sharing data between medical institutions. Its main objective is the creation of a Virtual Electronic Patient Record. The VEPR is the collection of all the medical documents referring a given patient, which can be located in any of the hospitals that take part in MedIGS. The system is quite complex, and its design is out of the scope of this work. Here we only provide a general view in order to introduce the problem and the proposal of the paper.

Each hospital in MedIGS has a front-end where the doctors can ask for the VEPR of an specific patient. When a request is submitted, the system launches a mobile agent, called *Collector Agent*, which will visit a specified set of hospitals looking for medical data related to the patient and storing its references. Moreover, in each hospital, the mobile agent collects new locations to visit where more medical documents may be available. Once there are no more hospitals to visit, the *Collector Agent* returns to the initial hospital and shows the data references to a *Document Broker Agent*, which makes the documents locally available. The mobile agent does not carry the documents it founds because of the overhead that means carrying a huge amount of data during the collecting itinerary.

As the system is treating medical data, security is a strong requirement. Self protected mobile agents [3] are used to prevent improper modifications on code and data carried with agents. Regarding access control, Role-Based Access Control is used to authorize the data access. Before submitting a request, the user must be identified in its own system by a previous login step. Once a user is properly authenticated, local policies specify if she is allowed to execute document retrieval requests. Nevertheless, a local system is not able to determine which actions a user can perform in a foreign system. This restriction lies on the independence of each domain, which has the authority to define their own structure and managing their own policies. It entails that a role in one system can have a different meaning in another system or just not being understood.

When the system generates a *Collector Agent*, it embeds the VEPR retrieval request into the agent. Information about user's identity, the origin domain and user's attributes must be also included. When a *Collector Agent* arrives to a foreign hospital, it has to interact with the *Remote Broker Agent*. However, a previous authentication step is required. Authentication can be done through identity certificates signed by the user's local domain. Once the user has been identified, the *Collector Agent* launches the request. At this moment, the system has to determine if the agent is allowed or not to access the data that it is asking for. The decision will be made in the RBAC Policy Decision Point (PDP) of the platform taking into account the user's information related to the request.

If a user who has launched a request owns the suitable set of attributes specified in the local policies, the action will be permitted. Currently, the *Remote Broker Agent* provides the *Collector Agent* with the local interfaces needed to discover the data. Note that there is a substantial difference between collectors agents and users: whilst users are supposed to read the data, *Collector Agents* are supposed to discover it. As discovering is a previous step of reading the data, a *Collector Agent* will be only allowed to discover it if the user who launched the request is allowed to read the data. Once the *Collector Agent* has been authorized, it will obtain the references to the medical documents and the system will log the action as a read action executed by the user. If the user has not been authorized to continue with the action there is the possibility, under some restrictions, to override this decision, knowing that the action will be properly investigated later. It allows the system to deal with emergency situations, where the flexibility of the system takes priority to the security itself.

3 Global Access Control through Attribute Conversion

When adding access control to a multi-domain environment, it appears the need of a global understanding of the attributes defined in each domain. However, the independence of the domains, which have been working in an isolated manner before the interoperability agreement, makes it hard. Moreover, in RBAC the attributes not only reflect a set of permissions but also the structure of the organization through the dominance relationship between attributes. In this section, we propose the basis of an access control system for MedIGS, which is a clear example of a multi-domain environment.

We consider the attribute conversion as a technique for enabling interoperability in multi-domain scenarios. The main idea is to establish an attribute mapping between the different domains. This way, agents acting in a foreign domain will be subjected to the well-tested local access control policies defined in this domain, avoiding the need of generating specific access control policies for the interoperability. However, it is not guaranteed that a crisp relation between different attribute sets exists because of the domains' independence. That is why we consider attribute conversion by making an assessed attribute conversion, associating a magnitude to each translated attribute. The magnitude can be understood as the user-attribute assignment's degree in the range $[0-1]$, where 0 means no attribute possession and 1 means the maximal degree of possession. Note that the users-attribute assignment can be seen as a fuzzy set membership function.

The conversion of attributes is performed through static conversion policies which specify an output attribute set for each input attribute set. This policies must be defined once by the domains' authority as a previous step towards the interoperability. The format of the policies is not specific. They can be simple one-to-one conversion rules, conversion matrices which allow one-to-many conversions or more sophisticated policies which allow many-to-many conversions. The only mandatory requirement is that conversion policies must specify for each translated attribute its associated magnitude. The magnitude can reflect the similarity between attributes or another such metric. For example, it can be computed based on a credit system. In this initial approach, each domain must have as much conversion policies as domains are in the scenario. However, more sophisticated conversion structures, with the aim of reducing the global number of conversion policies, will be considered as future work.

As MedIGS is a middleware, which can be used with low impact in the hospital's systems, the integration of the access control system cannot involve the modification of the local access control policies of each domain. The magnitude of the converted attributes is only used to determine when an attribute is applicable or not. This is the way by which the systems adapts to external situations, being more permissive if there is an emergency or more restrictive if there is not. For this purpose, each domain has a dynamic *security threshold*, denoted as δ. An attribute is only applicable if its magnitude is equal or greater than δ. If not, it is possible to *override* this restriction if the attribute's magnitude is greater than a second threshold called *override threshold*, denoted by β. The *override threshold* gives the users the possibility to access the data, at their own responsibility, under situations which they wouldn't be able to access. It can be helpful in an emergency situation. Every override will be submitted to auditing post-processes to validate its legitimacy. Note that both thresholds define three regions where an attribute's magnitude can be located: the *applicable* region, if the magnitude is equal or greater than δ; the *applicable under override* region, if the magnitude is somewhere between β and δ; and the *denied* region if the magnitude is less than β. To disallow the overrides β and δ must take the same value. The duty of adjusting the thresholds relies on the managing staff of the hospitals.

MedIGS is supposed to work over RBAC-based systems. In RBAC there are two kinds of assignments. The former is the user-role assignments, which specifies which attributes (roles) a user can activate. The latter is the role-permission assignment, which specifies which actions can perform the members of the role. The attribute conversion must take into account the role-permission assignments to determine the similarity

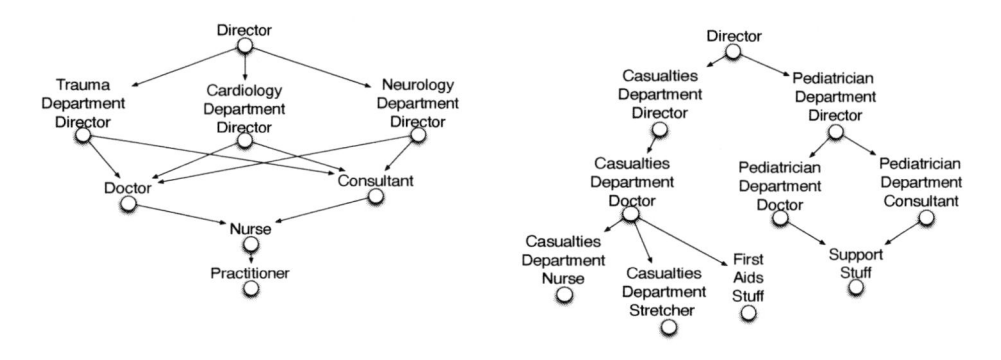

Fig. 1. H1 (left) and H2 (right) internal structure represented by roles

between roles if the system is performing a similarity-based mapping. Once the roles have been bestowed to the users, for each action the system will determine which is the user's minimum set of roles needed to be activated to perform the action. This is the way the system enforces the least-privilege principle [14]. Enforcing the thresholds' restrictions at the role activation has a minimum impact on the hospitals' access control system because there is no need to modify the well-tested local access control policies of each domain to specify not only the required attributes but its minimum magnitude for the action.

3.1 Example

Imagine that there are two hospitals involved in a medical data sharing agreement through MedIGS. The hospitals are named H1 and H2, and they present the internal structure, represented by roles, depicted in Figure 1. Based on similarity, it has been decided that the mapping of management roles between H1's and H2's is the one shown in Table 1.

Imagine that a user holding the Cardiology Department Director attribute of H1 wants to recover the VEPR of a given patient. With this purpose, when she submits the request, MedIGS launches a mobile agent which will migrate to H2 and ask there for the patient data. The Cardiology Department Director attribute in H1 ($H1\_CDD$) is equivalent to the set $\{[H2\_D, 0'05], [H2\_CDD, 0'7], [H2\_PDD, 0'7]\}$ in H2. H2 has a

Table 1. H1 to H2 role mapping, where H1_D represents H1's director, H2_CDD represents H2's Casualties Department Director and so on. Note that the relation is unidirectional from H1 to H2.

	H2_D	H2_CDD	H2_PDD
H1_D	0'95	0'5	0'5
H1_TDD	0'05	0'9	0'6
H1_CDD	0'05	0'7	0'7
H1_NDD	-	0'3	0'3

Table 2. Access control policy in H2 for direction staff. Rows represent actions and columns represent sufficient attributes to execute the action specified by an "x".

	H2_D	H2_CDD	H2_PDD
Create record	x	-	-
Modify record	-	-	-
Read record	x	x	x
Delete record	x	-	-

security threshold δ equal to $0'6$, and the override threshold β equal to $0'1$, in this moment the user will be able to activate the roles $H2\_CDD, H2\_PDD$ as their magnitude are greater or equal to δ, this is, they are in the *applicable* region.

H2 has the access control policies for management staff shown in Table 2. With δ equal to $0'6$, the user, activating the roles $H2\_CDD$ or $H2\_PDD$, will be able to read records. Moreover, she would have the possibility to activate the role $H2\_D$ if it were in the *applicable under override* region. As β is equal to $0'1$ and the $H2\_D$'s magnitude is $0'05$, the override on the role activation is not possible.

Imagine now that a disaster occurs and an emergency is declared. Adapting the security thresholds δ and β to $0'2$ and $0'03$ respectively will not only allow the users who own the H1's Cardiology Department Director attribute to activate the $H2\_D$ but also the users owning the H1's Neurology Department Director attribute to activate the $H2\_CDD$ and $H2\_PDD$ without the need to apply the attributes in the *override region*.

4 System's Architecture

The MedIGS' access control system has two main parts, shown in Figure 2. The former is the hospital's access control system, which is supposed to have at least a Policy Enforcement Point (PEP), a Policy Decision Point (PDP) and a set of access control policies (Ap). The latter part is the MedIGS extension which consists of an Attribute Management Service (AMS), the attribute conversion policies (Cp) and a local repository of certificates (Crep). We can see the integration of these two parts as a single system.

The PEP acts as an interface to the resources. Each action on the system must be queried to the PEP which will ask to the PDP if the action can be permitted or not. If the action is permitted, the PEP allows the user to execute it. The PDP acts as the decision engine of the system. When a request is submitted, the PDP checks in the access policies if the attributes presented by the user are enough to execute the given

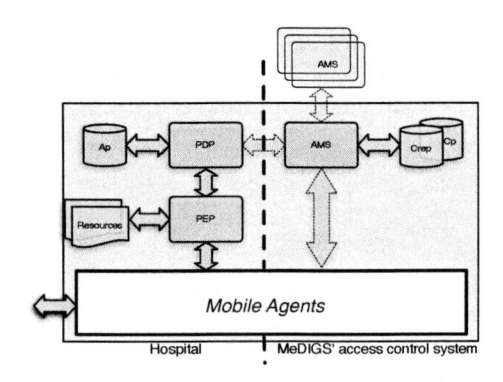

Fig. 2. Scheme of the access control system placed in each hospital. The left hand side represents the hospital part while the right hand represents the MedIGS extension. The interactions between the modules which depend on the design alternative are represented by dotted arrows.

action on the given resources. However, the PDP will not be able to understand the foreign users' attributes without a previous conversion into local attributes. This is the duty of the AMS which receives foreign attribute certificates and issues local attribute certificates understandable in the local domain. The way that these components interact has an impact on the integration of the middleware with the hospital system. It depends on different factors which are explained in the following section.

4.1 Integration Analysis

There are three factors influencing the interactions between the hospital's and the MedIGS' part of the access control system which determines the impact of the middleware integration. The first one is given by the hospital's access control system. If it works in a *push* manner, the users have the duty to attach their attribute certificate to each access request. This alternative removes the necessity of the hospital's systems to interact between them for the foreign user's attribute certificates recovering. However, if the system is working under a *pull* model, it has the duty to recover the attribute certificates of the users. In case of recovering a foreign user certificate the hospital system is obliged to contact the AMS which establishes a communication with the AMS of the origin domain for the certificate recovery.

The second factor depends on when the attribute conversion is performed. It can be done as a previous step for the access queries. This alternative breaks the user's interaction with the system in two parts. First of all, the user has to obtain a valid attribute certificate for the domain. For that purpose, the user has to ask the AMS for an attribute conversion. Once the user has a valid attribute certificate, it is allowed to launch access queries. However, this way of working can commit in non necessary attribute conversions. To prevent that, the *on demand* attribute conversion is performed only when the required attributes are known, that is, when an access request has been submitted. Nevertheless, it implies the interaction between the PDP of the hospital's access control system and the AMS to perform the ad-hoc attribute certificate conversion.

Finally, the third factor refers to where the attribute conversion is done. It can be in the local domain or in the user's origin domain. If the conversion is performed in the local domain there is no need of interaction between the hospital's access control system and the MedIGS' extension as the whole attribute set associated to the user is converted. However, it implies the disclosure of more access control information about a user in its origin domain than necessary. The origin attribute conversion solves that problem performing the attribute conversion in the user's origin domain. To do that, the conversion must be ad-hoc for each access request. It implies that the system has to perform an *on demand* attribute conversion, hence the interaction between the hospital access control system and the MedIGS extension is required.

Note that while the first factor is fixed by the working manner of each hospital's access control system, the second and the third factors can be considered as design alternatives. Choosing one of them depends on the integration's impact level that the hospitals are willing to tolerate.

5 Related Work

Most of the current work about interoperability of security systems [15, 13, 16] is focused towards the interoperability of security policies from different domains. In our case, we provide a low level interoperability by converting the attributes that will be used with existing policies in a similar way [9, 8], but instead of providing an absolute conversion we allow to relax the security requirements in order to provide a more realistic conversion. This idea comes from previous work on imprecise security systems. For instance, Foley [5] considers the use of similarity measures between attributes in order to use them in *trust management* systems. The same idea was applied to SAML-based decision engines [10].

Obviously, while our proposal introduces more flexibility and an approach more closely related to real situations, it forgoes the security of the whole system by relaxing it. Nevertheless, this approach is already considered a good option for corporate environments, where flexibility is preferred over high security requirements [11]. Following these research line we find a proposal where Alqatawna et. al. [2] introduce the notion of *override* in access control policies. An access request can be denied with the possibility of override. If the user agrees, the system allows the access under the audit of an authority. Our proposal differs in providing more flexibility when it comes to define how access control rules can be bypassed.

Finally, it is noteworthy to say that the way we convert attributes between domains has some parallelism with fuzzy set theory [18], as far as we can consider the attributes as fuzzy sets where the users have a membership degree and conversion policies as fuzzy relations. Fuzzy sets, although in different context, have been applied to security systems [7, 1, 12].

6 Conclusions

MedIGS is a multi-agent based application deployed over a multi-domain environment made up of independent heterogeneous medical institutions. The spirit of cooperation between different organizations through a multi-agent system generates the necessity, not only to regulate the actions that a user can perform, but the necessity to regulate the actions that previously unknown users can perform with the only information of where they come from and which roles do they hold in their own domains. As a middleware, MedIGS is supposed to have the minimum impact on the hospital's systems where it is integrated. The attribute conversion technique avoids the necessity of generating new access control policies for the domains, which have a big impact on them. On the contrary, the technique is based on finding a relation between attributes in different domains, and apply the foreign users to the local access control policies of each domain.

In this document, an attribute conversion technique has been presented. Furthermore, it has been described an access control system for the MedIGS scenario, and its integration analysis, which can be easily adapted to any multi-agent system over a multi-domain environment.

Acknowledgments

Partial support by the Spanish MEC (project TSI2006-03481 and project ARES - CONSOLIDER INGENIO 2010 CSD2007-00004) and by the Autonomous University of Barcelona (PIF grant 472-01-1/07).

References

1. Alo, R., Berrached, A., De Korvin, A., Beheshti, M.: Using fuzzy relation equations for adaptive acess control in distributed systems. In: Advances In Infrastructure For e-Bussiness And Education On The Internet, pp. 176–184 (2000)
2. Alqatawna, J., Rissanen, E., Sadighi, B.: Overriding of access control in xacml. In: POLICY 2007: Proceedings of the Eighth IEEE International Workshop on Policies for Distributed Systems and Networks, pp. 87–95. IEEE Computer Society, Los Alamitos (2007)
3. Ametller, J., Robles, S., Ortega-Ruiz, J.A.: An implementation of self-protected mobile agents. In: Eleventh IEEE International Conference and Workshop on the Engineering of Computer-Based Systems, Brno, Czech Republic, pp. 544–549. IEEE Computer Society Press, Los Alamitos (2004)
4. Ferraiolo, D., Kuhn, R.: Role-based access controls. In: 15th NIST-NCSC National Computer Security Conference, pp. 554–563 (1992)
5. Foley, S.N.: Supporting imprecise delegation in keynote using similarity measures. In: Sixth Nordic Workshop on Secure IT Systems (2001)
6. Gong, L., Qian, X.: Computational issues in secure interoperation. Software Engineering 22(1), 43–52 (1996)
7. Hosmer, H.H.: Security is fuzzy!: applying the fuzzy logic paradigm to the multipolicy paradigm. In: NSPW 1992-1993: Proceedings on the 1992-1993 workshop on New security paradigms, pp. 175–184. ACM, New York (1993)
8. López, G., Cánovas-Reverte, O., Gómez-Skarmeta, A.F.: Use of xacml policies for a network access control service. In: 4th International Workshop for Appiled PKI, IWAP 2005 (September 2005)
9. López, G., Cánovas, Ó., Gómez-Skarmeta, A.F., Otenko, S., Chadwick, D.W.: A heterogeneous network access service based on PERMIS and SAML. In: Chadwick, D., Zhao, G. (eds.) EuroPKI 2005. LNCS, vol. 3545, pp. 55–72. Springer, Heidelberg (2005)
10. Navarro-Arribas, G., Foley, S.: Approximating SAML using similarity based imprecision. Intelligence in Communication Systems (January 2005)
11. Odlyzko, A.: Economics, psychology, and sociology of security. In: Wright, R.N. (ed.) FC 2003. LNCS, vol. 2742, pp. 182–189. Springer, Heidelberg (2003)
12. Ovchinnikov, S.: Fuzzy sets and secure computer systems. In: NSPW 1994: Proceedings of the 1994 workshop on New security paradigms, pp. 54–62. IEEE Computer Soceity Press, Los Alamitos (1994)
13. Pearlman, L., Welch, V., Foster, I., Kesselman, C., Tuecke, S.: A community authorization service for group collaboration. In: POLICY 2002: Proceedings of the 3rd International Workshop on Policies for Distributed Systems and Networks (POLICY 2002), Washington, DC, USA, p. 50. IEEE Computer Soceity, Los Alamitos (2002)
14. Samarati, P., di Vimercati, S.d.C.: Access control: Policies, models, and mechanisms. In: Focardi, R., Gorrieri, R. (eds.) FOSAD 2000. LNCS, vol. 2171, pp. 137–196. Springer, Heidelberg (2001)
15. Shafiq, B., Joshi, J.B.D., Bertino, E., Ghafoor, A.: Secure interoperation in a multidomain environment employing rbac policies. IEEE Transactions on Knowledge and Data Engineering 17(11), 1557–1577 (2005)

16. Sun, Y., Pan, P., Leung, H., Shi, B.: Ontology based hybrid access control for automatic interoperation. In: Automatic and Trusted Computing. LNCS, pp. 323–332. Springer, Heidelberg (2007)
17. Vieira-Marques, P., Robles, S., Cucurull, J., Cruz-Correia, R., Navarro-Arribas, G., Martí, R.: Secure integration of distributed medical data using mobile agents. IEEE Intelligent Systems 21(6) (November-December 2006)
18. Zadeh, L.A.: Fuzzy sets. Information and Control 8(3), 338–353 (1965)

Bridging the Gap between the Logical and the Physical Worlds

Francisco García-Sánchez[1], Renato Vidoni[2], Rodrigo Martínez-Béjar[1], Alessandro Gasparetto[2], Rafael Valencia-García[1], and Jesualdo T. Fernández-Breis[1]

[1] Facultad de Informática, Universidad de Murcia, 30100 Espinardo (Murcia), Spain
{frgarcia,rodrigo,valencia,jfernand}@um.es
[2] Universita' degli Studi di Udine, Via delle Scienze, 208, 33100, Udine, Italy
{renato.vidoni,gasparetto}@uniud.it

Abstract. An implicit frontier have traditionally existed between services available online and those that can be offered by mechanical devices such as robots. However, many applications can benefit from the convergence of these two worlds. In this paper, we present BRIDGE, a knowledge technologies-based multi-agent system for joining together Web and robot-provided services. The fundamentals of the framework and its architecture are explained and a traffic control-related use case scenario described.

1 Introduction

Until few years ago, the Web was seen as a mere repository of information. With the emergence of Web Services [2], the Web naturally evolved into a distributed source of functionality. Web Services technology is based on a set of standard protocols and, thus, facilitates the remote execution of functionality regardless of the operating system the service is hosted in and the programming language used to implement the service. However, Web services cannot cover all the spectrum of possible functionality required by a user. End users might need to fulfil a goal that involves (physically) interacting with the real world. In such situations, mechanical devices capable of carrying out these interactions are needed. Robots are perfectly fit for this task. They can be used to gather information from the environment through sensors, and to manipulate and modify the environment by means of actuators.

The Intelligent Agents [20] topic has been broadly studied over the last 30 years and it is currently being revisited due to its relation to the Semantic Web [1]. Before the emergence of the Semantic Web, agents had to face the problems derived from the lack of structure characterizing much of the information published on the web. Nowadays, the semantically annotated information of the Semantic Web can be automatically processed by agents, so that new powerful opportunities open up for both application developers and users. The usefulness of agent technology in a service-intensive environment is two-fold. Firstly, they can act as autonomous software entities that discover, compose, invoke and monitor services without human intervention, thus preventing users from carrying out this kind of arduous, time-consuming tasks. Secondly, agents possess the ability to adapt to changing situations and handle the dynamism of these environments.

In this paper, we present BRIDGE, an agent framework that combines Web Services and robot-provided functionality to provide end users with added-value services. Our contribution is an overall solution based on a fully-fledged, loosely-coupled architecture that serves as a meeting point for services coming from both the logical and the physical worlds. The rest of the paper is organized as follows. Section 2 contains an overview of other related approaches in which agents are employed to interface the communication between humans and robots. The main features of the proposed framework are pointed out in Section 3. In Section 4, the application of the framework in an example scenario is described. Finally, conclusions are put forward in Section 5.

2 Related Work: Combining Reality with Virtuality

Human-Agent-Robot interaction focuses on cognitive, physical and social interaction between agents, robots [13, 12]) and people to provide for collaborative intelligence and extend human capabilities. Robots (and/or agents) can be useful in different contexts and environments, such as in disaster recovery, search and rescue tasks, delivering health care, assisting the elderly and increasing productivity in the workplace.

In literature, different approaches and applications for developing multi-agent teamwork and creating cooperation between human, agents and robots, can be found (see [17, 7, 11, 10] and the references there are in). Successful applications have been developed by integrating and extending KAoS (Knowledgeable Agent-oriented System), a collection of componentized agent services compatible with several popular agent frameworks [3], in order to create a single environment for human-agent work systems (e.g. with Brahms and Nomads as in [18, 6, 4]. A more recent evolution of this network infrastructure is presented in [11] where the KAoS HART (Human-Agent-Robot Teamwork) has been adapted for providing dynamic regulation between agents, robots, traditional computing platforms and Grid computing. Grid computing is the application of resources from many networked computers to a single problem at the same time. Its evolution led to the so-called socio-cognitive grid concept and applications in the human-agent domain can be found in [5, 16]. The idea behind the socio-cognitive grids is to provide cognitive and social resources accessible on electronic devices in support of common activities. This leads to transform the Net into a human resource, ideally accessible by anyone at anytime and anywhere for solving a particular problem.

Taking into account these human-artificial intelligence interactions, in our attempt an alternative integration between humans, software and physical agents is studied and realized. Thus, our goal with the BRIDGE system is to further facilitate the three-party communication between humans, the logical world and the physical world. In order to reconcile possible semantic mismatches while interacting, ontologies [19] have been used for knowledge representation. Besides, we face the dynamism of the logical and physical worlds by making use of agent technology. Intelligent agents' properties of reactivity, pro-activeness and social ability enable the framework to deal with the ever-changing environment. Furthermore, while semantically-described Web services operates in the logical world, robots are responsible for manipulating the physical world in the ways agents, which act on behalf of human users, dictate.

3 The BRIDGE System

BRIDGE (Bridge for Robots and Intelligent agents in Distributed, Geographically-dispersed Environments) is a platform that assists in developing complex applications in which services from both the physical and the logical worlds are necessary. The platform has been built on the basis of SEMMAS [8, 9], a framework for seamlessly integrating IA and SWS. Thus, the applications built on top of SEMMAS benefit from the autonomy, dynamism and goal-oriented behaviour agents provide and the high degree of interoperability across platforms and operating systems WS advocate. BRIDGE aims to extend SEMMAS to incorporate a new source of functionality, that provided by robots placed in different environments in the real world.

BRIDGE is composed of three loosely-coupled components (see Fig. 1): a Multi-Agent System (MAS) with eight different types of agents, an ontology repository in which several ontologies are stored, and a bunch of services divided into two distinguished groups: Web services and robot-provided services. Each component is described in detail separately in the following sub-sections.

3.1 Multi-agent System

The core of the proposed framework is a MAS composed of eight different types of agents. These agents operate as intermediaries between users that have needs and services capable of fulfilling those needs. Depending on their functionality, agents are grouped into three main categories: (1) service-representative agents, such as the "Provider Agent", the "Service Agent" and the "Robot Agent": (2) user-representative agents, such as the "Customer Agent", the "Discovery Agent" and the "Selection Agent"; and (3) system-management agents, such as the "Framework Agent" and the "Broker Agent". The agents that function as service representatives are responsible for

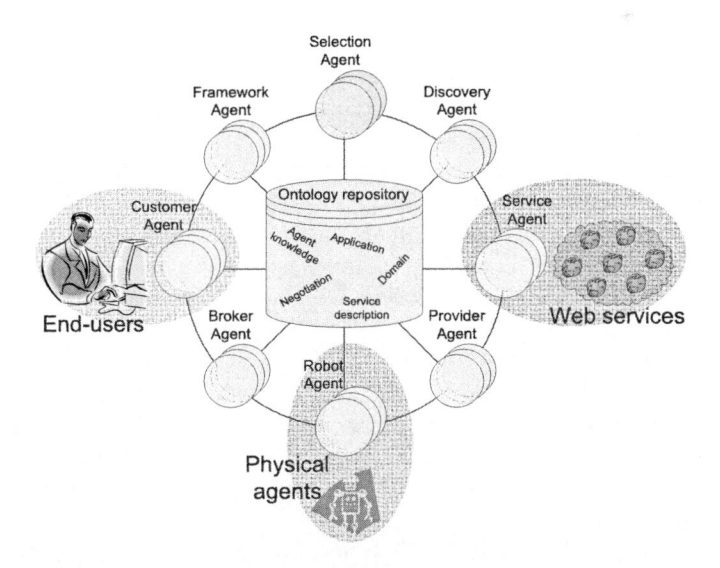

Fig. 1. The system architecture

Table 1. Description of the agents in the system

Agent	Brief description
Broker	It is responsible for solving interoperability issues. Three different levels are considered: data mediation, process mediation and functional interoperability
Customer	It acts as a user representative. First, users indicate their preferences and specify the goal to be achieved. Then, the goal is carried out and the results given back to the user. The whole process happens transparently to the user
Discovery	It is in charge of searching in the Semantic Web Services repository for the service or set of services (i.e. composition) that satisfy the requisites established by the users
Framework	It is responsible for monitoring and ensuring a correct functioning of the platform. This type of agent also controls and balances the workload
Provider	It acts as a service provider representative. The entities set their preferences regarding service execution and these are taken into account during the negotiation process with the service consumers
Robot	It acts as a robot representative and includes the machinery to make robots carry out a particular action and to retrieve the data gathered by them
Selection	It is in charge of selecting the most appropriate (single or compound) service from the set of services found by the discoverer according to the users' preferences. For that purpose, a negotiation process is carried out
Service	It acts as a service representative. The service provider establishes a concrete set of preferences regarding a particular service and these are taken into account when negotiating with service consumers

managing the access to services and ensuring that the conditions agreed for the service execution are satisfied. Agents that act as user representatives are in charge of locating the appropriate services, agreeing on contracts with their providers, and receiving and presenting the results of their execution. Management agents have a twofold purpose: to avoid system resources becoming overloaded and to monitor the status of all the interactions. This should not be understood as a centralized control mechanism that hampers the decentralized vision of MAS. Instead, our aim was to enable the system to both manage unexpected errors and improve its performance. Thus, the "Framework Agent" does not control the activity of the remaining agents, but makes sure no errors block the system functioning.

The system functionality is present in the form of roles. Roles are encapsulations of dynamic behaviour and properties that can be played by agents [21]. Therefore, the behaviour an agent shows at a particular point in time depends on the roles it plays. However, for most agents to keep their identity, they must mandatorily assume some particular roles. In Table 1, we show a summary with the basic functionality each agent type must provide.

The major difference with SEMMAS is the presence of the "Robot Agent", which interfaces the software agents' environment with the real world. For this, it undertakes the 'robot' role, which includes the means to communicate with the physical robot. Hence, this "Robot Agent" encloses the standard capabilities of the software agents and special purpose abilities: it is charged to open and close the communication with the correct autonomous robot, translate tasks information or requests into robot commands, and manage the results and the required information that come from the different available sensors. Wireless radio (i.e. Wi-Fi and Bluetooth) or infrared communication devices can be mounted and easily integrated into the robotic systems creating autonomous robots that can be reached, questioned and used in order to share, discover and use information.

3.2 Ontology Repository

Ontologies are the knowledge representation mechanism utilized in BRIDGE. Various ontologies must be considered for the system to properly function. These ontologies are categorized into five groups: domain ontologies, application ontologies, agent local knowledge ontologies, semantic web services ontologies and negotiation ontologies. Prior to describing the desirable contents of these ontologies, it is worth to note that when dealing with different ontologies it is necessary to account for the semantic heterogeneity issue [15]. The "Broker Agent" possesses the mediation mechanisms necessary to overcome this problem. A more detailed discussion of the way the framework faces interoperability issues can be found in [9] and [8].

A domain ontology models a specific domain knowledge and represents the particular meanings of terms as they apply to that domain. In our approach, the domain ontology represents a conceptualization of the specific domain the framework is going to be applied in. This ontology supports the communication among the components in the framework without misinterpretations. An application ontology may involve several domain ontologies to describe a certain application. For the purposes of this work, the application ontology embraces the knowledge entities (i.e. concepts, attributes, relationships, and axioms) that model the application in which the framework is to be used.

The small piece of ontology used by a single service for service description is defined as a service ontology. We assume the existence of one or various remote ontology repositories containing the semantic description of the available services. As stated before, the framework does not impose any restriction in terms of the kind of SWS specification (i.e. OWLS, WSMO, SWSF, WSDL-S or SAWSDL) to be used.

When a group of individual agents form a MAS, the presence of a mechanism to coordinate such a group becomes necessary. Negotiation is one of the means to achieving such coordination. Negotiation mechanisms in MAS are mainly composed of two elements: negotiation protocols and negotiation strategies. Until now, there have been conceived lots of negotiation mechanisms. However, the appropriateness of these in a particular situation highly depends on the problem under question and the application domain. For all that, we have provided the framework with a negotiation ontology, which comprises both negotiation protocols and strategies, so that agents can choose the best mechanism at run-time.

Finally, an agent local knowledge ontology generally includes knowledge about the assigned tasks as well as the mechanisms and resources available to achieve those tasks.

Thus, for example, the knowledge ontology for the "Broker Agent" may contain the mapping rules it has to apply to resolve the interoperability mismatches that might occur during the system execution. While setting up the framework, a developer can, by providing appropriate ontologies, customize it and make it work in a concrete domain to solve a specific kind of problem.

3.3 Services

Services are the entities that provide all the functionality that the system can show off. In BRIDGE, services come in two flavours: Web services (WS) and robot-provided services (RS). WS account for some of the companies' internal business processes. RS, on the other hand, refer to the features physical agents all around the world posses and may be of use and provide utility for end users.

WS can be implemented in almost any programming language and can be deployed in any platform. WSDL is the standard language to syntactically describe services capabilities. Through its WSDL interface, any client can have access to the functionality provided by a given WS. However, given that millions of services can be accessible worldwide at the same time, it is necessary a mechanism to automate the discovery and management of those services that are needed at a particular point in time. Semantic Web Services [14], that is, the semantic description of the services capabilities is meant to solve that particular issue. By formalizing the description of the functionality that WS offer, we enable machines (i.e., software programs) to autonomously process that information and interact with services without the need of human intervention. There is no standard way to define the services semantics. On the contrary, various approaches have been proposed and are being examined by experts in the field (e.g., OWL-S, WSMO or SAWSDL).

Robots can alter the world by means of the so called actuators and perceive changes in the environment through their sensors. These functions can be announced as services that the system consumes through the robots' APIs (Application Program Interface). Indeed, physical robots can be used to facilitate different kinds of services, ranging from the simple and direct knowledge acquisition (through sensors of vision, distance, sound and touch) to more complex behaviours and actions, including searching, finding, counting, manipulating and moving objects as well as discovering and analyzing information by means of the merge of intelligent capabilities, sensors and actuators. It has to be said that all these capabilities offer data and knowledge directly from the real world, so creating a source of real-time updated information. Moreover, the active manipulation of the environment and the possible subsequent direct sensing offer important capabilities that cannot be reached by solely using usual web services.

4 BRIDGE in Traffic Control

An application domain in which the use of BRIDGE can be significantly advantageous is that of traffic control. Various governmental bureaus keep traffic information (both in regional and national scopes) stored into databases being regularly updated. Commonly, this information is to some extent public and accessible to all citizens. Traffic information is generally more precise and up-to-date for big cities and severely congested roads.

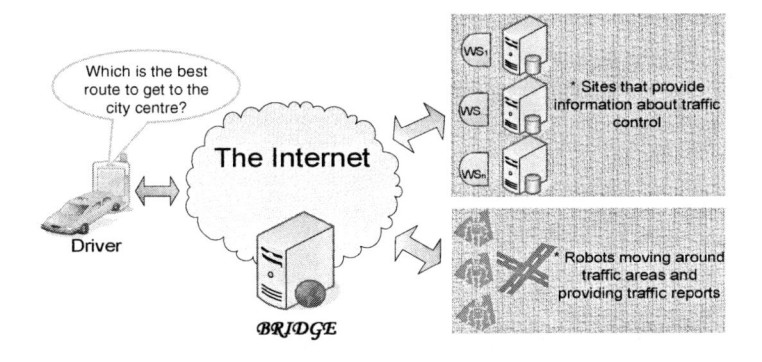

Fig. 2. Sample scenario

However, when it comes to side and small streets or less-travelled roads, there is little or even none information available. This lack of complete information makes it difficult for machines to automatically calculate an optimal route from a location to another on the basis of traffic conditions. In order to fill this gap, robots functionality can be very helpful. More precisely, given the mobility and flexibility that robots may possess, they can cover the streets -from which public databases do not have any information- gathering traffic data. BRIDGE users would have access to both the information coming from governmental databases and the live traffic data robots provide.

A sample scenario is shown in Fig. 2. In this scenario, a citizen is willing to get to the city centre through the optimal path. This goal is sent to BRIDGE, which first obtains all the possible routes to get to the target, and then finds out the traffic conditions in all the streets involved in the proposed routes. With all the information, BRIDGE can determine a route containing the less-travelled streets. We now describe this process step-by-step.

Query input: the citizen, by means of a web browser, sends BRIDGE the query in structured language. The query is received by an instance of the "Customer Agent" that will represent and act on behalf of that particular citizen within the system. The "Customer Agent" incorporates a (natural) language processing tool capable of translating the query in natural language into an ontology (a.k.a. goal ontology).

Service discovery, composition and selection: the "Customer Agent" sends the goal ontology to an instance of the "Selection Agent", which starts looking for services that can fulfil the goal. In principle, none of the available services can achieve the goal by itself. Given these circumstances, the "Selection Agent" decomposes the goal into finer-grained subgoals. The resulting plan involves to, first, get different possible routes to get to the target address and, second, get information about the traffic conditions in the streets that form part of those routes. At this point, the system discovers several services capable of fulfilling each subgoal and selects, by means of an instance of the "Selection Agent" the ones that better suits the user's needs (in terms of price, dispatch time, trustability, etc.). The discovery process is done based on the matching between the goal ontology and the WS semantic description, and the detection of robots that provide information about relevant areas. For service selection, a negotiation process takes place that tries to reconcile user requirements with service providers' conditions.

Service execution: once selected, the services are orchestrated for their execution. First, the "Customer Agent" has to notify the pertinent instances of the "Service Agent" and the "Robot Agent" (those that are the representatives of the WS and robots involved in the goal resolution, respectively) that they must initiate the service execution process. A "Service Agent" is basically a WS client that has the means to draw, from the semantic description of a service, the methods it has to invoke and the parameters that are necessary for the invocation. A "Robot Agent", on the other side, has access to the robot API and exploits it to both send commands and receive information.

Presenting the results: the "Customer Agent" is responsible for integrating the results of the execution of all the services. When the execution of all services has finalized, the corresponding agents send the results to the "Customer Agent". By using the goal ontology, the domain ontology and the services semantic descriptions, this agent is able to compose a unique response to be sent to the citizen.

System management: during this process, the "Framework Agent" is monitoring all the activities taking place in the system. If it finds out that the system has gotten blocked, the "Framework Agent" initiates the procedure to unblock it. On the other hand, when any agent receives a message it cannot interpret, it sends the message under question to the "Broker Agent", which is in charge of translating the contents of the message into the terms understood by the corresponding agent. For this, the "Broker Agent" has access to the mapping rules between the ontologies in the system that have been defined.

5 Conclusions

The approach proposed here aims to facilitate the convergence of the functionality coming from the services available on the Internet on the one hand and those solely reachable by direct physical interaction with the real world on the other. For it, the proposed platform seamlessly integrates four main ingredients: Web and robot-provided services, intelligent agents and ontologies. Ontologies are the true key to platform feasibility, as they enable the effective communication between the agents in the platform and the available services. With our approach, software applications can benefit from the autonomy, pro-activeness, dynamism and goal-oriented behavior agents provide, the high degree of interoperability across platforms and operating systems Web Services advocate, and the wide spectrum of functionality robots provide.

Several issues remain to be addressed. First, as it was pointed out in Section 3, when dealing with several disparate ontologies a mediation mechanism is needed. So far, ad-hoc solutions have been implemented into the "Broker Agent" and test case scenarios have been simplified to delimit this problem. Similarly, other service-related functions (e.g., discovery, selection, composition, etc.) have been implemented ad-hoc. The roles-based approach makes easier the inclusion, in the form of plugins, of already tested, more sophisticated implementations for all these components. Concerning robots, issues such as fault tolerance, communication and real-time requirements must be further analyzed. Finally, to fully exploit the potential benefits of ontologies, various ontology reasoners and inference engines will be examined and the most effective ones added to the framework.

Acknowledgement. This work has been partially supported by the Spanish Ministry for Industry, Tourism and Commerce under projects TSI-020100-2008-564 and TSI-020100-2008-665.

References

1. Berners-Lee, T., Hendler, J., Lassila, O.: The semantic web. Scientific American, 34–43 (May 2001)
2. Booth, D., Haas, H., McCabe, F., Newcomer, E., Champion, M., Ferris, C., Orchard, D.: Web services architecture. W3c working group note. In: World Wide Web Consortium (2004)
3. Bradshaw, J., Suri, N., Cañas, A., Davis, R., Ford, K., Huffman, R., Jeffers, R., Reichherzer, T.: Terraforming cyberspace. IEEE Computer (2001)
4. Bradshaw, J., Uszok, A., Jeffers, R., Suri, N., Hayes, P., Burstein, M., Acquisti, A., Benyo, B., Breedy, M., Carvalho, M., Diller, D., Johnson, M., Kulkarni, S., Lott, J., Sierhuis, M., Hoof, R.V.: Representation and reasoning for daml-based policy and domain services in kaos and nomads. In: Proc. of the AAMAS 2003 (2003)
5. Bruijn, O.D., Stathis, K.: Socio-cognitive grids: The net as a universal human resource. In: Proc. of Tales of the Disappearing Computer (2003)
6. Clancey, W., Sierhuis, M., Kaskiris, C., van Hoof, R.: Advantages of brahms for specifying and implementing a multiagent human-robotic exploration system. In: Proc. of the FLAIRS 2003 (2003)
7. Freedy, A., Sert, O., Freedy, E., Weltman, G., Mcdonough, J., Tambe, M., Gupta, T.: Multiagent adjustable autonomy framework (maaf) for multirobot, multihuman teams. In: International symposium on collaborative technologies (CTS) (2008)
8. García-Sánchez, F., Fernández-Breis, J.T., Valencia-García, R., Gómez, J.M., Martínez-Béjar, R.: Combining semantic web technologies with multi-agent systems for integrated access to biological resources. Journal of Biomedical Biomedical Informatics 41(5), 848–859 (2008); Special issue on Semantic Mashup of Biomedical Data
9. García-Sánchez, F., Martínez-Béjar, R., Valencia-García, R., Fernández-Breis, J.T.: An ontology, intelligent agent-based framework for the provision of semantic web services. Expert Systems with Applications 36(2P2), 3167–3187 (2009)
10. Innocenti, B., López, B., Salvi, J.: Resource coordination deployment for physical agents. In: From Agent Theory to Agent Implementation, 6th Int. Workshop AAMAS (2008) (2009)
11. Johnson, M., Feltovich, P., Bradshaw, J., Bunch, L.: Human-Robot Coordination through Dynamic Regulation. In: IEEE ICRA 2008, Pasadena, CA, USA (2008)
12. Kaminka, G.: Robots are agents, too! AgentLink News (2004)
13. Kaminka, G.: Robots are agents, too! In: Proc. of the 6th International Conference on Autonomous Agents and Multiagent Systems - Invited talk (2007)
14. McIlraith, S., Son, T.C., Zeng, H.: Semantic web services. IEEE Intelligent Systems 16(2), 46–53 (2001)
15. Noy, N.F., Doan, A., Halevy, A.Y.: Semantic integration. AI Magazine 26(1), 7–10 (2005)
16. Ryutov, T.: A socio-cognitive approach to modeling policies in open environments. In: Proc. of the 8th IEEE International Workshop on Policies for Distributed Systems and Networks (2007)
17. Scerri, P., Johnson, L., Pynadath, D., Rosenbloom, P., Si, M., Schurr, N., Tambe, M.: An automated teamwork infrastructure for heterogeneous software agents and humans. In: Proc. of the 2nd International joint conference on agents and multiagent systems (AAMAS) (2003)

18. Sierhuis, M., Bradshaw, J., Acquisti, A., van Hoof, R., Jeffers, R., Uszok, A.: Human-agent teamwork and adjustable autonomy in practice. In: Proc. of the i-SAIRAS 2003 (2003)
19. Studer, R., Benjamins, R., Fensel, D.: Knowledge engineering: Principles and methods. Data and Knowledge Engineering 25(1-2), 161–197 (1998)
20. Wooldridge, M.: An introduction to MultiAgent Systems. John Wiley & Sons Ltd., Chichester (2002)
21. Zhao, L., Mehandjiev, N., Macaulay, L.: Agent roles and patterns for supporting dynamic behavior of web services applications. In: Proc. of the 3rd International Conference on Autonomous Agents and Multi-Agent Systems, New York, USA (2004)

Building Service-Based Applications for the iPhone Using RDF: A Tourism Application

Javier Palanca, Gustavo Aranda, and Ana García-Fornes

DSIC - Universidad Politecnica de Valencia
jpalanca@dsic.upv.es, garanda@dsic.upv.es, agarcia@dsic.upv.es

Abstract. The Resource Description Framework (RDF) language is a widely used tool for information representation among several agent-based frameworks and applications. In this paper, one of these applications is introduced: a tourism application that allows the user to find touristic hotspots and venues matching specific criteria, an application that is used via an interface designed specially for the Apple iPhone^{TM}. Besides, this application uses a modern and efficient agent platform with RDF support to power the whole architecture.

1 Introduction

The *Resource Description Framework* (RDF) is a language for representing information that is widely used in web service technologies. Over the last few years, it has also been used on the Agent and MAS research field. Some agent platforms provide support for RDF to some extent, but in these cases RDF is usually only offered as a Content Language. However, the expressive power of RDF can be used not only in agent communication but also as a general form of managing information for MAS. RDF supports ontologies created by the user, according to its specific needs; it also supports system ontologies which can be specified to support flexible organization schemes, semantic querying and collaboration. Also, RDF provides a semantically rich and uniform user interface for building applications.

The development of one of these agent-based semantically rich applications is presented in this paper. Such application is a tourism application that allows a user to find a restaurant that matches with its particular needs, be it regarding price, quality or even details of the dishes and food present in the restaurant. As part of the process of building this application, a suitable agent platform with semantic capabilities should be chosen. Also, this application has to be designed to support multiple user interfaces, but its main one has to be a user interface designed and optimised for the Apple iPhone^{TM}.

Throughout this paper, the development of this application will be presented, along with a small review of the use of RDF in Multi-Agent environments. In section 2, such review is introduced. In section 3, the development of the tourism application is revealed. Following, in section 4, a more detailed look at the data flow within the application is seen, along with some implementation details in section 4.1. Finally, in section 5, some conclusions and leads to future works are presented.

2 Using RDF as a Knowledge Base Representation Language

RDF is a language for representing information about resources present on the World Wide Web. By generalizing the concept of a Resource on the Web, RDF can also be used to represent information about elements which are not directly located on the Web. RDF is based on the idea of identifying things using Web identiers (called Uniform Resource Identiers, or URIs), and describing resources in terms of simple properties and property values. The underlying structure of any expression in RDF is a collection of triples, each consisting of:

- **A Subject.** The subject can be any resource,
- **A Predicate.** The predicate is a named property of the subject, and
- **An Object.** The object denotes the value of that property.

A set of such triples is called a RDF graph (see figure 1). A RDF graph can be expressed by means of a XML-based format (called RDF/XML) which allows to storage and process these graphs with previously existing technologies (such as XML parsers). RDF is intended for situations in which the information needs to be processed by applications, rather than only being human-readable.

RDF was initially designed for the Semantic Web, but it can also be used in agent technology for storing information representation, exchange it and make queries. It is

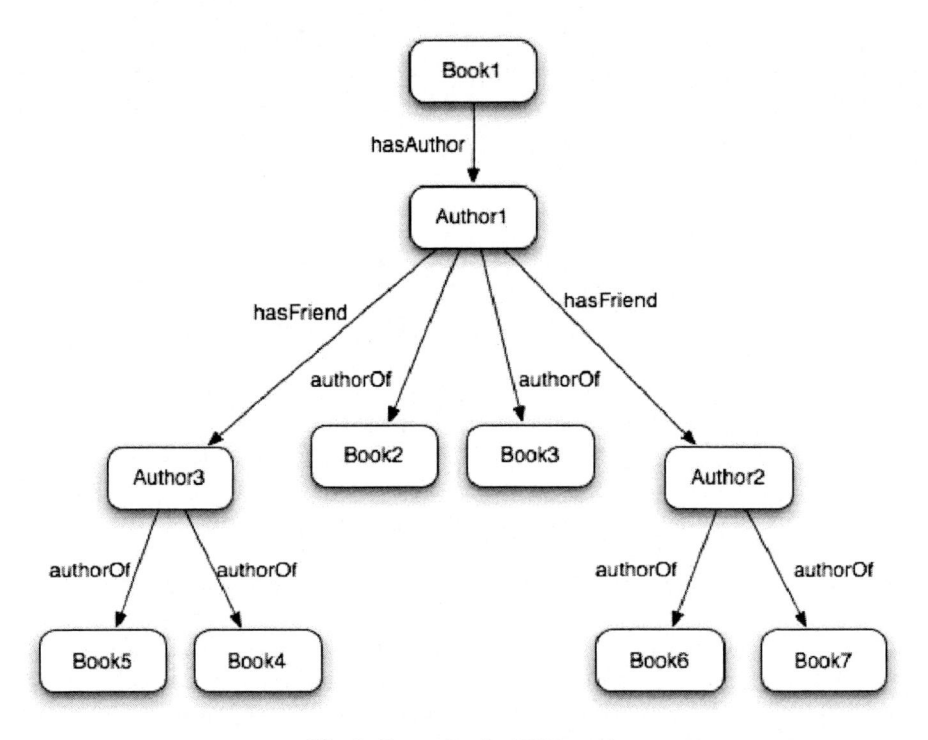

Fig. 1. Example of a RDF graph

no surprise that, over the last few years, the use of RDF in agent technology has been widely increasing. With the growth of the Semantic Web, RDF has become of the main information carriers for agents. Agents can now retrieve RDF descriptions from semantic web pages to extract relevant information. Agents can then understand and reason about that information and use it to meet the needs of the users. Moreover, one of the motivations for the development of RDF was the automated processing of Web information by software agents.

Besides, a derivative of RDF, the *Web Ontology Language* (OWL)[4], is used when greater expressiveness is needed. It is used to describe complete ontologies. Every OWL ontology is a valid RDF document, but not all RDF documents are valid OWL documents. From a Multi-Agent platform point of view, supporting RDF is the first step towards supporting OWL. Also, OWL-S[11] is an ontology built on top of OWL for describing Semantic Web Services. Ontologies provide knowledge representations. By means of the OWL-S ontology, users and software agents can automatically discover, invoke, compose, and monitor Web resources offering services.

Probably, the most common use of RDF in MAS technologies is as a Content Language. This use is specified as a standard by the Foundation for Intelligent Physical Agents (FIPA)[7] in its FIPA RDF Content Language Specification[8]. There are many agent platforms that support RDF as a content language for the agent messages. In these platforms, RDF/XML is usually the allowed serialization of the message contents.

2.1 RDF Applications

There are many example of the use of RDF in some way for communication purposes among agents. In some business domains such as e-commerce, chemistry, industry, etc... where dynamic environments and interoperability are required, RDF is used in interactions between software agents. For example, the work of Bauwens[3] in the telecommunication domain uses RDF to allow agents to communicate. This paper presents a case study where RDF is used in scenarios dealing with the provisioning of Virtual Private Networks. In [5], an infrastructure for the semantic integration of FIPA compliant platforms is given. That paper presents a platform component-matching engine for finding out the most appropriate platforms for cooperation among themselves. These kind of queries and its corresponding results are specified using RDF as a content language.

As previously stated, the expressive power of RDF is more far-reaching than just agent communication. According to[2], RDF seems to be well positioned to become the standard in the representation of ontologies in the future. Indeed, there are many studies in MAS that use RDF for other purposes apart from communication, like knowledge representation, interaction with Semantic Web, storage purposes, etc...

The Haystack project[10] is a repository which uses RDF for modeling and storing information. The project proposes and demonstrates a personal information management system that employs RDF as its primary data model. In addition, since RDF provides a standard, platform-neutral means for metadata exchange, the project claims that it naturally facilitates sophisticated features such as annotation and collaboration. Other uses of RDF as a content language include PASIBC[13], [6], [9] and [12] in a marketplace environment.

3 The Tourism Application

The properties of the RDF framework allow the use of this language as a great way to transport and store knowledge. RDF is a language which allows agents to represent and manage the knowledge described in their ontologies. Thanks to RDF properties, it can be used in, but not limited to, the following fields:

- Knowledge management internally to the agents;
- Storage in knowledge bases;
- Data description made by the user;
- and of course, Content language for the agent messages.

Such flexibility of RDF immediately brings one clear advantage: data conversion between formats is no longer needed, and such, one of the classic problems of knowledge management in multi-agent system disappears. The presence of RDF at a global level and using a single representation language allows for an efficient management of knowledge without giving up the expressivity of a content language as powerful as RDF.

This work describes how this design philosophy has been applied to a particular MAS application. This application works on an efficient agent platform using *exclusively* RDF as the knowledge representation language in each and everyone of the application layers.

Such developed application presents a tourist service search architecture based on MAS. This application allows the user to search for different kinds of tourism services (in particular, restaurants) using some search parameters based on the user's preferences. A MAS has been designed in which the agents play different predefined roles, such as:

SightAgent: The SightAgents are the ones who manage and publish the information of each tourism service. There is a SightAgent for every hotspot in the city that offers a tourism service. Such agents are maintained by the owners of the hotspot and, not only they show information about its place, but they also manage possible reservations and activities within the venue.

UserAgent: The UserAgent is the one agent which presents the application interface to the users of the system. It offers a search service and a reservations service in different formats, adapted to fit the actual device where the user sees the interface (e.g. a mobile phone, a desktop browser, etc...). The main user interface has been designed specifically for the Apple iPhone^{TM}.

PlanAgent: The PlanAgent is responsible for scheduling the activities of a single day for a given user (using the user's preferences). This agent is capable of building a schedule that fits what the user likes and, accordingly, make the necessary reservations if the user desires so. It also can modify certain aspects of the schedule to respond to fine-tuning made by the user (e.g. changes in the timetable of some events). This agent falls outside of the scope of this article.

BrokerAgent: The BrokerAgent is the mediator between the UserAgent and the SightAgent. This agent is responsible of receiving the search queries of the UserAgent and redirect them to the corresponding SightAgent. As soon as the BrokerAgent

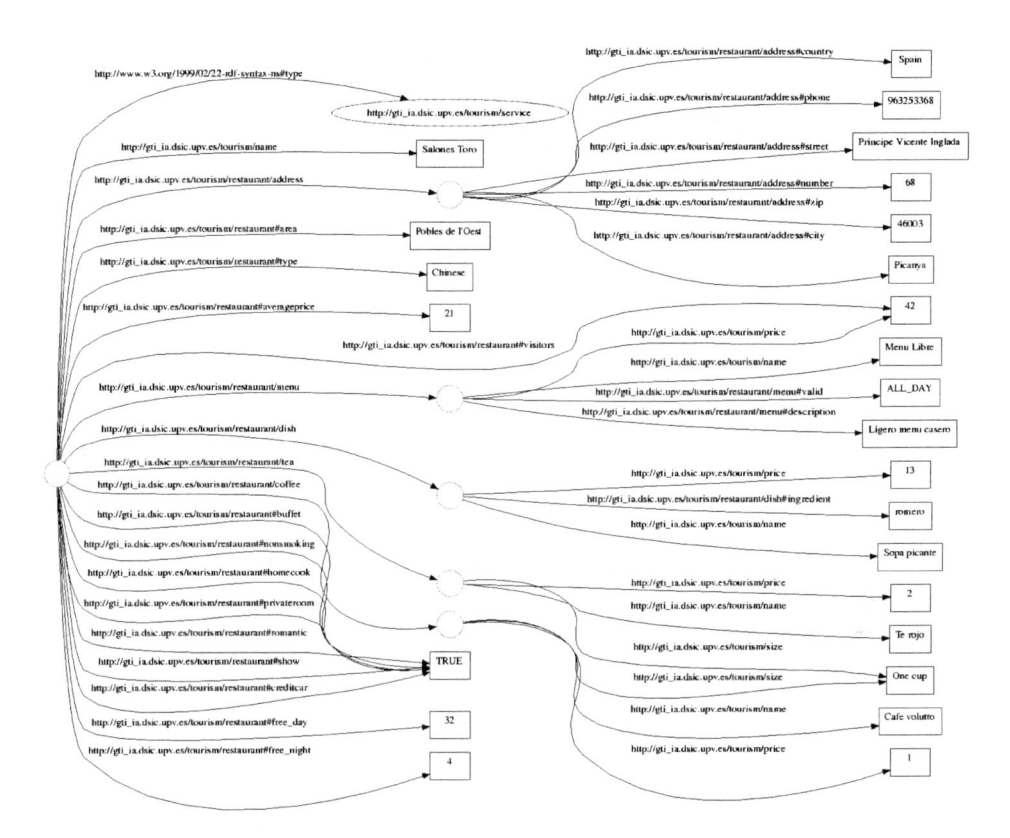

Fig. 2. Example of RDF restaurant representation

has all the information from the SightAgent gathered, it forwards the results of the query to the corresponding UserAgent for it to present the necessary choices to the user.

This application makes use of an ontology which has been specifically designed for this domain. In its current state, the application successfully works with restaurant venues, and so, the ontology is limited to define the concepts related with restaurants, such as the name of the restaurant, its address, the menus, the dishes that compose each menu, the ingredients that compose each dish, etc...

An actual example of a restaurant in RDF-triples representation can be seen following in the Figure 2.

As seen, RDF represents the relevant restaurant information using a graph. The User-Agent reads its information from a file in RDF format and uploads it to its knowledge base, which is in the same format and, thus, the information requires zero processing. This way, thanks to the platform also using the RDF language to transmit messages through the network, there is no overhead for processing and converting the information to any other formats. The whole system gains efficiency and coherence.

4 Data Flow through the Application

There are three distinct types of operations within the application: registering and unregistering a SightAgent in the BrokerAgent's *'address book'* and search for a restaurant. Registering or unregistering a SightAgent is a simple operation as it only implies sending one message from the SightAgent, with its id, to the BrokerAgent and the reply (affirmative or negative) to that message. From that moment on, the BrokerAgent will be aware of such SightAgent until it performs an unregister operation.

The search operation, however, is more complex. It starts with the sending of a message from the UserAgent with the search parameters and, after the relevant search is performed with the registered SightAgents, the BrokerAgent returns the results to the UserAgent. The whole process comprises:

1. The UserAgent sends a message to the BrokerAgent with the search parameters.
2. The Brokeragent forwards the message to all the registered SightAgents.
3. Each SightAgent compares the received message with its own RDF graph representing a restaurant. If it raises a match, the SightAgent sends the full graph to the BrokerAgent.

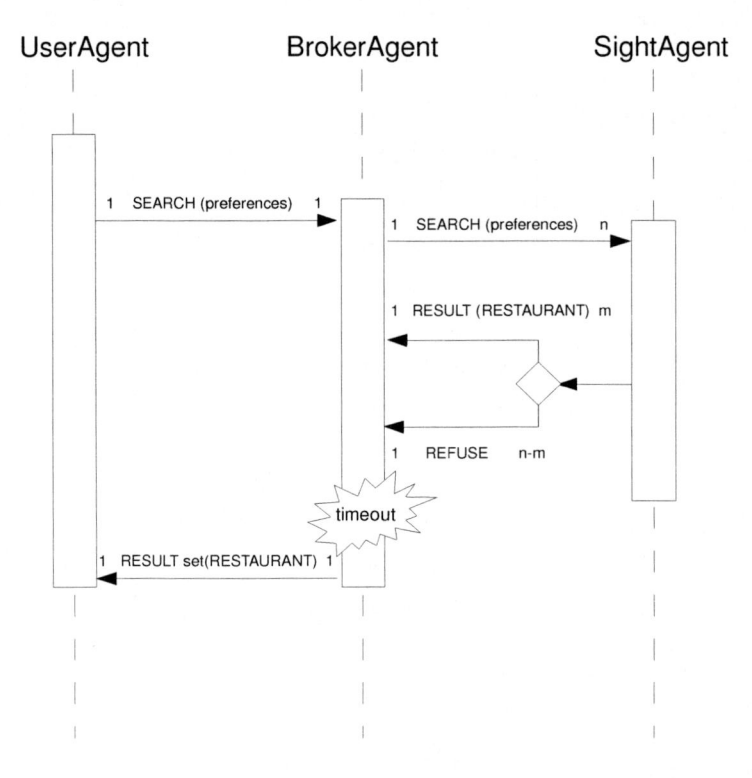

Fig. 3. Interactions Diagram

Fig. 4. iPhone User Interface

4. After a timeout, the BrokerAgent gathers all the RDF graphs that it has received from the different restaurants that match the criteria set by the initial search and forwards them all in a single message towards the UserAgent.
5. Finally, the UserAgent gets the reply message with the set of restaurants that match the criteria and presents them to the user through the interface.

In the figure 3, it can be seen the diagram that shows the interactions among the agents of this system.

As one can see, the great advantage of using RDF in this application is the almost inexistent necessity of transforming the data. All the information that is managed can be represented as RDF graphs. Besides, the ontology has been designed in such a way that it is not necessary to process the data. Along the application data flow, only copies and search processes are performed with the information stored in the RDF graphs. The only data conversion is made by the UserAgent when it presents data from the RDF graphs in a more user-friendly (more readable) fashion through the interface.

4.1 Implementation Details

For implementing this application it was obvious that an agent platform with heavy RDF support was needed. The chosen platform was the MAGENTIX agent platform[1]. MAGENTIX uses RDF not only as a the content language for the message, but also as the language in which the message itself is expressed, blurring the division between *container* and *content*, and such, it is perfectly suited for the needs of this application.

In order to develop the user interface, a web interface has been built using directly the services offered by the *mod_python* module of the Apache web server. Using python server-side scripting, the interface processes the RDF graph received from the User-Agent. This agent and the web server communicate through a custom MAGENTIX-HTTP bridge designed for this application.

Finally, the main interface of this application was conceived as a web application for the Apple iPhone^{TM}(Figure 4). For this to work, a custom cascade stylesheet (CSS) that provides the *look and feel* of iPhone^{TM} applications was used, along with some additional development in Javascript and modern HTML in order to be able to generate and animate dynamic web contents on the iPhone^{TM} web browser, as well as taking advantage of some features specific to the device. For instance, whenever a telephone number of a given restaurant is presented to the user, it acts as a telephone hyperlink, and so the user can simply *tap* the number and the device will call it. Another interesting feature is Google Maps integration: all the restaurant addresses are presented as Google Maps hyperlinks, so the user can *tap* the address and the device will open its native Google Maps application showing the actual address of the restaurant on the map.

A free version of the iPhone^{TM} web application can be accessed through the web address `http://itourism.gti-ia.dsic.upv.es`.

5 Conclusions and Future Work

The prominent role of RDF in the MAS scene has been reviewed through this paper. As a content language, a foundation for building consistent ontologies, a data model, or even as the communication language itself for the agents. It is clear that RDF plays a key role in the current and future developments of MAS applications.

One of these applications that takes advantage of RDF is a tourism application that allows a user to find a restaurant that matches with the particular needs of that user, be it regarding price, quality or even details of the dishes and food present in the restaurant. For deploying this application, a new MAS has been developed with the help of the MAGENTIX agent platform, which fully embraces RDF as its message communication language.

The application has been designed to be able to have many web interfaces. The main one, however, is a web interface for the Apple iPhone^{TM}, a specific interface that takes advantage of the features of the device, such as telephone hyperlinks and Google Maps integration.

The next evolution of this application will progress in at least a couple of areas: first, support for more types of venues (such as museums) is already underway, expanding the ontology but keeping the overall graph structure; second, the main interface will become a full-fledged iPhone^{TM}application that will take advantage of even more features of the device, such as adding user positioning (i.e. GPS) data to the queries to be able to determine which venues are closer to the user.

References

1. Alberola, J.M., Such, J.M., Espinosa, A., Botti, V., Garca-Fornes, A.: Magentix: a multiagent platform integrated in linux (2008)
2. Alexaki, S.: Managing rdf metadata for community webs. In: 2nd International Workshop on the WWW and Conceptual Modelling (WCM 2000), pp. 140–151 (2000)
3. Bauwens, B.: Xml-based agent communication: Vpn provisioning as a case study. In: XML Europe 1999 (1999)
4. Bechhofer, S., van Harmelen, F., Hendler, J., Horrocks, I., McGuinness, D., Patel-Schneider, P., Stein, L., et al.: OWL Web Ontology Language Reference. W3C Recommendation 10, 2006–01 (2004)
5. Cenk, R., Dikenelli, O., Seylan, I., Gurcan, O.: An infrastructure for the semantic integration of pa compliant agent platforms. In: AAMAS, pp. 1316–1317 (2004)
6. Delgado, J., Gallego, I., Garca, R., Gil, R.: An architecture for negotiation with mobile agents, pp. 21–32 (2002)
7. FIPA: FIPA (the foundation for intelligent physical agents), http://www.fipa.org/
8. FIPA: FIPA RDF Content Language Specification. Technical report, FIPA (2001)
9. Hui, K., Chalmers, S., Gray, P., Preece, A.: Experience in using rdf in agent-mediated knowledge architectures, pp. 82–89 (2003)
10. Huynh, D., Karger, D.R., Quan, D.: Haystack: A platform for creating, organizing and visualizing information using rdf. In: Eleventh World Wide Web Conference Semantic Web Workshop (2002)
11. Martin, D., Burstein, M., Hobbs, J., Lassila, O., McDermott, D., McIlraith, S., Narayanan, S., Paolucci, M., Parsia, B., Payne, T., et al.: OWL-S: Semantic Markup for Web Services. W3C Member Submission 22 (2004)
12. Patkos, T., Plexousakis, D.: A semantic marketplace of peers hosting negotiating intelligent agents. In: CAiSE Forum, pp. 82–89 (2005)
13. Sova, I.: Pasibc - a novel.net agent platform for knowledge based information systems, pp. 743–746. IEEE Computer Society, Los Alamitos (2005)

Designing a Visual Sensor Network Using a Multi-agent Architecture*

Federico Castanedo, Jesús García, Miguel A. Patricio, and José M. Molina

University Carlos III of Madrid
Computer Science Department
Grupo de Inteligencia Artificial Aplicada (GIIA)
{fcastane,jgherrer,mpatrici}@inf.uc3m.es, molina@ia.uc3m.es
http://www.giaa.inf.uc3m.es

Abstract. An intelligent Visual Sensor Network (VSN) should consist of autonomous visual sensors, which exchange information with each other and have reasoning capabilities. The information exchanged must be fused and delivered to the end user as one unit. In this paper, we investigate the use of the Multi-Agent paradigm to enhance the fusion process in a VSN. A key issue in a VSN is to determine which information to exchange between nodes, what data to fuse and what information to present to the final user. These issues are investigated and reported in this paper and the benefits of an agent based VSN are also presented. The aim of the paper is to report how the multi-agent architecture contributes to solving VSNs problems. A real prototype of an intelligent VSN using the Multi-Agent paradigm has been implemented with the objective to enhance the data fusion process.

1 Introduction

The main objective of a VSN is to be able to monitor a large geographic area and to receive information of what is happening there [1]. In a VSN with several sensors monitoring an area, data fusion is a main element. The visual sensors could be deployed with overlapped field of views, therefore the information must be fused to provide to the final user a single view [2]. With sensor data correlation and fusion, the VSN can detect targets and track movements in the whole monitoring area. In order to achieve target tracking in a big area, one single visual sensor is not sufficient, the use of several sensors is necessary. Also, some redundancy is needed to carry out the fusion process among all updating sensors.

Multi-sensor data fusion [3] is about combining and relating data from several visual sensors to achieve a better quality than could be achieved by a single visual sensor. Fused data represents an entity in greater detail and with less uncertainty than what is obtainable from any of the individual sources. Combining additional independent and redundant data usually provides an improvement of the results.

A common situation in data fusion is that data sources are spatially or temporally distributed. In complex distributed systems it is impossible, at the design time of the

---

* This work was supported in part by Projects CICYT TIN2008-06742-C02-02/TSI, CI-CYT TEC2008-06732-C02-02/TEC, SINPROB, CAM MADRINET S-0505/TIC/0255 and DPS2008-07029-C02-02.

system, to know all the possible interactions between the components of the system. Because communication will occur at unpredictable times and for unpredictable reasons. If we use a classical mathematical modelling approach, in the design phase of the system, all the possible combinations should be programmed. This task could be very complex and requires modelling the error distributions of the visual sensors. Therefore, it is better, from a design point of view, to let the agents at run time decide when to communicate and what to communicate.

VSN are naturally distributed: data sources are spatially distributed, data processing is performed in a distributed manner and the users interested in the result could also be distributed. For these characteristics, VSN definitely falls into the class of potential multi-agent applications.

The multi-agent paradigm has been applied to computer vision in several different works. Zhou et. al [4] use an agent-based approach that uses an utility optimization technique to guarantee that vision tasks are fulfilled. In the embedded multi camera tracking context, the multi-agent paradigm has been also applied. Quaritsch et. al. [5] present an agent-based autonomous multi camera tracking system, providing a soft hand-over in tracking tasks. The basic idea is that agents are instantiated in the embedded device as the tracking object is moving. Several papers using multi-agent systems in distributed sensor networks (without visual capabilities) have been published [6]. However, multi-agent systems applied to visual sensor networks are less explored. VSNs are different from other type of sensor networks because it is necessary to perform image processing tasks in order to obtain valuable information. Berge-Cherfaoui [7] proposed a Multi-Agent approach based on the blackboard model. The blackboard model is one of the first multi-agent communication models and is less flexible than current communication models based on FIPA-ACL [8]. The work of Pavón et. al [9] is very related and provides an useful information on how a multi-sensor system is modelled using agents from a software engineering point of view.

In contrast to other works, this paper focuses on how the multi-agent paradigm is used to develop a VSN enhancing the fusion process. Agents behavior should involve perception, reasoning, communicating, acting and learning in complex environments. In multi-agent VSNs, visual sensing is the mechanism or process whereby the system can be influenced by the environment around it. Therefore, digital image processing techniques must be carried on the agents with the purpose of obtain the information of the object (position, size, velocity and so on). Instead of exchange the raw images between the agents, which implies a lot of communication bandwidth, only relevant information is sent. The main objectives of the proposed multi-agent VSN architecture are:

- To build an open architecture which is easy to scale. We could easily add new agents (with different or same goals) in the multi-agent VSN system.
- A standard based architecture, which would allow us to inter-operate with third part developments.
- Improving the visual sensor processes through reasoning and coordination of the involved visual sensors.

The rest of the paper is organized as follows. In the next section the requirements and objectives of a VSN are presented. Section 3 introduces the multi-agent systems and how they could solve VSN problems. Then, the approach used to implemented the multi-agent VSN is presented. Finally, section 5 concludes the paper with a summary of the major benefits on using a multi-agent VSN.

2 Visual Sensor Networks: Requirements and Objectives

The work on VSN can be divided into two categories, based on visual configuration: (1) spatially overlapping field of view and (2) non-overlapping field of view. In non-overlapping field of view configuration, since there might be significant gaps between the field of views of the cameras, the precision in sensor data fusion may be suffer. Most sensor fusion algorithms assume an overlapping field of view. In this work, an overlapping field of view is also assumed.

Implementing and developing a VSN is a challenging task which involves solving several problems. Firstly, VSN applications must cope with a dynamic and not defined number of detected objects. Therefore, it is mandatory to use algorithms that performs detection, tracking and removing objects from the captured images. In VSN, movement detection and object tracking is performed using several visual sensors connected in a network. VSN are extremely useful because they provide an extended area coverage and could exploit the redundancy of the information. In order to integrate the different parts of a VSN it is necessary to establish: (1) the objectives of each module and (2) how the different modules interact to reason about the context in the global scene.

Some of the open problems to fuse data from sensor in a VSN are: (1) visual sensor configuration, (2) camera calibration, (3) object matching, (4) hand-over, (5) state estimation and (6) occlusion management.

(1) The physical installation of the visual sensors provides a big impact in the performance and economic cost of the system. On one hand, many redundant sensors increase the economic cost. On the other hand, if some areas are not being covered, it could be a risk. Therefore, cover all the area with the minimum visual sensors is an important task. Pavlidis et al. [10] presented an algorithm to cover a parking area with the minimum visual sensors. The basic idea is to increase the visual sensors number with the constraint of having 25%-50% of covered area.

(2) The calibration process (or common referencing) involves projecting the coordinates from the local plane of each visual sensor to global coordinates. The calibration process uses a geometrical projection known as the pinhole model or perspective projection model [11]. The algorithms that reconstruct the 3-D structure from an image, or calculate the object's position in the space need equations to match the 3-D points with their corresponding 2-D projections. Therefore, we have to calculate what is known as the *extrinsic parameters* (those which relate both 3-D reference systems) and the *intrinsic ones* (which relate the reference system of the camera to the image). This task is known in computer vision as camera calibration process [11]. The pinhole model establishes the mathematical relation between the spatial points and their equivalents in the camera image.

Stein [12] and Lee et al. [13] use the tracking positions of an object and the ground plane restriction to determine the projection's matrix transformation. A different approach is used by Intille and Bobick [14], they propose to use the football field mark lines for calibrate the environment. In our work, the constraint in planar motion (ground

Fig. 1. Local projection on the ground plane (white point)

plane restriction) is used to estimate a homograph correspondence between the different views and avoid the use of more complex stereo-vision processes. In order to obtain the global coordinates of the local tracks, visual sensors compute the inverse transformation of the middle-bottom point of each object (see Figure Fig. 1).

(3) Object matching involves associating sensor measurements with target tracks. It is known as data association. Data association could be applied at different levels in the sensor environment. In a visual sensor network, the identity of each object must be maintained over the environment being monitored. The most common algorithms used for data association are: nearest neighbor, Multiple Hypothesis Tracking (MHT) [15], Probabilistic Multiple Hypothesis Tracking (PMHT) [16], Probabilistic Data Fusion (PDA) [17], Joint Probabilistic Data Fusion (JPDA) [18] and all their variants.

(4) In VSN, hand-over takes place when the tracked object moves from one camera's field of view to another camera's field of view. The main reason for doing a hand-over is because the object is going to disappear from the camera's field of view. VSN environments must cope with hand-over in order to allow a global coherence of the objects being monitored in the environment. Quaritsch et. al [5] propose a mobile multi-agent approach in which an agent migrates from one smart sensor to another for solving the hand-over problem. Other authors use a color based approach for matching the detected objects. However this technique is not very robust and fails with illumination changes and with people wearing clothes with similar colors.

(5) State combination from multiple visual sensors is an important problem which many potential applications.

The information exchanged by the visual sensors in the network must be *aligned in time* in the fusion process. Therefore, the information should be time-stamped as soon as possible. The internal clocks of each visual agent are different and perhaps not synchronized. We use the Network Time Protocol algorithm [19] in order to maintain each clock synchronized in a common time basis.

When the information is going to be fused on the fusion node, it must be aligned in time. For this task, an algorithm that interpolates the information to the fused instant is used.

The data fusion techniques involve the combination of estimates obtained with different visual sensors. As we said before, instead of sending the raw images, which imply a lot of communication bandwidth, only tracking information is sending. In the proposed system, each visual sensor performs tracking of the detected persons on their field of view. The tracks of the different overlapped sensor should be fused to obtain a global track. The internal tracking process provides for each detected object X_{T_j}, an associated track vector of features $\hat{X}_{T_j}^{S_i}[n]$, containing the numeric description of their features and state their location, speed, dimensions, etc. as well as the associated error covariance matrix, $\hat{P}_{T_j}^{S_i}[n]$.

Other important problems to solve in a VSN are: (1) data fusion coherence, (3) remove inconsistent information in the fusion process and (3) the coordination of the visual sensors to coverage the area being monitored. Data fusion coherence aims to solve situations in which the detected objects in an overlapped area are different. Thus, it is mandatory a reasoning process between the sensors in order to clarify the situation. Inconsistent information in the fusion process is also related with data fusion coherence and it is carried out by statistical tests between the information [2]. Sensor coordination is necessary in order to provide a mechanism for monitoring an extended area.

3 Multi-agent Systems and Their Application to Visual Sensor Networks

There are different types of multi-agent architectures and can be grouped into: (1) reactive, (2) deliberative and (3) hybrid. We choose the Belief-Desire-Intention agent paradigm (deliberative type), however there are also different ways of modelling multi-agent systems;for example, Corchado and Laza [20] proposed to construct agents using case based reasoning technology. Focusing on the Belief-Desire-Intention (BDI) model of agency [21], an agent is formally specified by its mental state, consisting of its beliefs, desires, intentions and ongoing interaction (with other agents and its environment). Therefore, each agent has its own set of beliefs, desires and intentions. The state of the agent at any given time is a triple (B, D, I), where $B \subseteq Beliefs$, $D \subseteq Desires$ and $I \subseteq Intentions$. An agent's *beliefs* correspond to the information the agent acquires from the environment and the other agents. It represents the knowledge of the state of the world. *Desires* capture the motivation of the agents. A desire represents the state of affairs that the agent would like to bring about. *Intentions* are the basic steps chosen by the agent to achieve its *Desires* and represent the desires which an agent has committed to achieve. *Intentions* constrain the reasoning that an agent requires to perform in order to select the action that will be performed. Practical reasoning involves two important processes: (1) deciding the goals to be achieved and (2) how to achieve them. In our proposed multi-agent VSN architecture, intentions are dropped as the environment evolves or the intention has been already achieved. For example, if the tracking objects disappear from the field of view, the tracking intention will not be achievable. The proposed multi-agent VSN architecture is based on the PRS systems computational model [22] [23], and specifically on the open source multi-agent framework JADEX [24].

The autonomous agents cooperate for the purpose of improving their local information, and cooperation is achieved by message exchange through a declarative high-level agent communication language. FIPA-ACL [8] is the standard language for the communication between the agents. This communication consist of high level speech act theory. Those speech acts can be informing, requesting, accepting, rejecting, believing, and so on. The common objectives of the agents in the multi-agent system are achieved with the help of their interactions. Agents communication takes place using FIPA ACL [8] messages in an asynchronous manner. Therefore, the sender and the receiver do not

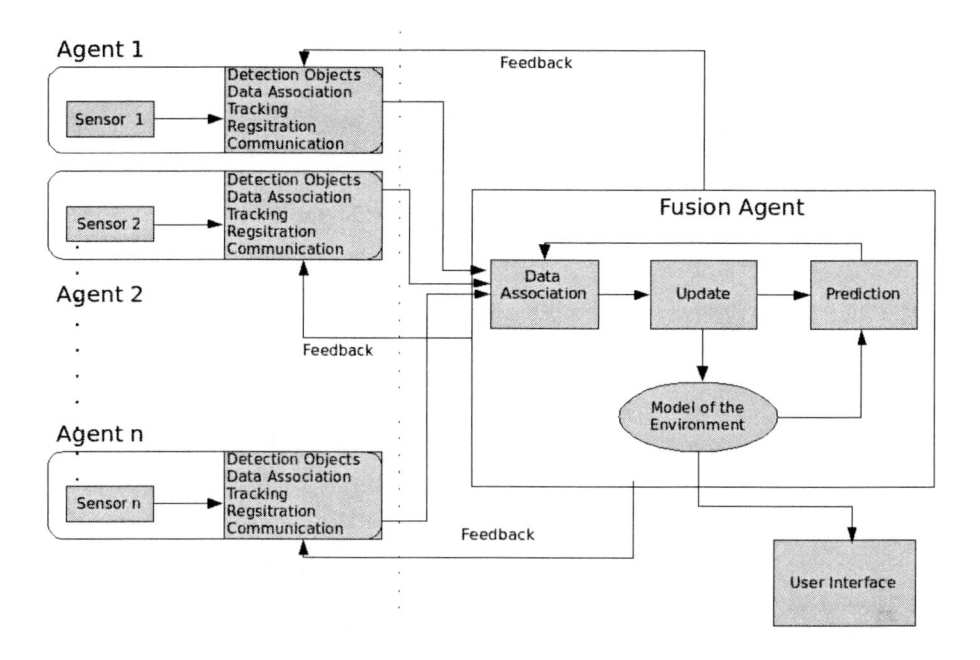

Fig. 2. Multi-Agent VSN architecture with feedback. The figure depicts n sensor agents. Each of them performs: object detection, data association, tracking, registration and communicates with their respective fusion agent. The fusion agent performs data association, updates the (global) model of the environment, make predictions and communicates the information to the interface agent.

need to be synchronized in order to communicate. The communications and interactions of the agents provide a natural way by which the information is exchanging in a VSN. Others works, related with agent communication and computer vision, are based on specific communication protocols: (1) the blackboard model [7], (2) the information subscription coordination model [25] and (3) the contract-net protocol [26]; and do not exploit the benefits of a standard communication language. However, FIPA ACL includes an implementation of the contract net protocol.

The use of a BDI multi-agent architecture in a VSN environment provides several advantages with respect to the traditional visual sensor systems:

1. The improvement of accuracy and performance due to cooperation. By means of cooperation agents exchange information. Information redundancy (given by overlapped field of views) could be exploited to enhance the accuracy and performance, i.e., a better tracking accuracy [27] [2].
2. The agents solve problems which are part of a global problem, i.e., most of them perform local tracking using local signal processing, but there is a global view of the tracking which is performed by all of them [28].
3. Sharing information between them, allow agents to correct errors by their ability to make an explicit coordination.

4. In a complex environment with several occlusions, lost of tracks usually occurs, therefore a fusion process is justified to avoid tracking inconsistency.

In figure Fig. 2 a global view of the data fusion architecture is showed. The data fusion algorithm consists on weighting each local sensor track according to the covariance matrix after performing a statistical test and is based on the algorithm of the reference [2]. There is a feedback loop between the fused tracks and each local sensor agent. The objective of the feedback information is to allow a reasoning process in each agent in order to correct the local information about the scene. The inverse transformation of the tracked positions is carried in each local agent and a disparity measure is obtained. Then, if the obtained information differs on a defined threshold, the state of the object is changed according to the fused state. Feedback information provides a way to improve the sensor agent's local quality information, using the data fused from all the involved sensor agents. This process is related with the local/global model of the environment. The communication is being performed by means of FIPA ACL messages and provides a communication middleware in the VSN.

In summary, agents in a VSN should reason about the quality of the information which is going to send and receive, negotiate about the targets that are tracking and decide if they take actions to correct or to delete information.

4 Implementation of a Multi-agent VSN

Using an existing multi-agent platform could provide all common agent tasks as message handling, message encoding/decoding, monitoring and so on. There are several development frameworks of multi-agent platforms, by which some of them are more mature than other. In this work, we choose JADEX [24] as the underlying platform by several reasons. First, it is FIPA compliant, secondly, it is open source and finally it is implemented in Java which allows an easy development, giving the opportunity of obtain experimental results quickly.

The system is divided in three different type of agents, however notice that it could be running N instances of each different type.

1. Sensor agent: this type of agent tracks all the objects and sends the data to a fusion agent. It acquires environmental information through a camera and performs the local signal processing task. The tasks carried out are: object detection, data association, estimation of state and communication with the fusion agent. These tasks are achieved entirely using local data and pixel coordinates and cope with the largest amount of data that can be captured from the video stream. The vision algorithms are based on the OpenCv [29] library. The agent accesses them through Java Native Invocation (JNI) [30]. Each surveillance agent can freely communicate with other agents by exchanging ACL (Agent Communication Language) messages to make the surveillance process more effective.
2. Fusion agent: It fuses the received data from each surveillance agent, which involves: data association, update the model of the environment and predict the evolution of the information. It also sends feedback information to the sensor agents and informs the interface agent.

3. Interface agent: This agent receives the fused data and shows it to the final user, which is also the user interface of the surveillance application. In this graphical user interface the fused tracking information is shown.

The main belief of the sensor agents is the knowledge about the detected objects in their field of view. Sensor agents have the following *desires*: the tracking desires which are performed continuously, the looking desire which allows the agent to observe the environment and the communication desires. Sensor agents have the following *Intentions* (which are generated from the previous desires): tracking, looking the environment, sending new track information (a new object is detected), updating track information (a tracked object change the position) and delete track information (an object disappear).

An indoor evaluation of the proposed architecture was undertaken in our laboratory. Three Sony EVI-100 Pan-Tilt-Zoom cameras were connected to a matrox morphis frame grabber. In this experiment they were used as static cameras. Each PTZ camera is controlled by a surveillance-sensor agent which runs on a dedicated personal computer. There is also a fusion agent which receives each surveillance-sensor agent's tracking information. A record of 300 frames (with a resolution of 768x576 pixels) from one person randomly moving (in a zone of 660cm x 880cm) is performed.

The image sequences of each camera were analyzed by a sensor agent. Inside each sensor agent, the looking interval parameter value was set to 10 milliseconds. The fusion agent data fusion frequency was set to 10 milliseconds, the spatial difference was set to 140 centimeters and the temporal difference was set to 20 milliseconds. For each detected object (track) the centroid was obtained. This position was projected on the ground plane, and then the information in terms of global coordinates was sent to the fusion agent.

5 Conclusions

In this paper the benefits of applying the multi-agent paradigm for modelling a VSN are presented. The specific nature of the multi-sensor visual network, logically and geographically distributed, fit well with the multi-agent paradigm. The use of a classical approach to deal with a visual sensor environment requires to model the sensor's errors which are unknown in most of the visual environments. In contrast, the use of a multi-agent approach allows a distributed reasoning of the information acquired by each visual sensor by means the social ability of each agent. In this work, the standard FIPA ACL communication language is used to share the agent's information. In addition, the use of a multi-agent architecture, based on the FIPA standard, allows the inter-operation with other systems.

References

1. Patricio, M.A., Carbó, J., Pérez, O., García, J., Molina, J.M.: Multi-agent framework in visual sensor networks. EURASIP Journal on Advances in Signal Processing 2007, Article ID 98639, 21 pages (2007), doi:10.1155/2007/98639
2. Castanedo, F., Patricio, M.A., Garcia, J., Molina, J.M.: Robust data fusion in a visual sensor multi-agent architecture. In: 10th International Conference on Information Fusion (July 2007)

3. Hall, D.L., Llinas, J.: Handbook of MultiSensor Data Fusion. CRC Press, Boca Raton (2001)
4. Zhou, Q., Parrott, D., Gillen, M., Chelberg, D.M., Welch, L.: Agent-based computer vision in a dynamic, real-time environment. Pattern Recognition 37(4), 691–705 (2004)
5. Quaritsch, M., Kreuzthaler, M., Rinner, B., Bischof, H., Strobl, B.: Autonomous Multicamera Tracking on Embedded Smart Cameras. EURASIP Journal on Embedded Systems 2007(1), 35–35 (2007)
6. Lesser, V., Ortiz, C.L., Tambe, M.: Distributed Sensor Networks: A Multiagent Perspective. Kluwer Academic Publishers, Dordrecht (2003)
7. Berge-Cherfaoui, V., Vachon, B.: A multi-agent approach of the multi-sensor fusion. In: Fifth International Conference on Advanced Robotics. Robots in Unstructured Environments, ICAR, pp. 1264–1269 (1991)
8. FIPA. Fipa communicative act library specification (2001)
9. Pavón, J., Gómez-Sanz, J., Fernándex-Caballero, A., Valencia-Jiménez, J.J.: Development of intelligent multisensor surveillance systems with agents. Robotics and Autonomous Systems 55(12), 892–903 (2007)
10. Pavlidis, I., Morellas, V., Tsiamyrtzis, P., Harp, S.: Urban surveillance systems: from the laboratory to the commercial world. Proceedings of the IEEE 89(10), 1478–1497 (2001)
11. Tsai, R.: A versatile camera calibration technique for high accuracy 3d machine vision metrology using off-the-shelf tv cameras and lenses. IEEE Journal of Robotics and Automation 3(4), 323–344 (1987)
12. Stein, G.P.: Tracking from multiple view points: Self-calibration of space and time. In: Proceedings of the IEEE Computer Society Conference on Computer Vision and Pattern Recognition, vol. 1, pp. 521–527 (1999)
13. Lee, L., Romano, R., Stein, G.: Monitoring Activities from Multiple Video Streams: Establishing a Common Coordinate Frame. IEEE Transactions on Pattern Analysis and Machine Intelligence 22(8), 758–767 (2000)
14. Intille, S.S., Bobick, A.F.: Closed-world tracking. In: Proceedings of the Fifth International Conference on Computer Vision, pp. 672–678 (1995)
15. Reid, D.: An algorithm for tracking multiple targets. IEEE Transactions on Automatic Control 24(6), 843–854 (1979)
16. Streit, R.L., Luginbuhl, T.E.: Maximum likelihood method for probabilistic multihypothesis tracking. In: Proceedings of SPIE, vol. 2235, p. 394 (1994)
17. Bar-Shalom, Y., Tse, E.: Tracking in a cluttered environment with probabilistic data association. Automatica 11, 451–460 (1975)
18. Fortmann, T.E., Bar-Shalom, Y., Scheffe, M.: Multi-target tracking using joint probabilistic data association. In: 19th IEEE Conference on Decision and Control including the Symposium on Adaptive Processes, vol. 19, pp. 807–812 (December 1980)
19. Mills, D.L.: Internet time synchronization: the network time protocol. IEEE Transactions on Communications 39(10), 1482–1493 (1991)
20. Corchado, J.M., Laza, R.: Constructing deliberative agents with case-based reasoning technology. International Journal of Intelligent Systems 18(12), 1227–1241 (2003)
21. Bratman, M.E.: Intentions, Plans and Practical Reasoning. Harvard University Press, Cambridge (1987)
22. Georgeff, M.P., Lansky, A.L.: Reactive reasoning and planning. In: Proceeedings of the Sixth National Conference on Artificial Intelligence (AAAI), pp. 677–682. AAAI Press, Menlo Park (1987)
23. Ingrand, F.F., Georgeff, M.P., Rao, A.S.: An architecture for real-time reasoning and system control. IEEE Expert, [see also IEEE Intelligent Systems and Their Applications] 7(6), 34–44 (1992)
24. Pokahr, A., Braubach, L., Lamersdorf, W.: Jadex: Implementing a bdi infraestructure for jade agents. Search of Innovation (Special Issue on JADE) 3(3), 76–85 (2003)

25. Multimedia Object Descriptions Extaction from Surveillance Types, http://www.tele.ucl.ac.be/PROJECTS/MODEST/index.html
26. Graf, T., Knoll, A.: A Multi-Agent System Architecture for Distributed Computer Vision. International Journal on Artificial Intelligence Tools 9(2), 305–319 (2000)
27. Castanedo, F., García, J., Patricio, M.A., Molina, J.M.: A multi-agent architecture to support active fusion in a visual sensor network. In: Second ACM/IEEE International Conference on Distributed Smart Cameras (September 2008)
28. Castanedo, F., Patricio, M.A., García, J., Molina, J.M.: Bottom-up/top-down coordination in a multiagent visual sensor network. In: IEEE Conference on Advanced Video and Signal Based Surveillance, AVSS, pp. 93–98, September 5-7 (2007)
29. OpenCV, intel.com/technology/computing/opencv/index.htm
30. JNI. Java native method invocation specification v1.1 (1997)

Designing Virtual Organizations

N. Criado, E. Argente, V. Julián, and V. Botti

DSIC Universidad Politécnica de Valencia
Camino de Vera s/n, 46022 Valencia, Spain
{ncriado,eargente,vinglada,vbotti}@dsic.upv.es

Abstract. Virtual Organizations are a suitable mechanism for enabling coordination of heterogeneous agents in open environments. Taking into account many concepts of the Human Organization Theory, a model for Virtual Organizations has been developed. This model describes the structural, functional, normative and environmental aspects of the system. It is based on four main concepts: organizational unit, service, norm and environment. All these concepts have been applied in a case-study example for the management of a travel agency system.

Keywords: Multi-Agent Systems, Norms, Virtual Organizations.

1 Introduction

The development of new technologies, such as Internet, e-commerce, Web Services, etc. has shown the need of building distributed applications in which autonomous and heterogeneous agents work together [1]. Both multi-agent systems (MAS) and service-oriented architectures (SOA) are particularly well suited as a support for this new tendency. In this sense, service standards provide an infrastructure for interactions among heterogeneous agents. On the other hand, agent features such as sociability, autonomy, etc. allows the development of dynamic and complex services [2].

Virtual Organizations are a set of individuals and institutions that need to coordinate resources and services across institutional boundaries [3, 4]. Thus, they are open systems formed by the grouping and collaboration of heterogeneous entities and there is a separation between form and function that requires defining how a behavior will take place. They have been employed as a paradigm for developing agent systems [5, 6]. Organizations allow modeling systems at a high level of abstraction. They include the integration of organizational and individual perspectives and also the dynamic adaptation of models to organizational and environmental changes [7] by forming groups with visibility boundaries [6].

Based on the Human Organization Theory [8], we have defined a *Virtual Organization Model* (VOM) that allows describing the main aspects of an organization: its structure, behavior, dynamics, norms and environment. VOM mainly defines which are the system entities, how they are related between them and their environment, which is their functionality, which services they need and/or they offer, how entities adopt some predefined functionality inside the system and which are their behavioral restrictions. The main contribution of VOM is that it is aimed at modeling open societies in which heterogeneous and autonomous agents work together, focused on the integration of both Web Services and MAS technologies. Thus, entities' functionality is described, published and accessed by means of services.

The *Virtual Organization Model* approach takes into account the four main aspects of a Virtual Organization: (i) Structural dimension, which describes components of the system and their relationships; (ii) Functional dimension, that details system functionality; (iii) Normative dimension, that defines mechanisms employed by a society in order to influence its member behavior; and (iv) Environmental dimension, which describes the environment in terms of its resources and how agents can perceive and act on them.

This paper is structured as follows: section 2 contains a detailed explanation of related works and motivations for the VOM model. Section 3 introduces the developed model, describing its different dimensions. Section 4 describes a case study example that has been designed using VOM. Finally, some conclusions and future work are detailed in section 5.

2 Modeling Open Systems

Several models and tools have been presented inside the MAS community in order to give support for the design of virtual organizations. Our proposal of organizational model takes these works as a starting point and increases them in order to overlap their drawbacks and thus ensuring behavior of dynamical societies, which are formed by heterogeneous and autonomous entities. Following, related proposals on organizational, functional, normative and environmental models are briefly commented.

Regarding organizational models, VOM tries to combine several organization modeling language proposals, especially AML [9], AGRE [10], MOISE+ [11], INGENIAS [12] and OMNI [13]. Thus, the Structural Dimension of VOM takes into account the agent-group-role concepts employed in AGRE; the group, roles and links notions employed in MOISE+; and also the *organizational unit* concept of AML and its related usage in the Human Organization Theory. In AML, an organizational unit is seen both as a global atomic entity and as an association of internal entities, which are related among them according to their roles, functionality, resources and environment. Therefore, the structural dimension allows the specification of a system at a high level of abstraction by means of role and organizational unit concepts.

The Functional Dimension is normally represented by means of tasks and goals pursued by agents. For example, in MOISE+ global goals are defined and decomposed into missions performed by agents. VOM Functional description extends previous proposals in three ways: (i) functionality in VOM is described employing the OWL-S standard, which allows defining functionality more expressively, i.e. representing service preconditions and effects; (ii) global functionalities are described as complex services that are composed of atomic services, so a complex service specification describes how agent behaviors are orchestrated; and (iii) functionality is detailed in two ways: services that entities perform and services that entities need. Therefore, our proposal focuses on expressing functionality of a system and its components by means of service descriptions. Thus, Service Oriented Computing (SOC) concepts such as ontologies, process models, choreography, facilitators, service level agreements and quality of service measures can be applied to MAS.

The Normative Dimension contains a set of mechanisms for ensuring social order and preventing self-interested behaviors. Our proposal makes use of the normative approach of both MOISE+ and OMNI (based on E-Institution framework [14]). They define norms as a description of the expected behavior. However, none deviation from the desired behavior is possible. In this sense, they assume the existence of a middle-ware that controls all agent interactions. Our proposal is not based on a centralized norm enforcer. Thus, BDI agents are free to decide to respect norms. The VOM normative dimension defines sanctions and rewards as a persuasive method for norm fulfillment.

Finally, the Environmental Dimension, which focuses on describing the elements of the environment, has been mainly considered in AGRE, AML and INGENIAS works. VOM Environmental Dimension describes the environment components in a standard way, integrating the main abstraction of these approaches. The *Resource* concept has been adopted from INGENIAS framework. This concept is similar to the *Body* abstraction of AGRE models, which indicates how agents perform actions on resources. Moreover, the *Port* concept of AML is also integrated in VOM, which represents an abstraction for accessing both system resources and published functionality. Therefore, VOM Environmental Dimension allows heterogeneous agents to access to diverse functionalities and resources.

As a result, the proposed VOM integrates different MAS modeling approaches, and it mainly focuses on the integration of SOA and MAS techniques for supporting dynamical and open societies. In the next section, a detailed description of the VOM dimensions is presented.

3 VOM Description

As explained before, the *Virtual Organization Model* is divided into four dimensions: structural, functional, environmental and normative. Related meta-models have also been defined using UML notation language and extending ANEMONA [15] and INGENIAS approaches [12], but they have not been included here due to lack of space. These VOM meta-models are deeply explained in [16]. Moreover, a specific ontology for Virtual Organizations has been specified, which includes the VOM elements and allows agents to describe the organization structure, its norms, services and how roles can be dynamically adopted.

Following subsections contain a more detailed description of the four dimensions covered with VOM.

3.1 Structural Dimension

The Structural dimension (Figure 1a) describes which are the elements of the system (agents and organizational units) and how they are related.

The proposed VOM defines an *Organizational Unit* (OU) as a basic social entity that represents the minimum set of agents that carry out some specific and differentiated activities or tasks, following a predefined pattern of cooperation and communication [16, 17]. This association can also be seen as a single entity at analysis and design phases, since it pursues goals, offers and requests services and it plays a specific role inside other units.

An OU is formed by different *entities* (*has_member* relationship) along its life cycle, which can be both single agents and other OUs viewed as a single entity. The Organizational Units present different topologies and communication relationships depending on their environment, the type of activities that they perform and their purpose. The basic topologies are: (i) *Simple Hierarchy*, in which a supervisor agent has control over other members; (ii) *Team*, which are groups of agents that share a common goal, collaborating and cooperating between them; and (iii) *Flat*, in which there is none agent with control over other members. Using these three basic social entities, more complex organizational structures can be modeled, such as federations, bureaucracies, congregations or coalitions, as it is deeply explained in [18].

An OU includes a set of roles that can be acquired by its members (*has_role*) and the sort of relationships among them (*has_ relationship*). The Role concept is defined by three attributes: *Visibility*, *Accessibility* and *Position*. *Visibility* indicates whether agents can obtain information of this role from outside the unit in which this role is defined (*public*) or from inside (*private*). *Accessibility*, based on [13], considers two types of roles: (a) *internal* roles, which are designed to enforce the social behavior of other agents and/or they require a fixed, trusted and reputated functionality, so they are assigned to internal agents of the system platform; and (b) *external* roles, which can be enacted by any agent, whose functionality might have not been implemented by the system designers, but it must be controlled by the system norms and services. Finally, *Position* determines its structural position inside the unit, such as supervisor or subordinate.

The *Relationship* concept, based on [19], represents social connections between *Roles*. An *information* relationship connects agents entitled to know each other and communicate relevant information. A *monitoring* relationship implies a monitoring and controlling process of agent activity. Finally, a *supervision* relationship implies that an agent has power over another, i.e. it delegates one or more objectives or tasks to its subordinate agents.

3.2 Functional Dimension

The functional dimension (Figure 1b) details the specific functionality of the system, based on services, tasks and goals. It also defines system interactions, which are activated by means of goals or service usage.

Our proposal is focused on the integration of both SOA and MAS technologies. Services represent the functionality that agents offer to other entities, independently

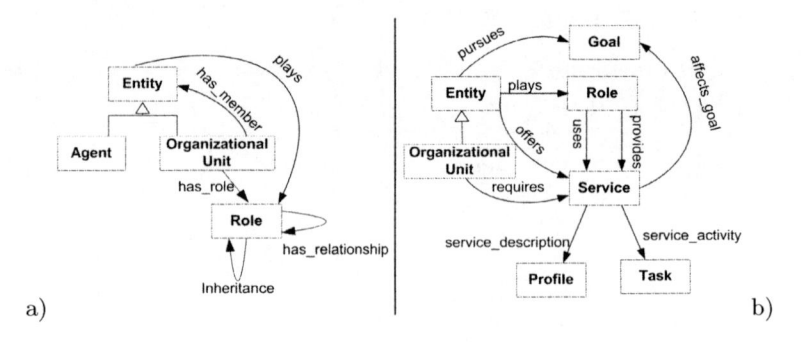

Fig. 1. a) Structural Dimension; b) Functional Dimension

of the concrete agent that makes use of it. Services can be atomic (simple task) or formed by several tasks. These tasks can be performed by the agent that offers the service or they can be delegated to other agents, by means of service invocation, composition and orchestration. VOM allows describing agent functionality in terms of services, by means of OWL-S [20] ontology, highly standardized and popular in the Web Service domain. Thus, VOM describes the functionality of a *Service* (*service_description*) and the way in which it is performed (*service_activity*) using OWL-S, standardizing the agent functionality and modeling MAS as a service oriented architecture.

An *Entity* is described by an identifier and a membership relation inside a unit in which it *plays* a specific role. It is also capable of *offering* some specific functionality to other entities. Its behavior is motivated by their *pursued goals*. Moreover, an organizational unit can also publish its requirements of services (*requires* relation), so then external agents can decide whether to participate inside, thus providing those services. Any service has one or more roles that are in charge of its provision (*provides*) and others that consume it (*uses*). Furthermore, any service obviously has influence over system goals (*affects_goal* relation).

3.3 Environmental Dimension

Taking into account the Human Organization Theory as a starting point [8], the environment should be considered from two different perspectives: (i) from a *structural* perspective, the environment must be described in terms of its components, i.e. resources and applications; (ii) from a *functional* perspective, the environment details how agents perceive and act on it.

VOM includes both functional and structural perspectives of the environment. Our purpose is focused on describing environmental components, perceptions and acts on these elements, and defining permissions for accessing them. The proposed Environmental Dimension (Figure 2a) defines each element of the environment as a *Resource*, which represents an environmental component. It belongs to an entity (*has_resource*), which can be a single agent or an organizational unit. In this last case, an entity in charge of managing the access permissions to this element is needed (*has_portControl*). The resource is accessed

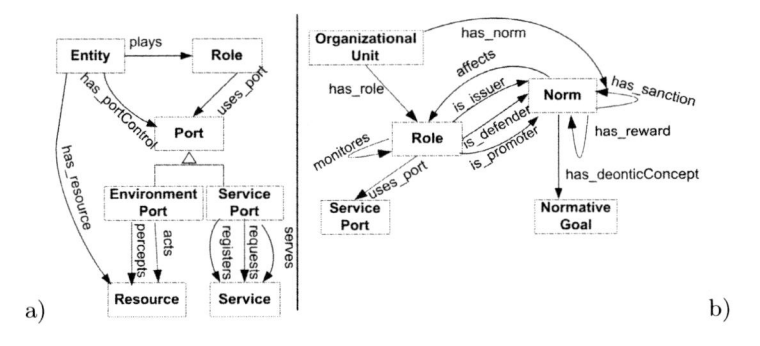

Fig. 2. a) Environmental Dimension; b) Normative Dimension

and perceived through an *EnvironmentPort*. On the other hand, the *ServicePort* concept details the registration of a service in a service directory (*registers*) or its consumption (*serves* or *requests*). Each port is *controlled* by an entity and it is employed by one or more roles (*uses_port*).

3.4 Normative Dimension

The Normative dimension (Figure 2b) describes the organizational norms and normative goals that agents must follow, including sanctions and rewards.

Open systems, in which heterogeneous and self-interested agents work together, need some mechanisms that restrict agent behaviors but maintain agent autonomy. The Normative Dimension of VOM is a coordination mechanism that attempts to: (i) promote behaviors that are satisfactory to the organization, i.e. actions that contribute to the achievement of global goals; and (ii) avoid harmful actions, i.e. actions that prompt the system to be unsatisfactory or unstable. Therefore, the normative definition allows specifying the desired behavior in terms of actions that agents should do or not. More specifically, norms determine the role functionality by restricting its actions and establishing its consequences.

Each OU has a set of norms that restricts its member behaviors (*has_norm*). A norm *affects* a role directly, which it is *obliged, forbidden* or *permitted* to perform the specified action (*has_deonticConcept*). Valid *actions* are service requesting, registering or providing. Sanctions and rewards are expressed by means of norms. The *issuer* role is in charge of controlling norm fulfillment, whereas *defender* and *promoter* roles are responsible for carrying out sanctions and rewards, respectively. All those roles can be played by the same agent or by different ones.

4 Case Study

In this section, an example of a travel agency system has been developed employing VOM. The system acts as a regulated meeting-point, in which providers and clients contact between them for consuming or producing touristic services. A similar example has also been modeled using electronic institutions in previous works [21, 22].

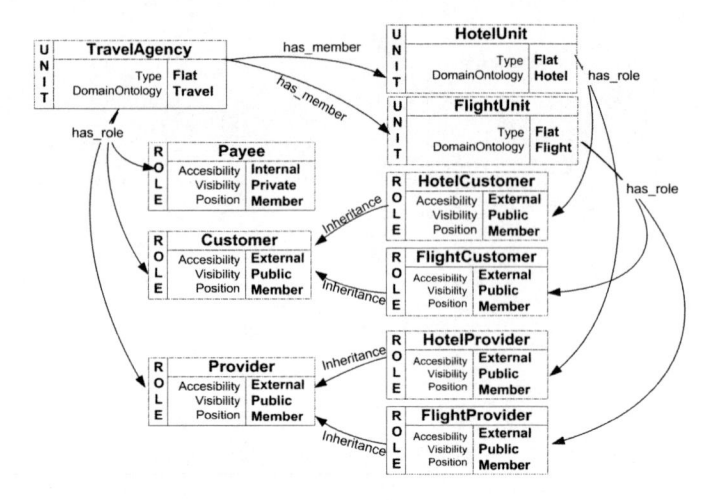

Fig. 3. Travel Agency structural description

In this system, agents can search for and make hotel and flight bookings. *TravelAgency* is an application that facilitates the interconnection between clients (individuals, companies, travel agencies) and providers (hotel chains, airlines). It is a service-oriented architecture. Therefore, it delimits services that each agent can offer and/or provide. However, the internal functionality of these services is responsibility of the provider agents. The definition of this system, employing the presented model, corresponds to the identification of structural, functional, normative and environmental components. Following this process is detailed.

- **Structural Description.** In the *TravelAgency* system, three kinds of roles can be identified:
    - **Customer:** This role requests services of the system. Being an accessible role, it can be played by external agents. It is specialized into *HotelCustomer* and *FlightCustomer* roles, according to each type of product.
    - **Provider:** This role offers and produces services. A *Provider* agent offers hotel or flight search services; and, in some cases, reservation of rooms or flight seats. It is also specialized into *HotelProvider* and *FlightProvider* roles.
    - **Payee:** It gives advanced payment service. It represents a bank institution through which reservation payment is carried out. It is an internal role, so external agents are not able to acquire it.

This case study is modelled as a main unit (*TravelUnit*) inside which there are two OUs (*HotelUnit* and *FlightUnit*), which are dedicated to flight and hotel services, respectively. These units act as members of the main unit, playing the *Provider* role. When an external agent is registered in the *Travel Agency* system, it is given a system description, employing the VOM ontology. Figure 3 shows a partial description of the system.

Table 1. TravelSearch Service Profiles

Service: TravelSearch	**ProfileID:** TravelSearchPF	**Description:** Search for travel information	
UnitID: TravelAgency	**ClientRole:** Customer	**ProviderRole:** Provider	
Inputs: city:string country:string	**Outputs:**	[city ok] company:string price:float	[not in city] error

- **Functional Description.** In addition to the system description, there is a service directory in which *TravelAgency* services are registered and published. In this example, services that can be requested or provided describe the domain-related functionality. Inside the main unit (*TravelAgency*) three services are offered: *TravelSearch*, *Reserve* and *Payment*. These services are specialized, inside each unit, according to each type of product. Table 1 shows a brief profile of the *TravelSearch* service. Due to lack of space, a more detailed description of all services has not been included.
- **Normative Description.** This example is regulated through different kinds of norms such as role incompatibility norms, role cardinality norms, functionality norms, etc. As a norm example, Norm 1 contains a role incompatibility norm that forbids the *Payee* role to play *Provider* and *Customer* roles, simultaneously. This norm has been expressed by means of a normative language detailed in [23].

$$FORBIDDEN\ Payee\ REQUEST\ AdoptRole$$
$$MESSAGE\ (\ CONTENT\ (role\ "Provider",\ unit\ "TravelAgency") \quad (1)$$
$$OR\ CONTENT\ (role\ "Customer",\ unit\ "TravelAgency"))$$

- **Environmental Description.** Agent Organizations are not isolated; on the contrary, they are situated in a concrete environment. In the *TravelAgency* example there is a database that contains geographical information. Accordingly to the environmental dimension, it can be modeled as a *resource* belonging to the *TravelAgency* unit that can be employed by all unit members.

5 Conclusions and Future Work

Virtual Organizations have been employed as a coordination mechanism in open societies. This work presents a new model for designing Virtual Organizations based on both Human Organization Theory and Service Oriented Architecture. This model makes possible the definition of open organizations by means of four dimensions: Structural, Functional, Normative and Environmental. Those dimensions describe which are the system entities, how they are related between them and their environment and which is their functionality and behavior, defined by means of services. VOM also describes restrictions on which services must be provided by each role, as well as how agents should interact for consuming or providing these services.

VOM concepts have been applied into a case study, implementing its organizational concepts in a multi-agent platform. In addition, both Normative and

Environmental dimensions of VOM have also been implemented. This implementation makes possible the integration VOM into a MAS platform that offers support for unit, role and norm management. All this work belongs to a higher project whose driving goal is to develop a new way of working with open agent organizations.

Acknowledgements. This work is supported by TIN2005-03395 and TIN2006-14630-C03-01 projects of the Spanish government, GVPRE/2008/070 project, FEDER funds and CONSOLIDER-INGENIO 2010 under grant CSD2007-00022, FPU grant AP-2007-01256 awarded to N.Criado.

References

1. Luck, M., McBurney, P., Shehory, O., Willmott, S.: Agent Technology: Computing as Interaction (A Roadmap for Agent Based Computing). In: AgentLink (2005)
2. Luck, M., McBurney, P.: Computing as Interaction: Agent and Agreement Technologies. In: Proc. IEEE Conf. Distributed Human-Machine Systems, pp. 1–6 (2008)
3. Foster, I., Kesselman, C., Tuecke, S.: The Anatomy of the Grid: Enabling Scalable Virtual Organizations. Int. J. High Perform. Comput. Appl. 15(3), 200–222 (2001)
4. Boella, G., Hulstijn, J., van der Torre, L.: Virtual organizations as normative multiagent systems. In: HICSS. IEEE Computer Society, Los Alamitos (2005)
5. Boissier, O., Padget, J., Dignum, V., et al. (eds.): ANIREM 2005 and OOOP 2005. LNCS (LNAI), vol. 3913. Springer, Heidelberg (2006)
6. Ferber, J., Gutknecht, O., Michel, F.: From agents to organizations: An organizational view of multi-agent systems. In: Giorgini, P., Müller, J.P., Odell, J.J. (eds.) AOSE 2003. LNCS, vol. 2935, pp. 214–230. Springer, Heidelberg (2004)
7. Dignum, V., Dignum, F.: A landscape of agent systems for the real world. Tech. Report 44-cs-2006-061, Institute of Information and Computing Sciences, Utrecht University (2006)
8. Daft, R.: Organization Theory and Design. South-Western College Pub. (2003)
9. Trencansky, I., Cervenka, R.: Agent modeling language (AML): A comprehensive approach to modeling MAS. Informatica 29(4), 391–400 (2005)
10. Ferber, J., Michel, F., Báez-Barranco, J.A.: AGRE: Integrating environments with organizations. In: Weyns, D., Van Dyke Parunak, H., Michel, F. (eds.) E4MAS 2004. LNCS, vol. 3374, pp. 48–56. Springer, Heidelberg (2005)
11. Hubner, J., Sichman, J., Boissier, O.: S-MOISE+: A middleware for developing organised multi-agent systems. In: Proc. Int. Workshop on Organizations in Multi-Agent Systems. LNCS, vol. 3913, pp. 64–78. Springer, Heidelberg (2006)
12. Pavón, J., Gómez-Sanz, J.J.: Agent oriented software engineering with INGENIAS. In: Mařík, V., Müller, J.P., Pěchouček, M. (eds.) CEEMAS 2003. LNCS, vol. 2691, pp. 394–403. Springer, Heidelberg (2003)
13. Dignum, V., Vázquez-Salceda, J., Dignum, F.: OMNI: Introducing social structure, norms and ontologies into agent organizations. In: Bordini, R.H., Dastani, M., Dix, J., El Fallah Seghrouchni, A. (eds.) PROMAS 2004. LNCS, vol. 3346, pp. 181–198. Springer, Heidelberg (2005)
14. Esteva, M., Rodriguez-Aguilar, J., Sierra, C., Arcos, J., Garcia, P.: On the formal specification of electronic institutions. In: Sierra, C., Dignum, F.P.M. (eds.) AgentLink 2000. LNCS (LNAI), vol. 1991, pp. 126–147. Springer, Heidelberg (2001)

15. Botti, V., Giret, A.: ANEMONA: A Multi-Agent Methodology for Holonic Manufacturing Systems. Springer, Heidelberg (2008)
16. Argente, E., Julian, V., Botti, V.: MAS modeling based on organizations. In: Proc. AOSE, pp. 1–12 (2008)
17. Argente, E., Palanca, J., Aranda, G., Julian, V., et al.: Supporting agent organizations. In: Burkhard, H.-D., Lindemann, G., Verbrugge, R., Varga, L.Z. (eds.) CEEMAS 2007. LNCS, vol. 4696, pp. 236–245. Springer, Heidelberg (2007)
18. Horling, B., Lesser, V.: A survey of multiagent organizational paradigms. The Knowledge Engineering Review 19, 281–316 (2004)
19. Lopez, F., Luck, M.: A model of normative multi-agent systems and dynamic relationships. In: Lindemann, G., Moldt, D., Paolucci, M. (eds.) RASTA 2002. LNCS, vol. 2934, pp. 259–280. Springer, Heidelberg (2004)
20. Service Profile. OWL-S applications and issues (March 08, 2004)
21. Sierra, C., Thangarajah, J., Padgham, L., Winikoff, M.: Designing institutional multi-agent systems. In: Padgham, L., Zambonelli, F. (eds.) AOSE VII / AOSE 2006. LNCS, vol. 4405, pp. 84–103. Springer, Heidelberg (2007)
22. Dignum, F., Dignum, V., Thangarajah, J., Padgham, L., Winikoff, M.: Open agent systems?? In: Luck, M., Padgham, L. (eds.) Agent-Oriented Software Engineering VIII. LNCS, vol. 4951, pp. 73–87. Springer, Heidelberg (2008)
23. Argente, E., Criado, N., Julián, V., Botti, V.: Designing norms for Virtual organizations. In: Proc. CCIA, pp. 16–24 (2008)

Dynamic Orchestration of Distributed Services on Interactive Community Displays: The ALIVE Approach

I. Gómez-Sebastià[1], Manel Palau[2], Juan Carlos Nieves[1], Javier Vázquez-Salceda[1], and Luigi Ceccaroni[2]

[1] Universitat Politècnica de Catalunya, C/Jordi Girona 1-3, E08034, Barcelona, Spain
{igomez, jcnieves, jvazquez}@lsi.upc.edu
[2] Tech Media Telecom Factory, C/Marina 16-18, Torre Mapfre 28 A E08005 Barcelona, Spain
{manel.palau, luigi}@tmtfactory.com

Abstract. Interconnected service providers constitute a highly dynamic, complex, distributed environment. Multi-agent system design-methodologies have been trying to address this kind of environments for a long time. The European project ALIVE presents a framework of three interconnected levels that tackles this issue relying on organisation and coordination techniques, as well as on developments in the Web-services world. This paper presents initial results focused on a high-tech, real use case: interactive community displays with touristic information and services, dynamically personalized according to user preferences and local laws.

1 Introduction

In our cities, urban information services are provided in ways which have not changed much in a century: bus-stop shelters, metro-stations panels and screens, maps, and urban furniture. This scenario brings up the opportunity to improve the information services that a city provides to citizens and tourists not only making it dynamic but also with the novel possibility of providing ubiquitous access to informational and other services. In this paper we will focus in a specific scenario, played in a dynamic and changing environment. The scenario proposed shows the difficulty to offer (on real time) personalized recommendations adapted to user profiles and contexts.

We propose a solution to cope with this scenario by means of intelligent agents, capable of reorganising themselves to compose and control the execution of dynamic and flexible services. The technologies and mechanisms concerning coordination and cooperation among agents (such as communication, negotiation and monitoring techniques) are key in order to get the right content with the right agent and in the right location. Furthermore, in open and complex agent societies [10], concepts such as organisational meta-models and self-organisation [7], trust, reputation, norms, obligations and joint-intentions [6] are required in order to effectively model uncertain and competitive environments.

The ALIVE framework (see Section 2) presents a new approach to the engineering of distributed software systems based on the adaptation of coordination and organisation mechanisms, often seen in human and other societies, to service-oriented architectures. In this paper we show how this framework facilitates the interaction between content and service providers (e.g., city administrators, advertisers, restaurants), and service integrators in the development and maintenance of distributed systems. Section 3 of the paper presents a scenario based on the dynamic orchestration, organisation and coordination of services: a personalised recommendation tool that brings city services closer to residents and tourists by interconnecting service providers, locations and people. Finally, in Section 4 the proposed system is compared with existing work.

2 The ALIVE Architecture

New generations of networked applications, based on the notion of software services that can be dynamically deployed, require profound changes in the way in which software systems are designed. New approaches are needed, which allow integrating new functionalities, and thus new services, into existing running systems of already active, distributed and interdependent processes.

The ALIVE framework combines *Model Driven Design*[1] (MDD) with coordination and organisational mechanisms, providing support for "live" (that is, highly dynamic) and open systems of services. ALIVE's multi-level approach helps the maintenance of distributed systems by reorganising and adapting services. This is achieved via a set of intelligent agents responsible of adding control and flexibility to service composition and execution, effectively providing intelligent service orchestration. These agents are fully aware of the organisation's structure and goals, and thus can reflect them on service composition and execution, such as serving with higher priority requests coming from an actor higher on the organisation's hierarchy.

2.1 ALIVE Levels

The ALIVE framework extends current trends in engineering by defining three levels (see Fig. 1) as follows.

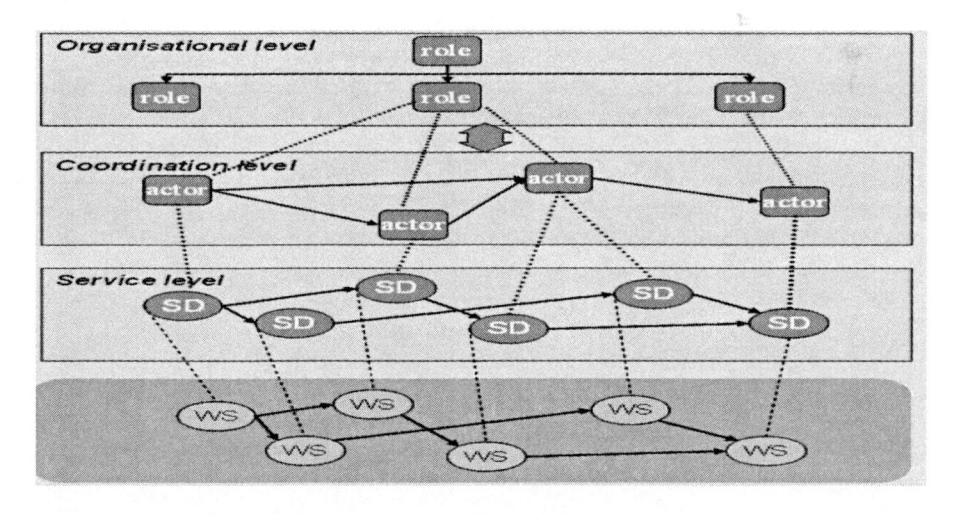

Fig. 1. The ALIVE architecture

---

[1] See [http://www.omg.org/mda/], visited on October 29, 2008.

A *Service Level* augments and extends existing service models in order to make components aware of their social context and of the rules of engagement with other services by making use of semantic Web technologies. This comes in handy when highly dynamic and always changing services (WS in figure) are present in the system, as the *meta-information* contained in each service description (SD in figure) eases the process of finding substitute services if a given service does not respond to invocations.

A *Coordination Level* provides the means to specify, at a high level, the patterns of interaction between services, using a variety of powerful coordination techniques from recent European research in the area [5, 11]. This is very useful when the system has to react to failures or exceptions inherent to the main processes (*e.g.*, failing payment or booking systems).

Finally, an *Organisational Level* provides context whether directly or not for the Coordination and the Service Levels, specifying the organisational rules that govern interaction and using recent developments in organisational dynamics [16] to allow the structural adaptation of systems over time. This is important when frequent changes on rules and restrictions are expected, as adapting the other two levels without this information would be a tough task.

The ALIVE multi-level framework is specially fit for scenarios where changes can occur at either abstract or concrete levels, and where services are expected to be continuously changing, with new services entering the system and existing services leaving it at run-time. For example, when there is a significant change on a high level (*e.g.*, a change in the organisational structure), the service orchestration at lower levels is automatically reorganised. Another example is the automatic adaptation of higher levels when lower ones suffer significant changes (*e.g.*, the continuous failure in some low-level services).

3 Application of the ALIVE Framework

In order to demonstrate the applicability of the ALIVE architecture, we describe a scenario about multimedia distribution of information and orchestration of services, where *interactive community displays* (ICDs) [3] are used by TMT Factory (a software company) to provide a hyperlocal Web[2] in urban environments. The services and information provided, and how users' information is stored, processed and distributed, are subjected to various municipal, national and European regulations. The challenge is to achieve the interaction among content providers, service providers and integrators, to provide personalized information and recommendations.

For this scenario we consider the case of an underage user interacting with an ICD in search of entertainment and cultural options around the city. In this case, the system identifies (if possible) the user; gathers ratings and reviews about available services, along with user preferences and location; and finally offers personalized recommendations. In the following sections details are given about the system's model and operation.

---

[2] The difference between the semantic Web and the hyperlocal Web is that the ontologies of the latter one contain not only semantic information, but also geographic coordinates.

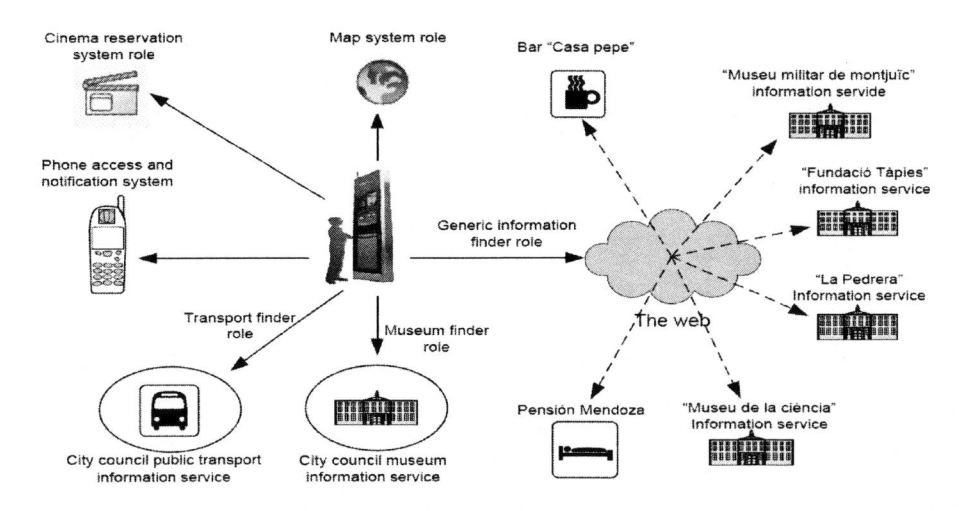

Fig. 2. System organisational model

3.1 System Model and Components

In the simplified model in Fig. 2, general components are shown together with components specific to the case, such as a map system, an information finder, museums, public transport information, the cinema reservation system and a public museum information service, provided by the city. It is important to notice the coexistence among local services and services offered via the Internet, and among public and commercial information.

Fig. 3 depicts the main components of the ALIVE architecture. In the organisation level, the *Domain Ontology* includes all domain-specific terms in the system, such as "Cinema" or "Museum". The *Organisational Model* (Org Model) describes the role each stakeholder in the system should fulfil. This model can be maintained by a set of tools, including a *Model Checker*, which can verify the consistency of the model (if consistency is important). The *Utils* component receives high-level exceptions and handles them at run-time, updating the organisational model as required.

In the coordination level, the *MAS Generator* component receives information from the *Utils* component to (re)generate the agents that will populate the *MAS*. In the scenario presented, entities such as "Cinema agent & transport" or "Map agent" will be generated. These entities are the agents responsible of monitoring service executions and composing available services when required. The *Workflows* component contains the workflows to be executed by the agents when performing individual tasks, including workflows such as "Cinema reservation" or "Museum information composition". The *Monitor* keeps track of the execution of the workflows to detect anomalies or deviations from expected behaviour as soon as possible, enabling the system to react to them.

In the service level the *Execution Engine* is used by agents to execute the workflows. The *Workflow Builder* allows dynamically (re)creating workflows that orchestrate the execution of the agents in order to reach some organisational goal. *Service*

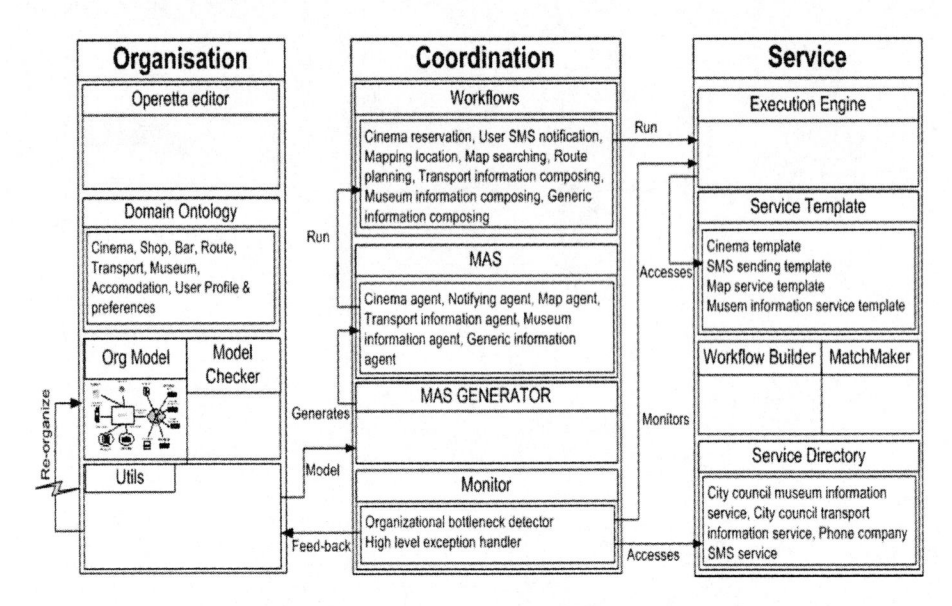

Fig. 3. Main components of the ALIVE architecture used in the scenario

Templates are made available for the prototypical services to be offered; in this scenario, this component contains, for instance, templates for the "Cinema" service that enable the system to make ticket reservations on different theatres, regardless the concrete booking service.

3.2 Use Case Scenario in Detail

In the scenario presented, if the user wants to visit some museums, she can ask the system for museum information, including a brief description and location on the map. In the current organisational structure of the system, this corresponds to the *museum information searching* task of the *museum information finder* role, which is performed by an agent, residing at the coordination level, which invokes the City council's museum information service through the service level. Let us suppose, however, that this service has become unavailable, so in order to be able to retrieve museum information upon the user request, the system has to find alternative ways of performing the task that was assigned to the *museum information finder*.

As there is no alternative service to be invoked in the current workflow in the service level, an exception is passed to the *Monitor* at the coordination level, which looks for alternate ways of enacting the workflow. In this case, there are no such alternatives, though. Consequently, the notification of the exception goes up to the organisational level, which reorganises the model in order to perform the task requested by the user, triggering changes in both coordination and service levels.

After these changes, the system responsible of retrieving museum information is an agent able to gather and compose museum information from data found on the Internet. As a result, the actor responsible of enacting this role has changed, and the new actor, residing at the coordination level is invoked. This actor requests the service level to find out concrete museum services, access them, and come back with the required information. Then, the agent is able to compose (i.e. put together and format) this information, providing opening times, prices and brief descriptions for each available museum. When the information is composed, it is returned to the user, who gets what she requested even if one of the core components of the system was not available (see Fig. 4).

Once the user has information about all museums, she requests the system to locate them on the map. The components at organisational level look for an actor responsible of performing this task: in this case, an agent which can generate queries to existing map servers. This agent, residing at the coordination level, requests a map of a given location to the service level. The service level searches for existing services which implement the mapping service template. Three services are found: one specialised in road maps, which does not have accurate maps of most museums' areas; a second one that requires a payment for the maps; and a third one which is suitable for the required tasks.

After checking the museums information, the user decides that she would rather leave museum visits for another day, and go and try the new 3D-cinema instead. Therefore, the user asks the system to book two tickets. As this entertainment facility is quite far from the user's actual location, the system suggests a route on public transportation to the cinema.

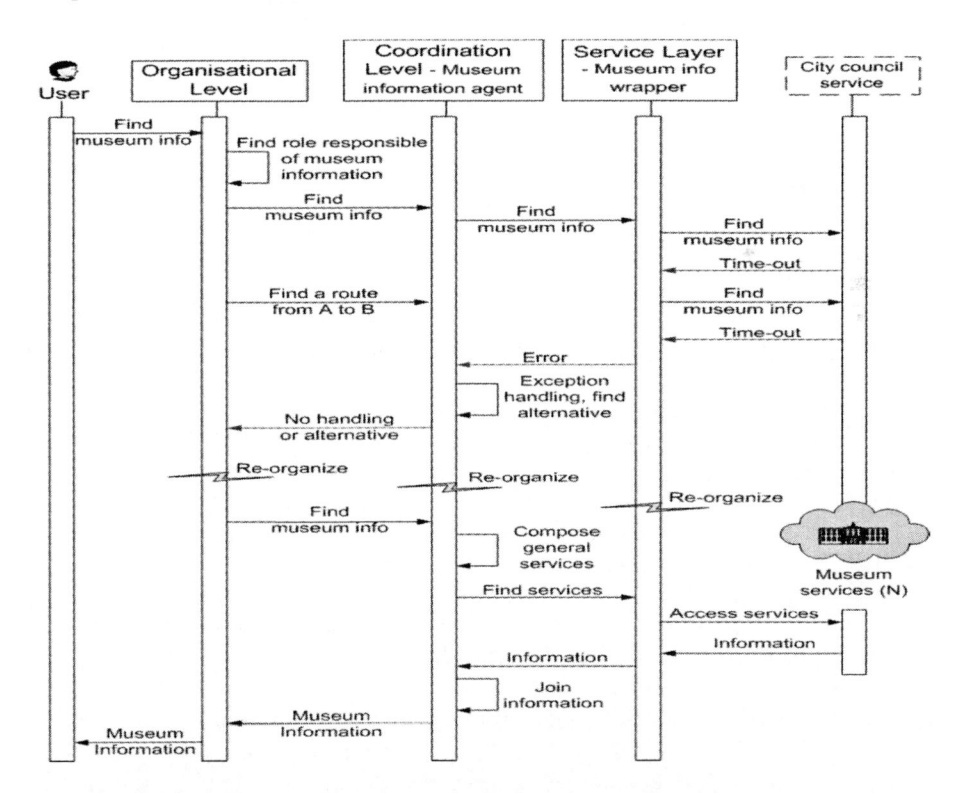

Fig. 4. Part of the scenario execution workflow showing reorganisation at all levels

For this, the organisational level finds the actor responsible of performing this task. In this case, it can be an agent for booking the cinema and finding a route. This agent finds out the actions required to enact the workflow associated to the requested task, and realizes that, for this, it needs to access a cinema booking service on the Internet, and request two tickets. Consequently, it commands the service level to perform this task. Once it is done, it needs to find a route from user's location to the cinema via a public transportation system, so the next task in the workflow is to ask the service level for information about such system. Once the service level provides this information, the agent is able to find out means of reaching the cinema. The user is presented this information on the ICD screen: the booked session, the bus to be taken, buses' route, and the nearest bus stop on a map of the zone. At this point, the user can log out of the system and head to the bus stop.

Later on, the system might realise that changed traffic condition would delay the bus long enough for the user to miss her session. If necessary, it proceeds in cancelling current cinema reservation, and informing the user about the situation via SMS. Then, the user can, for instance, reply to the SMS choosing to book a ticket for the following session and ask for some bars or shops near the cinema to pass the time before the newly booked session starts, receiving all this information on her cell-phone.

As shown in the scenario description, the coordination level, via its monitoring system, can detect that a given workflow is continuously failing or been executed in a rather inefficient way due, for instance, to the presence of one agent responsible of too many tasks which is becoming a bottleneck. In this case, the coordination level can suggest changing the organisational level in order to reallocate task responsibilities to avoid such bottleneck.

4 Related Work

Main lines of research about incorporating agent-based methods and techniques into Web-service contexts [14] include the extension of communication with more flexible patterns or protocols [1], which can support service-to-service negotiation [4, 11] and dynamic service composition. For the latter, most of the proposed approaches are based on either pre-defined workflow models or in AI planning techniques.

In the case of the pre-defined workflow models approach, a composite service is pre-defined by the designer by means of abstract workflows where nodes are not bounded directly to services but to search recipes [2, 15]. At runtime these recipes are used to find and bind to concrete services. In our solution we use a similar approach by means of the Service Templates, which are used to dynamically bind the concrete services at runtime. The difference is that workflows do not need to be predefined, but can be dynamically generated from the organisational objectives.

In the case of the AI planning approach, existing methods tend to be based on the assumption that each Web service method is an action or task which alters the state of the world. Typically, input and output parameters in service methods act as preconditions and effects. Examples of this approach include the use of existing planning languages, such as Golog [12] or SHOP2 [18], or the synthesis of services based on theorem proving methods [17]. An important remark to be done is that most approaches use centralised planning, where a specific component (the planner or the theorem prover) is the one that dynamically generates the full plan based on its (centralised) perceptions of the state of the world. In the ALIVE approach we make use of distributed planning, based on the TAEMS framework [9], where the agents at the coordination level build partial plans that are then shared to generate a partial global plan.

An extra distinction of our approach over existing ones [13] is that it provides organisational context (mainly organisation's objectives and structure) that can be used when selecting and invoking services, providing an organisation awareness to some components (such as agents on coordination level or matchmaker on service level), which can direct the decision making in order to move towards the higher-level organisational objectives. One of the effects is that exceptions can be managed at the low-level, searching for a service (or composition of services) to substitute services that happen to be non-operational. This kind of exception handling is common in other SOA architectures. However, exceptions can also be managed at the high-level, looking for alternative ways to fulfil a workflow, or executing some tasks when certain states are reached. This high-level exception handling is not commonly seen in other SOA architectures.

5 Conclusions

The ALIVE framework aims to design and develop distributed systems suitable for highly dynamic environments, based on model-driven engineering, and three interconnected levels: service, coordination and organisation. The project fulfils the needs of TMT Factory (a software company) for more flexible, adaptive technologies to be used in interactive community displays, focussing on providing an easy to maintain system.

Due to the connection among levels, a change in the service level can trigger changes on both coordination and organisation levels; whereas a change in the organisation level can also trigger changes in the other two. These changes can be automatically carried out (with the assistance of ALIVE tools). This provides the system with high dynamicity, making it capable to react to changes on the environment.

Through the monitoring and exception system in the coordination level, work-flow errors can be detected and treated properly. As shown in a use case, this provides the system with high stability, enabling it to fulfil its goals when a core component fails. The fact that service templates can abstract the service level and organisation level from concrete Web services makes workflow specifications more stable through time, whereas service discovery provides continuous system adaptation to a highly dynamic environment.

Finally, the high-level organisation, connected to the coordination and, indirectly, to the service level, ensures system's compliance with organisational structure, as available services will be composed and used according to organisation objectives, roles, norms and restrictions.

Acknowledgments. This work has been partially supported by the FP7 project ALIVE IST-215890, which is funded by the European Commission. The authors would like to acknowledge the contributions of their colleagues from the ALIVE consortium (http://www.ist-alive.eu). The views expressed in this paper are not necessarily those of the ALIVE consortium.

References

1. Ardissono, L., Goy, A., Petrone, G.: Enabling Conversations with Web Services. In: Proceedings of the Second International Joint Conference on Autonomous Agents and Multi-agent Systems, Melbourne, pp. 819–826 (2003)
2. Casati, F., Ilnicki, S., Jin, L.: Adaptive and Dynamic Service Composition in eFlow. In: Wangler, B., Bergman, L.D. (eds.) CAiSE 2000. LNCS, vol. 1789, p. 13. Springer, Heidelberg (2000)
3. Ceccaroni, L., Codina, V., Palau, M., Pous, M.: PaTac: Urban, ubiquitous, personalized services for citizens and tourists. In: Proceedings of the Third International Conference of Digital Society, Cancun (2009) (to appear)
4. Ermolayev, V., et al.: Towards a Framework for Agent-Enabled Semantic Web Service Composition. Int. J. of Web Services Research 1(3), 63–87 (2004)
5. Ghijsen, M., Jansweijer, W., Wielinga, B.B.: Towards a framework for agent coordination and reorganization, agentCoRe. In: Sichman, J.S., Padget, J., Ossowski, S., Noriega, P. (eds.) COIN 2007. LNCS, vol. 4870, pp. 1–14. Springer, Heidelberg (2008)
6. Jennings, N.R.: Controlling Cooperative Problem Solving in Industrial Multi-Agent Systems using Joint Intentions. Artificial Intelligence 75(2), 195–240 (1995)
7. Juan, T., Sterling, L.: The ROADMAP meta-model for intelligent adaptive multi-agent systems in open environments. In: Giorgini, P., Müller, J.P., Odell, J.J. (eds.) AOSE 2003. LNCS, vol. 2935, pp. 231–253. Springer, Heidelberg (2004)
8. Klusch, M.: Semantic Web Service Coordination. In: Helin (ed.) CASCOM - Intelligent Service Coordination in the Semantic Web. WSSAT, pt. I, pp. 59–104. Birkhäuser Basel (2008)
9. Lesser, V., et al.: Evolution of the GPGP/TÆMS Domain-Independent Coordination Framework. In: Proceedings of the First International Conference on Autonomous Agents and Multiagent Systems, vol. 9, pp. 87–143. Kluwer Academic Publishers, Dordrecht (2004)
10. Luck, M., McBurney, P., Shehory, O., Willmott, S.: Agent Technology: Computing as Interaction. In: A Roadmap for Agent Based Computing. AgentLink III (2006)
11. Matskin, M., et al.: Enabling Web Services Composition with Software Agents. In: Proceedings of the Conference on Internet and Multimedia Systems, and Applications, Honolulu (2005)
12. McIlraith, S., Cao Son, T.: Adapting Golog for Composition of Semantic Web Services. In: Proceedings of the 8th International Conference on Knowledge Representation and Reasoning, Toulouse (2002)
13. Papazoglou, M.P., van den Heuvel, W.J.: Service oriented architectures: approaches, technologies and research issues. J. VLDB 16, 389–415 (2007)
14. Singh, P., Huhns, N.: Service-Oriented Computing Semantics, Processes, Agents. Wiley, Chichester (2004)

15. Thandar, M., Edmond, D.: The use of patterns in service composition. In: Bussler, C.J., McIlraith, S.A., Orlowska, M.E., Pernici, B., Yang, J. (eds.) CAiSE 2002 and WES 2002. LNCS, vol. 2512, pp. 28–40. Springer, Heidelberg (2002)
16. van der Vecht, B., Dignum, F., Jules, J., Meyer, C., Dignum, V.: Organizations and autonomous agents: Bottom-up dynamics of coordination mechanisms. In: The Fifth Workshop on Coordination, Organizations, Institutions, and Norms in Agent Systems, Estoril (2008)
17. Waldinger, R.: Web agents cooperating deductively. In: Rash, J.L., Rouff, C.A., Truszkowski, W., Gordon, D.F., Hinchey, M.G. (eds.) FAABS 2000. LNCS, vol. 1871, pp. 50–262. Springer, Heidelberg (2001)
18. Wu, D., et al.: Automating DAML-S web services composition using SHOP2. In: Fensel, D., Sycara, K.P., Mylopoulos, J. (eds.) ISWC 2003. LNCS, vol. 2870, pp. 195–210. Springer, Heidelberg (2003)

Efficiency in Electrical Heating Systems: An MAS Real World Application

José R. Villar, Roberto Pérez, Enrique de la Cal, and Javier Sedano

Departamento Informática y Automática
Universidad de Oviedo
Campus de Viesques s/n 33204 Gijón, Spain
`villarjose@uniovi.es`, `UO24411@uniovi.es`, `delacal@uniovi.es`,
`jsedano@ubu.es`

Abstract. In electrical heating systems, the electrical power consumption should be lower than the Contracted Power Limit. Energy distribution devices are used to solve this problem, but they are only concerned with the electrical energy. We claim that this energy distribution must also consider the comfort level in the building. In this work, an electrical energy distribution MAS for coordinating the electrical heaters is proposed. The MAS is responsible for both objectives: the electrical power must be lower than the Contracted Power Limit and the comfort level in the building must be maintained if sufficient power is available. The MAS is now being implemented in a real world application of an electrical heating system marketed by a local company.

Keywords: MAS applications, Software Agents in Industry, Energy distribution, Electrical heating systems.

1 Introduction

In Spain, a local company marketed a new catalogue of dry electrical heaters in the middle of the year 2008. The total power installed in a domestic electrical heating system in Spain easily surpasses 7 kW. When electrical heaters are used, the electrical energy consumption could be higher than the common contracted power limit (for short, CPL). Typically, an energy distribution device shares the electrical energy between the electrical heaters, avoiding surpassing the CPL. The main drawback is that it does not consider the comfort level in the house.

To improve the efficiency of the electrical heating system, the local company desires to design a Central Control Unit (CCU) in order to save energy and distribute it among the heaters in a building while maintaining the comfort level in the house. In previous works, the development of such a device has been analyzed, and a multi agent hybrid fuzzy system has been proposed [15, 14, 16, 13]. The solution integrates the electric heaters and the CCU. The CCU is responsible for distributing the available electric energy to the heaters based on the energy balance concept and with the objective of keeping the user defined comfort level in the house. Although the proposal has proved successful in the suitable distribution of the available electric energy, it suffers from some deficiencies. Specifically, some bias was found in the steady state. Also, there was a lack of stability in the electric energy output due to fluctuations that must be faced.

In this work, a solution to the electrical energy distribution for the electrical heating system is described. An MAS architecture has been adopted, with the CCU and the heaters collaborating to reach both objectives: to keep the power consumption lower than the CPL and to maintain the comfort level in the building. The energy distribution algorithm, which makes use of a Fuzzy Rule Base System (FRBS), and the MAS design are detailed.

This work is organized as follows. The next section deals with the Spanish regulations and the problem definition. Section 3 details the MAS proposed in this work. Implementation aspects and experimentation results are given in Section 4. Finally, conclusions and future work are presented.

2 Preliminaries

In the first quarter of 2007, the Spanish parliament approved a new building regulation [5]. As a result, building methods have been updated [6]. This new regulation had many consequences, as it determined how new buildings must be accomplished [4]: materials, isolation, energy efficiency, ventilation rates, etc. In Spain, the LIDER software [7] has been developed and should be used to calculate the heating installation in a building: the number of heaters and their nominal power are fixed.

The RITE establishes 5 climate winter zones, named with a letter from A to E, where E represents the maximum in weather severity. A peculiar fact [5] is that in Spain only 3 of the 5 winter zones defined in the RITE are considered. Moreover, a number between 1 and 5, related with the summer weather severity, is also given. The combination of winter and summer severities determines the climate zone for each location in Spain.

Furthermore, the constructors build many different kinds of buildings: condominiums -each apartment includes 2, 3 or 4 bedrooms-, detached and terraced houses, etc. All of them can have an electrical heating system installed, so the design of an energy distribution device must consider all the possible cases. The term building topologies refers to all of the building parameters that influence the heating system. Such parameters include the type of house, the geometrical aspects, the inner partition, the materials, etc. For example, the building envelope could help in reducing the heating losses. The building topologies have been extracted from the analysis of the building market, and have been reported in [13, 16]. The proposed building topologies should be used to validate any proposal of an electrical energy distribution for electrical heating systems.

2.1 The Problem Description

The main goal is the design of a heating system to distribute the available power without exceeding the CPL. The predefined comfort level in the house must also be reached. The heating system comprises the electrical heaters installed in all of the rooms of a house and the CCU. The range of the nominal powers for the heaters are 500, 1000, 1200 and 1500 Watts. The houses must comply with the Spanish regulations.

The inhabitants establish the comfort level in the house or in a room by setting the predefined suitable environmental variables. In the case of this project, only the temperature in each room is considered due to economic reasons. Future works should consider other measurements, such as the humidity percentage. The different room temperatures are to be measured using the temperature sensors included in each heater. Although the temperatures measured in the heaters are noisy, a pre-process module is carried out in order to improve the measurement quality; the heaters designers have introduced this pre-process module.

Even though there are devices for energy distribution –specifically, energy rationalizers–, these devices distribute the energy consumption but do not consider the comfort level in the house. Consequently, it is necessary to introduce a new device, called CCU, which will be responsible for the saving and distribution of energy to the installed heaters; all of the devices will collaborate to achieve the goal. A measure of the current power consumption is needed to maintain the electrical power consumption below the CPL.

Some hints about costs are to be considered when deciding the system architecture. Firstly, the communications between devices should be wireless in order to reduce the cost of the installation, specifically, the communications must be managed by a Zigbee wireless network [10]. Secondly, the cost of configuring the system should also be low: the installer must configure the whole system in a limited period of time. The configuration must be as simple as possible, with a reduced set of parameters.

3 The MAS Approach for Electrical Energy Distribution

An MAS architecture based on a Zigbee network [10] has been proposed as a solution for the electrical energy distribution. In this approach, both the heaters and the CCU act as agents, collaborating to reach the goals. This MAS solution is based on several reasons. Firstly, the use of a wireless network such as Zigbee reduces the installation cost. Moreover, the robustness of the solution is improved as the electrical heaters agents can behave in a stand-alone mode when the distributed system collapses. Nevertheless, the CCU is responsible for the electrical energy distribution when the distributed system is up and the heaters collaborate with the CCU carrying out several necessary tasks. Finally, the system must be implemented in language C for microcontrollers, so the modules should not be of high computational complexity.

In order to describe the whole solution this section is organized as follows. First, the MAS system is outlined, and then a short description of the energy distribution algorithm (EDA) is presented, concluding with the MAS architecture and the agents' descriptions.

3.1 The Electrical Energy Distribution System

The concept of energy balance is used in the multi-agent system outlined in Figure 1. In short, the heaters send the CCU the temperature of the room and the heating energy error. The heating energy error is calculated as the difference between the estimated required energy to reach the comfort level in the room and the energy that has been spent in heating the room (heating energy). The CCU measures the instantaneous consumed current in the house and estimates the power that can be spent on heating the house (available power). The CCU also stores the set point temperature profiles for all the rooms in the house, and the association between rooms and heaters. Finally, the CCU carries out the Energy Distribution Algorithm (EDA) to distribute the instantaneous power (heating power) for each heater. The heating power is the fraction of the nominal power that a heater is allowed to spend; the heating energy is the heating power by time unit.

The EDA makes use of an FRBS (a fuzzy controller –FRBS-1–) to calculate the percentage of nominal power for a heater given the room temperature error and the energy error. The temperature error is calculated as the difference between the temperature set point and the room temperature. Fuzzy logic is chosen to manage the wide variety of topologies that the system must work with: the same algorithm must ensure the power distribution and the energy saving for many different cases. An FRBS is a Fuzzy inference system that allows to model with human interpretability [2]. An FRBS includes a Knowledge Base with two main parts: the fuzzy rule base and the database. The former includes the rules and the relationships between the fuzzy partitions of each variable. The latter comprises the description of each membership function for each fuzzy partition.

A block diagram of the whole process is shown in Figure 2. There are two stages in the solution: the Design Stage and the Run Stage. In the Design Stage, an FRBS-1 is generated for each pair of climate zone and building topology (configuration). To generate the learning datasets for the FRBS-1 the simulation software tool HTB2 [11] has been used.

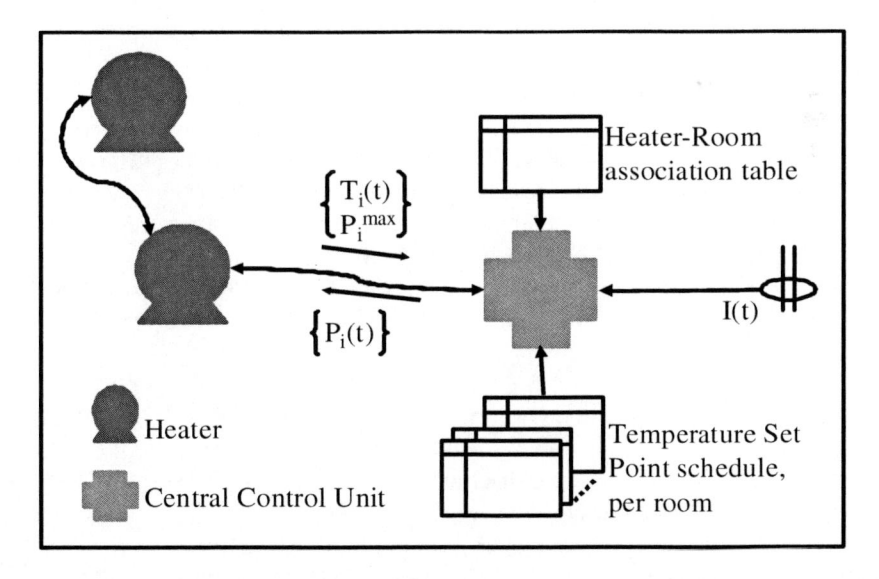

Fig. 1. The multi-agent system: the involved agents

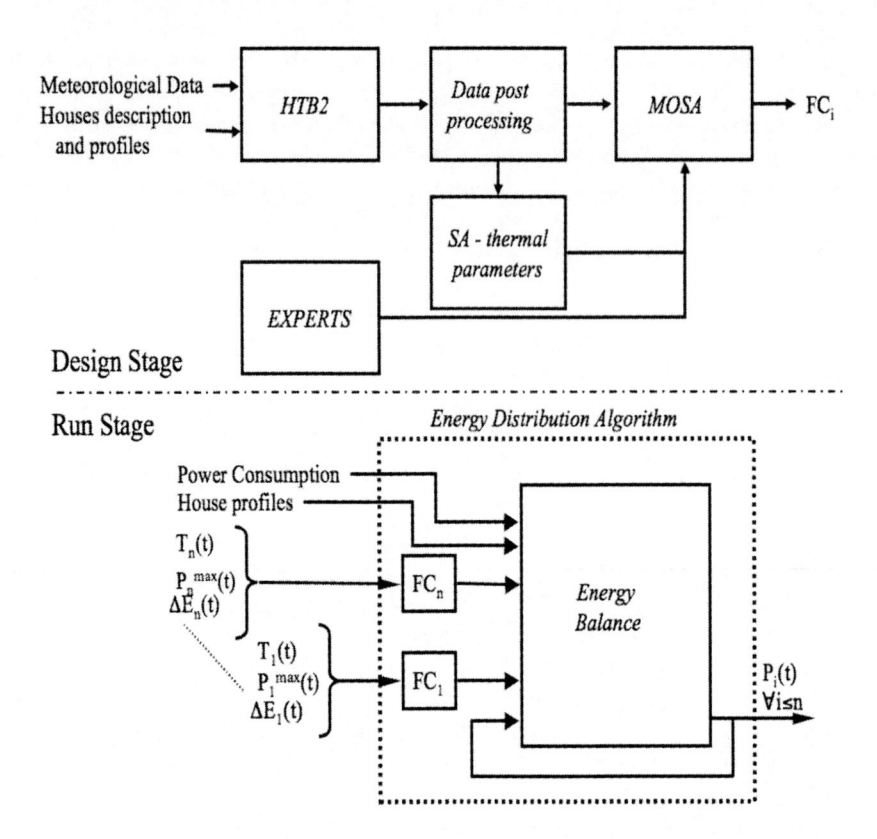

Fig. 2. The two steps procedure for distributing the power between the heaters

In the Run Stage the heating power for each heater is calculated using the concept of the balance of energy: the heating energy must equal the required energy in order to reach the comfort in the house. The available power is distributed between the heaters according to their percentages of nominal power. The heater agents calculate the energy error and the temperature in the room and send them to the CCU. The CCU uses that information, the power consumption in the house and the temperature set point profiles to distribute the available power. Also, the CCU uses an FRBS-1 for each room: according to the configuration, an initial FRBS-1 is chosen. The CCU can modify an FRBS-1 in order to improve the energy distribution.

In the Design Stage two optimization algorithms are used. A simulated annealing algorithm is used to identify the thermal dynamics of each room in order to train the FRBS-1. A multi-objective simulated annealing (MOSA) is used to train the FRBS-1 attending to the main objectives: to avoid surpassing the CPL in the house and to maintain the comfort level in the house if there is enough available power. A Michigan approach is used [8], and the initial individual is given by the company experts. Future work must include testing different FRBS learning approaches in the literature [1].

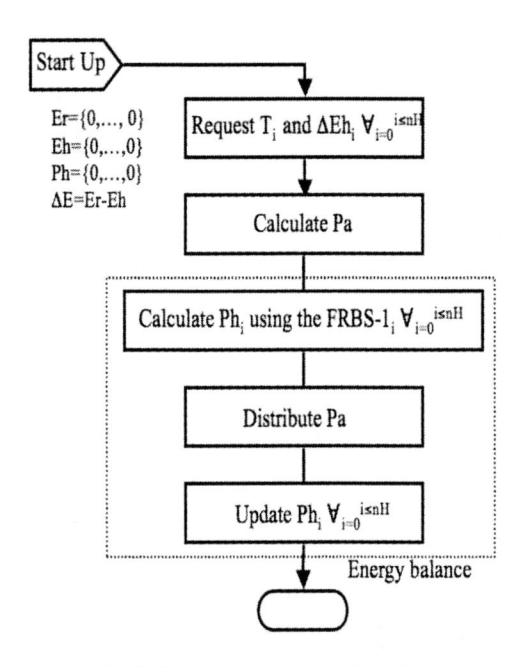

```
              ┌─────────┐
              │ Start Up│
              └────┬────┘
Er={0,..., 0}      │
Eh={0,...,0}   ┌───▼─────────────────────┐
Ph={0,...,0}   │ Request T_i and ΔEh_i ∀_{i=0}^{i≤nH} │
ΔE=Er-Eh       └───┬─────────────────────┘
                   │
              ┌────▼────┐
              │Calculate Pa│
              └────┬────┘
                   │
       ┌───────────▼─────────────────────┐
       │Calculate Ph_i using the FRBS-1_i ∀_{i=0}^{i≤nH}│
       └───────────┬─────────────────────┘
                   │
              ┌────▼────┐
              │Distribute Pa│
              └────┬────┘
                   │
              ┌────▼─────────────┐
              │Update Ph_i ∀_{i=0}^{i≤nH}│
              └────┬─────────────┘
                   │  Energy balance
                 (End)
```

Fig. 3. The EDA iteration flowchart

3.2 The EDA Description

The power distribution among the heaters is carried out using the concept of the balance of energy. This means that the energy spent on heating the house must equal the energy estimated as required. Each time the energy distribution is needed the algorithm in Figure 3 is run. The required energy is estimated locally by the heaters, which send it to the CCU. Also, the CCU receives from each heater the temperature in the room. The CCU then aggregates both measures for each room in the building according to the number of heaters installed in it.

The CCU calculates the available power (Pa) as the difference between the CPL and the current consumption due to the small power devices and lighting. The FRBS-1 from room i is then used to propose the heaters heating power (Phi). Finally, the Pa is distributed between the heaters according to their Phi, which is updated. This Phi is the heating power that the heater i is allowed to spend on the next duty cycle. The last step in this algorithm is to join the switch ON time slots assigned to each heater to avoid fluctuation in the power output.

3.3 The MAS Architecture

The MAS architecture is based on two kinds of agents: the CCU agent –which acts as a coordinator– and the heater agent –which acts as data provider and actuation unit–. Both types of agents collaborate to distribute the power without surpassing the CPL while maintaining the comfort level in the house if there is sufficient available power.

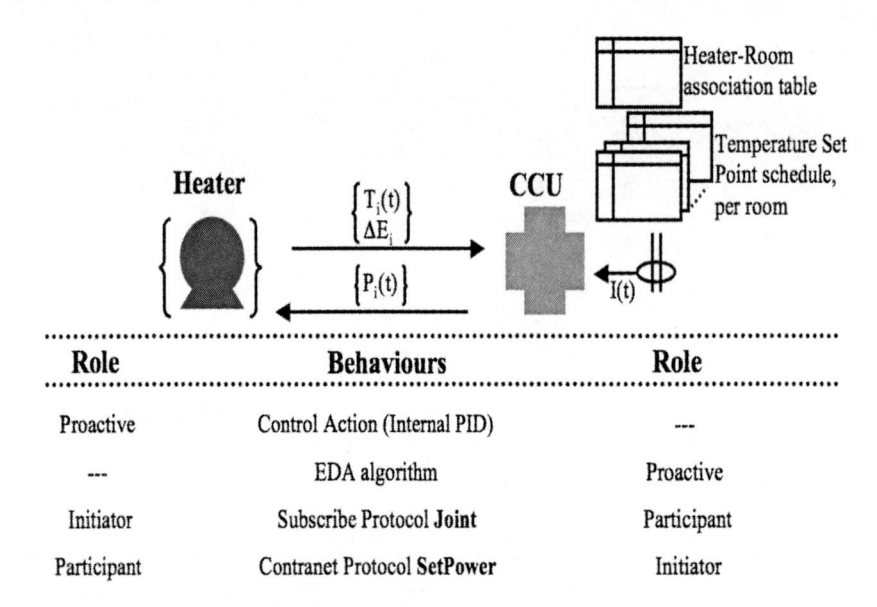

Fig. 4. The MAS architecture and some of the behaviours: the role of each type of agents is included

An MAS paradigm is considered to provide the heating system with a robust behavior, as proposed in [3]. The MAS methodology used is that resembled in [9] [12]. The MAS runs over a Zigbee wireless network. The EDA –distributed between the CCU and the heaters– directs the power output of all the heaters when the network is up. When the network is down, or when the CCU is out of service, all the heaters must act as normal stand-alone heaters until the system recovers. Figure 4 shows the schema of the MAS, where the data flows between heaters and CCU and some of the agent behaviors are presented.

3.4 The CCU Agent

The CCU agent is responsible for managing the building information: the building topology, the climate zone, the association table between heaters and rooms and the temperature set point profiles for each room. Also, the CCU measures the power consumption in the building in order to calculate the available power.

In the Run Stage, the CCU runs the EDA, distributing the power and synchronizing all the heaters in the building.

3.5 The Electrical Heater Agent

The electrical heater agent is responsible for heating the room where its respective heater is installed. This agent has the ability to detect the collapse of the distributed system (the CCU breaks down, the wireless network is OFF, etc.). In this case, each heater acts autonomously in an isolated installation and no overall objectives are to be accomplished. Also, it will automatically change to the collaborative work if the system recovers.

This agent is responsible for some important tasks, such as calculating the temperature in each room or estimating the required power. Also, some behaviors like open window detection or the initiation of the association to a CCU correspond to this agent.

4 The Implementation of the MAS Approach

The MAS described in the previous section is now being implemented in microcontroller devices. The CCU includes the FRBS-1, the EDA, the rest of the pre-defined behaviours and a human-machine interface (HMI) to set up the system. This HMI is text based to simplify the code needed and to reduce the cost of the CCU unit in the market.

Also, the heater agents have an HMI, which is also text based. This HMI allows users to configure the heater in the stand-alone mode and to set the network address in case of being integrated within the energy distribution system.

The CCU needs a small set of parameters to be configured: the climate zone and the building topology. Also, the temperature set up profile for each room and for the building must be given, but this is common to all heating system controllers. Finally, a procedure of discovering the heaters from the CCU and associating them to a room of the building must be executed.

In Figure 5 the photo of the prototype being developed is included. The CCU is the micro-controller device on the left; the remaining devices are the heaters. The centred heater includes the HMI. The Atmel ATAVRRZ200 Zigbee demonstration kit is being used to develop the prototype.

Fig. 5. The MAS prototype: the left-most micro-device is the CCU, while the remaining devices are heaters

5 Conclusions and Future Works

A valid approach to distribute the available electrical power between the heaters in an electrical heating system has been developed. An MAS approach has proven as valid for these purposes, providing robustness and simplicity to the distributed heating system. The HAIS proposed for the design stage is able to extract the information and to successfully train the FRBS-1.

Future work includes validating the prototype in real situations previous to manufacturing a marketable product. Also, the HAIS presented in this work must be tested against different FRBS learning approaches available in the literature in order to obtain better fuzzy controllers.

Acknowledgements. This research work has been funded by Gonzalez Soriano, S.A. –by means of the CN-08-028-IE07-60 FICYT research project– and by the Spanish Min. of Education, under the grant TIN2005-08386-C05-05.

References

[1] Alcalá-Fdez, J., Sánchez, L., García, S., del Jesus, M.J., Ventura, S., Garrell, J.M., Otero, J., Romero, C., Bacardit, J., Rivas, V.M., Fernández, J.C., Herrera, F.: Keel: A software tool to assess evolutionary algorithms to data mining problems. Soft Computing (in press, 2007), doi: 10.1007/s00500-008-0323-y
[2] Casillas, J., Cordón, O., Herrera, F., Villar, P.: A hybrid learning process for the knowledge base of a fuzzy rule-based system. In: Proceedings of the Information Processing and Management of Uncertainty in Knowledge-Based Systems IPMU 2004, Perugia, Portugal (2004)
[3] Davidsson, P., Boman, M.: Distributed monitoring and control of office buildings by embedded agents. Information Sciences 171, 293–307 (2005)
[4] Consejo Superior de Colegios de Arquitectos de España. CTE-HE ahorro de energía, Aplicación a edificios de uso residencial vivienda-DAV. Consejo Superior de Colegios de Arquitectos de España (2006) ISBN: CTE 84-934051-7-5
[5] Ministerio de la Presidencia. Real Decreto 1027/2007, de 20 de julio, por el que se aprueba el Reglamento de Instalaciones Térmicas en edificios. BOE num 207, Agosto (2007), http://www.boe.es
[6] Ministerio de la Vivienda. Real Decreto 314/2006, de 17 de marzo, por el que se aprueba el Código Técnico de la Edificación (2006) ISBN: 84-340-1641-9
[7] Ministerio de Vivienda. Documento Básico de Ahorro de Energia. Limitación de la demanda de Energía. Dirección General de Arquitectura y Politítica de Vivienda (2005)
[8] Holland, J.H.: Escaping Brittleness: The Possibilities of General-Purpose Learning Algorithms Applied to Parallel Rule-Based Systems, vol. 2. Morgan Kaufmann, Los Altos (1986)
[9] Julián, V., Botti, V.: Developing real-time multi-agent systems. Integrated Computer-Aided Engineering 11(2), 135–149 (2004)
[10] Kinney, P.: Zigbee technology: Wireless control that simply works. Technical report, The ZigBee Alliance (2007), http://www.zigbee.org/
[11] Lewis, P.T., Alexander, D.K.: Htb2: A flexible model for dynamic building simulation. Building and Environment (1), 7–16 (1990)

[12] The Foundation of Intelligent Physical Agents. The FIPA official site (2008),
http://www.fipa.org/
[13] Villar, J.R., de la Cal, E.A., Sedano, J.: A fuzzy logic based efficient energy saving approach for domestic heating systems. Integrated Computer-Aided Engineering (submitted, 2008)
[14] Villar, J.R., de la Cal, E., Sedano, J.: Energy saving by means of fuzzy systems. In: Yin, H., Tino, P., Corchado, E., Byrne, W., Yao, X. (eds.) IDEAL 2007. LNCS, vol. 4881, pp. 155–161. Springer, Heidelberg (2007)
[15] Villar, J.R., de la Cal, E.A., Sedano, J.: Energy savings by means of multi agent systems and fuzzy systems. In: Proceedings of the 6th International Workshop on Practical Applications of Agents and Multiagent Systems IWPAAMS 2007, Salamanca, Spain, pp. 119–128 (2007)
[16] Villar, J.R., de la Cal, E.A., Sedano, J.: Energy saving by means of fuzzy systems. In: Corchado, E., Abraham, A., Pedrycz, W. (eds.) HAIS 2008. LNCS (LNAI), vol. 5271, pp. 583–590. Springer, Heidelberg (2008)

Hardware Protection of Agents in Ubiquitous and Ambient Intelligence Environments[*]

Antonio Maña, Antonio Muñoz, and Daniel Serrano

University of Malaga
{amunoz,amg,serrano}@lcc.uma.es

Summary. Agent-based computing represents a promising paradigm for distributed computing. Unfortunately the lack of security is hindering the application of this paradigm in real world applications. We focus on a new agent migration protocol that takes advantage of TPM technology. The protocol has been validated using AVISPA model checking toolsuite. In order to facilitate its adoption, we have developed a software library to access TPM functionality from agents and to support their secure migration. This paper presents hardware-based system to protect agent systems. Concretely our work is based on trusted platform module (TPM), which is the cornerstone to build the solution. In order to build our solution on a robust basis, we validate this protocol by means of a model checking tool called AVISPA. Then as final result we provide a library to access to TPM (Trusted Platform Module) functionality from software agents. Along this paper we detail more relevant aspects of this library both in the development stage of it and while we use it to develop a system based agent.

1 Introduction

A well accepted definition of software agent is found in [16]: "An agent is an encapsulated computer system that is situated in some environment and that is capable of flexible, autonomous action in that environment in order to meet its design objectives". More important characteristics of an agent are that are highly agree upon to include: Autonomy (behaves independent according to the state it encapsulates), proactiveity (able to initiate without external order), reactivity (react in timely fashion to direct environment stimulus), situatedness (ability to settle in an environment that might contain other agents), directedness (agents live in a society of other collaborative or competitive groups of agents), and the social ability to interact with other agents, possibly motivated by collaborative problem solving.

Then an agent is supposed to live in a society of agents; multi-agent systems (MAS). A MAS is known as a system composed of several agents collectively capable of reaching goals that are difficult to achieve by an individual agent of monolithic system. Also a MAS represents a natural way of decentralization,

---

[*] Work partially supported by E.U. through projects SERENITY (IST-027587) and OKKAM (IST- 215032) and DESEOS project funded by the Regional Government of andalusia.

where there are autonomous agents working as peers, or in teams, with their own behaviour and control. Mobile agents are a specific form of mobile code and software agent paradigms. However, in contrast to the Remote evaluation and Code on demand paradigms, mobile agents are active in that they may choose to migrate between computers at any time during their execution. This makes them a powerful tool for implementing distributed applications in a computer network. Development process required some considerations, such as a JADE system is composed for a platform that contains a main container in which agents are deployed.

Additional containers can be added to this platform, some of them can be remote containers, and different platform can interacts among them, allowing the migration of the agents among them. Them containers play agencies roles in JADE on which agents will be deployed. For this purpose we based our work on JADE, the main reason to choose this platform was its interplatform migration mechanism. Taking into account JADE structure, we conclude that two different kinds of migration exists, migration among containers from different platforms and migration from containers from the same platform. In the case that the migration is from containers from different platforms, the agent migrates from a container from source agency to the destination agency main container. In such a case that destination agency is not a JADE built-on platform architecture can be different, depending on the platform. In the other case, the agent migrates from a container to another one but in the same platform. Both migration processes imply some security concerns. Platform migration is not secure because main container from source platform can be untrusted. Also containers migration has the same problem, it is, if destination container is not trusted then the migration is not secure. Secure migration library solves arisen problems.

Agent migration consists on a mechanism to continue the execution of an agent on another location [5]. This process includes the transport of agent code, execution state and data of the agent. In an agent system, the migration is initiated on behalf of the agent and not by the system. The main motivation for this migration is to move the computation to a data server or a communication partner in order to reduce network load by accessing a data server a communication partner by local communication. Then migration is performed from a source agency where agent is running to a destination agency, which is the next stage in agent execution. This migration can be performed by two different ways. First way is by moving, that is, the agent moves from the source agency to the destination one. And the second way is by cloning. In this case, the agent is copied to the destination agency. From this moment on, the two copies of the agent coexist executing in different places. Henceforth we use the term migration to refer to both agent cloning and agent moving. Software agents are a promising computing paradigm. It has been shown [19] that scientific community has devoted important efforts to this field. Indeed, many applications exist based on this technology. However, all this efforts are not useful in practice due to the lack of a secure robust basis to develop new applications. In [19] two different solutions are presented to provide a security framework for agents building. The first of these is a software solution based on the protected computing approach [11]. The second one consists on a hardware based solution that takes advantage of the TPM technology. In this paper we use that theoretical and abstract solution to build a solution for the secure migration of software agents.

2 State of the Art

The hardware-based protection infrastructure described in the present paper takes advantage of the recent advances in trusted hardware. The basic idea behind the concept of Trusted Computing is the creation of a chain of trust between all elements in the computing system, starting from the most basic ones. Consequently, platform boot processes are modified to allow the TPM to measure each of the components in the system and securely store the results of the measurements in Platform Configuration Registers (PCR) within the TPM. This mechanism is used to extend the root of trust to the different elements in the computing platform. Therefore, the chain of trust starts with the aforementioned TPM, which analyses whether the BIOS of the computer is to be trusted and, in that case, passes control to it. This process is repeated for the master boot record, the OS loader, the OS, the hardware devices and finally the applications. In a Trusted Computing scenario a trusted application runs exclusively on top of trusted and pre-approved supporting software and hardware. Additionally the TC technology provides mechanisms for the measurement (obtaining a cryptographic hash) of the configuration of remote platforms. If this configuration is altered or modified, a new hash value must be generated and sent to the requester in a certificate. These certificates attest the current state of the remote platform.

Several mechanisms for secure execution of agents have been proposed in the literature with the objective of securing the execution of agents. Most of these mechanisms are designed to provide some type of protection or some specific security property. In this section, we focus on solutions that are specifically tailored or especially well-suited for agent scenarios. More extensive reviews of the state of the art in general issues of software protection can be found in [11] [12]. Some protection mechanisms are oriented to the protection of the host against malicious agents. Among these, SandBoxing and proof-carrying code [13]. One of the most important problems of these techniques is the difficulty of identifying which operations (or sequences of them) can be permitted without compromising the local security policy. Other mechanisms are oriented toward protecting agents against malicious servers. Among them the concept of sanctuaries [14] was proposed. Several techniques can be applied to an agent in order to verify self-integrity and avoid that the code or the data of the agent is inadvertently manipulated. Anti-tamper techniques, such as encryption, checksumming, anti-debugging, anti-emulation and some others [15] [17] share the same goal, but they are also oriented toward the prevention of the analysis of the function that the agent implements. Additionally, some protection schemes are based on self-modifying code, and code obfuscation [18]. Finally there are techniques that create a two-way protection. Some of these are based on the aforementioned protected computing approach [11].

3 Trusted Computing Technology in the Agent Protection Road

Previously we argued that is essential to integrate strong security mechanisms in agent systems world to achieve a reasonable security level to support real world applications. For this reason we propose a hardware-based mechanism to provide security to agent systems. The TPM provides mechanisms, such as cryptographic algorithms, secure key storage and remote attestation that provides important tools to achieve a high level of security. Unfortunately, the TPM technology is very complex and so are the procedures and action sequences to use it. The access to the device and the use of TPM functionality is not an easy task, especially for average software developers. Therefore, we have developed a library that provides access to the TPM from software agents. The main advantage for this approach is that developers of agent systems do not need to become security experts, and can access the security mechanisms without taking care of low level details of the TPM technology. Hardware-based solutions can be built on the basis of different devices. We have selected the TPM for our solution because it provides all necessary features. Concretely, it provides cryptography capabilities, secure storage, intimate relation to the platform hardware, remote attestation, etc. Remote attestation is perhaps the most important feature; it allows the computer to recognize any unauthorized/authorized changes to software. If certain requirements are met then the computer allows the system to continue its operations. This same technology that TPM utilizes can also be applied in agent-bases systems to scan machines for changes to their environments made by any entity before the system boots up and accesses the network.

Additional reasons are the support of a wide range of industries and the availability of computers equipped with TPM In summary, this element is the cornerstone of our approach and the security of our system is focused on it. The use of TPM in these systems is as follows. Each agency takes measures of system parameters while booting to determine its security, such as BIOS, keys modules from Operating System, active processes and services in the system. Through these parameters an estimation of the secure state of the agency can be done. Values taken are stored in a secure way inside the trusted device, preventing unauthorized access and modification. Agencies have the ability to report previously stored configuration values to other agencies in order to prove their security.

Our system takes advantage of the remote attestation procedure provided by the TPM in order to verify the security of the agencies. In this way agents can verify the security of the destination agency before migration.

4 Secure Migration Library for Agents

This Section introduces the Secure Migration Library for Agents. It aims to serve as a platform for the agent migration on top of the JADE platform. In order to provide a secure environment this library makes use of a TPM. In this way, the migrations are supported by a trusted platform. The SMLA is based on the one hand on the JADE platform and on the other hand on the TPM4Java library. TPM4Java provides access to the TPM functions from the Java programming language. However, as it is, it is not possible to use TPM4Java in agents. For this reason, we have extended TPM4Java.

Among the central objectives for the development of this library we highlight the following, providing a secure environment in which agents can be securely executed and migrated, to facilitate the integration of TPM technology with JADE, to provide mechanisms that are simple to use from the programmers point of view, to provide a library that is easy to adapt to be based on other security devices such as smartcards, to provide a library with a generic and flexible interface. Throughout this mechanism users are able to use the library to solve a wide range of problems. It is important to mention the fact that is important to provide a library that can easily be extended in the future. Finally, the most important function provided by the library is the secure migration mechanism. This secure mechanism is based on the remote attestation of the destination agency, before migration actually takes place, in order to guarantee that agents are always executed in a secure environment. Additionally we provide a secure environment for agents to run, in such a way that malicious agents are not able to modify the host agency.

Our library is designed according to the following set of requirements. On the one hand, among the functional requirements we considered that our main objective was to provide a mechanism for secure agent migration in JADE platform. The goal was to make possible for agents to extend their trusted computing base to the next destination agency before migrating to it. It is important to mention that each agency must provide local functionality to perform the secure migration mechanism, which requires the presence of the TPM for remote attestation. Similarly each agency must provide the functionality to allow to other agencies to take remote integrity measures to determine whether its configuration is secure. In order to support the previous processes, the library includes a protocol that allows two agencies to exchange information about their configurations. This exchange establishes the bases of the mutual trust between agencies. Concerning the non-functional requirements, the library must seamlessly integrate in the JADE platform, in such a way that its use does not imply modifications in the platform. Additionally, the operation of the library must be transparent to the user. In order to facilitate the adaptation of the library to new devices we decided to create an additional abstraction layer that provides generic services. In this way the generic services remain unchanged. One final requirement has to support both types of migration (moving and cloning). Concerning the architecture each agency can host different agents and is supported by TPM functionalities. TPM plays the role of trusted element of each agency, able to certify to other agencies its trustworthiness.

4.1 Secure Migration Protocol

Following we study all the protocols that allowed that our library provides the security to the migration. Firstly we analyse the different attestation protocols as well as secure migration protocols, studying their benefits, so that we can build our own secure migration protocol. From this point we use the concept of migration both to agent cloning and agent moving, because from protocol point of view there is no difference between them. A first approach of a secure migration protocol is in [11]. This protocol provides some important ideas to take into account during the design process of the final protocol, we highlight that the agent trusts in its platform to check the migration security. The necessity of the use of TPM to obtain and report configuration data by a secure way is a key issue. Other protocol that provides interesting ideas to consider in the development of a secure migration system is in [15]. The most relevant ideas provided are; a protocol shows how an agent from the agency requests to TPM the signed values from PCRs. As well as the protocol shows how the agent obtain platform credentials. These credentials, together with PCRs signed values allow to determine whether the destination configuration is secure. This is performed by using the quote function provided by the TPM. TPM Quote function allows remotely verifying the contents of the integrity registers. Finally, we analyse the protocol from [15]. The most relevant ideas from this protocol are; the use of an AIK to sign the PCR values, the use of a CA that validates the AIK and the use of configurations to compare received results from remote agency. We designed a new protocol based on the study of these three protocols, in such a way that we took advantage of the appeals provided by each of them. Our protocol has some characteristics. The agency provides to the agent the capacity to migrate by a secure way. Also the agency uses a TPM that provides configuration values stored in PCRs. TPM signs PCRs values using a specific AIK for the destination agency. In such a way that data receiver knows securely TPM identity which signed. A Certification Authority generates needed credentials to correctly verify the AIK identity. Together with signed PCRs values the agency provides AIK credentials in such a way that the requester can correctly verify that data comes from agency TPM. Following we define the 18 steps protocol, used to perform secure migration.

We can observe that the protocol fulfils the five main characteristics we mentioned previously. Then we have a clear idea of the different components of the system as well as the interaction between them to provide the security in the migration.

4.2 Verification of Secure Migration Protocol with AVISPA

The Secure Migration protocol previously described is the basis of this research. Thus we want to build a robust solution, for this purpose a validation of this protocol is the next step in this research. Among different alternatives we selected a model checking tool called AVISPA.

Algorithm 1. Secure migration protocol

1. Agent Ag requests to his agency the migration to Agency B.
2. Agency A (source agency) sends to agency B (destination agency) a request for attestation.
3. Agency B accept the request for attestation and send a nonce (this value is composed by random bits used to avoid repetition attacks) and indexes of PCRs values that needs.
4. Agency A request to its TPM PCR values requested by agency B together with the nonce all signed.
5. TPM returns requested data.
6. Agency A obtains AIK credentials from its credentials repository.
7. Agency A requests agency B for PCRs values requested and nonce all signed. Then it sends AIK credentials, which contains the public key corresponding with the private key used to sign data. Additionally, it sends a nonce and the indexes of PCRs that wants to know.
8. Agency B validates the authenticity of received key verifying the credentials by means of the CA public key which was used to generate those credentials.
9. Agency B verifies the PCRs values signature and the nonce received using the AIK public key.
10. Agency B verifies that PCRs values received belongs to the set of accepted values and them the agent configuration is valid.
11. Agency B requests to its TPM the PCR values requested by the agency A together with the nonce signed.
12. TPM returns requested data.
13. Agency B obtains AIK credentials from its credentials repository.
14. Agency B sends to agency A PCR values requested and the nonce signed. Also it sends AIK credentials, which contains the public key corresponding to the private key used to encrypt the data.
15. Agency A validates the authenticity of received key verifying the credentials by means of CA public key that generated those credentials.
16. Agency A verifies the PCR values signature and the nonce received using the AIK public key.
17. Agency A verifies that PCR values received belongs to the set of accepted values and then the agency B configuration is secure. From this point trustworthy between agencies A & B exists.
18. Then Agency A allows to the agent Ag the migration to agency B.

AVISPA is an automatic push-button formal validation tool for Internet security protocols, developed in a project sponsored by the European Union. It encompasses all security protocols in the first five OSI layers for more than twenty security services and mechanisms. Furthermore this tool covers (that is verifiable by it) more than 85 of IETF security specifications. AVISPA library available on-line has in it verified with code about hundred problems derived from more than two dozen security protocols. AVISPA uses a High Level Protocol Specification Language (HLPSL) to feed a protocol in it; HLPSL is an extremely expressive and intuitive language to model a protocol for AVISPA. Its operational semantic is based on the work of Lamport on Temporal logic of Actions. Communication using HLPSL is always synchronous. Once a protocol is fed in AVISPA and modelled in HLPSL, it is translated into Intermediate Format (IF). IF is an intermediate step where re-write rules are applied in order to further process a given protocol by back-end analyzer tools. A protocol, written in IF, is executed over a finite number of iterations, or entirely if no loop is involved. Eventually, either ran attack is found, or the protocol is considered safe over the given number of sessions.

System behaviour in HLPSL is modelled as a 'state'. Each state has variables which are responsible for the state transitions; that is, when variables change, a state takes a new form. The communicating entities are called 'roles' which own variables. These variables can be local or global. Apart from initiator and receiver, environment and session of protocol execution are also roles in HLPSL.

Roles can be basic or composed depending on if they are constituent of one agent or more. Each honest participant or principal plays one role. It can be parallel, sequential or composite. All communication between roles and the intruder are synchronous. Communication channels are also represented by the variables carrying different properties of a particular environment. The language used in AVISPA is very expressive allowing great flexibility to express fine details. This makes it a bit more complex than Hermes to convert a protocol into HLPSL. Further, defining implementation environment of the protocol and user-defined intrusion model may increase the complexity. Results in AVISPA are detailed and explicitly given with reachable number of states. Therefore regarding result interpretation, AVISPA requires no expertise or skills in mathematics contrary to other tools like HERMES[20] where a great deal of experience is at least necessary to get meaningful conclusions. Of the four available AVISPA Back-Ends we chose the OFMC Model, which is the unique that uses fresh values to generate nonce's. However, this alternative requires a limit value for the search. The results of our research are the following:

```
SUMMARY
  SAFE
DETAILS
  BOUNDED_NUMBER_OF_SESSIONS
PROTOCOL
  /home/anto/avispa/avispa-1.1/testsuite/results/protocolo_2.txt.if
GOAL
  as_specified
BACKEND
  OFMC
COMMENTS
STATISTICS
  parseTime: 0.00s
  searchTime: 3.23s
  visitedNodes: 4 nodes
  depth: 200 plies
environment()
```

These results show that the summary of the protocol validation is safe. Also some statistics are shown among them depth line indicates 200 plies, but this process has been performed for 250, 300, 400, 500, 1000 and 2000 of depth values with similar results. Appendix A depicts the HLPSL code to verify the Secure Migration Protocol in AVISPA. For this purpose we check the authentication on pcra(pcrib), pcrb(pcria), aikab and aikba, due to these are exchanges values among agent A and B.

5 Conclusions and Future Work

In this paper we have presented a new approach for the agent based systems protection based on specific hardware. To validates our approach we provide a proof of concept by means of the 'secure migration library for agents'. Along this paper we describe in details this library and all protocols which it relies on. Especially we discuss the 'secure migration protocol' as the basis to our library. Moreover, we provide the methodology to validate this protocol by means of a validation tools called AVISPA. Concerning the future work, an important issue is the implementation of the Anonymous Direct Attestation protocol. We mentioned that our library is based on the attestation protocol based on the Certification Authority protocol described in section 3.1. Our solution is based on this protocol due to the fact that TPM4Java only implements this. Certification Authority protocol provides important advantages as previous sections described but this entails some disadvantages. Among these disadvantages we highlight the needed to generate a certificate for each used key in every attestation, which implies that many requests might be performed to the CA and this provokes a bottleneck for the system. Besides, we found that whether verification authority and CA act together the security of the attestation protocol can be violated. That is, the use of the anonymous direct attestation protocol entails some advantages. Let us see through an example, certify issuer entity and verifier entity can not collaborate to violate the system security. Therefore, a unique entity can be verifier and issuer of certificates. Last but not least of the found advantages of this approach is that certificates only need to be issued once which solves the aforementioned bottleneck. These advantages prove that the anonymous attestation protocol is an interesting option for attesting. Possible future lines of research are the improvement of the keys management system of the library. Our library uses RSA keys for attestation protocol that must be loaded in TPM. However the size for key storage in the TPM is very limited, then it must be carefully managed to avoid arisen space problems. Our library handles the keys in such a way that only one key is loaded in TPM simultaneously, therefore keys are loaded when will use and downloaded when used. This procedure is not very efficient due to the continuous loading and downloading of keys. For this problem we propose the possibility of downloading the same key that we will use in next step. Of course this entails more issues. A different improvement in the key management lies on caching these keys. Then, several keys can be loaded simultaneously in TPM and that these can be change when more available space is needed. This solution entails to develop a key replace policy to determine which key delete from TPM to load a new key in the cache, but this task is out of the scope of this paper. Another future work is to extend the library with new functionalities to secure migration services in order to provide concurrency. Secure migration services implemented in the library provide secure migration to a remote container. However they have capacity to handle a unique request simultaneously, then migration requests arriving while migration is performed are refused. This happens due to TPM management of the problem together with the aforementioned key management problem. A possible improvement of the library is to provide of secure migration service with the capability handle simultaneously several requests.

References

1. Mouratidis, H., Kolp, M., Faulkner, S., Giorgini, P.: A Secure Architectural Description Language for Agent Systems. In: AAMAS 2005, Utrecht, Netherlands, July 25-29 (2005)
2. Wang, H., Wang, C.: Intelligent Agents in the Nuclear Industry. IEEE Computer 30(11), 2834 (1997)

3. Schwuttke, U.M., Quan, A.G.: Enhancing Performance of Cooperating Agents in Real-Time Diagnostic Systems. In: Proceedings of the Thirteenth International Joint Conference on Artificial Intelligence (IJCAI 1993), pp. 332–337. International Joint Conferences on Artificial Intelligence, Menlo Park (1993)
4. Clements, P., Papaioannou, T., Edwards, J.: Aglets: Enabling the Virtual Enterprise. In: The Proc. Of Managing Enterprises - Stakeholders, Engineering, Logistics and Achievement (ME-SELA 1997), p. 425 (1997) ISBN 1 86058 066 1
5. General Magic, Inc. The Telescript Language Reference (1996), http://www.genmagic.comTelescript.TDE.TDEDOCSHTMLtelescript.html
6. http://cougaar.org/
7. Shepherdson, D.: The JACK Usage Report. In: The Proc. Of Autonomous Agents and Multi Agents Systems (AAMAS 2003) (2003)
8. http://jade.tilab.com
9. http://www.irit.fr/recherches/ISPR/IAM/JavAct.html
10. Alechina, N., Alechina, R., Habner, J., Jago, M., Logan, B.: Belief revision for AgentSpeak agents. In: The Proc. Of Autonomous Agents and Multi Agents Systems 2006, Hakodate, Japan, pp. 1288–1290 (2006) ISBN:1-59593-303-4
11. Maña, A.: Protección de Software Basada en Tarjetas Inteligentes. PhD Thesis. University of Málaga (2003)
12. Hachez, G.: A Comparative Study of Software Protection Tools Suited for E-Commerce with Contributions to Software Watermarking and Smart Cards. PhD Thesis. Universite Catholique de Louvain (2003), http://www.dice.ucl.ac.be/~hachez/thesis_gael_hachez.pdf
13. Necula, G.: Proof-Carrying Code. In: Proceedings of 24th Annual Symposium on Principles of Programming Languages (1997)
14. Yee, B.S.: A Sanctuary for Mobile Agents. Secure Internet Programming (1999)
15. Trusted Computing Group: TCG Specifications (2005), https://www.trustedcomputinggroup.org/specs/
16. Wooldrigde, M.: Agent-based Software Engineering. In: IEE Proceedings on Software Engineering, vol. 144(1), pp. 26–37 (February 1997)
17. Stern, J.P., Hachez, G., Koeune, F., Quisquater, J.J.: Robust Object Watermarking: Application to Code. In: Pfitzmann, A. (ed.) IH 1999. LNCS, vol. 1768, pp. 368–378. Springer, Heidelberg (2000), http://www.dice.ucl.ac.be/crypto/publications/1999/codemark.pdf
18. Collberg, C., Thomborson, C.: Watermarking, Tamper-Proofing, and Obfuscation - Tools for Software Protection. University of Auckland Technical Report 170 (2000)
19. Maña, A., Muñoz, A., Serrano, D.: Towards secure agent computing for ubiquitous computing and ambient intelligence. In: Indulska, J., Ma, J., Yang, L.T., Ungerer, T., Cao, J. (eds.) UIC 2007. LNCS, vol. 4611, pp. 1201–1212. Springer, Heidelberg (2007)
20. Hussain, M.: Dominique Seret. A Comparative study of Security Protocols Validation Tools: HERMES vs. AVISPA. In: Proceecings of International Conference on Advanced Communication Technology (ICACT). IEEE Communication Society, Los Alamitos (published, 2006)

Management System for Manufacturing Components Aligned with the Organisation IT Systems

Diego Marcos-Jorquera, Francisco Maciá-Pérez, Virgilio Gilart-Iglesias, Jorge Gea-Martínez, and Antonio Ferrándiz-Colmeiro

University of Alicante, Technology and Computer Science Department
P.O. Box 99, 03080, Alicante, Spain
{dmarcos, pmacia, vgilart, jgea, aferrandiz}@dtic.ua.es

Abstract. This paper proposes a management system for industrial production elements based on a multi-agent system which introduces a set of methods, models and architectures designed to provide a framework for specifying, implementing and establishing maintenance systems within the global business model. The main contribution of this proposal is that, as well as providing an adequate approach for the management of this type of infrastructure, it is possible to evolve from mass production models towards mass customisation models, increasingly bringing organisation objectives closer to the interests and requirements of customers.

1 Introduction

The market is increasingly demanding personalised products in order to satisfy customer requirements and, specifically, tailored manufacturing is shown to be one of the most effective channels for lowering costs and amortising the high investment required by these new approaches.

Internet, and the use made of it by organisations and customers, encourages this trend, enabling rapid, global and personalised access to an increasingly wide and varied catalogue of products.

The solution is found in new business models such as agile manufacturing systems with a high capacity for flexibility and dynamism which enables a rapid adaptation of production based on market demand.

IT may also be a valuable tool for achieving these objectives, putting it to the service of business and providing it with the requisite functionality, yet without increasing the complexity of processes and the system maintenance. In particular, the use of Multi-Agents Systems (MAS) together with the Service Oriented Architecture paradigm (SOA) is presented as one of suitable approaches in order to resolve the problems introduce by the new production models [1].

This paper proposes a component manufacturing management system (sections 3 and 4) based on information technologies that incorporate the concepts of high availability and business continuity into the low manufacturing levels of the organisation, providing toughness and, above all, the flexibility required for implementing new business models. A review of the state of the art is also presented, together with the model and its implementation (section 2), the system architecture (section 5), a functional prototype of the system (section 6) and the main conclusions arising from the work such as future lines of work (section 7).

2 Background

Either through models or techniques such as *flexible manufacturing*, *mass customisation* or *lean production* it is necessary to seek systems with greater flexibility for adapting in a dynamic way to changes and which systems enable them to be integrated with existing elements of the organisation in order to achieve global management. *Agile manufacturing* is one type which is best adapted to new requirements deriving mainly from Internet and IT [2]. *Agile manufacturing* may be defined as a production model integrating technology, human resources and organisation through IT providing flexibility, speed, quality and efficiency and enabling a deliberate, effective and coordinated response to changes in the environment. The literature in this area has provided numerous contributions as well as very different approaches, such as the use of intelligent control systems [3], the advantages of agent technologies, [4] or effects deriving from the development of IT [5], all of which indicate the considerable interest shown by the scientific community in these fields.

Flexibility is an important feature of various sectors within a manufacturing organisation. We can find flexibility in: machinery, production, mixing, product range, production volumes, delivery, routing, expansion and response [6]. The common denominator in this type of approach is the need for models which will provide adaptability, agility, capacity for integration, dynamism, fault tolerance, business continuity and scalability.

One of the main fronts open in this field is the integration of production processes with other business processes. One proposal were based on traditional models of automation based on proprietary protocols at eBusiness model level of resources, as systems external to business processes (Modbus, Profibus, AS-I, FIPIO, DeviceNET, Interbus or Ethernet industrial), and consisted of initial attempts to facilitate integration with business components [7]. In other cases, as in Schneider's proposal [8], embedded devices were used under concepts such as transparent factory. In [9] it is proposed to benefit from these same techniques in order to raise the level of abstraction of the production elements to the business level in such a way that the integration of resources, processes and in general business logic is produced in a natural and transparent manner within the existing business models. Finally, also within the framework of European research projects, there are some important initiatives which support the interest along these lines, with significant results which have advanced towards Service-oriented architecture (SOA) and embedded devices in industrial machinery as valid technologies [10].

The incorporation of management systems in industry began during the nineties with the use of more open and flexible control mechanisms which improved the benefit of automata [7]. Although the power and connectivity of these new components have been facilitating integration at all levels of the manufacturing process [11], their indiscriminate or poorly planned application, lack of standards and their inherent complexity have generated a new problem, namely their management.

The first open standards which attempted to address the management of devices in a generic manner were SNMP and CMIP [12], specified by the IETF; both protocols are oriented mainly towards network supervision and control.

The numerous tasks associated with network management, as well as the fact that it is extremely diverse and complex, means that maintenance work on these systems presupposes a high cost to organisations both in terms of time and personnel resources. Many companies use administration systems which incorporate the characteristics of self management and self configuration which facilitate network management [12]. Examples of these systems are: Solstice, Sun Enterprise Manager and NESTOR.

The use of multi-agent systems for computer network management provides a series of characteristics which favour automation and self-regulation in maintenance processes [13]. The creation of projects such as AgentLink III, the first Coordinated Action on Agent Based Computation financed by the *European Commission 6th Framework Programme* is a clear indicator of the considerable degree of interest currently manifested in software agents. In manufacturing and industrial automation field the MAS systems also are presented as a suitable solution in order to approach the requirements of new production models. Nowadays, the majority of the industrial automation and control approaches (Bionic Manufacturing Systems, Reconfigurable Manufacturing Systems, Holonic Manufacturing Systems, Balanced Automation systems, Evolvable Production Systems) propose the use of intelligent distributed systems where MAS system can supply a lot of advantages, instead of focused on traditional centralized systems [14]. Production management in physically distributed environments involves the necessity to take decisions at a local level, and therefore it is necessary to transfer part of the control and supervision logic to industrial machinery [15]. In addition, management process should be self-organised, establishing a reactive and proactive behaviour [16] in order to resolve disruptions and unexpected changes in the short term and in order to anticipate and prepare for critical situations, establishing machinery maintenance policies [17] or process quality control [18].

All these features are described in [1], where is presented a general overview about emergent approaches in the industrial automation field, highlight the advantages of use of MAS systems together with SOA paradigm.

Among existing maintenance systems, high availability, disaster recovery and self management systems apply techniques which aim to avoid or at least reduce down time caused by failures or incidents in the services offered [19], which permit full restoration of equipment information within time constraints. With these mechanisms problems arising from degradation or loss of stored information can be resolved and they also facilitate the start up of new equipment replicating the information obtained on the basis of a similar model.

There are currently various regeneration systems in the commercial sector which include: Ghost from Symantec and REMBO; as open code projects: Clonezilla, G4L, Linbox y UDP Cast [21]. The main disadvantage of all these systems is their high dependency on technology to be recovered. This means that multiple solutions need to be applied to cover the range of technology in the organisation. The Department of Computer Technology of the University of Alicante has developed Gaia [21], a multi-platform regeneration system based on open code. The system is designed in a modular manner, with an agent based system, which enables it to be adapted for our purposes to this environment.

3 Management System Approach

Our proposal focuses on a management system for industrial production elements which permits installation, maintenance, updating and reconfiguration of these production elements, acting on either embedded or external computer control systems which regulate industrial machinery.

The main tasks that the management system will accomplish are: software maintenance, setting up of new devices, device software regeneration, device reconfiguration, monitoring and inventory

The main advantages of these proposals are:

- The application of IT methods to achieve management and self management of production elements.
- Reduction of configuration times and set up of industrial machinery avoiding manual configurations.
- Dynamic adaptation to production changes.
- Reduction of set up times for new devices in the event of substitution through failure or extension.
- Detection and proactive correction of errors and damages in the manufacturing components.

The management system proposed is developed as a set of methods, models and architectures designed to provide a reference framework which permits specifying, implementing and establishing management IT systems of the production elements so that these can be aligned with the global business model of the organisation, as well as processes, services, resources and present technologies.

The system is designed for the management of industrial machinery and other auxiliary intermediate elements. In order to enable manufacturing elements to interact with the remainder of the system we should ensure that they have network communication capacity as well as processing capacity to execute the required software for carrying out the reconfiguration tasks.

In the case of low level machinery, in order to achieve the proposal objectives, it is possible to incorporate embedded systems which provide a minimum computation and communication platform making it possible to interact with the reconfiguration system. This same strategy could be applied in the case of industrial machinery which has a software platform based on the closed proprietary system.

4 Management Model Based on a Multi-Agent System

The management model has been proposed as an agent based system where agents represent the entities responsible for executing control and management activities, not only of production elements, but also, of IT services placed at their disposal for this purpose.

In this agent based system the set of tasks are defined on the basis of the processes, sub-processes and activities identified and defined.

Table 1. Management service associated tasks

Name	Description
wake_up	Wake up a manufacturing component starting its boot process by sending a special network datagram like WoL
get_configuration	Obtains the stored configuration of a manufacturing component
put_configuration	Stores the configuration of a manufacturing component
get_agent	Transfers an agent from the agent farm to the manufacture component
configure	Configures a manufacturing component
get_management_plan	Obtains the maintenance plan for a manufacturing component
init_agent	Starts the functionality of an agent
get_next_production_order	Gets the next production order to process
put_logs	Stores the activity logs
shutdown	Shutdown or reboot a manufacturing component

Table 2. Roles of agents implicated in management services

ID	Role	Goal
RCN	Reconfiguration	Reconfiguring a manufacturing component
INV	Inventory	Storing and providing access to the system information
MOV	Mobility	Transferring the agent for the system
INT	Interaction	Interacting with administrators
PLN	Planning	Planning and triggering the reconfiguration process
GST	Management	Managing the parameters of the system
CNT	Context	Establishing the context for an agent

Table 1 provides a brief example with the main tasks of a software restoration process and reconfiguration of a manufacturing component.

In order to determine the agents that the system needs and their tasks, the first step requires identification of the various roles. Table 2 compiles the main roles identified for the management process of industrial components.

Having identified roles, a set of agents are proposed which are required to develop the tasks relating to each process which are characterised through the assignment of specific roles.

Table 3 provides a list of the principal agents implicated in the management process, along with their respective roles and their relations with other agents and with the principal IT services that agents use to achieve their goals.

Table 3. Agents implicated in management system

Id	Agent	Roles	Relation with other agents	Access to services	MB	MEM	AUT	LC
RA	Reconfiguration	RCN MOV	IA CNT	Reconfiguration Log	▲	▼	▲	▼
SA	Schedule	PLN	IA CNT	Inventory Configuration	▼	■	▲	▲
IA	Inventory	INV MOV	CNT	Inventory Repository Configuration	■	▲	■	■
MA	Management	GST MOV INT	SA IA CNT	Repository Configuration	▲	▼	▼	■
CA	Context	CNT MOV		Context Configuration	▲	■	■	▼

MB – Mobility, MEM – Memory, AUT – Autonomy, LC – Life cycle
▲ – High, ■ – Midway, ▼ – Low

The agent based system, by itself, is sufficiently complex to merit attention in the form of more specific works.

Although the management system is largely supported by the system's general services, some of these in particular stand out: the planning service which is responsible for registering and programming the tasks of different agents involved in the management services; the inventory service providing all the information on the hardware and software configuration of all the equipment to be managed; the management services which provide all the utilities for developing its function.

5 System Architecture

The proposed service architecture has a layered structure (see figure 1a). The first layer contains the information system as a central core of the system. It is responsible for compiling all the information relating to the services provided as well as any additional information on the different organisation resources, enabling all types of maintenance processes to be automated. This system acts as a *Configuration-Management Database* (CMDB), offering a unified vision of all the organisation information.

The two following layers are service layers: *basic services layer* and *management services layer*. The first one provides distributed services of a general nature, destined to provide support to the whole system: directory, register or database management services. The *management services layer* consists of management services. Some of these are more generic in nature such as configuration and system management services, whereas others provide a specific functionality which needs to be incorporated in manufacturing elements, such as the dynamic reconfiguration of the machinery.

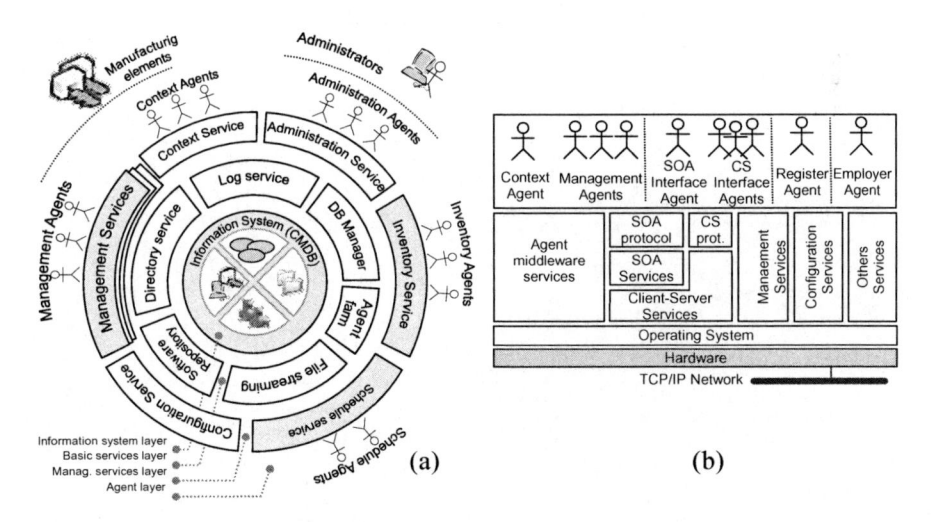

Fig. 1. Service and management device architecture

The fourth and final layer (*agent layer*) contains the software agents responsible for executing machinery maintenance tasks. The agents provide a normalised service interface and facilitate the incorporation of characteristics such as pro-activity, autonomy, scalability, ubiquity and support for heterogeneity.

The management device architecture (figure 1b) is organised in a layer-based manner and has a widely accepted structure for embedded devices. In the lower layer the device hardware has been defined, based on a computational system including microprocessor, volatile memory (for the execution), non volatile memory (for the stored system) and communication module for its connection to the network.

The middleware layer comprise different network services, commonly used in the management field, under the client-server model (CS) and services oriented architectures (SOA). These modules give basic services so the device can communicate, at application level, with other external components. In order to adapt this communication to the syntax of management service instructions, the management protocol is implemented on these modules (in both SOA and CS versions). One of the other important services placed in this layer is the core, which contains all the procedures offered by the management services. The middleware platform is also located in the same layer in order to provide support to the service software agents. These agents are placed in the last layer (application layer) and in fact, they undertake to provide the service: acting as interface with other services and applications (*SOA/CS Agents*), registering the service in a discovery service (*Register Agents*), or, simply, manage selected services (*Management Agents*).

6 Test Scenario

For the test scenario we used an industrial prototype in scale made with the standard models of Staudinger GmbH. The prototype is composed of: a high level storage warehouse, a flexible process line, a production unit, a turntable and conveyor belts (see figure 2a).

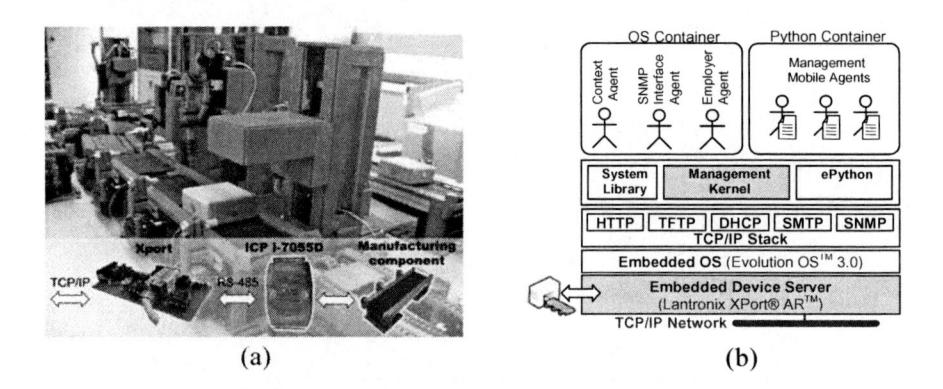

Fig. 2. Test scenario and device prototype

Furthermore, in this section the implementation of a prototype device is presented, taking into account the general architecture described in the previous section and specifying the different structural blocks according to the available technologies. In figure 2b the resulting architecture has been given graphic shape.

The manufacturing devices have for the input-output interface a 8-channel isolated digital input-output module ICP I-7055D.

The computational hardware platform chosen for the prototype development is a *Lantronix Xport® AR*™ device which has a 16 bit *DSTni-EX*™ processor with 120MHz frequency reaching 30MIPS, SRAM (1,25MB), ROM (16KB) and flash memory (4MB). These capacities are sufficient for the memory requirements of the software developed for implementing the protocol.

In the service layer, the implementation process has been conditioned by the characteristics of *XPort* device. Although, with the current hardware miniaturisation level their computational capacities have been increased, the devices continue to present considerable limitations in their resources. In this layer, three service blocks are implemented: the middleware that provides the communication mechanisms of the service, the management service kernel and the middleware platform that provides the execution of software agents.

The communication service middleware is upheld by standard protocols and technologies included in the *Evolution OS*. For this prototype a Client-Server SNMP architecture has been selected. For its implementation, a C module has been developed to provide a protocol syntactic analyser.

The management service kernel has been implemented as a functions library written in C language and offered as API for the others device modules.

In order to implement service agents, a division has been made in the implementation process between static and mobile agents. In the first case, an ad hoc implementation for the XPort device has been developed in C language, using an operative system such as the agents' container. In the second case, in order to establish an execution framework for the mobile agents (the management mobile agents), a Python embedded engine (ePython version 2.5) has been adapted to the XPort features. These mobile agents are implemented as Python text scripts.

7 Conclusions

In this paper we have discussed the advantages of integrating production elements in the global organisation model in order to achieve viable new agile manufacturing models.

The research centred on proposing a model which would enable sophisticated IT services to be fully integrated at production levels with other technological elements, and in alignment with the existing business model. On the basis of this model a management system has been specified which facilitates the self regulating management of these elements and the associated processes, incorporating a high degree of flexibility and robustness at the levels involved.

The next step consists of incorporating semantics in the definition of services in a manner which will secure high levels of self configuration and self management for production elements.

Acknowledgments. This work was supported by the Spanish Ministry of Education and Science with Grant TIN2006-04081.

References

1. Ribeiro, L., Barata, J., Mendes, P.: MAS and SOA: Complementary Automation Paradigms. In: Innovation in Manufacturing Networks, vol. 266, pp. 259–268. Springer, Boston (2008)
2. Avella, L., Vázquez, D.: Is Agile Manufacturing a New Production Paradigm? Universia Business Review (6), 94–107 (2005)
3. Brennan, R.W., Fletcher y, M., Norrie, D.H.: An Agent-Based Approach to Reconfiguration of Real-Time Distributed Control Systems. IEEE Transactions on Robotics and Automation 18(4) (2002)
4. Giret, A., Julián, V., Botti, V.J.: Agentes Software y Sistemas Multi-Agente: Conceptos, arquitecturas y aplicaciones. In: Aplicaciones Industriales de los sistemas Multiagentes. Pearson Prentice Hall, London (2005) ISBN : 84-205-4367-5
5. McFarlane, D.C., Bussmann, S.: Developments in Holonic Production Planning and Control. Int. Journal of Production Planning and Control 11(6), 522–536 (2000)
6. Slack, N.: The flexibility of manufacturing systems. International Journal of Operations & Production Management 25(12), 1190–1200 (2005)
7. Moreno, R.P.: Ingeniería de la automatización industrial. Ra-Ma, Madrid, Spain (2004)
8. Transparent Factory. Manual de usuario y planificación (2001), http://www.modicon.com
9. Gilart-Iglesias, V., Maciá-Pérez, F., Gil-Martínez-Abarca, J.A., Capella-D'alton, A.: Industrial Machines as a Service: A model based on embedded devices and Web Services. In: 4th International IEEE Conference on Industrial Informatics (INDIN 2006), pp. 630–635 (2006)
10. Jammes, F., Smit, H.: Service-Oriented paradigms in industrial automation. IEEE Transaction on industrial informatics 1(1), 62–70 (2005)
11. Chang, H.: A Model of Computerization of Manufacturing Systems: an International Study. Information and Management 39(7), 605–624 (2002)
12. Kim, M., Choi, M., Hong, J.W.: A load cluster management system using SNMP and web. International Journal of Network Management 12(6), 367–378 (2002)
13. Guo, J., Liao, Y., Parviz, B.: An Agent-based Network Management System. In: Hamza, H.M. (ed.) Internet and Multimedia Systems and Applications (IMSA 2005), Honolulu, pp. 20–88 (2005)
14. Marik, V., Mcfarlane, D.C.: Industrial adoption of agent-based technologies. Intelligent Systems 20(1), 27–35 (2005)
15. Lee, S.-M., et al.: A component-based distributed control system for assembly automation. In: Proceedings of 2nd International Conference on Industrial Informatics, INDIN 2004 (2004)
16. McFarlane, D.C., Bussmann, S.: Developments in Holonic Production Planning and Control. Intenational Journal of Production Planning and Control 11(6), 522–536 (2000)
17. Carnero, M.C.: An evaluation system of the setting up of predictive maintenance programmes. Reliability Engineering and System Safety 91, 945–963 (2005)
18. Cianfrani, C.A., West, J.E.: ISO 9001:2000 aplicada a la fabricación. In: AENOR (2004)

19. Ivinskis, K.: High availability of commercial applications. In: Carey, M., Schneider, D. (eds.) Proceedings of the 1995 ACM SIGMOD international conference on Management of data, pp. 433–434. ACM Press, New York (1995)
20. Cuff, J.A., Coates, G.M.P., Cutts, T.J.R., Rae, M.: The Ensembl Computing Architecture. In: Genome Research, vol. 14, pp. 971–975. Cold Spring Harbor Laboratory Press (2004)
21. Marcos Jorquera, D., Maciá Pérez, F., Gilart Iglesias, V., Gil Martínez-Abarca, J.A.: High Availability for Manufacturing Components. In: IEEE International Conference on Industrial Informatics, pp. 474–479 (2006)

MASITS – A Tool for Multi-Agent Based Intelligent Tutoring System Development

Egons Lavendelis and Janis Grundspenkis

Department of Systems Theory and Design
Riga Technical University
Kalku 1, LV-1658, Riga, Latvia
egons.lavendelis@cs.rtu.lv, janis.grundspenkis@cs.rtu.lv

Abstract. Intelligent Tutoring Systems (ITS) have some specific characteristics that must be taken into consideration during the development. However, there are no specific tools for agent based ITS development. This paper proposes such tool named MASITS for multi-agent based ITS development. The tool supports the whole life-cycle of ITS development. It provides an environment to create models needed in all phases of the development. During the analysis phase a goal diagram and a use case diagram is used. The design is divided into two stages, namely external and internal design. The tool provides code generation from the diagrams created during the design. Source code of JADE agents, behaviours and ontology are generated.

Keywords: Agent Oriented Software Engineering, Agent Development Tool, Intelligent Tutoring System.

1 Introduction

Extensive research in the agent oriented software engineering field is ongoing. Several agent oriented software engineering methods and methodologies have been proposed, for example, Gaia [1], Prometheus [2], PASSI [3], MaSE [4], and Tropos [5]. However, only few of them provide a CASE tool to be used in agent oriented software development. A few examples of existing tools are the following: agentTool [4], Prometheus Design Tool [6] and Goal Net Designer [7]. Agent implementation environments like JADE [8], JACK [9], MADE [7] exist at their own. One of the main tasks that an agent development tool has to accomplish is to generate implementation code from the design. Prometheus Design Tool generates JACK code and Goal Net Designer generates MADE agents. However, many other tools fail to provide sufficient code generation.

Intelligent agents and multi-agent systems are widely used in Intelligent Tutoring System (ITS) development, for an overview see [10]. Additionally a few multi-agent architectures for ITS development have been built [11, 12, 13]. At the same time, specific tools for multi-agent based ITS development do not exist. However, ITSs have some specific characteristics that must be taken into consideration during the development. Firstly, ITSs are hardly integrated into organisation and they have very few actors. So, organisational and actor modelling can hardly be used and requirements come only from the system's goals and functionality. Thus, methodologies and tools that use organisational modelling as one of the main techniques are not applicable to ITS. Secondly, ITSs consist of known set of agents and have a well established architecture [10, 13, 14], that should be taken into consideration during the design. However, agent (or agent type or role) definition is major activity in the above mentioned tools. Prometheus Design Tool provides agent types' definition by grouping functionality. AgentTool provides role definition and derives agents and multi-agent architecture from previously defined roles. Thus, a specific ITS development tool can have advantages over general purpose tools by supporting appropriate activities.

Additionally, CASE tool usage during the agent based system development process has the following advantages: (1) the tool enhances diagram drawing by providing the appropriate elements; (2) relationships among different diagrams can be created, which are used for consistency checking and crosscheck; (3) some diagrams can be partly generated from previously created diagrams; (4) source code for agents and other system's elements can be generated from the design. Thus, there are significant advantages in design of ITSs using a specific tool. This paper contains description of a MASITS (Multi-Agent System based Intelligent Tutoring System) tool for ITS development.

All diagrams used in MASITS tool are described in details in Section 3. Section 4 includes an overview of interdiagram links used for consistency checks and crosscheck. Finally, Section 5 gives conclusions and outlines the future work.

2 The MASITS Tool

The MASITS tool has been built for multi-agent based intelligent tutoring system development. The tool supports full ITS development life cycle from requirements analysis to implementation. The purpose of the MASITS tool prescribes that the main characteristics of ITSs are taken into consideration. Appropriate requirements analysis techniques have been chosen. ITSs are hardly integrated into any organization and there are very few actors involved. Thus, requirements analysis has to be done using techniques that focus on system's goals and functionality, but not organizational or role analysis. Similarly, during the design phase the results of agent based ITS research are included. The MASITS tool is intended to design multi-agent systems where agents communicate by sending simple messages without any complicated protocols. Besides, the set of agents that build up an ITS is known [10]. The set of agents can be adjusted to meet specific needs, but there is no need for any agent definition activities during the design. Additionally, ITSs can be built using a holonic agent architecture [13]. Thus, a support for holons and their hierarchy is included in the MASITS tool.

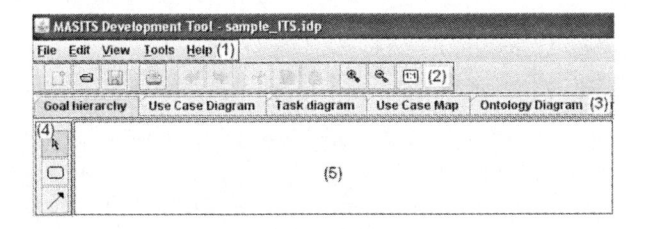

Fig. 1. Main parts of the interface

Interface of the MASITS tool consists of the following main parts, as it is shown in Figure 1. Main menu (1) contains all functions of the tool. The most frequently used features are included in the main toolbar, too (2). All diagrams created during the development of the system are included in tabs (3). Each tab of the diagram contains a drawing toolbar (4) and a page for diagram drawing (5).

The first phase that MASITS tool supports is requirements analysis. It is done using goal modelling and use case modelling. The goal modelling is done first, because goals are used in use case creation. The second phase is design that is divided into two stages, namely, external and internal design of agents. During the external design agent functionality and interactions among agents are specified. During the internal design the internal structure of agents is specified, i.e., it is defined how agents achieve functionality specified during the external design.

Of course, design tools which can be used to specify systems that are implemented on different platforms have wider usage. However, transformation from design concepts to agent implementation platform concepts is almost unique for each combination of design concept set and agent platform [14]. Thus, we have chosen to use a set of JADE agent development framework's concepts [8] already during the design phase. JADE was chosen, because it provides simple way to create Java agents and organise their interactions. Moreover, Java classes of JADE agents can be easily generated from the design elements. Agent communication in JADE is organised using predicates from the domain ontology. Agents are Java classes and their actions are defined as behaviours. So, main implementation elements are ontology, agent and behaviour classes. Additionally to these elements, a batch file to start the system is needed. The batch file includes deployment details of the system, which are specified in the deployment diagram. The batch file is generated automatically from the deployment diagram. So, the MASITS tool supports implementation and deployment phases, too.

3 Diagrams Used in MASITS Tool

The ITS development using MASITS tool consists of creation of a set of diagrams. The following diagrams are included: a goal diagram, a system level use case diagram, a task-agent diagram, a use case map, an interaction diagram, a ontology diagram, an agent's internal view and a holon hierarchy diagram. All diagrams included in the MASITS tool and dependencies among them are shown in Figure 2. Sequential creation of diagrams is denoted with an arrow, crosscheck is denoted with a dashed line.

3.1 Goal Diagram

The goal diagram depicts goals and hierarchical relationships among them. Two types of relationships are distinguished: (1) AND decomposition (simple line), meaning that all subgoals have to be achieved to achieve the higher level goal. (2) OR decomposition (black circle and connecting lines), meaning that at least one of the subgoals has to be achieved to achieve the higher level goal. An example of a goal diagram is shown in Figure 3.

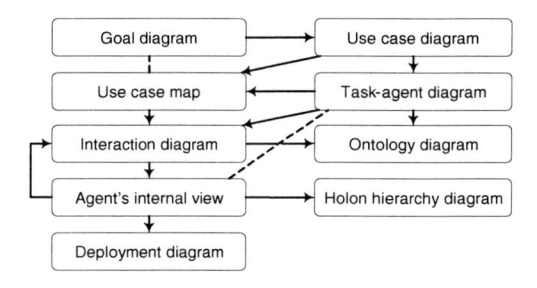

Fig. 2. Diagrams included in MASITS tool and dependencies among them

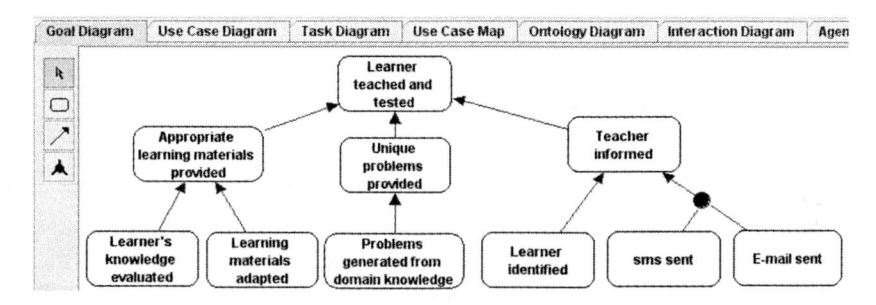

Fig. 3. Goal diagram

3.2 Use Case Diagram

The second diagram included in MASITS tool is well known use case diagram. It includes system level use cases and actors interacting with the system. At first, all actors are identified and included in the diagram. Then use cases and their descriptions are created corresponding to actions that have to be done to accomplish the lower level goals of the goal diagram. Interdiagram links to goals that are supported by use cases are created. A well known UML use case notation is used [16]. An example of a use case diagram is shown in Figure 4.

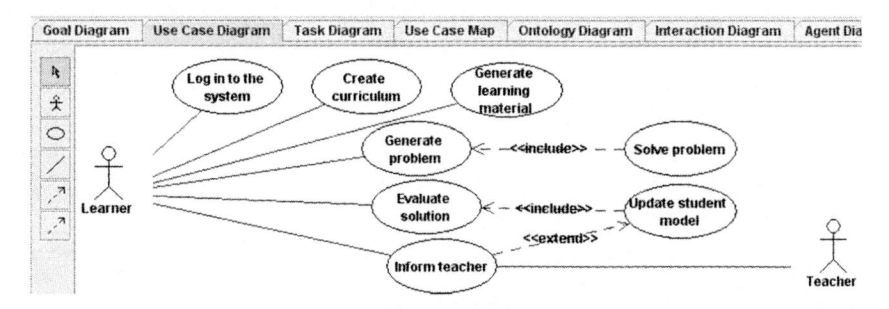

Fig. 4. Use case diagram

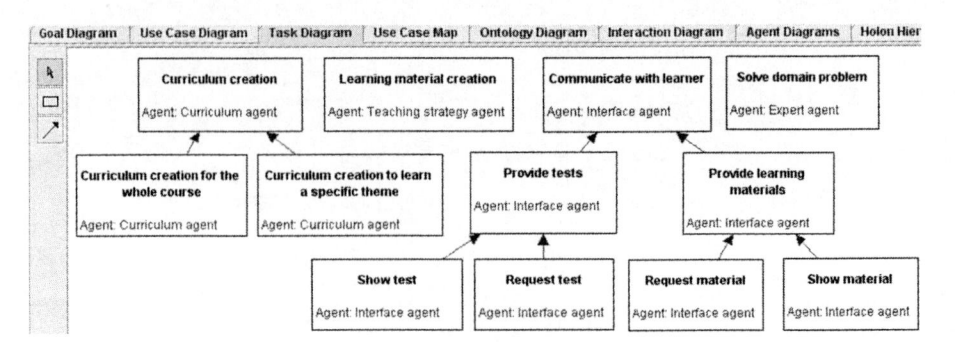

Fig. 5. Task-agent diagram

3.3 Task-Agent Diagram

The task-agent diagram is a hierarchy of tasks that the system has to accomplish. The diagram contains information about task allocation to agents, too. Name of agent that is responsible for each task is added to the task's node.

The first step of task-agent diagram creation is task decomposition resulting in a task hierarchy. The task-agent diagram consists of one or more task hierarchies. The task hierarchy is created by defining tasks corresponding to use case scenario steps, i.e., a task is created for each step of the use case scenario. Task hierarchies are created by linking up tasks that can be assigned to the same agents not by corresponding use cases or goals. So, the structure of task hierarchy is different from the goal hierarchy. After finishing task decomposition, tasks are allocated to agents using basic principles of ITS architecture [10, 13, 17]. An example of a task-agent diagram is shown in Figure 5.

3.4 Use Case Map

In object-oriented approach use case maps are used to model the control passing path during the execution of use case (for details, see [18]). In multi-agent design use case maps include agents, their tasks and message paths among them. A use case map represents the use case scenario explicitly showing interactions among agents. Each link between two tasks can be considered as a message between corresponding agents. Thus, after creating the use case map interactions among agents can be easily specified just adding message content.

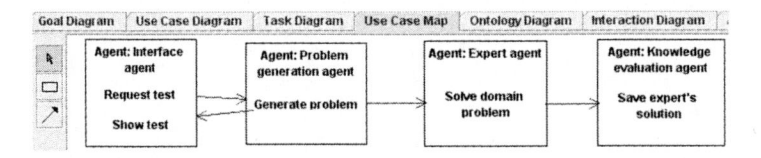

Fig. 6. Use case map for use case "Generate problem"

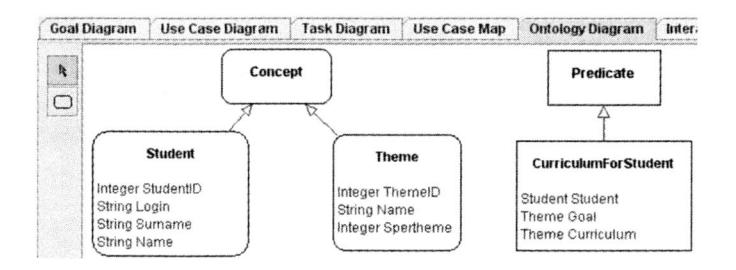

Fig. 7. Ontology diagram

Use case map creation is the first step of agent interaction design. Interaction is designed for each pair of interacting agents. Use case maps are created for pairs of agents whose interaction is too complicated to be specified directly in interaction diagram. An example of a use case map is shown in Figure 6.

3.5 Ontology Diagram

The ontology diagram is a model of the problem domain. Concepts of the problem domain and predicates used in agent interaction are described. The ontology diagram is a modified version of the class diagram. In fact, the ontology diagram is a class hierarchy. Two superclasses are used: (1) Concept – subclasses of this class are domain concepts; (2) Predicate – subclasses of this class are predicates used in agent interactions as message contents.

All other classes are subclasses of the above-mentioned two superclasses. Each class has attributes which have name, cardinality and type. Attribute's type may be a primitive type, Java class or a concept defined in the ontology diagram. Predicates are not allowed as attribute types. Initial ontology diagram usually is created during interaction design by adding predicates used in agent interaction specification. Concepts needed to define these predicates are added during interaction design, too. During agents' internal design the ontology diagram is refined by adding concepts and predicates used in agents' internal view. An example of the ontology diagram is shown in Figure 7.

3.6 Interaction Diagram

The interaction diagram specifies interactions among agents. The MASITS tool allows specifying interactions by messages sent among agents. Protocols are not included, because the tool is designed for agent based ITS development and our research has shown that complicated protocols are not widely used in this domain.

The diagram consists of two main elements, namely, agents and messages. Agents are denoted as rounded rectangles. Messages are denoted as labelled links with full arrows between agent vertexes. Labels of message links are names of predicates. The diagram includes interaction between an interface agent and a user, too. Thus, a few additional elements are added: a user (denoted as an actor), events monitored by an agent (dashed line) and methods of interface called by interface agent (line with simple arrow).

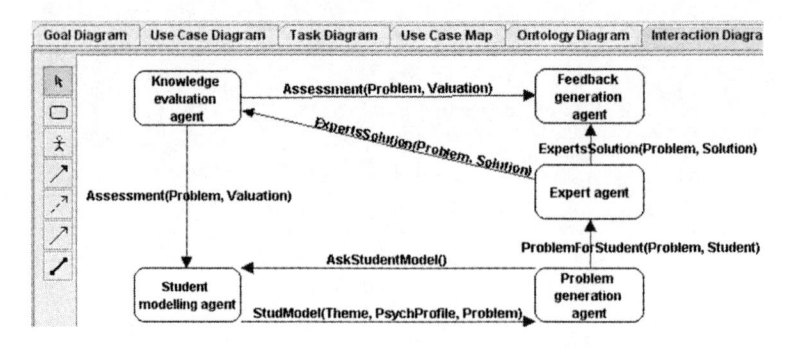

Fig. 8. Fragment of the interaction diagram

The main interaction diagram is created to specify interactions among higher level agents. Additionally, the interaction diagram is created for each holon. Interaction diagrams for holons may include sets of typical agents [17] and directory facilitator agents. These elements are introduced to allow specifying open holons. An example of an interaction diagram is shown in Figure 8.

3.7 Agent's Internal View

The agent's internal view specifies design of agent's internal structure. An internal view is created for each agent. It consists of the following elements:

- An agent diagram, including agent's perceptions, messages sent by an agent, agent's actions and links among these elements.
- Perception (message and event) processing rules.
- Startup rules describing actions that are performed by an agent during the startup.
- A list of agent's beliefs. Agents' beliefs are specified as pairs of belief's type and name: <Type>, <Name>. Type can be either primitive type used in Java language, Java class or ontology class (predicate or concept) from ontology diagram.
- A list of agent's actions specifying implementation details. Actions are described in table containing the following columns: name of the action, name of corresponding behaviour class, type of the action (one-shot, cyclic, timed, etc.), inner (implemented as inner class in agent's class) or outer (implemented as a class in the same package as agent) behaviour class.

The agent diagram includes the following elements: (1) all messages sent and received by agent. The agent diagram of holon's head includes messages designed in higher level interaction diagram and also messages sent to and received from body agents of the holon; (2) agent's actions and interactions among them. So, agent's plans are modelled; (3) perceived events from the environment; (4) agent's actions to user interface.

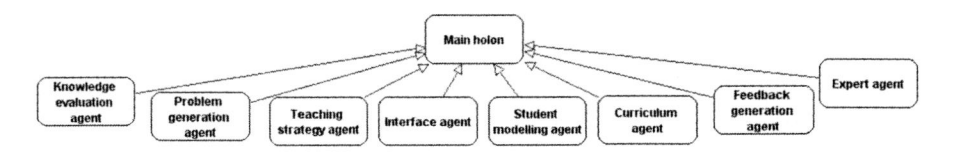

Fig. 9. The holon hierarchy diagram

The agent diagram specifies an agent as its actions and interactions among them. Actions mainly are initiated reacting on received messages or perceptions. Additionally actions initiated by time can be used (denoted by a ticker). Each social agent has at least two actions – message sending and receiving. These actions are not depicted in the agent diagram. Instead, received and sent messages have message contacts. Agent can have four types of contacts: message receiving, message sending, event receiving and action to external environment contacts. Each message, event or action is added to its own contact.

Perception processing and startup rules are designed as IF-THEN rules: IF <Condition> THEN <Action> ELSE <Action>. The following templates can be used as condition: Received (check if particular predicate is received), Compare and True. Additionally, Boolean operators (AND, OR, NOT) can be applied to conditions. Templates Action, Belief Set (sets value of one of the agent's beliefs), Action conjunction (sequence of actions) and IF-THEN rule can be used as action.

It is advisable to perform internal design of each agent iteratively. Each iteration can include design elements corresponding to either a set of perceptions or a set of actions. For details and an example of agent's internal view, see [17].

3.8 Holon Hierarchy Diagram

The holon hierarchy diagram shows holons that exist in the system and a hierarchy among them. This diagram is created automatically. Holon nodes are added to the hierarchy when the user creates a holon to implement an agent. The only purpose of the diagram is to summarize the structure of the multi-agent system. An example of the holon hierarchy diagram is shown in Figure 9.

3.9 Deployment Diagram

The deployment diagram specifies how the designed agents are used in the system. The diagram consists of containers that are used to run the system and agent instances that are launched during the system startup. Each container is a JADE container that is started during the startup of the system. One of the containers is the main JADE container, included in all deployment diagrams. Interactions among containers are specified, too. Agents are defined in the JADE containers. Each agent has a name and agent class defined in the agent diagram. In fact, agents designed in the interaction diagrams and agent internal views are agent classes that can have multiple instances. Concrete instances are specified in the deployment diagram.

4 Interdiagram Links

One of the main advantages of MASITS tool is the introduced concept of interdiagram links that are links among elements of two different diagrams. Firstly, interdiagram links are used to ensure the diagram consistency. Majority of them are created if an element is used as a part of another element. For example, an interdiagram link is created from predicate vertex to communication links, where the predicate is used as message content.

MASITS tool allows editing each element only in one place. All other occurrences of the element are changed automatically to ensure consistency. For example, agent's name can be changed only in the interaction diagram. If it is changed, the tool changes the name of the agent in the agent diagram, the holon hierarchy diagram, the task diagram and other interaction diagrams. Elements can be deleted in the same way as changed. However, if an element is deleted, significant parts of other models can be affected. Two choices are available. The tool just deletes dependant elements if the deleted element has the same semantics as the dependant elements. Deletion of the element is restricted if the dependant part is a significant part of other diagrams with different semantics. Due to the space limitations, further details about consistency checking links are omitted.

Secondly, interdiagram links are used to specify semantic dependencies between elements of different diagrams, too. Such links are created by the user and are used during the crosscheck among diagrams. The MASITS tool contains interface to define such interdiagram links and to see defined links and unlinked elements. The unlinked elements can be used to check, if all elements of one diagram are supported by elements in another diagram. Such approach is used in the MASITS tool four times. Firstly, links between use cases and goals are created. Unlinked goals show which goals need use cases to be created. Secondly, links between goals and tasks are created to show which tasks support which goals. This kind of links is used during the crosscheck between goal and task diagrams. Any unsupported goal indicates that defined tasks are not sufficient to achieve system's goals. Thirdly, paths from use case maps are linked to messages in the main interaction diagram to ensure that every link from paths in use case maps have corresponding messages in the main interaction diagrams. Finally, each agent must have actions to realize all tasks assigned to it in the task diagram. Thus, interdiagram links are created among tasks and agents' actions.

5 Conclusions and Future Work

A specific tool for multi-agent based ITS development is proposed. The main advantages of the proposed tool are the following. The tool supports the whole life cycle of ITS development, thus there is no need for any additional tool. The very popular agent development environment (JADE) is used for implementation. The tool provides consistency checking and important crosschecks during the development of system, helping to find possible errors. Finally, code generation is done automatically from the diagrams created during the design phase.

Regardless of the MASITS tool is developed for specific purpose, it can be used to develop other multi-agent systems, especially those that have similar characteristics with multi-agent based ITSs.

Our future work is to develop a full case study of ITS development using the MASITS tool. After finishing the case study it will be possible to evaluate the tool in more details. However, the abovementioned advantages make the tool to be a promising one. The main direction of our research is to formulate a full lifecycle methodology for agent based ITS development supported by the MASITS tool.

References

1. Wooldridge, M., Jennings, N.R., Kinny, D.: The Gaia methodology for agent-oriented analysis and design. Journal of Autonomous Agents and Multi-Agent Systems (2000)
2. Padgham, L., Winikoff, M.: Prometheus: A Methodology for Developing Intelligent Agents. In: Proceedings of the Third International Workshop on AgentOriented Software Engineering, AAMAS 2002 (2002)
3. Burrafato, P., Cossentino, M.: Designing a multi-agent solution for a bookstore with the PASSI methodology. In: Fourth International Bi-Conference Workshop on Agent-Oriented Information Systems at CAiSE 2002 (2002)
4. DeLoach, S.: Analysis and Design Using MaSE and agentTool. In: Proceedings of the 12th Midwest Artificial Intelligence and Cognitive Science Conference, pp. 1–7 (2001)
5. Giunchiglia, F., Mylopoulos, J., Perini, A.: The Tropos Software Development Methodology: Processes, Models and Diagrams. In: Proceedings of the First International Joint Conference on Autonomous Agents and Multiagent Systems, pp. 35–36 (2002)
6. Padgham, L., Thangarajah, J., Winikoff, M.: Tool Support for Agent Development Using the Prometheus Methodology. In: Fifth International Conference on Quality Software (QSIC 2005), pp. 383–388 (2005)
7. Yu, H., Shen, Z., Miao, C.: Intelligent Software Agent Design Tool Using Goal Net Methodology. In: Proceedings of the 2007 IEEE/WIC/ACM International Conference on Intelligent Agent Technology, pp. 43–46 (2007)
8. JADE Home Page (last visited: 10.06.07), http://jade.tilab.com/
9. Howden, N., Rönnquist, R., Hodgson, A., Lucas, A.: JACK Intelligent Agents - Summary of an Agent Infrastructure. In: Proceedings of the Fifth International Conference on Autonomous Agents (2001)
10. Grundspenkis, J., Anohina, A.: Agents in Intelligent Tutoring Systems: State of the Art. In: Scientific Proceedings of Riga Technical University Computer Science, Riga. Applied Computer Systems, 5th series, vol. 22, pp. 110–121 (2005)
11. Capuano, N., et al.: A Multi-Agent Architecture for Intelligent Tutoring. In: Proceedings of the International Conference on Advances in Infrastructure for Electronic Business, Science, and Education on the Internet, SSGRR 2000 (2000)
12. Webber, C., Pesty, S.: A two-level multi-agent architecture for a distance learning environment. In: de Barros Costa, E. (ed.) ITS 2002 Workshop on Architectures and Methodologies for Building Agent-based Learning Environments, pp. 26–38 (2002)
13. Lavendelis, E., Grundspenkis, J.: Open Holonic Multi-Agent Architecture for Intelligent Tutoring System Development. In: Proceedings of IADIS International Conference Intelligent Systems and Agents 2008, pp. 100–108 (2008)

14. Smith, A.: Intelligent Tutoring Systems: personal notes (1998) (last visited 18.04.2005), http://www.cs.mdx.ac.uk/staffpages/serengul/table.of.contents.htm
15. Massonet, P., Deville, Y., Neve, C.: From AOSE methodology to agent implementation. In: Proceedings of the First International Joint Conference on Autonomous Agents and Multiagent Systems, Bologna, Italy, pp. 27–34 (2002)
16. OMG UML Superstructure 2.1.2 (last visited: 03.10.2008), http://www.omg.org/docs/formal/07-11-02.pdf
17. Lavendelis, E., Grundspenkis, J.: Design of Multi-Agent Based Intelligent Tutoring Systems. In: Scientific Proceedings of Riga Technical University Computer Science. Applied Computer Systems. RTU Publishing, Riga (2009) (accepted for publishing)
18. Buhr, R.J.A., Elammari, M., Gray, T., Mankovski, S.: Applying Use Case Maps to Multi-Agent Systems: A Feature Interaction Example. In: Proceedings of the Thirty-First Hawaii International Conference on System Sciences, pp. 171–179 (1998)

Multi-agent Reasoning Based on Distributed CSP Using Sessions: DBS

Pierre Monier, Sylvain Piechowiak, and René Mandiau

LAMIH UMR CNRS 8530
Université de Valenciennes et du Hainaut Cambrésis
59313 Valenciennes Cedex 9, France
{pierre.monier,sylvain.piecowiak,
rene.mandiau}@univ-valenciennes.fr

Abstract. Early researches on Constraints Satisfaction Problems (CSP) in Artificial Intelligence began in the 1970s. The CSP formalism addresses many problems in a simple and efficient way. However, it is not possible to solve some of these problems in a classical and centralized way, for various reasons such as prohibitive computation time or unsafe security data. To solve these naturally distributed problems, the Distributed CSP (DisCSP) have been proposed. In this paper, we present an algorithm called *DBS* for DisCSP solving and we discuss about the performances obtained with a random DisCSP generator.

1 Preliminaries

In a multi-agent context, important issues concern both the study of interactions between agents and the reasonning performed by these agents. In this paper, we assume that the reasoning is defined by a set of relations between variables.

In such a context, the first researches on Constraints Satisfaction Problems (CSP) in Artificial intelligence began in the 1970s. A CSP makes it possible to describe, in a simple formalism, a large variety of problems such as truth maintenance [4].

Distributed CSPs (DisCSP) were proposed in order to solve naturally distributed problems. They are studied in the domain of Distributed Artificial intelligence (DAI). The problem to be solved is thus distributed on all agents. These agents interact with each other in order to find a global solution based on each agent's local solutions (figure 1). DisCSP are usually applied to solve distributed problems such as timetabling [10], road traffic management[5] and manufacturing control [3]. More generally:

Definition 1. *A Distributed CSP is a quadruplet (X,D,C,A) where:*

- X *is a finite set of p variables:* $\{x_1, x_2, ..., x_p\}$.
- D *is a set of domains associated with these variables:*
  $D = \{Dom(x_1), Dom(x_2), ..., Dom(x_p)\}$.
- C *is a finite set of m constraints:* $\{c_1, c_2, ..., c_m\}$.
- A *is finite set of N agents:* $\{A_1, A_2, ..., A_n\}$ *where each agent is given a subset of X.*

The set of constraints C can be decomposed into two subsets called inter-agent constraints (C_{inter}) and intra-agent constraints (C_{intra}). Intra-agent constraints represent the

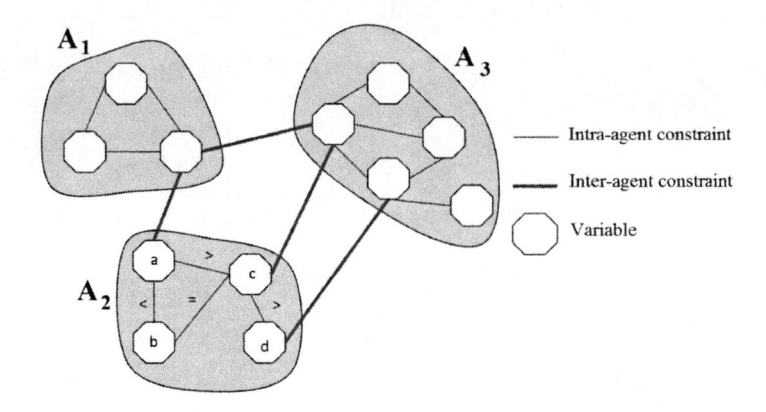

Fig. 1. Agent's local decision based on sub-CSP : illustration

relations between variables of a same agent. Inter-agent constraints correspond to the relations between variables assigned to different agents.

Variables appearing in the inter-agent constraints are called *interface variables*. In this paper, we assume that there exists only one *interface variable* per agent, in order to better understand the behavior of our algorithm.

After having defined DisCSP formalism, we will solve them. Many algorithms for DisCSP resolution exist such as *ABT* [11], *DIBT* [7] and *DDB* [2].

The remainder of this paper is organized as follows. The next section presents an improvement of *DBS* (Distributed Backtracking with Sessions) algorithm [6]. Section 3 describes some experiments and comparisons of *DBS* with *ABT*, a classical algorithm.

2 DBS Algorithm

2.1 Assumptions and Notations

A sub-CSP is assigned to each agent. The CSP of different agents are linked by binary inter-agent constraints. As previously mentionned, each agent possesses a single *interface variable*, called x_{Self}.

A total order, noted $\succ$, is established between the different agents (a priority is assigned to each agent), to avoid infinite loop problems. For example, a change for A_1 implies a change for A_2 which involves a change for A_1. Given two agents A_1 and A_2, $A_1 \succeq A_2$ means that A_1 has a higher or equal priority compared to A_2. For each agent, two sets are defined: Acc^+ and Acc^- respectively representing the higher and lower *accointances*. Given a generic agent, called $self$, if Acc_{Self} represents the set of $self$'s accointances, we have:

- $Acc^+_{Self} = \{A_i \in Acc_{self} \mid A_i \succ Self\}$
- $Acc^-_{Self} = Acc_{Self} - Acc^+_{Self}$
- $DirectAcc^-_{Self} = \{A_i \in Acc^-_{Self} \mid \exists C_{A_i,Self} \in C_{inter}\}$

Agents are ordered according to an adaptation of the *First − Fail*[1] heuristic [8]. The smaller an agent's interface variable size is, the higher priority this agent receives. In case of equality between two agents, the lexicographic order is used.

Parallel to global research, each agent carries out its local search, and stores the solutions in a list called *my_solutions*. This list contains the solutions which propose an alternative for the *interface variable*. When the local search is finished, a boolean called *endLocalSearch* (initially put on *false*) is set to *true*.

Each agent assigns (in a concurrent way) a value for its interface variable then sends it to its $DirectAcc^-_{self}$ with a submission message. Each agent possesses a set called $agent\_view_{self}$ containing the different values suggested by higher priority agents. If an agent cannot find a possible instanciation which respects the received assignments (contained in $agent\_view_{self}$), it informs an agent in Acc^+_{self} with a backtrack message.

Each agent uses a list of triplet (x,v,s) called $TotalBTList_{self}$ (x is an agent, v is a value and s is a number of session (integer)). When *self* must transmit a backtrack message, following the reception of a backtrack message, this list is used to determine the appropriate receiver. Given A and B two sets of triplet (x,v,s), we note $A \uplus B = A \cup \{(x,v,s) \in B \mid (x,\_,\_) \notin A\}$.

DBS associates to each solution from the list *my_solutions* a tag called *propose* which makes it possible to know if this solution has already been proposed in the current *self*'s session (Cf. definition 2). Agent *self* tries to find a solution which has not been already proposed in this current session and which satisfies the constraints C. If there is no solution which respects simultaneously these two conditions, agent *self* verifies whether its local search is over. Indeed, if it is not the case, new possible solutions satisfying these two conditions can be added to the *my_solutions* set. The *global* search for *self* is then interrupted for T seconds, and after this delay, *self* checks again if new solutions have been added to the list *my_solutions*.

2.2 Asynchronism Management

In order for an agent to determine whether a received message is obsolete or not, the concept of session is used: each message is associated with a session number.

Definition 2. *Given an agent self and Acc^-_{self} the set of accointances whose priority is lower than self's priority. A work session between self and Acc^-_{self} is an integer indicating for each element of Acc^-_{self} the state of global search from self's point of view.*

A backtrack message m_b is valid if the session attached to m_b is equals to the current session of the agent which receives this message m_b.

When *self* sends a submission message, its current session is attached to this message. The design of a backtrack message proceeds according to two possibilities. Agent *self* transmits a backtrack message, following the reception of a submission message (section 2.2.1) or a backtrack message (section 2.2.2).

---
[1] This heuristic was chosen because of its effectiveness and its relatively simple implementation.

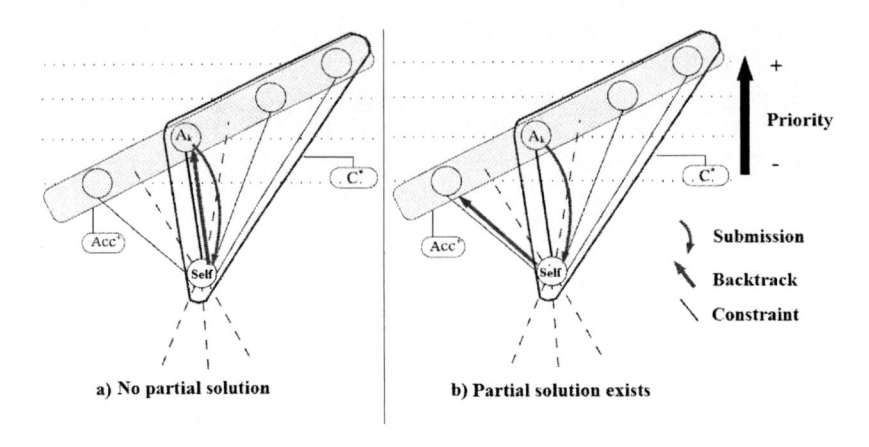

Fig. 2. Backtrack in DBS

2.2.1 Reception of a Submission Message

As shown in figure 2. *Self* has just received a submission message from agent A_k. Unfortunately, *self* does not find a value for x_{Self} satisfying all the inter-agent constraints which connect *self* to the agents contained in *agent_view*$_{Self}$. In this situation, *Self* must send a backtrack message.

In order to determine the receiver of this message, *self* tries to find a partial solution by taking into account only the received assignments from the agents of higher or equal priority than A_k (in figure 2, this set is called C^*). There are two possibilities.

- There is no partial solution and then *self* sends a backtrack message to A_k (figure 2a).
- A partial solution exists and then the backtrack message is sent to the lowest priority agent contained in *agent_view*$_{Self}$(figure 2b).

In both cases, a backtrack message for an agent A_i is defined by:

(backtrack, (A_i, v_i, s_i), *BTList*) where v_i and s_i are respectively the value and the session of A_i (contained in *agent_view*$_{self}$) and $BTList^2 = \{(x,v,s) \in agent\_view_{self} \uplus TotalBTList_{self} \mid x \succ A_i\}$.

2.2.2 Reception of a Backtrack Message

We now consider the case where agent *self* must transmit a backtrack message following the reception of a backtrack message m_b. The *BTList* contained in the message m_b is noted $BTList_{m_b}$.

The backtrack message is transmitted to the lowest priority agent, noted A_i, contained in *agent_view*$_{self} \uplus BTList_{m_b}$. If A_i belongs to $BTList_{m_b}$, this message will be: (backtrack, (A_i, v_i, s_i), *BTList*) with $BTList = agent\_view_{self} \uplus \{BTList_{m_b} \uplus TotalBTList_{self}\} - \{(A_i, v_i, s_i)\}$. The details of *DBS* algorithm are given in the appendix of this paper.

---

[2] *BTList* allows the agent who receives this backtrack message to continue backtracking (if necessary).

3 Experiments

The proposed *DBS* algorithm is now numerically evaluated under two different situations: (i) Scenario with 3 agents, (ii) Experiments with a random DisCSP generator.

3.1 Scenario with 3 Agents

The DisCSP describes in figure 3 is distributed over three agents A_1, A_2 and A_3. To better illustrate the behaviour of the *DBS* algorithm, the *first – fail* heuristic was not used to order the agents in this example. During the initialization step, each agent chooses a value (called *currentValue*) for its interface variable and sends it with a submission message. In this example, notations are simplified so that no difference are made between variables and agents. A possible order of execution is given below:

1. A_1 chooses value 1 then sends the message m_1 (submission, $(A_1, 1, 0)$) to A_3.
2. A_2 chooses the value 2 and sends the message m_2 (submission, $(A_2, 2, 0)$) to A_3.
3. A_3 chooses value 1. Since $DirectAcc^-_{A_3} = 0$, A_3 does not send a submission message. A_3 receives m_1. The current value of A_3 is set to 2 in order to satisfy the constraint $(A_1 \neq A_3)$. $Agent\_view_{A_3} = (A_1, 1, 0)$. A_3 receives m_2. $agent\_view_{A_3} = \{(A_1, 1, 0), (A_2, 2, 0)\}$. Since A_3 cannot find any solution respecting $agent\_view_{A_3}$, it sends a backtrack message m_3 (backtrack, $(A_2, 2, 0)$, $BTList = \{(A_1, 1, 0)\}$) to A_2. The triplet $(A_2, 2, 0)$ is then removed from $agent\_view_{A_3}$.
4. A_2 receives m_3. Since there is no possible solution, A_2 performs backtracking until the lowest priority agent in $agent\_view_{A_2} \uplus \{BTList_{m_3} \uplus TotalBTList_{A_2}\}$. A_2 sends the message m_4 (backtrack, $(A_1, 1, 0)$, $BTList = \{\}$) to A_1. Since m_4 was sent to an agent not in $agent\_view_{A_2}$, A_2 sends its current value again. It sends the message m_5 (submission, $(A_2, 2, 1)$) to A_3.
5. A_1 receives the message m_4. It modifies its current value and then A_1 sends the message m_6 (submission, $(A_1, 2, 0)$) to A_3.
6. A_3 receives the messages m_5 and m_6. $Agent\_view_{A_3}$ becomes $\{(A_1, 2, 0), (A_2, 2, 0)\}$. Value 1 is coherent with $agent\_view_{A_3}$. There is no more exchanged messages. *DBS* algorithm is in a stable state which means that a solution has been found. The global solution is built from the local solutions of each agent: $\{(A_1 = 2), (A_2 = 2), (A_3 = 1)\}$.

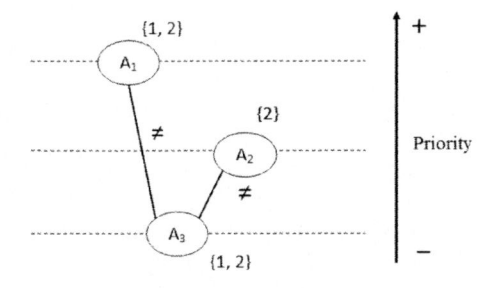

Fig. 3. Simple DisCSP

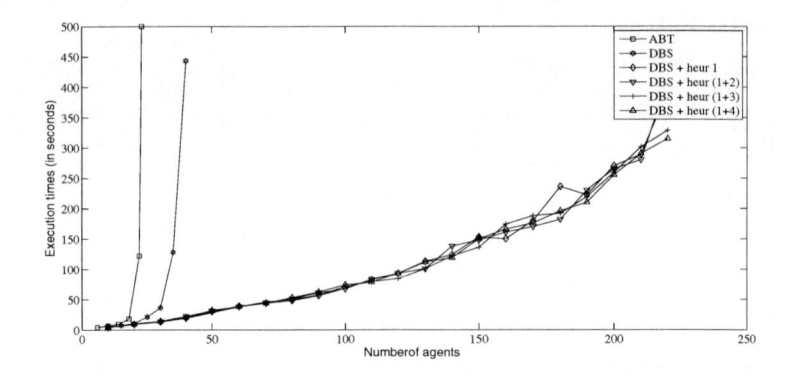

Fig. 4. Comparison between *ABT* and *DBS*: numerical results (execution times)

3.2 Evaluation

To evaluate the proposed *DBS* algorithm, we have implemented a particular DisCSP generator: connections between sub-CSP rely on a single variable (the *interface variable*). This particular feature makes it possible to obtain a DisCSP which can be solved by single-variable DisCSP resolution algorithms per agent, such as *ABT* or *DDB*. Our generator is an adaptation of a centralised CSP generator proposed by C. Bessiere[3].

In order to generate random DisCSP instances, the following parameters are used: number of variables n, size of the domain of each variable d, C_{intra} connectivity (a C_{intra} connectivity equals to X represents the total possible number of C_{intra} multiplied by $X/100$), C_{intra} hardness, number of agents, C_{inter} connectivity and C_{inter} hardness.

Each agent is given n variables all having a same domain of size d. Then, we add intra agents constraints between those variables. Their number is equal to the total number of possible constraints $(n*(n-1)/2)$ multiplied by C_{intra}. Then, for each constraint (binary), we define a list of forbidden couples using D_{intra}. The size of this list is: $d^d * D_{intra}$.

We implemented both the *ABT* and *DBS* algorithms, as well as many versions of *DBS* algorithm where several heuristics are used to reduce the number of messages without eliminating any solution. Those heuristics are listed below:

- *heur. 1* : If the reception box of a given agent *self* contains several submission messages coming from the same agent A_k, then only the last submission message from A_k is preserved.
- *heur. 2* : If the reception box of a given agent *self* contains several backtrack messages and at least one submission message, then all backtrack messages are removed from its reception box.
- *heur. 3* : If a given agent *self* has sent a backtrack message to an agent A_k in $agent\_view_{self}$, then while *self* does not receives a submission message from A_k, *self* removes all backtrack messages from its reception box.
- *heur. 4* : If the reception box of a given agent *self* contains several submission messages, *self* treats in priority those coming from higher priority agents.

---
[3] http://www.lirmm.fr/~bessiere/generator.html

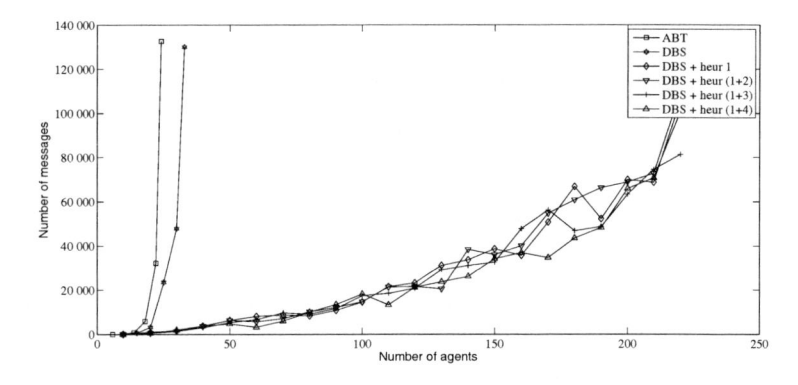

Fig. 5. Comparison between *ABT* and *DBS*: numerical results (number of messages)

We decided to evaluate *DBS* by comparing its performances with the *ABT* algorithm. The latter was chosen because it presents many similarities with *DBS*. Moreover, *ABT* is used as a reference to compare DisCSP algorithms in many publications such as [9].

In general, assigning a local CSP per agent having only one *interface variable* is the same as adding values to the domain of the interface variables during the global search. So the number of variables per agent, the connectivity and the hardness of C_{intra} (that we arbitrarily set to 40%) have little influence. They only modify the time required by an agent to find all possible values for its *interface variable*.

We arbitrarily set both the size of the domain and the number of variables to 6. A value of 6 corresponds to a local CSP containing 6^6 (approximately 47000) leaves in its search tree.

We chose to use the multi-agent platform JADE [1] to implement *ABT* and *DBS*. Figures 4 and 5 give respectively the execution time and the number of exchanged messages required for a DisCSP resolution with a C_{inter} connectivity and a C_{inter} hardness both equal to 40%. The number of agents ranges from 10 to 230 for each algorithm (*ABT*, *DBS* and *DBS* with additionnal heuristics). For a given number of agents, results were obtained by averaging ten different experiments.

For a DisCSP composed of 40 agents, *DBS* finds a solution within 40 seconds whereas *ABT* requires the same computation times to solve a DisCSP composed only of 24 agents. When *DBS* uses heuristic 1, it obtains very good results : a DisCSP composed of 230 agents can be solved within 300 seconds. Only heuristic 1 makes it possible to decrease the execution time and the number of exchanged messages. The other heuristics make it possible to remove obsolete messages but use too much CPU time. So the results obtained by *heur.* $(1+2)$, *heur.* $(1+3)$ and *heur.* $(1+4)$ are similar to the results obtained with *heur.* 1.

We carried out an additionnal series of experiments to study the influence of C_{inter} connectivity and C_{inter} hardness parameters. The higher C_{inter} connectivity is, the more time *DBS* takes to solve a DisCSP. For a C_{inter} connectivity set to 40%, if C_{inter} hardness increases from 0% to about 15%, execution time increases and if C_{inter} hardness increases from about 15% to 100%, execution time are reduces.

4 General Conclusion

In this paper, we have proposed a complete DisCSP algorithm called *DBS* to solve DisCSP. The assumptions, the required notations and the management of asynchronism were given.

Performances of the proposed *DBS* algorithm, in terms of computation times and number of exchanged messages, were given in section 3. *DBS* was implemented with the JADE multi-agent platform. We have also developed a random DisCSP generator to compare several variants of the proposed algorithm with *ABT*. When *DBS* used *heur.* 1, it achieves very good performances. Without using this heuristic, *DBS* requires 400 seconds to solve a 40 agents DisCSP whereas *ABT* already requires more than 500 seconds for a DisCSP of only 24 agents.

We are planning on generalizing the *DBS* algorithm to solve DisCSP where each agent possesses several *interface variables*. Moreover, we will improve the work exposed in [5] to apply the *DBS* algorithm to a road traffic application. In this context, each vehicle is considered as an agent and represents its perception using a CSP and tries to find its own solution while maintaining a consistent collective behavior.

References

1. Bellifemine, F., Giovani, C., Tiziana, T., Rimassa, G.: Jade programmer's guide. Tech. rep. (2000)
2. Bessiere, C., Maestre, A., Meseguer, P.: Distributed dynamic backtracking. In: Silaghi, M.C. (ed.) IJCAI 2001 workshop on Distributed Constraint Reasoning, Seattle, WA, pp. 9–16 (2001)
3. Clair, G., Gleizes, M.P., Kaddoum, E., Picard, G.: Self-Regulation in Self-Organising Multi-Agent Systems for Adaptive and Intelligent Manufacturing Control. In: Second IEEE International Conference on Self-Adaption and Self-Organization (SASO 2008), Venice, Italy. IEEE Computer Society, Los Alamitos (2008)
4. Dechter, R.: A constraint-network approach to truth maintenance. Tech. rep., Cognitive Systems Laboratory, Computer Science Dept., Univ. of California, Los Angeles (1987)
5. Doniec, A., Mandiau, R., Piechowiak, S., Espié, S.: Anticipation based on constraint processing in a multi-agent context. Journal of Autonomous Agents and Multi-Agent Systems (JAAMAS) 17, 339–361 (2008)
6. Doniec, A., Piechowiak, S., Mandiau, R.: A discsp solving algorithm based on sessions. In: Russell, I., Markov, Z. (eds.) FLAIRS 2005: Recent advances in artificial intelligence: Proceedings of the eighteenth International Florida Artificial Intelligence Research Society Conference, pp. 666–670. AAAI Press, Menlo Park (2005)
7. Hamadi, Y.: Traitement des problèmes de satisfaction de contraintes distribués. Ph.D. thesis, Université de Montpellier II, France (1999)
8. Haralick, R.M., Elliott, G.L.: Increasing tree search efficiency for constraint satisfaction problems. Artificial Intelligence 14(3), 263–313 (1980)
9. Meisels, A., Zivan, R.: Asynchronous forward-checking for discsps. Constraints 12(1), 131 (2007)
10. Tsuruta, T., Shintani, T.: Scheduling meetings using distributed valued constraint satisfaction algorithm. In: 14th European Conference on Artificial Intelligence (ECAI), pp. 383–387 (2000)
11. Yokoo, M.: Distributed Constraint Satisfaction: Foundation of Cooperation in Multi-agent Systems. Springer, Heidelberg (2000)

Appendix

procedure **dbs**
1. run processus localSearch()
2. **if** searchSolution(C_{inter}) **then**
3.   sendSolution()
4.   analyzeMessage()
5. **else**
6.   **for all** $A \in Acc_{self}$ **do**
7.   send (end) to A
8.   **end for**
9. **end if**

procedure **searchSolution**
Require: C : set of constraints
Ensure: *boolean*
1. $sol \leftarrow \{s \in my\_solutions \mid state[s] = \neg propose\}$
2. $sol^* \leftarrow \{s \in sol \mid consistant(s,C)\}$
3. **if** $sol^* \neq \emptyset$ **then**
4.   choose $\widetilde{sol}$ in sol^*
5. $state[s] \leftarrow propose$
6. $currentValue \leftarrow value(x_{self}, \widetilde{sol})$
7.   return (*true*)
8. **else**
9.   **if** endLocalSearch **then**
10.   return (*false*)
11.   **else**
12.   wait t seconds
13.   return searchSolution(C)
14.   **end if**
15. **end if**

procedure **sendSolution**
Require:
Ensure:
1. **for all** $A \in DirectAcc_{self}^-$ **do**
2.   send (submission, $(self, currentValue, currentSession)$) to A
3. **end for**

procedure **analyzeMessage**
Require:
Ensure:
1. **while** $message \neq end$ **do**
2.   **if** $message = (submission, (sender, assign, session))$ **then**
3. $agent\_view_{self} \leftarrow \{(sender, assign, session)\} \uplus agent\_view_{self}$
4.   closeSession()
5.   checkSubmission(*sender*)
6.   **end if**
7.   **if** $message = (backtrack, (x,v,s), BTList)$ **then**
8.   **if** $s = currentSession$ **then**
9.   **if** $v \notin ValBTReceive$ **then**
10. $ValBTReceive \leftarrow ValBTReceive \cup \{v\}$
11.   backtrack($BTList$)
12.   **end if**
13.   **end if**
14.   **end if**
15. **end while**

procedure **closeSession**
Require:
Ensure:
1. $currentSession \leftarrow currentSession + 1$
2. $ValBTReceive \leftarrow \emptyset$
3. **for all** $s \in my\_solutions$ **do**
4. $state[s] \leftarrow \neg propose$
5. **end for**

procedure **checkSubmission**
Require: $Agent$: Agent
Ensure:
1. **if** searchSolution(C_{inter}) **then**
2.   sendSolution()
3. **else**
4.   **if** searchPartialSolution($C_{inter}, Agent$) **then**
5. $(x,v,s) \leftarrow \{(x',v',s') \in agent\_view_{self} \mid \forall A \in agent\_view_{self}, A \succeq x\}$
6.   **else**
7. $(x,v,s) \leftarrow \{(x',v',s') \in agent\_view_{self} \mid x' = Agent\}$
8.   **end if**
9. $BTList \leftarrow \{(x',v',s') \in agent\_view_{self} \uplus TotalBTList \mid x' \succ x\}$
10. $agent\_view_{self} \leftarrow agent\_view_{self} - \{(x,v,s)\}$
11.   send ((backtrack, $(x,v,s), BTList$)) to x
12. **end if**

procedure **searchPartialSolution**
Require: C : set of constraints, A : Agent
Ensure: *boolean*
1. $C^* \leftarrow \{c_{Self,Y} \in C \mid Y \succeq A\}$
2. return searchSolution(C^*)

procedure **backtrack**
Require: $BTList$: set of triplets $(Agent, value, session)$
Ensure:
1. $TotalBTList \leftarrow BTList \uplus TotalBTList$
2. $BTList^* \leftarrow agent\_view_{self} \uplus TotalBTList$
3. **if** searchSolution(C_{inter}) **then**
4.   sendSolution()
5. **else**
6.   **if** $BTList^* = \emptyset$ **then**
7.   **for all** $A \in Acc_{self}$ **do**
8.   send (end) à A
9.   **end for**
10.   **else**
11. $(x,v,s) \leftarrow \{(x',v',s') \in BTList^* \mid \forall A \in BTList^*, A \succeq x'\}$
12. $BTList^* \leftarrow BTList^* - \{(x,v,s)\}$
13.   send (backtrack, $(x,v,s), BTList^*$) à x
14.   **if** $(x,v,s) \in agent\_view_{self}$ **then**
15. $agent\_view_{self} \leftarrow agent\_view_{self} - \{(x,v,s)\}$
16.   **else**
17.   closeSession()
18.   sendSolution()
19.   **end if**
20.   **end if**
21. **end if**

Alg. 1. Distributed Backtracking with Sessions (*DBS*)

Natural Interface for Sketch Recognition

D.G. Fernández-Pacheco[1], N. Aleixos[2], J. Conesa[1], and M. Contero[2]

[1] DEG. Universidad Politécnica de Cartagena, 30202 Cartagena, España
 daniel.garcia@upct.es, julian.conesa@upct.es
[2] Instituto de Investigación e Innovación en Bioingeniería (Universidad Politécnica de Valencia), España
 naleixos@dig.upv.es, mcontero@dig.upv.es

Abstract. New interfaces for CAD applications are coming up due to new available devices endowed with alternate technology. The use of this kind of interfaces is not extended at all because of the lack of robustness and reliability in recognition and interpretation of user inputs. In this work we develop a first approach of an interpreter of user sketches suitable for working on-line in CAD environments. An agent based architecture for the recogniser is presented. The recognition process uses invariant descriptors as the fast Fourier transform of the radius and arc length versus cumulative turning angle signatures, the HU moments and other invariant shape factors. The final classification is carried out by statistical learning using non-linear discriminant analysis. The method proposed in this article is not dependent on the number of strokes, neither on the sketching sequence order powered by the user. Finally, a validation of the recogniser is done.

Keywords: Natural interfaces, statistical learning, agent-based systems.

1 Introduction

New mainstream portable computing devices endowed with a digitising tablet and a pen, such as Tablet-PCs or personal digital assistants (PDAs), are emerging as a standard drawing tool allowing the development of new user interfaces –the so-called Calligraphic Interfaces [1]– created to substitute the traditional pen and paper. The recognition of hand-sketched symbols is a very active research field, but automatic gesture recognition is a complex task since different users can draw the same symbols with a different shape, size or orientation.

Due to these inconveniences, many recogniser algorithms are not robust or present ambiguity in their decisions. To minimise the impact of these problems, techniques based on digital image processing, such as convolution filters or morphological operations, can be applied. Moreover, the problems related to the sketch scale or orientation has to be solved by means of techniques that remain invariant in the face of these features [2]. Shape description from morphological features as length, width, perimeter, area, inertial moments, bounding box, etc., presents the drawback of similar results for different shapes and is dependent on the size or rotation, which increases the percentage of mistakes in the classification of the algorithm [3]. Fourier descriptors are an example of invariant ones [4], and they have been largely used as a general technique for image information compression or classification of regular shapes [5] or handwriting characters [6]. Apart from Fourier descriptors, there are the set of regular moment invariants as one of the most popular and widely used contour-based shape descriptors. In this set we can find the descriptors derived by Hu [7]. These geometrical moment invariants have been then extended to larger sets (Wong and Siu [8]) and to other forms (i.e. Dudani et al. [9] and Liao & Pawlak [10]).

Along the time, some works are carried out using these techniques in sketch recognition to detect symbols, diagrams, geometric shapes and other user command gestures. Some of them are scale and rotation dependent, others use invariant features related to shape factor ratios, Fourier descriptors or invariant moments, and others can accept multiple strokes but in a strict input order ([11-16]. For classification, fuzzy logic, distances from ideal shape, linear and non-linear discriminant analysis, etc. are used ([17-20]).

Technology based on agents has been widely used for process simulation, process control, traffic control and so on, but their use is being extended more and more to recognition process for supporting natural interfaces. For instance Juchmes et al. [21] base their freehand-sketch environment for architectural design on a multi-agent system. Also Achten and Jessurum [22] use agents for recognition task in technical drawings. So do Mackenzie et al. [23] in order to classify sketches of animals. Other examples can be found in Azar et al. [24], who base their system for sketch interpretation on agents, or in Casella et al. [25], who interpret sketched symbols using an agent-based system.

The goal of this paper is to develop an agent based recogniser for hand-drawn sketches, intended to be integrated in a CAD application. The proposed recognition is based on the extraction of invariant shape descriptors, as the Hu moments and the fast Fourier transform (FFT) of the shape of the sketched form. The recogniser we propose for the automatic recognition consists of four steps. In the first step, we use preprocessing image analysis techniques to smooth and remove noise from the sketch. In the second step its contour is extracted. In the third step the descriptors are calculated. As a final step, a Bayesian non-linear discriminant analysis is used to classify the sketch by attending to the chosen descriptors. To design the recogniser, an agent-based architecture has been implemented, and in order to evaluate the recogniser, an alphabet that includes geometrical and dimensional constraints for 2D sections and commands for basic modelling operations to create solid models in a CAD environment has been defined. The model was tested on classifying 2590 CAD gestures, collected by means of a Tablet-PC from 9 different users. The results showed an average success rate of 95%.

The paper is organised in the following way: first the recognition process is explained (including the noise suppression, the shape descriptors and the statistical model for classification), second the recogniser structure is shown (including the algorithmic representation and the agent structure), and, finally an evaluation and discussion of the recogniser are given.

2 Recognition Process

The implemented gesture alphabet is presented in table 1, in which constraint and command gestures are distinguished. These gestures have been chosen as basic ones

Table 1. Alphabet of constraint and command gestures

Constraint gestures	Class	Constraint gestures	Class	Command gestures	Class
	Concentric		Vertical		Extrusion
	Dimension		Horizontal		Revolve
	Diametral dimension		Parallel		Cross-out (erase)
	Tangency		Perpendicular		

for modelling task to allow the users to construct parametric geometry from sketches and basic solid models. In general, gestures can be composed of several strokes where each stroke is a set of points that are digitised by the device between consecutives pen down and pen-up events. The last stroke for a gesture occurs when a timeout is reached from the last pen-up event.

2.1 Smoothing and Noise Suppression

Ideally, when a sketch is introduced, its points have to be uniformly distributed along the gesture, but the faster the speed a gesture is introduced at, the fewer points are digitized, what causes different concentration of points in different parts of the gesture. An algorithm that estimates the distance between each point and their neighbors is capable of detecting and removing isolated points. A smoothing mask is then passed along the points of each stroke to avoid the effect of tremors caused by a bad drawing. For each point, the sum of the X coordinate of two previous points, two subsequent points and twice its value are calculated and divided by six (the number of points involved in the operation). Then the X coordinate of the point is substituted by the result of the operation. The same process is done with the Y coordinate. Lastly, the line equation between two consecutive digitised points is calculated, and the gaps are filled in with new points.

2.2 Contour Extraction

To avoid the dependence of the sequence of introduction and the number of strokes used, the contour of the shape is extracted. As the subsequent FFT analysis needs the input normalised to a fixed number of points, previous to extract the contour, the shape is cut out by its bounding box and fitted into a canvas of fixed dimensions. Then the contour is extracted and its points' number normalised, in such a way the shape is not altered and the original information is preserved. The FFT algorithm requires an input with 2^n points, so the number of points of the extracted contours is normalised to a number power of 2.

2.3 Shape Descriptors Utilised

There are amounts of descriptors to use for describing the shape, but we have chosen some of those that are independent for translocation and rotations, since sketched shapes can appear in any position or orientation in the sketchy space. In this way, the descriptors used for classification are Hu moments, perimeter, circularity and the fast Fourier transform.

Hu derived a set of moments using algebraic invariants. In particular, Hu defines seven values, computed by normalizing central moments through order three, that are invariant to object scale, position, and orientation. The moments used in here are the first six ones, since the seventh one if for symmetry. Equations detailed in [7].

Also perimeter and circularity have been chosen for describing the shape, since they are independent for translocation and rotations [26]. For the calculation of the perimeter an 8-connectivity neighborhood has been followed.

Finally, the Fast Fourier Transform of two signatures of the shape of the gesture is calculated. The two signatures are: 1) the distance of each point from the shape centroid, and 2) the changes in the direction of consecutive points. For the first signature, the centroid of the gesture is calculated as the average X,Y coordinates of all the points. Then, the Euclidean distances from the centroid to each point in the gesture are also calculated and stored in an array of distances, thus giving a characteristic spectrum for the gesture. This is the radius signature. In order to extract the second signature, we use the definition proposed by Yu in [27], where direction gives the angle between two consecutive points of the shape in the range $[-\pi,\pi]$. When a shape keeps its direction, points are aligned and there is no angle increment, otherwise, increment is non-zero. This is the arc length versus cumulative turning angle signature.

In this work, the FFT is calculated for both earlier signatures. The number of harmonics used in the classification is 10, since Tao et al. [28] demonstrated that the most relevant shape information of objects can be reconstructed from their first 10 harmonics. The signature arrays of radius distance and direction are used separately as inputs to the FFT algorithm. As a result, 20 values corresponding to the fundamental frequencies are obtained.

2.4 Statistical Model

One of the most used approaches from the statistical point of view for pattern recognition is the Bayesian analysis. This estimates all the probabilities of a pattern to pertain to any of the classes present in the problem, assuming that it belongs to the class with the highest probability. The probabilities *a posteriori* are estimated through the Bayes theorem, which allows to estimate the *a posteriori* probability $P(\omega_i|x)$, of a pattern x formed by a set of j features $(x_1...x_j)$ to belong to any of the N classes ω_i, from the *a priory* $P(\omega_i)$ and the conditional $P(x|\omega_i)$ probabilities. In our case, the patterns are the user inputs and the features are the shape descriptors above mentioned.

In order to design the classificatory function, a training set of the gestures described in Table 1 were collected from 9 CAD users. They were previously shown how to use the sketcher in a Tablet PC and how to sketch each gesture. To fill the training set, the users were asked to sketch several occurrences of each gesture in different sizes and orientations, and in two different days. A standard non-linear Bayesian discriminatory analysis was used to determine the classification functions. Thus, a classification function was obtained for each class which maximises its value when a gesture is assigned to its corresponding class. Assuming the cost of a classification error to be equal for each class, the maximisation function is defined as (2.1):

$$P(w_i \mid x) = \frac{p(x \mid w_i)P(w_i)}{\sum_{j=1}^{m} p(x \mid w_j)P(w_j)} \quad ; \quad i=1,\dots m \qquad (2.1)$$

Where m is the number of classes, x is the n-dimensional observed vector of a gesture, w_i $(i=1..m)$ is one of the m different classes and $P(w_i)$ is the 'a priori' probability of a gesture to pertain to a specific class (without knowing it's observed vector). In our algorithm this probability is the same for each class. For the statistical model, the six Hu moments, the number of perimeter points, the circularity and Fourier descriptors have been used. The model was implemented on a computer application in order to automatically classify the kind of gesture the user had sketched. The model was validated with separated sets of data, one set to train the model (550 samples, 50 for each class) and other different set to test it (1948 different samples).

3 Recogniser Implementation

The proposed algorithm consists of two different parts. The first part is the offline process to collect the training set of the gesture alphabet (Figure 1 in red colour). Then, the classification model for the training set is created and used in the online process, where users sketch gestures from the earlier alphabet in our CAD application on a Tablet PC, and classification and corresponding action is carried out. Figure 1 shows the recogniser structure in a programming sequential environment.

In a traditional environment, the recognition process is carried out sequentially following the flow of the Figure 1. In here, we have determined the specific tasks of the recognition process and have assigned them to different agents that can act independently, avoiding the sequential process and arranging the recognition process to execute in a more logical way. Regarding to the pre-processing stage (see Figure 1), it is

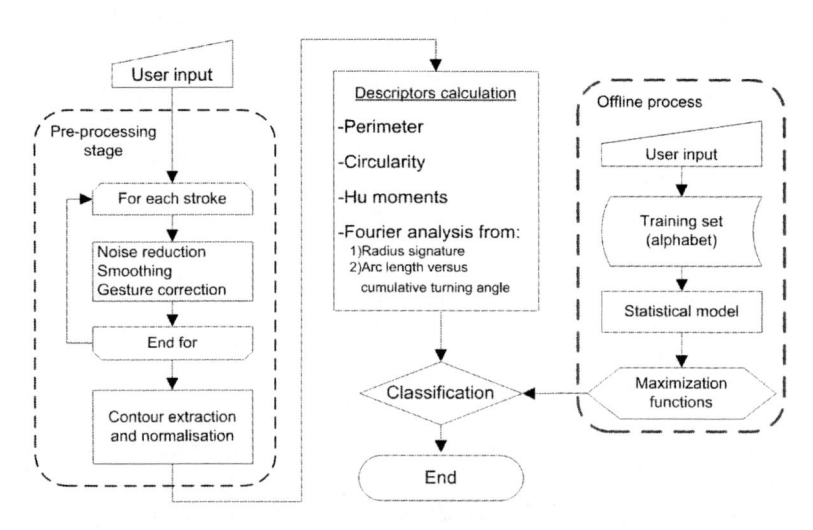

Fig. 1. Recogniser structure

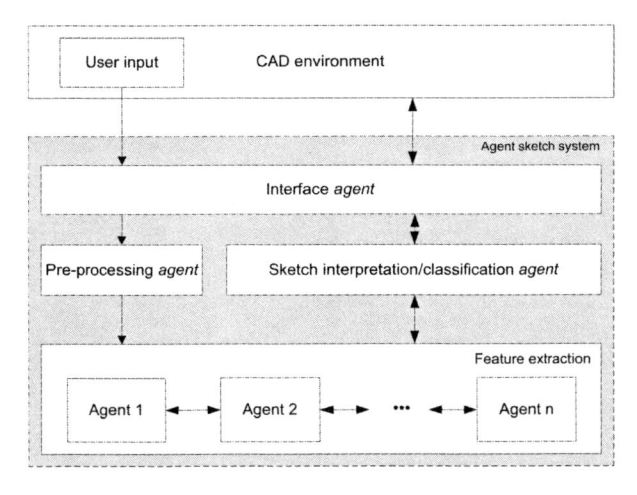

Fig. 2. Agent-based sketch scheme for the recogniser

a set of different tasks that have to be carried out in a sequential order, since one task begins when the previous one has finished. A single agent with different behaviours is dedicated to perform the pre-processing stage. We refer to pre-processing agent as PA. Figure 2 outlines the agent-based architecture proposed.

Once the pre-processing stage has finished, the features' extraction begins. These features can be extracted in parallel, since they are independent each other. Therefore, different agents are assigned to extract different features, and there will be so many agents as features of the sketch to be calculated. The extracted features are described in section 2.3. The feature agents (FA) will start their processes as soon as the pre-processing agent PA notifies them.

A feature agent can communicate with other feature agents if needs some external information. When a feature agent finishes, it notifies the superior level agent responsible for classifying the sketch, that we call the sketch interpretation/classification agent (SICA). This agent waits for the feature agents to finish. With the information it gets from feature agents, the SICA agent uses the Bayes theorem to classify the sketch by means of a non-linear discriminant analysis, as described in section 2.4. On the top of the agent-based scheme resides the interface agent (IA), which manages the interaction with the user. This agent gets the user input and passes it to agents on the lower level, later gets the result from the sketch interpretation agent and passes it to the CAD application.

3.1 The Agent Sketch System

The architecture of the agent sketch system for recognizing free hand-drawn sketches is described in Figure 3. It has been implemented on top of the Jade agent-based platform [29] using Java 1.6. For communication, messages are encoded in FIPA-ACL messages [30], a communication language natively supported by Jade. FIPA-ACL specifies both the message fields and the message performatives, which consist of communicative acts such as requesting, informing, making a proposal, accepting or refusing a proposal, and so on. Jade offers the means for sending and receiving

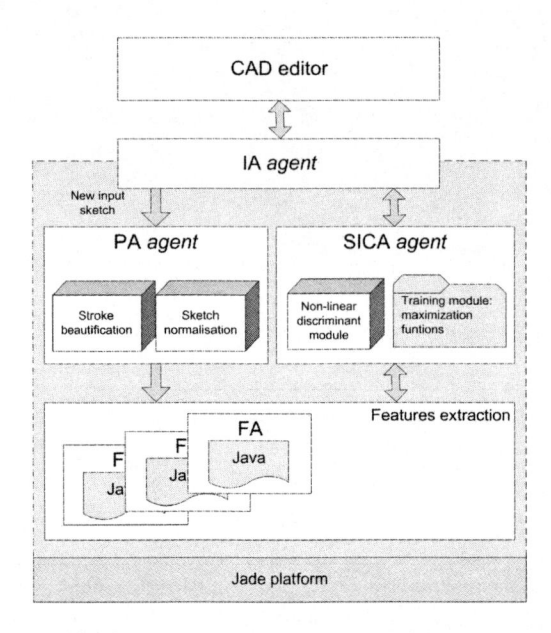

Fig. 3. The agent sketch system architecture

messages, also to and from remote computers, in a way that is transparent to the agent developer. For each agent, it maintains a private queue of incoming ACL messages that agents access in a blocking mode. Jade also allows the MAS developer to monitor the exchange of messages using the ''sniffer'' built-in agent.

3.2 The Sketch Interpretation/Classification Agent

The goal of the sketch interpretation agent (SICA) is to give an interpretation of the sketch drawn by the user, every time he/she requests it. In order to build the correct sketch interpretation/classification, the SICA requests the information from different FAs, and by means of a non-linear discriminant analysis, SICA classifies the sketch, giving the solution to the CAD application.

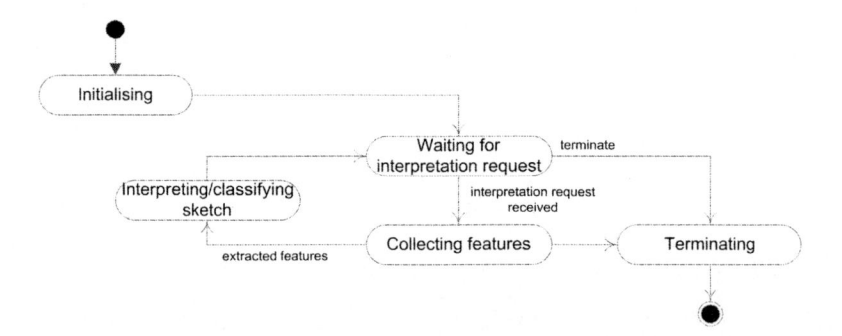

Fig. 4. Diagram describing the behaviour of sketch interpretation agent

The SICA behaviour is detailed in the diagram of Figure 4. In the Initializing state the SICA initializes itself and reads some configuration files to learn how to communicate with FAs (i.e., addresses, supported communication protocols, etc.). When the SICA is in the *waiting for interpretation request* state, it just puts itself in waiting state until a request message is received. On the other hand, if a sketch interpretation request is received while the SICA is not in this waiting state, then the message is stored in its incoming message queue.

4 Evaluation and Discussion

The tests to evaluate the recogniser were conducted with 9 CAD users. Each user introduced several occurrences of each of the different classes of gestures (a total of 2590 gestures) with different orientations and sizes, and each gesture was recorded.

Table 2. Confusion matrix for constraints and commands for a CAD application (values in %)

Recognised gestures \ Sketched gestures	Concentric	Dimension	Diametral	Tangency	Vertical	Horizontal	Parallel	Perpendicular	Extrusion	Revolve	Cross-out
Concentric	100.0	0.00	0.00	0.00	0.00	0.00	0.00	0.00	0.00	0.00	0.00
Dimension	0.00	92.65	0.00	0.00	0.00	0.00	0.00	0.00	0.00	4.41	2.94
Diametral	0.00	3.33	92.67	1.33	0.00	0.67	0.00	0.00	0.67	0.67	0.67
Tangency	0.00	0.00	0.00	96.59	0.00	0.98	0.00	0.00	0.00	0.49	1.95
Vertical	0.00	0.00	0.00	0.00	97.74	0.00	0.00	0.90	0.45	0.45	0.45
Horizontal	0.00	0.56	0.00	0.00	4.44	94.44	0.00	0.00	0.00	0.00	0.56
Parallel	0.00	0.00	0.00	0.00	0.00	0.00	100.0	0.00	0.00	0.00	0.00
Perpendicular	0.00	0.00	0.00	0.00	0.00	0.00	0.65	94.19	0.65	3.23	1.29
Extrusion	0.00	0.00	5.59	0.00	0.00	0.00	0.00	0.00	94.4	0.00	0.00
Revolve	0.00	14.63	0.00	0.00	0.00	0.00	0.00	0.00	0.00	85.37	0.00
Cross-out	0.00	0.00	1.33	0.00	0.67	1.33	0.00	0.00	0.00	0.00	96.6

The results of the classification are shown in Table 2, which shows success ratio in classification for each class in the diagonal cells (highlighted in black), and the percentage of misclassification. The average success ratio achieved with the algorithm proposed was of 94.87% with all descriptors used altogether, that is, HU moments, perimeter, elongation and Fourier descriptors.

5 Summary

A recogniser for gestures from users' sketches in a sketch-based environment has been proposed. The recognition process is supported by feature agents (FAs) that collect the features of the sketched gesture and by the sketch interpretation/classification agent (SICA) that manages data from FAs and uses it as parameters for the classification functions to interpret the sketch. Our agent-based sketch system offers a rapid solution for an operating online application, but the ambiguity for gesture shapes still remains since it does not consider contextual information. The proposed recogniser has the advantage to be easily extensible, so external information can be incorporated in the system.

It has been tested within a modelling system application. It is based on the shape description of the gestures using radius and arc length versus cumulative turning angle signatures analysed by means of the fast Fourier transform, and on other invariant descriptors as HU moments, perimeter and circularity. The final classification is carried out by statistical learning using non-linear discriminant analysis, and the validation of the model was made with separated sets of data, one for the training set and a different one for the test set.

The method proposed is not dependent on the number of strokes and neither on the sketching sequence order carried on by the user. Regarding to the off-line process, it is quite easy to add new gestures to the alphabet, and to train the system to recognise them. The averaged success ratio for a sketched-based modelling environment was nearly 95% (94.87%) what is a good result in an on-line application.

Acknowledgments. The Spanish Ministry of Science and Education and the FEDER Funds, through the CUESKETCH project (Ref. DPI2007-66755-C02-01) partially supported this work.

References

1. Contero, M., Naya, F., Jorge, J., Conesa, J.: CIGRO: A minimal instruction set calligraphic interface for sketch-based modeling. In: Kumar, V., Gavrilova, M.L., Tan, C.J.K., L'Ecuyer, P. (eds.) ICCSA 2003. LNCS, vol. 2669, pp. 549–558. Springer, Heidelberg (2003)
2. Kan, C., Srinath, M.D.: Invariant character recognition with Zerkine and orthogonal Fourier-Mellin moments. Pattern Recognition 35, 143–154 (2002)
3. Blasco, J., Aleixos, N., Moltó, E.: Machine vision system for automatic quality grading of fruit. Biosystems Engineering 85(4), 415–423 (2003)
4. Gonzalez, R.C., Woods, R.E.: Digital Image Processing, 2nd edn. Prentice Hall, Upper-Saddle River (2002)
5. Mokhtarian, F., Abbasi, S.: Robust automatic selection of optimal views in multi-view free-form object recognition. Pattern Recognition 38, 1021–1031 (2005)
6. Chen, G.Y., Bui, T.D., Krzyzak, A.: Rotation invariant pattern recognition using ridgelets, wavelet cycle-spinning and Fourier features. Pattern Recognition 38, 2314–2322 (2005)
7. Hu, M.: Visual pattern recognition by moment invariants. IRE Trans. Inf. Theor. IT-8, 179–187 (1962)
8. Wong, W.H., Siu, W.C.: Improved digital filter structure for fast moment computation. IEE Proc. Vision, Image Signal Process 46, 73–79 (1999)
9. Dudani, S.A., Breeding, K.J., Mcghee, R.B.: Aircraft identification by moment invariants. IEEE Trans. Comput. C-26, 39–46 (1977)
10. Liao, S.X., Pawlak, M.: On the accuracy of Zernike moments for image analysis. IEEE Trans. Pattern Anal. Mach. Intell. 20, 1358–1364 (1998)
11. Rubine, D.H.: Specifying Gestures by Example. Computer Graphics 25(4); Proceedings of the SIGGRAPH 1991, 329–337 (1991)

12. Ajay, A., Vo, V., Kimura, T.D.: Recognising Multistroke Shapes: An Experimental Evaluation. In: Proceedings of the ACM (UIST 1993), Atlanta, Georgia, pp. 121–128 (1993)
13. Gross, M.D.: Recognising and Interpreting Diagrams in Design. In: Proceedings of ACM (AVI 1994), Bari, Italy, pp. 88–94 (1994)
14. Zang, D., Lu, G.: A Comparative Study of Fourier Descriptors for Shape Representation and Retrieval. In: The 5th Asian Conference on Computer Vision, ACCV 2002, Melbourne, Australia, pp. 1–6 (2002)
15. Harding, P.R.G., Ellis, T.J.: Recognising Hand Gesture Using Fourier Descriptors. In: Proceedings of the 17th International Conference on Pattern Recognition, vol. 3, pp. 286–289 (2004)
16. Zion, B., Shklyar, A., Karplus, I.: Sorting fish by computer vision. Comput. Electron. Agric. 23, 175–187 (1999)
17. Fonseca, M.J., Jorge, J.: Using Fuzzy Logic to Recognise Geometric Shapes Interactively. In: Proceedings of 9th IEEE Conference on Fuzzy Systems, vol. 1, pp. 291–296 (2000)
18. Xiangyu, J., Wenyin, L., Jianyong, S., Sun, Z.: On-Line Graphics Recognition. In: Pacific Conference on Computer Graphics and Applications, pp. 256–264 (2002)
19. Zhengxing, S., Liu, W., Binbin, P., Bin, Z., Jianyong, S.: User adaptation for online sketchy shape recognition. In: Lladós, J., Kwon, Y.-B. (eds.) GREC 2003. LNCS, vol. 3088, pp. 305–316. Springer, Heidelberg (2004)
20. Park, C.H., Park, H.: Fingerprint classification using fast Fourier transform and non-linear discriminant analysis. Pattern Recognition 38, 495–503 (2005)
21. Juchmes, R., Leclercq, P., Azar, S.: A freehand-sketch environment for architectural design supported by a multi-agent system. Computers & Graphics 29(6), 905–915 (2005)
22. Achten, H.H., Jessurun, A.J.: An agent framework for recognition of graphic units in drawings. In: Proceedings of 20th International Conference on Education and Research in Computer Aided Architectural Design in Europe (eCAADe 2002), Warsaw, pp. 246–253 (2002)
23. Mackenzie, G., Alechina, N.: Classifying sketches of animals using an agent-based system. In: Petkov, N., Westenberg, M.A. (eds.) CAIP 2003. LNCS, vol. 2756, pp. 521–529. Springer, Heidelberg (2003)
24. Azar, S., Couvreury, L., Delfosse, V., Jaspartz, B., Boulanger, C.: An agent-based multimodal interface for sketch interpretation. In: Proceedings of International Workshop on Multimedia Signal Processing (MMSP 2006), British Columbia, Canada (2006)
25. Casella, G., Deufemia, V., Mascardi, V., Costagliola, G., Martelli, M.: An agent-based framework for sketched symbol interpretation. Journal of Visual Languages and Computing 19, 225–257 (2008)
26. Wojnar, L., Kurzydłowski, K.J.: Practical Guide to Image Analysis. In: ASM International, pp. 157–160 (2000) ISBN 0-87170-688-1
27. Yu, B.: Recognition of freehand sketches using mean shift. In: Proc. of IUI 2003, pp. 204–210 (2003)
28. Tao, Y., Morrow, C.T., Heinemann, P.H., Sommer, H.J.: Fourier-Based Separation Technique for Shape Grading of Potatoes Using Machine Vision. Transactions of the ASAE 38(3), 949–957
29. Bellifemine, F., Poggi, A., Rimassa, G.: Developing multi-agent systems with JADE. In: Castelfranchi, C., Lespérance, Y. (eds.) ATAL 2000. LNCS, vol. 1986, pp. 89–103. Springer, Heidelberg (2001)
30. FIPA ORG, FIPA ACL Message Structure Specification, Document no. SC00061G (2002)

Performance of an Open Multi-Agent Remote Sensing Architecture Based on XML-RPC in Low-Profile Embedded Systems

Guillermo Glez. de Rivera, Ricardo Ribalda, Angel de Castro, and Javier Garrido

Escuela Politécnica Superior, UAM, Francisco Tomás y Valiente, 11. 28049, Madrid, Spain
{guillermo.gdrivera, ricardo.ribalda, angel.decastro,
javier.garrido}@uam.es

Abstract. GdRBot is a platform for controlling a set of collaborative agents through XML-RPC. It was originally designed and tested for using PCs as robots and servers. Unfortunately, common computers can't be used on every scenario. On this paper the requirements of the architecture and the performance of two low-performs devices implementing the architecture are discussed. The first device is a Gumstix, a platform based on an Intel PXA microprocessor and the second is a recycled router by Linksys, the NSLU2. Both powered by Linux as operating system.

1 Introduction

In recent years, multi-agent systems (MAS) have received increasing attention in the artificial intelligence community. Research in multi-agent systems involves the investigation of autonomous, rational and flexible behaviour of entities such as software programs or robots. MAS researchers develop communications languages, interaction protocols, and agent architectures that facilitate the development of multiagent systems. The paradigm of agents and multi agent systems is one of the most promising approaches to develop and control individual robots and teams of robots. In the field of remote sensing the use of robots is increasing constantly [10], [6], [8]. Robots can reach places where common sensors are unable to get. Because of the many different items involved in robotics (from AI to electronics), an heterogeneous group of people is needed to reach good results, and usually, heterogeneous people means heterogeneous languages and ways of working.

The GdRBot [3] platform provides a simple but powerful platform for developing robots. Heterogeneous groups can work together with pre-defined interfaces and roles, reducing the coordination problems. Also, everything developed in the platform can be easily reused.

The GdRBot platform (Fig. 1[1]) consists on an extendible superset of XML-RPC calls on an heterogeneous network. Robots behave like servers, providing access to their sensors through a simple interface, accessible from almost any programming language or operating system. Robots can also behave as clients to access other robots, supporting collaborative acts or shared knowledge. All the devices in the architecture can be interconnected using any interface: Ethernet, Wifi, Bluetooth, GPRS, USB,...

---
[1] Cliparts obtained from OpenClipArt.

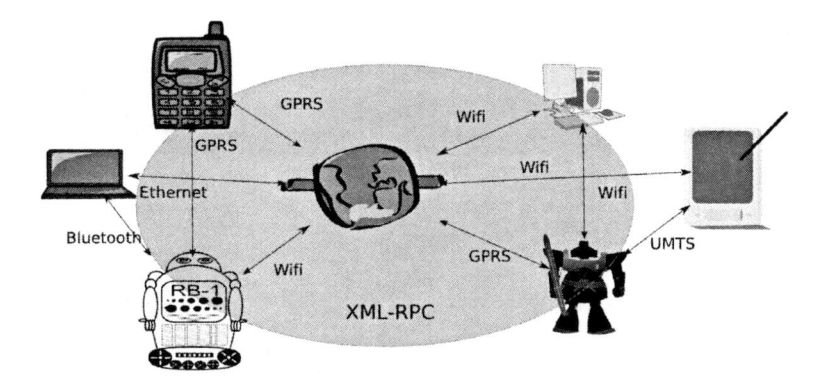

Performance of an Open Multi-Agent Remote Sensing Architecture 521

Fig. 1. GdRBot Platform

1.1 Legancy System

The first system used to support GdRBot has been an EPIA mother board [20] based robot (Fig. 2). All the system were controlled by a Debian [13] Linux distribution. The server routines were implemented as Apache's CGIs. Debug and development tools could be easily installed and used inside and outside the robot. Also, EPIA mother boards present many ports to attach sensors and communication interfaces to the robot. These are the complete specifications of the system:

- EPIA ML1000 motherboard.
- 7AH plumb battery.
- 2 step by step motors.
- Wifi card by USB.
- GPBot [2]: A development board with a micro-controller, attached by serial port.
- USB camera.

Fig. 2. EPIA Architecture monted on a motorized structure

Unfortunately, EPIA based robots are not the best solution for all the applications. They are very big (17x17 cm, area size: 289 cm^2) and consume a lot of power, needing a big and heavy battery. In order to be able to move all this weight, they need very powerful motors, which even waste more energy. The price of all its components is also considerable.

2 New Architecture

GdRBot doesn't depend on any hardware or software, it only depends on a protocol specification that can be easily approached by any device. Therefore, alternative architectures have been developed trying to minimize their cost and size.

2.1 Software Aspects

Target devices will need at least one network interface and a TCP-IP stack to place an XML-RPC server on the top of it. Instead of developing a full network stack, an operating system (OS) will be used. The OS will also provide drivers for some common sensors and some applications as web servers which reduce drastically the porting effort. Because the first robot used Linux, which can run on almost any architecture, it has been chosen as the target OS on the next platforms. Some particular aspects will be analyzed.

Compiler&LibC: If it is planned to port software to an architecture different than the host machine, a compiler is needed. A cross-compiler allows compiling applications for other architectures from the one the compiler is running on. GNU GCC provides this facility. Apart from the compiler, the LibC [16] library is needed in order to access all system functionality from the applications. Both GCC and LibC can be easily obtained from their site and automatically compiled on the host machine using crosstool [12].

Kernel: Once the cross-compiler is obtained, the next step is to obtain a kernel that will manage all the hardware in our robot, and provide a network interface to it. The kernel source code can be obtained from its official web site [19], being the same for every architecture. After it is downloaded it has to be configured to fit in the final hardware. All the possible information about the hardware should be collected previously. After the configuration, the kernel should be compiled with a cross-compiler to the image format needed by the boot-loader.

Boot-Loader: The kernel is just a program for our target processor. It should be loaded into memory and the micro should start running its entry point. Each architecture has its own peculiarities about its booting process. Very often a simple program like UBoot [18] is executed before the kernel to load it into memory, uncompress it (if needed), set up some registers and run it.

Web-Server: As explained before, the whole platform is based on XML-RPC. This is the reason why an implementation of HTTP is needed. It is a very simple implementation and it can be easily developed or a mature web-server can be installed in the robot.

Fig. 3. Gumstix Architecture

The most common web server is Apache. It has many options for security, caching, etc... but need a lot of resources. If the architecture is not very powerful other options should be studied like thttpd, cherokee or boa [11].

2.2 Hardware Platforms

Two different embedded architectures have been used as hardware platforms to support the multi-agent architecture.

GdRBot in Gumstix: The Gumstix [15] (Fig. 3) is an extremely embedded device based on the Intel XScale PXA255 Microprocessor. This platform is characterized by its reduced size, 8x2x0.6 cm, (area size: 16 cm^2 the size of a stick of gum), its reduced comsuption (it draws less than 250 mA) and its wireless connectivity via Bluetooth. It has several expansions packs including Ethernet, and a "Robostix" board adapter which is a micro-controller (Atmel ATMega128) used to control many sensors by a GPIO, SPI and I2C.

The target device was composed by a connex 400xm-bt board extended with a Robostix board and a netcf board. The total price of the device has been about 250$.

The Gumstix is already provided with a Linux running inside it [21]. So the only big effort done has been the installation of the cross-compiler in the host machine, and the recompilation of the server components.

The Gumstix doesn't have a very powerful processor, so if complex algorithms are needed, they should run in the client machine. Also it can't handle many queries per second and it doesn't have a big permanent memory.

Any sensor/actuator can be controlled by this device, thanks to the Robostix board. It provides: A/D Converters, PWM output, Serial Lines, Digital IO, etc...

The Gumstix has a relatively good connectivity. It can be connected using Ethernet, WIFI or Bluetooth. Also it has a quite good autonomy with a small battery.

Fig. 4. NSLU2 Architecture

Gumstix has been used with success by other research groups in robotics [1], [9], [7].

GdrRBot in Linksys NSLU2: The Linksys NSLU2 [17] (Fig. 4) was firstly designed by Linksys as a device for sharing hard drives through a network. But it can also be considered as a multipurpose computer based on an PXA microprocessor.

The Linux community has released distributions designed specifically for this device, like OpenSlug [5] or DebianSlug [14] (a coplete Debian port).

The main benefits of this device is that it is very cheap (about 100$) and can be bought almost everywhere. It provides two USB ports and an Ethernet port. Installing DebianSlug in the device is quite straightforward. Firmwares with DebianSlug can be obtained in the Internet. A disadvantage is that it has only USB ports to extend it. To avoid this, an USB micro-controller, like Cypress EZ-USB, can be used to control external sensors/actuators. Unfortunatelly, drivers for the network chipset are closed-source. In order to avoid using them, a WIFI USB dongle can be used for networking. It can be used in the same scenarios as the Gumstix, but is much easier to obtain, or to get support and it is cheaper. On the other hand it is bigger (area size: 108 cm^2) and less flexible than the Gumstix, because only USB-based sensors can be used with it. There some groups working on this device, but less than in the Gumstix [4].

As resume, each architecture (old and news) has its own restriction as shown on Table 1.

Table 1. Main Restrictions

Architecture	Restrictions
EPIA	No A/D lines, high consumption
Gumstix	Reduced computation power
NSLU2	Only USB sensors/actuators

3 Verification and Results

Some tests were done in order to check the performance of the different systems and software configurations. All the experiments consist on a set of calls where the robot always acts as the server but with different clients. If the client is an external system, it is a remote call but if the client is locaded in the robot itself it is a local call. A typical call could be reading GPIOs or changing a motor parameter.

3.1 Different Architectures Performance

In this experiment, a set of calls have been made to the EPIA, the NSLU2 and the Gumstix platforms. All were running a lightweight webserver (Boa), because it is more appropriate for embedded systems. The calls have been made from a client allocated in a different machine (remote call). As it can be seen in Fig. 5, EPIA platform presents the best performance as expected, because it has a more powerful CPU and then, it can handle a higher number of calls per second (approx. EPIA: 180 calls/sec, Gumstix: 50 calls/sec and NSLU2: 30 calls/sec).

3.2 Different Web Servers Architectures

In this experiment, a set of calls have been made to a CGI server allocated in the top of two different web servers from a different computer. Apache (complex) and Boa (lightweight) have been used as web servers. Because Apache web server is not natively supported by Gumstix, this experiment has only been run on EPIA and NSLU2 both connected via Fast Ethernet.

As it can be seen in Fig. 6, Boa works much better in both architectures. Boa has been designed with speed and size restrictions in mind. On the other hand Apache was designed to provide a huge amount of services not taking care about size or speed

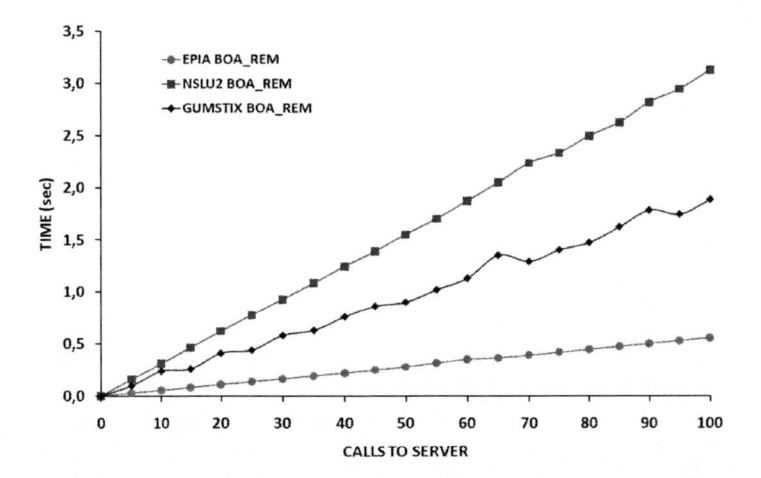

Fig. 5. Architecture Performance

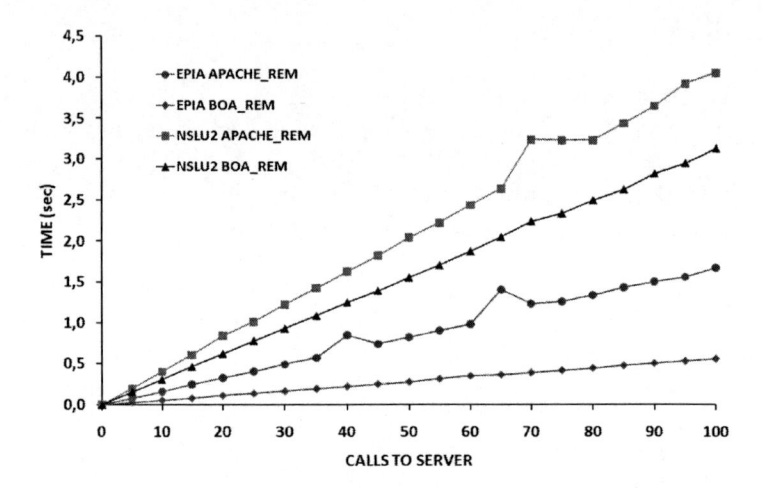

Fig. 6. Web-Server Performance

restrictions. As we need just a small set of services the use of lightweight web servers is highly recommended.

3.3 Different Allocation of the Client

To test the importance of the possible network latency, in this experiment a set of calls has been made from a client allocated in the robot (local) and outside it (remote). This experiment has been run in two architectures: EPIA and NSLU2, using boa as web server.

As it can be seen in Fig. 7, the remote processing of the call only adds an insignificant overhead, in both EPIA and NSLU2 platforms.

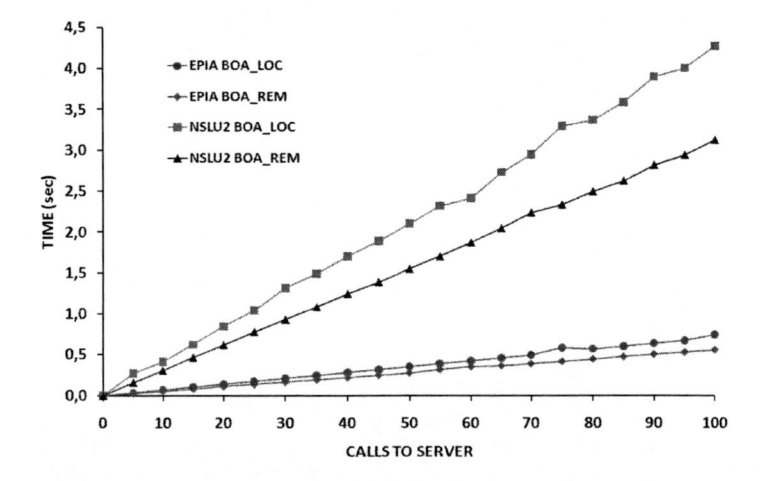

Fig. 7. Client-Allocation Performance

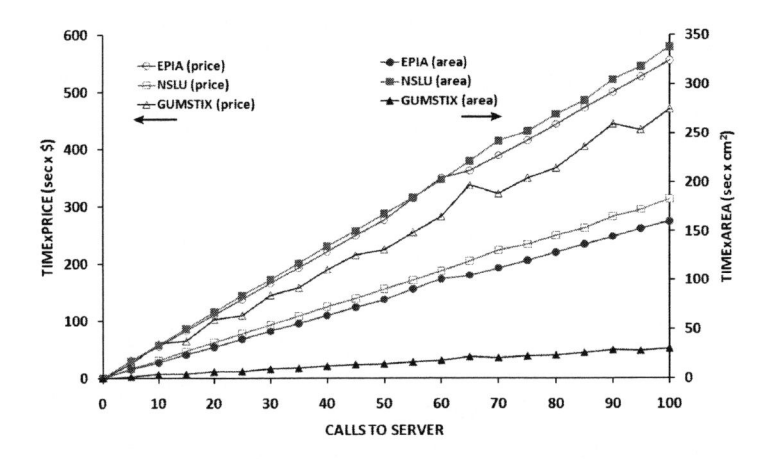

Fig. 8. Performance vs HW Platform Area and Price

3.4 HW Platform Price and Size Versus Performance

In Fig. 8, the performance of the system normalized with the platform area size and platform price is shown. The performance has been measured with the same web-server (boa) and client allocation (remote).

All the systems are not very different in performance per dollar (open dots). However, the Gumstix platform has the best outstanding performance per cm^2 (solid dots).

4 Conclusions

Thanks to the platform specification, porting an application to another architecture is very easy, it is transparent for the client the involved hardware or software of the robotic agent it wants to control.

Embedded devices can provide a performance comparable to PCs with better performance/price and performance/size relation.

Open Source operating systems provide a great flexibility, security and network facilities.

The use of lightweight web servers in platforms based on XML-RPC improve the total performance.

Acknowledgements

This work has been supported by the TEC2006-13141-C03-03 project of the Spanish Ministry of Science and Technology. Ricardo Ribalda is supported by a FPU Fellowship from the Ministerio de Educacion y Ciencia (Spanish Ministry of Science).

References

1. De Nardi, R., Holland, O.: UltraSwarm: A further step towards a flock of miniature helicopters. In: Şahin, E., Spears, W.M., Winfield, A.F.T. (eds.) SAB 2006 Ws 2007. LNCS, vol. 4433, pp. 116–128. Springer, Heidelberg (2007)
2. Glez. de Rivera, G., Lopez-Buedo, S., González, I., Venegas, C., Garrido, J., Boemo, E.: GPBOT: Plataforma Hardware para la enseñanza de Robótica en Ingeniería Informática. In: Tecnologías Aplicadas a la Enseñanza de la Electrónica (TAEE 2002), pp. 67–70 (February 2002)
3. Glez. de Rivera, G., Ribalda, R., Colás, J., Garrido, J.: A Generic Software Platform for Controlling Collaborative Robotic System using XML-RPC. In: IEEE/ASME International Conference on Advanced Intelligent Mechatronics, pp. 1336–1341 (July 2005)
4. Handziski, V., Kopke, A., Willig, A., Wolisz, A.: TWIST: A Scalable and Reconfigurable Wireless Sensor Network Testbed for Indoor Deployments. TKN Technical Reports Series (2005)
5. (Web server): Slug-Firmware.net (cited October 27, 2008), http://www.openslug.org/
6. Posadas, J.L., Poza, J.L., Simó, J.E., Benet, G., Blanes, F.: Agent-based distributed architecture for mobile robot control. Engineering Applications of Artificial Intelligence (2007), doi: 10.1016/j.engappai.2007.07.008
7. Rusu, R.B., Robotin, R., Lazea, G., Marcu, C.: Towards Open Architectures for Mobile Robots: ZeeRO. In: IEEE International Conference on Automation, Quality and Testing, Robotics, pp. 260–265 (2006)
8. Sanfeliu, A., Hagita, N., Saffiotti, A.: Network robot systems. Robotics and Autonomous Systems (208), doi: 10.1016/j.robot.2008.06.007
9. Sauze, C., Neal, M.: An Autonomous Sailing Robot for Ocean Observation. In: TAROS 2006, pp. 190–197 (September 2006)
10. Suri, D., Hotel, A., Schmidt, D., Biswas, G., Kinnebrew, J., Otte, W., Shankaran, N.: A Multi-Agent Architecture provides Smart Sensing for the NASA Sensor Web. In: IEEE Aerospace Conference, paper. 1198 (March 2007)
11. Boa Web Server (cited October 27, 2008), http://www.boa.org/
12. Building and Testing gcc/glibc cross toolchains (cited October 27, 2008), http://www.kegel.com/crosstool/
13. Debian GNU/Linux (cited October 27, 2008), http://www.debian.org/
14. Debian/NSLU2 Downloads (cited October 27, 2008), http://www.slug-firmware.net/d-dls.php
15. Gumstix Home Page (cited October 27, 2008), http://www.gumstix.com/
16. GNU C Library (cited October 27, 2008), http://www.gnu.org/software/libc/
17. Linksys Home Page (cited October 27, 2008), http://www1.linksys.com/products/product.asp?prid=640
18. U-boot (cited October 27, 2008), http://www.denx.de/wiki/u-boot/
19. The Linux Kernel Archives (cited October 27, 2008), http://www.kernel.org/
20. Via Home Page (cited October 27, 2008), http://www.via.com.tw/en/products/mainboards/
21. Wolfe, A.: Tiny Linux Computer Has High Hopes For Robotics Apps. In: TechWeb News (2005)

Privacy Preservation in a Decentralized Calendar System

Ludivine Crépin[1,3], Yves Demazeau[1], Olivier Boissier[2], and François Jacquenet[3]

[1] Laboratoire d'Informatique de Grenoble, CNRS, Maison Jean Kuntzmann - 110 av. de la Chimie, Domaine Universitaire de Saint Martin d'Heres, 38041 Grenoble cedex 9, France
Ludivine.Crepin@imag.fr, Yves.Demazeau@imag

[2] Ecole Nationale Supérieure des Mines de Saint-Etienne, Centre G2I, Equipe SMA, 158 Cours Fauriel, 42000 Saint-Etienne, France
Olivier.Boissier@emse.fr

[3] Université de Lyon, Université Jean Monnet, Laboratoire Hubert Curien, CNRS, 18 rue Benoit Lauras, 42000 Saint-Etienne, France
Francois.Jacquenet@univ-st-etienne.fr

Abstract. Privacy perservation, in terms of sensitive data access and management, is an important feature of decentralized calendar systems. Indeed, when users delegate their sensitive data such as timetables to an autonomous agent, this one executes many automatic data processing without their intervention: users lost a part of their data control. To tackle this problem, we propose in this article to extend a concrete application of calendars management multi-agent system by implementing a specific protocol for sensitive data transactions that represents the first step of privacy preservation in multi-agent systems.

Keywords: Calendar Management, Multi-Agent system, Privacy, Sensitive Data Transaction, Interaction Protocol.

1 Introduction

In applications such as calendar management systems, users's sensitive information may be disclosed to other ones for instance while trying to schedule meeting. Considering the sensitiveness in such systems is a difficult and time consuming task. This is an even more important issue when considering management realized by a multi-agent system. That leads us to consider the problem of privacy preservation in terms of data management and access, like Deswarte and Melchor define it in [7].

We propose to extend a decentralized calendar multi-agent system [6] with the model of Hippocratic Multi-Agent Systems (HiMAS) [5] that takes into account this data sensitivity regarding moral issues and users' wishes.

In fact, privacy preservation must be considered during three critical steps. The first one is the storage of data that requires security. The second one is the transaction of sensitive data that requires security and many constraints in relation to users' desires (in terms of disclosure, use and retention of information for example). The last one concerns the becoming of data after a communication: we need to guarantee its protection. In this article we focus on the second critical step by implementing a specific protocol for the transaction of sensitive data [4] in a decentralized calendar application [6]. In fact, we will see that this protocol also proposes the basis for the realisation of the third step.

The next section presents the basic calendar multi-agent system [6]. Then we present the context of the extension that is the transformation of this application into a HiMAS. The fourth section defines the basis of the sensitive data protocol and the fifth section presents the implementation of this protocol into our agenda management system. We finish this article by some conclusions and some considerations for future works.

2 Decentralized Calendars Application

This article focuses on a concrete user-centred application presented in [6]. This application is an multi-agent approach for decentralized calendars management. The architecture (see Figure 1) proposes to represent each user by an agent in charge of the user's timetable (event scheduling, tasks, meetings).

Each timetable contains events, including tasks and meetings, that are characterized by two subjective attributes: the importance and the urgency. These attributes aim to give a priority level for each event. By default, the importance takes priority over the urgency but an agent can choose the contradictory, according to the user's wishes.

The agents in charge of the users' timetables have two possibilities to interact. The first one is the meeting negotiation based on GeNCA [8], a general negotiation API based on a contract protocol. When an user wants to fix a meeting with another user, the corresponding agent sends to the other one a proposition represented by a contract for the meeting date. The agent that receives this proposition chooses to accept or to modify this contract according to its strategy. The strategy is related to the importance and the urgency of the event and depends on the sender. Two agents that have different definition of importance and urgency do not propose same slots of times for the meeting negotiation.

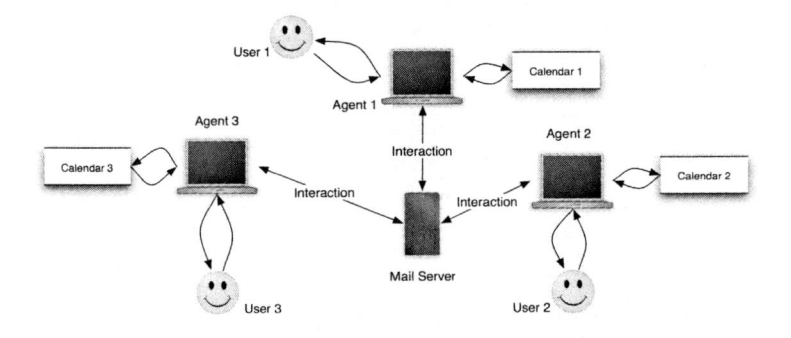

Fig. 1. Application architecture

The second possibility to fix a meeting is the calendars sharing. This kind of interaction is based on trust. When an agent asks another agent for its calendar, this last one chooses to send it according to its trust relationships with the first one. The trust model we implemented [6] is in direct relation with users: they determine what are the trusted agents according to their believes.

In this article, our proposition focuses only on this second kind of interaction, the calendar sharing. We propose now to introduce privacy preservation by extending this application to a Hippocratic Multi-Agent System (HiMAS) [5] presented in the next section. We consider that the sensitive data are the slots of time for each timetable, in particular the attributes of importance and urgency for each slot.

3 Foundations: Hippocratic Multi-Agent Systems

3.1 Required Definitions

The **private sphere** contains data that an agent[1] considers as sensitive and all the associated management rules defining the conditions of its disclosure, its use or its sharing for example.

To represent the possible positions of an agent with respect to the private sphere, we define two roles in relation with the sensitive data transaction. The **consumer** role characterizes the agent which asks for sensitive data and uses it. The **provider** role characterizes the agent which discloses sensitive data. The provider defines a **policy** and the consumer a **preference** to define their desires regarding the sensitive data manipulations (use, disclosure...).

The consumer's policy and the provider's preference are defined in a similar way to the policies and the preferences defined in [12]: purpose specification, retention time and possible use. They are composed of the transaction objectives[2], the retention time of collected data, the broadcasting list and the data format (required references).

We can notice that we install a provider-centered view on the management of sensitive data, the opposite vision of the service- centered vision like for example in [9], regarding the terms of consumer and provider. This vision defines the user as the provider of information and the service as the consumer of information, it is the service and not the user that asks for data to the user and uses it. This is due to the fact that we mainly have a user-centered view on privacy preservation: users should be confident in the management of the sensitive data they delegate to their personal agent.

3.2 Nine Principles for HiMAS

In order to respect the private sphere, a HiMAS must respect the nine principles inspired by the Hippocratic Databases [1] described below.

---

[1] In this approach, we consider users' as agents.
[2] The objectives are close to the concept of goal, like for example in BDI model [2] or [10].

1. **Purpose specification:** The provider must know the objectives of the sensitive data transaction. Therefore it can evaluate the transaction consequences.
2. **Consent:** Each sensitive data transaction requires the provider's consent.
3. **Limited collection:** The consumer commits to cutting down to a minimum the amount of data for realizing its objectives.
4. **Limited use:** The consumer commits to only use sensitive provider's data to satisfy the objectives that it has specified and nothing more.
5. **Limited disclosure:** The consumer commits to only disclose sensitive data to reach its objectives.
6. **Limited retention:** The consumer commits to retain sensitive data only for the minimum amount of time it takes to realize its objectives.
7. **Safety:** The system must guarantee sensitive data safety during storage and transactions.
8. **Openness:** The transmitted sensitive data must remain accessible to the provider during the retention time.
9. **Compliance:** Each agent should be able to check the obedience to the previous principles.

4 Interaction Protocol for Sensitive Data Sharing

4.1 Content Language

For each principle of a HiMAS (and for the notion of format[3] that is required in our approach) we define an associated concept in a conceptual graph [11] (refer to Figure 2) implemented in an OWL file [13]. To formalize this conceptual graph, we use an existential positive conjunctive fragment of the first order logic in order to obtain no contradictory logical information. Each principle and the notion of format is represented by a concept linked to another according to a semantic relationship.

To take the domain of the application into account, the conceptual graph must be instantiated by all the possible values of each concept according to the domain (refer to doted set in Figure 2).

To ensure the syntax, we define all the required elements of sensitive data transaction in an XSD schema. Then the HiMAS agents build an XML file that validates the XSD schema and where all the values are present in the OWL file to ensure the semantics. This system allows HiMAS agents to create sensitive data transaction with regards to privacy preservation at a semantic and syntactic level. A preference and a policy are based on the same concepts. Therefore we represent the provider's preference by the modifications that the provider induces from the consumer's policy if the policy does not match with the preference. Each consumer and each provider validate their policy and their preference using the content language in order to build and to process a sensitive data transaction with respect to the private sphere.

---

[3] All the required references for the asked data like the urgency or the importance for instance.

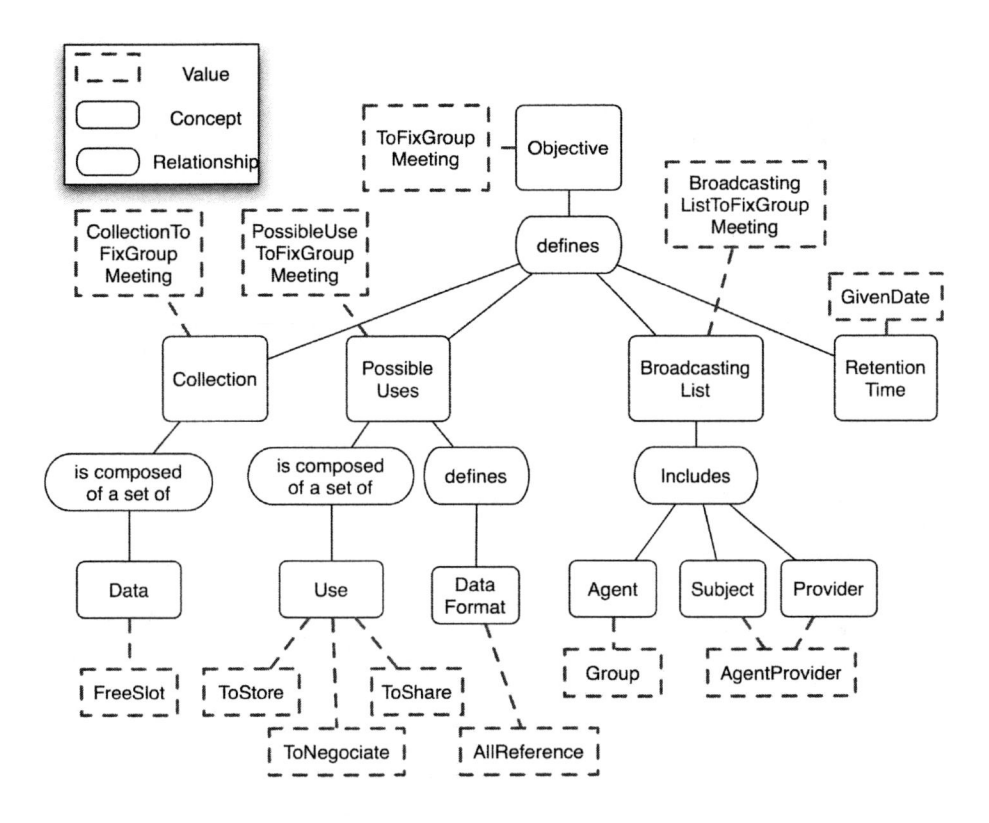

Fig. 2. Principles formalization

4.2 Sensitive Data Transaction Protocol

We now present in a chronological order the three steps of the interaction protocol represented in Figure 3: the design of the policy, the sensitive data transaction and the design of the preference.

The first step of this protocol is the **design of the consumer's policy**. This agent builds its policy according to its objectives thanks to the content language in order to preserve privacy. Afterwards it executes the first interaction: the consumer includes its policy in a **sensitive data transaction** and sends it to the provider. The constraint of this step is that the transaction file must syntactically and semantically validate the content language (see Subsection 4.1) in order to respect agents' privacy.

Afterwards the provider begins to check the validity of the received sensitive data transaction. Then, from the management rules of its private sphere, the provider **designs its preference** thanks to the content language and tries to map this preference with the received policy. If the policy matches its preference, the provider sends the consumer the asked sensitive data. In the other way, the provider proposes some modifications to the policy in order to find an

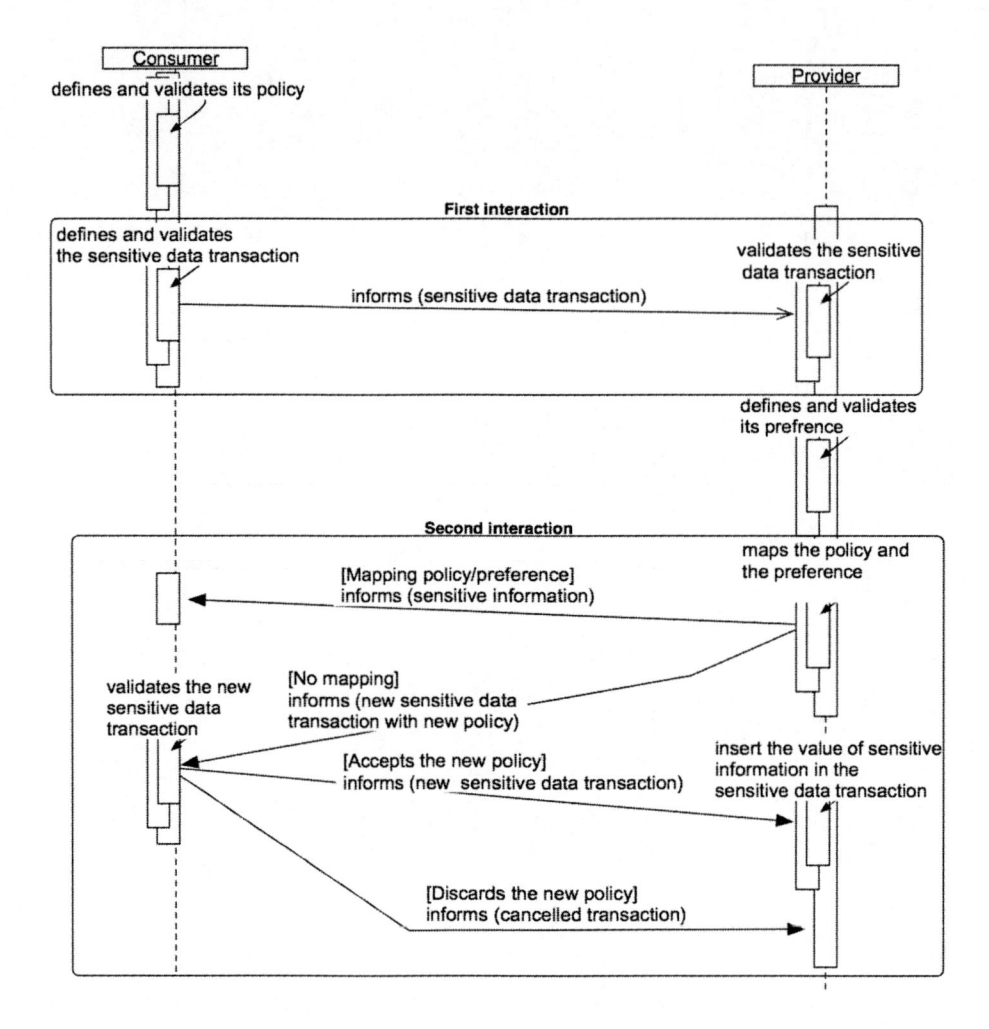

Fig. 3. Sensitive data transaction protocol

agreement. For instance, the provider can change the broadcasting list if the proposed list contains some agent that it does not trust. If the consumer accepts these adaptations, the provider sends it the data, else the consumer cancels the transaction.

This approach allows HiMAS agents to verify the constraints defined by the principles of the HiMAS thanks to the content language and so gives the basis for the compliance principle. The consumer (resp. provider) can design its policy (resp. preference) with respect to the constraints defined by HiMAS principles. This obedience is made by the semantic links between the concepts representing the HiMAS principles.

5 Implementation into the Calendar Application

5.1 Current Implementation

In this application [6], each agent manages a calendar of one user and is registered with a specific agent, the server agent. This agent aims to deliver the message to the given agent and to prevent the agency about the subscription of a new agent. The calendar is constituted by a set of resources that represent slot of time. The negotiated contract is about these resources: each agent proposes a finite set of free resources and asks to GeNCA to find a free slot of time for each agent with regards to the importance and the urgency of the meeting. During the sharing, an agent sends to an other the asked resources.

To communicate, each message is transmitted using a GMAIL server with the Jabber protocol. This protocol allows to ensure the security, the confidentiality and the decentralization of the application. This server is represented by the server agent. At a high level, each message is defined according to a specific kind of interaction: registration, negotiation and sharing.

The graphical user interface proposes to the user one tab for each possible action: visualization of the calendar, to ask the agenda of an other user, management of the trust, informations about canceled, accomplish and current contract. The presentation of the calendar is made thanks to mig calendar that is a Java component allowing to create events visualization.

To extend this application, we propose to implement three objectives for the meeting sharing: to inform, to fix a meeting and to fix a group meeting. To give an example of our work, we present in this article only the implementation of one objective: a consumer wants to fix a group meeting with a provider and other agents (group G) in a given period of time (interval between two slots of time).

5.2 Content Language Interpretation

The first point of this implementation is the content language interpretation. To instantiate the classes representing the HiMAS principles, we determine the maximal set of values for each class according to the semantics relationships. For example, in order to fix a group meeting, we define the following constraints for the content language (see the doted set in Figure 2):

- The sensitive data that the consumer can collect is the free slots of time for a given period.
- The consumer can disclose this sensitive data to the group G and it must guarantee that the provider is able to access this data.
- If the sensitive data has been disclosed, all the possible references (urgency and importance) can be disclosed.
- The consumer cannot retain collected data after a given time.
- The possible uses of the collected sensitive data are storage, negotiation and sharing.

5.3 Agents Reasoning

The second important point of our implementation concerns the agents reasoning. For each calendar sharing, the consumer builds its policy by parsing the content language, according to the objective chosen by the user (in the example, to fix a group meeting). The parsing of the OWL file is based on the same technique than the XML parsing. Once the consumer finds its objectives, it creates its policy in a XML file by including every possible values for each class in its policy. Afterwards this agent validates its policy: it checks that all values are present in the content language and that the XML validates the XSD file to ensure the syntax. Now the consumer can create the sensitive data transaction file, including its policy, and validates this file in the same way that the policy.

However the calendar sharing interaction is started by the user and timetables that are managed by agents represent sensitive data of users, so we need to take the user intervention into account in the implementation. Indeed, we allow users to personalize the consumer's policy. When an user wants to access another calendar, he indicates to his agent his objectives and also each policy element thanks to a form. The agent checks the validity of this policy and rejects all policy that is not valid at a syntactic and/or a semantic level. If this policy is valid, the consumer creates the sensitive data transaction and sends it to the provider. In this way, a user can choose to send an automatic or a personalized policy to the provider.

To accept the sensitive data transaction, the first condition is that the provider trusts the consumer, else the transaction is canceled. The provider begins to check the validity of the sensitive data transaction and of the policy in order to verify the consumer's intentions. If a semantic or syntactic error occurs, the provider rejects the transaction. As for the policy, the preference may then be personalized by the user thanks to a form. So, after the content language validation, the provider verifies if the received policy agrees with the user's preference. If an agreement is found, the provider sends the required sensitive data to the consumer, else the provider proposes a new policy based on the user' wishes to the consumer.

When the consumer received a new version of the policy, it accepts this one if the user has given his agreement and changed in its reasoning the terms of its policy, else the consumer cancels the sensitive data transaction.

6 Conclusion

The extension of the agenda system [6] that we propose allows users to manage their sensitive data access according to the domain of the application. To introduce this privacy preservation, we have implemented a specific protocol [4] for all the sensitive data transactions in relation to the calendar sharing interactions. Our contribution allows a dynamic management of privacy thanks to the content language and allows also users to personalize this management. Users can delegate their private sphere to the agents that respect their preferences and the private sphere thanks to the semantic and syntactic validation.

In order to ensure a complete privacy preservation in the future, we will need to implement a secure media of communication to prevent attacks. Moreover we plan to use the trust model presented in [6] to pass a judgment on the agents in relation with the past sensitive data transactions in order to prevent users from malicious behaviors. For this purpose, we need to implement the compliance principle, in particular the detection of policy violation. With this detection and the trust model, we will be able to establish a social order [3] for privacy preservation.

Acknowledgments. This work is supported by Web Intelligence project, financed by the ISLE cluster of Rhône-Alpes region.

References

1. Agrawal, R., Kiernan, J., Srikant, R., Xu, Y.: Hippocratic databases. In: Proceedings of the International Conference Very Large Data Bases, pp. 143–154. Morgan Kaufmann, San Francisco (2002)
2. Bratman, M.E.: Intention, plans, and practical reason. O'Reilly, Harvard University Press, Cambridge (1987)
3. Castelfranchi, C.: Engineering social order. In: Omicini, A., Tolksdorf, R., Zambonelli, F. (eds.) ESAW 2000. LNCS, vol. 1972, pp. 1–18. Springer, Heidelberg (2000)
4. Crépin, L., Demazeau, Y., Boissier, O., Jaquenet, F.: Sensitive data transaction in hippocratic multi-agent systems. In: 9th International Workshop Engineering Societies in the Agents World (2008)
5. Crépin, L., Vercouter, L., Jaquenet, F., Demazeau, Y., Boissier, O.: Hippocratic multi-agent systems. In: Proceedings of the 10th International Conference of Entreprise Information Systems, pp. 301–308 (2008)
6. Demazeau, Y., Melaye, D., Verrons, M.-H.: A decentralized calendar system featuring sharing, trusting and negotiating. In: Ali, M., Dapoigny, R. (eds.) IEA/AIE 2006. LNCS, vol. 4031, pp. 731–740. Springer, Heidelberg (2006)
7. Deswarte, Y., Melchor, C.A.: Current and future privacy enhancing technologies for the internet. Annales des Télécommunications 61(3-4), 399–417 (2006)
8. Mathieu, P., Verrons, M.-H.: A general negotiation model using XML. Artificial Intelligence and Simulation of Behaviour Journal 1, 523–542 (2005)
9. Rezgui, A., Ouzzani, M., Bouguettaya, A., Medjahed, B.: Preserving privacy in web services. In: Chiang, R.H.L., Lim, E.-P. (eds.) Proceedings of the Workshop on Web Information and Data Management, pp. 56–62. ACM, New York (2002)
10. Sichman, J.S., Demazeau, Y.: Exploiting social reasoning to deal with agency level inconsistency. In: Proceedings of the First International Conference on Multiagent Systems, pp. 352–359. MIT Press, Cambridge (1995)
11. Sowa, J.F.: Conceptual Structures: Information Processing in Mind and Machine. Addison-Wesley, Reading (1984)
12. W3C. Plateform for privacy preferences (2002), http://www.w3.org/p3p/
13. W3C. Owl web ontology language (2004), http://www.w3.org/tr/owl-features/

Protected Computing Approach: Towards the Mutual Protection of Agent Computing[*]

Antonio Maña, Antonio Muñoz, and Daniel Serrano

University of Malaga
{amunoz,amg,serrano}@lcc.uma.es

Abstract. After a first phase of great activity in the field of multi-agent systems, researchers seemed to loose interest in the paradigm, mainly due to the lack of scenarios where the highly distributed nature of these systems could be appropriate. However, recent computing models such as ubiquitous computing and ambient intelligence have introduced the need for this type of highly distributed, autonomous and asynchronous computing mechanisms. The agent paradigm can play an important role and can suit the needs of many applications in these scenarios. In this paper we argue that the main obstacle for the practical application of multi-agent systems is the lack of appropriate security mechanisms. Moreover, we show that as a result of recent advances in security technologies, it is now possible to solve the most important security problems of agent-based systems.

1 Introduction

Mobile agents are software entities with the ability to migrate from node to node in a network acting autonomously and in cooperation with other agents in order to accomplish a variety of tasks. Currently there are different agent-based applications in numerous computer environments such as peer-to-peer networks, Web crawlers, and surveillance of local area networks, just to mention a few. Multi-agent systems (MAS) represent a promising architectural approach for building distributed Internet-based applications. Agent-systems, and in particular multi-agent systems, can bring important benefits especially in application scenarios where highly distributed, autonomous, intelligent, self organizing and robust systems are required. Furthermore, the high levels of autonomy and self-organization of agent systems provide excellent support for the development of systems in which dependability is essential. Ubiquitous Computing and Ambient Intelligence scenarios belong to this category. However, despite the attention given to the field by the research community, the agent technology has failed to gain a wide acceptance and has been applied only in a few specific real world scenarios. Security issues play an important role in the development of multi-agent

---

[*] Work partially supported by E.U. through projects SERENITY (IST-027587) and OKKAM (IST- 215032) and DESEOS project funded by the Regional Government of andalusia.

systems and are considered to be one of the main issues to solve before agent technology is ready to be widely used outside the research community. But for the provision of appropriate security in the context of multiagent systems it is not enough that the agent platform provides a set of standard security mechanisms such as sandboxing, encryption and digital signatures. It is necessary to design the whole infrastructure with security in mind. Furthermore, the security of an agent platform needs to be tailored to the specific characteristics of these systems. This paper describes some current technologies that can be used to build secure agent systems suitable for applications in ubiquitous computing and ambient intelligence scenarios. We must note that in these scenarios the computational infrastructure is composed of a very large number of computing devices with heterogeneous capabilities and under the control of different owners. This heterogeneity introduces the need for agents and agencies to learn about the capabilities and needs of each other. This problem was addressed in a previous paper [1] based on the use of agent profiles, where we also introduced a protection approach based on the Trusted Computing paradigm. We introduced the application of the Protected Computing paradigm to agents in [2]. In this paper we further develop this latter approach. In particular we describe in detail the development process of agent systems using the protected computing approach, including the use of different automated tools to support it. The organization of the paper is as follows. Section 2 describes the background on the security (or lack of) in current and past agent platforms. Section 3 describes how to apply our approach, called Protected Computing, focusing on the development cycle and the tools used. Section 4 summarizes conclusions and describes some ongoing work.

2 Background

We have mentioned that the focus of our paper is the description of suitable mechanisms for the security of agent systems in ubiquitous computing and ambient intelligence scenarios. The purpose of this section is to provide a view about the main agent-based systems and agent oriented tools, focusing on their security mechanisms. This review covers a wide range of applications from the first applications to the more recent ones. The objective of this analysis is to draw attention to the fact that these systems have traditionally neglected the need of a secure underlying infrastructure. The first MAS applications appeared in the middle 80s. These pioneer systems covered a wide variety of environments (manufacturing systems, process control, air traffic control, information management...), but almost all of them were built upon non secure infrastructures [3, 4, 5]. At that time, considering the foreseen scenarios and threats, agent technology developers assumed that the underlying infrastructure was secure, but now it is obvious that it is not. Some other agent-based applications lacking a security infrastructure were even proposed for nuclear plants [6] and aircraft control [7] applications. If we focus not on the applications, but on the infrastructures for agent-based systems, the situation is quite similar. Unfortunately, most of the platforms for agents, like Aglet [8], Cougaar [9], JACK [10], the popular JADE [11], JAVACT [12], and AgentSpeak(L) [13], share a negative common point, which is their insufficient security.

3 Mutual Protection for Secure Agent Computing

The fact that the current situation is far from satisfactory does not mean that the problems do not have a solution. In fact, the goal of this paper, and in particular of this section, is to present several viable technologies that can be used in multi-agent systems in order to provide a secure infrastructure. Protected Computing approaches are based on the remote execution of part of the code of an application (an agent, in our case). We present two schemes for providing security on multi-agent systems based on the Protected Computing paradigm. The Trusted Computing paradigm uses a specific hardware architecture containing trusted hardware elements. We show how this paradigm can enhance the trust on the configuration of the agencies. Finally, this section is closed by presenting the Proof-Carrying Code technology and showing how this technology contributes to the protection against malicious agents.

3.1 Protected Computing

The Protected Computing approach is based on the partitioning of the software elements into two or more parts. The basic idea is to divide the application code into two or more mutually dependent parts. Some of these parts (which we will call private parts) are executed in a secure processor, while others (public parts) are executed in any processor even if it is not trusted. A detailed description of this technology is presented in reference [2]. As mentioned, the approach uses a trusted processor to enforce the correct execution of the private parts of the program. Therefore, these parts must be carefully selected in order to obtain the best protection. In general, the Protected Computing model requires the use of secure coprocessors that have asymmetric cryptography capabilities; secure storage to contain a key pair generated inside the coprocessor and ensure that the private key never leaves it. Depending on the scenario some of these requirements may be relaxed. An important advantage of this scheme is the fact that different types of coprocessors can be used for different applications, as well as for different partitions of the same application. It is possible to apply the protected computing model in order to protect agent societies in a multi-agent setting, where several agents are sent to different (untrusted) agencies in order to perform some collaborative task. Because agents run in potentially malicious hosts, the goal in this scenario is to protect agents from the attacks of malicious hosts. The basic idea is to make agents collaborate, not only in the specific tasks they are designed to perform, but also in the protection of other agents. In this way each agent acts as secure coprocessor for other agents. Therefore, using the protected computing model, the code of each agent is divided into public and private parts. For the sake of simplicity, and without loss of generality, we will consider the simplest case where the code of each agent is divided in two parts: a public one and a protected one. From this description, it is easy to derive the possibilities that the division of the code into more parts opens. In particular the inclusion of multiple private parts, which could even be designed to be executed in different coprocessors, is especially relevant for the scenarios that we are considering. In the general case, the private part of each agent must be executed by some other agent running in a different host. This scheme is suitable for protecting a set of several mutually dependent agents. Consequently, in this general case, a conspiracy of all hosts is necessary in order to attack the system. Fig. 1 depicts the development cycle of the Protection of multiagent system following this approach (note that we have divided the complete development cycle into two parts, development and deployment, here we present the development and in the next sections we will review deployment). It begins with the usual development of the complete multiagent system code by the developer. Then, in order to split the code into parts the developer has to use the "Automatic Tool for Code Partitioning" (CPT). Since the code partition is a difficult task, and specialized expertise is required for performing it, this tool eases the code partitioning (creating the public and private parts) according to a set of rules that we call "protection profiles". The product of the operation of this tool is a set of public parts and a set of private parts. Then we obtain private parts and public parts. These parts will be used in a different way depending on the mutual protection scheme applied (static or dynamic). Fig. 1 depicts two

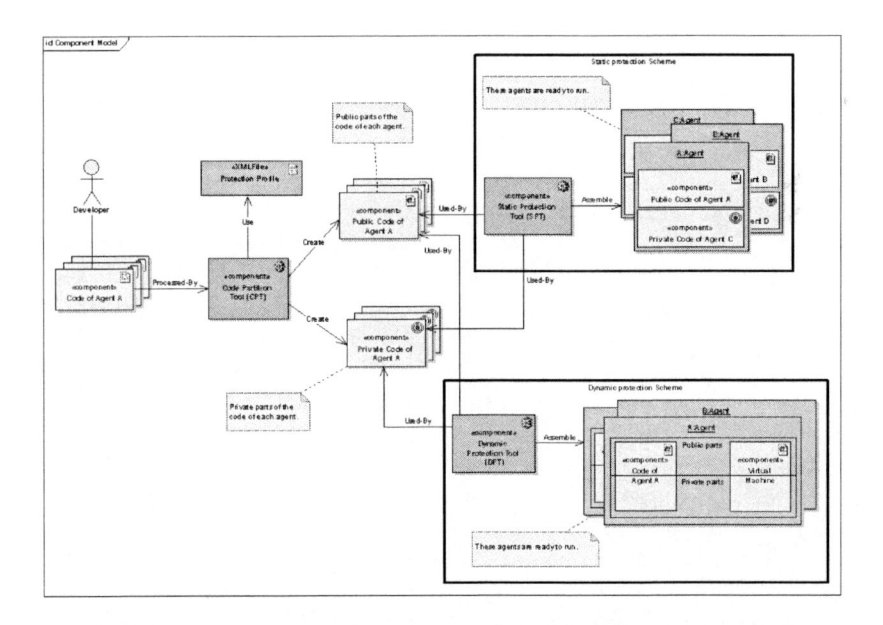

Fig. 1. Development time of a multi-agent system using a protected computing approach

different protection schemes like two different ways. A detailed description of these approaches can be found in the subsequent subsections.

A new element appeared in this scheme is the Protection Profile. A Protection Profile consists on a set of rules used to apply the Protected Computing approach. These rules describe the way in which the partitioning is carried out, it establishes the number of partitions, the kind of instructions to be protected, the sizes of these partitions, etc. Next table shows an example of how the partition method is defined by a set of rules in a XML file. These rules have parameters which can be set from CPT. These settings ease the protection of a Multi Agent System to developers. The process is easy to use since the developer is provided by a tool that allows by means of a graphic user interface set up the values to establish a concrete partitioning method.

```
<?xml version="1.0" encoding="iso-8859-1"?>
<agent xmlns:xsi="http://www.w3.org/2001/XMLSchema-instance"
xsi:noNamespaceSchemaLocation="D:\informacion\ProtectionProfile.xsd">
    <settings>
        <MaxNumberOfPartitions> Number </MaxNumberOfPartitions>
        <SetOfProtectedIns>
            <Ins> ControlFlow </Ins>
            <Ins> Load&Store </Ins>
            <Ins> MethodInvocations </Ins>
        </SetOfProtectedIns>
        <MaxPartitionSize> Size</MaxPartitionSize>
        <PriorityOrder>
            <Instructions>
                <Ins> Load&Store </Ins>
                <Ins> MethodInvocations </Ins>
                <Ins> ControlFlow </Ins>
            </Instructions>
            <Data>  key </Data>
</PriorityOrder>
    </settings>
</agent>
```

This code shows an example of a Protection Profile. This example of Protection Profile in a XML file shows how to set up is performed by the developer to protect a MAS. It provides of a higher degree of priority to instructions that data, at the same time that establish an order for instructions priority according to the necessity of the final system. In the case that developer needs a faster system he will indicate that protected part must be as small as possible achieving a lower security level. However in the case that the developer needs a high security level system, then he will set up Protection Profile tool to generate a system with a high computation activity in Trusted processors. Therefore the implementation of the protection can be done following to schemes. We can protect by means of data, it is marking the data to be protected. For this purpose the tool allocate the instructions to protect taking into account java labels such as final, static, etc. Also we can protect by means of instructions. In order to do that the tool has a classification of instructions and we mark the type of instructions to be protected (or a group of them). And finally a combination of both techniques can be implemented. Developer can set up the Protection Profile, adjusting it to get the more suitable configuration for his requirements. An example of this combination is shown in the protection profile code shown. Another important approach is marking the code to be protected while developing, instead of perform the configuration by means of a Protection Profile,. More concrete is the selection of parts of code to protect. Then Java compiler marks these byte codes, by developer command, adding some annotations which will be protected in next steps. This approach can be used both to protect instructions, data and a combination of both of them. Data Protection must be commanded by developer according to the code and context restrictions. However Instructions Protection is suitable to be automates. For this purpose we grouped the byte code instruction set currently consists of 212 instructions. The instruction set can be roughly grouped as follows:

- Stack operations: Constants can be pushed onto the stack either by loading them from the constant pool with the ldc instruction or with special "shortcut" instructions where the operand is encoded into the instructions, e.g. iconst 0 or bipush (push byte value).
- Arithmetic operations: The instruction set of the Java Virtual Machine distinguishes its operand types using different instructions to operate on values of specific type. Arithmetic operations starting with i, for example, denote an integer operation. E.g., iadd that adds two 5 integers and pushes the result back on the stack. The Java types boolean, byte, short, and char are handled as integers by the JVM.
- Control flow: There are branch instructions like goto and if icmpeq, which compares two integers for equality. There is also a jsr (jump sub-routine) and ret pair of instructions that is used to implement the finally clause of try-catch blocks. Exceptions may be thrown with the athrow instruction. Branch targets are coded as offsets from the current byte code position, i.e. with an integer number.
- Load and store operations for local variables like iload and istore. There are also array operations like iastore which stores an integer value into an array.
- Field access: The value of an instance field may be retrieved with getfield and written with putfield. For static fields, there are getstatic and putstatic counterparts.
- Method invocation: Methods may either be called via static references with invokestatic or be bound virtually with the invokevirtual instruction. Super class methods and private methods are invoked with invokespecial.
- Object allocation: Class instances are allocated with the new instruction, arrays of basic type like int[] with newarray, arrays of references like String[][] with anewarray or multianewarray.
- Conversion and type checking: For stack operands of basic type there exist casting operations like f2i which converts a float value into an integer. The validity of a type cast may be checked with checkcast and the instanceof operator can be directly mapped to the equally named instruction.

This general scheme can be implemented in different ways. However, we can distinguish two different strategies. In the first one, the collaboration between agents is predefined. This means that every agent has the private parts of the code of one or more of the other agents that are collaborating. We call this strategy Static Mutual Protection. On the other hand the Dynamic Mutual Protection strategy makes it possible for any of the collaborating agents to serve as a secure coprocessor to any other agent. Therefore, in this case, the interactions between the agents are not predefined. This latter strategy is more powerful and flexible, but it also entails more complexity and reduced performance.

Static Mutual Protection

The Protected Computing scheme can be applied in order to protect a society of collaborating agents by making every agent collaborate with one or more remote agents running in different hosts. These agents act as secure coprocessors for the first one. Likewise, these agents are in turn protected by other agents as shown in 2. In this setting, an attack requires the cooperation of all agencies.

For this specific strategy, it is possible that the protected parts of an agent are directly included in the other agents as shown in 2. This strategy increases the performance by avoiding the transmission of the protected code sections over the network. In contrast, it is only suitable in those scenarios where the set of agents to be protected is static and can be determined before their actual execution. Fig. 1 depicted a tool used to generates new agents with the structured based on Static Protection technology named SPT (Static Protection Tool). This tool receives as input two different set of agent codes. First set consists on public code of each agent generated by CPT. Similarly the second set contains the private code of each agent. SPT will chain these agents according to the philosophy of this strategy. An example of the possible applications of this scheme is that of a competitive bidding. In this scenario a client requests bids from several contractors to provide a good or a service. It is important that the bidding takes place simultaneously, so that none of the contractors can access the offer from the other contractors, because this would give it advantage over the others. The client can use several single-hop agents to collect the offers from the contractors. Each agent will be protected, using the Static Protection strategy, by other agents. This is possible since the client generates the set of agents, which is static and

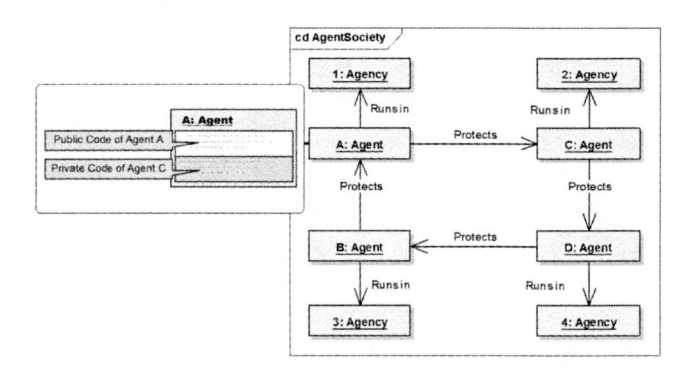

Fig. 2. Mutual Static Protection between four agents

known a priori. We can also safely assume that a coalition of all contractors is will not happen. In fact, no technological solution can prevent all contractors to reach an external agreement. Because each agent is protected by other agents running in the hosts of the competitors, and because the protected computing model ensures that it is neither possible to discover nor to alter the function that the agents perform and it is also impossible to impersonate the agents, we know that all agents will be able to safely collect the bids, guaranteeing the fairness of the process.

Dynamic Mutual Protection

The Static Mutual Protection strategy can be successfully applied to many different scenarios. However, there will be scenarios where will not possible to foresee the possible interactions between the agents, where the agents will be generated by different parts, or that will involve very dynamic multi-hop agents. In these cases the Static Mutual Protection strategy will be difficult or impossible to apply. Therefore we propose a new strategy named Dynamic Protection where each agent will be able to execute arbitrary code sections on behalf of other agents in the society. This strategy is currently under development. As shown in 3, each agent will include a public part, an encrypted private part and a specific virtual machine similar to the one described in [14]. This virtual machine will allow agents to execute on-the-fly the code sections (corresponding to the private parts) will be received from other agents. The Dynamic Protection Strategy process will be as illustrated in next figure. In the first exchange ag1 will act as the protected agent while ag2 will act as protecting agent (secure coprocessor) for the first one. In the exchange, ag1 will send a private code section to the virtual machine of ag2. This virtual machine will process the private section and then will return some results (results1). Subsequent will exchange illustrate ag3 acting as protecting agent for ag2 (in this case protected agent), and finally ag1 protecting ag3. The scalability of this scheme will be very good since only a few agents (one in most cases) will be involved in the protection of any other agent.

Additionally, agents will have to be able to recognise other trusted agents and to decrypt the private code sections of other agents that will be sent to them for execution previously. For these reasons a mutual authentication mechanism will have to be in place. In the simplest case, we can make all agents in the society to share a symmetric key. In this case all private sections will be encrypted using this key. In a more complex setting, each agent will have an asymmetric key pair, a certificate of the public key issued by an appropriate authority and the public key of all trusted authorities. In this case, agent will mutually authenticate themselves by way of digital signatures and will be able to receive specific licenses for the execution of the protected sections of other agents. In the simplest case the license is the symmetric key used to encrypt the private part of the protected agent encrypted with the public part of the protecting agent. The Fig. 1 depicted how the deployment of the agents following this strategy will be. It is important to remark that this deployment could be done by an automatic way. As in the previous scenario there will be a Dynamic Protection Tool (DPT). This tool will

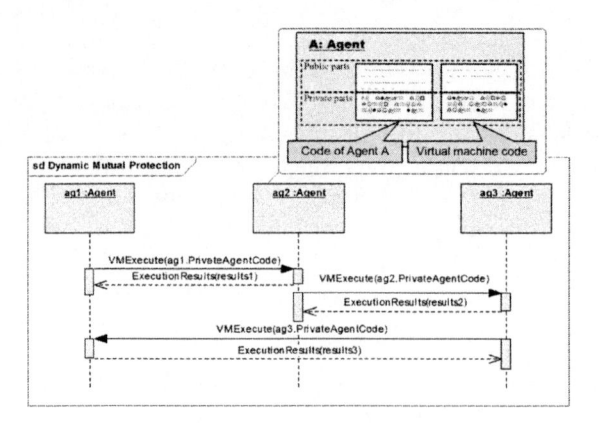

Fig. 3. Sequence diagram showing a Mutual Dynamic Protection between three agents

be able to allocate into every agent a little virtual machine code. This virtual machine will be able to execute public and private code from other agents on the fly. Doing this, there will be not necessity of fixed assignations between agents, because every agent will be a potential secure processor for the rest of the agents in the system.

4 Conclusions

We have shown in this paper the potential for agent systems in ubiquitous computing and ambient intelligence scenarios. We have described some current technologies that can be used to build secure agent systems suitable for applications in these scenarios, where the computational infrastructure is composed of a very large number of computing devices with heterogeneous capabilities and under the control of different owners. In this paper we have extended and detailed our preliminary work on the application of the Protected Computing paradigm to agent systems. Finally, we have described in detail the development process of agent systems using the protected computing approach, including the use of different automated tools to support it. Our ongoing work is focused on two ways. On one hand, we are working on the development of the automatic tools that support the development process. Actually, we count on prototypes of the tools that process the agents. This tools work with a predefined policy. We are working on the flexibility and adaptability of those tools to different parameters in the policies. This first line also includes the description of such policies. On the other hand, we are working on the dynamic mutual protection approach. The main task here is to identify problems raised when we put together concepts like dynamism, virtual machine and agent private parts. This is because possible deadlocks or other problems could rise from this paradigm. After this task we are going to work in our first prototypes of the Dynamic Protection Tool (DPT).

References

1. Lopez, J., Maña, A., Muñoz, A.: A Secure and Auto-configurable Environment for Mobile Agents in Ubiquitous Computing Scenarios. In: Ma, J., Jin, H., Yang, L.T., Tsai, J.J.-P. (eds.) UIC 2006. LNCS, vol. 4159, pp. 977–987. Springer, Heidelberg (2006)
2. Maña, A., Muñoz, A.: Mutual Protection for Multiagent Systems. In: Proceedings of the Third International 3rd International Workshop on Safety and Security in Multiagent Systems (SASEMAS 2006) in Autonomous Agents and Multi-agents Systems (AAMAS 2006) (2006)
3. Jennings, N., Sycara, K., Wooldridge, M.: A Roadmap for Agent Research and Development. Autonomous Agents and Multiagent Systems 1(1) (1998) (forthcoming)
4. Chaib-draa, B.: Industrial Applications of Distributed AI. Communications of the ACM 38(11), 49–53 (1995)
5. Jennings, N.R., Corera, J.M., Laresgoiti, I.: Developing Industrial Multiagent Systems. In: Proceedings of the First International Conference on Multiagent Systems, pp. 423–430. AAAI Press, Menlo Park (1995)
6. Wang, H., Wang, C.: Intelligent Agents in the Nuclear Industry. IEEE Computer 30(11), 28–34 (1997)
7. Schwuttke, U.M., Quan, A.G.: Enhancing Performance of Cooperating Agents in Real-Time Diagnostic Systems. In: Proceedings of the Thirteenth International Joint Conference on Artificial Intelligence (IJCAI 1993), pp. 332–337. International Joint Conferences on Artificial Intelligence, Menlo Park (1993)
8. Clements, P., Papaioannou, T., Edwards, J.: Aglets: Enabling the Virtual Enterprise. In: The Proc. of Managing Enterprises - Stakeholders, Engineering, Logistics and Achievement (ME-SELA 1997), p. 425 (1997) ISBN 1 86058 066 1
9. http://cougaar.org/
10. Shepherdson, D.: The JACK Usage Report. In: The Proc of. Autonomous Agents and Multi Agents Systems (AAMAS 2003) (2003)
11. http://jade.tilab.com/
12. http://www.irit.fr/recherches/ISPR/IAM/JavAct.html
13. Alechina, N., Alechina, R., Hübner, J., Jago, M., Logan, B.: Belief revision for AgentSpeak agents. In: The Proc. of Autonomous Agents and Multi Agents Systems, Hakodate, Japan, pp. 1288–1290 (2006) ISBN:1-59593-303-4
14. Maña, A., López, J., Ortega, J., Pimentel, E., Troya, J.M.: A Framework for Secure Execution of Software. International Journal of Information Security 3(2) (2004)
15. Trusted Computing Group: TCG Specifications. Workshop on Safety and Security in Multiagent Systems (SASEMAS 2006) (2005),
 https://www.trustedcomputinggroup.org/specs/
16. Necula, G.: Proof-Carrying Code. In: Proceedings of 24th Annual Symposium on Principles of Programming Languages (1997)

Toward a Conceptual Framework for Multi-points of View Analysis in Complex System Modeling: OREA Model

Mahamadou Belem[1] and Jean-Pierre Müller[2]

[1] CIRAD-UPR GREEN TA C-47 Campus International de Baillarguet
 34398 Montpellier cedex 5-France
 `mahamadou.belem@mpl.ird.fr`
[2] CIRAD-UPR GREEN TA C-47 Campus International de Baillarguet
 34398 Montpellier cedex 5-France
 `jean-pierre.muller@cirad.fr`

Abstract. Complex system design requires a multi-disciplinarily approach deploying a multiplicity of point of views. The OREA (Organization-Role-Entity-Aspect) model presented in this paper proposes a conceptual framework allowing to represent a complex system in different points of view and to integrate them in order to form a coherent system. Using an Agent-Group-Role (AGR) extension, OREA allows not only to represent a system from various points of view through the roles played by entities within organizations, but also to allow various entities to play a same role in various ways through the notion of aspects. Unlike most roles based models, OREA provides a clear distinction between (1) the external properties of an entity described by the roles and (2) the internal properties described by the aspects. The OREA model increases organization reuse and the adaptability of a system. OREA model have been implemented in Mimosa and used to implement a simulator for carbon dynamics analysis at village territory level.

Keywords: Complex system, Design, Multi-point of views, Multi-agents System, Organization, AGR, OREA, MIMOSA.

1 Introduction

Today, there is an increase in complex system modelling by the researchers of different disciplines having to collaborate for a better comprehension of the system under study. Müller [1] defines a complex system as (1) a set of entities with non-linear behaviour, interacting with each other and evolving at many scales of time and space and (2) that the behaviour cannot be resumed to the summation of the components behaviour. The analysis of such systems requires taking into account several factors. These factors can concern one or several disciplines. The different disciplines have not necessarily the same points of view on the objects of the system. However these points of view may be complementary and require to be taken into account to deal efficiently with the system complexity. Then, considering the properties of a complex system, its description requires (1) multi-scales description including both (a) individual and global levels and (b) the explicit representation of environment and (2) multiplicity of points of view at individual, global and environmental levels.

However, at the individual level, two kinds of points of view can be observed and must be explicitly represented: (1) the external points of view which describe the relations of the component with its world and (2) the internal points of view which describe the intrinsic properties of the component and how it behaves and change its properties.

The objective of this study is to propose a conceptual framework allowing multi-points of view description of complex systems. This framework (1) intends to the knowledge representation of the design of complex systems and (2) is a tool for the dialogue between scientists from different disciplines for a multidisciplinary analysis of a complex system. This conceptual framework must:

- Allow a multipoint of view analysis and description of a complex system both at individual and global levels while taking into account the integration of environment;
- Allow an explicit representation and distinction of the external and internal points of view of the components;

Using multi-agent system (MAS) organizations-centred approach, we proposed Organization-Role-Entity-Aspect (OREA) model. OREA is an extension of AGR [2]. OREA allows representing of a complex system in different points of view from the **roles** the entities play in **organization** and allows different types of **entities** to play a same role in different ways through the concept of **aspect**.

This paper is organized as following. In the second section, we present the context of our study. In the thirst section, we present a review of roles based models and to underline the contributions of our model. The OREA meta-model and dynamics are described in fourth section.

2 Context of the Study

Multi-agents system (MAS) provide sustainable framework to handle simultaneously the individual and the global levels of a complex system. MAS treat a system as a set of agents interacting at different scales of time and space, evolving in an environment that they can perceive and modify through their actions. Two approaches are currently used for building complex systems with MAS: the Agent-Centred Multi-Agent System (ACMAS) [3] and the Organization-Centred Multi-Agent System (OCMAS) [2].

The ACMAS approach is based on agent-oriented point of view. *"In that view, the designer of a multi-agent system is only concerned with agents' individual actions, and it is supposed that social structures come from patterns of actions that arise as a result of interactions"* [2]. The structure of agent-oriented models considers agents as the first class entity where each agent is defined by its internal state and set of operations defining its behaviour. Providing one level of description (individual level), the ACMAS approach does not enable an explicit multi-point of view description of a system.

As to OCMAS models, they assume that the social structure must exist a priori and constraint the agents behaviour. Unlike ACMAS models, OCMAS models treat MAS as organizations interacting through agents playing roles. However, OCMAS introduce the notion of organization, role and agent to represent the structure of a system.

An organization defines a collection of roles and their relationship. A role is an agent function; it is closed to an organization and can be played by several agents. Using organization, it is possible to describe explicitly the organizational relationship and reduce the system complexity. It can be used to limit the scope of interactions, provide strength in number, reduce or manage uncertain ... [4]. Several OCMAS models have been proposed. They can be classified according the problem they aim to resolve and their structure. According to their objective, we have tree categories of models intending (1) to the decomposition of a system into sub-systems [5; 6], (2) to the co-ordination and the collective tasks execution [7; 8] and (3) to the design of opened system [9; 10]. According to their structure, we have tree categories of models: (1) roles based models intending to propose a social level and to take into account heterogeneity and modularity in the MAS [2; 5], (2) componental based models very closed to software engineering intending to increase the reuse and modularity in MAS specification and (3) the models coupling the two first approaches [11].

The model proposed in this paper aims to the decomposition of a complex system into sub-systems and uses a structure based on roles based approach. But the model is focussed on knowledge representation and analysis of complex system and not only to MAS implementation. Then, in OREA model, we use the notion of organization to define a point of view at global level, the notion of role to define the external point of view on a component, the notion of entity to define the objects of the system and the notion of aspect to describe the internal point of view of the entities and how they behave.

3 Review of Roles Based Models

In this section we describe some existing roles based models and underline the contributions of our model.

Gaia Model

Gaia [5] intends to allow an analyst to go systematically from statement of requirements to a design that is sufficiently detailed that it can be implemented directly. Gaia uses two categories of concept: abstract entities used in analysis stage and concrete entities used in design stage. The Gaia organization structure takes into account both the macro level and micro level (agent). But the Gaia organization structure is static. Agent cannot enter and leave dynamically the groups. Also, organization is not represented explicitly. The model concerns the system where agents pursue one goal. Unlike Gaia, OREA concerns dynamic structure, the organization representation is explicit and takes into account a multi-organizational structure.

AGR Model

The Agent-Group-Role (AGR) model [2] is a generic meta-model using the concepts of agents, groups and roles to describe the structure of a MAS. A group is a set of agents interacting through their roles. A role defines agent behaviour; it is closed to a group and can be played by several agents. Only agents sharing same groups can communicate. Unlike Gaia, AGR supports simultaneously several groups and the agents can enter and leave dynamically the roles.

As limitations, the AGR model has a very limited expressiveness and offers few concepts for the design of complex systems. The roles in AGR are represented by labels and a group is just a collection of agents. In addition, the structure of the role is strongly linked to the structure of agent decreasing the organization modularity and reuse. The structure of an agent in AGR describes both the organizational properties and the individual properties. In OREA, the notion of group and role are represented as first class entities and OREA makes a clear distinction between the external and internal properties of an entity. The external properties are defined in the structure of role and the internal properties in the architecture of agent through the notion of **aspect.**

MOCA Model

MOCA [11] deals with how agents can play multiple roles while guarantying system and agents coherence. MOCA defines group, role as first class entities. The structure of an agent in MOCA is a set of components that can be added or removed dynamically according to the contexts. MACO model increases the reusability and modularity of organization and allows implementation of dynamic and evolving systems

MOCA sets strict constraints on roles implementation. Role defines an effective behaviour of an agent and its dynamics is specified by a statechart. A role in MOCA defines how an agent behaves. MOCA does not respect the first principle of organization centered general principles [12]. In addition, in MOCA, a role has two properties: (1) external properties allowing an agent to be in relation with others agents and (2) specifies in part the internal behaviour of the agent. In consequence, an agent structure is linked to the role implementation in MOCA. This approach decreases role reusability. In addition, it is impossible for two types of agents to play the same role in different ways. For example, in the dynamics of carbon resources at village territory level, soil, plant and the atmosphere play the role of storage. But they do not store the same elements and do not have the same dynamic. Thus, the function of storage is not necessary the same for the plant, soil or atmosphere. A role in MOCA does not take into account this specification.

In OREA model, we assume that a role can be played in different ways by different types of entities. A role describes only the organization properties and the internal properties are described in the internal structure of the agents. Unlike MOCA, a role does not describe how it is played. We use the notion of **Aspect** to describe the internal properties of an agent and define how roles are played. With OREA, it is possible to define several behaviours for the same type of role by implementing several aspects. As MOCA, we use the concept of competence to define the behaviour of the agent and roles. But a role in OREA does not use the competences of the agents. Then, in OREA, the verification of roles acquisition does not depend on agents competences. It depends only on the global structure of the system: types of agent, types of role, cardinalities, etc.

4 OREA Model

4.1 Meta-model of OREA

The OREA model provides two levels of analysis. The first level concerns the **abstract level** characterized by the concepts of organization, role type, entity type and aspect type, competence type, influence type. The second level concerns the **concrete level**. The concrete level is an instance of the abstract level and characterized by the concepts of group, role, entity, aspect, competence and influence concepts used to create a particular model.

4.1.1 Abstract Level

The abstract level provides concepts for the description of the structure of a system. It is composed of:

Organization: an organization provides an abstract description of a group. It provides the description of the structure of groups. It is defined by a set of role types, their cardinalities and their relationships. Organization defines some constraints to guarantee the group coherence. The constraints define the entities acceptance conditions in a group, a group status (minimum and maximum member), etc.

Role Type: a role type is an abstraction of an entity function, an external point of view on a type of entity. A role type is characterized by its cardinalities and the type of provided competences. A role type describes only the extrinsic properties of the entity types from the point of view of an organization. In addition, a role type defines the conditions an entity must obtain before playing its instances.

Entity type: it provides an abstract description of a category of entities which have the same properties. An entity type describes a type of passive or active object. An entity type defines the architecture of the entities instancing it. Thus, the structure of an entity is characterized by (1) the types of aspects which describe the intrinsic properties, (2) the types of roles which describe its extrinsic properties and (3) the types of relationship between the types of aspects and types of roles.

Aspect Type: an aspect type provides an abstract description of aspects. It is the description of a behaviour which can possibly be played by an entity type. The relationship between an entity type and an aspect type is one-one relationship. An aspect type can be linked to several role types. Then an instance of an aspect type can control several roles.

4.1.2 Concrete Level

The concrete level is an instance of the abstract level. It defines **how** a system behaves to fulfil its goal. The components of the concrete level are:

Group: a group is a set of roles interacting among them. Concretely a group is an instance of an organization. It is represented by an object class. Then, a group can be created and removed dynamically.

Role: A role is an instance of a role type. A role defines the external behaviour of an entity and the membership of an entity to a group. It allows an entity to interact with others entities by sending and receiving influences. A role is defined by a set of competences it can provide. A role in OREA does not use the competences of entities. Also, a role in OREA doesn't implement how it is played. A role provides two interfaces of communication: (1) internal interface allowing to role to interact with its player and (2) the external interface allowing interactions with others roles in the same group.

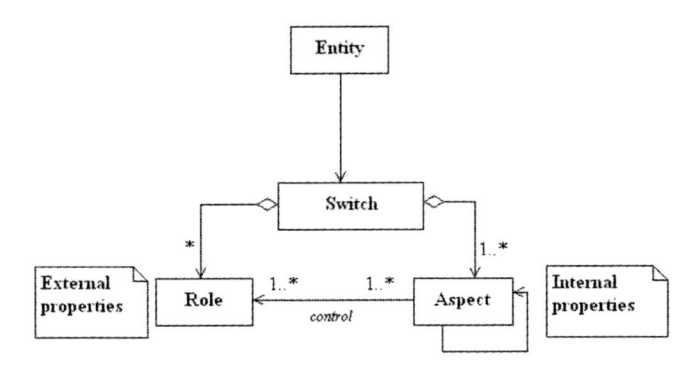

Fig. 1. Entity internal structure

Entity: An entity is an active or a passive object playing roles in the system through its aspects. An entity can play several roles in several groups simultaneously. As noticed previously, roles do not use the competences of entities. In this case, entity may have its own behaviour with allows it to carry out its competences according the internal and external contexts. Then, the dynamic of agent has been divided into sub-dynamics (**Aspect**) to specify the internal dynamics according the context. Each subdivision is represented by an aspect. Aspect describes separately entities mental issues and how entities behave according the internal and external contexts. If an entity handles a role, it must have aspects controlling this role. The internal structure of an entity is represented trough a compound object: *Switch* (Fig. 1) managing (adding and removing) a set of components (roles, aspects, competences). The components are the aspects, the competences and the roles played by the entity.

Aspect: In OREA a role specifies *what* entity has to do in an organization and an aspect specifies *how* and in what context an entity behaves and achieves its roles. The behaviour of an aspect is based on the competences of the entity with handles it and those one provided by the controlled roles. To handle an aspect, an entity must provide the competences required by the aspect. If an entity leaves all roles controlled by an aspect, it leaves also this aspect. Two aspects of a same entity do not describe the same internal properties but they can control a same role. The aspects of two types of entity can implement differently how a same role type will be played. This approach allows to different types of entities to play the same role type in different way.

In addition to these concepts, OREA use the notion of influence [13] competence [11] are used (Fig. 2).

4.2 Dynamics with OREA

4.2.1 The Organizational Dynamics

As noticed previously, a group structure is described by an organization. In OREA model, the organizations manage group creation and deletion. A group manages the role handling by the entities. Thus, the roles are created and assigned to entities by the groups.

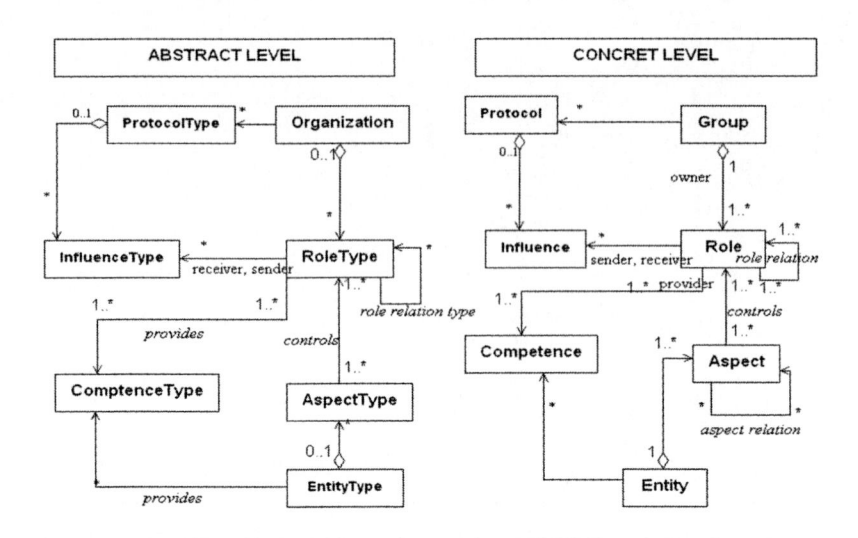

Fig. 2. OREA Meta-model

4.2.2 The Dynamics of Entities

Two entities cannot interact directly between them. They interact only through their roles. All influences are received and sent by the roles. When a role receives an external influence, this influence is sent to its player. When entity receives an influence, a specific aspect is selected and executed by the *Switch* by using the description of the entities for that. The description of an entity (Entity type) provides the description of aspects handled by an entity. Each type of aspect describes its interface (*AspectInterface*) defined as set of influence types:

$$AspectInterface = < InfluenceType^* >$$

The aspects can interact among them to share information or to perform some tasks. To interact among them, the aspects can send internal influences or call a method of others aspects. The execution of an aspect depends on the roles played by the entity. An aspect can be executed if only if the entity plays all roles required by the aspects.

4.2.3 Protocol of Communication

In OREA, the protocols of communication are represented as reuse components. Each type of role describes locally a part of a protocol. The set of protocols used by a role defines its external used for the verification of interactions. The interfaces of roles are defined as following:

$$ExternalInterface = < ProtocolDescription^* >$$
$$ProtocolDescription = < name, InfluenceType^* >$$

5 Conclusion and Future Work

This paper presented OREA model. The aim of the study concerns the proposition of a conceptual framework based on OCMAS approach for complex system analysis. OREA is an extension of AGR integrating the notion of Aspect. OREA model allows to define separately and explicitly the internal and external properties of entities and permit various types of entities to play differently the same role. While roles describe the external properties, the aspects describe the internal properties of entities. Roles in OREA do not define how entities behave in organization. But they allow entities to interact and provide competences for the entities behaviour. The behaviour of an entity is based on the dynamics of a set of aspects. An aspect defines a part of entity behaviour and uses the roles futures. Aspects describe entities mental issues and define how a role is played.

The OREA model has been used in the design and implementation of a model dedicated to the dynamic resource-scale carbon land West African. For future works, it will define a set of graphics formalisms to help the designer in design stage.

References

[1] Müller, J.P.: Emergence of collective behaviour and problem solving. In: Omicini, A., Petta, P., Pitt, J. (eds.) Engineering Societies in the Agents World: 4th International Workshops, pp. 1–21. Springer, London (2004)
[2] Ferber, J., Gutknecht, O.: A meta-model for the analysis and design of organisation in multi-agents. In: Proceedings of Third international Conference on Multi-Agent Systems, pp. 128–135. I.C.S. Press Ed., Paris (1998)
[3] Drogoul, A., Ferber, J., Cambier, C.: Multi-agent Simulation as a Tool for Analysing Emergent Processes in Societies. In: Simulating Societies Symposium, University of Surrey, Guildford (1992)
[4] Horling, B., Lesser, V.: A Survey of Multi-Agent Organizational Paradigms. The Knowledge Engineering Review 19, 281–316 (2005)
[5] Wooldridge, M., Jennings, N., Kinny, D.: The Gaia Methodology for Agent-Oriented Analysis and Design. In: Autonomous Agents and Multi-Agent Systems, vol. 3, pp. 285–312 (2000)
[6] Ricordel, P.-M.: Programmation Orientée Multi-Agents: Développement et Déploiement de Systèmes Multi-Agents Voyelles. In: Mathématiques, Sciences et technologies de l'information,Informatique INPG (2001)
[7] Barbuceanu, M., Gray, T., Mankovski, S.: Coordinating with obligations. In: Proceedings of the Second International Conference on Autonomous Agents (AGENTS 1999), pp. 62–69. ACM Press, New York (1998)
[8] Hannoun, M., Boissau, O., Sichman, J.S., Carron, T., Sayettat, C.: Conception et mise en oeuvre du modèle organisationnel MOISE. In: Actes des 8èmes Journées Francophones Intelligence Artificielle Distribuée & Systèmes Multi-Agents (JFIADSMA 2000), Saint-Jean-la-Vêtre, France (2000)
[9] Esteva, M., Padget, J., Sierra, C.: Formalizing a language for institutions and norms. In: Meyer, J.-J.C., Tambe, M. (eds.) ATAL 2001. LNCS, vol. 2333, pp. 348–366. Springer, Heidelberg (2002)

[10] Vazquez-Salceda, J., Padget, J.A., Lopez-Navidadc, A., Caballeroc, F.: Formalizing an electronic institution for the distribution of human tissues. Artificial Intelligence in Medicine 27, 233–258 (2003)

[11] Amiguet, M., Müller, J.P., Baez, J., Nagy, A.: La plate-forme MOCA: conception de SMA organisationnels à structure dynamique. In: JFIADSMA Paris, pp. 151–164. Hermès, Lille, France (2002)

[12] Bàez, J., Stratulat, T., Ferber, J.: Un modèle institutionnel pour SMA organisationel. Journée Francopohone des Systèmes Multi-Agents (2005)

[13] Ferber, J., Müller, J.-P.: Influences and Reaction: a Model of Situated Multiagent Systems. In: Proceedings of ICMAS 1996, pp. 72–79. AAAI Press, Kyoto (1996)

Using Hitchhiker Mobile Agents for Environment Monitoring

Oscar Urra[1], Sergio Ilarri[1], Eduardo Mena[1], and Thierry Delot[2]

[1] Department of Computer Science and Systems Engineering, University of Zaragoza, María de Luna 1, 50018, Zaragoza, Spain
ourra@ita.es, {silarri,emena}@unizar.es
[2] Université de Valenciennes Bat. ISTV2, Le Mont Houy 59313 Valenciennes Cedex 9, France
Thierry.Delot@univ-valenciennes.fr

Abstract. Recently, there has been a great interest in the development of protocols and data management techniques for vehicular networks (VANETs). In a VANET, the vehicles form a wireless ad hoc network where different types of useful data can be exchanged by using the dynamic links that a vehicle can establish with its neighboring vehicles. While this offers opportunities to develop useful applications, many research challenges arise from the point of view of data management.

In this paper, we propose the use of cars equipped with sensors in a VANET for environment monitoring. Our approach is based on mobile agents, which jump from car to car as necessary to reach the area of interest and keep themselves in that area. Thus, relying on an expensive fixed infrastructure of sensors is avoided. Instead, any area can be monitored with low cost as long as there are enough vehicles traversing it. We present experiments that compare different traveling strategies for the agents.

1 Introduction

Vehicular ad hoc networks (VANETs) are attracting a great interest, both in research and in industry. One of the most interesting features is the possibility to use a spontaneous and inexpensive wireless ad hoc network between the vehicles to exchange interesting information (e.g., to warn the driver of an accident or a danger).

On the other hand, the relevance of environmental issues has grown considerably, and there are many areas of study on this subject. In many of them, it is important to have environmental data collected in the field, such as CO_2 or other gas concentration levels, the presence of harmful substances, or meteorological parameters such as the temperature, the humidity, and many others. The usual way of collecting these measures may be problematic. Thus, fixed measurement instruments can be expensive to maintain and they require an infrastructure to operate them, a protected location, power, and communication lines. An alternative could be the use of mobile equipment operated by a person who travels in the area of interest while sampling the required environment parameters, which is also a slow and expensive process. To avoid these drawbacks, we can benefit from regular vehicles traveling along the roads within the geographical area of interest, as long as those vehicles are equipped with the appropriate measurement device; to encourage participation in the monitoring among sensor-enabled vehicles, different techniques can be applied (e.g., based on the concept of *virtual currency*, as in [1]).

In this paper, we advocate the use of mobile agent technology [8, 2] (programs that can move between computers) as the ideal candidate to implement such a system in an efficient and flexible manner. In our proposal, mobile agents jump from vehicle to vehicle as necessary to reach the area of interest and keep themselves within that area. As each vehicle follows its own route, which may be different from the optimal route or even be unsuitable for the monitoring task, the mobile agents may need to change to a different vehicle frequently. Thus, we can compare a monitoring mobile agent in our proposal with a hitchhiker, who may use several vehicles to reach the intended destination. The main difference is that a monitoring mobile agent cannot live outside the execution environment provided by the cars (i.e., outside a mobile agent platform, as explained in Section 3); therefore, once the agent arrives in the area to be monitored, it must jump from car to car to keep itself within such area. With a mobile agent-based strategy, the required environment data can be collected quickly on a wide area (as long as there are enough vehicles). Moreover, the cost of a support infrastructure is avoided, as the idle resources of regular vehicles are used instead.

As far as we know, no other work proposes taking advantage of mobile agents' features to perform monitoring tasks in a vehicular network. Indeed, [3] is the only work that uses this technology in a vehicular field; however, its goal is different (traffic control and management) and it does not face the research issues appearing in our context (agents that must move from car to car to perform the monitoring and transferring data without the need of a dedicated network). Other works that focus on monitoring using vehicles are MobEyes [4] and CarTel [5]. In MobEyes it is not possible to define specific monitoring tasks; instead, the vehicles diffuse data summaries, which are collected by nearby vehicles such as police patrols. CarTel assumes the existence of open Wi-Fi access points to send the sensor readings directly to a central server. Neither of these works benefit from mobile agents to perform a flexible and inexpensive monitoring.

The rest of this paper is structured as follows. In Section 2, we describe how a VANET can be used for environment monitoring. Based on that general proposal, in Section 3 we describe our monitoring approach based on the use of mobile agents. In Section 4, we present some tests that compare different *hitchhiking strategies* for the agents. Finally, in Section 5 we summarize our conclusions and present some lines of future work.

2 Using VANETs for Monitoring

A *Vehicular Ad hoc Network* (VANET) is a mobile, ad hoc, communication network which is dynamically established between vehicles traveling along roads in a geographical area. The vehicles use only short-range networks (100-200 meters), like IEEE 802.11 or based on Ultra Wide Band (UWB) standards, in order to establish temporary communication links to exchange information between vehicles in a mobile P2P fashion [6]. Hence, it is possible that there exists no direct connection between two vehicles in the network, in which case the use of some multi-hop communication protocol [7] is necessary. These protocols are usually complex and it is difficult to limit the maximum time needed to deliver a message to a recipient, due to the fact that the existing links change constantly. However, using short-range networks has three important advantages: 1) there is no need of a dedicated support infrastructure (expensive to deploy and maintain), 2) the users do not need to pay for the use of these networks, and 3) it allows a very quick exchange of information between two vehicles that are within range of each other. Moreover, many application scenarios do not need to communicate with a specific target vehicle but with all the vehicles within a certain area. Although we do not rely on a fixed network infrastructure, we can benefit from the existence of some *relaying devices* on the roads: static devices, deployed along the roadside, which provide Internet-wide coverage to nearby vehicles by using a fixed network (thus enabling vehicle-to-infrastructure communications).

We argue in this paper that a VANET can be used for monitoring purposes. Thus, vehicles can measure certain environmental parameters in a specific area by means of different types of sensors installed on the vehicles. For example, we may think of devices that measure the CO_2 or the pollen concentration, the temperature, or even the coverage level of a cell phone company. As another example, monitoring data such as the number of available parking spaces or the average speed of vehicles in an area are also interesting to provide useful information to drivers. Using a VANET for monitoring implies a process of five steps:

1. *Determining the goal of the monitoring task.* The coordinates of the *monitored area*, the environmental parameter to measure and the monitoring period.
2. *Allocating vehicles for the monitoring.* Vehicles equipped with the required sensors must be assigned the task to measure the required environmental parameter.
3. *Collecting the data of interest.* The data sources will be sensors installed on the vehicles, which measure the required parameters from the environment.
4. *Routing the collected data.* The acquired data are sent to a predefined place using on-board short-range wireless devices to transfer the data to other nearby cars.
5. *Processing the data retrieved.* The collected information is gathered and stored in an information system for later analysis and processing.

In the next section, we describe the mobile agent-based approach that we propose to perform the monitoring indicating how these steps are realized with mobile agents.

3 Environment Monitoring Using Mobile Agents

Mobile agents are software components that run on an execution environment (traditionally called *place*) provided by a certain *mobile agent platform*, and can autonomously travel from *place* to *place* (within the same computer or between different computers) [8, 2]. A mobile agent platform provides services such as transportation of agents to other computers, communication with other agents, security, etc., in a transparent way to the programmer. Mobile agents provide some benefits (e.g., autonomy, flexibility, and effective usage of the network [8]) that make them very attractive for distributed computing and wireless environments (e.g., see [9]).

A mobile agent can be seen as a program that has the ability to pause its execution, move to another *place*, and resume its execution there, maintaining the values of its data structures (the state of the agent). Thanks to this capability, it is easy to build complex distributed applications that are at the same time flexible: If the task executed by an agent must be changed in the future, a new version of the agent (a new agent implementation) can be delivered. Thus, there is no need to keep specialized software installed on the computers/devices composing the distributed system: Only the generic mobile agent platform software is needed and an agent implementing the required behavior can be sent there at any time.

Mobile agent systems and monitoring VANETs bear several similarities. Thus, in a monitoring VANET there are many vehicles, distributed on a wide geographic area, that obtain data (measured by sensors) which must be moved from vehicle to vehicle based on certain conditions (e.g., location and direction) to try to reach their target. The existing similarity with a situation where some software agents move from one computer/device to another makes mobile agents a very suitable option to implement a monitoring solution for VANETs. The five steps of the monitoring process described in Section 2 can be implemented using mobile agents as follows:

1. *Determining the goal of the monitoring task.* A number of monitoring parameters must be provided to a mobile agent implementation, such as: the type of environmental parameter to measure, a definition of the monitored area (e.g. given by the GPS coordinates of its perimeter), the monitoring precision required (see step 3), and the monitoring period (given by a time limit after which the agent will end the monitoring task and will return the collected data). All these parameters are determined before the monitoring agent deployment, which is initiated from the agent platform hosted on a *monitoring computer*.
2. *Allocating vehicles for the monitoring.* The monitoring agent moves to the *relaying device* (see Section 2) that is the closest to the area of interest. Once there, the agent waits for a suitable car passing by and hops there. Then, as it travels in the car, the agent will constantly assess the possibility to jump to a different car if it considers that it may be a better alternative to reach the target area.
3. *Collecting the data of interest.* The target area may be too large to be monitored by a single agent. Thus, we divide the area in sub-areas (cells), according to the monitoring precision required (the larger the number of cells the higher the precision, as samples in more locations within the area will be taken), and allocate one clone of the agent (a *cell monitoring agent*) to each sub-area. They will need to move to a different car whenever its current car leaves the cell, or if the required sensor type is not available. When the agent reaches its cell in a car with suitable sensors, it will take data samples and store them in its data structures. This process is performed autonomously by each agent, without the collaboration of any other agent.
4. *Routing the collected data.* Once the monitoring period has elapsed, the cell monitoring agents return to the monitoring computer with the collected data. If the monitoring computer is attached to the fixed network, they jump from car to car trying to reach the closest relaying device from which they travel to the monitoring computer directly (using the fixed network). However, some application scenarios require the monitoring computer to be mobile. For example, the driver of a car could automatically receive information about the traffic ahead or about the availability of parking spaces in areas near his/her destination. In this case, the agent

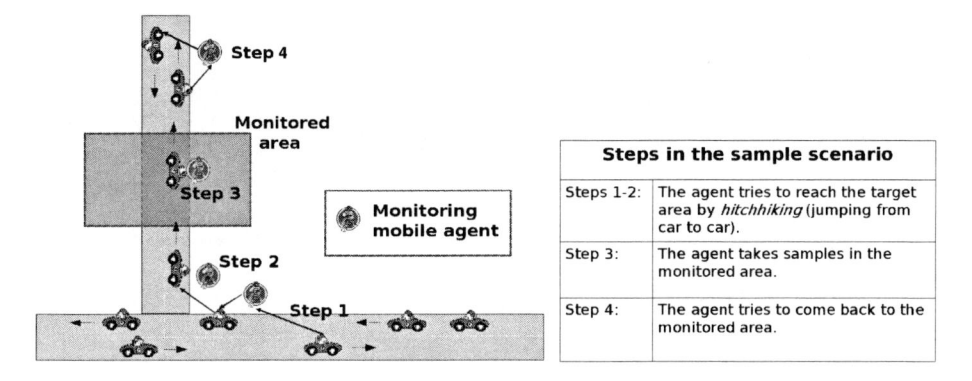

Fig. 1. Example scenario: a hitchhiker agent in action

Steps in the sample scenario	
Steps 1-2:	The agent tries to reach the target area by *hitchhiking* (jumping from car to car).
Step 3:	The agent takes samples in the monitored area.
Step 4:	The agent tries to come back to the monitored area.

jumps from car to car to reach the area where the monitoring device is (this area can be computed from the initial location of the device, its maximum speed, and the time elapsed), and then it broadcasts itself within that area.

5. *Processing the data retrieved.* The monitoring computer gathers the data transported by the incoming agents and stores these data (e.g., in a relational database) for further processing. The arriving agents can then finish their execution.

Figure 1 shows a scenario where an agent reaches the monitored area and later has to "come back" with another vehicle because its current vehicle leaves the area.

In the rest of this section, we first describe the technology required to implement the proposed approach. Then, we emphasize the benefits of this approach based on mobile agents. Finally, we enumerate some difficulties and how we solve them.

3.1 Technological Elements

Apart from the existence of certain relaying devices on the roads (as mentioned at the end of the first paragraph in Section 2), vehicles taking part in the approach described are required to be equipped with several hardware components and run certain software:

- They must be equipped with *sensors* that measure values of the type required in the monitoring task. Different vehicles with different types of sensors may participate in different monitoring tasks. These sensors will probably not be installed by car manufacturers but by voluntary users willing to take part in the distributed monitoring. Since these devices usually operate in a passive and non-intrusive way, the users' driving experience will not be altered.
- They must have a *computing device* with enough resources to execute an agent platform and manage the sensors (e.g., a PDA or an ultra mobile PC). This computing device must provide:
  - A *wireless communication device*, that allows the vehicle to communicate with its neighbors.
  - A *GPS receiver*, which can be queried by the monitoring agents to know if they are within the intended geographic area.

In this sense, any wireless-enabled PDA with a working navigation system (e.g., TomTom, see http://www.tomtom.com) would be enough.
- They must execute a (lightweight) *mobile agent platform* that offers suitable services to the monitoring agents, such as a wireless transportation service to other devices and an interface to query the available sensors and the GPS receiver.

It should be noted that most of the elements indicated above are interesting for a variety of applications, not only for our monitoring purposes. Thus, for example, many vehicles will have a GPS receiver as part of a navigation system. Moreover, we can envision that a wide variety of applications could be deployed in a VANET if the vehicles execute a mobile agent platform. Cars not providing the features described simply cannot cooperate in the monitoring task.

3.2 Benefits of Using Mobile Agents in Monitoring VANETs

The use of mobile agents for environment monitoring in vehicular networks has a number of advantages, such as:

- *Flexibility regarding how the monitoring task is deployed and performed.* A VANET can be very heterogeneous and dynamic. Thus, there are different types of sensors that may be available on the vehicles, very different road infrastructures (e.g., urban/rural roads or highways) with different traffic density, etc. Depending on the context, different traveling strategies could be considered by the agents. Thanks to the flexibility provided by an approach based on mobile agents, if a better *traveling strategy* is found or a new class of sensors is introduced, a new version of the monitoring agents with the needed enhancements can be deployed in the network without altering the ongoing VANET operations: A mobile agent can implement the behavior required and carry it to any vehicle which hosts a mobile agent platform (without any extra software installation in the vehicle).
- *Cost minimization.* As sensors in vehicles are constantly "moving", a small number of them are needed to cover a certain area. Instead of deploying an expensive fixed infrastructure of static sensors, an approach based on agents that travel in a vehicular network benefit from existing resources available on regular vehicles.
- *Global coverage.* Any geographic area can be monitored, as long as there are suitable vehicles traveling nearby. Mobile agents carry the monitoring task wherever it is needed. For example, if there is a traffic accident involving a lorry carrying dangerous substances, mobile agents can travel there to monitor the scene.
- *Good performance.* Mobile agents exhibit a good performance in comparison with other alternative approaches, such as traditional client/server architectures (e.g., [9] is one of several studies showing this).
- *Natural implementation.* Routing the collected data between the vehicles can be implemented naturally using mobile agents. In general, mobile agents allow a convenient implementation of the monitoring steps described in Section 2.

For all the above reasons, mobile agents are a suitable technology for monitoring in VANETs.

3.3 Challenges and Solutions

However, there are some challenges to consider to perform an efficient monitoring:

- *Size of the monitored area*. The monitored area could be very large, and so using a single monitoring agent would be inefficient. Thus, the agent should move within the area to sample the environmental data at several locations within the area, making it very difficult to obtain all the samples of the data with a high sampling frequency. Instead, as mentioned in the description of step 3 in Section 3, we propose to divide the monitored area in several sub-areas (*cells*) and allocating a different *cell monitoring agent* to each of those cells.
- *Routing the monitoring agent to the target area*. To reach the target area, a monitoring agent must jump from car to car[1] until it finds one car that moves into that area (see steps 1-2 in Figure 1). For this, the agent tries to find a suitable vehicle that can physically transport it closer to the area that must be monitored.
- *Keeping an agent within its assigned cell*. Another important question is how to keep an agent inside its cell while it is collecting data. Thus, if the vehicle carrying the agent leaves the cell, then the agent will need to come back (using a different vehicle) to continue the monitoring task (e.g., see step 4 in Figure 1).
- *Returning to the monitoring computer*. Once the monitoring task has finished, the agent must return to the monitoring computer (probably via a relaying device).

Regarding the last three issues, different traveling strategies (that an agent can apply to try to reach a certain location, such as the center of its target cell) can be considered, such as:

- *Random jump (RND)*. The agent jumps to another car with a 50% probability.
- *Basic Encounter Probability (BEP)*[2]. The angle between the movement vector of the vehicle and a straight line to the destination is considered, in order to estimate the probability that the vehicle will move towards the destination. The agent jumps if, by jumping, its BEP increases.
- *Distance (DST)*. The agent jumps whenever the distance between the target car and the agent's destination decreases along time.
- *Frontal angle (ANG)*. The angle of direction of the target car regarding the agent's target location is considered. The agent jumps if this angle is less than 90°. The difference with the BEP strategy is that the decision is taken independently of the status of the current vehicle carrying the agent.

With some of these strategies the decision is based on information that must be obtained by querying the target car. Therefore, a traveling protocol for mobile agents where a trip succeeds only if certain conditions hold at the destination would be useful. These strategies will be evaluated experimentally in the next section.

---

[1] The *target car* could move out of range at any time. A mobile agent platform ensures the reliability of agents' movements: Either a trip succeeds or the agent has the opportunity to re-try (traveling to the same car or to a different car).

[2] This measure is inspired by the concept of *Encounter Probability (EP)* presented in [10], that estimates the probability that a vehicle will meet an *event* (e.g., an accident) on a road.

4 Experimental Evaluation

As stated in the previous section, defining a suitable hitchhiking strategy for the agents is an important issue. Therefore, we have evaluated the four strategies proposed by simulating vehicles moving within a graph network. The simulation is run on a road network represented by the graph shown in Figure 2.a, extracted from a real map, which corresponds to an area of four squared kilometers in the region of Valenciennes (France). The area to monitor is divided in six cells. A monitoring agent is created on a fixed computer at node S, and then this agent travels to a relaying device R. When a suitable vehicle passes within range of R, the agent jumps in the vehicle to try to reach the target area. Once in the target area, this agent transforms itself into six cell monitoring agents, one for each cell within the monitored area. The simulated vehicles move along the edges of the graph with (random) speeds between 50 and 100 km/h, taking a random turn at each intersection. The range of the wireless communications is between 140 and 200 meters, and each agent takes one second to perform a jump to another car within range.

To compare the different traveling strategies, we measure the total number of samples taken by the agents during a 50-minute monitoring task with each strategy: The longer an agent is able to remain within its cell, the higher the number of samples it will be able to take and, therefore, the monitoring will be more accurate. Each test is repeated 10 times and the average results are reported in Figure 2.b, for scenarios with different numbers of vehicles. As expected, the worst strategy is RND because with this strategy the status of the cars is not considered in the decision process. The best strategy is DST, which is also quite simple and intuitive. Next in performance is ANG, and then BEP. These last two strategies are similar but the second one takes into account both the current and the potential target car; as a consequence, the number of jumps performed by the agents with the second strategy is smaller. As shown in the figure, all the proposed strategies behave better with a higher number of vehicles, as this offers the agents more transportation means and alternative paths to reach their target areas. Moreover, with enough vehicles, a sufficiently high sampling frequency can be maintained

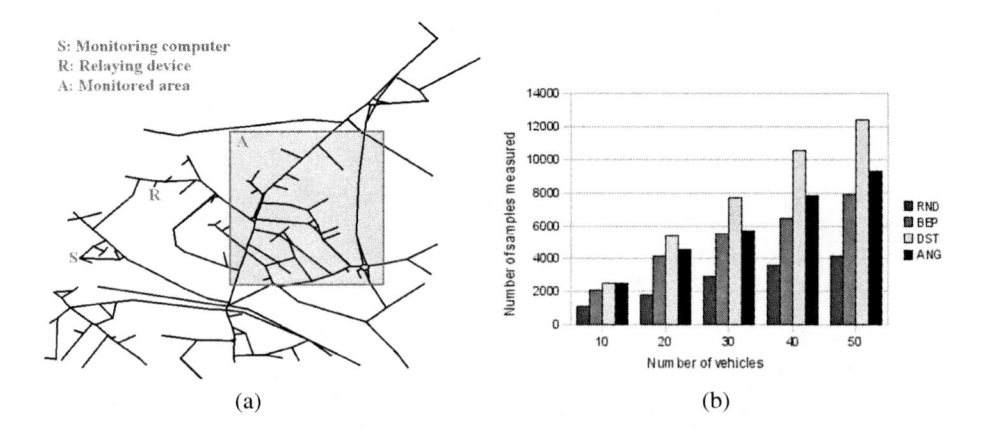

Fig. 2. Comparing traveling strategies: (a) scenario for evaluation and (b) samples measured

(e.g., about 40 samples per minute and cell with the DST strategy in a scenario with 50 vehicles). It is expected that the best strategy will depend on a number of factors, such as the traffic density or the speed of the vehicles. We plan to perform more experiments in a wide variety of scenarios.

5 Conclusions and Future Work

In this paper, we have presented a novel approach that combines vehicular networks with mobile agent technology for environment monitoring. In our approach, the mobile agents jump from car to car to arrive to the target geographic area and to keep themselves there to perform the monitoring task. We have analyzed different research issues and proposed and evaluated different routing strategies for the agents. Our initial experimental results are promising. However, there are some factors that can challenge the system, such as a low number of equipped vehicles or the existence of poor wireless communications. More work is needed to analyze the limitations of our current proposal in those circumstances.

As future work, we plan to perform more experiments in other scenarios and with different experimental settings. We will also study other strategies (e.g., using replicas of the monitoring agents as a form of redundancy to perform the monitoring). Finally, we will also analyze the suitability (and perform some adaptations) of the mobile agent platform SPRINGS [2] to implement a prototype; some experiments with this platform have already been performed in wireless environments [11].

Acknowledgements. The authors acknowledge the support of the CICYT project TIN2007-68091-C02-02 and the following institutions: the International Campus on Safety and Intermodality in Transportation, the Nord-Pas-de-Calais Region, the European Community, the Regional Delegation for Research and Technology, the Ministry of Higher Education and Research, the Aragón Institute of Technology, and the National Center for Scientific Research.

References

1. Xu, B., Wolfson, O., Rishe, N.: Benefit and Pricing of Spatio-temporal Information in Mobile Peer-to-Peer Networks. In: 39th Hawaii International Conference on System Sciences (HICSS-39), p. 223 (2006)
2. Trillo, R., Ilarri, S., Mena, E.: Comparison and Performance Evaluation of Mobile Agent Platforms. In: 3rd International Conference on Autonomic and Autonomous Systems (ICAS 2007), p. 41. IEEE Computer Society, Los Alamitos (2007)
3. Chen, B., Cheng, H.H., Palen, J.: Integrating mobile agent technology with multi-agent systems for distributed traffic detection and management systems. Transportation Research Part C (June 3, 2008) (in press), doi:10.1016/j.trc.2008.04.003
4. Lee, U., Magistretti, E., Zhou, B., Gerla, M., Bellavista, P., Corradi, A.: MobEyes: Smart Mobs for Urban Monitoring with a Vehicular Sensor Network. IEEE Wireless Communications 13(5), 52–57 (2006)

5. Hull, B., et al.: CarTel: A Distributed Mobile Sensor Computing System. In: 4th International Conference on Embedded Networked Sensor Systems (SenSys 2006), pp. 125–138. ACM, New York (2006)
6. Luo, J., Hubaux, J.-P.: A survey of research in inter-vehicle communications. In: Embedded Security in Cars - Securing Current and Future Automotive IT Applications, pp. 111–122. Springer, Heidelberg (2005)
7. Zhao, J., Cao, G.: VADD: Vehicle-assisted data delivery in vehicular ad hoc networks. IEEE Transactions on Vehicular Technology 57(3), 1910–1922 (2008)
8. Lange, D., Oshima, M.: Seven good reasons for mobile agents. Communications of the ACM 42(3), 88–89 (1999)
9. Spyrou, C., Samaras, G., Pitoura, E., Evripidou, P.: Mobile agents for wireless computing: the convergence of wireless computational models with mobile-agent technologies. Mobile Networks and Applications 9(5), 517–528 (2004)
10. Cenerario, N., Delot, T., Ilarri, S.: Dissemination of Information in Inter-Vehicle Ad Hoc Networks. In: 2008 IEEE Intelligent Vehicles Symposium (IV 2008), pp. 763–768. IEEE Computer Society, Los Alamitos (2008)
11. Urra, O., Ilarri, S., Mena, E.: Testing Mobile Agent Platforms Over the Air. In: 1st Workshop on Data and Services Management in Mobile Environments (DS2ME 2008), pp. 152–159. IEEE Computer Society, Los Alamitos (2008)

Using Multiagent Systems and Genetic Algorithms to Deal with Problems of Staggering

Arnoldo Uber Junior and Ricardo Azambuja Silveira

Dept. of Computing and Statistics
UFSC - Federal University of Santa Catarina
Florianópolis, Santa Catarina, Brazil
arnoldo.u.jr@gmail.com, silveira@inf.ufsc.br

Abstract. Production systems need information in real time to deal with diagnosis of problems and decision taking. For that, accuracy, processing agility, and mainly agility in managing information, is necessary. Those characteristics are found in the approach of the Multi Agents System. Processes Staggering is a complex task, which depends on many variables and demands real time in a satisfactory solution where the search space is extremely complex and, thus, Genetic Algorithms techniques can help to solve these situations. This article presents a case study where a hybrid modeling for the implementation of a process staggering system in a production system.

Keywords: Multi Agent System, MAS, Genetic Algorithms, Processes Staggering.

1 Introduction

The demand for information in real time becomes every moment more real, with production systems in need of agile and continuous information flow, seeking the improvement of manufacturing processes to attend to the great variety of configurations that your products can compose, having flexibility, lowering costs, improving quality and handing the products on due time, in the appropriate quantity and according to what was requested.

Sacile [6] says that industrialization flexibility is the ability to organize and reorganize production resources efficiently according to: pricing, quality, response time to environmental changes and mainly, technological demand changes.

According to Soares [7], nowadays it is not sufficient work with a good production plan and control. To be competitive it is necessary to optimize the levels of stock, the use of resources, the production costs, the change time and the unattended requirements. For this reason, many techniques and computational paradigms have appeared to solve the problem of production plans or programs creation under the concept of optimization. Some examples are the use of such strategies as: Linear and non Linear Programming, Simulated Annealing, Neural Nets, Taboo Search, and Genetic Algorithms in optimization [8].

As it is a problem which tends to be NP-Complete and enables a vast application, the JSS problem was chosen to evaluate the framework proposed in this article. The tools that enable the use of these techniques independently, or as a group, forming new hybrid technologies, allow satisfactory solutions to complex problems through other perspectives, turning into what before was unimaginable: real solutions.

This task proposes the modeling of a framework for staggering of processes distributed using a collaborative and self organizable multi agent system and the search and optimization techniques: Genetic Algorithms and Taboo search[8].

The study of processes staggering techniques refers to the creation of the first operational systems (SO) with the staggerer's processes algorithms without pre-emption (FCFS – first to arrive first to execute, SJF – first the shortest, etc.), (Round – Robin, Priority – SFTF – first the process that takes less time, Multilevel Lines, etc.) However the use of staggerer's processes extends to other areas besides OS: it affects all the problems where there is a group of tasks to be executed and a group of executing units, and the time of final execution of the tasks is directly affected by the sequence of execution adopted. Such problems are found in: Projects sequence, Job-Shop Scheduling (JSS), production planning, etc. As it is a problem that tends to be NP – complete and enables a vast application, the JSS problem was chosen for the application of the framework proposed in this article.

The majority of programming problems studied is applied to the environment known as Job Shop. The traditional Job Shop is characterized by allowing different flows from the orders between the machines and the different operation numbers in order, which are processed just once in each machine according to Oliveira [4].

Oliveira [4] affirms that the JSS problem is a combinatorial problem, which then turns into NP- complete in certain situations (explicit or implicit enumeration of all the possible alternatives to guarantee an optimal solution) as exemplified in Fig. 1. In this way, optimizing algorithms are computationally possible when applied to small real problems, with limited objectives. For problems of an importance similar to those found in the real environment, it is customary to sacrifice the obtainment of a solution optimal for heuristic methods, which result into a suboptimal solution with acceptable computational time.

Its complexity can be observed in Jain [3], where even an analogy with the Traveling Salesman NP complete problem is made.

According to Borges [1] in a manufacturing environment, a continuous spectrum of differentiation with a flow shop at one edge and a job shop at the other can also be seen.

Pinedo [5] affirms that in multi-operation shops, jobs often have different routes. Such an environment is referred to as a job shop, which is a generalization of a flow shop (a flow shop is a job shop in which each and every job has the same route).

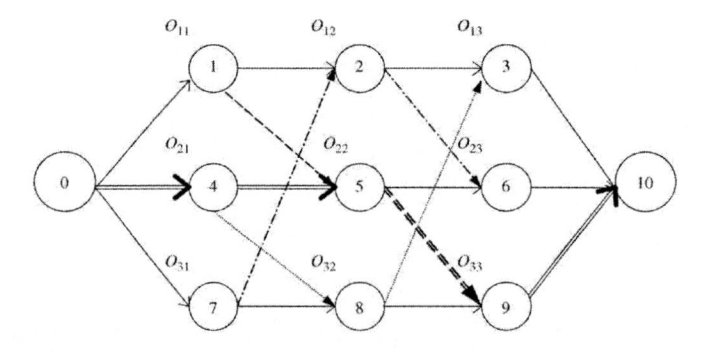

Fig. 1. Job Shop scheme [3]

A Job Shop type environment is that where the materials dislocate in the factory, in routes depending on the type of job to be executed; whereas the Flow Shop type is characterized by the fact that the materials and pieces dislocate in the factory in constant routes. The real production situations fit between these two types or as a combination of both [1]. A better way of differentiating both can be seen in Table 1 below.

Table 1. Comparison between Job Shop and Flow Shop [1]

Flow Shop	Job Shop
Fixed routes	Variable routes
Layout by product	Layout by process
Specialized equipment	Flexible equipment
Production to stock in large quantities	Production for orders in small quantities
Longer lead time to increase the capacity	Shorter lead time to increase the capacity
Well defined capacity	Capacity difficult to define
Intensive regarding capital	Intensive regarding workforce
Low intermediate stocks	Significant intermediate stocks
Low material coordination	High material coordination
Highly trained and specialized operators	Operators possess certain ability in some kind of function to monitor and control the processing equipment and / or machine which manufactures the products.
Overlapping in the operations	No existence of overlapping in the operations
Failure in the equipment can stop the plant	Failure in the equipment can stop the production of some items
Delay in receiving materials	Delay in receiving materials and pieces can stop the plant and delay the production of items
More energy consumption	Less energy consumption

The solution proposed in this article takes into account a problem composed of Job Shop and Flow Shop characteristics, which will be mentioned here just as JSS.

2 Problem

A hypothetical production scenario was developed to represent the problem, as shown below in Fig. 2.

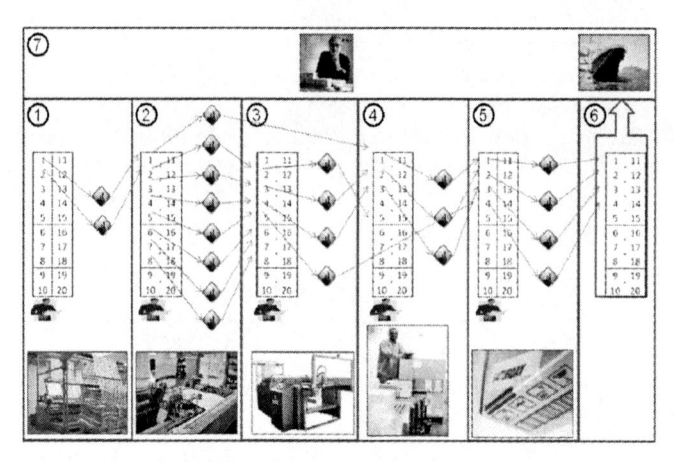

Fig. 2. Production Scenario

The production scenario represents a productive flow composed of 6 different phases, each of them presenting a transformation process of the material, however it does not necessarily have a sequence. Each phase has a group of processes, also known as intermediate stock for the supply of resources. Once the process in the resources is executed, they feed another group of processes to be executed in the next phase, which may or may not be in the sequence.

Depending on the size of this group of processes, the number of alternatives for choosing the process and the sequence to be executed can turn out to be an operation which demands considerable processing power. This situation is multiplied a considerable number of times if we take into account that the local choice of the phase will cause a sequence change in all the subsequent phases.

The sequence of a phase or resource determined by the imposition of other process generally causes a loss of local efficiency, making the objectives to be fulfilled detrimental to the global efficiency of the system.

Thus there is a need for a tool which optimizes the communication between processing phases, seeking to reach equilibrium between local efficiency and global efficiency, through communication, cooperation and self organization of the agents of that organization.

3 Proposed Tool

The proposed tool was modeled using the Multi Agent System Engineering Methodology (MaSE) proposed by De Loach [2].

The rules diagram presented in Fig. 3 shows the interaction between the advisor agent and the phase's agents, the monitor agent and the phase agents, as well as the interaction of the phase's agents with other phase's agents.

We defined three rules to carry out the processes staggering: Advisor, Phase Agent, and Monitor.

These rules allow the development of the following operations:

• To monitor staggering: It allows the Monitor rule to coordinate the sequence of actual processes to the phase agents' rules;

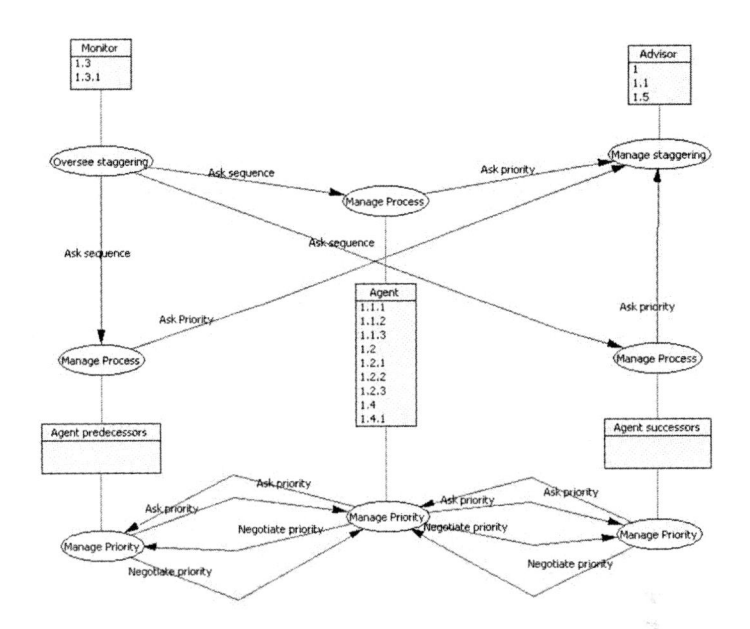

Fig. 3. SMA rules diagram

- To manage processes: The phases agents look for the priority of the processes with the advisor agent;
- To manage priorities: The phases agents also ask for the priority of the phase agents which are their predecessors and successors.

The class agents diagram is shown in Fig. 4, where the three rules must be registered in the framework JADE and each one of the phase agent requests from the framework the existence of the agents: Advisor, Monitor, predecessor and successor.

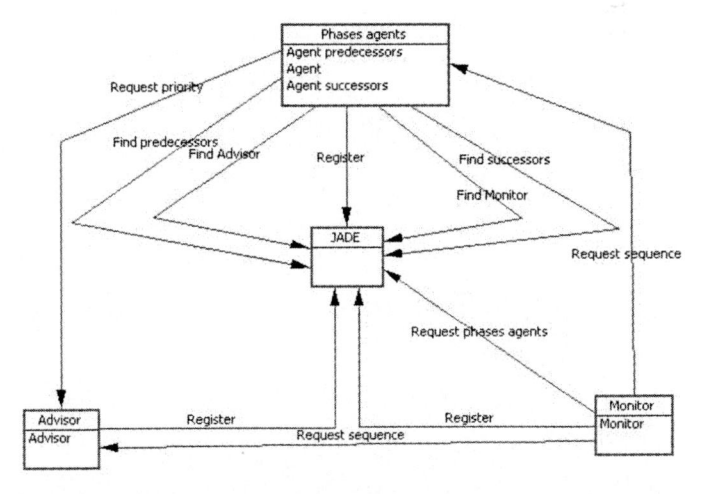

Fig. 4. Agents class diagram

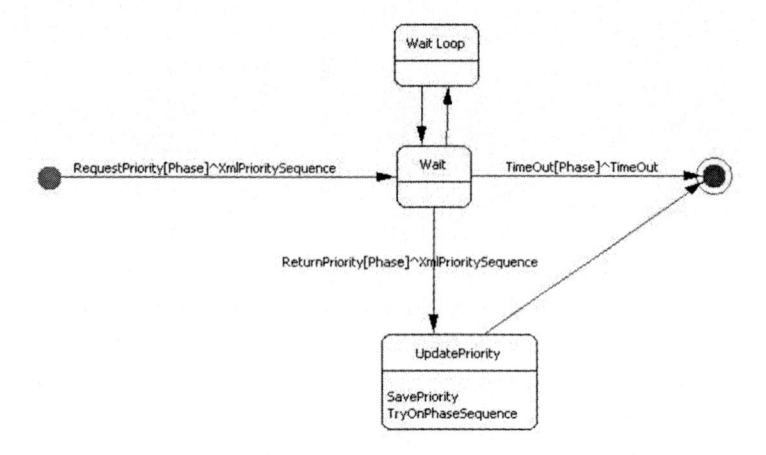

Fig. 5. Communication classes agents

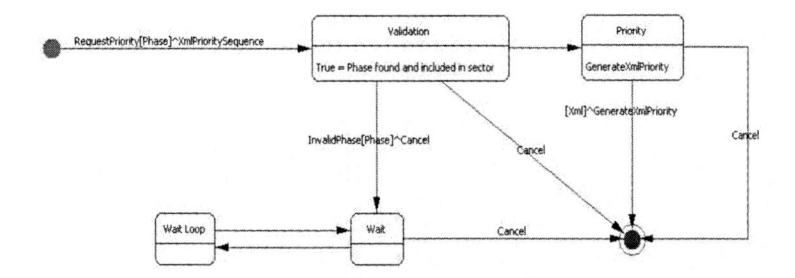

Fig. 6. Communication classes diagram

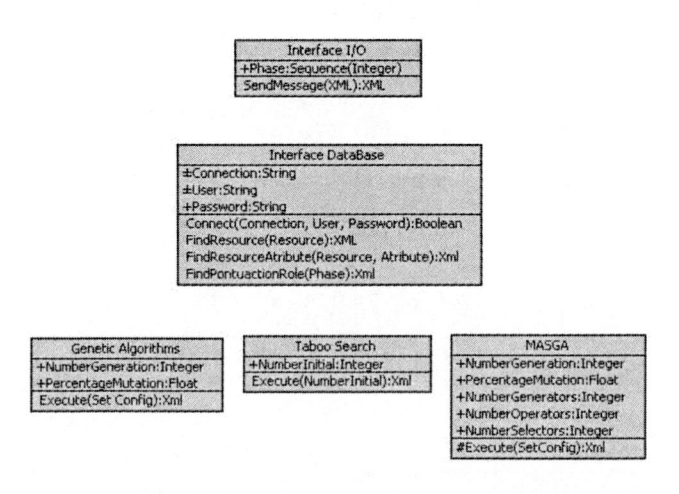

Fig. 7. Phase agent architecture diagram

One of the communication class diagrams developed is shown in Fig. 5 and Fig. 6. This diagram establishes the states of the Priority Requisition operation between the phase agents and the advisor agent. The first diagram shows the question (Fig. 5) and the second, the answer (Fig. 6).

In the mounting of agents' classes, according to the MaSE methodology, the agents' internal structures are defined. The phase agent structure is shown in Fig. 7.

The implantation diagram defined for this article is shown in Fig. 8, where there are two groups: managing and phases, both being part of the environment together with the Jade framework.

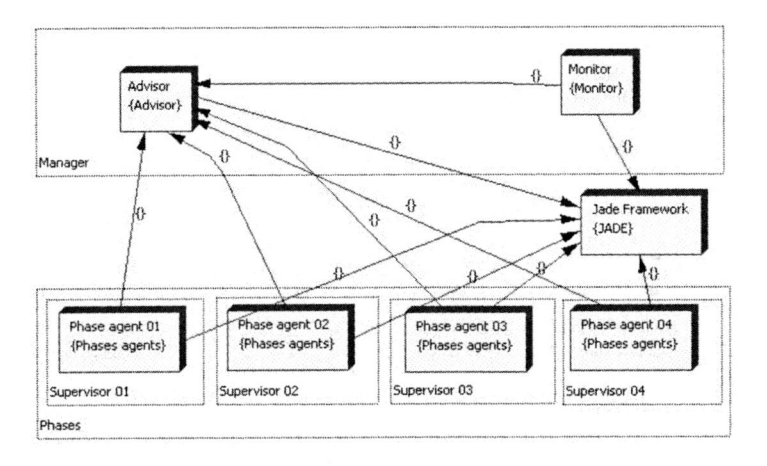

Fig. 8. Implantation diagram

4 Life Cycle

The proposed SMA's life cycle is commenced with the instance of the Monitor and Advisor agents, and later of the phase agents necessary for the situation.

The phase agents seek the information of the processes to be staggered from their inter phase with the data bank (DB). The staggering priorities are asked to the advisor agent and to the other phase agents. It is necessary to know the production priority of the phase agents that precede and succeed so as to create a sequence between the phases, and in this way maintain a continuous flow.

Having ready all the necessary information, the optimization and sequence procedures of the processes are executed through the optimization inter phase, using the techniques GA and Taboo search [8].

The techniques foreseen in the optimization inter phase, using the available resources and details of the processes to be staggered, look for the best local sequence of the processes through the obtainment of the makespan (length of the longest route, that is to say, longest process time) decrease. The rules for evaluation of the makespan are through punctuation, defined by the user who interacts with the framework.

Having all the processes staggered, acting in an automatic way, they are carried out in the execution list and wait for new processes to enter the staggering line. In case it acts in manual mode, the interaction with the user is required for analysis and to make the staggered sequence effective.

5 Conclusion

This article achieved its objective when it modeled a framework for processes distributed staggering using search heuristics techniques.

Through the described modeling, it is possible to develop the framework and apply it in processes staggering problems, where the number of variables for analysis is essentially great and hence allowing their division into smaller parts.

By solving each part individually, with local optimizations but taking into account the problem as a whole, a solution of the original problem is obtained in an optimized way and with less efficiency loss in the individual parts.

The theory of Multi Agent System showed itself to be fundamental in developing this research task, as it made the division of the problem into smaller parts possible, with distributed processing, apprenticeship and local optimization of the information.

The choice of the heuristic search techniques GA or Taboo search [8], allows gains in the choice of a local satisfactory solution, avoiding the analysis of a whole group of possibilities which are continually altered during the SMA life cycle.

Improvements can be made in the inclusion of negotiation techniques between the phase agents and the advisor to change the priorities. Increasing the search options by adding techniques such as: Simulated Annealing, Ant Colonies and hybrid methods, also allows advances in the sense of optimization quality and processing time.

Acknowledgements. We would like to thank the Operacional Têxtil Consultoria e Sistemas Ltda for the support and the opportunity to do research in your environment. Comments by anonymous reviewers improved the quality of this paper.

References

1. Borges, F.H., Dalcol, P.R.T.: Indústria de processos: comparações e caracterizações. In: ENEGEP. XXII Encontro Nacional de Engenharia de Produção. Proceedings. ABEPRO, Rio de Janeiro (2002)
2. DeLoach, S.A., Wood, M.: Developing Muiltiagent Systems with agentTool. LNCS (LNAI). Springer, Berlin (2000)
3. Jain, A.S., Meeran, S.: Deterministic Job-Shop Scheduling: Past, Present and Future. European Journal of Operational Research 113, 390–434 (1999)
4. Oliveira, R.L., Walter, C.: Escalonamento de um Job-Shop: um algoritmo com regras heurísticas. Thesis UFRGS, Porto Alegre (2000)
5. Pinedo, M.L.: Planning and Scheduling in Manufaturing and Services. Springer+Business Media Inc., New York (2005)

6. Sacile, R., Paolucci, M.: Agent-Based Manufacturing and Control Systems. CRC Press LLC, Flórida (2005)
7. Soares, M.M., et al.: Otimização do planejamento mestre da produção através de algoritmos genéticos. In: ENEGEP. XXII Encontro Nacional de Engenharia de Produção. Proceedings. ABEPRO, Rio de Janeiro (2002)
8. Wilson, R.A., Keil, F.C.: The MIT Encyclopedia of the Cognitive Sciences. MIT Press, London (1999)

VisualChord: A Personal Tutor for Guitar Learners

Alberto Romero, Ana-Belén Gil, and Ana de Luis

University of Salamanca,
Department of Computer Science and Automation – Sciences Faculty
Plaza de la Merced s/n, 37008, Salamanca, Spain
romero.rume@gmail.com, {abg,adeluis}@usal.es

Abstract. This paper describes VisualChord, a Web application, as personal tutor of initiation to the guitar that based on agents architecture that extracts files, tablatures and songs from Internet repositories, normalizing by rules and with a disambiguation algorithm to be stored in the internal repository with semantic tagging including a difficulty measure for each piece. This allows the user to training with a personalized music pieces selection with his/her guitar. There tries to offer a small personalized and flexible tutor who adapts to the tastes and aptitudes of the user. We describe the VisualChord platform, discuss the architecture, and describe some usability test and results with information about acquiring and making use of this develop.

1 Introduction

The development of the current Web has transformed our way of working, buying, communicating, learning and teaching, etc. The environment is socialized in the called virtual communities and there turn out to be a big amount of new possibilities to share and recollect web information to consume with all kind of technologies and perspectives. On the other hand, the irruption of the paradigm of the Web 2.0 supposes a democratization of the tools of access to the information and elaboration of contents [1]. Jointly with the materialization of the Semantic Web, it will allow improve the engines of search and to create personal agents much more advanced allowing to integer all this web information. With this perspective the agents' technologies acquires a fundamental aspect to extract, and filter all this information to consume.

Along all kind of diverse information in web there exist diverse types of computer applications and files related to the art of playing the guitar. The majority of the applications work with files called tablatures (*Guitar Tablature File*). These applications generate and emulate songs in MIDI or any other format, with all kind of extra functionalities. There exists a great traffic of this type of tablature files that circulate along Internet. Such it is their volume, which already has specified a not formal definition for these files, with extension ".tab", see Fig. 1. This tab files contains not only the music transcription for the guitar but also the song letter accomplish. Tablatures files contain dense information because serve to define riffs, alones, arpeggios and accompaniments musical pieces, etc. There are important Web communities and portals (lacuerda.net, ultimate-guitar.com, etc.) that contain and allow exchanging tablatures and files for the any applications to play the guitar.

```
Start note to sing: B
Guitar: E G D B d

 //:                                        Every
breath you - /Every single-
d----------------------
B-----------0------------
F------2---------2-------
D----0---0----0---0------           (x2)
G--0------------------
D----------------------

:take  /day
```

Fig. 1. Tablature Document (tab)

The facility of diffusion that offer Internet and the technological possibilities allows us to develop an application that extract the needed information from the repositories or web pages in tab format with musical information and once treated would be consumed by the developed platform. This application Web, called VisualChord, is a personal tutor of initiation to the guitar that we will detail in the following points. The second paragraph starts by analyzing other software to support guitar learners, the stage of work from which we will define our tool. The third paragraph makes a selection of the main modules of the developed application, analyzing aspects of the realized development as well as its functionality. The paragraph 4 details the analyses of usability of the platform interface and about its facility of use. The article finishes with a few conclusions for the application, including a few lines to improve.

2 Motivations and Related Work

There exist diverse types of computer applications that of a way or other one give support to everything that one that wants to begin to play the guitar. The same programs usually offer tools to modify and to create the tablatures. These programs proper and facilitate the composition to authors who have not a musical education. Many of these programs also transport automatically the notation of the coding used in the tablatures, to the own notation on a score. There exist also computer programs that follow the classical linear methods for playing guitar. They provide one interface for the computer with lessons they teach to play or improve the skill with the instrument. These programs usually are not configurable and the lessons are in general predefined.

The principal software references for this kind of work are:

Power Tab Editor[1]: It is highly used by its great potential, facility of use and reliability, beside this is also a free tool. The program works with files .tab, has a player for MIDI to emulate the song, shows the equivalence of the coding on a stave, allows to edit and to create scores and tablatures, it adjusts to tablatures of guitar and bass, etc.

Guitar Pro (GP)[2]: It's considered the best program of the genre. It has absolutely all the utilities of the Power Tab Editor but in addition it complements

---

[1] http://www.power-tab.net/
[2] http://www.guitar-pro.com

these utilities by adding more instruments. It does not only centre on the guitar; also it might add any type of instrument. It includes the technology RSE (*Realistic Sound Engine* - Engine of Real Sound) providing with a quality of interpretation to the reproduction almost equivalent to the instruments real sound.

GuitarVision[3]: It's specializes exclusively in the guitar. The interface is based on the neck of a guitar, where the chords appear positioned according to the user selects. Also it has predefined songs in which it is marking the changes of chord as advancing the song.

Guitar Guru[4]: Tool very similar to the previous one.

We observe that the studied programs can qualify according to the files of the songs that treat in two blocks. The first one formed by programs type Power Tab Editor that work with tabs files and that contain the transcription to the coding of the tablature of the song. They are much extended in the Internet area and are easy to find. But the final result would not be alike the accompaniment of a guitar. This aim is obtained by the second type of tools among which the Guitar Guru is, but we meet that the files of the songs that the program uses are proprietary and there is needed extra payment.

Another principal difference is the type of user whom they are directed. The first block is dedicated to a user who already plays the guitar with a certain skill and fluency can like to dare with riffs solos and accompaniments, whereas the second block is more orientated to users who are starting to play the guitar, to know the chords and to practise the changes of the diverse rhythms. In this second block also there might be included the digital methods of guitar, where the guitar can connect to the computer and the program monitors the execution of the pupil, which they do not stop being a classic method adapted for a computer, but the inflexibility from these can come to discourage the people who only wants to play the guitar in a basic way without technical notions nor knowledge of music theory.

As result of there we generated a classification of the existing applications:

1. According to the user that is directed: 'starter users' attended by the applications GuitarVision and GuitarGuru and 'initiated-advanced users' with Guitar Pro and Publishing Power Editor.
2. According to the used methodology:'Digital methodologies' included in Guitar Pro and Publishing Power Tab, and 'Classical methodologies' though they do not use a classic strict method, it is possible to think that the applications are based on this type of methods: GuitarVision and Guitar Guru.

Another important aspect to stand out is that the emulations that produce the applications of the treated songs. Only the applications destined for new users have an engine capable of emulating a song played as a real song, not with notes MIDI, and given the simplicity of these applications, and that the files of support of the songs are of payment, the potential of this utility diminishes considerably. Nevertheless, the applications that have a great free repository of songs have not this so beneficial emulation for the user. So the desired application need for free repositories and real song emulation along a flexible and personalized method to work with tabs files and all the functionalities to proper a learning context for any person who wants to learn to play guitar.

---

[3] http://www.guitarvision.com
[4] http://www.musicnotes.com/guitarguru/

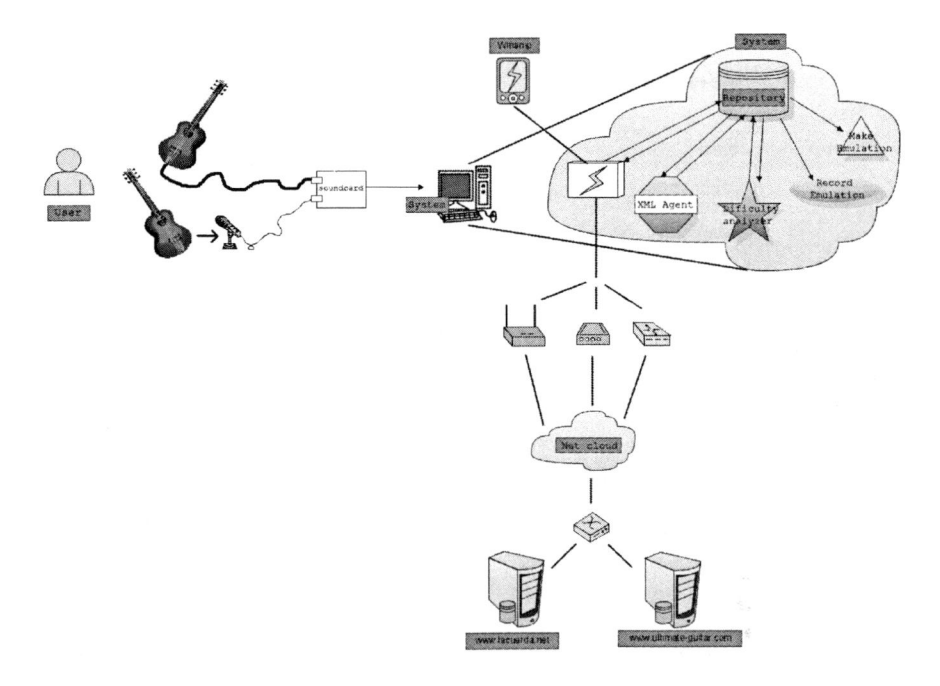

Fig. 2. Visual-Chord Architecture

3 VisualChord Overview

VisualChord allows in connection with the free web repositories of musical resources, generates a personal tutor of Spanish guitar with the requirement of emulation and other aspects that the following points detail. The architecture of the tool, see Fig. 2, shows how there exists a module that extracts the resources from Internet repositories with technologies of web recovery information [2].

They allow to shape the musical resources in a way completely disconnected, that is to say, there are not orientated to restrict the structure or the format of the resources, but just to write annotations that allow us to classify and to describe their properties. Across a wrapper, which normalizes the heterogeneous extracted resources and the Agent XML that tagged adequately enriching the content with semantic labels based on a musical ontology and it stores them in the internal repository of the system. These resources stored in an XML format contains information about the song, the notes and the letters of the songs, an indicator of the difficulty that the agent measures (analyzer of difficulty) and any other information that the user wants to be added. Once stored in the repositories of the system emulate the song music by any program the user establish (e.g. Winamp); then the user can record, linked his/her interpretation and to store it, etc.

3.1 Wrapper for Searching Sources

Wrappers are specialized program routines that automatically extract data from Internet and convert the information into a structured format. There are lot of these

```
Begin:
        WinampInfo=GetWinampInfo()
        Adecuate(WinampInf)
        SongInf = FindInFile(WinampInfo)
        if (SongInf == true)
        Begin:
                ShowInfo(SongInf)
        end if
        else
        begin:
                SongInf = FindInServer(WinampInfo)
                if (SongInf == true)
                begin:
                        SaveInRepository(SongInf,WinampInfo)
                        ShowInfo(SongInf)
                end if
                else
                begin:
                SongInf=FindInOtherServer(WinampInfo,ChordFormat)
                        if (SongInf == true)
                        begin:
        SaveInRepository(SongInf,WinampInfo)
                                ShowInfo(SongInf)
                        end if
                        else
                        begin:
SongInf=BuscarSegundoServer(WinampInfo,TAB)
                                if (SongInf == true)
                                begin:
        SaveInRepository(SongInf,WinampInfo)
                                        ShowInfo(SongInf)
                                end if
                        end else
                end else
        end else
end
```

Fig. 3. Wrapper Algorithm for extraction of song item

develops that exploits Web data sources using reconfigurable Web wrapper agents [3]. The wrapper in VisualChord works according the algorithm in fig. 3.

It has three main steps. Firstly, it must be able to recover information of the net cloud, as HTML pages, tabs files from several repositories of songs and web pages (e.g. www.lacuerda.net, www. ultimate-guitar.com). Secondly, search for, recognize and extract specified data. Thirdly, save this data in a suitably structured format to enable further manipulation by the platform. This work is made by the XML Agent.

3.2 Normaliser and Exportation Agent to XML

The files download from Internet repositories are any kind of flat text, and web pages where the content settles down it isolating of the etiquettes HTML. This supposes a problem at the moment of recognizing what part of the file is a chord, and what is a song letter, especially with certain chords that could be sensitive to mistakes, as SOL chord in Spanish (that means also sun) or MI chord that in Spanish means 'my'. Also the notation could be different (Spanish: Do Re Mi Fa Sol La Si; French: Ut Re Mi Fa Sol La Si; English: C D E F G A B; German: C D E F G A H).

VisualChord: A Personal Tutor for Guitar Learners   581

```
37   A·················E······················F#m
38   Si te vas para que regresaste
39   D·A··········E···················F#m
40   y además solo quise besarte
41   D········A···········E
42   ay cuando mi vida cuando
43            ····F#m··············D
44   va a ser el día que tu pared desaparezca
45
46   A·········E······················F#m
47   Fabriqué un millón de ilusiones
48   D·A··········E···················F#m
49   prisioneras que se hicieron canciones
50   D·····A···············E
51   ay cuando mi vida cuando
52            ····F#m··············D
53   vas a cerrar tus ojos por mi
54
55   D···E·····F#m······D
56   oooooooooooooooooooo
57   D·················E
58   antes que ver el sol
59      ····A·················D
Ln 1 : 84  Col 1  Sel 0                    2,75 KB    ANSI
```

Fig. 4. Some detail from a song download from the web

Appear then the serious problem of the disambiguation in the automatic recovery of information.

If we notice in fig 4, we will see that this extract of a song download. Already in the first line (marked like 37), and the first character of the same one, we meet a delicate situation. This "A" comes to be the Spanish preposition or the chord LA in the English notation? In the following line (line 38) we meet that the first word of the phrase, A, shows ambiguity. Is a chord or is the conjunction? We can see that in the successive lines of the song these cases continue appearing, since it is obvious, in a song there is an average between 4 and 10 different chords that are repeating themselves along the same one. Once a note has appeared, there are many possibilities that this one repeats itself several times along the song.

Since we have seen in this small chunk of song, the ambiguities are abundant along a file and can also proceed from English and Spanish musical notation and it depends on the song letter idiom also. To solve it we have generated a disambiguation algorithm. Doing searches for arguments and for dependences just only for Spanish song letters, looking if it can be or not a musical note. A state machine that expresses the conditions how it works the algorithm would be the following one, Fig. 5:

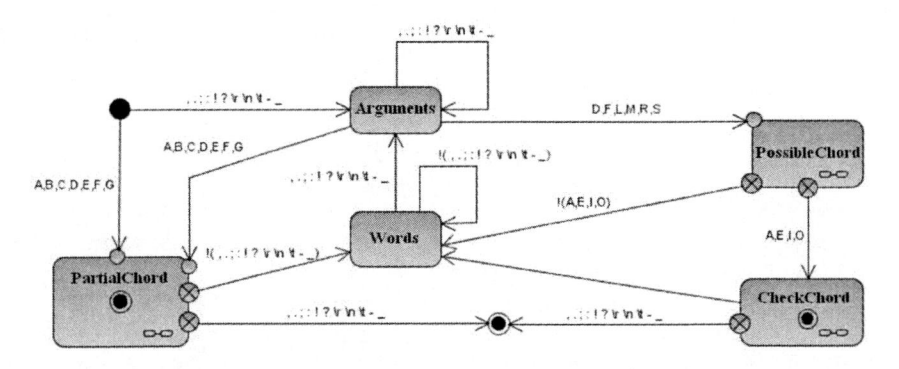

Fig. 5. State machine for disambiguation

Firstly the algorithm get the song text character by character and analyzes if it could be or not a chord. When the algorithm finds a possible match, it begins to make some checks by analyzing the possible chord's arguments (the characters before and after it) to resolve if it is or not a chord.

The agent does the same from the first up to the last character of the song to find any possible chord. When the algorithm finishes its execution all the chords in the song should have been identified.

Once settled this problem transforms the tab file to a XML file, with the following DTD:

```
<!ELEMENT song (info, ACORDES)>
<!ELEMENT info (mod, nota)>
<!ELEMENT mod (#PCDATA) >
<!ELEMENT diff (#PCDATA) >
<!ELEMENT CHORDS (chord*) >
<!ELEMENT chord (#PCDATA)>
```

It is a file that relies on information of the alterations of the chords (*mod*). Then the element CHORDS is where the real part of the song resides, that is to say, the letter of the same one and the chords. The chords will turn out to be delimited by a label of type chord. The labels *diff* is the value of the difficulty of the song, returned by the agent assessor of difficulty of the song, which we will see in following points.

Also the Agent of exportation to XML is creating a repository with the files that are generated to improve and to relieve processes in future executions on the same song. It is necessary to emphasize that before giving for the exportation finished and creation of the XML it appears to the user la view for approval and the possibility to add new annotations. If it does not agree with the obtained results, always he/she will modify the information directly in XML's file before store into the repository.

3.3 Agent Assessor for Musical Execution Difficulty

The application is a tutor personalized for the initiation to the guitar, for this reason there is implemented an algorithm that evaluates the difficulty of any piece. The level of difficulty is grade in low, average, high and very high. The evaluation it applies some rules that an agent measures and tagged with the appropriate labels the song in the repository file and alerting the user. When the song is openned by the application, the Difficulty Agent analyzes the song according to some parameters and show the user the results. The parameters that the agent evaluates to identify the difficulty are:

- **Notation of the chords:** since the principal user of the system is going to be a person who is starting playing the guitar, it is probable enough that he/she does not know the English notation (taken as international standard) or German one. For this motive, the system will add difficulty to the song. The dificulty degree goes from 0 up to 1 of 10 points.
- **Alterations of the notes** (flats and supported)**:** it is proportional to the number of alterations of the song. In the guitar to play the alterations it is necessary to use the forefinger as capo, and this has a high difficulty in beginners so relies on an increment in the algorithm. The dificulty degree goes from 1 up to 3.4 of 10 points.

- **Density of chords:** the difficulty of the piece bears in mind the number of chords that composes the song for unit of time, to major density, major difficulty. The dificulty degree goes from 1 up to 3.6 of 10 points.
- **Number of different chords in the song:** The difficulty grows with the major number of different chords. The dificulty degree goes from 0 up to 2 of 10 points.

3.4 Emulator of Interpretation

Other characteristic of the application is their emulation engine. This one uses the file XML to extract in a tail all the chords of the song, and it generates a song in real time, by connecting a chord after other one. The program create an object of the class CWave for each of the chords of the song and by concatenating (simultaneously that reproducing) one with another and counts the duration of the notes formed by the user.

4 Usability Requirements Achieved in VISUALCHORD

The structure of the information and the accesses across the interface it is organized attending to the needs of the final users of the tool, following the cycle of design of interface centred on the user ([4], [5]). Its utilization is simple and the degree of learning of every task is also rapid quantified by means of test of usability carried out with a group of 10 persons foreign to the development. This stage has great importance in the phase of development of the software. It serves to detect mistakes produced during the phase of implementation and even in the phase of design. It is in addition an effective way of checking the functionality of the system, verifying if the proposed aims were reached and if the previous requirements are fulfilled.

The first battery was carrying out to value the simplicity of the application task for the user. The users executed diverse, simple and clearly-defined task without knowing the system. If the user was not managing to finish the task in a reasonable time it took as a failure of the design. This way it was necessary to modify several times the design of the interface up to obtaining a few results adapted to the proposed aims. The most important test was the one that had to refine the XML exporter until it was managed to do usable and intuitively to the user.

As we observe in the Fig. 6(A), which it is the result of the study, at the end of the design of the interface, the percentage of success at the moment of the functionality of

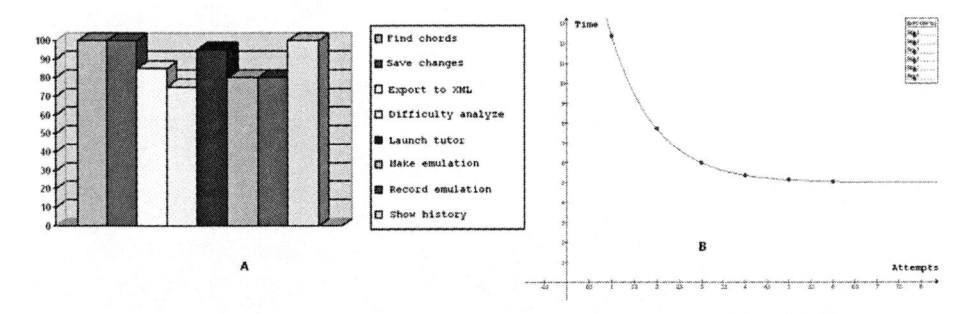

Fig. 6. VisualChord's usability test (A) By functionality (B) Learning Curve

the system finds the first time is at worst 75%, 100% of success being obtained in other occasions.

Once reached the final design of interface was analyzed the learning curve. We see in Fig. 6 (B), the graph generated and we can conclude that the result of the analysis reveals that the VisualChord interface is intuitive. In few executions, from the second one, already big changes are observed in the time used for the accomplishment of the same task. Also it is possible to observe how in a minimum of 4 executions the time used for the accomplishment of the task assigned (to generate the emulation of the song) becomes stable in 5 seconds approximately.

The tests in all the operating systems for that there had to be suitable the application (Ms. Windows 98/98 SE/ME/2000/XP), working with internal files of the repository and without them. The usability was evaluated in the system, guaranteeing that the system is intuitive and simple to use and therefore the application.

5 Conclusion and Further Aspects

The article presents an application orientated to facilitating and helping the guitarists amateurs to start playing guitar. There tries to offer a small personalized and flexible tutor who adapts to the tastes and aptitudes of the user. The application is sustained in a desk application that connects with the Web repositories to obtain the resources and songs that the application will handle by normalizing a tagged to store in an internal repository. The systems of wrapping are an effective solution of engineering that in our case helps to reduce the costs of development. VisualChord uses the advantage that offers the new tool's generation based on the web 2.0 as new concept of sharing and interchange information by using specialized resources in forums of Internet user's communities. The tool across agents normalizes the extracted information by tagging all the information and allows measuring the degree of difficulty of the pieces, to generate record of songs, connection with the sound card of the computer and some other modules, some of them presented in this article.

The developed system will be extended and improved in the near future. The first one is to share the internal repository of the system to be accessed by the specializing community. This will allow to generate the complete flow of information input and to share the improved and annotated in XML. The immediately aim is to translate the semantic Web is to change the annotations to describe the resources from XML to RDF or OWL to clearly jump into a semantic web languages.

Between other improvements in addition we aim that it is studied to apply the sound recognition in the application to provide the system as an assessor of the interpretation of the user. This might facilitate to the user an objective evaluation of his/her interpretations.

It is evident that the new way of understanding Internet promotes an increasing diversity and information flow. This allows an access much easier and decentralized to the contents. The generation of these kinds of tools will increase, by using and generate information specializing in the Web.

References

1. Musser, J., O'Reilly, T., The O'Reilly Radar Team.: Web 2.0 Principles and Best Practices. An O'Reilly Radar Report (November 2006)
2. Baeza-Yates, R., Ribeiro-Neto, B.: Modern Information Retrieval. Addison-Wesley, Wokingham (1999)
3. Chang, C.-H., Siek, H., Lu, J.-J., Hsu, C.-N., Chiou, J.-J.: Reconfigurable Web Wrapper Agents. IEEE Intelligent Systems 18(5), 34–40 (2003)
4. Constantine, L., Lockwood, L.: Software for Use: A Practical Guide to the Models and Methods of Usage Centered Design. Addison Wesley, Boston (1999)
5. Hassan, Y., Martín, F.J., Fernández, Iazza, G.: Diseño Web Centrado en el Usuario: Usabilidad y Arquitectura de la Información. Hipertext.net, núm. 2 (2004), http://www.hipertext.net

Author Index

Agüero, Jorge 60
Aleixos, N. 510
Alonso, Luis 207
Alonso-Betanzos, Amparo 339
Alonso-Ríos, David 339
Aranda, Gustavo 421
Arauzo, José Alberto 293
Argente, Estefania 319, 440

Bajo, Javier 20, 217
Barbucha, Dariusz 169
Belem, Mahamadou 548
Boissier, Olivier 529
Bonura, S. 227
Borrego, Carlos 150
Borrell, J. 401
Botía, Juan A. 197
Botti, V. 440
Brasser, Russell 356

Caarls, Jurjen 1
Cammarata, G. 227
Cano, Rosa 246
Carbo, Javier 266
Carrascosa, Carlos 60
Castanedo, Federico 430
Castillo, Andrés G. 237
Castro, Antonio J.M. 159
Ceccaroni, Luigi 450
Conesa, J. 510
Contero, M. 510
Corchado, Juan M. 20
Crépin, Ludivine 529
Criado, Natalia 319, 440

de Castro, Angel 520
de la Cal, Enrique 460
del Castillo-Mussot, M. 310
Delot, Thierry 557
de Luis, Ana 576
Demazeau, Yves 188, 529
De Paz, Juan F. 217
de Paz, Yanira 246
de Rivera, Guillermo Glez. 520
Díaz, F. 140
Duque, Néstor D. 237

Egyhazy, Csaba 356
Espinosa, Agustin 130

Fdez-Riverola, Florentino 50
Fernández-Breis, Jesualdo T. 411
Fernández-Lorenzo, Santiago 339
Fernández-Pacheco, D.G. 510
Ferrándiz-Colmeiro, Antonio 480
Francaviglia, G. 227
Fuentes-Fernández, Rubén 40, 70

Galán, José Manuel 293
García, Jesús 430
García-Fornes, Ana 130, 421
García-Magariño, Iván 40, 70
García-Sánchez, Francisco 411
Garrido, Javier 520
Gasparetto, Alessandro 411
Gea-Martínez, Jorge 480
Georgé, Jean-Pierre 302
Gil, Ana-Belén 576
Gilart-Iglesias, Virgilio 480
Gleizes, Marie-Pierre 70

Glez-Dopazo, Julia 50
Glez-Peña, Daniel 50, 140
Glize, Pierre 302
Gómez-Rodríguez, Alma 284
Gómez-Sanz, Jorge 40, 70
Gómez-Sebastià, I. 450
González, Angélica 274
González-Moreno, Juan C. 284
Grabska, Ewa 364
Grundspenkis, Janis 490
Guijarro-Berdiñas, Bertha 339
Gutiérrez, M.E. Beato 11

Hernandez, Luis 130
Herrero, Pilar 120

Ilarri, Sergio 557
Isaza, Gustavo A. 237

Jacquenet, François 529
Jaumard, B. 374
Jędrzejowicz, Piotr 169
Joumaa, Hussein 188
Julián, Vicente 60, 319, 440
Junior, Arnoldo Uber 567

Kameas, A.D. 383
Kemeny, Andras 110
Kodia, Zahra 90
Kubera, Yoann 100

Lacroix, Benoit 110
Lancho, B. Pérez 11
Lavendelis, Egons 490
Laza, Rosalía 50, 140
López, Vivian F. 207
López-López, Silvia 339
López-Paredes, Adolfo 293
Lueiro-Astray, Loxo 284
Luzón, M.V. 140

Maciá-Pérez, Francisco 480
Maña, Antonio 256, 470, 538
Mancini, Toni 329
Mandiau, René 501
Marcos-Jorquera, Diego 480
Marguglio, A. 227
Martí, Ramon 30
Martín-Campillo, Abraham 30, 401
Martínez-Béjar, Rodrigo 411

Martínez-García, Carles 30, 401
Martínez-Miranda, Juan 80
Martin, Pablo 266
Mata, Aitor 274
Mathieu, Philippe 100, 110, 374
Matos-Franco, Juan C. 392
Mena, Eduardo 557
Migeon, Frédéric 70
Molina, José M. 430
Moneva, Hristina 1
Monier, Pierre 501
Morais, A. Jorge 349
Moreno, María 207
Morreale, V. 227
Müller, Jean-Pierre 548
Muñoz, Andrés 197
Muñoz, Antonio 256, 470, 538

Navarro-Arribas, G. 401
Nieto, J.A. Fraile 11
Nieves, Juan Carlos 450
Nongaillard, A. 374

Oliveira, Eugenio 159
Orfila, Agustin 266

Pajares, Javier 293
Palanca, Javier 421
Palau, Manel 450
Paletta, Mauricio 120
Patricio, Miguel A. 430
Pavón, Juan 80
Pavón, Reyes 50, 140
Pérez, Belén 274
Pérez-Delgado, María Luisa 179, 392
Pérez, Roberto 460
Peyruqueou, Sylvain 302
Picault, Sébastien 100
Piechowiak, Sylvain 501
Pinzón, Cristian 246
Puccio, M. 227

Rebollo, Miguel 60
Régis, Christine 302
Ribalda, Ricardo 520
Robles, Sergi 30, 150
Rodríguez, Arezky H. 310
Rodríguez, Sara 217
Romero, Alberto 576
Romero-González, Rubén 284
Rougemaille, Sylvain 70
Rubio, Manuel P. 246

Said, Lamjed Ben 90
Sánchez, Ana 197
Sanchez-Anguix, Victor 130
Sedano, Javier 460
Serrano, Daniel 256, 470, 538
Silveira, Ricardo Azambuja 567
Ślusarczyk, Grażyna 364
Stamatis, P.N. 383
Strug, Barbara 364

Tapia, Dante I. 20, 274

Urra, Oscar 557

Valencia-García, Rafael 411
Vázquez, G.J. 310
Vázquez-Salceda, Javier 450
Verriet, Jacques 1
Vidoni, Renato 411
Villar, José R. 460
Vincent, Jean-Marc 188

Zaharakis, I.D. 383